# A CONCISE COURSE IN ADVANCED LEVEL STATISTICS

## With Worked Examples

## Fourth Edition

**J CRAWSHAW** BSc

Former Head of Mathematics and Deputy Head
Clifton High School, Bristol

**J CHAMBERS** MA

Former Head of Mathematics
Sutton High School, Surrey

First published in 1984 by:
Stanley Thornes (Publishers) Ltd

Second Edition 1990
Third Edition 1994
Fourth Edition 2001

Reprinted in 2002 by:
Nelson Thornes Ltd
Delta Place
27 Bath Road
CHELTENHAM
GL53 7TH
United Kingdom

02  03  04  05  /  10  9  8  7  6  5  4  3  2

A catalogue record of this book is available from the British Library

ISBN 0 7487 5475-X

Page make-up by Mathematical Composition Setters Ltd

Printed and bound in Spain by Mateu Cromo

# Contents

# Preface

## Introduction

This fully revised and updated edition of **A Concise Course in Advanced Level Statistics** is a comprehensive text for use primarily by students and teachers of Advanced Level Mathematics, both at AS and A2 level. It also provides a useful support for those studying statistics as part of science, social science and humanities courses.

## Features

- Points of theory are explained concisely and illustrated clearly by worked examples, many taken from Advanced Level papers.
- Carefully graded exercises help you to consolidate ideas and gain experience in applying theory to different situations.
- Frequent hints pinpoint common misunderstandings and reinforce ideas.
- Key concepts and formulae are highlighted in colour to increase clarity. Frequent summaries provide a quick reference.
- Extensive miscellaneous exercises and end-of-chapter tests provide practice in tackling examination questions, providing essential examination preparation.
- Answers to all exercises are provided.
- An ICT supplement explores the use of ICT in the study of statistics.

## Specifications

The text covers the main theory required in the specifications of all the examination boards for the statistics sections of AS and A2 Mathematics.

## Examination Questions

We are grateful to the following Awarding Bodies for permission to reproduce questions from their past examinations:

- Assessment and Qualifications Alliance (AQA), including Northern Examinations and Assessment Board (NEAB/JMB) and Associated Examining Board (AEB)
- The Edexcel Foundation including University of London Examinations and Assessment Councils (L)
- Mathematics in Education and Industry (MEI)
- Oxford, Cambridge and RSA (OCR) including University of Cambridge Local Examinations Syndicate (C), Oxford & Cambridge Schools Examination Board (O & C) and Oxford Delegacy of Local Examinations (O)
- Welsh Joint Education Committee (WJEC)

All answers and worked solutions provided for examination questions are the responsibility of the authors.

We hope that you will enjoy using this text and that it will enhance your understanding of statistics and give you confidence to succeed.

J Crawshaw & J Chambers
2001

# 1

# Representation and summary of data

*In this chapter you will learn about*

- discrete and continuous data
- stem and leaf diagrams (stemplots)
- histograms, frequency polygons and the shape of a distribution
- pie charts
- means and weighted means
- standard deviation and variance
- cumulative frequency
- medians, quartiles and inter-percentile ranges
- skewness, including Pearson's coefficient and quartile coefficient
- the shape of the normal distribution
- box-and-whisker diagrams (boxplots) and outliers

## DISCRETE DATA

In a survey of 1m quadrats in a field the number of snails in each of 30 quadrats was recorded as follows:

| 1 | 2 | 4 | 0 | 2 | 3 | 1 | 4 | 2 | 3 | 5 | 2 | 2 | 3 | 2 |
| 2 | 3 | 1 | 2 | 3 | 2 | 0 | 1 | 1 | 2 | 0 | 3 | 2 | 3 | 3 |

This is an example of **discrete raw data**.

Discrete data can take only exact values, for example

the number of cars passing a checkpoint in 30 minutes,
the shoe sizes of children in a class,
the number of tomatoes on each plant in a greenhouse.

The data are known as raw because they have not been ordered in any way.

## Frequency distribution for discrete data

To illustrate the data more concisely, count the number of times each value occurs and summarise these in a table, known as a **frequency distribution**.

| Number of snails | 0 | 1 | 2 | 3 | 4 | 5 | |
|---|---|---|---|---|---|---|---|
| Frequency | 3 | 5 | 11 | 8 | 2 | 1 | Total 30 |

The frequency distribution can be represented diagrammatically by a vertical line graph or a bar chart. The height of the line or bar represents the frequency.

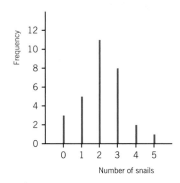

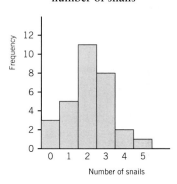

Notice that

- in the vertical line graph the distinct lines reinforce the discrete nature of the variable,
- in the bar chart the bars are all the same width and they are labelled in the middle of the bar on the horizontal axis.

## The mode

The mode is the value that occurs most often.

The mode is the most popular value, deriving from the French 'a la mode' meaning fashionable. It is easy to see from the diagrams above that the mode is 2 snails per quadrat.

## CONTINUOUS DATA

The following data were obtained in a survey of the heights of 20 children in a sports club. Each height was measured to the nearest centimetre.

| 133 | 136 | 120 | 138 | 133 | 131 | 127 | 141 | 127 | 143 |
|---|---|---|---|---|---|---|---|---|---|
| 130 | 131 | 125 | 144 | 128 | 134 | 135 | 137 | 133 | 129 |

This is an example of **continuous raw data**.

Continuous data cannot take exact values but can be given only within a specified range or measured to a specified degree of accuracy.

For example, the measurement 144 cm (given to the nearest cm) could have arisen from any value in the interval 143.5 cm $\leq h < 144.5$ cm.

Other examples of continuous data are

the speed of a vehicle as it passes a checkpoint,
the mass of a cooking apple,
the time taken by a volunteer to perform a task.

## Frequency distribution for continuous data

To form a frequency distribution of the heights of the 20 children, group the information into **classes** or **intervals**. Here are three different ways of writing the same set of intervals.

| Height (cm) |
| --- |
| $119.5 \leq h < 124.5$ |
| $124.5 \leq h < 129.5$ |
| $129.5 \leq h < 134.5$ |
| $134.5 \leq h < 139.5$ |
| $139.5 \leq h < 144.5$ |

| Height (cm) |
| --- |
| 119.5–124.5 |
| 124.5–129.5 |
| 129.5–134.5 |
| 134.5–139.5 |
| 139.5–144.5 |

| Height (to the nearest cm) |
| --- |
| 120–124 |
| 125–129 |
| 130–134 |
| 135–139 |
| 140–144 |

The values 119.5, 124.5, 129.5, ... are called the **class boundaries** or the **interval boundaries**. The upper class boundary (u.c.b.) of one interval is the lower class boundary (l.c.b.) of the next interval.

## Width of an interval

The width of an interval is the difference between the boundaries.

Width of an interval = upper class boundary – lower class boundary

Often intervals with equal widths are chosen, as in the above illustrations in which each width is 5 cm.

To group the heights it helps to use a tally column, entering the numbers in the first row 133, 136, 120, ... etc. and then the second row. It is a good idea to cross off each number in the list as you enter it. The frequency distribution for the above data should read:

| Height (cm) | Tally | Frequency |
| --- | --- | --- |
| $119.5 \leq h < 124.5$ | I | 1 |
| $124.5 \leq h < 129.5$ | ЖHt | 5 |
| $129.5 \leq h < 134.5$ | ЖHt II | 7 |
| $134.5 \leq h < 139.5$ | IIII | 4 |
| $139.5 \leq h < 144.5$ | III | 3 |
| | | Total 20 |

It is important to note that when the data are presented only in the form of a grouped frequency distribution, the original information has been lost. For example you would know that there was one item in the first interval, but you would not know what it was. You would know only that it was between 119.5 cm and 124.5 cm.

# STEM AND LEAF DIAGRAMS (STEMPLOTS)

A very useful way of grouping data into classes while still retaining the original data is to draw a **stem and leaf diagram**, also known as a **stemplot**.

These are the marks of 20 students in an assignment:

| 84 | 17 | 38 | 45 | 47 | 53 | 76 | 54 | 75 | 22 |
| 66 | 65 | 55 | 54 | 51 | 44 | 39 | 19 | 54 | 72 |

Notice that the lowest mark is 17 and the highest mark is 84.
In stem and leaf diagrams, all the intervals must be of equal width, so it seems sensible to choose intervals 10–19, 20–29, 30–39, ..., 80–89 for this data.

Take the **stem** to represent the tens and the **leaf** to represent the units.

The first five entries
84, 17, 38, 45 and 47
are represented like this:

| Stem (tens) | Leaf (units) |
|---|---|
| 1 | 7 |
| 2 | |
| 3 | 8 |
| 4 | 5 7 |
| 5 | |
| 6 | |
| 7 | |
| 8 | 4 |

When all the numbers have been entered the diagram looks like this:

| Stem | Leaf |
|---|---|
| 1 | 7 9 |
| 2 | 2 |
| 3 | 8 3 9 |
| 4 | 5 7 |
| 5 | 3 4 5 4 1 4 |
| 6 | 6 5 |
| 7 | 6 5 2 |
| 8 | 4 |

The entries in each leaf are now arranged in numerical order and a key is given to explain the stem and leaf. The final diagram looks like this:

**Stem and leaf diagram to show assignment marks**

| Stem | Leaf |
|---|---|
| 1 | 7 9 |
| 2 | 2 |
| 3 | 3 8 9 |
| 4 | 5 7 |
| 5 | 1 3 4 4 4 5 |
| 6 | 5 6 |
| 7 | 2 5 6 |
| 8 | 4 |

*Key* 1 | 7 means 17 marks

The stemplot gives a good idea at a glance of the shape of the distribution. It is easy to pick out the smallest and largest values and to see that the mode is 54. It is also obvious that the **modal class** is 50–59.

## Example 1.1

The maximum temperature in °C, measured to the nearest degree, was recorded each day during June in Sutton with the following results:

| 19 | 23 | 19 | 19 | 20 | 12 | 19 | 22 | 22 | 16 | 18 | 16 | 19 | 20 | 17 |
|----|----|----|----|----|----|----|----|----|----|----|----|----|----|----|
| 13 | 14 | 12 | 15 | 17 | 16 | 17 | 19 | 22 | 22 | 20 | 19 | 19 | 20 | 20 |

Draw a stem and leaf diagram to illustrate the temperatures and write down the modal temperature.

## Solution 1.1

The smallest value is 12 and the highest value is 23. Grouping the data into intervals 10–19, 20–29, ... would give you very little information.

Choose a sensible number of intervals; usually between 5 and 10. Since you must use intervals with equal width, you could use intervals of 2 °C and consider 12–13, 14–15, 16–17, 18–19, 20–21, 22–23.

First do a preliminary plot and then arrange the entries in each leaf in order.

Preliminary plot:

| Stem | Leaf |
|------|------|
| 1 | 2 3 2 |
| 1 | 4 5 |
| 1 | 6 6 7 7 6 7 |
| 1 | 9 9 9 9 8 9 9 9 9 |
| 2 | 0 0 0 0 0 |
| 2 | 3 2 2 2 2 |

The modal temperature is 19 °C.

Final diagram:

| Stem | Leaf |
|------|------|
| 1 | 2 2 3 |
| 1 | 4 5 |
| 1 | 6 6 6 7 7 7 |
| 1 | 8 9 9 9 9 9 9 9 9 |
| 2 | 0 0 0 0 0 |
| 2 | 2 2 2 2 3 |

*Key* 1|2 means 12 °C

Stemplot to show maximum temperatures

NOTE: The stem does not necessarily represent the tens digit. For example, suppose you want to use intervals 12–14, 15–17, 18–20, 21–23. The interval 18–20 cannot be represented by a stem of 1, since the tens digit changes during the interval. For the stem you can use 12, 15, 18 and 21. The leaf is then given as the number that is added to the stem.

| Stem | Leaf |
|------|------|
| 12 | 0 0 1 2 |
| 15 | 0 1 1 1 2 2 2 |
| 18 | 0 1 1 1 1 1 1 1 2 2 2 2 2 |
| 21 | 1 1 1 1 2 |

*Key* 15|2 means 17 °C
18|0 means 18 °C

NOTE: The key is essential in explaining how the stemplot has been formed.

In a stem and leaf diagram, or stemplot

(a) equal intervals must be chosen,
(b) a key is essential.

## Example 1.2

The table gives the number of days on which rain fell in 36 consecutive intervals of 30 days.

| 21 | 19 | 6 | 12 | 8 | 18 | 9 | 8 | 11 | 17 | 15 | 13 |
|----|----|----|----|----|----|----|----|----|----|----|----|
| 16 | 9 | 17 | 18 | 9 | 24 | 17 | 7 | 8 | 17 | 17 | 8 |
| 7 | 11 | 16 | 17 | 8 | 5 | 13 | 22 | 20 | 16 | 20 | 13 |

Draw stem and leaf diagrams with the following class intervals:

(a) 5–9, 10–14, 15–19, 20–24
(b) 4–6, 7–9, 10–12, 13–15, 16–18, 19–21, 22–24.

## Solution 1.2

(a) Using intervals 5–9, 10–14, 15–19, 20–24 the completed stem and leaf diagram is:

| Stem | Leaf |
|------|------|
| 0 | 5 6 7 7 8 8 8 8 8 9 9 9 |
| 1 | 1 1 2 3 3 3 |
| 1 | 5 6 6 6 7 7 7 7 7 7 8 8 9 |
| 2 | 0 0 1 2 4 |

Key 1 | 6 means 16

NOTE: The stem and leaf diagram could have been written differently, as follows:

| Stem | Leaf |
|------|------|
| 5 | 0 1 2 2 3 3 3 3 3 4 4 4 |
| 10 | 1 1 2 3 3 3 |
| 15 | 0 1 1 1 2 2 2 2 2 2 3 3 4 |
| 20 | 0 0 1 2 4 |

Key 15 | 1 means 16
5 | 3 means 8

(b) Using intervals 4–6, 7–9, 10–12, ... the completed diagram, arranged in order is:

| Stem | Leaf |
|------|------|
| 4 | 1 2 |
| 7 | 0 0 1 1 1 1 1 2 2 2 |
| 10 | 1 1 2 |
| 13 | 0 0 0 2 |
| 16 | 0 0 0 1 1 1 1 1 1 2 2 |
| 19 | 0 1 1 2 |
| 22 | 0 2 |

Key 13 | 2 means 15

Both diagrams show that the mode is 17 rainy days, but the seven intervals used in (b) show more clearly the two peaks, illustrating that the distribution is approximately **bi-modal**, with modal classes 7–9 and 16–18.

## Example 1.3

Look at this stem and leaf diagram and for each of the three keys provided, give

(a) the value ringed,
(b) the width of the interval containing the ringed value.

| Stem | Leaf |
|------|------|
| 0 | 7 |
| 0 | 9 |
| 1 | 0 1 |
| 1 | 2 2 |
| 1 | 4 4 4 5 5 |
| 1 | 6 6 7 7 7 |
| 1 | 8 8 8 8 9 9 ⑨ |
| 2 | 0 0 1 1 |
| 2 | 2 3 |
| 2 | 4 |

(i)   the widths of 30 metal components

*Key* 1 | 2 means 1.2 cm

(ii)  the reaction times of 30 volunteers

*Key* 1 | 2 means 12 hundredths of a second

(iii) the attendance at 30 matches

*Key* 1 | 2 means 1200 people

## Solution 1.3

(i)   (a)  1 | 9 means 1.9 cm.
       (b)  The interval is 1.8 cm–1.9 cm. Since width is a continuous variable, and assuming that widths have been measured to the nearest tenth of a centimetre, then 1.75 cm ⩽ width < 1.95 cm and the class width is 2 mm.

(ii)  (a)  1 | 9 means 19 hundredths of a second, i.e. 0.19 seconds.
       (b)  The interval is 0.18 sec–0.19 sec, i.e. 0.175 ⩽ time < 0.195, so the class width is 0.02 seconds.

(iii) (a)  1 | 9 means 1900 people.
       (b)  The interval is 1800 people–1900 people. Assuming that the number has been given to the nearest hundred, then 1750 ⩽ number < 1950, so the class width is 200 people.

## Back-to-back stemplots

Stem and leaf diagrams can be used to compare two samples by showing the results together on a **back-to-back stemplot**.

## Example 1.4

Use a stem and leaf diagram to compare the examination marks in French and English for a class of 20 pupils.

| French | 75 | 69 | 58 | 58 | 46 | 44 | 32 | 50 | 53 | 78 |
|--------|----|----|----|----|----|----|----|----|----|----|
|        | 81 | 61 | 61 | 45 | 31 | 44 | 53 | 66 | 47 | 57 |
| English | 52 | 58 | 68 | 77 | 38 | 85 | 43 | 44 | 56 | 65 |
|        | 65 | 79 | 44 | 71 | 84 | 72 | 63 | 69 | 72 | 79 |

**Solution 1.4**

The first four entries for French (75, 69, 58, 58) and for English (52, 58, 68, 77) are entered into a back-to-back stemplot as follows:

| Key (French) 9\|6 means 69 | French | | English | Key (English) 5\|2 means 52 |
|---|---|---|---|---|
| | | 3 | | |
| | | 4 | | |
| | 8 8 | 5 | 2 8 | |
| | 9 | 6 | 8 | |
| | 5 | 7 | 7 | |
| | | 8 | | |

The completed diagram, before rearranging, is:

| French | | English |
|---|---|---|
| 1 2 | 3 | 8 |
| 7 4 5 4 6 | 4 | 3 4 4 |
| 7 3 3 0 8 8 | 5 | 2 8 6 |
| 6 1 1 9 | 6 | 8 5 5 3 9 |
| 8 5 | 7 | 7 9 1 2 2 9 |
| 1 | 8 | 5 4 |

The final diagram, arranged in order:

| Key (French) 8\|5 means 58 | French | | English | Key (English) 6\|3 means 63 |
|---|---|---|---|---|
| | 2 1 | 3 | 8 | |
| | 7 6 5 4 4 | 4 | 3 4 4 | |
| | 8 8 7 3 3 0 | 5 | 2 6 8 | |
| | 9 6 1 1 | 6 | 3 5 5 8 9 | |
| | 8 5 | 7 | 1 2 2 7 9 9 | |
| | 1 | 8 | 4 5 | |

From the diagram it is clear that the class had higher marks in English than in French and it appears that they performed better in English. This would, however, depend on the standards of marking used in the two examinations.

---

## Exercise 1a   Stemplots

1. (a) Draw a stemplot to show the masses, correct to the nearest kilogram, of 30 men.
   Use intervals 50–54, 55–59, 60–64, ...
   (b) Write down the modal mass.

   | 74 | 52 | 67 | 68 | 71 | 76 | 86 | 81 | 73 |
   | 68 | 64 | 75 | 71 | 61 | 63 | 57 | 67 | 57 |
   | 59 | 72 | 79 | 64 | 70 | 74 | 77 | 79 | 65 |
   | 68 | 76 | 83 |

2. A teacher recorded the times taken by 20 boys to swim one length of the pool.

   The times are given to the nearest second.

   Using intervals 24–25, 26–27, ..., draw a stem and leaf diagram to illustrate the results.

   | 32 | 31 | 26 | 27 | 27 | 32 | 29 | 26 | 25 | 25 |
   | 29 | 31 | 32 | 26 | 30 | 24 | 32 | 27 | 26 | 31 |

3. A group of adults took part in an experiment which measured their reaction times. The results were given to the nearest hundredth of a second.

   | 0.14 | 0.17 | 0.21 | 0.20 | 0.20 | 0.22 |
   | 0.14 | 0.24 | 0.26 | 0.17 | 0.14 | 0.17 |
   | 0.21 | 0.20 | 0.22 | 0.14 | 0.24 | 0.26 |
   | 0.17 | 0.18 | 0.17 | 0.21 | 0.20 | 0.23 |
   | 0.17 | 0.23 | 0.21 | 0.23 | 0.24 | 0.23 |

   Use intervals 0.14–0.15, 0.16–0.17, 0.18–0.19, ... to draw a stemplot to illustrate the results. Comment on your diagram.

4. In a lesson on measurement, 30 pupils estimated the length of a line in centimetres and wrote down their value correct to the nearest mm. Using intervals 3.0–3.9, 4.0–4.9, ..., draw a stemplot.

9.2  7.3  7.0  6.5  5.4  5.3  10.1  8.4
8.8  7.1  7.6  7.9  6.7  9.6  5.5  7.4
7.0  8.2  5.5  7.8  8.2  7.5  6.1  6.1
3.9  6.8  7.6  8.1  8.0  10.0

5.  The daily hours of sunshine in London during August were

7.0  7.6  12.5  12.9  8.3  9.7  8.4  11.1
7.5  7.5  9.8  10.4  11.6  11.3  7.3  7.8
6.8  6.2  6.1  5.6  5.6  5.8  4.8  4.3
0.0  0.6  0.8  1.6  0.2  2.4  2.6

Illustrate these data on a stem and leaf diagram and comment.

6.  A stemplot is given below but it does not have a key.

| Stem | Leaf |
|------|------|
| 5 | 9 |
| 6 | 1 4 |
| 6 | 7 8 9 |
| 7 | 2 3 3 ④ |
| 7 | 5 6 6 6 7 8 |
| 8 | 0 3 4 |
| 8 | 5 |

State the value ringed and the width of the interval that it is in when the diagram illustrates
(a)  the times taken for a journey, where 6 | 8 represents 6.8 hours,
(b)  the masses, in g to three decimal places, of components, where 6 | 8 represents 0.068 g.

7.  Draw back-to-back stemplots for the following data. What conclusions can you draw?

(a)  The pulse rates of 30 company directors were measured before and after taking exercise.
*Before*: 110, 93, 81, 75, 73, 73, 48, 53, 69, 69, 66, 111, 105, 93, 90, 50, 57, 64, 90, 111, 91, 70, 70, 51, 79, 93, 105, 51, 66, 93.
*After*: 117, 81, 77, 108, 130, 69, 77, 84, 84, 86, 95, 125, 96, 104, 104, 137, 143, 70, 80, 131, 145, 106, 130, 109, 137, 75, 104, 75, 97, 80.
(Use class intervals 40–49, 50–59, 60–69, …)

(b)  The ages of teachers in two schools:
*School A*: 51, 45, 33, 37, 37, 27, 28, 54, 54, 61, 34, 31, 39, 23, 53, 59, 40, 46, 48, 48, 39, 33, 25, 31, 48, 40, 53, 51, 46, 45, 45, 48, 39, 29, 23, 37.
*School B*: 59, 56, 40, 43, 46, 38, 29, 52, 54, 34, 23, 41, 42, 52, 50, 58, 60, 45, 45, 56, 59, 49, 44, 36, 38, 25, 56, 36, 42, 47, 50, 54, 59, 47, 58, 57.
(Use class intervals 20–29, 30–39, 40–49, …)

(c)  20 boys and 20 girls took part in a reaction-timing experiment. Their results were measured to the nearest hundredth of a second.
*Girls*: 0.22, 0.21, 0.18, 0.18, 0.16, 0.19, 0.25, 0.22, 0.17, 0.19, 0.16, 0.21, 0.24, 0.22, 0.19, 0.22, 0.25, 0.22, 0.17, 0.22.
*Boys*: 0.14, 0.20, 0.22, 0.16, 0.19, 0.16, 0.15, 0.23, 0.23, 0.19, 0.16, 0.15, 0.09, 0.23, 0.11, 0.21, 0.22, 0.18, 0.18, 0.16.
(Use class intervals 0.08–0.09, 0.10–0.11, 0.12–0.13, …)

# WAYS OF GROUPING DATA

The following frequency distributions show some of the ways that data can be grouped. The information is more concise than the raw data, but the disadvantage is that the original information has been lost.

(i)  **Frequency distribution to show the lengths, to the nearest millimetre, of 30 rods**

| Length (mm) | 27–31 | 32–36 | 37–46 | 47–51 |
|-------------|-------|-------|-------|-------|
| Frequency | 4 | 11 | 12 | 3 |

The interval 27–31 means 26.5 mm ⩽ length < 31.5 mm.

The class boundaries are  26.5,  31.5,  36.5,  46.5,  51.5

The class widths are       5,    5,    10,   5

(ii)  **Frequency distribution to show the marks in a test of 100 students**

| Mark | 30–39 | 40–49 | 50–59 | 60–69 | 70–79 | 80–99 |
|---|---|---|---|---|---|---|
| Frequency | 10 | 14 | 26 | 20 | 18 | 12 |

This distribution can be interpreted in two ways:

(a)  As discrete data, the interval 30–39 represents $30 \leqslant mark < 40$.
The class boundaries are    30,   40,   50,   60,   70,   80,   100
The class widths are           10,   10,   10,   10,   10,   20

(b)  As continuous data, assuming marks are to the nearest integer, 30–39 would represent $29.5 \leqslant mark < 39.5$.
The class boundaries are    29.5,   39.5,   49.5,   59.5,   69.5,   79.5,   99.5
The class widths are              10,      10,      10,      10,      10,      20

(iii)  **Frequency distribution to show the lengths of 50 telephone calls**

| Length of call (min) | 0– | 3– | 6– | 9– | 12– | 18– |
|---|---|---|---|---|---|---|
| Frequency | 9 | 12 | 15 | 10 | 4 | 0 |

The interval '3–' means $3\ minutes \leqslant time < 6\ minutes$, so any time including 3 minutes and up to (but not including) 6 minutes comes into this interval.

The class boundaries are    0,   3,   6,   9,   12,   18

The class widths are           3,   3,   3,   3,   6

(iv)  **Frequency distribution to show the masses of 40 packages brought to a particular counter at a post office**

| Mass (g) | –100 | –250 | –500 | –800 |
|---|---|---|---|---|
| Frequency | 8 | 10 | 16 | 6 |

The interval '–250' means $100\ g < mass \leqslant 250\ g$, so any mass over 100 grams up to and including 250 grams comes into this interval.

The class boundaries are    0,   100,   250,   500,   800

The class widths are           100,   150,   250,   300

(v)  **Frequency distribution to show the speeds of 50 cars passing a checkpoint**

| Speed (km/h) | 20–30 | 30–40 | 40–60 | 60–80 | 80–100 |
|---|---|---|---|---|---|
| Frequency | 2 | 7 | 20 | 16 | 5 |

The interval 30–40 means $30\ km/h \leqslant speed < 40\ km/h$.

The class boundaries are    20,   30,   40,   60,   80,   100

The class widths are           10,   10,   20,   20,   20

(vi) **Frequency distribution to show ages (in completed years) of applicants for a teaching post**

| Age (years) | 21–24 | 25–28 | 29–32 | 33–40 | 41–52 |
|---|---|---|---|---|---|
| Frequency | 4 | 2 | 2 | 1 | 1 |

Since the ages are given in completed years (not to the nearest year) then '21–24' means $21 \leqslant \text{age} < 25$. Someone who is 24 years and 11 months would come into this category. Sometimes this interval is written '21–' and the next is '25–', etc.

The class boundaries are    21,   25,   29,   33,   41,   53

The class widths are         4,    4,    4,    8,    12

## HISTOGRAMS

Grouped data can be displayed in a **histogram** as in the following diagram.

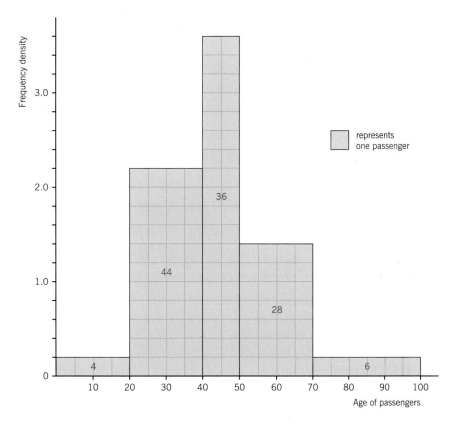

This histogram represents the following table for the distribution of ages of passengers on a shuttle flight from Denver, Colorado to Salt Lake City, Utah.

| Age, $x$ years | $0 \leqslant x < 20$ | $20 \leqslant x < 40$ | $40 \leqslant x < 50$ | $50 \leqslant x < 70$ | $70 \leqslant x < 100$ |
|---|---|---|---|---|---|
| Frequency | 4 | 44 | 36 | 28 | 6 |

Histograms resemble bar charts, but there are two important differences.

In a histogram

- there are no gaps between the bars,
- the area of each bar is proportional to the frequency that it represents. This means that

total area $\propto$ total frequency.

Histograms often have bars of varying widths, so the height of the bar must be adjusted in accordance with the width of the bar.

The vertical axis is not labelled frequency but **frequency density** where

$$\text{frequency density} = \frac{\text{frequency}}{\text{interval width}}$$

Consider the interval $20 \leqslant x < 40$ in the frequency table above.

Frequency = 44, interval width = 20, so frequency density = $\frac{44}{20} = 2.2$

The complete table looks like this:

| Ages | Interval width | Frequency | Frequency density |
|---|---|---|---|
| $0 \leqslant x < 20$ | 20 | 4 | 0.2 |
| $20 \leqslant x < 40$ | 20 | 44 | 2.2 |
| $40 \leqslant x < 50$ | 10 | 36 | 3.6 |
| $50 \leqslant x < 70$ | 20 | 28 | 1.4 |
| $70 \leqslant x < 100$ | 30 | 6 | 0.2 |

## Modal class

The highest bar in the histogram represents the interval $40 \leqslant x < 50$. This is the **modal class**. Notice that in the table this interval does not have the greatest frequency, but it does have the greatest frequency density.

In a grouped frequency distribution, the modal class is the interval with the greatest frequency density, i.e. the interval represented by the highest bar in the histogram.

### Example 1.5

The grouped frequency distribution records the masses, to the nearest gram, of 84 letters delivered by the postman.

| Mass (g) | 1–20 | 21–40 | 41–60 | 61–80 | 81–100 |
|---|---|---|---|---|---|
| Number of letters | 10 | 18 | 24 | 14 | 18 |

Draw a histogram to illustrate these data.

## Solution 1.5

The data are continuous.
The class boundaries are 0.5,   20.5,   40.5,   60.5,   80.5,   100.5
The interval widths are        20,    20,    20,    20,    20

In this example all the intervals are of equal width and you could use the frequency for the height of the bar. It is, however, a good idea to use the frequency density for the height of the bar. The resulting histogram will then have a **total area** which represents the **total frequency**.

| Mass (g) | Interval width | Frequency | Frequency density |
|---|---|---|---|
| $0.5 \leqslant x < 20.5$ | 20 | 10 | 0.5 |
| $20.5 \leqslant x < 40.5$ | 20 | 18 | 0.9 |
| $40.5 \leqslant x < 60.5$ | 20 | 24 | 1.2 |
| $60.5 \leqslant x < 80.5$ | 20 | 14 | 0.7 |
| $80.5 \leqslant x < 100.5$ | 20 | 18 | 0.9 |

**Histogram to show the masses of letters**

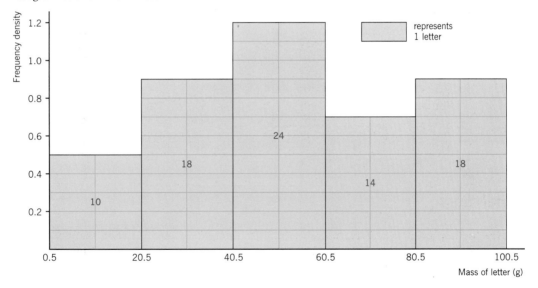

The main purpose of histograms is to illustrate grouped continuous data, but they can also be used to illustrate grouped discrete data.

## Example 1.6

These are the examination marks for a group of 120 first year statistics students.

| Mark | 0–9 | 10–19 | 20–29 | 30–49 | 50–79 |
|---|---|---|---|---|---|
| Frequency | 8 | 21 | 53 | 28 | 10 |

Represent the data in a histogram and comment on the shape of the distribution.

## Solution 1.6

The data are discrete, so, to avoid gaps in the histogram, use class boundaries 9.5, 19.5, 29.5, 49.5. This leads to −0.5 and 79.5 as the remaining two boundaries, even though these marks are outside the range of the discrete data.

The class boundaries are   −0.5,  9.5,  19.5,  29.5,  49.5,  79.5
The interval widths are           10,  10,    10,    20,    30

| Mark | Class Width | Frequency | Frequency Density |
|------|-------------|-----------|-------------------|
| 0–9 | 10 | 8 | 0.8 |
| 10–19 | 10 | 21 | 2.1 |
| 20–29 | 10 | 53 | 5.3 |
| 30–49 | 20 | 28 | 1.4 |
| 50–79 | 30 | 10 | 0.3 |

**Histogram to show examination marks**

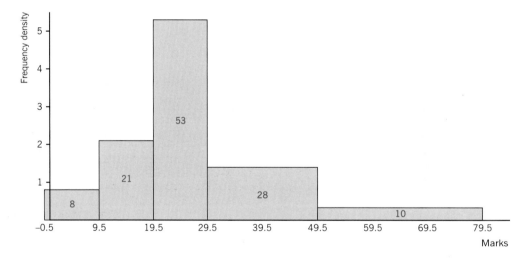

The distribution has a long tail of values to the right. It is said to be positively skewed.

*HINT*: when drawing the histogram you will find it easier to mark out the horizontal axis −0.5, 9.5, 19.5, ... using the lines of your squared paper. Then draw in the vertical frequency density axis in a suitable position. Anywhere will do for this; it does not have to go through (0, 0), but could be to the left of −0.5, for example

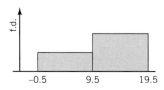

# Finding the frequencies from a histogram

To find the frequency in each interval, use

frequency = interval width × frequency density

## Example 1.7

A Passengers' Association conducted a survey on the punctuality of trains using a particular station. The histogram illustrates the results.

(a) Construct the frequency distribution.
(b) How many trains were there in the survey?

**Histogram to show lateness of trains**

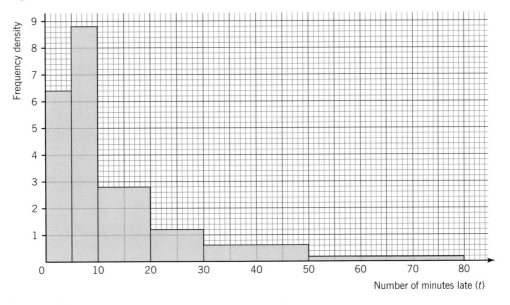

## Solution 1.7

(a) To find the frequency in each interval, use frequency = interval width × frequency density

| Number of ($t$) minutes late | $0 \leqslant t < 5$ | $5 \leqslant t < 10$ | $10 \leqslant t < 20$ | $20 \leqslant t < 30$ | $30 \leqslant t < 50$ | $50 \leqslant t < 80$ |
|---|---|---|---|---|---|---|
| Frequency | $5 \times 6.4$ <br> $= 32$ | $5 \times 8.8$ <br> $= 44$ | $10 \times 2.8$ <br> $= 28$ | $10 \times 1.2$ <br> $= 12$ | $20 \times 0.6$ <br> $= 12$ | $30 \times 0.2$ <br> $= 6$ |

(b) Number of trains = 32 + 44 + 28 + 12 + 12 + 6 = **134**

## Example 1.8

The number of letters delivered to the houses in Distribution Street is illustrated in the histogram. Given that 13 houses received three or four letters, how many houses are there in the street? Explain the scale on the vertical axis.

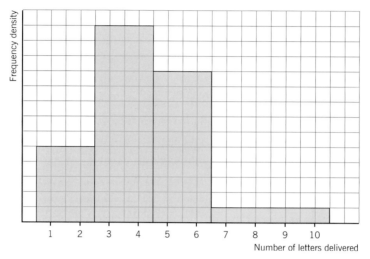

## Solution 1.8

The scale on the frequency density axis has not been marked but since you are given that there are 13 houses in the interval 3–4 it is easy to see the area of four small squares represents one house.

represents 1 house

The frequencies can be deduced directly from this, for example, the interval 7–10 contains two houses.

Total frequency = 5 + 13 + 10 + 2 = 30
**There are 30 houses in the street.**

To work out the scale on the frequency density axis, note that the interval 3–4 has frequency 13 and is of width 2, therefore frequency density = $13 \div 2 = 6.5$.

Since the bar is 13 squares high, each square on the vertical axis represents a frequency density of 0.5.

---

Although it is easier to use frequency density for the vertical scale in the histogram, other scales can be used, provided that area is proportional to frequency. This is illustrated in the following example.

## Example 1.9

A teacher recorded the time, to the nearest minute, spent reading during a particular day by each child in a group. The times were summarised in a grouped frequency distribution and represented by a histogram. The first class in the grouped frequency distribution was 10–19 and its associated frequency was eight children. On the histogram the height of the rectangle representing the class was 2.4 cm and the width was 2 cm. The total area under the histogram was $53.4 \text{ cm}^2$.

Find the number of children in the group.

(L)

## Solution 1.9

Rectangle representing 10–19 interval:

$$\text{Area of rectangle} = 2 \times 2.4$$
$$= 4.8 \text{ cm}^2$$

8 children, 2.4 cm, 2 cm

$$\text{Area} \propto \text{frequency}$$
$$\therefore \quad \text{Area} = k \times \text{frequency}$$
$$4.8 = k \times 8$$
$$k = 0.6$$
$$\text{Total area} = k \times \text{total frequency}$$
$$53.4 = 0.6 \times \text{total frequency}$$
$$\therefore \quad \text{Total frequency} = \frac{53.4}{0.6} = 89$$

There were 89 children in the group.

# FREQUENCY POLYGONS

A grouped frequency distribution can be displayed as a frequency polygon.

To construct a frequency polygon, for each interval plot frequency density against the mid-interval value, where

mid-interval value = $\frac{1}{2}$ (lower class boundary + upper class boundary)

Then join the points with straight lines.

## Example 1.10

Draw a frequency polygon to illustrate this frequency distribution which gives the times taken by 31 competitors to complete a cross-country run.

| Time $t$ (min) | $25 \leqslant t < 30$ | $30 \leqslant t < 35$ | $35 \leqslant t < 40$ | $40 \leqslant t < 50$ | $50 \leqslant t < 65$ |
|---|---|---|---|---|---|
| Frequency | 4 | 12 | 8 | 4 | 3 |

## Solution 1.10

| Time | Mid-interval value | Interval width | Frequency | Frequency density |
|---|---|---|---|---|
| $25 \leqslant t < 30$ | 27.5 | 5 | 4 | $\frac{4}{5} = 0.8$ |
| $30 \leqslant t < 35$ | 32.5 | 5 | 12 | $\frac{12}{5} = 2.4$ |
| $35 \leqslant t < 40$ | 37.5 | 5 | 8 | $\frac{8}{5} = 1.6$ |
| $40 \leqslant t < 50$ | 45 | 10 | 4 | $\frac{4}{10} = 0.4$ |
| $50 \leqslant t < 65$ | 57.5 | 15 | 3 | $\frac{3}{15} = 0.2$ |

Frequency polygon to show times taken to complete a cross-country run

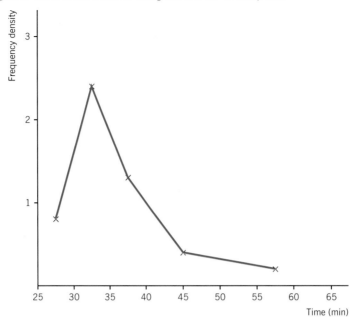

Note that this distribution is skewed with a tail at the right hand end, i.e. it is positively skewed.

You could of course construct the histogram first and then join the mid-points of the tops of the rectangles to give the frequency polygon.

# Comparative frequency polygons

Frequency polygons are very useful when comparing sets of data.

## Example 1.11

Draw frequency polygons to compare the age distribution of the teachers in two sixth form colleges:

| Age | 20– | 25– | 30– | 35– | 40– | 45– | 50– | 55– | 60– | 65– |
|---|---|---|---|---|---|---|---|---|---|---|
| College A | 4 | 6 | 11 | 14 | 9 | 5 | 5 | 3 | 0 | 0 |
| College B | 0 | 2 | 4 | 7 | 11 | 12 | 11 | 8 | 5 | 0 |

## Solution 1.11

Work out the mid-interval value for each interval, for example in the interval '20–' the lower boundary is 20 and the upper boundary is 25, so mid-interval value = $\frac{1}{2}(20 + 25) = 22.5$

The width of each interval is 5, so work out the frequency densities for each college by dividing the frequencies by 5.

| Mid-interval value | Frequency density College A | Frequency density College B |
|---|---|---|
| 22.5 | 0.8 | 0 |
| 27.5 | 1.2 | 0.4 |
| 32.5 | 2.2 | 0.8 |
| 37.5 | 2.8 | 1.4 |
| 42.5 | 1.8 | 2.2 |
| 47.5 | 1 | 2.4 |
| 52.5 | 1 | 2.2 |
| 57.5 | 0.6 | 1.6 |
| 62.5 | 0 | 1 |
| 67.5 | 0 | 0 |

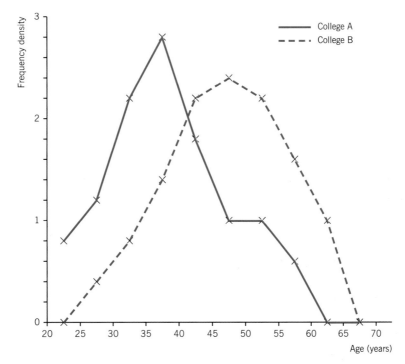

The bulk of the distribution for College A is further to the left than College B. This indicates that College A has a much younger staff than College B.

Notice that in this example, since all the intervals are of equal width, frequency could have been used on the vertical axis.

# FREQUENCY CURVES

When the number of intervals is large the frequency polygon consists of a large number of line segments. The frequency polygon approaches a smooth curve, known as a frequency curve.

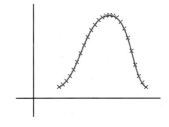

# The shape of a distribution

If distributions represented by a vertical line graph or a histogram are illustrated using a frequency curve, it is easier to see the general 'shape' of the distribution. For example:

A positively skewed distribution could occur when considering, for example,

- the number of children in a family,
- the age at which women marry,
- the distribution of wages in a firm.

(a) **Positive skew**

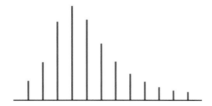   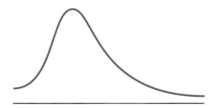

In a **positively skewed** distribution, there is a long tail at the *positive* end of the distribution.

(b) **Negative skew**

A negatively skewed distribution could occur when considering, for example,

- reaction times for an experiment,
- daily maximum temperatures for a month in the summer.

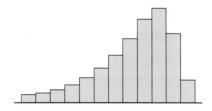   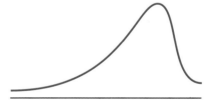

In a **negatively skewed** distribution, there is a long tail at the *negative* end of the distribution.

(c) **Reverse J-shape**

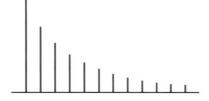   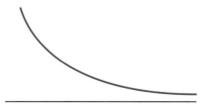

In a **J-shaped (reverse)** distribution an initial 'bulge' is followed by a long tail.

(d) **Uniform or rectangular**

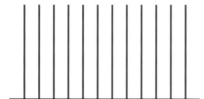 

In a **uniform** or **rectangular** distribution the data are evenly spread throughout the range.

(e) **The normal distribution**

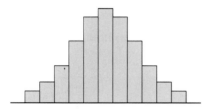 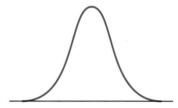

This symmetrical, bell-shaped distribution is known as a **normal** distribution.

An approximately normal distribution occurs when measuring quantities such as heights, masses, examination marks.

## Exercise 1b   Histograms and frequency polygons

1. A researcher timed how long it took for each of 38 volunteers to perform a simple task. The results are shown in the table.

| Time (seconds) | 5– | 10– | 20– | 25– | 40– | 45– |
|---|---|---|---|---|---|---|
| Frequency | 2 | 12 | 7 | 15 | 2 | 0 |

Draw a histogram to illustrate the data.

2. In a survey the masses of 50 apples were noted and recorded in the following table. Each value was given to the nearest gram.

```
86   101   114   118    87    92    93   116
105   102    97    93   101   111    96   117
100   106   118   101   107    96   101   102
104    92    99   107    98   105   113   100
103   108    92   109    95   100   103   110
113    99   106   116   101   105    86    88
108    92
```

(a) Construct a frequency distribution, using equal class intervals of width 5 g and taking the first interval as 85–89.
(b) Draw a histogram to illustrate the data and write down the modal class.
(c) Draw a stemplot to illustrate the data and write down the mode.

3. On a particular day the length of stay of each car at a city car park was recorded:

| Length of stay (min) | Frequency |
|---|---|
| $t < 25$ | 62 |
| $25 \leqslant t < 60$ | 70 |
| $60 \leqslant t < 80$ | 88 |
| $80 \leqslant t < 150$ | 280 |
| $150 \leqslant t < 300$ | 30 |

Represent the data by a histogram and state the modal class.

4. Draw a histogram to show the masses, measured to the nearest kilogram, of 200 girls.

| Mass (kg) | 41–50 | 51–55 | 56–60 | 61–70 | 71–75 |
|---|---|---|---|---|---|
| Frequency | 21 | 62 | 55 | 50 | 12 |

5. This histogram represents the speeds of cars passing a 30 miles per hour sign. Write out the frequency distribution.

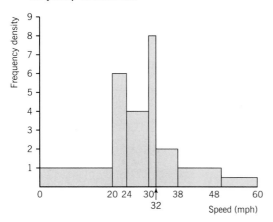

6. In a competition to grow the tallest hollyhock, the heights recorded by 50 primary school children were as follows. Heights were measured to the nearest centimetre.

| Height (cm) | Frequency |
|---|---|
| 177–186 | 12 |
| 187–191 | 8 |
| 192–196 | 8 |
| 197–201 | 9 |
| 202–206 | 7 |
| 207–216 | 6 |

Draw a histogram and superimpose a frequency polygon.

7. The table shows the duration, in minutes, of 64 telephone calls made from a High Street call box in a day.

| Length of call (min) | Frequency |
|---|---|
| 0– | 3 |
| $1\frac{1}{2}$– | 7 |
| 3– | 22 |
| 6– | 20 |
| 12– | 6 |
| 15– | 6 |
| 21– | 0 |

Draw a frequency polygon to illustrate the data.

8. These are the number of times the letter 'e' appears in each sentence in an article called 'My Kind of Day'. Make a grouped frequency distribution and draw a histogram.

15  12  8  12   3  10  14  17  5   3  8  11
 7  16  5  13  12  11   6   7  4  17  8   1

9. The table shows the ages, in completed years, of women who gave birth to a child at Anytown Maternity Hospital during a particular year. Without drawing a histogram first, draw a frequency polygon to illustrate the information. Describe the distribution.

| Age (years) | Number of births |
|---|---|
| 16– | 70 |
| 20– | 470 |
| 25– | 535 |
| 30– | 280 |
| 35– | 118 |
| 45– | 0 |

10. The patients at a chest clinic were asked to keep a record of the number of cigarettes they smoked each day.

| Number of cigarettes smoked per day | Frequency |
|---|---|
| 0–9 | 5 |
| 10–14 | 8 |
| 15–19 | 32 |
| 20–29 | 41 |
| 30–39 | 16 |
| 40 and over | 2 |

Draw a histogram to represent this data.

11. The marks awarded to 136 students in an examination are summarised in the table. Draw a histogram to illustrate the data.

| Marks | Frequency |
|---|---|
| 10–29 | 22 |
| 30–39 | 18 |
| 40–49 | 22 |
| 50–59 | 24 |
| 60–64 | 14 |
| 65–69 | 12 |
| 70–84 | 24 |

12.

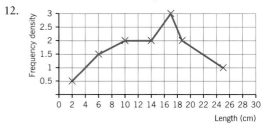

Complete the frequency distribution represented by the frequency polygon above.

| Length (cm) | Frequency |
|---|---|
| $0 \leqslant x < 4$ | 2 |
| $4 \leqslant x < 8$ | |
| $8 \leqslant x < 12$ | |
| $12 \leqslant x < 16$ | |
| $16 \leqslant x < 18$ | |
| $18 \leqslant x < 20$ | |
| $20 \leqslant x < 30$ | |

13. Lucy and Jack play a computer game every day and keep a record of their scores. Lucy's scores are shown in the table. Draw a frequency polygon to represent her scores.

| Lucy's scores | 50–99 | 100–149 | 150–199 | 200–249 | 250–299 |
|---|---|---|---|---|---|
| Frequency | 6 | 14 | 10 | 6 | 4 |

Jack's scores are as follows:

| Jack's scores | 50–99 | 100–149 | 150–199 | 200–249 | 250–299 |
|---|---|---|---|---|---|
| Frequency | 2 | 6 | 10 | 16 | 6 |

Draw a frequency polygon for Jack's scores on the same set of axes as Lucy's and use it to compare the two sets of scores.

14. Students were investigating the effects of a growth hormone placed on the growing tip of a maize seedling. The hormone was used in two different concentrations and distilled water was used as a control on a third set of seedlings. After three weeks the heights of the plants were measured to the nearest centimetre. They are shown in the table. Draw frequency polygons to represent the data and compare the results.

**Control**

| Height (cm) | Frequency |
|---|---|
| 45 | 0 |
| 46 | 7 |
| 47 | 11 |
| 48 | 12 |
| 49 | 14 |
| 50 | 14 |
| 51 | 18 |
| 52 | 12 |
| 53 | 8 |
| 54 | 3 |
| 55 | 1 |
| 56 | 0 |

**20% solution**

| Height (cm) | Frequency |
|---|---|
| 50 | 0 |
| 51 | 1 |
| 52 | 0 |
| 53 | 2 |
| 54 | 5 |
| 55 | 9 |
| 56 | 17 |
| 57 | 25 |
| 58 | 20 |
| 59 | 12 |
| 60 | 9 |
| 61 | 0 |

**40% solution**

| Height (cm) | Frequency |
|---|---|
| 54 | 0 |
| 55 | 2 |
| 56 | 2 |
| 57 | 2 |
| 58 | 7 |
| 59 | 10 |
| 60 | 11 |
| 61 | 18 |
| 62 | 18 |
| 63 | 16 |
| 64 | 9 |
| 65 | 5 |
| 66 | 0 |

15. In one month, a student recorded the length, to the nearest minute, of each of the lectures she attended. The table below shows her data and the calculations she made before drawing a histogram to illustrate these data.

| Length of lecture (minutes) | 50–53 | 54–55 | 56–59 | 60–67 |
|---|---|---|---|---|
| Number of lectures | $a$ | $b$ | 30 | $c$ |
| Frequency density | 5 | 13 | 7.5 | 1.5 |

Calculate
(a) the value of $a$, of $b$ and of $c$,
(b) the total number of lectures attended during the month. (C Additional)

# CIRCULAR DIAGRAMS OR PIE CHARTS

Pie charts are so called because they look like an apple pie! The areas of the slices or sectors of the pie are in proportion to the quantities being represented.

## Example 1.12

The pie chart, which is not drawn to scale, shows the distribution of various types of land and water in a certain county. Calculate

(a) the area of woodland,
(b) the angle of the urban sector,
(c) the total area of the county.                    (C)

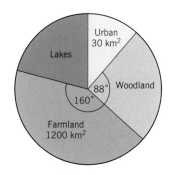

## Solution 1.12

(a) $160°$ represents $1200 \text{ km}^2$, $\therefore$ $88°$ represents $\frac{1200}{160} \times 88 = 660 \text{ km}^2$
**Area of woodland = 660 km$^2$**

(b) $1200 \text{ km}^2$ is represented by $160°$, $30 \text{ km}^2$ is represented by $\frac{160}{1200} \times 30 = 4°$
**Angle for the urban sector = 4°**

(c) $160°$ represents $1200 \text{ km}^2$, $360°$ represents $\frac{1200}{160} \times 360 = 2700 \text{ km}^2$
**Total area of county = 2700 km$^2$**

---

# Comparison pie charts

Pie charts of different sizes are useful when comparing two or more populations. The area of each pie will be in proportion to the different population sizes, so if the pies are drawn with radii $r_1$ and $r_2$ and represent total population sizes $F_1$ and $F_2$, then

$$\pi r_1^2 : \pi r_2^2 = F_1 : F_2$$

Dividing by $\pi$ $\qquad\qquad r_1^2 : r_2^2 = F_1 : F_2$

Taking square roots $\qquad r_1 : r_2 = \sqrt{F_1} : \sqrt{F_2}$

Radii should be chosen so that $\dfrac{r_1}{r_2} = \dfrac{\sqrt{F_1}}{\sqrt{F_2}}$.

## Example 1.13

The table shows, in millions of pounds, the sales of a company in two successive years.

| Year | Africa | America | Asia | Europe |
|---|---|---|---|---|
| First | 5.5 | 6.7 | 13.2 | 19.6 |
| Second | 5.8 | 15.2 | 9.2 | 29.8 |

Draw two pie charts which allow the total annual sales to be compared.          (*C Additional*)

## Solution 1.13

First calculate the total sales for each year and the angles in the pie charts.

Total sales (in millions of pounds):

First year $F_1 = 5.5 + 6.7 + 13.2 + 19.6 = 45$

Second year $F_2 = 5.8 + 15.2 + 9.2 + 29.8 = 60$

Angles:

|  | Africa | America | Asia | Europe |
|---|---|---|---|---|
| First year | $\dfrac{5.5}{45} \times 360° = 44°$ | $\dfrac{6.7}{45} \times 360° = 53.6°$ | $\dfrac{13.2}{45} \times 360° = 105.6°$ | $\dfrac{19.6}{45} \times 360° = 156.8°$ Total 360° |
| Second year | $\dfrac{5.8}{60} \times 360° = 34.8°$ | $\dfrac{15.2}{60} \times 360° = 91.2°$ | $\dfrac{9.2}{60} \times 360° = 55.2°$ | $\dfrac{29.8}{60} \times 360° = 178.8°$ Total 360° |

Work out the ratio of the radii using

$$r_1^2 : r_2^2 = F_1 : F_2 = 45 : 60 = 3 : 4$$
$$r_1 : r_2 = \sqrt{3} : \sqrt{4} = 1.73 \cdots : 2$$

So you could take $r_1 = 1.7$ cm, $r_2 = 2$ cm, or multiples of these e.g. $r_1 = 3.4$ cm, $r_2 = 4$ cm.

### Sales in first year

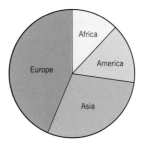

### Sales in second year

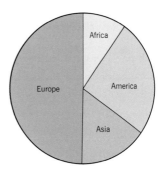

## Example 1.14

On a particular Wednesday the sales of sugar from a supermarket consisted of 250 large packets, 210 medium packets and 225 small packets. The mass of sugar in a large packet is $1\frac{1}{2}$ times that in a medium packet and $2\frac{1}{2}$ times that in a small packet. Calculate the angles needed to draw a pie chart representing the total masses of sugar sold in large, medium and small packets.

The radius of the corresponding pie chart for the following Saturday's sales of sugar was double that for the Wednesday's sales. On the Saturday 900 large packets and 900 medium packets were sold. Calculate the number of small packets sold on the Saturday.  (C)

## Solution 1.14

Let the mass of a small packet be $x$.

Then the mass of a large packet is $2\frac{1}{2}x$.

Also, you are given that

$$\text{mass of a large packet} = 1\frac{1}{2} \times \text{mass of medium packet}$$

so

$$2\frac{1}{2}x = 1\frac{1}{2} \times \text{mass of medium packet}$$

$$\therefore \quad \text{mass of a medium packet} = \frac{2\frac{1}{2}x}{1\frac{1}{2}} = \frac{5}{3}x.$$

Mass of 225 small packets $= 225x$

Mass of 210 medium packets $= 210 \times \frac{5}{3}x = 350x$

Mass of 250 large packets $= 250 \times \frac{5}{2}x = 625x$

$$\therefore \quad \text{total mass} = 1200x$$

Angle representing mass of small packets $= \dfrac{225}{1200} \times 360° = \mathbf{67.5°}$

Angle representing mass of medium packets $= \dfrac{350}{1200} \times 360° = \mathbf{105°}$

Angle representing mass of large packets $= \dfrac{625}{1200} \times 360° = \mathbf{187.5°}$

Let $F_W$ denote total number of packets sold on Wednesday.

Let $F_S$ denote total number of packets sold on Saturday.

Then $F_W = 250 + 210 + 225 = 685$.

Also $\quad r_S : r_W = 2 : 1$

$\therefore \qquad F_S : F_W = r_S^2 : r_W^2 = 4 : 1$

$\therefore \qquad F_S = 4 \times 685 = 2740$

Number of small packets sold on Saturday $= 2740 - (900 + 900)$

$$= 940$$

**940 small packets were sold on Saturday.**

---

## Exercise 1c   Pie charts

1. There are 28 pupils in Peter's class. He carried out a survey of how the pupils in his class travelled to school. His results are shown in the table below.

| Method of travel | Number of pupils |
|---|---|
| Bus | 12 |
| Car | 2 |
| Bicycle | 5 |
| Walking | 9 |

The data are to be illustrated by a pie chart.
   (a) Calculate, to the nearest degree, the sector angles of the pie chart.
   (b) Draw the pie chart using a circle of radius 5 cm, labelling each sector with the method of travel it represents.

There are 34 pupils in Shumilla's class. For these pupils she carried out the same kind of survey and drew a pie chart to show her results.

   (c) Calculate, giving your answer to three significant figures, the radius of a comparable pie chart which could be used to represent the results of Shumilla's survey. (C)

2. The following data summarise the expenditure by a county council during a particular year.

| Service | Expenditure (£m) |
|---|---|
| Education | 160.2 |
| Highways & Public Transport | 35.7 |
| Police | 28.9 |
| Social Services | 27.9 |
| Other | 24.5 |

These data are to be represented by a pie chart of radius 5 cm. Calculate, to the nearest degree, the angle corresponding to each of the five classifications. (**Do not draw the pie chart.**)

The following year the county council spent £305.2 m.

Find the radius of a comparable pie chart which could be used to represent this second set of data. (*L*)

3. Five companies form a group. The sales of each company during the year ending 5 April, 1988, are shown in the table below.

| Company | *A* | *B* | *C* | *D* | *E* |
|---|---|---|---|---|---|
| Sales (in £1000s) | 55 | 130 | 20 | 35 | 60 |

Draw a pie chart of radius 5 cm to illustrate this information.

For the year ending 5 April, 1989, the total sales of the group increased by 20%, and this growth was maintained for the year ending 5 April, 1990.

If pie charts were drawn to compare the total sales for each of these years with the total sales for the year ending 5 April, 1988, what would be the radius of each of these pie charts?

If the sales of company *E* for the year ending 5th April, 1990, were again £60 000, what would be the angle of the sector representing them? (*C*)

4. A charity obtains its income from various sources. The table below shows these sources and the corresponding amounts of income for 1993.

| Source | Income (£) |
|---|---|
| Advertising | 30 000 |
| Donations | *x* |
| Fees | 9 000 |
| Investments | 3 000 |
| Sponsorship | 10 000 |

A pie chart was drawn to illustrate the data. Given that the angle of the sector representing Donations was 204°, calculate
(a) the total income for 1993,
(b) the value of *x*,
(c) the angle of each of the remaining sectors.

A second pie chart was drawn to compare the corresponding 1996 data with that of 1993. In 1996 the income from Sponsorship had increased to £28 800 and this was represented by a sector of angle 60° in the pie chart for 1996. Given that the radius of the 1996 pie chart was 9 cm, calculate the radius of the 1993 pie chart. (*C*)

5. A golf club has four categories of membership: men, women, juniors and social members. The pie chart shown, which is not drawn to scale, illustrates the distribution of membership in 1995. Given that there were 147 men and 35 social members, calculate
(a) the number of junior members,
(b) the angle of the sector representing the social members,
(c) the number of women.

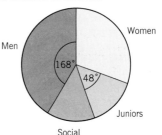

The corresponding pie chart for 2000 indicated that the number of men had increased by 49 although the angle of the corresponding sector remained the same. Calculate the total number of members in 2000.

Given that the radius of the 1995 pie chart was 26 cm, calculate the radius of the 2000 pie chart. (*C Additional*)

6. During a particular fortnight a family spends £52.27 on meat, £23.10 on fruit and vegetables, £19.72 on drink, £12.41 on toiletries, £102.68 on groceries and £9.82 on miscellaneous items.

These data are to be represented by a pie chart of radius 5 cm.
(a) Calculate, to the nearest degree, the angle corresponding to each of the above classifications. (**Do not draw the pie chart.**)

The following fortnight the family spends 20% more in total.
(b) Find the radius of a comparable pie chart to represent the data on this occasion. (*L*)

7. Pie charts *A*, *B* and *C* are drawn to compare, over a given period, the total value of the sales of certain items in each of three branches of a multiple store. The radii of the charts are 20 cm, 30 cm and 40 cm, respectively.
(a) If the total sales value represented by chart *B* is £4500, calculate the total sales value represented by each of charts *A* and *C*.
(b) The angle of the sector representing a particular item in chart *A* is 72°. Calculate the sales value of this item.
(c) The sales value represented by a sector in chart *C* is £600. Calculate the angle of the sector.
(d) One item occupies one quarter of chart *A*, and the sales value for this item is one half of that for the same item on chart *B*. Calculate the angle of the sector for this item on chart *B*. (*C Additional*)

8. On a certain day, 125 people, each buying one newspaper, were asked which newspaper they had bought. The results of the survey are shown in the table below.

| Newspaper | Number bought |
|---|---|
| The Times | 10 |
| The Telegraph | 25 |
| The Express | 40 |
| Some other paper | 50 |

Calculate the angles of the sectors of a pie chart of radius 5 cm which would illustrate these data.

The following day a similar survey was carried out and the radius of the pie chart necessary to compare the new set of data with the previous set was 6 cm. Calculate the number of people in the second survey. (C Additional)

9. A householder keeps an annual account of four items of expenditure. The figures for the year 1991 are shown in the table below.

| Item | Expenditure (£) |
|---|---|
| Taxes | $x$ |
| Travel | 1000 |
| Light/Heat | $y$ |
| Telephone | 300 |

A pie chart was drawn to illustrate these data. Given that the angles of the sectors representing Taxes and Travel were 124° and 80° respectively, calculate
(a) the total expenditure for the year,
(b) the value of $x$ and of $y$,
(c) the angle of each of the remaining sectors.

In 1992, the total expenditure on the same items was £8000. Given that the radius of the pie chart for 1991 was 6 cm, calculate the radius of the pie chart for 1992 in order that the two sets of data may be compared. (C Additional)

# THE MEAN

A typical or average value is useful when interpreting data. One such average is the mean.

Consider the five numbers

0.9,   1.4,   2.8,   3.1,   5.6.

The mean is $\dfrac{0.9 + 1.4 + 2.8 + 3.1 + 5.6}{5} = \dfrac{13.8}{5} = 2.76$

In general, for the $n$ numbers $x_1, x_2, x_3, ..., x_n$,

the mean is $\dfrac{x_1 + x_2 + x_3 + \cdots + x_n}{n}$.

## Example 1.15

To obtain Grade A, Ben must achieve an average of at least 70 in five tests. If his average mark for the first four tests is 68, what is the lowest mark he can get in his fifth test and still obtain Grade A?

## Solution 1.15

For the first four tests, $\dfrac{x_1 + x_2 + x_3 + x_4}{4} = 68$

$\therefore \qquad x_1 + x_2 + x_3 + x_4 = 68 \times 4 = 272$

For five tests, Ben wants his mean mark to be at least 70.

$$\frac{x_1 + x_2 + x_3 + x_4 + x_5}{5} \geqslant 70$$

$$\frac{272 + x_5}{5} \geqslant 70$$

$$272 + x_5 \geqslant 350$$

$$x_5 \geqslant 350 - 272$$

$$x_5 \geqslant 78$$

**To obtain Grade A, Ben must get at least 78 marks in his fifth test.**

---

A shorthand way of writing $x_1 + x_2 + x_3 + x_4$ is $\sum_{i=1}^{4} x_i$.

The symbol $\Sigma$ (the Greek capital letter 'sigma') is used to denote 'the sum of'. So for $x_1 + x_2 + x_3 + \cdots + x_n$ you could write $\sum_{x=1}^{n} x_i$.

The mean is often denoted by $\bar{x}$, so $\bar{x} = \dfrac{x_1 + x_2 + \cdots + x_n}{n} = \dfrac{\sum_{i=1}^{n} x_i}{n}$

This is rather cumbersome, so usually the subscript $i$ is omitted.

For discrete raw data:

$$\bar{x} = \frac{\Sigma x}{n}$$

## Example 1.16

The members of an orchestra were asked how many instruments each could play. Here are their results.

```
2  5  2  4  1  1  1  2  1  3
3  2  1  2  1  1  2  4  3  2
1  2  3  1  4  2  3  1  1  2
```

Find the mean number of instruments played.

## Solution 1.16

$n = 30$, $\Sigma x = 2 + 5 + 2 + \cdots + 1 + 2 = 63$

$$\bar{x} = \frac{\Sigma x}{n} = \frac{63}{30} = 2.1$$

**The mean number of instruments played is 2.1.**

---

In the above example, the data could have been arranged in a frequency distribution:

| Number of instruments, $x$ | 1 | 2 | 3 | 4 | 5 |
|---|---|---|---|---|---|
| Frequency, $f$ | 11 | 10 | 5 | 3 | 1 |

The total number of instruments played can be calculated in an organised way as follows:

| $x$ | $f$ | $f \times x$ |
|---|---|---|
| 1 | 11 | 11 |
| 2 | 10 | 20 |
| 3 | 5 | 15 |
| 4 | 3 | 12 |
| 5 | 1 | 5 |
| | $\Sigma f = 30$ | $\Sigma fx = 63$ |

↑ total number of people   ↑ total number of instruments played

$$\bar{x} = \frac{\text{total number of instruments}}{\text{total number of people}}$$

$$= \frac{\Sigma fx}{\Sigma f}$$

$$= \frac{63}{30}$$

$$= 2.1$$

**The mean number of instruments played is 2.1**

Note that $\Sigma fx$ is sometimes written $\Sigma xf$ and remember that $x$ and $f$ are multiplied.

In general, for data in an ungrouped frequency distribution

$$\bar{x} = \frac{\Sigma fx}{\Sigma f}$$

When the data have been grouped into intervals, the actual values of the readings are not known. You can only make an estimate of the mean. To do this, take the mid-interval value as representative of the interval.

Remember that mid-interval value = $\frac{1}{2}$ (lower class boundary + upper class boundary)

## Example 1.17

The speeds, to the nearest mile per hour, of 120 vehicles passing a check point were recorded and are grouped in the table below.

| Speed (m.p.h.) | 21–25 | 26–30 | 31–35 | 36–45 | 46–60 |
|---|---|---|---|---|---|
| Number of vehicles | 22 | 48 | 25 | 16 | 9 |

Estimate the mean of this distribution.

(C Additional)

## Solution 1.17

Work out the mid-interval value for the first interval 21–25, using lower class boundary = 20.5, upper class boundary = 25.5.

So mid-interval value = $\frac{1}{2}$ (20.5 + 25.5) = 23.

You then assume that all the values in the interval 21–25 are in fact 23.

Find the other mid-interval values and form a table:

| Speed (m.p.h.) | Mid-interval value, $x$ | $f$ | $fx$ |
|---|---|---|---|
| 21–25 | 23 | 22 | 506 |
| 26–30 | 28 | 48 | 1344 |
| 31–35 | 33 | 25 | 825 |
| 36–45 | 40.5 | 16 | 648 |
| 46–60 | 53 | 9 | 477 |
| | | $\Sigma f = 120$ | $\Sigma fx = 3800$ |

$$\bar{x} = \frac{\Sigma fx}{\Sigma f}$$
$$= \frac{3800}{120}$$
$$= 31\tfrac{2}{3}$$

The mean speed was $31\tfrac{2}{3}$ m.p.h.

## Using the calculator to find the mean

You can use your calculator in ordinary computation mode to calculate the total and also do the division. It is more useful, however, to work in the statistical mode, known as SD or STAT mode. Your calculator may operate as in one of the examples below. If yours does not appear to follow one of the patterns, you will need to consult your calculator manual.

Notice that once you put in the data you have access not only to the value for the mean, but also to $n$ and $\Sigma x$.

### Example 1.18

Find the mean of the numbers 33, 28, 26, 35, 38.

### Solution 1.18

| | Casio 570W/85W/85WA | | Sharp |
|---|---|---|---|
| Set SD mode | MODE MODE 1 or MODE 2 | | MODE 1 |
| Clear memories | SHIFT Scl = | | 2nd F CA |
| Input data | 33 DT | | 33 DATA |
| | 28 DT | | 28 DATA |
| | 26 DT | | 26 DATA |
| | 35 DT | | 35 DATA |
| | 38 DT | | 38 DATA |
| To obtain | | | |
| $\bar{x} = 32$ | SHIFT 1 = | | 2nd F ( |
| $n = 5$ | RCL C | Red letters on third | 2nd F ) |
| $\Sigma x = 160$ | RCL B | row of calculator | 2nd F + |
| To clear SD mode | MODE 1 | | MODE 0 |

From the calculator, **the mean is 32.**

## Example 1.19

Find the mean number of children per family for the following frequency distribution.

| Number of children per family, $x$ | 1 | 2 | 3 | 4 | 5 |
|---|---|---|---|---|---|
| Frequency, $f$ | 3 | 4 | 8 | 2 | 3 |

## Solution 1.19

| | Casio 570W/85W/85WA | Sharp |
|---|---|---|
| Set SD mode | MODE MODE 1 or MODE 2 | MODE 1 |
| Clear memories | SHIFT Scl = | 2nd F CA |
| Input data in the order $x \times f$ | 1 SHIFT ; 3 DT | 1 × 3 DATA |
| | 2 SHIFT ; 4 DT | 2 × 4 DATA |
| | 3 SHIFT ; 8 DT | 3 × 8 DATA |
| | 4 SHIFT ; 2 DT | 4 × 2 DATA |
| | 5 SHIFT ; 3 DT | 5 × 3 DATA |
| To obtain | | |
| $\bar{x} = 2.9$ | SHIFT 1 = | 2nd F ( |
| $\Sigma f = 20$ | RCL C (Red letters on third row of calculator) | 2nd F ) |
| $\Sigma fx = 58$ | RCL B | 2nd F + |
| To clear SD mode | MODE 1 | MODE 0 |

From the calculator, **the mean is 2.9 children per family.**

Make sure that you input the data in the order $x \times f$. Remember that $x$ usually comes first in the frequency table.

## Example 1.20

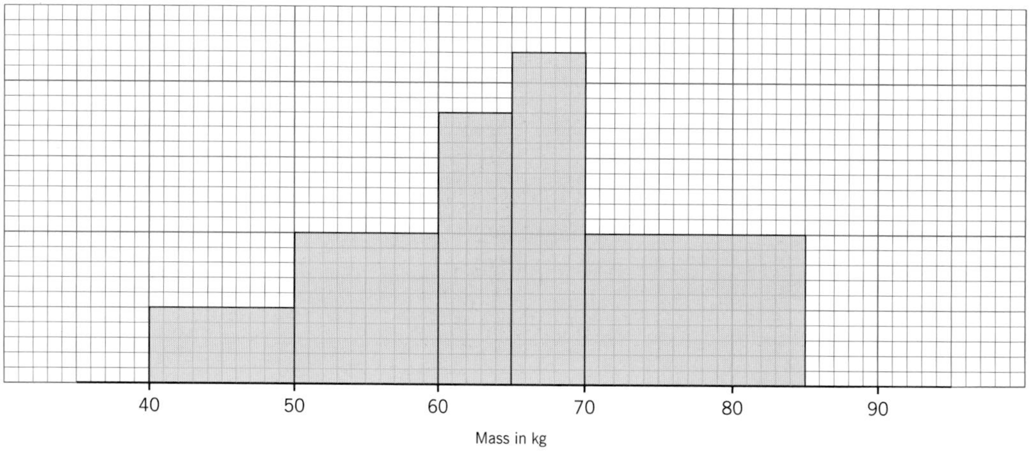

The diagram shows a histogram of the distribution of masses of 50 first-year University students. All the rectangles are there but the vertical axis has been torn off.

(a) Compile a grouped frequency table for the distribution.
(b) Use the values in your frequency table to find an approximate value for the mean mass of the students.

## Solution 1.20

Let one small square be $h$ on the vertical axis.

Remember that in a histogram, the area of each rectangle is proportional to the frequency.

The areas are

$5h \times 10$, $10h \times 10$, $18h \times 5$, $22h \times 5$, $10h \times 15$
i.e. $50h$, $100h$, $90h$, $110h$, $150h$.

So the total area $= 500h$.

But total frequency = the number of students = 50

$$\therefore \quad 500h = 50$$
$$h = 0.1$$

This means that the frequencies are 5, 10, 9, 11, 15, giving a total of 50.

(a) The frequency distribution is

| Mass (kg) | Frequency, $f$ |
|---|---|
| $40 \leqslant m < 50$ | 5 |
| $50 \leqslant m < 60$ | 10 |
| $60 \leqslant m < 65$ | 9 |
| $65 \leqslant m < 70$ | 11 |
| $70 \leqslant m < 85$ | 15 |

(b) Take the mid-point of each interval to represent that interval. For example, the mid-point of the interval $60 \leqslant m < 65$ is $\frac{1}{2}(60 + 65) = 62.5$.

| mid-point, $x$ | frequency, $f$, | $fx$ |
|---|---|---|
| 45 | 5 | 225 |
| 55 | 10 | 550 |
| 62.5 | 9 | 562.5 |
| 67.5 | 11 | 742.5 |
| 77.5 | 15 | 1162.5 |
| | $\Sigma f = 50$ | $\Sigma fx = 3242.5$ |

i.e. $\bar{x} = \dfrac{\Sigma fx}{\Sigma f}$

$= \dfrac{3242.5}{50}$

$= 64.85$ kg

**The mean mass is 64.85 kg.**

Using the calculator:

| | Casio 570W/85W/85WA | Sharp |
|---|---|---|
| Set SD Mode | MODE MODE 1 or MODE 2 | MODE 1 |
| Clear memories | SHIFT Scl = | 2nd F CA |
| Input Data in the order $x \times f$ | 45 SHIFT ; 5 DT | 45 × 5 DATA |
| | 55 SHIFT ; 10 DT | 55 × 10 DATA |
| | 62.5 SHIFT ; 9 DT | 62.5 × 9 DATA |
| | 67.5 SHIFT ; 11 DT | 67.5 × 11 DATA |
| | 77.5 SHIFT ; 15 DT | 77.5 × 15 DATA |

| To obtain | | |
|---|---|---|
| $\bar{x} = 64.85$ | SHIFT 1 = | 2nd F ( |
| $\Sigma f = 50$ | RCL C | 2nd F ) |
| $\Sigma fx = 3242.5$ | RCL B | 2nd F + |
| To clear SD Mode | MODE 1 | MODE 0 |

Red letters on third row of calculator

Practise this yourself. Make sure that you are familiar with the method on **your** calculator

## Exercise 1d   The mean

1. Find the mean of each of the following sets of numbers,
   (i)   not using SD mode,
   (ii)  using SD mode.
   Compare your answers.

   (a)  5, 6, 6, 8, 8, 9, 11, 13, 14, 17
   (b)  148, 153, 156, 157, 160
   (c)  $44\frac{1}{2}$, $47\frac{1}{2}$, $48\frac{1}{2}$, $51\frac{1}{2}$, $52\frac{1}{2}$, $54\frac{1}{2}$, $55\frac{1}{2}$, $56\frac{1}{2}$
   (d)  1769, 1771, 1772, 1775, 1778, 1781, 1784
   (e)  0.85, 0.88, 0.89, 0.93, 0.94, 0.96
   (f)

   | $x$ | 1 | 2 | 3 | 4 | 5 | 6 | 7 |
   |---|---|---|---|---|---|---|---|
   | $f$ | 4 | 5 | 8 | 10 | 17 | 5 | 1 |

   (g)

   | $x$ | 27 | 28 | 29 | 30 | 31 | 32 |
   |---|---|---|---|---|---|---|
   | $f$ | 30 | 43 | 51 | 49 | 42 | 35 |

   (h)

   | $x$ | 121 | 122 | 123 | 124 | 125 |
   |---|---|---|---|---|---|
   | $f$ | 14 | 25 | 32 | 23 | 6 |

2. A sample of 100 boxes of matches was taken and a record made of the number of matches per box. The results were as follows:

   | Number of matches per box | 47 | 48 | 49 | 50 | 51 |
   |---|---|---|---|---|---|
   | Frequency | 4 | 20 | 35 | 24 | 17 |

   Calculate the mean number of matches per box.

3. On a certain day the numbers of books on 40 shelves in a library were noted and grouped as shown. Find the mean number of books on a shelf.

   | Number of books | Number of shelves |
   |---|---|
   | 31–35 | 4 |
   | 36–40 | 6 |
   | 41–45 | 10 |
   | 46–50 | 13 |
   | 51–55 | 5 |
   | 56–60 | 2 |

4. The amounts spent by 120 motorists at a petrol station were recorded.

| Amount spent, £x | Number of motorists |
|---|---|
| $x < 5$ | 12 |
| $5 \leqslant x < 10$ | 38 |
| $10 \leqslant x < 15$ | 42 |
| $15 \leqslant x < 20$ | 20 |
| $20 \leqslant x < 40$ | 8 |

(a) Draw a histogram to represent the data.
(b) Estimate the mean amount spent.

5. The age distribution of the population of a small village is recorded in the table below.

| Age (years) | Number of people |
|---|---|
| 0– | 54 |
| 15– | 78 |
| 30– | 120 |
| 50– | 88 |
| 70– | 60 |
| 100– | 0 |

Draw, on graph paper, a histogram to represent these data.

Estimate the mean of this distribution.
(C Additional)

6. Find the mean length for the data represented by the stem and leaf diagram.

*Key* 15 | 1 means 16 cm

| Stem | Leaf |
|---|---|
| 12 | 0 0 |
| 15 | 0 1 1 |
| 18 | 1 1 2 |
| 21 | 0 1 1 2 2 2 |
| 24 | 0 0 1 2 |
| 27 | 1 1 |
| 30 | 2 |

7. The height, correct to the nearest metre, was recorded for each of the 59 birch trees in an area of woodland. The heights are summarised in the following table.

| Height (m) | 5–9 | 10–12 | 13–15 | 16–18 | 19–28 |
|---|---|---|---|---|---|
| Number of trees | 14 | 18 | 15 | 4 | 8 |

(a) A student was asked to draw a histogram to illustrate the data and produced the following diagram.

**A histogram to illustrate the heights of birch trees**

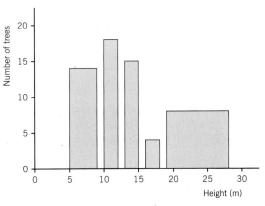

Give two critical comments on this attempt at a histogram.
(b) Using graph paper, draw a correct histogram to illustrate the above data.
(c) Calculate an estimate of the mean height of the birch trees, giving your answer correct to three significant figures. (C)

8. Telephone calls arriving at a switchboard are answered by the telephonist. The following table shows the time, to the nearest second, recorded as being taken by the telephonist to answer the calls received during one day.

| Time to answer (to nearest second) | Number of calls |
|---|---|
| 10–19 | 20 |
| 20–24 | 20 |
| 25–29 | 15 |
| 30 | 14 |
| 31–34 | 16 |
| 35–39 | 10 |
| 40–59 | 10 |

(a) Represent these data by a histogram. Give a reason to justify the use of a histogram to represent these data.
(b) Calculate an estimate of the mean time taken to answer the calls. (L)

## Weighted means

In some situations it may not be suitable to calculate an ordinary mean. There may be times when you wish to place greater emphasis on some of the values, as illustrated in the following example.

### Example 1.21

A candidate obtained the following results in her GCSE mathematics examination:

Paper 1: 72%, Paper 2: 64%, Coursework: 73%

The regulations state that the two written papers have equal weighting and count for 80% of the final result, whereas the coursework counts for 20%. What was the candidate's final mark?

### Solution 1.21

The results are in the following ratio:

$40\% : 40\% : 20\% = 4 : 4 : 2 = 2 : 2 : 1$.

For the final result, you have to take this weighting into account:

$$\text{weighted mean} = \frac{2(72) + 2(64) + 1(73)}{2 + 2 + 1} = \frac{345}{5} = 69$$

**Therefore the final mark is 69%.**

In general, if $x_1, x_2, ..., x_n$ are given weightings $w_1, w_2, ..., w_n$ then

$$\text{weighted mean} = \frac{w_1 x_1 + w_2 x_2 + \cdots + w_n x_n}{w_1 + w_2 + \cdots + w_n} = \frac{\Sigma w_i x_i}{\Sigma w_i} \quad \text{for} \quad i = 1, 2, ..., n$$

## Exercise 1e   Weighted means

1. Find the weighted mean of the numbers 8 and 12, if they are given the weights 2 and 3 respectively.

2. The final mark allocated to a student is calculated from her mark in each subject.
   (a) The class teacher worked out an ordinary mean.
   (b) The headteacher decided to weight the subjects in proportion to the number of lessons per week, as shown in the table.

| Subject | Mark | Number of lessons per week |
|---|---|---|
| Mathematics | 64% | 5 |
| English | 52% | 4 |
| Science | 71% | 6 |
| French | 75% | 3 |
| History | 82% | 2 |

   Which method gave the higher mark and by how much?

3. The prices of articles A, B and C are £30, £42 and £65. Find the mean price, if the three articles are given weights of 5, 3 and 2 respectively.

4. The weighted mean of the two numbers 30 and 15 is 20. If the weightings are 2 and $x$ respectively, find $x$.

5. Two students, Jack and Jill, take an examination in French, German and English. The table below shows the marks for each student and the weight to be applied to each subject.

| Subject | French | German | English |
|---|---|---|---|
| Marks for Jack | 80 | 72 | 46 |
| Marks for Jill | 64 | 82 | 40 |
| Weight | 2 | $x$ | 3 |

   Calculate the value of $x$ for which Jack and Jill have the same weighted mean mark and find the value of this mean.   (C Additional)

## VARIABILITY OF DATA

Each of these sets of numbers has a mean of 7 but the spread of each is set is different:

(a) 7, 7, 7, 7, 7
(b) 4, 6, 6.5, 7.2, 11.3
(c) −193, −46, 28, 69, 177

There is no variability in set (a), but the numbers in set (c) are obviously much more spread out than those in set (b).

There are various ways of measuring the variability or spread of a distribution, two of which are described here.

## The range

The range is based entirely on the extreme values of the distribution.

Range = highest value − lowest value

In (a) the range = 7 − 7 = 0
In (b) the range = 11.3 − 4 = 7.3
In (c) the range = 177 − (−193) = 370

Note that there are also ranges based on particular observations within the data and these **percentile** and **quartile ranges** are considered on page 68.

## THE STANDARD DEVIATION, $s$, AND THE VARIANCE, $s^2$

The standard deviation, $s$, is a very important and useful measure of spread. It gives a measure of the deviations of the readings from the mean, $\bar{x}$. It is calculated using all the values in the distribution. To calculate $s$:

- for each reading $x$, calculate $x - \bar{x}$, its deviation from the mean,
- square this deviation to give $(x - \bar{x})^2$ and note that, irrespective of whether the deviation was positive or negative, this is now positive,
- find $\Sigma(x - \bar{x})^2$, the sum of all these values,
- find the average by dividing the sum by $n$, the number of readings;
  this gives $\dfrac{\Sigma(x - \bar{x})^2}{n}$ and is known as the **variance**,
- finally take the positive square root of the variance to obtain the standard deviation, $s$.

The standard deviation, $s$, of a set of $n$ numbers, with mean $\bar{x}$, is given by

$$s = \sqrt{\frac{\Sigma(x - \bar{x})^2}{n}}$$

Each of the three sets of numbers on the previous page has mean 7, i.e. $\bar{x} = 7$.

(a) For the set 7, 7, 7, 7, 7

Since $x - \bar{x} = 7 - 7 = 0$ for every reading, $s = 0$, indicating that there is no deviation from the mean.

(b) For the set 4, 6, 6.5, 7.2, 11.3

$$\Sigma(x - \bar{x})^2 = (4 - 7)^2 + (6 - 7)^2 + (6.5 - 7)^2 + (7.2 - 7)^2 + (11.3 - 7)^2 = 28.78$$

$$s = \sqrt{\frac{\Sigma(x - \bar{x})^2}{n}} = \sqrt{\frac{28.78}{5}} = 2.4 \ (1 \ \text{d.p.})$$

(c) For the set −193, −46, 28, 69, 177

$$\Sigma(x - \bar{x})^2 = (-193 - 7)^2 + (-46 - 7)^2 + (28 - 7)^2 + (69 - 7)^2 + (177 - 7)^2 = 75\ 994$$

$$s = \sqrt{\frac{\Sigma(x - \bar{x})^2}{n}} = \sqrt{\frac{75\ 994}{5}} = 123.3 \ (1 \ \text{d.p.})$$

Notice that set (c) has a much higher standard deviation than set (b), confirming that it is much more spread about the mean.

Remember that

Standard deviation = $\sqrt{\text{variance}}$

Variance = (standard deviation)$^2$

*NOTE:*

- The standard deviation gives an indication of the lowest and highest values of the data as follows. In most distributions, the bulk of the distribution lies within two standard deviations of the mean, i.e. within the interval $\bar{x} \pm 2s$ or $(\bar{x} - 2s, \bar{x} + 2s)$. This helps to give an idea of the spread of the data.
- The units of standard deviation are the same as the units of the data.
- Standard deviations are useful when comparing sets of data; the higher the standard deviation, the greater the variability in the data.

## Example 1.22

Two machines, A and B, are used to pack biscuits. A random sample of ten packets was taken from each machine and the mass of each packet was measured to the nearest gram and noted. Find the standard deviation of the masses of the packets taken in the sample from each machine. Comment on your answer.

| Machine A (mass in g) | 196, 198, 198, 199, 200, 200, 201, 201, 202, 205 |
|---|---|
| Machine B (mass in g) | 192, 194, 195, 198, 200, 201, 203, 204, 206, 207 |

## Solution 1.22

Machine A $\quad \bar{x} = \dfrac{\Sigma x}{n} = \dfrac{2000}{10} = 200 \qquad$ Machine B $\quad \bar{x} = \dfrac{\Sigma x}{n} = \dfrac{2000}{10} = 200$

Since the mean mass for each machine is 200, $x - \bar{x} = x - 200$

To calculate $s$, put the data into a table:

| Machine A | | |
|---|---|---|
| $x$ | $x - 200$ | $(x - 200)^2$ |
| 196 | −4 | 16 |
| 198 | −2 | 4 |
| 198 | −2 | 4 |
| 199 | −1 | 1 |
| 200 | 0 | 0 |
| 200 | 0 | 0 |
| 201 | 1 | 1 |
| 201 | 1 | 1 |
| 202 | 2 | 4 |
| 205 | 5 | 25 |
| | | 56 |

| Machine B | | |
|---|---|---|
| $x$ | $x - 200$ | $(x - 200)^2$ |
| 192 | −8 | 64 |
| 194 | −6 | 36 |
| 195 | −5 | 25 |
| 198 | −2 | 4 |
| 200 | 0 | 0 |
| 201 | 1 | 1 |
| 203 | 3 | 9 |
| 204 | 4 | 16 |
| 206 | 6 | 36 |
| 207 | 7 | 49 |
| | | 240 |

$$s^2 = \frac{\Sigma(x - 200)^2}{10}$$
$$= 5.6$$
$$s = \sqrt{5.6}$$
$$= 2.37 \text{ (2 d.p.)}$$

$$s^2 = \frac{\Sigma(x - 200)^2}{10}$$
$$= 24$$
$$s = \sqrt{24}$$
$$= 4.90 \text{ (2 d.p.)}$$

Machine A:   s.d. = 2.37 g (2 d.p.)        Machine B:   s.d. = 4.90 g (2 d.p.)

**Machine A has less variation, indicating that it is more reliable than machine B.**

# Alternative form of the formula for standard deviation

The formula given above is sometimes difficult to use, especially when $\bar{x}$ is not an integer, so an alternative form is often used. This is derived as follows:

$$s^2 = \frac{1}{n} \Sigma(x - \bar{x})^2$$
$$= \frac{1}{n} \Sigma(x^2 - 2\bar{x}x + \bar{x}^2)$$
$$= \frac{1}{n} (\Sigma x^2 - 2\bar{x}\Sigma x + \Sigma\bar{x}^2)$$
$$= \frac{\Sigma x^2}{n} - 2\bar{x}\frac{\Sigma x}{n} + \frac{n\bar{x}^2}{n}$$
$$= \frac{\Sigma x^2}{n} - 2\bar{x}(\bar{x}) + \bar{x}^2 \quad \text{since} \quad \frac{\Sigma x}{n} = \bar{x}$$
$$= \frac{\Sigma x^2}{n} - \bar{x}^2$$

Alternative format for standard deviation
$$s = \sqrt{\frac{\Sigma x^2}{n} - \bar{x}^2}$$

NOTE: It is useful to remember that $\dfrac{\Sigma x^2}{n} - \bar{x}^2$ can be thought of as 'the mean of the squares minus the square of the mean'.

## Example 1.23

The mean of the five numbers 2, 3, 5, 6, 8 is 4.8. Calculate the standard deviation.

## Solution 1.23

**Method 1** using $\quad s = \sqrt{\dfrac{\Sigma(x - \bar{x})^2}{n}}$ $\qquad$ **Method 2** using $\quad s = \sqrt{\dfrac{\Sigma x^2}{n} - \bar{x}^2}$

| $x$ | $x - \bar{x}$ | $(x - \bar{x})^2$ |
|---|---|---|
| 2 | −2.8 | 7.84 |
| 3 | −1.8 | 3.24 |
| 5 | 0.2 | 0.04 |
| 6 | 1.2 | 1.44 |
| 8 | 3.2 | 10.24 |
| | | 22.80 |

| $x$ | $x^2$ |
|---|---|
| 2 | 4 |
| 3 | 9 |
| 5 | 25 |
| 6 | 36 |
| 8 | 64 |
| | 138 |

$$s^2 = \frac{22.80}{5}$$
$$= 4.56$$

$$s^2 = \frac{138}{5} - (4.8)^2$$
$$= 4.56$$

$$s = \sqrt{4.56}$$
$$= 2.14 \text{ (2 d.p.)}$$

$$s = \sqrt{4.56}$$
$$= 2.14 \text{ (2 d.p.)}$$

The working for method 2 is less involved.

# Using the calculator to find the standard deviation

The standard deviation can be found directly using the calculator in SD mode. The numbers are entered in the same way as when you are finding the mean.

To find the standard deviation of the five numbers 2, 3, 5, 6, 8 used in Example 1.23:

| | Casio 570W/85W/85WA | | Sharp |
|---|---|---|---|
| Set SD mode | MODE MODE 1 | or MODE 2 | MODE 1 |
| Clear memories | SHIFT Scl = | | 2nd F CA |
| Input data | 2 DT | | 2 DATA |
| | 3 DT | | 3 DATA |
| | 5 DT | | 5 DATA |
| | 6 DT | | 6 DATA |
| | 8 DT | | 8 DATA |
| To obtain $s = 2.135 \ldots$ | SHIFT 2 = | | 2nd F ÷ |
| You can check $\bar{x} = 4.8$ | SHIFT 1 = | | 2nd F ( |
| $\Sigma x = 24$ | RCL B | | 2nd F + |
| $\Sigma x^2 = 138$ | RCL A | Red letters on third | 2nd F − |
| $n = 5$ | RCL C | row of calculator | 2nd F ) |
| To clear SD mode | MODE 1 | | MODE 0 |

When data are in the form of a frequency distribution, the formula for $s$ is

$$s = \sqrt{\frac{\Sigma f(x - \bar{x})^2}{\Sigma f}}$$

or in the alternative form

$$s = \sqrt{\frac{\Sigma f x^2}{\Sigma f} - \bar{x}^2} \quad \text{where } \bar{x} \text{ is the mean.}$$

Consider again the data given in Example 1.19, on page 32, which shows the number of children in 20 families. The mean is 2.9.

| Number of children per family, $x$ | 1 | 2 | 3 | 4 | 5 |
|---|---|---|---|---|---|
| Frequency, $f$ | 3 | 4 | 8 | 2 | 3 |

You could use one of these three methods for finding the standard deviation. Method 2 is more popular than Method 1.

**Method 1** – using $\quad s = \sqrt{\dfrac{\Sigma f(x - \bar{x})^2}{\Sigma f}}$

| $x$ | $x - 2.9$ | $(x - 2.9)^2$ | $f$ | $f(x - 2.9)^2$ |
|---|---|---|---|---|
| 1 | −1.9 | 3.61 | 3 | 10.83 |
| 2 | −0.9 | 0.81 | 4 | 3.24 |
| 3 | 0.1 | 0.01 | 8 | 0.08 |
| 4 | 1.1 | 1.21 | 2 | 2.42 |
| 5 | 2.1 | 4.41 | 3 | 13.23 |
| | | | $\Sigma f = 20$ | $\Sigma f(x - \bar{x})^2 = 29.80$ |

$$s^2 = \frac{\Sigma f(x - 2.9)^2}{\Sigma f}$$
$$= \frac{29.80}{20}$$
$$= 1.49$$
$$s = \sqrt{1.49}$$
$$= 1.22 \text{ (2 d.p.)}$$

**The standard deviation of the number of children per family is 1.22 (2 d.p.).**

**Method 2** – using $\quad s = \sqrt{\dfrac{\Sigma f x^2}{\Sigma f} - \bar{x}^2}$

| $x$ | $f$ | $x^2$ | $fx^2$ |
|---|---|---|---|
| 1 | 3 | 1 | 3 |
| 2 | 4 | 4 | 16 |
| 3 | 8 | 9 | 72 |
| 4 | 2 | 16 | 32 |
| 5 | 3 | 25 | 75 |
| | $\Sigma f = 20$ | | $\Sigma f x^2 = 198$ |

$$s^2 = \frac{\Sigma fx^2}{\Sigma f} - (2.9)^2$$
$$= \tfrac{198}{20} - (2.9)^2$$
$$= 1.49$$
$$s = \sqrt{1.49}$$
$$= 1.22 \text{ (2 d.p.)}$$

**The standard deviation is 1.22 (2 d.p.), as before.**

**Method 3** – using the calculator in SD mode.

This time you need to take account of the frequencies, and this is done in exactly the same way as when finding the mean:

| | Casio 570W/85W/85WA | | Sharp |
|---|---|---|---|
| Set SD mode | MODE MODE 1 or MODE 2 | | MODE 1 |
| Clear memories | SHIFT Scl = | | 2nd F CA |
| Input data<br>Do this in the<br>order $x \times f$ | 1 SHIFT ; 3 DT<br>2 SHIFT ; 4 DT<br>3 SHIFT ; 8 DT<br>4 SHIFT ; 2 DT<br>5 SHIFT ; 3 DT | | 1 × 3 DATA<br>2 × 4 DATA<br>3 × 8 DATA<br>4 × 2 DATA<br>5 × 3 DATA |
| To obtain<br>$\bar{x} = 2.9$<br>$s = 1.220...$<br>$\Sigma f = 20$<br>$\Sigma fx = 58$<br>$\Sigma fx^2 = 198$ | SHIFT 1 =<br>SHIFT 2 =<br>RCL C<br>RCL B<br>RCL A | Red letters on third<br>row of calculator | 2nd F (<br>2nd F ÷<br>2nd F )<br>2nd F +<br>2nd F − |
| To clear<br>SD mode | MODE 1 | | MODE 0 |

**Therefore the standard deviation is 1.22 (2 d.p.), as before.**

In a grouped frequency distribution, the **mid-interval value** is taken as **representative of the interval,** as in the following example.

## Example 1.24

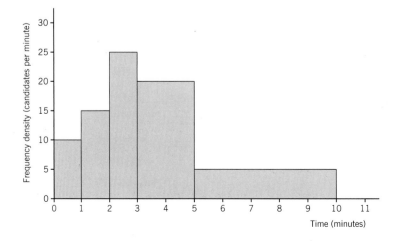

An intelligence test was taken by 115 candidates. For each candidate the time taken to complete the test was recorded, and the times were summarised in a histogram (see diagram). Write down the frequency for each of the class intervals 0–1, 1–2, 2–3, 3–5 and 5–10 minutes.

Calculate estimates of the mean and standard deviation of the times taken to complete the test. (C)

## Solution 1.24

Frequency = frequency density × interval width. Note that the interval 2–3, for example, represents $2 \leqslant \text{time} < 3$.

| Time (min) | 0–1 | 1–2 | 2–3 | 3–5 | 5–10 |
|---|---|---|---|---|---|
| Frequency | 10 | 15 | 25 | 40 | 25 |

To calculate estimates for the mean and standard deviation, use mid-interval values, $x$.

| Time (min) | $x$ | $f$ | $fx$ | $fx^2$ |
|---|---|---|---|---|
| 0–1 | 0.5 | 10 | 5 | 2.5 |
| 1–2 | 1.5 | 15 | 22.5 | 33.75 |
| 2–3 | 2.5 | 25 | 62.5 | 156.25 |
| 3–5 | 4 | 40 | 160 | 640 |
| 5–10 | 7.5 | 25 | 187.5 | 1406.25 |
| | | $\Sigma f = 115$ | $\Sigma fx = 437.5$ | $\Sigma fx^2 = 2238.75$ |

$$\bar{x} = \frac{\Sigma fx}{\Sigma f} = \frac{437.5}{115} = 3.8 \text{ (2 s.f.)}$$

$$s = \sqrt{\frac{\Sigma fx^2}{\Sigma f} - \bar{x}^2} = \sqrt{\frac{2238.75}{115} - 3.80\ldots^2} = 2.2 \text{ (2 s.f.)}$$

**The mean time is 3.8 minutes and the standard deviation is 2.2 minutes.**

[You could have calculated these directly using the calculator in SD mode. Check them yourself.]

If you are given summary information, rather than the raw data or frequency distribution, you cannot use the calculator in SD mode. You will have to use the formulae to calculate the mean and standard deviation, as in the following example.

## Example 1.25

(a) Cartons of orange juice are advertised as containing 1 litre. A random sample of 100 cartons gave the following results for the volume, $x$.

$$\Sigma x = 101.4, \quad \Sigma x^2 = 102.83$$

Calculate the mean and the standard deviation of the volume of orange juice in these 100 cartons.

(b) A machine is supposed to cut lengths of rod 50 cm long.

A sample of 20 rods gave the following results for the length, $x$.

$$\Sigma fx = 997, \quad \Sigma fx^2 = 49\ 711$$

(i) Calculate, the mean length of the 20 rods.
(ii) Calculate the variance of the lengths of the 20 rods.

State the units of the variance in your answer.

**Solution 1.25**

(a) $\Sigma x = 101.4, \Sigma x^2 = 102.83, n = 100$

$$\therefore \quad \bar{x} = \frac{\Sigma x}{n} = \frac{101.4}{100} = 1.014$$

The mean volume is **1.014 litres.**

$$s = \sqrt{\frac{\Sigma x^2}{n} - \bar{x}^2} = \sqrt{\frac{102.83}{100} - 1.014^2} = 0.0101\ \ldots$$

The standard deviation of the volume is **0.010 litres (2 s.f.)**

(b) $\Sigma fx = 997, \Sigma fx^2 = 49\ 711, \Sigma f = 20$

(i) $\bar{x} = \dfrac{\Sigma fx}{\Sigma f} = \dfrac{997}{20} = 49.85$

The mean length of the rods is **49.85 cm.**

(ii) Variance $= \dfrac{\Sigma fx^2}{\Sigma f} - \bar{x}^2 = \dfrac{49\ 711}{20} - 49.85^2 = 0.5275$

The variance is **0.5275 cm$^2$.**

---

## Exercise 1f   Mean and standard deviation

1. *Do not use the statistical program on your calculator for this question.*
   (i) For each of the following sets of numbers, calculate the mean and the standard deviation. Try using both forms of the formula for the standard deviation in parts (a) to (c). In parts (d) to (f) choose one of the methods.

   (a)  2, 4, 5, 6, 8
   (b)  6, 8, 9, 11
   (c)  11, 14, 17, 23, 29
   (d)  5, 13, 7, 9, 16, 15
   (e)  4.6, 2.7, 3.1, 0.5, 6.2
   (f)  200, 203, 206, 207, 209

   (ii) Now check your answers using your calculator in SD (STAT) mode.

2. The table shows the weekly wages in £ of each of 100 factory workers.
   (a) Draw a histogram to illustrate this information.
   (b) Calculate the mean wage and the standard deviation.

| Wage £ | Number of workers |
|---|---|
| $200 \leqslant x < 250$ | 10 |
| $250 \leqslant x < 300$ | 16 |
| $300 \leqslant x < 375$ | 40 |
| $375 \leqslant x < 400$ | 26 |
| $400 \leqslant x < 500$ | 8 |

3. Do this question
   (a) without using SD mode,
   (b) using SD mode on your calculator.

   The score for a round of golf for each of 50 club members was noted. Find the mean score for a round and the standard deviation.

   | Score, $x$ | Frequency, $f$ |
   | --- | --- |
   | 66 | 2 |
   | 67 | 5 |
   | 68 | 10 |
   | 69 | 12 |
   | 70 | 9 |
   | 71 | 6 |
   | 72 | 4 |
   | 73 | 2 |

4. The scores in an IQ test for 60 candidates are shown in the table. Find the mean score and the standard deviation.

   | Score | Frequency |
   | --- | --- |
   | 100–106 | 8 |
   | 107–113 | 13 |
   | 114–120 | 24 |
   | 121–127 | 11 |
   | 128–134 | 4 |

5. The stemplot shows the times, recorded to the nearest second, of 12 people in a race.

   Calculate the mean time and the standard deviation.

   | Stem | Leaf | |
   | --- | --- | --- |
   | 1 | 2 3 | *Key* 1 \| 5 means 15 seconds |
   | 1 | 5 5 6 6 | |
   | 1 | 7 9 9 | |
   | 2 | 0 1 | |

6. A vertical line graph for a set of data is shown below. Calculate the mean and standard deviation of the data.

   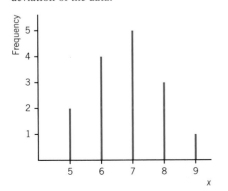

7. The following table shows the duration of 40 telephone calls from an office via the switchboard.
   (a) Obtain an estimate of the mean length of a telephone call and the standard deviation.
   (b) Illustrate the data graphically.

   | Duration in minutes | Number of calls |
   | --- | --- |
   | $\leqslant 1$ | 6 |
   | 1–2 | 10 |
   | 2–3 | 15 |
   | 3–5 | 5 |
   | 5–10 | 4 |
   | $\geqslant 10$ | 0 |

   *(O&C)*

8. For a set of ten numbers $\Sigma x = 290$ and $\Sigma x^2 = 8469$. Find the mean and the variance.

9. For a set of nine numbers $\Sigma(x - \bar{x})^2 = 234$. Find the standard deviation of the numbers.

10. For a set of nine numbers $\Sigma(x - \bar{x})^2 = 60$ and $\Sigma x^2 = 285$. Find the mean of the numbers.

11. A group of 20 people played a game. The table below shows the frequency distribution of their scores.

    | Score | 1 | 2 | 4 | $x$ |
    | --- | --- | --- | --- | --- |
    | Number of people | 2 | 5 | 7 | 6 |

    Given that the mean score is 5, find
    (a) the value of $x$,
    (b) the variance of the distribution.
    *(C Additional)*

12. From the information given about each of the following sets of data, work out the missing values in the table:

    | | $n$ | $\Sigma x$ | $\Sigma x^2$ | $\bar{x}$ | $s$ |
    | --- | --- | --- | --- | --- | --- |
    | (a) | 63 | 7623 | 924 800 | | |
    | (b) | | 152.6 | | 10.9 | 1.7 |
    | (c) | 52 | | 57 300 | 33 | |
    | (d) | 18 | | | 57 | 4 |

13. At a bird observatory, migrating willow warblers are caught, measured and ringed before being released. The histogram below illustrates the lengths, in millimetres, of the willow warblers caught during one migration season.

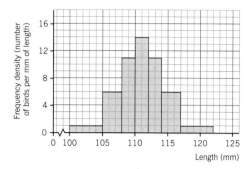

(b) State briefly how it may be deduced from the histogram (without any calculation) that an estimate of the mean length is 111 mm. Explain briefly why this value may not be the true mean length of the willow warblers caught.

(c) Given that the lengths, $x$ mm, of the willow warblers caught during this migration season were such that $\Sigma x = 13\,099$ and $\Sigma x^2 = 1\,455\,506$, calculate the standard deviation of the lengths. (C)

(a) Explain how the histogram shows that the total number of willow warblers caught at the observatory during the migration season is 118.

14. For a particular set of observations $\Sigma f = 20$, $\Sigma f x^2 = 16\,143$, $\Sigma f x = 563$. Find the values of the mean and the standard deviation.

15. For a given frequency distribution $\Sigma f(x - \bar{x})^2 = 182.3$, $\Sigma f x^2 = 1025$, $\Sigma f = 30$. Find the mean of the distribution.

16. The speeds of cars passing a speed camera are shown in the histogram.
Calculate estimates of the mean speed and the standard deviation.

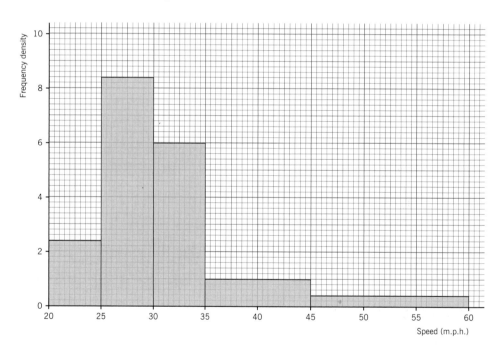

# Calculations involving the mean and standard deviation

## Example 1.26

(a) Calculate the mean and the standard deviation of the four numbers 2, 3, 6, 9.

(b) Two numbers, $a$ and $b$, are to be added to this set of four numbers, such that the mean is increased by 1 and the variance is increased by 2.5. Find $a$ and $b$. (L Additional)

**Solution 1.26**

(a) $\bar{x} = \dfrac{2+3+6+9}{4} = 5$

$s^2 = \dfrac{\Sigma x^2}{n} - \bar{x}^2 = \dfrac{4+9+36+81}{4} - 5^2 = 7.5, \quad s = \sqrt{7.5} = 2.7 \text{ (2 s.f.)}$

(b) New mean $= 5 + 1 = 6$

$$6 = \dfrac{2+3+6+9+a+b}{6}$$

$$6 = \dfrac{20+a+b}{6}$$

$20 + a + b = 36$

$a + b = 16 \ \dots\dots \text{①}$

Variance of original set $= s^2 = 7.5$. So new variance $= 7.5 + 2.5 = 10$

$$10 = \dfrac{4+9+36+81+a^2+b^2}{6} - 6^2$$

$$10 = \dfrac{130+a^2+b^2}{6} - 36$$

$\dfrac{130+a^2+b^2}{6} = 46$

$130 + a^2 + b^2 = 276$

$a^2 + b^2 = 146 \ \dots\dots \text{②}$

From ① $b = 16 - a$. Substituting in ②

$a^2 + (16-a)^2 = 146$

$a^2 + 256 - 32a + a^2 = 146$

$2a^2 - 32a + 110 = 0$

$a^2 - 16a + 55 = 0$

$(a-11)(a-5) = 0$

$\therefore \quad a = 11, a = 5$

If $a = 11$, $b = 16 - 11 = 5$

If $a = 5$, $b = 16 - 5 = 11$

**So the two numbers are 5 and 11.**

## COMBINING SETS OF DATA

### Example 1.27

The number of errors, $x$, on each of 200 pages of typescript was monitored. The results when summarised showed that

$\Sigma x = 920 \qquad \Sigma x^2 = 5032$

(a) Calculate the mean and the standard deviation of the number of errors per page. A further 50 pages were monitored and it was found that the mean was 4.4 errors and the standard deviation was 2.2 errors.

(b) Find the mean and the standard deviation of the number of errors per page for the 250 pages.

$(L)$

## Solution 1.27

(a) $\bar{x} = \dfrac{\Sigma x}{n} = \dfrac{920}{200} = 4.6$

$s^2 = \dfrac{\Sigma x^2}{n} - \bar{x}^2 = \dfrac{5032}{200} - 4.6^2 = 4$

$s = \sqrt{4} = 2$

**The mean is 4.6 errors per page and the standard deviation is 2 errors.**

(b) For the errors, $y$, on the further 50 pages

Mean = 4.4

$\therefore \quad 4.4 = \dfrac{\Sigma y}{50}$

$\Sigma y = 50 \times 4.4 = 220$

The standard deviation = 2.2

$\therefore 2.2^2 = \dfrac{\Sigma y^2}{50} - 4.4^2$

$\Sigma y^2 = 50(2.2^2 + 4.4^2) = \mathbf{1210}$

For the combined set of 250 pages:

Total number of errors $= \Sigma x + \Sigma y = 920 + 220 = 1140$

$\text{Mean} = \tfrac{1140}{250} = \mathbf{4.56}$

$(\text{Standard deviation})^2 = \dfrac{\Sigma x^2 + \Sigma y^2}{250} - 4.56^2$

$= \dfrac{5032 + 1210}{250} - 4.56^2$

$= 4.1744$

Standard deviation $= \sqrt{4.1744} = \mathbf{2.04}$ **(3 s.f.)**

---

In general, for a combined set of numbers

$$\text{mean} = \dfrac{\Sigma x + \Sigma y}{n_1 + n_2} \qquad \text{and} \qquad \text{variance} = \dfrac{\Sigma x^2 + \Sigma y^2}{n_1 + n_2} - (\text{mean})^2$$

Remember that standard deviation = $\sqrt{\text{variance}}$

## Example 1.28

Three statistics students, Ali, Les and Sam, spent the day fishing. They caught three different types of fish and recorded the type and mass (correct to the nearest 0.01 kg) of each fish caught. At 4 p.m., they summarised the results as follows.

| | Number of fish by type | | | All fish caught | |
| | Perch | Tench | Roach | Mean mass (kg) | Standard deviation (kg) |
|---|---|---|---|---|---|
| Ali | 2 | 3 | 7 | 1.07 | 0.42 |
| Les | 6 | 2 | 8 | 0.76 | 0.27 |
| Sam | 1 | 0 | 1 | 1.00 | 0 |

(a) State how it may be deduced from the data that the mass of each fish caught by Sam was 1.00 kg.

(b) The winner was the person who had caught the greatest total mass of fish by 4 p.m. Determine who was the winner, showing your working.

(c) Before leaving the waterside, Sam catches one more fish and weighs it. He then announces that, if this extra fish is included with the other two fish he caught, the standard deviation is 1.00 kg. Find the mass of this extra fish. (C)

## Solution 1.28

(a) If the standard deviation is 0, there is no deviation from the mean. All the readings must be exactly the same as the mean.

Since the mean is 1.00 kg, both fish must have weighed 1.00 kg.

(b)

|  | Number of fish | Mean | Total mass |
|---|---|---|---|
| Ali | 12 | 1.07 kg | $12 \times 1.07 = 12.84$ kg |
| Les | 16 | 0.76 kg | $16 \times 0.76 = 12.16$ kg |
| Sam | 2 | 1.00 kg | $2 \times 1.00 = 2.00$ kg |

**The winner was Ali.**

(c) Sam: let mass of extra fish be $x$, so masses of his three fish are 1, 1, $x$.

$$\bar{x} = \frac{2+x}{3}$$

$$s = \sqrt{\frac{1^2 + 1^2 + x^2}{3} - \bar{x}^2}$$

$$\therefore \quad 1.00 = \sqrt{\frac{2+x^2}{3} - \left(\frac{2+x}{3}\right)^2}$$

$$1 = \frac{2+x^2}{3} - \left(\frac{2+x}{3}\right)^2 \quad \text{(squaring both sides)}$$

$$1 = \frac{2+x^2}{3} - \left(\frac{4 + 4x + x^2}{9}\right)$$

$$9 = 3(2 + x^2) - (4 + 4x + x^2) \quad \text{(multiplying by 9)}$$
$$9 = 6 + 3x^2 - 4 - 4x - x^2$$
$$0 = 2x^2 - 4x - 7$$

$$x = \frac{4 \pm \sqrt{16 - 4(2)(-7)}}{4}$$

$$= \frac{4 \pm \sqrt{72}}{4}$$

$$x = 3.121 \ldots \quad \text{(ignoring negative value for } x\text{)}$$

**Mass of Sam's extra fish is 3.12 kg (2 d.p.)**

# Exercise 1g   Mean and standard deviation

1. The mean of ten numbers is 8. If an eleventh number is now included in the results, the mean becomes 9. What is the value of the eleventh number?

2. The mean of four numbers is 5, and the mean of three different numbers is 12. What is the mean of the seven numbers together?

3. The mean of $n$ numbers is 5. If the number 13 is now included with the $n$ numbers, the new mean is 6. Find the value of $n$.

4. The mean of the numbers 3, 6, 7, $a$, 14, is 8. Find the standard deviation of the set of numbers.

5. The numbers $a$, $b$, 8, 5, 7 have mean 6 and variance 2. Find the values of $a$ and $b$, if $a > b$.

6. For a set of 20 numbers $\Sigma x = 300$ and $\Sigma x^2 = 5500$. For a second set of 30 numbers $\Sigma x = 480$ and $\Sigma x^2 = 9600$. Find the mean and the standard deviation of the combined set of 50 numbers.

7. If the mean of the following frequency distribution is 3.66, find the value of $a$.

| $x$ | 1 | 2 | 3 | 4 | 5 | 6 |
|-----|---|---|---|---|---|---|
| $f$ | 3 | 9 | $a$ | 11 | 8 | 7 |

8. A bag contained five balls each bearing one of the numbers 1, 2, 3, 4, 5. A ball was drawn from the bag, its number noted, and then replaced. This was done 50 times in all and the table below shows the resulting frequency distribution.

| Number | 1 | 2 | 3 | 4 | 5 |
|--------|---|---|---|---|---|
| Frequency | $x$ | 11 | $y$ | 8 | 9 |

If the mean is 2.7, determine the values of $x$ and $y$.

9. Parplan Opinion Polls Ltd conducted a nationwide survey into the attitudes of teenage girls. One of the questions asked was 'What is the ideal age for a girl to have her first baby?' In reply, the sample of 165 girls from the Northern zone gave a mean of 23.4 years and a standard deviation of 1.6 years. Subsequently, the overall sample of 384 girls (Northern plus Southern zones) gave a mean of 24.8 years and a standard deviation of 2.2 years.

Assuming that no girl was consulted twice, calculate the mean and standard deviation for the 219 girls from the Southern zone.      (AEB)

10. The manager of a car showroom monitored the numbers of cars sold during two successive five-day periods. During the first five days the numbers of cars sold per day had mean 1.8 and variance 0.56. During the next five days the numbers of cars sold per day had mean 2.8 and variance 1.76. Find the mean and variance of the numbers of cars sold per day during the full ten days.      (NEAB)

11. Prior to the start of delicate wage negotiations in a large company, the unions and the management take independent samples of the work force and ask them at what percentage level they believe a settlement should be made. The results are as follows:

| Sample | Size | Mean | Standard deviation |
|--------|------|------|--------------------|
| 'management' | 350 | 12.4% | 2.1% |
| 'union' | 237 | 10.7% | 1.8% |

Assuming that no individual was consulted by both sides, calculate the mean and standard deviation for these 587 workers.      (AEB)

12. In a germination experiment, 200 rows of seeds, with ten seeds per row, were incubated. The frequency distribution of the number of seeds which germinated per row is shown below.

| Number of seeds germinated | Frequency |
|----------------------------|-----------|
| 0 | 4 |
| 1 | 10 |
| 2 | 16 |
| 3 | 28 |
| 4 | 34 |
| 5 | 44 |
| 6 | 32 |
| 7 | 16 |
| 8 | 10 |
| 9 | 6 |
| 10 | 0 |

(a) Calculate the mean and the standard deviation of the number of seeds germinating per row.

For another 50 rows an analysis shows that the mean is 4.4 seeds and the standard deviation is 2.2 seeds.

(b) Determine the mean and, to two decimal places, the standard deviation for the 250 rows.      (L)

13. The figures in the table below are the ages, to the nearest year, of a random sample of 30 people negotiating a mortgage with a bank.

| 29 | 26 | 31 | 42 | 38 |
|----|----|----|----|----|
| 45 | 35 | 37 | 38 | 38 |
| 36 | 39 | 49 | 40 | 32 |
| 32 | 34 | 27 | 61 | 29 |
| 33 | 31 | 33 | 52 | 44 |
| 32 | 30 | 38 | 42 | 33 |

Copy and complete the following stem and leaf diagram. Use the diagram to identify two features of the shape of the distribution.

| 25 | 4  1 |
|----|------|
| 30 | 1 |
| 35 | |

Find the mean age of the 30 people. Given that 18 of them are men and that the mean age of the men is 37.72, find the mean age of the 12 women. (*MEI*)

14. A travel agency has two shops, R and S. The number of holidays purchased in a particular week and the mean and standard deviation of the costs of these holidays at each shop are shown in the following table.
Calculate the mean, and, to the nearest penny, the standard deviation of the costs of all the 56 holidays purchased.

|        | Number of holidays | Mean cost (£) | S.D. (£) |
|--------|--------------------|---------------|----------|
| Shop R | 32                 | 190.35        | 10.4     |
| Shop S | 24                 | 202.25        | 15.5     |

(*L*)

15. Three random samples of 50, 30 and 20 bags respectively are taken from the production line of '12 kg bags' of cat litter. The contents of each bag are then weighed. A summary of the results is shown in the table.

| Sample | Size | Mean wt. (kg) | S.D. (kg) |
|--------|------|---------------|-----------|
| 1      | 50   | 11.8          | 0.5       |
| 2      | 30   | 12.1          | 0.9       |
| 3      | 20   | 11.7          | 1.1       |

Find, in kilograms to two decimal places, the mean weight per bag and the standard deviation for the 100 bags. (*L*)

16. The average height of 20 boys is 160 cm, with a standard deviation of 4 cm. The average height of 30 girls is 155 cm, with a standard deviation of 3.5 cm. Find the standard deviation of the whole group of 50 children.

# SCALING SETS OF DATA

## Example 1.29

Sweets are packed into bags with a nominal mass of 75 g. Ten bags are picked at random from the production line and weighed. Their masses, in grams, are

76, 74.2, 75.1, 73.7, 72, 74.3, 75.4, 74, 73.1, 72.8

(a) Use your calculator to find the mean mass and the standard deviation.

It was later discovered that the scales were reading 3.2 g below the correct weight.

(b) What was the correct mean mass of the ten bags and the correct standard deviation?

(c) Compare your answers to (a) and (b) and comment.

## Solution 1.29

(a) According to the scales with measurements being given in grams

$\bar{x} = 74.06, s = 1.166 \ldots = 1.17$ (2 d.p.)

(b) The correct readings are:
79.2, 77.4, 78.3, 76.9, 75.2, 77.5, 78.6, 77.2, 76.3, 76
$\bar{x} = 77.26$, $s = 1.166 \ldots = 1.17$ (2 d.p.)

(c) Notice that $77.26 - 74.06 = 3.2$ i.e. correct mean – original mean = 3.2
**So correct mean = original mean + 3.2: correct s.d. = original s.d.**

**If each reading is increased by 3.2, then the mean is increased by 3.2. The standard deviation, however, remains unaltered.**

Showing the two sets of readings on a graph helps to show that although the mean increased, the spread of the data about the mean remained the same.

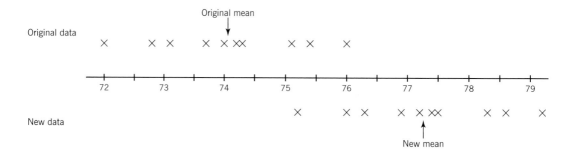

In general, if each number is increased by a constant $c$

- the mean is increased by $c$,
- the standard deviation remains unaltered,

i.e.   if $y = x + c$,   then $\bar{y} = \bar{x} + c$
and $s_y = s_x$

Now consider what happens when each number in a set of readings is *multiplied by a constant*.

For the four numbers 2, 3.5, 5, 6   $\bar{x} = 4.125$,   $s_x = 1.515 \ldots$

Multiplying each number by 3 to obtain $y$, where $y = 3x$
gives the numbers 6, 10.5, 15, 18.

For these, $\bar{y} = 12.375$,   $s_y = 4.546 \ldots$

Now $12.375 \div 4.125 = 3$,       so $\bar{y} = 3\bar{x}$
and $4.546 \cdots \div 1.515 \cdots = 3$,   so $s_y = 3s_x$

You can see from the diagram that the new set of data is much more spread out.

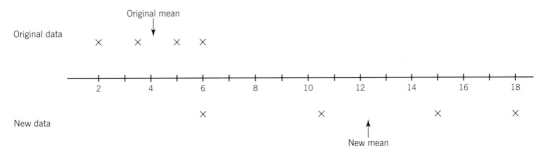

In general, if each number is multiplied by a constant $k$

- the mean is multiplied by $k$,
- the standard deviation is multiplied by $|k|$ where $|k|$ is the positive value of $k$.

i.e.   if $y = kx$,   then $\bar{y} = k\bar{x}$
$$s_y = |k|\, s_x$$

For example, if $y = -\frac{1}{2}x$,   then $\bar{y} = -\frac{1}{2}\bar{x}$ and $s_y = \frac{1}{2}s_x$   since $\left|-\frac{1}{2}\right| = \frac{1}{2}$.

Combining these two results,

if $\qquad y = ax + b$   where $a$ and $b$ are constants
then $\quad \bar{y} = a\bar{x} + b$
and $\quad s_y = |a|\, s_x$

## Example 1.30

Joe's mean mark for the physics tests for the term was 72. His teacher decided to scale all the marks according to the formula $y = 2x - 6$, where $y$ is the new mark and $x$ the original mark.

Find Joe's new mean mark.

## Solution 1.30

$$y = 2x - 6$$
$$\therefore \qquad \bar{y} = 2\bar{x} - 6$$
$$= 2 \times 72 - 6$$
$$= 84$$

**Joe's new mean mark is 84.**

---

## Example 1.31

The standard deviation of three numbers $a$, $b$, $c$ is 3.2.

(a) State the standard deviation of the three numbers $3a$, $3b$, $3c$.
(b) State the standard deviation of the three numbers $a + 2$, $b + 2$, $c + 2$.
(c) State the standard deviation of the three numbers $2a + 5$, $2b + 5$, $2c + 5$. (C)

## Solution 1.31

(a) If $y = 3x$, then $s_y = 3s_x = 3 \times 3.2 = 9.6$

(b) If $y = x + 2$, then $s_y = s_x = 3.2$

(c) If $y = 2x + 5$, then $s_y = 2s_x = 2 \times 3.2 = 6.4$

---

# Comparing data by scaling

If you wish to compare two sets of data, for example examination marks in two papers, you can scale one of the sets of data so that the two means are the same and the two standard deviations are the same.

## Example 1.32

For students on an Electronics course the assessment consists of two components: a written examination paper and a project. The marks for the examination paper are distributed with a mean of 62 and a standard deviation of 16. Those for the project have a mean of 37 and a standard deviation of 6. Anna, a student on the course, scored 80 marks on the examination paper and 46 marks for her project.

(a) Transform each of Anna's marks into a standardised score, such that, for each component, the mean and standard deviation for all students on the course are 50 and 20, respectively.

(b) Hence compare Anna's relative performance in the two assessment components.    (*NEAB*)

## Solution 1.32

(a) Standardised values: $\bar{y} = 50$, $s_y = 20$

Examination    $\bar{x} = 62$, $s_x = 16$

Let                      $y = ax + b$

then                    $\bar{y} = a\bar{x} + b$

$50 = 62a + b$ ......①

Now                   $s_y = as_x$

∴                        $20 = a \times 16$

$a = 1.25$

Substituting in ①         $50 = 62 \times 1.25 + b$

$b = -27.5$

The transformation for the examination paper is $y = 1.25x - 27.5$

When $x = 80$, $y = 1.25 \times 80 - 27.5 = 72.5$

**Anna's standardised mark for the examination is 72.5.**

Project $\quad \bar{x} = 37, s_x = 6$

Let $\qquad y = cx + d$

then $\qquad \bar{y} = c\bar{x} + d$

$$50 = 37c + d \ \dots\dots ②$$

Now $\qquad s_y = cs_x$

$\therefore \qquad 20 = c \times 6$

$$c = 3\tfrac{1}{3}$$

Substituting in ② $\qquad 50 = 37 \times 3\tfrac{1}{3} + d$

$$d = -73\tfrac{1}{3}$$

The transformation for the project is $y = 3\tfrac{1}{3}x - 73\tfrac{1}{3}$.

When $x = 46$, $y = 3\tfrac{1}{3} \times 46 - 73\tfrac{1}{3} = 80$

**Anna's standardised mark for the project is 80.**

(b) Relatively, Anna performed **better on the project** than in the examination.

---

# Exercise 1h   Scaling sets of data

1. (a) Find the mean and the standard deviation of the set of numbers 4, 6, 9, 3, 5, 6, 9.
   (b) Deduce the mean and the standard deviation of the set of numbers 514, 516, 519, 513, 515, 516, 519.
   (c) Deduce the mean and the standard deviation of the set of numbers 52, 78, 117, 39, 65, 78, 117.

2. A set of numbers has a mean of 22 and a standard deviation of 6. If 3 is added to each number of the set, and each resulting number is then doubled, find the mean and standard deviation of the new set. (*C Additional*)

3. A set of values of a variable $X$ has a mean $\mu$ and a standard deviation $\sigma$. State the new value of the mean and of the standard deviation when each of the variables is (a) increased by $k$, (b) multiplied by $p$. Values of a new variable $Y$ are obtained by using the formula $Y = 3X + 5$. Find the mean and the standard deviation of the set of values of $Y$. (*C Additional*)

4. Show that the standard deviation of the integers 1, 2, 3, 4, 5, 6, 7 is 2.

   Using this result find the standard deviation of the numbers
   (a) 101, 102, 103, 104, 105, 106, 107.
   (b) 100, 200, 300, 400, 500, 600, 700.
   (c) 2.01, 3.02, 4.03, 5.04, 6.05, 7.06, 8.07.
   (d) Write down seven integers which have mean 5 and standard deviation 6. (*L Additional*)

5. It is proposed to convert a set of marks whose mean is 52 and standard deviation is 4 to a set of marks with mean 61 and standard deviation 3. The equation for the transformation necessary to convert the marks is $y = ax + b$. Find

   (a) the values of $a$ and $b$,
   (b) the value of the scaled mark which corresponds to a mark of 64 in the original data,
   (c) the value in the original data if the scaled mark is 79.

6. The marks of five students in a mathematics test were 27, 31, 35, 47, 50.
   (a) Calculate the mean mark and the standard deviation.
   (b) The marks are scaled so that the mean and standard deviation become 50 and 20 respectively. Calculate, to the nearest whole number, the new marks corresponding to the original marks of 31 and 50. (*C Additional*)

7. It is proposed to convert a set of values of a variable $X$, whose mean and standard deviation are 20 and 5 respectively, to a set of values of a variable $Y$ whose mean and standard deviation are 42 and 8 respectively. If the conversion formula is $Y = aX + b$, calculate the values of $a$ and of $b$. (*C Additional*)

8. In order to compare the performances of candidates in two schools a test was given. The mean mark at school $A$ was 45, and the mean mark at school $B$ was 31 with a standard deviation of 5. The marks of school $A$ are scaled so that the mean and standard deviation are the same as school $B$ and a mark of 85 at school $A$ becomes 63. Find the values of $a$ and $b$ if the transformation used is $y = ax + b$. Find also the original standard deviation of the marks from school $A$.

9.  The following is a set of 109 examination marks ordered for convenience.

| 6 | 11 | 11 | 12 | 13 | 14 | 16 | 17 | 18 | 20 |
|---|----|----|----|----|----|----|----|----|----|
| 21 | 21 | 23 | 24 | 25 | 25 | 25 | 25 | 26 | 26 |
| 27 | 27 | 28 | 28 | 28 | 29 | 29 | 29 | 30 | 31 |
| 31 | 32 | 32 | 32 | 33 | 33 | 34 | 34 | 35 | 36 |
| 36 | 37 | 37 | 37 | 37 | 38 | 38 | 38 | 39 | 39 |
| 39 | 39 | 39 | 39 | 39 | 39 | 40 | 40 | 40 | 40 |
| 40 | 40 | 41 | 41 | 41 | 42 | 42 | 42 | 42 | 43 |
| 43 | 43 | 44 | 45 | 46 | 46 | 47 | 47 | 47 | 47 |
| 48 | 50 | 50 | 51 | 51 | 52 | 52 | 52 | 53 | 53 |
| 54 | 54 | 55 | 57 | 58 | 58 | 59 | 59 | 61 | 62 |
| 63 | 64 | 66 | 66 | 67 | 70 | 76 | 77 | 82 | |

(a) Construct a grouped frequency distribution using a class width of 10 and starting with 0–9.

(b) Draw a histogram and comment on the shape of the distribution.

(c) Using the frequency table estimate the mean and standard deviation of the marks.

(d) The marks are to be scaled linearly by the relation $Y = a + bX$ where $X$ is the old mark and $Y$ the new mark. The new mean and standard deviation are to be 50 and 10 respectively. Using your estimates in (c) calculate suitable values for $a$ and $b$.

10. The mean of the marks scored by candidates in an examination is 45. These marks are scaled linearly to give a mean of 50 and a standard deviation of 15. Given that the scaled mark of 80 corresponds to an original mark of 70, calculate
(a) the standard deviation of the original marks,
(b) the mark which is unchanged by the scaling.

Given that the greatest and least scaled marks are 92 and 2 respectively, calculate the corresponding original marks.      (C *Additional*)

# USING A METHOD OF CODING TO FIND THE MEAN AND STANDARD DEVIATION

## Example 1.33

Salt is packed in bags which the manufacturer claims contain 25 kg each. Eighty bags are examined and the mass, $x$ kg, of each is found. The results are $\Sigma(x - 25) = 27.2$, $\Sigma(x - 25)^2 = 85.1$. Find the mean and the standard deviation of the masses.

## Solution 1.33

You do not know the actual masses and a coding has been used to summarise the results. The coding is $y = x - 25$, where $\Sigma y = 27.2$ and $\Sigma y^2 = 85.1$

Therefore
$$\bar{y} = \frac{\Sigma y}{n}$$
$$= \frac{27.2}{80}$$
$$= 0.34$$

$$s_y{}^2 = \frac{\Sigma y^2}{n} - \bar{y}^2$$
$$= \frac{85.1}{80} - 0.34^2$$
$$= 0.948\ 15$$
$$s_y = 0.9737 \ldots$$

Now if $y = x - 25$, then     $x = y + 25$

So                                     $\bar{x} = \bar{y} + 25$

Therefore                         $\bar{x} = 0.34 + 25$

$= 25.34$

Also                                  $s_x = s_y$

so                                      $s_x = 0.9737 \ldots$

**The mean mass is 25.34 kg and the standard deviation is 0.97 kg (2 d.p.).**

NOTE: The value 25 used here is sometimes known as the **assumed mean**.

## Example 1.34

Use the coding $y = \dfrac{x - 200\,000}{25\,000}$ to find the mean and standard deviation of the following:

| $x$ | 125 000 | 150 000 | 175 000 | 200 000 | 225 000 | 250 000 | 275 000 |
|-----|---------|---------|---------|---------|---------|---------|---------|
| $f$ | 5 | 19 | 27 | 35 | 24 | 12 | 3 |

## Solution 1.34

$$y = \frac{x - 200\,000}{25\,000}$$

so $\qquad 25\,000y = x - 200\,000$

i.e. $\qquad x = 25\,000y + 200\,000$

$\qquad \bar{x} = 25\,000\bar{y} + 200\,000 \quad$ and $\quad s_x = 25\,000s_y$

| $x$ | $y = \dfrac{x - 200\,000}{25\,000}$ | $f$ | $fy$ | $fy^2$ |
|-----|-----|-----|-----|-----|
| 125 000 | −3 | 5 | −15 | 45 |
| 150 000 | −2 | 19 | −38 | 76 |
| 175 000 | −1 | 27 | −27 | 27 |
| 200 000 | 0 | 35 | 0 | 0 |
| 225 000 | 1 | 24 | 24 | 24 |
| 250 000 | 2 | 12 | 24 | 48 |
| 275 000 | 3 | 3 | 9 | 27 |
| | | $\Sigma f = 125$ | $\Sigma fy = -23$ | $\Sigma fy^2 = 247$ |

$\bar{y} = \dfrac{\Sigma fy}{\Sigma f} = -\dfrac{23}{125} = -0.184$

$s_y^2 = \dfrac{\Sigma fy^2}{\Sigma f} - \bar{y}^2$

$\quad = \dfrac{247}{125} - (-0.184)^2$

$\quad = 1.942 \ldots$

$s_y = 1.393 \ldots$

$\bar{x} = 25\,000 \times (-0.184) + 200\,000$
$\quad = 195\,400$

$s_x = 25\,000s_y$
$\quad = 25\,000 \times 1.393 \ldots$
$\quad = 34\,840.207 \ldots$
$\quad = 34\,800 \; (3 \text{ s.f.})$

**The mean is 195 400 and standard deviation 34 800 (3 s.f.).**

---

In general, if the set of numbers $x_1, x_2, \ldots, x_n$ is transformed to the set of numbers $y_1, y_2, \ldots, y_n$ by means of the coding

$$y = \frac{x - a}{b}$$

then $\qquad x = a + by$

so $\qquad \bar{x} = a + b\bar{y}$

and $\qquad s_x = bs_y$

# Exercise 1i   Coding

1. Find the mean and the standard deviation of the following sets of data, using the coding indicated:

(a)

| $x$ | $f$ |
|-----|-----|
| 304 | 1 |
| 308 | 5 |
| 312 | 9 |
| 316 | 4 |
| 320 | 4 |
| 324 | 2 |

$y = \dfrac{x - 312}{4}$

(b)

| Interval | $f$ |
|----------|-----|
| $100 \leqslant x < 200$ | 3 |
| $200 \leqslant x < 300$ | 7 |
| $300 \leqslant x < 400$ | 12 |
| $400 \leqslant x < 500$ | 18 |
| $500 \leqslant x < 600$ | 12 |
| $600 \leqslant x < 700$ | 6 |

$y = \dfrac{x - 450}{100}$

(c)

| Interval | $f$ |
|----------|-----|
| 0– | 5 |
| 0.005– | 10 |
| 0.01– | 13 |
| 0.015– | 18 |
| 0.02– | 12 |
| 0.025– | 6 |
| 0.03– | 6 |
| 0.035– | 0 |

$y = \dfrac{x - 0.0225}{0.005}$

2. For a particular set of data
$$n = 100, \quad \Sigma(x - 50) = 123.5, \quad \Sigma(x - 50)^2 = 238.4$$
Find the mean and the standard deviation of $x$.

3. Find the variance of $x$ if
$$\Sigma f(x - 100) = 127, \quad \Sigma f(x - 100)^2 = 2593,$$
$$\Sigma f = 20$$

4. Each morning for a month the owner of a smallholding timed how long it took to feed the animals. The results were as shown:

| Time (min) | Frequency |
|------------|-----------|
| –15 | 0 |
| –20 | 3 |
| –25 | 2 |
| –30 | 6 |
| –35 | 10 |
| –40 | 7 |
| –45 | 2 |
| –50 | 1 |

Calculate the mean time taken to feed the animals, using a method of coding.

5. The table shows the times taken on 30 consecutive days for a coach to complete one journey on a particular route. Times have been given to the nearest minute. Find the mean time for the journey and the standard deviation, using a method of coding.

| Time (min) | Frequency |
|------------|-----------|
| 60–63 | 1 |
| 64–67 | 3 |
| 68–71 | 12 |
| 72–75 | 10 |
| 76–79 | 4 |

6. In a practical class students timed how long it took for a sample of their saliva to break down a 2% starch solution. The times, to the nearest second are shown in the table below. Find the mean time, using a method of coding.

| Time (seconds) | Frequency |
|----------------|-----------|
| 11–20 | 1 |
| 21–30 | 2 |
| 31–40 | 5 |
| 41–50 | 11 |
| 51–60 | 8 |
| 61–70 | 2 |
| 71–90 | 1 |

# CUMULATIVE FREQUENCY

The **cumulative frequency** is the total frequency up to a particular item. A cumulative frequency distribution can be obtained from a frequency distribution and can be illustrated

(a) when the data are discrete and ungrouped – by drawing a step diagram,

(b) when the data are continuous or in the form of a grouped discrete distribution – by drawing a cumulative frequency polygon or curve.

## (a) Cumulative frequency – step diagrams for discrete ungrouped data

The table shows the number of attempts needed to pass the driving test by 100 candidates at a particular test centre.

| Number of attempts | 1 | 2 | 3 | 4 | 5 | 6 |
|---|---|---|---|---|---|---|
| Frequency (Number of candidates) | 33 | 42 | 13 | 6 | 4 | 2 |

The cumulative frequency distribution is formed as follows:

| Number of attempts | $\leqslant 1$ | $\leqslant 2$ | $\leqslant 3$ | $\leqslant 4$ | $\leqslant 5$ | $\leqslant 6$ |
|---|---|---|---|---|---|---|
| Cumulative frequency | 33 | 75 | 88 | 94 | 98 | 100 |

$\uparrow$     $\uparrow$         $\uparrow$

$33 + 42$    $33 + 42 + 13$        total number of candidates

Plot the cumulative frequency against the number of attempts and decide how to join the points.

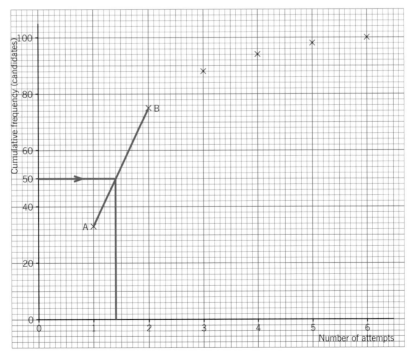

Plotting the values $(1, 33), (2, 75), (3, 88), \ldots$ tells you that 33 people took 1 attempt, 75 people took $\leqslant 2$ attempts, 88 people took $\leqslant 3$ attempts.

If you join the points with straight lines, such as from A to B, and consider a cumulative frequency of 50, this would suggest that 50 people took $\leqslant 1.4$ attempts which is nonsense.

Clearly it is not sensible to join the points directly.

When data are discrete (and usually integer) values and also are ungrouped, the points can be joined by a series of **steps** as shown:

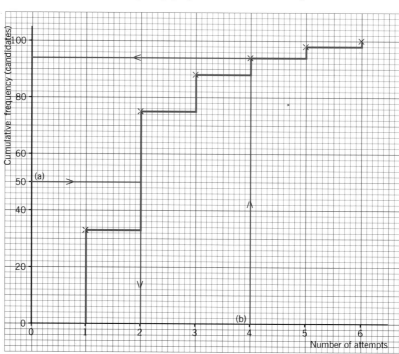

**Cumulative frequency graph of number of attempts**

The step diagram is necessary because the values are not distributed evenly throughout the intervals 1 to 2, 2 to 3, ... but 'jump' or 'step up' from 1 to 2, then 2 to 3 and so on.

(a) Consider a cumulative frequency of 50. From the graph, the number of attempts is two. This means that when the data are placed in ascending order of size, the 50th item is 2, i.e. the 50th person took two attempts.

(b) To find how many took up to four attempts, go up vertically from 4 on the horizontal axis, to the **top** of the step, then go left to the cumulative frequency axis. This shows that 94 candidates took up to four attempts. If you go to the bottom of the step, this tells you the number of candidates who took fewer than four attempts (88 in this case).

Notice that it only makes sense when you read from the discrete values on the horizontal axis. It would be silly to consider 3.6 attempts, for example.

Note that in a step diagram, the **mode** is given by the value of the variable that gives the 'steepest' step.

From the graph above, the mode is **two**.

## (b) Cumulative frequency polygons and curves for grouped data

Consider this situation:

Six weeks after planting, the heights of 30 broad bean plants were measured and the frequency distribution formed as shown.

| Height, $x$ cm | $3 \leqslant x < 6$ | $6 \leqslant x < 9$ | $9 \leqslant x < 12$ | $12 \leqslant x < 15$ | $15 \leqslant x < 18$ | $18 \leqslant x < 21$ |
|---|---|---|---|---|---|---|
| Frequency | 1 | 2 | 11 | 10 | 5 | 1 |

The cumulative frequency is calculated up to each **upper class boundary.**

The upper class boundaries are 6, 9, 12, 15, 18, 21.

The lower boundary of the first class is 3. This is inserted for completeness.

| Height, $x$ cm | $< 3$ | $< 6$ | $< 9$ | $< 12$ | $< 15$ | $< 18$ | $< 21$ |
|---|---|---|---|---|---|---|---|
| Cumulative frequency | 0 | 1 | 3 | 14 | 24 | 29 | 30 |

$$\underset{1+2}{\uparrow} \qquad \underset{1+2+11}{\uparrow} \qquad\qquad \underset{\substack{\text{total number}\\\text{of plants.}}}{\uparrow}$$

This can be shown diagrammatically in a **cumulative frequency graph.**

The cumulative frequencies are plotted against the **upper class boundaries** and the points are joined as follows:

(i)  for a **cumulative frequency polygon,**
   – join the points with straight lines, indicating that you are assuming that the readings are evenly distributed throughout the interval. This ties in with the fact that you draw horizontal lines at the tops of the blocks in a histogram.
(ii) for a **cumulative frequency curve,**
   – join the points with a smooth curve. In this case you are assuming a distribution of readings throughout the interval which might not be even.

**Cumulative frequency curve to show the heights of 30 broad bean plants**

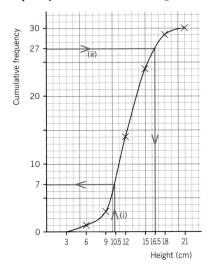

Values can be estimated from the graph. Note that the graph can be read in either direction.

(i)  To find the number of plants that were less than 10.5 cm tall:
- Find the height 10.5 cm on the horizontal axis
- Draw a vertical line up from 10.5 to meet the curve
- Draw a horizontal line to the cumulative frequency axis and read the value

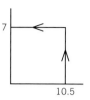

**From the graph, seven plants were less than 10.5 cm tall**

(ii)  To find $x$ where 90% of the plants were less than $x$ cm tall:
- 90% of 30 = 27
- Find 27 on the vertical axis and draw a horizontal line to meet the curve
- Draw a vertical line to the horizontal or 'height' axis and read the value

**From the graph, 27 plants were less than 16.5 cm tall, so $x = 16.5$**

## Example 1.35

A survey is carried out to determine the numbers of pupils in various age groups who are attending nurseries, schools and colleges within a certain area. The results are summarised in the following histogram.

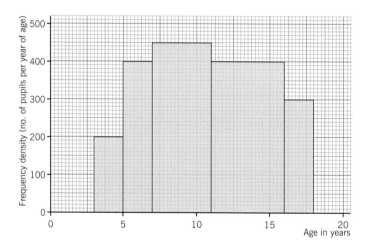

(a)  Copy and complete the following table showing the ages of the pupils and the corresponding cumulative frequencies.

| Age in years up to | 3 | 5 | 7 | 11 | 16 | 18 |
|---|---|---|---|---|---|---|
| Cumulative frequency | 0 | | | | | 5600 |

(b)  Draw a cumulative frequency diagram for the distribution.
(c)  Use your cumulative frequency diagram to estimate the age exceeded by 30% of the pupils in the survey.

(NEAB)

### Solution 1.35

Frequency = frequency density × width, so for $3 \leqslant \text{age} < 5$, $f = 200 \times 2 = 400$

Calculating the other frequencies gives

| Age | $3 \leqslant x < 5$ | $5 \leqslant x < 7$ | $7 \leqslant x < 11$ | $11 \leqslant x < 16$ | $16 \leqslant x < 18$ |
|---|---|---|---|---|---|
| Frequency (Number of pupils) | 400 | 800 | 1800 | 2000 | 600 |

(a) The cumulative frequency table is

| Age in years up to | 3 | 5 | 7 | 11 | 16 | 18 |
|---|---|---|---|---|---|---|
| Cumulative frequency | 0 | 400 | 1200 | 3000 | 5000 | 5600 |

(b) **Cumulative frequency polygon showing numbers in various age groups**

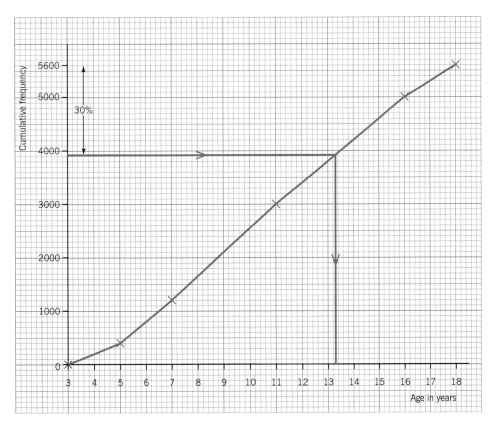

(c) If 30% of the pupils exceed a certain age, then 70% of pupils are younger than this age.
Now 70 % of 5600 = 3920
From the graph, 3920 pupils have an age up to 13.3 years, i.e. approx 13 years 4 months

so **30% of the pupils are older than 13 years 4 months.**

## Example 1.36

Students were asked how long it took them to travel to college on a particular morning. A cumulative frequency distribution was formed:

| Time taken (minutes) | Cumulative frequency |
|:---:|:---:|
| <5 | 28 |
| <10 | 45 |
| <15 | 81 |
| <20 | 143 |
| <25 | 280 |
| <30 | 349 |
| <35 | 374 |
| <40 | 395 |
| <45 | 400 |

(a) Draw a cumulative frequency polygon.
(b) Estimate how many students took less than 18 minutes.
(c) Taking equal class intervals of 0–, 5–, 10–, ..., construct a frequency distribution and draw a histogram.

## Solution 1.36

(a) **Cumulative frequency polygon to show the times taken to travel to college**

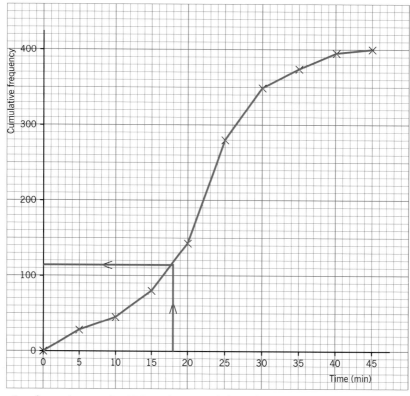

(b) Estimating from the graph, **114 students took less than 18 minutes.**

(c) Form the frequency distribution and calculate the frequency density for each interval, where frequency density = frequency ÷ interval width.
Note that the width of each interval is 5.

| Upper boundary | Cumulative frequency | Time (min) | Frequency | Frequency density |
|---|---|---|---|---|
| 5 | 28 | 0– | 28 | 5.6 |
| 10 | 45 | 5– | 45–28 = 17 | 3.4 |
| 15 | 81 | 10– | 81–45 = 36 | 7.2 |
| 20 | 143 | 15– | 143–81 = 62 | 12.4 |
| 25 | 280 | 20– | 280–143 = 137 | 27.4 |
| 30 | 349 | 25– | 349–280 = 69 | 13.8 |
| 35 | 374 | 30– | 374–349 = 25 | 5 |
| 40 | 395 | 35– | 395–374 = 21 | 4.2 |
| 45 | 400 | 40–(45) | 400–395 = 5 | 1 |
| | | | Total = 400 | |

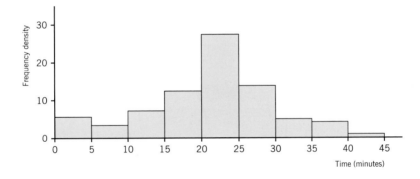

# CUMULATIVE PERCENTAGE FREQUENCY DIAGRAMS

These are particularly useful when two or more distributions are to be compared. For example, suppose we have the examination marks of 200 boys and 300 girls in Year 8.

| Mark | Cumulative frequency (boys) | Cumulative frequency (girls) |
|---|---|---|
| <10 | 6 | 0 |
| <20 | 22 | 6 |
| <30 | 60 | 12 |
| <40 | 140 | 24 |
| <50 | 172 | 42 |
| <60 | 188 | 75 |
| <70 | 196 | 120 |
| <80 | 198 | 246 |
| <90 | 200 | 294 |
| <100 | 200 | 300 |

Obtain the cumulative percentage frequencies as follows:

| Mark | Boys (total 200) | | Girls (total 300) | |
| --- | --- | --- | --- | --- |
| | Cumulative frequency | Cumulative % frequency | Cumulative frequency | Cumulative % frequency |
| <10 | 6 | $\frac{6}{200} = 3\%$ | 0 | $\frac{0}{300} = 0\%$ |
| <20 | 22 | $\frac{22}{200} = 11\%$ | 6 | $\frac{6}{300} = 2\%$ |
| <30 | 60 | $\frac{60}{200} = 30\%$ | 12 | $\frac{12}{300} = 4\%$ |
| <40 | 140 | $\frac{140}{200} = 70\%$ | 24 | $\frac{24}{300} = 8\%$ |
| <50 | 172 | $\frac{172}{200} = 86\%$ | 42 | $\frac{42}{300} = 14\%$ |
| <60 | 188 | $\frac{188}{200} = 94\%$ | 75 | $\frac{75}{300} = 25\%$ |
| <70 | 196 | $\frac{196}{200} = 98\%$ | 120 | $\frac{120}{300} = 40\%$ |
| <80 | 198 | $\frac{198}{200} = 99\%$ | 246 | $\frac{246}{300} = 82\%$ |
| <90 | 200 | $\frac{200}{200} = 100\%$ | 294 | $\frac{294}{300} = 98\%$ |
| <100 | 200 | $\frac{200}{200} = 100\%$ | 300 | $\frac{300}{300} = 100\%$ |

**Cumulative percentage frequency curves**

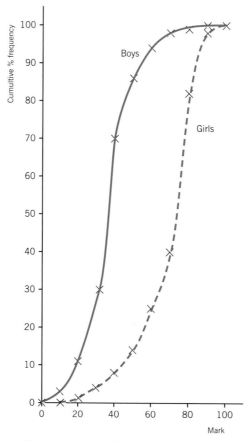

Great care must be taken when comparing these curves. A common mistake is to say that the boys have done better than the girls because the boys' graph is above that of the girls. If you calculate the corresponding percentage frequencies and draw the histograms you will see that this is not the case.

| Boys | | | | |
|---|---|---|---|---|
| Mark | Cumulative % frequency | Mark | % Frequency | Frequency density |
| <10 | 3% | 0– | 3% | 0.3 |
| <20 | 11% | 10– | 8% | 0.8 |
| <30 | 30% | 20– | 19% | 1.9 |
| <40 | 70% | 30– | 40% | 4.0 |
| <50 | 86% | 40– | 16% | 1.6 |
| <60 | 94% | 50– | 8% | 0.8 |
| <70 | 98% | 60– | 4% | 0.4 |
| <80 | 99% | 70– | 1% | 0.1 |
| <90 | 100% | 80– | 1% | 0.1 |
| <100 | 100% | 90– | 0% | 0 |
| | | | Total 100% | |

| Girls | | | | |
|---|---|---|---|---|
| Mark | Cumulative % frequency | Mark | % Frequency | Frequency density |
| <10 | 0% | 0– | 0% | 0 |
| <20 | 2% | 10– | 2% | 0.2 |
| <30 | 4% | 20– | 2% | 0.2 |
| <40 | 8% | 30– | 4% | 0.4 |
| <50 | 14% | 40– | 6% | 0.6 |
| <60 | 25% | 50– | 11% | 1.1 |
| <70 | 40% | 60– | 15% | 1.5 |
| <80 | 82% | 70– | 42% | 4.2 |
| <90 | 98% | 80– | 16% | 1.6 |
| <100 | 100% | 90– | 2% | 0 2 |
| | | | Total 100% | |

**Boys' results**

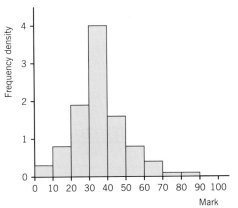

**Girls' results**

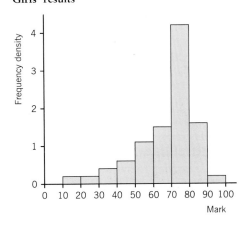

It is easy to see that the girls have done better than the boys. The modal class for the boys' marks is 30–39, whereas the modal class for the girls' marks is 70–79. The type of distribution for the boys' marks is said to be positively skewed and that for the girls' marks is said to be negatively skewed.

# MEDIAN, QUARTILES AND PERCENTILES

The median is an average that is unaffected by extreme values. It is often described as the middle value.

For a set of observations arranged in order of size, the median is the value 50% of the way through the distribution.

Other quantities that are unaffected by extreme values are the **quartiles** and **percentiles**. These are useful in giving an idea of the variability or spread of the data.

For data arranged in order of size:

- the lower quartile $Q_1$ is the value 25% of the way through the distribution,
- the upper quartile $Q_3$ is the value 75% of the way through the distribution,
- the $n$th percentile, $P_n$ is the value $n$% of the way through the distribution.

Therefore the median (sometimes called $Q_2$) is the 50th percentile, the lower quartile is the 25th percentile and the upper quartile is the 75th percentile. The quartiles, together with the median, split the distribution into four equal parts.

# Interquartile range

The difference between the quartiles, $Q_3 - Q_1$ is known as the **interquartile range**. It tells you the range of the middle 50% of the distribution and so is also unaffected by extreme values.

Interquartile range = upper quartile – lower quartile = $Q_3 - Q_1$

# Interpercentile range

Ranges between various percentiles can be found. For example, the range giving the middle 80% of the readings is found by subtracting the 10th percentile from the 90th percentile, i.e. $P_{90} - P_{10}$.

When finding the median and percentiles, it is important to take note of whether the data are grouped or ungrouped.

# Ungrouped data – median

For ungrouped data consisting of $n$ observations in order of size, the median is the $\frac{1}{2}(n + 1)$th observation.

(a) Consider this set of numbers: 7, 7, 2, 3, 4, 2, 7, 9, 31.
There are nine numbers, so the median is the $\frac{1}{2}(9 + 1)$th observation, i.e. the 5th observation.
Arranging them in order gives 2, 2, 3, 4, $\boxed{7}$ 7, 7, 9, 31

↑<br>5th observation

The median is 7.

(b) Consider this set of numbers: 36, 41, 27, 32, 29, 39, 39, 43.

There are eight numbers, so the median is the $\frac{1}{2}(8 + 1)$th observation, i.e. the 4.5th observation. As this does not exist, find the value that is half-way between the 4th and 5th observations.

Arranging the numbers in order of size gives 27, 29, 32, $\boxed{36, 39}$ 39, 41, 43

↑<br>4.5th observation

The median is half-way between 36 and 39, so median = $\frac{1}{2}(36 + 39) = 37.5$

Note that
- if there is an odd number of observations, the median is the middle value,
- if there is an even number of observations, there are two middle values. If these are $c$ and $d$, the median is halfway between them, i.e. $\frac{1}{2}(c + d)$.

## Ungrouped data – quartiles

The quartiles should divide in half the two distributions either side of the median, for example:

(a) 3  3  ⑤  6  8  ⑨  12  14  ⑲  20  24
$\qquad$ ↑ $\qquad\qquad$ ↑ $\qquad\qquad\qquad$ ↑
$\qquad$ $Q_1$ (lower $\quad$ $Q_2$ (median) $\quad$ $Q_3$ (upper
$\qquad$ quartile) $\qquad\qquad\qquad\qquad$ quartile)

$Q_1 = 5$
$Q_2 = 9$
$Q_3 = 19$

Interquartile range = $Q_3 - Q_1 = 19 - 5 = 14$

(b) 20  ㉓  23 ↑ 26  ㉗  28
$\qquad$ ↑ $\qquad$ ↑ $\qquad$ ↑
$\qquad$ $Q_1$ $\quad$ $Q_2$ $\quad$ $Q_3$

$Q_1 = 23$
$Q_2 = \frac{1}{2}(23 + 26) = 24.5$
$Q_3 = 27$

Interquartile range = $Q_3 - Q_1 = 27 - 23 = 4$

(c) 147  150 ↑ 154  158 ↑ 159  162 ↑ 164  165
$\qquad\qquad$ $Q_1$ $\qquad\quad$ $Q_2$ $\qquad\quad$ $Q_3$

$Q_1 = \frac{1}{2}(150 + 154) = 152$
$Q_2 = \frac{1}{2}(158 + 159) = 158.5$
$Q_3 = \frac{1}{2}(162 + 164) = 163$

Interquartile range = $Q_3 - Q_1 = 163 - 152 = 11$

(d) 10  12 ↑ 13  15  ⑲  19  24 ↑ 26  26
$\qquad\qquad$ $Q_1$ $\qquad\quad$ $Q_2$ $\qquad\quad$ $Q_3$

$Q_1 = \frac{1}{2}(12 + 13) = 12.5$
$Q_2 = 19$
$Q_3 = \frac{1}{2}(24 + 26) = 25$

Interquartile range = $Q_3 - Q_1 = 25 - 12.5 = 12.5$

Sometimes the following rule is used to find the quartiles:

$Q_1 = \frac{1}{4}(n + 1)$th value, $\qquad$ $Q_3 = \frac{3}{4}(n + 1)$th value.

This rule agrees with the above method when $n$ is odd, but there is a discrepancy when $n$ is even. It does not, however, make a great deal of difference which method is used.

## Example 1.37

A reaction time experiment was performed first with 21 girls and then with 24 boys. The results are shown on the stem and leaf diagram.

**Reaction times**

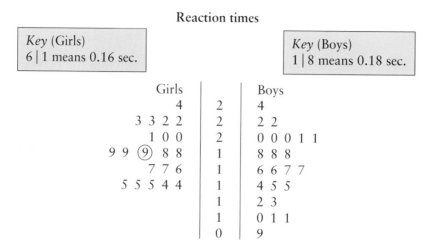

Find the median and the interquartile range for both sets of reaction times. Comment on your answers.

## Solution 1.37

*For the girls:*

There are 21 girls, so the median is the $\frac{1}{2}(21 + 1)$th value, i.e. the 11th value. This value has been ringed on the diagram and is obtained by counting up from the bottom, 14, 14, 15, 15, ... or counting down from the top, 24, 23, 23, 22, ... until the 11th value is reached.

**The median is 0.19 seconds.**

Find the quartiles by dividing in half the two distributions either side of $Q_2$.

So $Q_1$ is the 5.5th value. This is halfway between the 5th and 6th values. On this stem and leaf diagram, count from the bottom up, so the 5th value is 0.15 and the 6th value is 0.16.

**Therefore $Q_1$ is 0.155 seconds.**

$Q_2$ is the 16.5th value. This is halfway between the 16th and 17th values, i.e. between 0.21 and 0.22.

**Therefore $Q_3$ is 0.215 seconds.**

The interquartile range $= Q_3 - Q_1 = 0.215 - 0.155 = \mathbf{0.06}$ **seconds.**

*For the boys:*

There are 24 boys, so the median is the $\frac{1}{2}(24 + 1)$th value, i.e. the 12.5th value. This is halfway between the 12th and 13th values which are both 0.17 seconds.

**So the median reaction time for the boys is 0.17 seconds.**

$Q_1$ is the 6.5th value. This is halfway between the 6th and 7th values which are 0.13 and 0.14.

Therefore $Q_1$ is 0.135 seconds.

$Q_3$ is the 18.5th value. This is halfway between the 18th and 19th values, which are both 0.2 seconds.

Therefore $Q_3$ is 0.2 seconds.

The interquartile range = $Q_3 - Q_1 = 0.2 - 0.135 = 0.065$ seconds

Summary of results

|  | Girls | Boys |
|---|---|---|
| Median | 0.19 s | 0.17 s |
| Interquartile range | 0.06 s | 0.065 s |

These results confirm what the stem and leaf diagram shows, that the girls generally are slower than the boys to react, but that there is more variability in the boys' results.

## Ungrouped data in a frequency distribution – median and quartiles

To find the median and quartiles of data in the form of an ungrouped frequency distribution, it is useful to find the **cumulative frequency** as this gives the total frequency up to a particular item.

### Example 1.38

The table shows the number of children in the family for 35 families in a certain area. Find the median number of children per family, and the interquartile range.

| Number of children | 0 | 1 | 2 | 3 | 4 | 5 |
|---|---|---|---|---|---|---|
| Frequency (number of families) | 3 | 5 | 12 | 9 | 4 | 2 |

### Solution 1.38

The cumulative frequency distribution is formed as follows:

| Number of children | 0 | $\leqslant 1$ | $\leqslant 2$ | $\leqslant 3$ | $\leqslant 4$ | $\leqslant 5$ |
|---|---|---|---|---|---|---|
| Cumulative frequency (families) | 3 | 8 | 20 | 29 | 33 | 35 |

$$\begin{array}{cc} \uparrow & \uparrow \\ 3+5 & 3+5+12 \end{array}$$

Since there are 35 values the median is the $\frac{1}{2}(35 + 1)$th value, i.e. the 18th value. Since there are eight families with $\leqslant 1$ child and 20 families with $\leqslant 2$ children, the 18th value must be 2.

**Therefore the median number of children per family is 2.**

Since $n$ is odd:

$Q_1 = \frac{1}{4}(35 + 1)$th value = 9th value = 2

$Q_3 = \frac{3}{4}(35 + 1)$th value = 27th value = 3.

Therefore, interquartile range = $3 - 2 = \mathbf{1}$ **child per family.**

Illustrating this on a 'step' diagram showing the cumulative frequency:

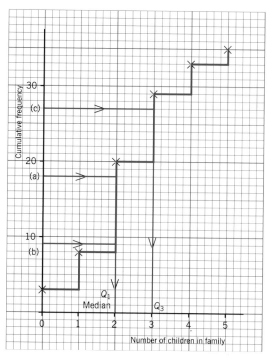

(a) To find the median, i.e. the 18th value, read across from 18 on the vertical axis, then down to 2.
**Median = 2**

(b) To find the lower quartile, i.e. the 9th value, read across from 9 on the vertical axis.
**Lower quartile = 2**

(c) To find the upper quartile, i.e. the 27th value, read across from 27 on the vertical axis.
Upper quartile = 3

**Therefore interquartile range = 3 − 2 = 1**

## Example 1.39

A teacher asked her class of thirty 14-year-olds how many novels they had read during the term. The results are illustrated in the cumulative frequency graph overleaf.

(a) Write down the mode.
(b) Find the median number of novels read.
(c) What percentage of the class read more than 5 novels?

Cumulative frequency graph of the number of novels read

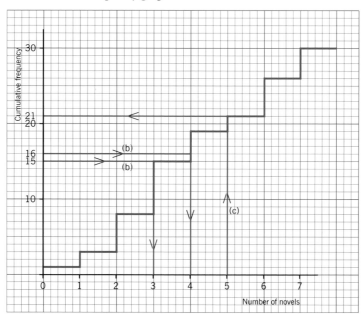

## Solution 1.39

(a) The 'steepest' step occurs when $x = 3$, so **mode = 3 novels**

(b) There are 30 pupils, so the median is the $\frac{1}{2}(30 + 1)$th value, i.e. the 15.5th value. This is half-way between the 15th value and the 16th value.
From graph,    15th value = 3,    16th value = 4
so **median = 3.5 novels**

(c) From the graph the number that read $\leqslant 5$ novels is 21. The 22nd person must have read 6 novels.

Number who read more than 5 novels = 30 − 21 = 9

Percentage that read more than 5 novels = $\frac{9}{30} \times 100\%$
$= \mathbf{30\%}$

## Exercise 1j   Cumulative frequency, median and quartiles – ungrouped data

1. Find the median of each of the following sets of numbers:
   (a)  4, 6, 18, 25, 9, 16, 22, 5, 20, 4, 8
   (b)  192, 217, 189, 210, 214, 204
   (c)  1267, 1896, 895, 3457, 2164
   (d)  0.7, 0.4, 0.65, 0.78, 0.45, 0.32, 1.9, 0.0078

2. The table shows the scores obtained when a die is thrown 60 times. Find the median score.

| Score, $x$ | 1 | 2 | 3 | 4 | 5 | 6 |
|---|---|---|---|---|---|---|
| Frequency, $f$ | 12 | 9 | 8 | 13 | 9 | 9 |

3.  These are the test marks of 11 students.

    52, 61, 78, 49, 47, 79, 54, 58, 62, 73, 72

    Find
    (a) the median
    (b) the lower quartile
    (c) the upper quartile
    (d) the interquartile range.

4.  Find the median and interquartile range of the
    following distributions:

    (a)

    | Stem | Leaf |
    |---|---|
    | 1 | 0 5 |
    | 2 | 3 4 4 |
    | 3 | 2 8 8 |
    | 4 | 1 5 6 6 7 |
    | 5 | 2 3 3 |
    | 6 | 5 7 8 8 |
    | 7 | 2 4 |
    | 8 | 0 |

    Key  5 | 2 means 52

    (b)

    | Stem | Leaf |
    |---|---|
    | 3 | 6 |
    | 3 | 1 2 |
    | 2 | 5 7 |
    | 2 | 0 3 4 4 |
    | 1 | 6 7 8 8 9 9 |
    | 1 | 2 2 3 4 |
    | 0 | 5 5 |
    | 0 | 1 3 3 |

    Key  1 | 2 means 1.2

    (c)

    | Stem | Leaf |
    |---|---|
    | 6 | 0 2 2 |
    | 10 | 1 1 2 3 |
    | 14 | 0 2 2 3 3 |
    | 18 | 0 2 3 3 3 3 3 3 |
    | 22 | 3 3 3 3 |
    | 26 | 0 0 2 |
    | 30 | 1 3 |

    Key  22 | 1 means 23

5.  Find the median and interquartile range of each
    of the following frequency distributions.

    (a)

    | x | 5 | 6 | 7 | 8 | 9 | 10 |
    |---|---|---|---|---|---|---|
    | f | 6 | 11 | 15 | 18 | 6 | 5 |

    (b)

    | x | 12 | 13 | 14 | 15 | 16 |
    |---|---|---|---|---|---|
    | f | 3 | 9 | 11 | 15 | 7 |

6.  The frequency table shows the number of goals
    scored in netball by Jemima in 25 games played.

    | Number of goals | 0 | 1 | 2 | 3 | 4 | 5 | 6 |
    |---|---|---|---|---|---|---|---|
    | Frequency | | 0 | 1 | 3 | 2 | 5 | 8 | 6 |

    (a) Construct a cumulative frequency table.
    (b) Draw a step diagram to illustrate the table.
    (c) Find the median number of goals.
    (d) Find the interquartile range.

7.  This cumulative frequency table shows the
    number of absences for each of a class of 32
    children during one term.

    | Times absent | 0 | ≤1 | ≤2 | ≤3 | ≤4 | ≤5 | ≤6 | ≤7 |
    |---|---|---|---|---|---|---|---|---|
    | Cumulative frequency | 5 | 11 | 20 | 23 | 27 | 28 | 31 | 32 |

    (a) Find the median number of absences.
    (b) Find the range of the middle 50% of the
        observations.
    (c) Calculate the mean number of absences.
    (d) Calculate the standard deviation.

8.  A researcher, studying the effectiveness of Family
    Income Supplement, carried out a survey of
    120 families receiving the benefit. As part of the
    survey the researcher recorded the number of
    children in each family. The results are illustrated
    in the cumulative frequency graph below.

    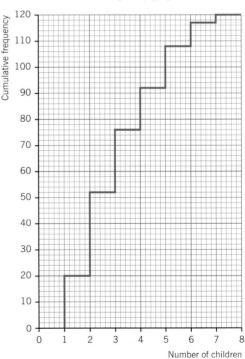

    (a) Write down the mode and the median of the
        number of children per family.
    (b) Find the interquartile range of the number
        of children per family.
    (c) Explain why the interquartile range is only a
        rough measure of spread for this type of
        distribution.                              (NEAB)

# Grouped data – median and quartiles

When data have been grouped into intervals, the original information has been lost, so it is only possible to make estimates of the median and quartiles. One way of doing this is to use a cumulative frequency graph, or cumulative percentage frequency graph as follows:

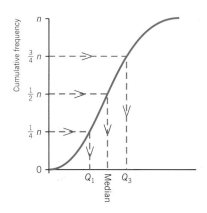

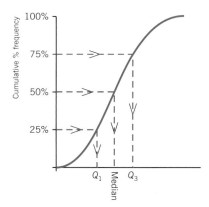

| Grouped data | Cumulative frequency curve | Cumulative percentage frequency curve |
|---|---|---|
| Lower quartile, $Q_1$ | $\frac{1}{4}n$th reading | 25% reading |
| Median, $Q_2$ | $\frac{1}{2}n$th reading | 50% reading |
| Upper quartile, $Q_3$ | $\frac{3}{4}n$th reading | 75% reading |

**Cumulative frequency curve**  **Cumulative % frequency curve**

Note that the $\frac{1}{2}(n + 1)$th reading is not used for the median. If you used this value you would not arrive at the same point on the cumulative frequency axis when you worked down from the top of the scale as you would when you worked up from the bottom of the scale. The $\frac{1}{2}n$th or 50% value is needed for the median.

Note also, that if preferred, a cumulative frequency polygon or cumulative percentage frequency polygon can be drawn. The values obtained for the median and quartiles will not vary greatly from those obtained from curves.

## Example 1.40

The table gives the cumulative distribution of the heights (in centimetres) of 400 children in a

| Height (cm) | <100 | <110 | <120 | <130 | <140 | <150 | <160 | <170 |
|---|---|---|---|---|---|---|---|---|
| Cumulative frequency | 0 | 27 | 85 | 215 | 320 | 370 | 395 | 400 |

certain school:
(a) Draw a cumulative frequency curve.
(b) Find an estimate of the median height.
(c) Determine the interquartile range.

(d) Determine the 10 to 90 percentile range.

## Solution 1.40

(a) **Cumulative frequency curve to show the heights of 400 children**

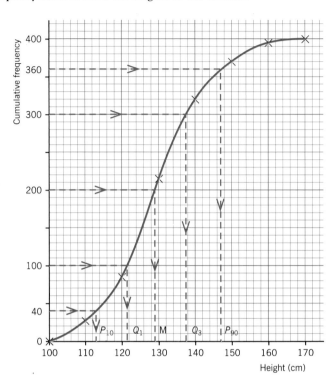

(b) For the median, find the $\frac{1}{2}(400)$th value, i.e. the 200th value.
From the graph, an estimate of the median is **129 cm**.

(c) For the lower quartile, $Q_1$, find the 100th value and for the upper quartile, $Q_3$, the 300th value.

From the graph, $Q_1 = 121.5$ cm and $Q_3 = 137.5$ cm
The interquartile range $= Q_3 - Q_1$
$$= 137.5 - 121.5$$
$$= \textbf{16 cm}$$

Note that this is the range of the middle 50% of the readings.

(d) For the 10th percentile (written, $P_{10}$) find the value which is 10% of the way through the readings, the $\frac{10}{100}(400)$th value, i.e. the 40th value. The 90th percentile is the $\frac{90}{100}(400)$th value, i.e. the 360th value.

$P_{90} = 147$ cm,   $P_{10} = 113$ cm

The 10 to 90 percentile range $= P_{90} - P_{10}$
$$= 147 - 113$$
$$= \textbf{34 cm.}$$

### Example 1.41

The masses, measured to the nearest kilogram, of 50 boys are noted and a cumulative percentage frequency distribution formed.

| mass (kg) | <59.5 | <64.5 | <69.5 | <74.5 | <79.5 | <84.5 | <89.5 |
|---|---|---|---|---|---|---|---|
| Cumulative % frequency | 0 | 4 | 16 | 40 | 68 | 88 | 100 |

Draw a cumulative percentage frequency curve and use it to estimate the median mass and the interquartile range.

### Solution 1.41

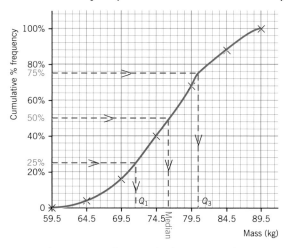

Cumulative % frequency curve to show masses of 50 boys

The median is the 50% reading. From the graph this is **76.3 kg.**

The lower quartile, $Q_1$ is the 25% reading, so $Q_1 = 71.5$ kg.

The upper quartile $Q_3$ is the 75% reading, so $Q_3 = 80.5$ kg.

The interquartile range $= Q_3 - Q_1 = 80.5 - 71.5 = $ **9 kg.**

It is interesting to note that if the data are represented by a histogram, the median divides the area exactly in half.

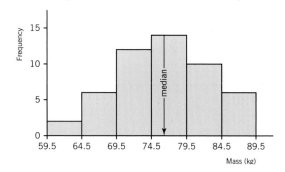

Histogram to show the masses of 50 boys

## Example 1.42

Examinations in English, Mathematics and Science were taken by 400 students. Each examination was marked out of 100 and the cumulative frequency graphs illustrating the results are shown below.

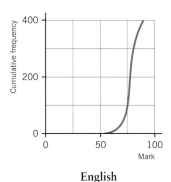

English

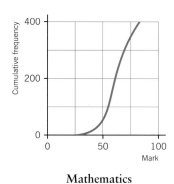

Mathematics

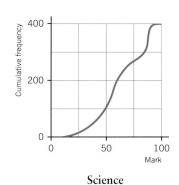

Science

(a) In which subject was the median mark the highest?
(b) In which subject was the interquartile range of the marks the greatest?
(c) In which subject did approximately 75% of the students score 50 marks or more?  (C)

## Solution 1.42

Showing the working on the diagrams:
Median $Q_2$ is 200th reading, lower quartile $Q_1$ is 100th, upper quartile $Q_3$ is the 300th reading.

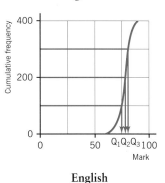

English

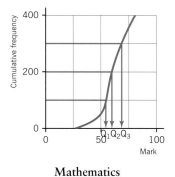

Mathematics

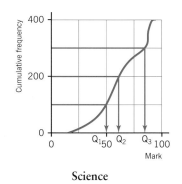

Science

(a) The median, $Q_2$, is the highest in English.
(b) The interquartile range, $Q_3 - Q_1$, is greatest for Science.
(c) The subject in which 300 students scored 50 or more, i.e. 75% scored 50 or more is Science.

# Using linear interpolation

It is possible to estimate the median, quartiles or other percentiles for grouped data without drawing the cumulative frequency graph. The method is known as **linear interpolation**.

## Example 1.43

The ages of 160 members of a bridge club are grouped as shown in the table.

| Age | 30– | 40– | 50– | 60– | 70– | 90– |
|---|---|---|---|---|---|---|
| Number of members | 5 | 42 | 61 | 37 | 15 | 0 |

Without drawing a cumulative frequency curve, estimate

(a) the median age,
(b) the number of members aged 67 or over,
(c) the 20th percentile.

## Solution 1.43

Form a cumulative frequency distribution.

| Age | <40 | <50 | <60 | <70 | <90 |
|---|---|---|---|---|---|
| Cumulative frequency | 5 | 47 | 108 | 145 | 160 |

(a) Since there are 160 observations, the median is the 80th observation.
From the table, 47 are under 50 and 108 are under 60, so the 80th person has an age in the interval 50–60.

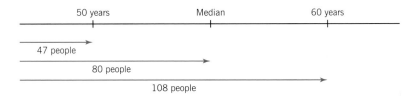

108 – 47 = 61, so there are 61 people in the interval 50–60.
80 – 47 = 33, so, assuming that the ages are evenly distributed, the median value will be $\frac{33}{61}$ of the way along the interval which has a width of ten years.
$\therefore$ median $= 50 + \frac{33}{61} \times 10 =$ **55.4 years**

(b) The age 67 is in the interval 60–70 which has a width of 10.
67 is located $\frac{7}{10}$ of the way through this interval.

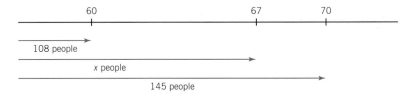

The number of people in the interval 60–67 is 145 – 108 = 37.
Using linear interpolation, $\frac{7}{10}$ of the 37 people will be under 67 years old.
Now $\frac{7}{10}$ of 73 = 25.9 $\approx$ 26
$\therefore$ number of people under 67 years old = 108 + 26 = 134
So **number of people 67 or over = 160 – 134 = 26**

(c) The 20th percentile is the value 20% of the way through the distribution.
There are 160 observations and 20% of 160 = 32, so the age of the 32nd person is needed.
Five people were under 40, 47 people were under 50, so the 32nd person is in the interval 40–50.
The number of people in this interval = 47 − 5 = 42.

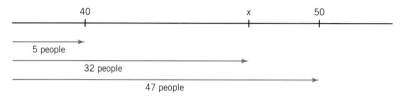

$32 - 5 = 27$, so $x$ will be $\frac{27}{42}$ of the way through the interval 40–50.
$x = 40 + \frac{27}{42} \times 10 = 46.4 \ldots$

**The 20th percentile is 46.4 years (1 d.p.)**

## Example 1.44

The distribution of the lengths of time of a large number of telephone calls made from an office in a given week was such that the median was 100 seconds and the 80th percentile was 190 seconds. Without drawing the cumulative frequency curve, estimate

(a) the upper quartile,
(b) the number of calls, out of 500, that lasted less than a minute.

## Solution 1.44

(a) Show the information on a diagram, denoting the upper quartile by $Q_3$.

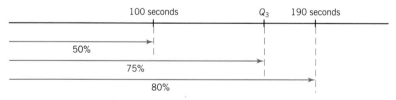

Percentage of distribution in interval 100–190 is 30%.
Percentage of distribution in interval 100–$Q_3$ is 25%.
So $Q_3$ is $\frac{25}{30}$ of the way through the interval 100–190.
This interval has a width of 90 so $Q_3 = 100 + \frac{25}{30} \times 90 = 175$

**The upper quartile is 175 seconds.**

(b)

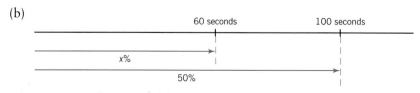

$x\%$ of calls lasted less than 60 seconds and 50% lasted less than 100 seconds.

Using ratios:

$$\frac{x}{50} = \frac{60}{100} \qquad \text{so } x = \frac{60}{100} \times 50 = 30$$

The number of calls = 30% of 500 = 150

**150 calls lasted less than a minute.**

---

## Exercise 1k   Cumulative frequency, median and quartiles – grouped data

1. The table below shows the frequency distribution of the masses of 52 women students at a college. Measurements have been recorded to the nearest kilogram.

| Mass (kg) | Frequency |
|---|---|
| 40–44 | 3 |
| 45–49 | 2 |
| 50–54 | 7 |
| 55–59 | 18 |
| 60–64 | 18 |
| 65–69 | 3 |
| 70–74 | 1 |

   (a) Construct a cumulative frequency table and draw a cumulative frequency curve.
   (b) How many students weighed less than 57 kg?
   (c) How many students weighed more than 61 kg?
   (d) 20% were heavier than x kg. Find the value of x.
   (e) Estimate the median.
   (f) Estimate the interquartile range.

2. Fifty soil samples were collected in an area of woodland, and the pH value for each sample was found. The cumulative frequency distribution was constructed as shown in the table.

| pH value | Cumulative frequency |
|---|---|
| <4.8 | 1 |
| <5.2 | 2 |
| <5.6 | 5 |
| <6.0 | 10 |
| <6.4 | 19 |
| <6.8 | 38 |
| <7.2 | 43 |
| <7.6 | 46 |
| <8.0 | 49 |
| <8.4 | 50 |

   (a) Draw a cumulative frequency curve.
   (b) What percentage of the samples had a pH value less than 7?

   (c) 50% of the samples had a pH value greater than x. Find x. What name is given to this value?
   (d) Taking equal class intervals of $4.4 \leqslant x < 4.8$ $4.8 \leqslant x < 5.2$, etc., construct the frequency distribution and draw a histogram. Show the median on the histogram.

3. Eggs laid at Hill Farm are weighed and the results grouped as shown:

| Mass (g) | Frequency |
|---|---|
| –50 | 3 |
| –54 | 2 |
| –58 | 5 |
| –62 | 12 |
| –66 | 10 |
| –70 | 6 |
| –74 | 2 |

   Construct a cumulative frequency table and draw a cumulative frequency curve. Use the curve to estimate the median mass.

4.

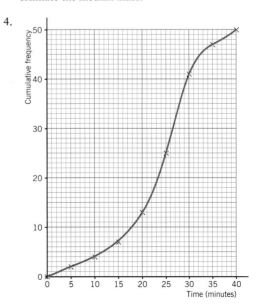

The cumulative frequency curve has been drawn from information about the amount of time spent by 50 people in a supermarket on a particular day.

(a) Construct the cumulative frequency table, taking boundaries ⩽5, ⩽10, ...
(b) How many people spent between 17 and 27 minutes in the supermarket?
(c) 60% of the people spent less than or equal to $t$ minutes. Find $t$.
(d) 60% of the people spent longer than $s$ minutes. Find $s$.
(e) Estimate the median.
(f) Find the interquartile range.

5. In a quality-control survey, the length of life, in hours, of 50 light bulbs is noted.
The results are summarised in the table.
Using linear interpolation, calculate estimates of
(a) the median,
(b) the interquartile range.

| Length of life ($h$) | Frequency |
|---|---|
| $650 \leqslant h < 670$ | 3 |
| $670 \leqslant h < 680$ | 7 |
| $680 \leqslant h < 690$ | 20 |
| $690 \leqslant h < 700$ | 17 |
| $700 \leqslant h < 702$ | 3 |

6. A factory produces a certain component. The masses of 500 of these components were measured to the nearest gram and are grouped in the following table.

| Mass (g) | 60–69 | 70–74 | 75–79 | 80–84 | 85–89 |
|---|---|---|---|---|---|
| Number of components | 38 | 93 | 120 | 196 | 53 |

Without drawing a cumulative frequency curve, estimate
(a) the 60th percentile,
(b) the number of components whose mass is less than 78 grams. (C)

7. The weekly maximum temperatures in a certain town were recorded, to the nearest degree Celsius, over a period of two years and grouped in the following table.

| Temperature (°C) | Number of weeks |
|---|---|
| −5 to −1 | 8 |
| 0 to 4 | 12 |
| 5 to 9 | 17 |
| 10 to 14 | 31 |
| 15 to 19 | 23 |
| 20 to 24 | 9 |
| 25 to 29 | 4 |

(a) Draw the cumulative frequency curve.
(b) Use your curve to estimate the median temperature.
(c) In a particular house it was found that the central heating was turned on when the weekly maximum temperature fell below 17 °C. Use your curve to estimate the number of weeks when the heating was turned on.
(d) A week is classified as *extremely warm* when the weekly maximum is greater than 21 °C.
Use your curve to estimate the percentage of weeks that are classified as *extremely warm*.
(C)

8. The times, to the nearest minute, taken by a group of 120 students to write a particular essay, were recorded and are grouped in the table below.

| Ten (minutes) | 40–44 | 45–49 | 50–54 | 55–59 | 60–64 |
|---|---|---|---|---|---|
| Number of students | 8 | 22 | 34 | 30 | 26 |

Construct the cumulative frequency table for this distribution and draw the cumulative frequency curve.
Use your curve to estimate
(a) the interquartile range of the times,
(b) the percentage of these students who spent over 62 minutes in writing the essay.

Another group of 30 students wrote the same essay and all took over 65 minutes to complete it.
Use your curve to estimate the median time of all 150 students. (C)

9. Each of 50 sportsmen was asked to state the distance, $x$ km, he needs to travel to obtain access to suitable training facilities. The results are summarised in the table below.

| Distance ($x$ km) | Number of sportsmen |
|---|---|
| $0 \leqslant x < 4$ | 1 |
| $4 \leqslant x < 10$ | 2 |
| $10 \leqslant x < 20$ | 6 |
| $20 \leqslant x < 35$ | 19 |
| $35 \leqslant x < 60$ | 12 |
| $60 \leqslant x < 100$ | 10 |

Construct the cumulative frequency table for the distribution and draw the cumulative frequency curve.

Use your curve to estimate
(a) the median distance,
(b) the interquartile range of the distances,
(c) the percentage of sportsmen who need to travel more than 30 km. (C)

10. The prices, on a particular day, of 53 stocks on the London Stock Exchange are summarised in the table below.

| Price £x | Number of stocks |
|---|---|
| $75 < x \leqslant 95$ | 6 |
| $95 < x \leqslant 100$ | 10 |
| $100 < x \leqslant 105$ | 12 |
| $105 < x \leqslant 110$ | 13 |
| $110 < x \leqslant 120$ | 7 |
| $120 < x \leqslant 135$ | 5 |

Construct the cumulative frequency table for this distribution and draw the cumulative frequency curve.
Use your curve to estimate
(a) the median price,
(b) the interquartile range,
(c) the number of stocks costing between £89 and £123. (C)

11. The masses, measured to the nearest gram, of 80 eggs were recorded and are grouped in the table below.

| Mass (g) | 50–59 | 60–64 | 65–69 | 70–79 |
|---|---|---|---|---|
| Number of eggs | 18 | 20 | $x$ | $y$ |

Assuming that the readings in each group are linearly distributed and given that 60% of all these eggs have actual masses below 66.5 g, calculate the value of $x$ and of $y$. (C)

12. 30 specimens of sheet steel are tested for tensile strength, measured in kN m$^{-2}$. The table below gives the distribution of the measurements.

| Tensile strength | Number of specimens |
|---|---|
| 405–415 | 4 |
| 415–425 | 3 |
| 425–435 | 6 |
| 435–445 | 10 |
| 445–455 | 5 |
| 455–465 | 2 |

Draw a cumulative frequency diagram of this distribution.
Estimate the median and the 10th and 90th percentiles. (O&C)

13. Every day at 08:28 a train departs from one city and travels to a second city. The times taken for the journey were recorded in minutes over a certain period and were grouped as shown in the table.

| Time | Frequency |
|---|---|
| –80 | 0 |
| –85 | 6 |
| –90 | 12 |
| –95 | 22 |
| –100 | 31 |
| –105 | 15 |
| –110 | 7 |
| –115 | 4 |
| –120 | 2 |
| –125 | 1 |
| Over 125 | 0 |

(The interval '–90' indicates all times greater than 85 minutes up to and including 90 minutes.)
From these figures draw a cumulative frequency curve and from this curve estimate
(a) the median time for the journey,
(b) the interquartile range,
(c) the number of trains which arrived at the second city between 10:00 and 10:15.
(C Additional)

14. Two hundred and fifty Army recruits have the following heights.

| Height (cm) | No. of recruits |
|---|---|
| 165– | 18 |
| 170– | 37 |
| 175– | 60 |
| 180– | 65 |
| 185– | 48 |
| 190–195 | 22 |

Plot the data in the form of a cumulative frequency curve. Use the curve to estimate
(a) the median height,
(b) the lower quartile height.
The tallest 40% of the recruits are to be formed into a special squad. Estimate
(c) the median,
(d) the upper quartile of the heights of the members of this squad.

15. The distribution of the times taken when a certain task was performed by each of a large number of people was such that its twentieth percentile was 25 minutes, its fortieth percentile was 50 minutes, its sixtieth percentile was 64 minutes and its eightieth percentile was 74 minutes.

Use linear interpolation to estimate (a) the median of the distribution, (b) the upper quartile of the distribution, (c) the percentage of persons who performed the task in forty minutes or less. (*NEAB*)

16. The times, correct to the nearest second, for 100 athletes to cover one lap of a running track were recorded and are shown in the table below.

| Recorded time (s) | Number of athletes |
|---|---|
| 65–69 | 0 |
| 70–74 | 8 |
| 75–79 | 20 |
| 80–84 | 25 |
| 85–89 | 31 |
| 90–94 | 10 |
| 95–99 | 6 |

Draw a cumulative frequency graph and hence determine the interquartile range.
To qualify for an athletics meeting, a runner needs to record a lap time of 78 seconds or under. Estimate the number of athletes who qualified and the median time for these qualifiers. (*C Additional*)

17. A group of 125 children raised money for a charity by sponsored activities. The amount raised by each child was recorded. These amounts, taken to the nearest £, are grouped in the table below.

| Amount raised, £ | Number of children |
|---|---|
| 1–5 | 70 |
| 6–10 | 36 |
| 11–15 | 19 |

State the smallest possible amount which may have been raised by one child. Without drawing a cumulative frequency curve, estimate the median amount raised.
Also estimate the mean amount raised and explain briefly why this is larger than the median. (*C Additional*)

18. In a borehole the thickness, in millimetres, of the 25 strata are shown in the table.

| Thickness (mm) | Number of strata |
|---|---|
| 0– | 2 |
| 20– | 5 |
| 30– | 9 |
| 40– | 8 |
| 50– | 1 |
| 60– | 0 |

Draw a histogram to illustrate these data. Construct a cumulative frequency table and draw a cumulative frequency polygon. Hence, or otherwise, estimate the median and the interquartile range for these data.
Find the proportion of the strata that are less than 28 mm thick. (*L*)

# SKEWNESS

On page 20 you considered the shape of various distributions.

There are mathematical ways of expressing the degree of skewness of a distribution:

In a **negatively-skewed** distribution the tail of the distribution is pulled in the **negative** direction.

mean < median < mode

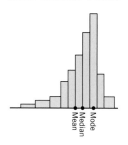

In a **symmetrical** distribution, mean = mode = median

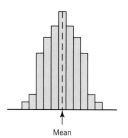

In a **positively-skewed** distribution the tail of the distribution is pulled in the **positive** direction.

mode < median < mean

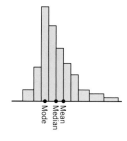

Notice that when distributions are skew, the median generally lies between the mode and the mean, and the following relationship is satisfied

mean − mode ≈ 3(mean − median)

One measure of skewness is given by Pearson's coefficient of skewness.

$$\text{Pearson's coefficient of skewness} = \frac{\text{mean} - \text{mode}}{\text{standard deviation}}$$

If mean > mode, the skew is positive.
If mean < mode, the skew is negative.
If mean = mode, the skew is zero and the distribution is symmetrical.

Alternatively

$$\text{Pearson's coefficient of skewness} = \frac{3(\text{mean} - \text{median})}{\text{standard deviation}}$$

Generally skewness can take any value between 3 and −3.

For example, the measure of skewness for these distributions might be as shown:

## Example 1.45

Electric fuses, nominally rated at 30 amperes (30A), are tested by passing a gradually increasing electric current through them and recording the current, $x$ amperes, at which they blow. The results of this test on a sample of 125 such fuses are shown in the following table.

| Current ($x$ A) | Number of fuses |
|---|---|
| $25 \leqslant x < 28$ | 6 |
| $28 \leqslant x < 29$ | 12 |
| $29 \leqslant x < 30$ | 27 |
| $30 \leqslant x < 31$ | 30 |
| $31 \leqslant x < 32$ | 18 |
| $32 \leqslant x < 33$ | 14 |
| $33 \leqslant x < 34$ | 9 |
| $34 \leqslant x < 35$ | 4 |
| $35 \leqslant x < 40$ | 5 |

Draw a histogram to represent these data.

For this sample calculate

(a) the median current,
(b) the mean current,
(c) the standard deviation of current.

A measure of the *skewness* (or asymmetry) of a distribution is given by

$$\frac{3(\text{mean} - \text{median})}{\text{standard deviation}}$$

Calculate the value of this measure of skewness for the above data.

Explain briefly how this skewness is apparent in the shape of your histogram. (L)

## Solution 1.45

| Current | Interval width | Frequency | Frequency density $= \dfrac{\text{frequency}}{\text{interval width}}$ |
|---|---|---|---|
| $25 \leqslant x < 28$ | 3 | 6 | 2 |
| $28 \leqslant x < 29$ | 1 | 12 | 12 |
| $29 \leqslant x < 30$ | 1 | 27 | 27 |
| $30 \leqslant x < 31$ | 1 | 30 | 30 |
| $31 \leqslant x < 32$ | 1 | 18 | 18 |
| $32 \leqslant x < 33$ | 1 | 14 | 14 |
| $33 \leqslant x < 34$ | 1 | 9 | 9 |
| $34 \leqslant x < 35$ | 1 | 4 | 4 |
| $35 \leqslant x < 40$ | 5 | 5 | 1 |

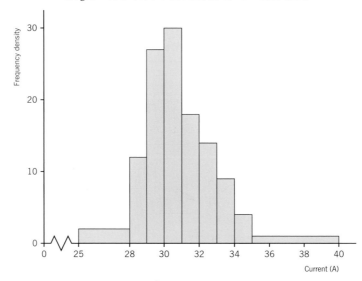

Histogram to show the current at which fuses blow

(a) For grouped data, the median is the $\frac{1}{2}n$th value.
   Since there are 125 observations, the median is the 62.5th. This can be found by linear interpolation as follows:

   Since 45 fuses blew at a current less than 30 A and 75 fuses blew at a current less than 31 A, the median lies in the interval, of width 1 A, from 30 A to 31 A.

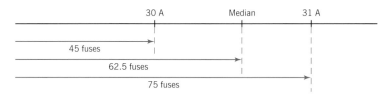

$$\text{Median} = 30 + \frac{17.5}{30} \times 1 = 30.58 \ldots = \textbf{30.6 A (1 d.p.)}$$

| Mid-point ($x$) | $f$ |
|---|---|
| 26.5 | 6 |
| 28.5 | 12 |
| 29.5 | 27 |
| 30.5 | 30 |
| 31.5 | 18 |
| 32.5 | 14 |
| 33.5 | 9 |
| 34.5 | 4 |
| 37.5 | 5 |
| | $\Sigma f = 125$ |

(b) $\bar{x} = \dfrac{\Sigma fx}{\Sigma f}$

$\phantom{(b)\ \bar{x}} = \dfrac{3861.5}{125}$

$\phantom{(b)\ \bar{x}} = 30.892$

(c) $s^2 = \dfrac{\Sigma f^2}{\Sigma f} - \bar{x}^2$

$\phantom{(c)\ s^2} = \dfrac{119\,905.25}{25} - 30.892^2$

$\phantom{(c)\ s^2} = 4.926 \ldots$

$\phantom{(c)\ } s = 2.219 \ldots$

[Check these on your calculator, using SD mode.]

**Therefore the mean is 30.892 A and the standard deviation is 2.22 A (2 d.p.)**

Now  skewness $= \dfrac{3(\text{mean} - \text{median})}{\text{standard deviation}}$

$\phantom{Now  skewness} = \dfrac{3(30.892 - 30.58\ldots)}{2.219\ldots}$

$\phantom{Now  skewness} = 0.42 \ (2 \text{ d.p.})$

Since skewness $> 0$, the **distribution is positively skewed.**

Note that the resulting frequency polygon confirms that the distribution is skewed to the right, i.e. positively skewed.

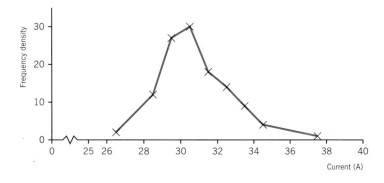

## Quartile coefficient of skewness

Another measure of skewness is defined in terms of the quartiles. Writing $Q_1$ for the lower quartile, $Q_2$ the median and $Q_3$ the upper quartile,

$$\text{Quartile coefficient of skewness} = \frac{(Q_3 - Q_2) - (Q_2 - Q_1)}{Q_3 - Q_1}$$

$$= \frac{Q_3 - 2Q_2 + Q_1}{Q_3 - Q_1}$$

| Symmetrical distribution | Positively skewed distribution | Negatively skewed distribution |
|---|---|---|
| 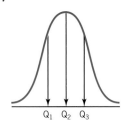 | 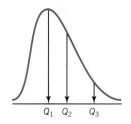 | 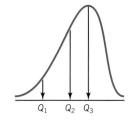 |
| $Q_3 - Q_2 = Q_2 - Q_1$ | $Q_3 - Q_2 > Q_2 - Q_1$ | $Q_3 - Q_2 < Q_2 - Q_1$ |
| Quartile skewness = 0 | Quartile skewness > 0 | Quartile skewness < 0 |

### Example 1.46

31 students tried to estimate the length of a line. The line was actually 60 mm long. These are their results, in millimetres.

```
61  70  46  44  26  23  30  83  52  44  38
37  49  59  58  63  31  29  37  48  76  61
46  31  38  41  49  52  56  75  61
```

Find the median and the quartiles of this distribution and use the quartiles to estimate the skewness.
Draw a histogram with equal class intervals $20 \leqslant l < 30$, $30 \leqslant l < 40$, …

### Solution 1.46

Arrange the results in order.

```
23   26  29  30  31  31  37  ③⑦   38   38  41  44  44  46  46
④⑧  49  49  52  52  56  58   59   ⑥①  61  61  63  70  75  76  83
```

There are 31 results, so the median, $Q_2$, is the $\frac{1}{2}(31 + 1)$th value, i.e. the 16th value.
So median = **48**.

To find the quartiles, since $n$ is odd (see page 69)

$Q_1 = \frac{1}{4}(31 + 1)$th values = 8th value = **37**

$Q_3 = \frac{3}{4}(31 + 1)$th values = 24th value = **61**

Now    $Q_3 - Q_2 = 61 - 48 = 13$
        $Q_2 - Q_1 = 48 - 37 = 11$

Since $Q_3 - Q_2 > Q_2 - Q_1$ the distribution is positively skewed.

Quartile coefficient of skewness
$$= \frac{(Q_3 - Q_2) - (Q_2 - Q_1)}{Q_3 - Q_1}$$
$$= \frac{13 - 11}{61 - 37}$$
$$= 0.083 \ldots$$

**This indicates a positive skew.**

The frequency distribution is as shown below, together with the histogram. Since each interval width is 10, frequency density = frequency ÷ 10.

| Length (mm) | Frequency | Frequency density |
|---|---|---|
| $20 \leqslant l < 30$ | 3 | 0.3 |
| $30 \leqslant l < 40$ | 7 | 0.7 |
| $40 \leqslant l < 50$ | 8 | 0.8 |
| $50 \leqslant l < 60$ | 5 | 0.5 |
| $60 \leqslant l < 70$ | 4 | 0.4 |
| $70 \leqslant l < 80$ | 3 | 0.3 |
| $80 \leqslant l < 90$ | 1 | 0.1 |

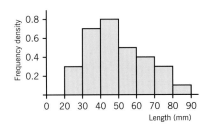

The histogram confirms the positive skew.

# THE NORMAL DISTRIBUTION

There is a special symmetrical distribution known as the **normal** distribution. This is bell-shaped, centred around the mean.
Here are two normal distributions with the same mean, but different standard deviations.

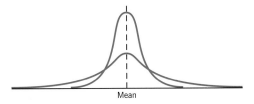

There are two normal distributions with the same standard deviation but with different means.

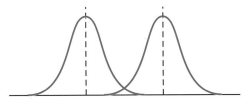

In a normal distribution:

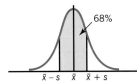

Approximately 68% of the distribution lies within one standard deviation (s) of the mean.

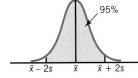

Approximately 95% of the distribution lies within two standard deviations (2s) of the mean.

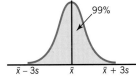

Over 99% of the distribution (nearly all!) lies within three standard deviations (3s) of the mean.

The quartiles are approximately $\frac{2}{3} \times$ standard deviation either side of the mean.

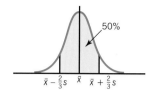

The normal distribution is studied in greater detail in Chapter 7.

## Exercise 1I  Skewness

1. Calculate Pearson's coefficient of skewness for the following frequency distributions where

$$\text{skewness} = \frac{\text{mean} - \text{mode}}{\text{standard deviation}}$$

(a)

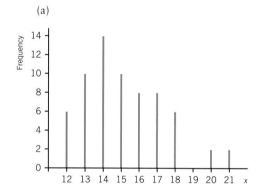

(b)

| $x$ | $f$ |
|-----|-----|
| 20 | 2 |
| 21 | 1 |
| 22 | 4 |
| 23 | 5 |
| 24 | 7 |
| 25 | 8 |
| 26 | 4 |
| 27 | 1 |

2. For a skewed distribution, the mean is 16, the median is 20 and the standard deviation is 5. Calculate Pearson's coefficient of skewness and sketch the curve.

3. For a skewed distribution, the mean is 86, the mode is 78 and the variance is 16. Calculate Pearson's coefficient of skewness and sketch the curve.

4. The following table shows the time, to the nearest minute, spent reading during a particular day by a group of school children.

| Time | Number of children |
|------|--------------------|
| 10–19 | 8 |
| 20–24 | 15 |
| 25–29 | 25 |
| 30–39 | 18 |
| 40–49 | 12 |
| 50–64 | 7 |
| 65–89 | 5 |

(a) Represent these data by a histogram.
(b) Comment on the shape of the distribution.

(L)

5. Over a period of four years a bank keeps a weekly record of the number of cheques with errors that are presented for payment. The results for the 200 accounting weeks are as follows.

| Number of cheques with errors ($x$) | Number of weeks ($f$) |
|-------------------------------------|------------------------|
| 0 | 5 |
| 1 | 22 |
| 2 | 46 |
| 3 | 38 |
| 4 | 31 |
| 5 | 23 |
| 6 | 16 |
| 7 | 11 |
| 8 | 6 |
| 9 | 2 |

$(\Sigma fx = 706, \quad \Sigma fx^2 = 3280)$

Construct a suitable pictorial representation of these data.
State the modal value and calculate the median, mean and standard deviation of the number of cheques with errors in a week.

Some textbooks measure the *skewness* (or asymmetry) of a distribution by

$$\frac{3(\text{mean} - \text{median})}{\text{standard deviation}}$$

and others measure it by

$$\frac{(\text{mean} - \text{mode})}{\text{standard deviation}}$$

Calculate and compare the values of these two measures of skewness for the above data. State how this skewness is reflected in the shape of your graph.           (*AEB*)

6. Find Pearson's coefficient of skewness for the distribution represented by this stem and leaf plot, which gives marks in an examination.

| Stem | Leaf |
|---|---|
| 1 | 9 |
| 2 | 2 8 |
| 3 | 3 7 |
| 4 | 5 |
| 5 | 2 5 5 7 |
| 6 | 1 1 6 6 8 8 8 |
| 7 | 3 5 5 |
| 8 | 2 9 |
| 9 | 1 |

*Key* 3 | 7 means 37

7.

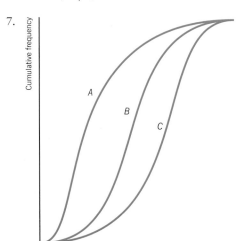

These are the three frequency curves associated with the cumulative frequency curves *A*, *B*, *C* above. Label each frequency curve with the appropriate letter.

(a)

(b)

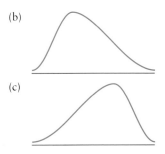

(c)

8. The following table gives the blood pressure of 60 students.

| Blood pressure | Frequency |
|---|---|
| 95– | 2 |
| 105– | 5 |
| 110– | 6 |
| 115– | 9 |
| 120– | 14 |
| 125– | 3 |
| 130– | 6 |
| 135– | 5 |
| 140– | 4 |
| 150–180 | 6 |

(a) Find
   (i)  Pearson's coefficient of skewness
   (ii) the quartile coefficient of skewness.

(b) Draw the histogram.

9. The following grouped frequency distribution summarises the time, to the nearest minute, spent waiting by a sample of patients in a doctor's surgery.

| Waiting time (to the nearest minute) | Number of patients |
|---|---|
| 3 or less | 6 |
| 4–6 | 15 |
| 7–8 | 27 |
| 9 | 49 |
| 10 | 52 |
| 11–12 | 29 |
| 13–15 | 13 |
| 16 or more | 9 |

The mean of the times was 9.63 minutes and the standard deviation was 3.03 minutes.

(a) Using interpolation, estimate the median and semi-interquartile range of these data.
   Semi-interquartile range = interquartile range ÷ 2

For a normal distribution of the ratio of the semi-interquartile range to the standard deviation would be approximately 0.67.

(b) Calculate the corresponding value for the above data. Comment on your result.

For a normal distribution, 90% of times would be expected to lie in the interval (mean ± 1.645 standard deviations).

(c) Find the theoretical limits for these data.
(d) Using appropriate percentiles, estimate comparable limits. Comment on your result.

(L)

10. Calculate the quartile coefficient of skewness for each of the following distributions:

(a)

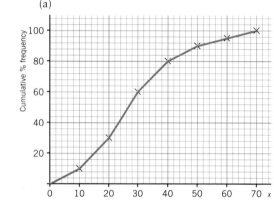

(b)

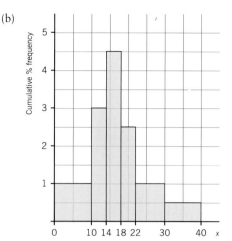

(c)

| Stem | Leaf |
|---|---|
| 2 | 0 0 |
| 5 | 0 1 1 |
| 8 | 1 1 2 2 2 |
| 11 | 0 0 1 2 2 2 |
| 14 | 1 1 2 2 |
| 17 | 0 1 |
| 20 | 1 |

Key   5 | 1 means 6

# BOX AND WHISKER DIAGRAMS (BOX PLOTS)

Consider the cumulative percentage frequency curves for girls' and boys' marks, drawn on page 66.

These are shown below, together with the median $Q_2$ and quartiles $Q_1$, $Q_3$.

Below each diagram is a box and whisker diagram, or box plot.

**Boys' marks**

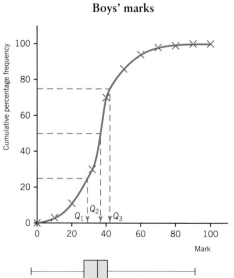

**Box plot for boys' marks**

**Girls' marks**

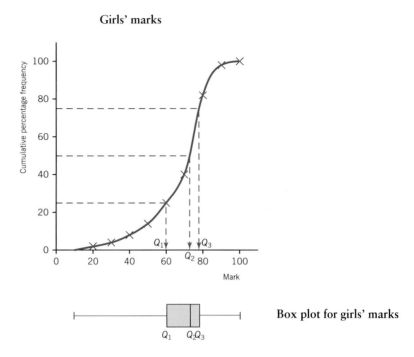

**Box plot for girls' marks**

The box plots can be drawn horizontally, as shown above, or vertically like this.

**Boys' marks    Girls' marks**

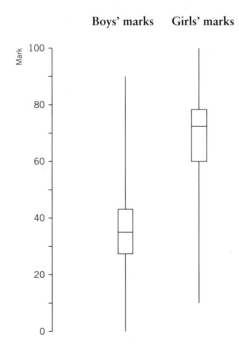

The **box and whisker diagram**, or **box plot**, illustrates the dispersion, or spread of the distribution. It uses the highest and lowest values of the data, the quartiles ($Q_1$ and $Q_3$) and the median ($Q_2$). For example:

*Vertically*

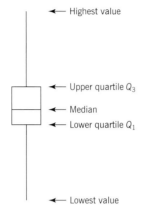

Notice that the 'box' extends from $Q_1$ to $Q_3$ and so encloses the middle 50% of the data.

The 'whiskers' extend from the box to the highest and lowest values and illustrate the range of the data.

*Horizontally*

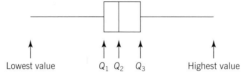

A box plot for a **symmetrical distribution** would look like this:

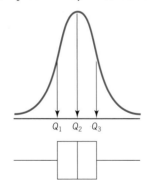

The whiskers are of equal length and the median is in the middle of the box.

For a **positively skewed distribution**:

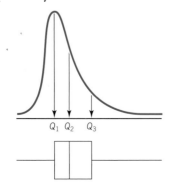

The right-hand whisker is longer and the median is nearer to the lower quartile.

For a **negatively skewed distribution:**

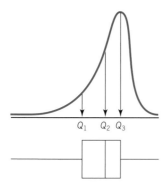

The left-hand whisker is longer and the median is nearer to the upper quartile.

## Example 1.47

A class of pupils played a computer game which tested how quickly they reacted to a visual instruction to press a particular key. The computer measured their reaction times in tenths of a second and stored a record of the sex and reaction time of each pupil. Finally it displayed the following summary statistics for the whole class.

|  | Median | Lower quartile | Upper quartile | Min | Max |
|---|---|---|---|---|---|
| Girls | 10 | 8 | 15 | 6 | 19 |
| Boys | 10 | 7 | 13 | 4 | 16 |

(a)  Draw two box plots suitable for comparing the reaction times of boys and girls.
(b)  Write a brief comparison of the performance of boys and girls in this game.        (*NEAB*)

## Solution 1.47

(a)

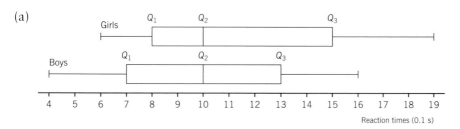

(b)  The 'typical' (median) reaction time for boys and girls is the same ($10 \times 0.1 = 1$ second). However the times for the boys are more evenly distributed, with a smaller range. There is a bigger spread of times for girls and their distribution is positively skewed.

In general, **the boys have the faster reaction time.**

## Example 1.48

A group of children carry out a survey of the numbers of sweets in each of 50 packets. Their results are shown in the following stem and leaf diagram.

> *Key* 20|4 means 24 sweets in a packet.

```
20 | 4  6  6  7  7  7  7  8  8  9  9  9  9
30 | 0  0  0  0  0  1  1  2  2  2  3  3  4  4  4
30 | 5  5  5  6  6  7  7  7  8  8  8  8  8  9  9
40 | 0  0  1  1  2  4  4
```

(a) Calculate the median and the quartiles of this distribution.
(b) Draw a box plot for the distribution.                                   (*NEAB*)

## Solution 1.48

(a) There are 50 items, so the median is the $\frac{1}{2}(50 + 1)$th value, i.e. the 25.5th value.
This is half-way between the 25th and 26th value which are 33 and 34.
So median, $Q_2 = $ **33.5 sweets**.

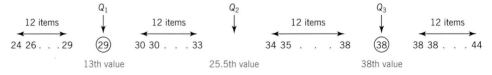

```
                                                Q₁
                                                ↓
20 | 4  6  6  7  7  7  7  8  8  9  9  9  ⑨                    (13 values)
30 | 0  0  0  0  0  1  1  2  2  2  3  [3  4]  4  4            (15 values)
30 | 5  5  5  6  6  7  7  7  ⑧  8  8  8↑8  9  9               (15 values)
40 | 0  0  1  1  2  4  4           ↑        |                (7 values)
                                   Q₃       Q₂
```

There are 25 items to the left of $Q_2$, so $Q_1$ is the 13th value, i.e. $Q_1 = $ **29 sweets**.
There are 25 items to the right of $Q_2$, so $Q_3$ is the 38th value, i.e. $Q_3 = $ **38 sweets**.

Remember that the quartiles divide in half the two distributions either side of the median.

Note that the pattern is easy to see:

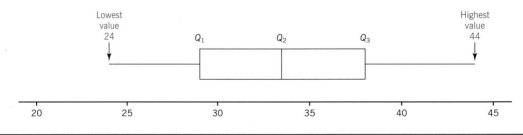

(b) Box plot

## Example 1.49

A group of athletes frequently run round a cross-country course in training. The box and whisker plots below represent the times taken by athletes A, B, C and D to complete the course.

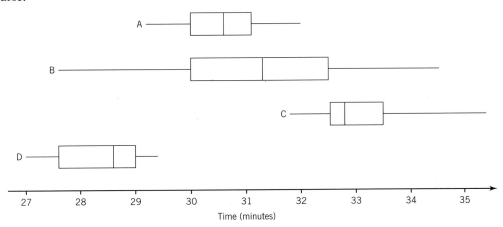

(a) Compare the times taken by athletes C and D.

Assume that the distributions shown above are representative of the times the athletes would take in a race over the same course.

(b) Which of the athletes A or B would you choose if you were asked to select one of them to win a race against

    (i)    C

    (ii)   D?

    Give a reason for **each** answer.

(c) Which athlete would be most likely to win a race between A and B?           (AEB)

## Solution 1.49

(a) D is always faster than C.
    C's times are more variable than D's.
    C's times are positively skewed.
    D's times are negatively skewed.

(b) (i)    B's median average time is faster than C's, but B's times are more variable. It is probable that B would win against C.
        Although A's slowest time of approximately 32 minutes appears to be slightly greater than C's fastest time. A will almost certainly win against C.
        **Therefore choose A to win a race against C.**

    (ii)   A has a small chance of winning against D, but B has a slightly greater chance of winning against D.
        **Therefore choose B to win against D.**

(c) A's average time is faster than B's and A's times are not as variable as B's.
    **Therefore choose A to win a race between A and B.**

# Outliers

Sometimes unusually high or low values occur in a set of data.
There may be good reason for these unusual results but quite often they occur because an error was made when the data were recorded.

To investigate extreme values you would use the mean and standard deviation, or the quartiles and interquartile range (IQR).

As a guide, the term 'outlier' can be applied to data which are

(a) at least 2 standard deviations from the mean, i.e. less than $\bar{x} - 2s$ or greater than $\bar{x} + 2s$, or

(b) at least $1\frac{1}{2} \times$ interquartile range beyond the nearer quartile.

Interquartile range $= Q_3 - Q_1$, so illustrating on a diagram gives:

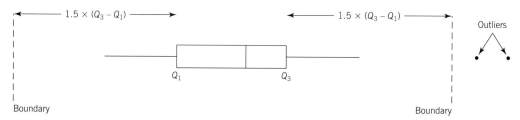

## Example 1.50

A class of 31 children recorded the maximum daily temperature for the month of July with the following results. The median and quartiles are shown on the stem and leaf diagram.

```
9 | 4
8 |
8 | 1  1
7 | 7  7  9  9
7 | 0  0  ⓪  0  2  2  2  3  3  3  ③
6 | ⑥  8  8  8  9  9
6 | 1  3  4  4  4  4
5 | 7
```

Key  6 | 8  means 68°F

Identify any outliers

(a) by using the values of the quartiles and illustrating your results on a boxplot,
(b) by using the mean and the standard deviation.

## Solution 1.50

(a) The values ringed are 66, 70 and 73, so $Q_1 = 66\,°F$,   $Q_2 = 70\,°F$,   $Q_3 = 73\,°F$.

Interquartile range $= Q_3 - Q_1 = 73° - 66° = 7\,°F$
Upper boundary    $= Q_3 + 1.5 \times 7° = 73° + 10.5° = 83.5°$
Lower boundary    $= Q_1 - 1.5 \times 7° = 66° - 10.5° = 55.5°$

Outliers therefore lie outside the interval 55.5° to 83.5°.

It would appear that the temperature recorded as 94 °F is an outlier. (It was probably recorded wrongly, since it is most unusual to have just one day with an extremely high temperature.) The temperature of 57 °F, however is not an outlier.

The whiskers are drawn down to 57 °F and up to 81 °F and the temperature of 94 °F is labelled as an outlier, as shown.

**Boxplot to show temperatures**

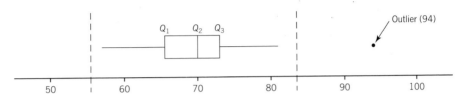

(b) Using calculator in SD mode:

$\bar{x} = 71$, and $s = 7.11$, so
$\bar{x} - 2s = 56.6$, $\bar{x} + 2s = 85.3$

Since outliers lie outside these values, 57 °F is not an outlier, but 94 °F is an outlier.

## Exercise 1m    Box plots

1. The table below gives the lengths, in minutes, of 50 telephone calls from a school office.

| Length of call (min) | Number of calls |
|---|---|
| ≤1 | 8 |
| 1–2 | 11 |
| 2–3 | 17 |
| 3–5 | 8 |
| 5–10 | 6 |
| ≥10 | 0 |

(a) Draw a cumulative frequency polygon.
(b) Estimate the median and the quartiles.
(c) Draw a box plot and comment on the distribution.

2. Two groups of people took part in a reaction-timing experiment. Their results, to the nearest hundredth of a second, are shown below. Construct box plots to represent the distributions, and comment.

| Key | 4\|2 means 24 hundredths of a second |
|---|---|

| Key | 2\|4 means 24 hundredths of a second |
|---|---|

| Group 1 | | Group 2 |
|---:|:---:|:---|
| 6 6 | 2 | |
| 5 4 4 | 2 | 4 5 |
| 3 3 3 3 2 2 | 2 | 2 2 2 3 3 |
| 1 1 0 0 0 | 2 | 0 0 1 |
| 8 | 1 | 8 9 9 9 9 |
| 7 6 6 | 1 | 6 6 7 7 |
| 5 4 4 | 1 | 4 4 |
| | 1 | 2 |
| | 1 | |
| | 0 | 9 |

3. Twenty-one girls estimated the length of a line, in millimetres. The results were

51  45  31  43  97  16  18  23  34  35  35
85  62  20  22  51  57  49  22  18  27

Draw the box plot and use it to identify any outliers.

4. Draw cumulative frequency polygons and then construct box and whisker diagrams to represent the following histograms.

   For each one, calculate $Q_3 - Q_2$ and $Q_2 - Q_1$. What do you notice?

   *Hint*:   remember that

   $$\text{frequency density} = \frac{\text{frequency}}{\text{class width}}$$

(a)
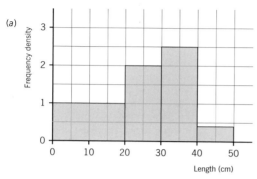

(b)
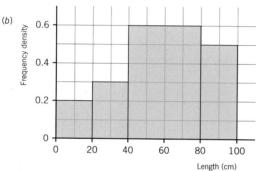

(c)
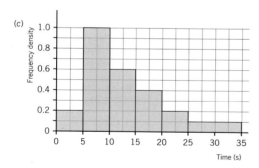

(d)
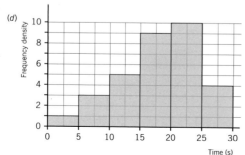

5. Thirty-one people completed a jigsaw in the following times (in minutes).

| | | | | | | | | | | |
|---|---|---|---|---|---|---|---|---|---|---|
| 11 | 53 | 72 | 48 | 48 | 49 | 39 | 87 | 73 | 23 | 120 |
| 24 | 61 | 36 | 66 | 67 | 86 | 79 | 65 | 47 | 36 | 133 |
| 78 | 81 | 70 | 75 | 53 | 42 | 42 | 72 | 144 | | |

   Using the mean and standard deviation, identify any outliers.

6. The box plots show the distributions of marks obtained by a class in English and in Mathematics. Comment on the distributions of marks.

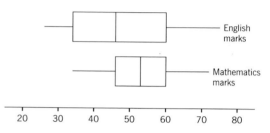

7.

| Key | 4 | 6 means 6.4 hours | | Key | 6 | 5 means 6.5 hours |

| December | | | July |
|---|---|---|---|
| | 12 | | 0  1 |
| | 11 | | 1  3  3  4 |
| | 10 | | |
| | 9 | | 2  2  8  8 |
| | 8 | | |
| | 7 | | 3  4 |
| 4  2  0 | 6 | | 5  5  6  6  6 |
| 1 | 5 | | 0  0  2  8  4 |
| 9  1 | 4 | | 1  3 |
| 1 | 3 | | 5  5 |
| 7  6  4  4  3  3 | 2 | | 6 |
| 9  8  8  6  0  0 | 1 | | 1  2 |
| 8  8  7  3  3  2  0  0  0  0  0  0 | 0 | | 3  4 |

   This back-to-back stemplot gives daily hours of sunshine in December and July

   Find the median and quartiles for each month and construct the box plots.

   Comment on the distributions.

8. These are the times of the postal delivery to my house over four successive weeks.

| | | | | | |
|---|---|---|---|---|---|
| 9:01 | 9:22 | 9:30 | 9:19 | 9:15 | 9:29 |
| 9:45 | 9:53 | 9:02 | 9:05 | 9:31 | 9:47 |
| 9:17 | 9:48 | 9:29 | 9:09 | 9:29 | 9:02 |
| 9:10 | 9:12 | 9:25 | 9:10 | 9:13 | 9:19 |

   (a) Draw a stem and leaf diagram.
   (b) Find the median time.
   (c) Find the quartiles.
   (d) Draw a box and whisker diagram.

9. Draw box plots to represent the following frequency distributions.

(a)

| $x$ | 0 | 1 | 2 | 3 | 4 |
|---|---|---|---|---|---|
| $f$ | 4 | 12 | 6 | 2 | 1 |

(b)

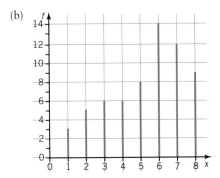

10. A frequency diagram for a set of data is shown below.

(a) Find the median and the mode of the data.
(b) Given that the mean is 5.95 and the standard deviation is 2.58, explain why the value 15 may be regarded as an outlier.
(c) Explain how you would treat the outlier if the diagram represents
   (i) the ages (in completed years) of children at a party,
   (ii) the sums of the scores obtained when throwing a pair of dice.
(d) Find the median and the mode of the data after the outlier is removed.
(e) *Without doing any calculations* state what effect, if any, removing the outlier would have on the mean and on the standard deviation.
(f) Does the diagram exhibit positive skewness, negative skewness or no skewness? How is the skewness affected by removing the outlier? (*MEI*)

11. In a test on the protein quality of a new strain of corn, a farmer fed 20 new born chicks with the new corn and observed how much weight they gained after three weeks. The results are given below.

Weight gain (grams)

360, 445, 403, 376, 434, 402, 397, 425, 407, 369

462, 399, 427, 420, 410, 391, 430, 369, 410, 397

(a) Make an ordered stem and leaf display of these data.

The farmer also fed a further 20 new-born chicks on the standard strain of corn he had previously used and he recorded their weight gains after three weeks. The results for this control group are given in the ordered stem and leaf display in the table below.

| Weight gain (grams) Unit is 1 gram | | | | |
|---|---|---|---|---|
| 32 | 1 | 5 | | |
| 33 | | | | |
| 34 | 5 | 9 | | |
| 35 | 0 | 6 | 6 | |
| 36 | 0 | 1 | 6 | |
| 37 | 1 | 2 | 2 | 3 |
| 38 | 0 | 6 | | |
| 39 | 9 | | | |
| 40 | 2 | | | |
| 41 | | | | |
| 42 | 1 | 3 | | |

(b) On a single diagram draw two box-and-whisker plots, one for the weight gains of the chicks fed the new strain of corn and the other for the weight gains of the control group fed the standard strain of corn.
(c) Use your box and whisker plots to compare and contrast the two distributions. (*O*)

12. A random sample of 51 people was asked to record the number of miles they travelled by car in a given week. The distances, to the nearest mile, are shown below.

| 67 | 76 | 85 | 42 | 93 | 48 | 93 | 46 | 52 |
|---|---|---|---|---|---|---|---|---|
| 72 | 77 | 53 | 41 | 48 | 86 | 78 | 56 | 80 |
| 70 | 70 | 66 | 62 | 54 | 85 | 60 | 58 | 43 |
| 58 | 74 | 44 | 52 | 74 | 52 | 82 | 78 | 47 |
| 66 | 50 | 67 | 87 | 78 | 86 | 94 | 63 | 72 |
| 63 | 44 | 47 | 57 | 68 | 81 | | | |

(a) Construct a stem and leaf diagram to represent these data.
(b) Find the median and the quartiles of this distribution.
(c) Draw a box plot to represent these data.
(d) Give one advantage of using
   (i) a stem and leaf diagram,
   (ii) a box plot,
   to illustrate data such as that given above. (*L*)

## Summary

- Vertical line graphs – ungrouped discrete data

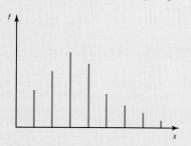

- Height represents frequency.
- Mode is denoted by the tallest line.

- Stem and leaf diagrams (stemplot)

Key  2 | 7  means 27

| Stem | Leaf |
|------|------|
| 1 | 0 4 5 9 |
| 2 | 2 2 3 5 6 7 7 |
| 3 | 1 2 2 7 8 |
| 4 | 3 3 4 6 |
| 5 | 2 3 7 |

- The stemplot must have a key.
- Intervals are 10–19, 20–29, 30–39, 40–49, 50–59
- Equal width intervals must be chosen.

- Histograms – grouped data

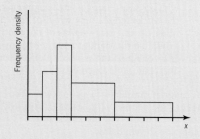

- Area ∝ frequency
- Frequency density = $\dfrac{\text{frequency}}{\text{interval width}}$
- Modal class is represented by tallest rectangle.
- Interval width = upper class boundary – lower class boundary.

- Frequency polygons – grouped data

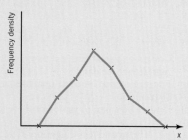

- Plot frequency density against the mid-interval value
- Join with straight lines

- Pie charts

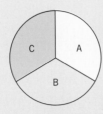

radius $r_1$        radius $r_2$

- Area $\propto$ frequency
- To compare sets of data with total frequencies $F_1$ and $F_2$, draw circles with radii in the ratio $\sqrt{F_1}:\sqrt{F_2}$.

- Mean, $\bar{x}$

  Raw data       $\bar{x} = \dfrac{\Sigma x}{n}$

  Frequency distributions    $\bar{x} = \dfrac{\Sigma fx}{\Sigma f}$

- When data are grouped, the mid-interval value, $\frac{1}{2}$(lower boundary + upper boundary) is taken to represent the interval.

- Standard deviation, $s$

  Raw data      $s = \sqrt{\dfrac{\Sigma(x - \bar{x})^2}{n}}$   or   $s = \sqrt{\dfrac{\Sigma x^2}{n} - \bar{x}^2}$

  Frequency distributions    $s = \sqrt{\dfrac{\Sigma f(x - \bar{x})^2}{\Sigma f}}$   or   $s = \sqrt{\dfrac{\Sigma fx^2}{\Sigma f} - \bar{x}^2}$

  variance $= s^2$

  $s \qquad = \sqrt{\text{variance}}$

- Scaling data

  If   $y = ax + b$, where $a$ and $b$ are constants

  then $\bar{y} = a\bar{x} + b$ and $s_y = |a| s_x$

- Coding data

  If $y = \dfrac{x - a}{b}$   then   $x = a + by$

  $$\bar{x} = a + b\bar{y}$$
  $$s_x = |b| s_y$$

- Combining sets of numbers, $x$ and $y$

  new mean $= \dfrac{\Sigma x + \Sigma y}{n_1 + n_2}$

  new variance $= \dfrac{\Sigma x^2 + \Sigma y^2}{n_1 + n_2} - (\text{new mean})^2$

- Weighted means

  If $x_1, x_2, ..., x_n$ are given weightings $w_1, w_2, ..., w_n$ then

  weighted mean $= \dfrac{w_1 x_1 + w_2 x_2 + \cdots + w_n x_n}{w_1 + w_2 + \cdots + w_n} = \dfrac{\Sigma w_i x_i}{\Sigma w_i}$

- Cumulative frequency is the total frequency up to a particular observation.

(a) Ungrouped data – step diagram.

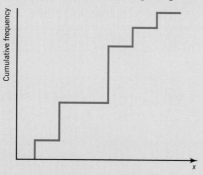

- steepest step denotes the mode.

(b) Grouped data – cumulative frequency curve or polygon.

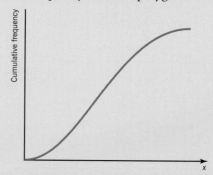

- plot cumulative frequency against upper class boundary.
- join with a curve (or with straight lines for a cumulative frequency polygon).

- Median, quartiles and percentiles

For $n$ observations arranged in order of size

- the median $Q_2$ is the value 50% of the way through the distribution,
- the lower quartile, $Q_1$, is the value 25% of the way through the distribution,
- the upper quartile, $Q_3$, is the value 75% of the way through the distribution,
- the $x$th percentile, $P_x$, is the value $x$% of the way through the distribution.

|  | Ungrouped date | Grouped data |
|---|---|---|
| $Q_2$ | $\frac{1}{2}(n + 1)$th value | $\frac{1}{2}n$th value = 50% value |
| $Q_1$ | Divides the distribution either side of the | $\frac{1}{4}n$th value = 25% value |
| $Q_3$ | median in half | $\frac{3}{4}n$th value = 75% value |

- Ranges

Range = highest value – lowest value

Interquartile range = upper quartile – lower quartile = $Q_3 - Q_1$

Middle 80% of readings = $P_{90} - P_{10}$

● Skewness

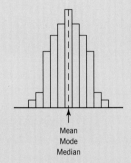

In a **symmetrical** distribution,
mean = mode = median
$$Q_3 - Q_2 = Q_2 - Q_1$$

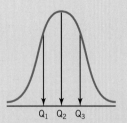

In a **positively-skewed** distribution the tail of the distribution is pulled in the **positive** direction.

mode < median < mean
$$Q_3 - Q_2 > Q_2 - Q_1$$

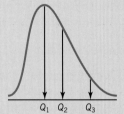

In a **negatively-skewed** distribution the tail of the distribution is pulled in the **negative** direction.

mean < median < mode
$$Q_3 - Q_2 < Q_2 - Q_1$$

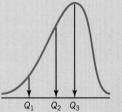

Pearson's coefficient of skewness $= \dfrac{\text{mean} - \text{mode}}{\text{standard deviation}} \approx \dfrac{3(\text{mean} - \text{median})}{\text{standard deviation}}$

Quartile coefficient of skewness $= \dfrac{(Q_3 - Q_2) - (Q_2 - Q_1)}{Q_3 - Q_1}$

● Box and whisker diagrams (boxplots)

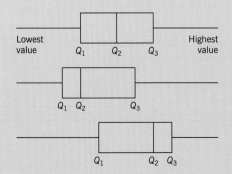

Symmetrical distribution

Positively skewed distribution

Negatively skewed distribution

> • Outliers – as a rough guide:
>
>   • Points at least two standard deviations from the mean.
>
>   • Points lying more than 1.5 times the interquartile range above $Q_3$ or below $Q_1$.

## Miscellaneous worked examples

### Example 1.51

The times $t$ (in seconds) taken by an athlete to run 400 metres on ten successive days were

53.2,   55.7,   54.2,   52.7,   53.6,   56.8,   54.0,   53.7,   59.3,   53.8.

[If required, you may use $\Sigma t = 547.0$, $\Sigma t^2 = 29\ 957.48$.]

(a)  Calculate the mean of the times.
(b)  Calculate the standard deviation of the times.
(c)  Determine the median of the times. (C)

### Solution 1.51

(a)  mean $\bar{t} = \dfrac{\Sigma t}{n} = \dfrac{547.0}{10} = 54.7$ seconds

(b)  $s = \sqrt{\dfrac{\Sigma t^2}{n} - \bar{t}^2} = \sqrt{\dfrac{29\ 957.48}{10} - 54.7^2} = \sqrt{3.658} = 1.9$ seconds (2 s.f.)

(c)  There are 10 values, so the median is the $\frac{1}{2}(10 + 1)$th value = 5.5th value.
Re-arranging the times in order of size

   52.7, 53.2, 53.6, 53.7, 53.8, 54.0, 54.2, 55.7, 56.8, 59.3
   ↑
   5.5th value

The median is half way between 53.8 and 54.0, i.e.

median = $\frac{1}{2}(53.8 + 54.0) =$ **53.9 seconds**

### Example 1.52

Applicants for an assembly job are required to take a test of manual dexterity. The times, in seconds, taken to complete the task by 19 applicants were as follows:

63, 229, 165, 77, 49, 74, 67, 59, 66, 102, 81, 72, 59, 74, 61, 82, 48, 70, 86.

For these data find

(a)  the median,
(b)  the upper and lower quartiles.

An outlier here is defined as any observation less than $Q_1 - 1.5(Q_3 - Q_1)$ or greater than $Q_3 + 1.5(Q_3 - Q_1)$, where $Q_1$ is the lower quartile and $Q_3$ is the upper quartile.

(c)  Identify any outliers in the data.

(d)  Illustrate the data by a box and whisker plot. Outliers, if any, should each be denoted by a '∗' and should not be included in the whiskers.    (*AEB*)

## Solution 1.52

Arranging the data in order

48  49  59  59  $\boxed{61}$  63  66  67  70  $\boxed{72}$  74  74  77  81  $\boxed{82}$  86  106  165  229

                $Q_1$                      $Q_2$                $Q_3$

(a)  $Q_2 = \frac{1}{2}(n + 1)$th item $= \frac{1}{2}(19 + 1)$th item $= 10$th item $= 72$ seconds

(b)  $Q_1 = 61$, $Q_3 = 82$

(c)  $Q_3 - Q_1 = 82 - 61 = 21$
$Q_1 - 1.5(Q_3 - Q_1) = 61 - 1.5 \times 21 = 29.5$
$Q_3 + 1.5(Q_3 - Q_1) = 82 + 1.5 \times 21 = 113.5$
So outliers are less than 29.5 or greater than 113.5
**Therefore the outliers are 165 and 229.**

(d)  Box and whisker plot to show times taken to complete the task.

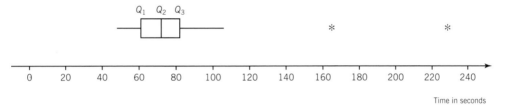

# Example 1.53

Whig and Penn, solicitors, monitored the time spent on consultations with a random sample of 120 of their clients. The times, to the nearest minute, are summarised in the following table.

| Time | Number of clients |
| --- | --- |
| 10–14 | 2 |
| 15–19 | 5 |
| 20–24 | 17 |
| 25–29 | 33 |
| 30–34 | 27 |
| 35–44 | 25 |
| 45–59 | 7 |
| 60–89 | 3 |
| 90–119 | 1 |
| Total | 120 |

(a)  By calculation, obtain estimates of the median and quartiles of this distribution.

(b)  Comment on the skewness of the distribution.

(c) Explain briefly why these data are consistent with the distribution of times you might expect in this situation.

(d) Calculate estimates of the mean and variance of the population of times from which these data were obtained.

The solicitors are undecided whether to use the median and quartiles, or the mean and standard deviation to summarise these data.

(e) State, giving a reason, which you would recommend them to use.

(f) Given that the least time spent with a client was 12 minutes and the longest time was 116 minutes, draw a box plot to represent these data. Use graph paper and show your scale clearly.

Law and Court, another group of solicitors monitored the times spent with a random sample of their clients. They found that the least time spent with a client was 20 minutes, the longest time was 40 minutes and the quartiles were 24, 30 and 36 minutes respectively.

(g) Using the same graph paper and the same scale draw a box plot to represent these data.

(h) Compare and contrast the two box plots.                                                     (L)

## Solution 1.53

(a)

| Time | Cumulative Frequency |
|------|----------------------|
| < 14.5 | 2 |
| < 19.5 | 7 |
| < 24.5 | 24 |
| < 29.5 | 57 |
| < 34.5 | 84 |
| < 44.5 | 109 |
| < 59.5 | 116 |
| < 89.5 | 119 |
| < 119.5 | 120 |

For grouped continuous data, with $n = 120$

$Q_1$ is the $\frac{1}{4}n$th value, i.e. the 30th value

$Q_2$ is the $\frac{1}{2}n$th value, i.e. the 60th value

$Q_3$ is the $\frac{3}{4}n$th value, i.e. the 90th value

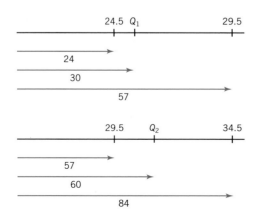

$Q_1$ lies in the interval 24.5–29.5 (width 5).

There are 33 items in this interval,

so $Q_1 = 24.5 + \frac{6}{33} \times 5 = \textbf{25.4 min}$

$Q_2$ lies in the interval 29.5–34.5 (width 5).

There are 27 items in this interval,

so $Q_2 = 29.5 + \frac{3}{27} \times 5 = \textbf{30 min}$

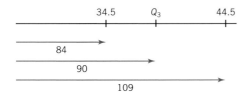

$Q_3$ lies in the interval 34.5–44.5 (width 10).

There are 25 items in this interval,

so $Q_3 = 34.5 + \frac{6}{25} \times 10 = $ **36.9 min**

(b) $Q_3 - Q_2 = 6.9$ min, $Q_2 - Q_1 = 4.6$ min

So $Q_3 - Q_2 > Q_2 - Q_1$. **This implies a positive skew.**

(c) Very few consultations take over an hour.
Most take just under half an hour.

(d)

| Mid-point $x$ | $f$ |
|---|---|
| 12 | 2 |
| 17 | 5 |
| 22 | 17 |
| 27 | 33 |
| 32 | 27 |
| 39.5 | 25 |
| 52 | 7 |
| 74.5 | 3 |
| 104.5 | 1 |

Using the calculator:

$\bar{x} = 32.6$ (1 d.p.)
$s^2 = 160.7$ (1 d.p.)

(e) It is better to use the median and quartiles because the distribution is skewed.

(g)

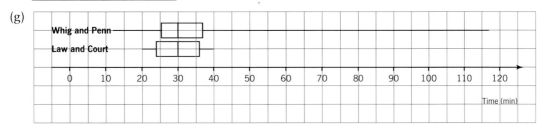

(h) Both have a small interquartile range i.e. small variability.

Both have the same median (30 min).

Whig and Penn times have a much greater range of values, so the Law and Court would appear more efficient.

# Miscellaneous exercise 1n

1. In 1798 the English scientist Henry Cavendish measured the specific gravity of the earth by careful work with a torsion balance. He obtained the 29 measurements given below.

   | | | | | | | | |
   |---|---|---|---|---|---|---|---|
   | 4.07 | 4.88 | 5.10 | 5.26 | 5.27 | 5.29 | 5.29 | 5.30 |
   | 5.34 | 5.34 | 5.36 | 5.39 | 5.42 | 5.44 | 5.46 | 5.47 |
   | 5.50 | 5.53 | 5.55 | 5.57 | 5.58 | 5.61 | 5.62 | 5.63 |
   | 5.65 | 5.75 | 5.79 | 5.85 | 5.86 | | | |

   The sum of these measurements is 157.17 and the sum of their squares is 855.0227.
   (a) Calculate the mean measurement and the standard deviation of the measurements. Obtain for these data the range, the median and the quartiles.
   Draw a box plot and use it to identify the outlier in Cavendish's data set.
   (b) If the data were analysed without this outlier, calculate the new values of
       (i) the median
       (ii) the mean
       (iii) the standard deviation. *(NEAB)*

2. The table shows the distribution of the lifetimes (measured to the nearest hour) of a sample of batteries.

   | Lifetime (to nearest hour) | Frequency |
   |---|---|
   | 690–709 | 3 |
   | 710–719 | 7 |
   | 720–729 | 15 |
   | 730–739 | 38 |
   | 740–744 | 41 |
   | 745–749 | 35 |
   | 750–754 | 21 |
   | 755–759 | 16 |
   | 760–769 | 14 |
   | 770–789 | 10 |

   (a) Draw a histogram to represent the data.
   (b) Draw a cumulative frequency polygon.
   (c) Calculate the mean and standard deviation.
   (d) Estimate the median and the quartiles.
   (e) Calculate Pearson's coefficient of skewness.
   (f) Calculate the quartile coefficient of skewness.
   (g) Draw a box and whisker diagram to illustrate the distribution.

3. The sum of 20 numbers is 320 and the sum of their squares is 5840. Calculate the mean of the 20 numbers and the standard deviation.
   (a) Another number is added to these 20 so that the mean is unchanged. Show that the standard deviation is decreased.

   (b) Another set of 10 numbers is such that their sum is 130 and the sum of their squares is 2380. This set is combined with the original 20 numbers. Calculate the mean and standard deviation of all 30 numbers. *(C Additional)*

4. A grouped frequency distribution of the ages of 358 employees in a factory is shown in the table below. Estimate, to the nearest month, the mean and the standard deviation of the ages of these employees.
   Graphically, or otherwise, estimate
   (a) the median and the interquartile range of the ages, each to the nearest month,
   (b) the percentage, to one decimal place, of the employees who are over 27 years old and under 55 years old. *(L)*

   | Age (last birthday) | Number of employees |
   |---|---|
   | 16–20 | 36 |
   | 21–25 | 56 |
   | 26–30 | 58 |
   | 31–35 | 52 |
   | 36–40 | 46 |
   | 41–45 | 38 |
   | 46–50 | 36 |
   | 51–60 | 36 |
   | 61– | 0 |

5. 200 candidates sat an examination and the distribution was obtained as shown in the table.
   (a) If the limits of class 40–49 are 39.5 to 49.5, what is the mid-interval value of this class?
   (b) Calculate the mean of the marks explaining any limitations of your calculation.
   (c) Plot a cumulative frequency curve and use it to estimate the upper and lower quartiles.
   (d) Assuming that your estimates are exact, find values for $a$ and $b$ correct to two significant figures, in order that the above marks can be scaled by the equation $y = ax + b$, where $y$ is the new mark, so that the mean becomes 45 and the lower quartile becomes 35.
   (e) State, with reason, whether the quartiles of the original marks will scale into the quartiles of the scaled marks.

   | Marks ($x$) | Frequency |
   |---|---|
   | 10–19 | 10 |
   | 20–29 | 18 |
   | 30–39 | 20 |
   | 40–49 | 30 |
   | 50–59 | 49 |
   | 60–69 | 46 |
   | 70–79 | 20 |
   | 80–89 | 5 |
   | 90–99 | 2 |

6. A school entered 88 students for an examination. The results of the examination are shown in the table below.

| Mark ($x$) | Frequency |
|---|---|
| $0 < x \leqslant 10$ | 3 |
| $10 < x \leqslant 20$ | 6 |
| $20 < x \leqslant 30$ | 9 |
| $30 < x \leqslant 40$ | 10 |
| $40 < x \leqslant 50$ | 12 |
| $50 < x \leqslant 60$ | 18 |
| $60 < x \leqslant 70$ | 14 |
| $70 < x \leqslant 80$ | 11 |
| $80 < x \leqslant 90$ | 5 |

(a) Calculate, showing your working and giving your answers correct to two decimal places, an estimate of
   (i)   the mean mark,
   (ii)  the variance,
   (iii) the standard deviation.

(b) Copy and complete the following cumulative frequency table.

| Mark ($x$) | Cumulative frequency |
|---|---|
| $x \leqslant 10$ | |
| $x \leqslant 20$ | |
| $x \leqslant 30$ | |
| $x \leqslant 40$ | |
| $x \leqslant 50$ | |
| $x \leqslant 60$ | |
| $x \leqslant 70$ | |
| $x \leqslant 80$ | |
| $x \leqslant 90$ | |

(c) Using 2 cm to represent 10 marks on the horizontal axis and 2 cm to represent 10 students on the vertical axis, draw, on graph paper, a cumulative frequency polygon to illustrate the distribution of the examination marks.

(d) Use your graph to estimate
   (i)   the median mark,
   (ii)  the interquartile range.
   The lowest mark required to obtain a grade A in the examination was 75.

(e) Estimate from your graph the number of students who were awarded a grade A for this examination. *(C)*

7. Data were collected from a survey of 150 students in a college canteen. Their daily expenditure on mid-day meals is summarised in the table opposite.
Illustrate the data by means of a cumulative frequency graph. Hence estimate the median value of the daily expenditure. *(C)*

| Daily expenditure (£) | Frequency |
|---|---|
| 0.66–0.90 | 11 |
| 0.91–1.15 | 28 |
| 1.16–1.30 | 38 |
| 1.31–1.45 | 34 |
| 1.46–1.70 | 27 |
| 1.71–2.00 | 12 |

8. The hourly wages, £$x$, of the 15 workers in a small factory are as follows:

£6.60,  £3.40,  £6.45,  £5.20,  £3.60,
£7.25,  £9.60,  £3.75,  £4.20,  £8.75,
£5.75,  £4.50,  £3.95,  £4.75,  £12.25.

(a) Illustrate the data in a stem and leaf diagram, using pounds for the stem and pence for the leaves. Clearly indicate the median wage. State the range.

(b) Given that $\Sigma x = 90.00$ and $\Sigma x^2 = 631.25$, calculate the mean and standard deviation of hourly wages of the workers.

After delicate wage negotiations, the workers are offered a choice of one of the following pay rises:

(A) an increase of 30 pence per hour,
(B) a 5% rise in hourly rates.

(c) Use your answers in part (b) to deduce the mean and standard deviation of the hourly wages of the 15 workers under both schemes.

(d) Explain why the management would not mind which scheme was implemented, but the workers might. *(MEI)*

9. The table below shows the length distribution of pebbles from the bed of a river.

| Length, $x$ millimetres | Frequency |
|---|---|
| $0 \leqslant x < 5$ | 10 |
| $5 \leqslant x < 10$ | 8 |
| $10 \leqslant x < 20$ | 12 |
| $20 \leqslant x < 50$ | 25 |
| $50 \leqslant x < 100$ | 30 |

(a) You are given that the frequency density for the class $0 \leqslant x < 5$ is 10. Write down the frequency densities for the other classes.

(b) Represent the data in a histogram.

(c) Use your histogram, or otherwise, to estimate the modal length of a pebble.

(d) Calculate estimates of
   (i)   the mean length of a pebble,
   (ii)  the median length of a pebble. *(O)*

10. An advertising campaign to promote electric showers consists of a mailshot which includes a pre-paid postcard requesting further details. Prospective customers who return the postcard are then contacted by one of five sales staff: Gideon, Magnus, Jemma, Pandora or Muruvet. The pie charts below represent the number of sales completed and the number of potential customers contacted during a one-month period.

### Sales completed

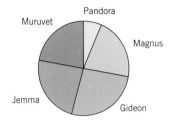

### Potential customers contacted

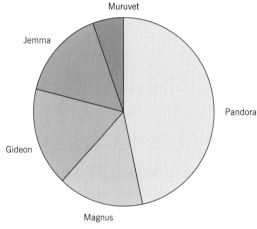

(a) The total number of potential customers contacted is 1100. Find, approximately, the total number of sales completed.
(b) Describe the main features of the data revealed by the pie charts.
(c) The manager wishes to compare the sales staff according to the number of sales completed. What type of diagram would you recommend, in place of a pie chart, so that this comparison could be made easily?
(AEB)

11. An examination is taken by two sets of candidates from the same school. The number of candidates in each set, the mean marks and the variances are shown below.

| | Number of candidates | Mean mark | Variance |
| --- | --- | --- | --- |
| Set A | 20 | 66 | 9 |
| Set B | 30 | 51 | 39 |

Calculate the mean mark for all 50 candidates and show that the standard deviation of all 50 marks is 9.
It is suggested that the original marks of the candidates from Set A should be linearly scaled so that their scaled marks would have a standard deviation of 9 and a mean mark equal to the mean mark of all 50 candidates.
(a) What effect would this have on an original mark of 60 obtained by a candidate from Set A?
(b) Given that the original marks of the candidates in Set A were all integers, explain why no mark would remain unchanged.
(C Additional)

12. Machine A is set to cut lengths of wood 100 mm long. To test the accuracy of the machine, a random sample is taken from the output. The sample size is denoted by $n$ and the length in millimetres of each piece of wood is denoted by $x$. The results are summarised by

$$n = 50, \quad \Sigma x = 5035, \quad \Sigma x^2 = 507\,033.$$

Calculate the mean and standard deviation of the lengths in the sample, giving your answers correct to one decimal place.

Machine B is also set to cut lengths of wood 100 mm long. A random sample of 50 items from this machine has mean 100.2 mm and standard deviation 1.1 mm. Giving your reasons, comment briefly on the accuracy of the two machines. (C)

13. A frequency diagram for a set of data is shown in the figure. No scale is given on the frequency axis, but summary statistics are given for the distribution:

$$\Sigma f = 50, \quad \Sigma fx = 100, \quad \Sigma fx^2 = 344.$$

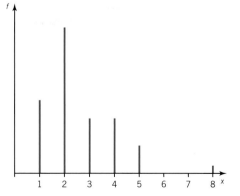

(a) State the mode and the mid-range value of the data.
(b) Identify two features of the distribution.
(c) Calculate the mean and standard deviation of the data and explain why the value 8, which occurs just once, may be regarded as an outlier.

(d) Explain how you would regard the outlier if the diagram represents
  (A) the difference of the scores obtained when throwing a pair of ordinary dice,
  (B) the number of children per household in a neighbourhood survey.
(e) Calculate new values for the mean and standard deviation if the single outlier is removed. (*MEI*)

14. Data collected from a survey of the cost of 4320 houses in a town are summarised in the table below.

| Cost (£) | Number of houses |
|---|---|
| 20 001–50 000 | 540 |
| 50 001–60 000 | 1150 |
| 60 001–70 000 | 1320 |
| 70 001–100 000 | 860 |
| 100 001–150 000 | 450 |

On graph paper, illustrate the data by means of a cumulative frequency graph, and use your graph to estimate the median cost and the interquartile range. (*C*)

15. The age distribution of the applicants for a job is recorded in the table below.

| Age (years) | 20– | 35– | 40– | 45– | 50– | 60– |
|---|---|---|---|---|---|---|
| Number of applicants | 14 | 12 | 7 | 8 | 9 | 0 |

Draw, on graph paper, a histogram to represent these data.

Estimate
(a) the mean of the distribution,
(b) the median age of the applicants who are less than 50 years of age. (*C*)

16. The cumulative frequency table below refers to the lengths, in minutes, of 400 telephone calls made from a certain household during a period of three months.

| Length of call in minutes | Number of calls |
|---|---|
| $\leqslant 1$ | 20 |
| $\leqslant 2$ | 67 |
| $\leqslant 2\frac{1}{2}$ | 118 |
| $\leqslant 3$ | 177 |
| $\leqslant 5$ | 315 |
| $\leqslant 10$ | 400 |

Construct the corresponding frequency table and draw a histogram to illustrate the data.
Use linear interpolation to estimate the median length of call and explain the geometrical significance of a vertical line drawn through the histogram at this value. (*C Additional*)

17. The figure shows a cumulative frequency curve for the length of telephone calls from my house during the first six months of last year.

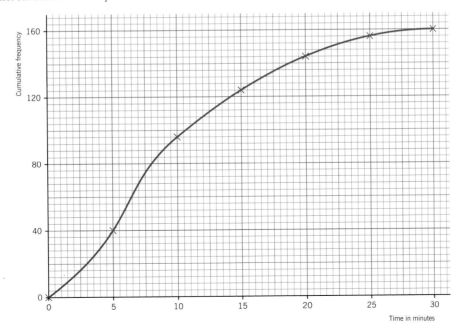

(a) Find the median and inter-quartile range.
(b) Construct a histogram with six equal intervals to illustrate the data.
(c) Use the frequency distribution associated with your histogram to estimate the mean length of call.
(d) State whether each of the following is true or false

(A) the distribution of these call times is negatively skewed,
(B) the majority of the calls last longer than 6 minutes,
(C) the majority of the calls last between 5 and 10 minutes,
(D) the majority of the calls are shorter than the mean length. (MEI)

# Mixed test 1A

1. One hundred runners competed in a half-marathon race. The table below shows $N$, the number of runners who completed the course within $T$ minutes of the start.

| $T$ | 65 | 85 | 95 | 105 | 115 | 155 |
|-----|-----|-----|-----|-----|-----|-----|
| $N$ | 0 | 25 | 53 | 73 | 90 | 100 |

Construct the corresponding frequency table and use this
(a) to draw a histogram to represent the data,
(b) to estimate the mean value of $T$. (C)

2. The following stem and leaf diagram summarises the blood glucose level, in mmol/l, of a patient, measured daily over a period of time.

Blood glucose level    5 | 0 means 5.0    Totals

```
5 | 0 0 1 1 1 2 2 3 3 3 4 4   (12)
5 | 5 5 6 6 7 8 8 9 9          ( 9)
6 | 0 1 1 1 2 3 4 4 4 4        (10)
6 | 5 5 6 7 8 9 9              (  )
7 | 1 1 2 2 2 3                (  )
7 | 5 7 9 9                    (  )
8 | 1 1 1 2 2 3 3 4            (  )
8 | 7 9 9                      ( 3)
9 | 0 1 1 2                    ( 4)
9 | 5 7 9                      ( 3)
```

(a) Write down the numbers required to complete the stem and leaf diagram.
(b) Find the median and quartiles of these data.
(c) On graph paper, construct a box plot to represent these data. Show your scale clearly.
(d) Comment on the skewness of the distribution. (L)

3. The 30 members of the *Darton* town orchestra each recorded the amount of individual practice, $x$ hours, they did in the first week of June. The results are summarised as follows:

$$\Sigma x = 225, \quad \Sigma x^2 = 1755.$$

The mean and standard deviation of the number of hours of practice undertaken by the members of the *Darton* orchestra in this week were $\mu$ and $\sigma$ respectively.
(a) Find $\mu$.
(b) Find $\sigma$.

Two new people joined the orchestra and the number of hours of individual practice they did in the first week of June were $\mu - 2\sigma$ and $\mu + 2\sigma$.
(c) State, giving your reasons, whether the effect of including these two members was to increase, decrease or leave unchanged the mean and standard deviation. (L)

4. A newsagent carried out a survey to gather general information about her customers, and the readability of her magazines. Table 1 shows a classification of the customers during one hour of trading and Table 2 shows the number of words per sentence for a sample of 100 sentences taken from a magazine.

Table 1

| | Child/ Student | Adult female | Adult male |
|---|---|---|---|
| Number of customers | 5 | 28 | 22 |

Table 2

| Words per sentence | 1–5 | 6–10 | 11–15 | 16–25 | 26–45 |
|---|---|---|---|---|---|
| Number of sentences | 18 | 32 | 22 | 14 | 14 |

(a) State a suitable type of diagram which could represent the data in Table 1.
(b) The survey was carried out on a Monday morning. Give one possible reason why conclusions based upon the results of Table 1 should be treated with caution.
(c) Represent the data in Table 2 by means of an accurately drawn histogram on graph paper.
(d) Use the figures in Table 2 to calculate, correct to three significant figures, estimates of the mean and standard deviation of the number of words per sentence in the sample.

(e) Data on the number of words per sentence for a sample of 100 sentences taken from a second magazine were presented in a table similar to Table 2 and with the same class intervals. From this new table, an estimate for the mean number of words per sentence was calculated to be 9.145. In fact, the only sentence in this sample with more than 25 words was one which was 32 words long. Calculate an improved estimate for the mean of this second sample. (C)

5. The following table shows data about the time taken, in seconds to the nearest second, for completing each one of a series of 75 similar chemical experiments.

| Time (s) | Number of experiments |
|---|---|
| 50–60 | 4 |
| 61–65 | 13 |
| 66–70 | 26 |
| 71–75 | 22 |
| 76–86 | 10 |

(a) State the type of diagram appropriate for illustrating the data.

(b) Calculations using the data in the table give estimates as follows

mean time of the experiments      69.64 s,
standard deviation                        6.37 s.

Explain why these are estimates rather than precise values.

(c) Estimate the median and the interquartile range of the times taken for completing the experiments.

(d) It was subsequently revealed that the four experiments in the 50–60 class had actually taken 57, 59, 59 and 60 seconds respectively. State, without further calculation, what effect (if any) there would be on the estimates of the median, interquartile range and mean if this information were taken into account. (C)

6. As part of a detailed study of its workforce, a large company selected a random sample of 100 male employees and recorded the length of time each employee had been with the company. The histogram illustrates the distribution produced.

**Time employed by the company for a random sample of 100 of its male employees**

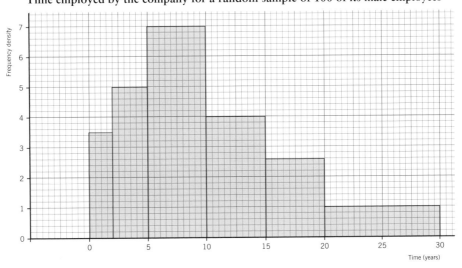

(a) Copy and complete the following table.

| Time (years) | 0– | 2– | 5– | 10– | 15– | 20–30 |
|---|---|---|---|---|---|---|
| Number of males in the sample | | | 35 | | | |

(b) Calculate estimates of the median and quartile times for the sample.

(c) An equivalent random sample of 100 female employees gave calculated estimates of 3.2 years for the median, and 1.8 years and 8.6 years for the quartile times. The longest serving woman in the sample had been with the company for 20 years. The sample also included a woman who had very recently joined the company. Draw adjacent box plots to compare the distributions of the lengths of time male employees and female employees had been with the company.

(d) List **three** differences between the two distributions as illustrated by the box plots.
(*NEAB*)

# Mixed test 1B

1. In a transport survey, the number of passengers in each of 523 cars travelling into a town centre on a particular morning was recorded. The results are summarised in the following table.

| Number of passengers in a car | 0 | 1 | 2 | 3 | 4 | 5 |
|---|---|---|---|---|---|---|
| Number of cars | 183 | 160 | 108 | 63 | 8 | 1 |

   (a) Calculate the mean number of passengers in a car, giving your answer correct to three significant figures.
   (b) State the mode of the number of people (i.e. passengers plus driver) in a car in the survey.
   (c) It is given that, correct to three significant figures, the standard deviation of the number of passengers in a car in the survey is 1.09. State the standard deviation of the number of people in a car in the survey. (C)

2. A student collected some data on the heights, $x$ cm, of plants of a particular species. She chose to represent the data in a stem and leaf display, as shown below.

                    Unit is 1 cm
   1 | 1  2  2  3  4  4  4  5  6  7  7  9
   2 | 1  1  1  2  5  5  7
   3 | 1  2  2  5  5  9
   4 | 1  3  4  5

   (a) (i) Explain why the data might be better represented by a two-part stem and leaf display.
       (ii) Rewrite the above data in such a display.
   (b) Calculate an estimate of the mean height, in centimetres, of plants of this species.
   (c) Calculate the median of the data given in the display.
   (d) State which of the mean and median would be a better measure of location for the heights of these 29 plants. Give a reason for your answer. (O)

3. A pie chart was drawn, for each of the years 1990 and 1995, to illustrate the amounts spent by a householder on electricity, gas, water and telephone, and to compare the total amounts spent in the two years.
   (a) Given that the radii of the 1990 and 1995 charts were 15 cm and 18 cm respectively, calculate the percentage increase in the total amount spent.

   The amount spent on water in 1995 was twice the amount spent in 1990. In the 1995 chart the amount spent on water was represented by an angle of 47°.

   (b) Calculate, to the nearest degree, the angle of the corresponding sector in the 1990 pie chart. (C)

4. A school cleaner is approaching pensionable age. She lives halfway between two post offices, A and B, and has to decide from which of the two she will arrange to collect her pension. For a few months she has deliberately used the two post offices alternately when she has required postal services. On each of these visits she has recorded the time taken between entering the post office and being served.

   The boxplots below show these waiting times for the two post offices. The symbol * represents an outlier.

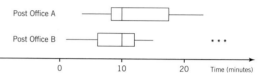

   (a) Compare, in words, the distributions of the waiting times in the two post offices.
   (b) Advise the cleaner which post office to use if the outliers were due to

       (i) a cable laying company having severed the electricity supply to the post office,
       (ii) the post office being short-staffed.
                                        (AEB)

5. Thirty children were given a task to perform and the times taken were recorded, each to the next whole number of minutes above the actual time. The results were as follows:

   12  20  14  17  17   8  19  13  27  13
   16  18  10   7  22  16  11  18  13   6
   16  12  14  23  15   8  10  17  16  19

   (a) Copy and complete the following stem and leaf diagram to illustrate the above data.

   | Key 10 \| 5 represents a time of 15 minutes |
   |---|

   | 00 | 6  7  8  8 |
   | 10 | 0  0  1  2  2 |
   | 10 | 5  6  6 |
   | 20 | 0  2  3 |
   | 20 | 7 |

   (b) Use your diagram to estimate the median and the quartiles of the distribution of times taken to complete the task.
   (c) Draw a box plot to illustrate the distribution. (NEAB)

6.

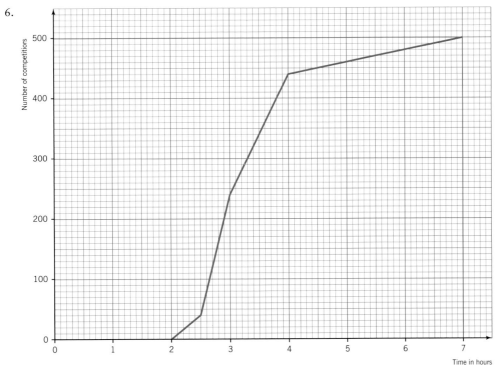

The diagram shows a cumulative frequency polygon for the numbers of competitors who completed a marathon within 2, $2\frac{1}{2}$, 3, 4 and 7 hours of the start.

(a) Use the diagram to estimate
    (i)   the median,
    (ii)  the quartiles,
of the times taken by the 500 competitors who completed the run.

(b) Copy and complete the following frequency table.

| Time in hours | $2-2\frac{1}{2}$ | $2\frac{1}{2}-3$ | 3-4 | 4-7 |
|---|---|---|---|---|
| No. of competitors | | 200 | | |

(c) Calculate an estimate of
    (i)   the mean,
    (ii)  the standard deviation,
of the 500 competitors' times.    (*NEAB*)

# 2

## Regression and correlation

*In this chapter you will learn how to*

- interpret scatter diagrams for bivariate data
- calculate the equations of the least squares regression lines and use them to estimate values
- calculate and interpret the value of the product-moment correlation coefficient
- calculate and interpret the value of Spearman's rank correlation coefficient

## SCATTER DIAGRAMS

Suppose you wish to investigate the relationship between two variables $x$ and $y$, for example

- the weight at the end of a spring $(x)$ and the length of the spring $(y)$,
- the number of hours spent studying for an examination $(x)$ and the mark achieved $(y)$
- a student's mark in a French test $(x)$ and the mark in a German test $(y)$,
- the diameter of the stem of a plant $(x)$ and the average length of leaf of the plant $(y)$.

Data connecting two variables are known as **bivariate data**. When pairs of values are plotted, a **scatter diagram** is produced. Here are some examples:

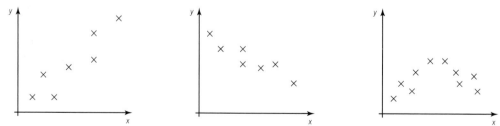

## Dependent and independent variables

If one of the variables has been controlled, it is called the **independent** or **explanatory** variable. The other variable is then the **dependent** or **response** variable.

For example, if you place weights of 10 g, 20 g, 25 g, 30 g, 35 g, 50 g, etc on the end of a spring and record the length of the spring for each weight, the weight is controlled so it is the independent variable. The length of the spring is the dependent variable.

## REGRESSION FUNCTION

Having drawn a scatter diagram, you can then look for a mathematical relationship between the variables, $y = f(x)$, where the function $f$, known as the **regression function**, is to be determined.

## LINEAR CORRELATION AND REGRESSION LINES

Consider the simplest type of regression function, where $y = f(x)$ is a straight line.
If the points on the scatter diagram appear to lie near a straight line, called a **regression line**, you would say that there is **linear correlation** between $x$ and $y$. Here are some examples:

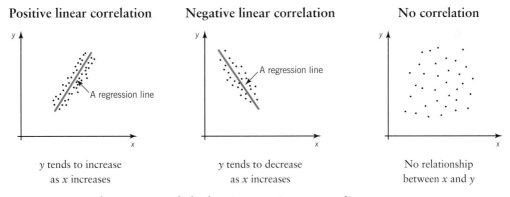

| Positive linear correlation | Negative linear correlation | No correlation |
| --- | --- | --- |
| $y$ tends to increase as $x$ increases | $y$ tends to decrease as $x$ increases | No relationship between $x$ and $y$ |

Common sense and care are needed when interpreting scatter diagrams.

- Mathematically, there may *appear* to be a relationship, but this does not imply that there is a relationship in reality. You might find, for example, that over a period of time in a particular city there has been an increase in the number of robberies and an increase in the number of health food shops. It would however be foolish to imply that there is a relationship between these two variables.
- The appearance of a mathematical relationship does not imply that there is a **causal** relationship. An increase in one variable does not necessarily cause an increase, or decrease, in the other variable.

If it appears from the scatter diagram that a linear relationship is a sensible interpretation, you may then attempt to find a **model** for the relationship in the form of a **regression line**.

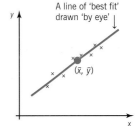

In previous work you may have drawn a **line of best** fit on the scatter diagram, attempting to draw it so that there are as many points above the line as below it, or as many points to the left of the line as to the right of it. The line should also go through the point $(\bar{x}, \bar{y})$, the means of the two sets of data.

This method, known as drawing 'by eye', is rather haphazard. There is a mathematical way of fitting the regression line, known as the **method of least squares** and this is illustrated in the following example.

Consider the situation in which the mass, $y$ g, of a chemical is related to the time, $x$ minutes, for which the chemical reaction has been taking place, according to the table:

| Time, $x$ min | 5 | 7 | 12 | 16 | 20 |
|---|---|---|---|---|---|
| Mass, $y$ g | 4 | 12 | 18 | 21 | 24 |

These results can be illustrated on a scatter diagram.

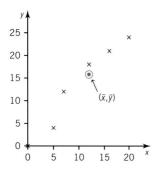

From the scatter diagram there seems to be a positive linear correlation between the mass and the time.

The line of best fit must pass through the means of both sets of data, i.e. the point $(\bar{x}, \bar{y})$. You should find, by calculation, that this is the point $(12, 15.8)$. It has been plotted on the diagram.

Diagrams 1 and 2 show attempts at drawing the line of best fit.

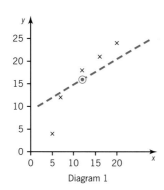

Diagram 1

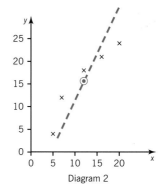

Diagram 2

In each of the attempts, the dotted line goes through $(\bar{x}, \bar{y})$ and there are three points above the line and two points below it. Yet neither of these lines is correct.

Diagram 3 shows the true line of best fit.
It has equation

$$y = 1.15 + 1.22x$$

This equation has been **calculated** by using the method of least squares and the calculations are shown on page 123.

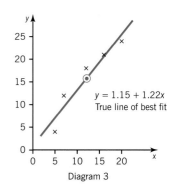

$y = 1.15 + 1.22x$
True line of best fit

Diagram 3

Note that the times, $x$, are chosen by the person holding the stopwatch, so $x$ is the independent variable. The values of the mass, $y$, depend on the results of the chemical process at these times, therefore $y$ is the dependent variable. If you were to repeat the experiment with the same values of $x$, you would almost certainly get a different set of values of $y$. So for a fixed value of $x$ you could have several different values of $y$, all in the same vertical line on the scatter diagram.

# Least squares regression line of $y$ on $x$

To find the equation of the least squares regression line of $y$ on $x$ for the chemical experiment data, consider **vertical** distances $m_1, m_2, m_3, m_4, m_5$ drawn from each point to the regression line. These distances will be positive or negative according to whether the points are above or below the line, so instead work with the **squares** of these values and consider their sum, $m_1^2 + m_2^2 + m_3^2 + m_4^2 + m_5^2$. A shorthand way of writing this is $\Sigma m_i^2$, where

$$\Sigma m_i^2 = m_1^2 + m_2^2 + m_3^2 + m_4^2 + m_5^2 \qquad \text{for } i = 1, 2, 3, 4, 5$$

A line that fits the data well is one that makes $\Sigma m_i^2$ as small as possible, i.e. it is drawn so that $\Sigma m_i^2$ **is minimised.**

This line is called the **least squares regression line of $y$ on $x$.**

Consider our three attempts at drawing the line of best fit. The vertical distances have been shown and you can see that $\Sigma m_i^2$ is least in diagram 3.

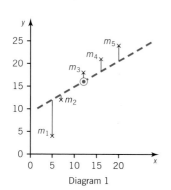
Diagram 1

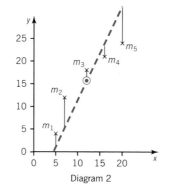
Diagram 2

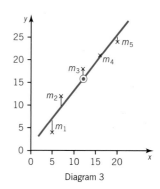
Diagram 3

# Useful formulae when calculating regression lines

Before looking at how to find the equation of the regression line, here is a reminder of the formulae for the mean (page 28) and the variance (page 37) of a set of data together with a new formula that connects the $x$ and $y$ data, the covariance.

*For the x data:*

The mean of the $x$ data is $\bar{x}$ where $\bar{x} = \dfrac{\Sigma x}{n}$ .

The variance is usually written $s^2$, but to distinguish that it is the variance of the $x$ data, you could write $s_x^2$. Usually, however, when working in the context of regression and correlation, the variance of the $x$ data is written $s_{xx}$.

Remember that there are alternative formats of the variance:

$$s_{xx} = \frac{1}{n}\Sigma(x-\bar{x})^2 = \frac{\Sigma(x-\bar{x})^2}{n} \qquad \text{or } s_{xx} = \frac{1}{n}\Sigma x^2 - \bar{x}^2 = \frac{\Sigma x^2}{n} - \bar{x}^2$$

*For the y data:*

$$\bar{y} = \frac{\Sigma y}{n}$$

$$s_{yy} = \frac{1}{n}\Sigma(y-\bar{y})^2 = \frac{\Sigma(y-\bar{y})^2}{n} \qquad \text{or } s_{yy} = \frac{1}{n}\Sigma y^2 - \bar{y}^2 = \frac{\Sigma y^2}{n} - \bar{y}^2$$

*For the x and y data:*
The **covariance**, $s_{xy}$, connects the $x$ and $y$ data and the formula is

$$s_{xy} = \frac{1}{n}\Sigma(x-\bar{x})(y-\bar{y}) = \frac{\Sigma(x-\bar{x})(y-\bar{y})}{n} \qquad \text{or } s_{xy} = \frac{1}{n}\Sigma xy - \bar{x}\bar{y} = \frac{\Sigma xy}{n} - \bar{x}\bar{y}$$

In some textbooks and formulae booklets you might see the notation $S_{xx}$, $S_{yy}$ and $S_{xy}$. These are known as the 'big $S$' formulae and are derived from the 'small $s$' formulae above as follows:

$$S_{xx} = ns_{xx} = \Sigma(x-\bar{x})^2 \qquad \text{or } S_{xx} = \Sigma x^2 - n\bar{x}^2 = \Sigma x^2 - \frac{(\Sigma x)^2}{n}$$

$$S_{yy} = ns_{yy} = \Sigma(y-\bar{y})^2 \qquad \text{or } S_{yy} = \Sigma y^2 - n\bar{y}^2 = \Sigma y^2 - \frac{(\Sigma y)^2}{n}$$

$$S_{xy} = ns_{xy} = \Sigma(x-\bar{x})(y-\bar{y}) \qquad \text{or } S_{xy} = \Sigma xy - n\bar{x}\bar{y} = \Sigma xy - \frac{\Sigma x \Sigma y}{n}$$

The big $S$ formulae are useful in calculations where the factor of $n$ cancels, but it should be remembered that they are not the formulae for the variance and covariance.

## The equation of the regression line *y* on *x*

You are probably familiar with the equation of a straight line in the form

$$y = mx + c$$

where $m$ is the gradient and $c$ is the $y$-intercept.

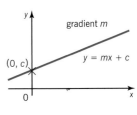

When writing the equation of the regression line, a slightly different format is usually used in which the constant term is written before the $x$-term and the letters used are $a$ and $b$. The format is

$$y = a + bx$$

where $b$ is the gradient and $a$ is the $y$-intercept.

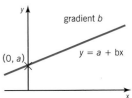

To define the regression line, you need to find the value of $a$ and $b$ for a particular set of data. This is done as follows:

For the regression line $y$ on $x$ written in the form

$$y = a + bx$$

the gradient, $b$, can be calculated as follows:

$$b = \frac{s_{xy}}{s_{xx}} \quad \text{or} \quad b = \frac{S_{xy}}{S_{xx}}$$

Note that $b$ is known as the **regression coefficient of $y$ on $x$**.

To find $a$, use the fact that $(\bar{x}, \bar{y})$ lies on the line,

If $y = a + bx$ then $\bar{y} = a + b\bar{x}$

Rearranging $a = \bar{y} - b\bar{x}$

To find the equation of the regression line $y$ on $x$ for the chemical experiment data on page 120:

| $x$ | $y$ | $x^2$ | $y^2$ | $xy$ |
|---|---|---|---|---|
| 5 | 4 | 25 | 16 | 20 |
| 7 | 12 | 49 | 144 | 84 |
| 12 | 18 | 144 | 324 | 216 |
| 16 | 21 | 256 | 441 | 336 |
| 20 | 24 | 400 | 576 | 480 |
| $\Sigma x = 60$ | $\Sigma y = 79$ | $\Sigma x^2 = 874$ | $\Sigma y^2 = 1501$ | $\Sigma xy = 1136$ |

There are five pairs of data, so $n = 5$.

$$\bar{x} = \frac{\Sigma x}{n} = \frac{60}{5} = 12 \text{ and } \bar{y} = \frac{\Sigma y}{n} = \frac{79}{5} = 15.8$$

$$s_{xy} = \frac{1}{n}\Sigma xy - \bar{x}\bar{y} = \frac{1}{5} \times 1136 - 12 \times 15.8 = 37.6$$

$$s_{xx} = \frac{1}{n}\Sigma x^2 - \bar{x}^2 = \frac{1}{5} \times 874 - 12^2 = 30.8$$

For the regression line $y$ on $x$ in the form $y = a + bx$:

$$b = \frac{s_{xy}}{s_{xx}} = \frac{37.6}{30.8} = 1.2207\ldots = 1.22 \text{ (2 d.p.)}$$

and $a = \bar{y} - b\bar{x} = 15.8 - 1.2207 \times 12 = 1.150\ldots = 1.15$ (2 d.p.)

**So the equation of the regression line $y$ on $x$ is $y = 1.15 + 1.22x$.**

If you use the big $S$ formulae:

$$S_{xy} = \Sigma xy - \frac{\Sigma x \Sigma y}{n} = 1136 - \frac{60 \times 79}{5} = 188$$

$$S_{xx} = \Sigma x^2 - \frac{(\Sigma x)^2}{n} = 874 - \frac{60^2}{5} = 154$$

$$b = \frac{S_{xy}}{S_{xx}} = \frac{188}{154} = 1.2207 \ldots = 1.22 \text{ (2 d.p.)}$$

and $a$ is calculated as above to give $a = 1.15$ (2 d.p.).

An alternative way of working out the equation is to use the following format for the equation of a straight line:

If $m$ is the gradient and the line goes through a fixed point $(h, k)$, the equation of the line can be written

$$y - k = m(x - h).$$

In the case of the regression line $y$ on $x$, the gradient is $b$ and the fixed point is $(\bar{x}, \bar{y})$, so the equation of the regression line $y$ on $x$ can be written

$$y - \bar{y} = b(x - \bar{x}) \qquad \text{where } b = \frac{s_{xy}}{s_{xx}} \text{ or } b = \frac{S_{xy}}{S_{xx}}$$

For the above data relating to the chemical experiment, the equation is

$$y - 15.8 = 1.2207(x - 12)$$
$$y - 15.8 = 1.2207x - 14.648$$
$$y = 1.2207x + 1.152$$
$$y = \mathbf{1.22}x + \mathbf{1.15} \text{ (2 d.p.) as before}$$

Summarising:

The least squares regression line $y$ on $x$ is

$$y = a + bx \qquad \text{where } a = \bar{y} - b\bar{x} \text{ and } b = \frac{s_{xy}}{s_{xx}} = \frac{S_{xy}}{S_{xx}}$$

Alternatively

$$y - \bar{y} = b(x - \bar{x})$$

$b$ is the gradient of the line and is known as the regression coefficient of $y$ on $x$.
$a$ is the y-intercept.

Note that any of the formats described above can be used to calculate the regression line. It is a good idea, however, to make sure that you are familiar with the one used in your examination formulae booklet, which may be one of these, or one written in a slightly different form.

## Making predictions using the regression line *y* on *x*

The regression line *y* on *x* gives you the *average* value of *y* for a given value of *x*, so in certain circumstances it can be used to predict or estimate missing values. This is known as *interpolating* from the given information.

The regression line *y* on *x* is used

- when *x* is the **independent** variable and you want to estimate *y* for a given value of *x*, or you want to estimate *x* for a given value of *y*.

- when neither variable is controlled and you want to estimate *y* for a given value of *x*.

For the chemical reaction data, in which *x* is the independent variable, you can use the regression line $y = 1.15 + 1.22x$ to estimate (a) *y* when $x = 10$, (b) *x* when $y = 20$, as follows:

(a) The estimate of *y* when $x = 10$, written $\hat{y}$, is given by

$$\hat{y} = 1.15 + 1.22 \times 10 = 13.35$$

(b) The estimate for *x* when $y = 20$, written $\hat{x}$, is given by

$$20 = 1.15 + 1.22\hat{x}$$
$$1.22\hat{x} = 18.85$$
$$\hat{x} = 15.4$$

*Warning*: you must take care, though, as estimating outside the range of your data is unreliable. For example, for the chemical reaction data, when the reactants have formed their product, the reaction ceases and the mass would not continue to increase. Going outside the range of data is known as *extrapolating* from the given information.

*Important note*: In the situation where neither variable is controlled and you want to estimate *x* for a given value of *y*, you would use a different regression line, the **least squares line *x* on *y***. You would also use the regression line *x* on *y* if *y* is the independent variable. This is described more fully on page 130.

## Using a calculator to find the regression line *y* on *x*

Linear regression (LR) mode on the calculator enables you to input the pairs of data $(x_i, y_i)$ and then obtain the values of *a* and *b* and also $\bar{x}$, $\bar{y}$, $\Sigma x$, $\Sigma x^2$, $\Sigma y$, $\Sigma y^2$, $\Sigma xy$ and *n*. On the calculator, the value of *a* is usually denoted by *A* and the value of *b* by *B*.

Your calculator may follow a similar procedure to that outlined below. If not, you should consult your calculator manual.

| Casio 85W/85WA/570W | | |
|---|---|---|
| Set LR mode | MODE 3 1 | |
| | or MODE MODE 2 1 | |
| Clear memories | SHIFT Scl = | |
| Input data | 5 , 4 DT | |
| | 7 , 12 DT | |
| | 12 , 18 DT | |
| | 16 , 21 DT | |
| | 20 , 24 DT | |
| You now have access to | | Equation of regression line: |
| $A = 1.1506 \ldots$ | SHIFT 7 = | $y = A + Bx$ |
| $B = 1.2207 \ldots$ | SHIFT 8 = | so $y = 1.15 + 1.22x$ |
| You can check the following | | |
| $\Sigma x^2 = 874$ | RCL A | |
| $\Sigma x = 60$ | RCL B | Red letters A, B, C, D, E and F on third row of calculator. |
| $n = 5$ | RCL C | |
| $\Sigma y^2 = 1501$ | RCL D | |
| $\Sigma y = 79$ | RCL E | |
| $\Sigma xy = 1136$ | RCL F | |
| $\bar{x} = 12$ | SHIFT 1 = | |
| $s_{xx} = 30.8$ | SHIFT 2 = $x^2$ = | |
| $\bar{y} = 15.8$ | SHIFT 4 = | |
| $s_{yy} = 50.56$ | SHIFT 5 = $x^2$ = | |
| To clear LR mode | MODE 1 | |

To estimate $y$ when $x = 10$, key in

10 SHIFT $\hat{y}$  to give 13.35 ...

To estimate $x$ when $y = 20$, key in

20 SHIFT $\hat{x}$  to give 15.44 ...

## Example 2.1

One measure of personal fitness is the time taken for an individual's pulse rate to return to normal after strenuous exercise; the greater the fitness, the shorter the time. Reg and Norman have the same normal pulse rates. Following a short programme of strenuous exercise they both recorded their pulse rates $P$ at time $t$ minutes after they had stopped exercising. Norman's results are given in the table below.

| $t$ | 0.5 | 1.0 | 1.5 | 2.0 | 3.0 | 4.0 | 5.0 |
|---|---|---|---|---|---|---|---|
| $P$ | 125 | 113 | 102 | 94 | 81 | 83 | 71 |

(a) Draw a scatter diagram to represent this information.

The equation of the regression line of $P$ on $t$ for Norman's data is

$P = 122.3 - 11.0t$.

(b) Use the above equation to estimate Norman's pulse rate 2.5 minutes after stopping the exercise programme.

Reg's pulse rate 2.5 minutes after after stopping the exercise was 100.

The full data for Reg are summarised by the following statistics:

$n = 8$,   $\Sigma t = 19.5$,   $\Sigma t^2 = 63.75$,   $\Sigma P = 829$,   $\Sigma Pt = 1867$

(c) Find the equation of the regression line of $P$ on $t$ for Reg's data.

(d) State, giving a reason, which of Reg or Norman you consider to be the fitter.   (L)

## Solution 2.1

(a) **Scatter diagram to show Norman's data**

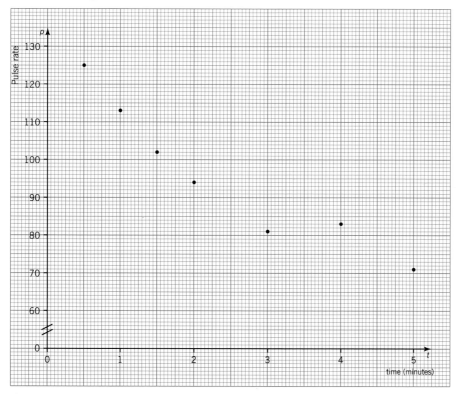

(b) $P = 122.3 - 11.0t$
so when $t = 2.5$, $P = 122.3 - 11.0 \times 2.5 = \mathbf{94.8}$

(c) Regression line of $P$ on $t$ for Reg's data is $P = a + bt$

where $a = \bar{P} - b\bar{t}$ and $b = \dfrac{s_{tp}}{s_{tt}} = \dfrac{S_{tp}}{S_{tt}}$

$\bar{P} = \dfrac{\Sigma P}{n} = \dfrac{829}{8} = 103.625$,   $\bar{t} = \dfrac{\Sigma t}{n} = \dfrac{19.5}{6} = 2.4375$

To find $b$ using the small $s$ format:

$$s_{tP} = \frac{1}{n}\Sigma tP - \bar{t}\bar{P} = \frac{1}{8} \times 1867 - 103.625 \times 2.4375 = -19.210\ldots$$

$$s_{tt} = \frac{1}{n}\Sigma t^2 - \bar{t}^2 = \frac{1}{8} \times 63.75 - 2.4375^2 = 2.027\ldots$$

$$b = \frac{s_{tP}}{s_{tt}} = \frac{-19.210}{2.027} = -9.4759\ldots$$

To find $b$ using the big $S$ format:

$$S_{tP} = \Sigma tP - \frac{\Sigma t \Sigma P}{n} = 1867 - \frac{19.5 \times 829}{8} = -153.6875$$

$$S_{tt} = \Sigma t^2 - \frac{(\Sigma t)^2}{n} = 63.75 - \frac{19.5^2}{8} = 16.21875$$

$$b = \frac{S_{tP}}{S_{tt}} = \frac{-153.6875}{16.21875} = -9.4759\ldots$$

To calculate $a$, use

$$a = \bar{P} - b\bar{t} = 103.625 - (-9.4759) \times 2.4375 = 126.72\ldots$$

**Regression line $P$ on $t$ for Reg is $P = 126.7 - 9.5t$.**

(d) Norman is fitter as his pulse rate decreases more rapidly. This can be seen from the gradients of the regression lines: the gradient for Norman is $-11.0$ and the gradient for Reg is $-9.5$.

---

## Drawing a regression line on a scatter diagram

### Example 2.2

The following data represent the lengths ($x$) and breadths ($y$) of 12 cuckoos' eggs measured in millimetres.

| $x$ | 22.3 | 23.6 | 24.2 | 22.6 | 22.3 | 22.3 | 22.1 | 23.3 | 22.2 | 22.2 | 21.8 | 23.2 |
|---|---|---|---|---|---|---|---|---|---|---|---|---|
| $y$ | 16.5 | 17.1 | 17.3 | 17.0 | 16.8 | 16.4 | 17.2 | 16.8 | 16.7 | 16.2 | 16.6 | 16.4 |

Draw a scatter diagram for the data.

Obtain the least squares regression line of $y$ on $x$ and plot this on the scatter diagram.

(NEAB)

### Solution 2.2

The scatter diagram is shown below together with the regression line.

To find the equation of the regression line use the formulae or find it directly on the calculator where you should find that $A = 11.473\,122\ldots$ and $B = 0.232\,717\,9\ldots$. Giving values to four significant figures, the equation of the least squares regression line of $y$ on $x$ is

$$y = 11.47 + 0.2327x$$

To plot the line on the scatter diagram, you need to work out *three points* on the line, including $(\bar{x}, \bar{y})$, the mean of each set of data.

From the calculator, $\bar{x} = 22.675$ and $\bar{y} = 16.75$, so plot $(\bar{x}, \bar{y})$ as accurately as you can.

Now choose two other $x$-coordinates and calculate the $y$ value for each. The $x$-coordinates should be within the range of data, perhaps at the extremities.

Choosing $x = 21.8$ and $x = 24.2$:

When $x = 21.8$, $y = 11.47 + 0.2327 \times 21.8 = 16.54 \ldots$, so plot $(21.8, 16.54)$.

To obtain this directly on the calculator key in

$\boxed{21.8}$ $\boxed{\text{SHIFT}}$ $\boxed{\hat{y}}$ to give 16.546 …

When $x = 24.2$, $y = 11.47 + 0.2327 \times 24.2 = 17.10 \ldots$, so plot $(24.2, 17.1)$.

Directly on the calculator: $\boxed{24.2}$ $\boxed{\text{SHIFT}}$ $\boxed{\hat{y}}$ gives 17.1048 …

Now draw the regression line, joining the three points, but do not take the line beyond the range of the data.

**Scatter diagram to show the lengths ($x$) and breadths ($y$) of 12 cuckoo eggs.**

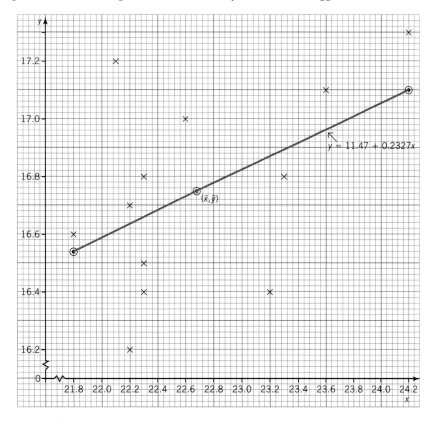

## Least squares regression line *x* on *y*

In Example 2.2, the regression line $y$ on $x$ would be used to estimate the breadth of a cuckoo's egg, $y$, for a given value of the width, $x$. Note that neither the length nor the breadth of the cuckoo's egg is controlled, so there is no independent variable. If you wanted to estimate the width, $x$, for a given value of the breadth, $y$, you would use a different line, the regression line $x$ on $y$.

The **least squares regression line *x* on *y*** is used

- when neither variable is controlled and you want to estimate $x$ for a given value of $y$.

- when $y$ is the controlled (independent) variable and you want to estimate $x$ for a given value of $y$, or $y$ for a given value of $x$.

This time the **horizontal** distances $n_1, n_2, n_3$ ... from the points to the line are considered.

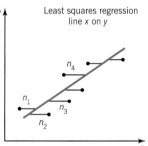
Least squares regression line *x* on *y*

The sum of their squares,

$$\Sigma n_i^2 = n_1^2 + n_2^2 + n_3^2 + ...$$

is made as small as possible, i.e. the line is drawn so that $\Sigma n_i^2$ **is a minimum.**

## The equation of the regression line *x* on *y*

The equation of the regression line $x$ on $y$ is often written in the form

$$x = c + dy \qquad \text{where } c = \bar{x} - d\bar{y}$$

and
$$d = \frac{s_{xy}}{s_{yy}} = \frac{S_{xy}}{S_{yy}}.$$

See page 122 for the formulae for $s_{xy}$, $s_{yy}$, $S_{xy}$ and $S_{yy}$.

Also, since the line goes through $(\bar{x}, \bar{y})$, the equation can be written

$$x - \bar{x} = d(y - \bar{y})$$

$d$ is known as the **regression coefficient** of $x$ on $y$.

Note, however, that $d$ is *not* the gradient of the line. This can be seen by rearranging the equation $x = c + dy$

$$dy = x - c$$
$$y = \left(\frac{1}{d}\right)x - \frac{c}{d}.$$

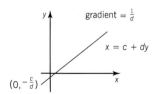

So the gradient of the regression line $x$ on $y$ is $\dfrac{1}{d}$ and the $y$-intercept is $-\dfrac{c}{d}$.

Considering the data of Example 2.2, the summary information is

$\Sigma x = 272.1$, $\Sigma x^2 = 6175.69$, $\Sigma y = 201$, $\Sigma y^2 = 3368.08$, $\Sigma xy = 4559.04$ and $n = 12$.

To find the equation of the regression line $x$ on $y$ in the form $x = cy + d$ calculate $c$ and $d$ as follows:

$$\text{Find } \bar{x} = \frac{\Sigma x}{n} = \frac{272.1}{12} = 22.675 \qquad\qquad \bar{y} = \frac{\Sigma y}{n} = \frac{201}{12} = 16.75$$

To calculate $d$ using the small $s$ format:

$$s_{xy} = \frac{1}{n}\Sigma xy - \bar{x}\bar{y} = \frac{1}{12} \times 4559.03 - 22.675 \times 16.75 = 0.11291 \ldots$$

$$s_{yy} = \frac{1}{n}\Sigma y^2 - \bar{y}^2 = \frac{1}{12} \times 3368.08 - 16.75^2 = 0.11083 \ldots$$

$$d = \frac{s_{xy}}{s_{yy}} = \frac{0.11291 \ldots}{0.11083 \ldots} = 1.0187 \ldots$$

To calculate $d$ using the big $S$ format:

$$S_{xy} = \Sigma xy - \frac{\Sigma x \Sigma y}{n} = 4559.03 - \frac{272.1 \times 201}{12} = 1.355$$

$$S_{yy} = \Sigma y^2 - \frac{(\Sigma y)^2}{n} = 3368.08 - \frac{201^2}{12} = 1.33$$

$$d = \frac{S_{xy}}{S_{yy}} = \frac{1.355}{1.33} = 1.0187 \ldots$$

To calculate $c$, use $c = \bar{x} - d\bar{y} = 22.675 - 1.0187 \times 16.75 = 5.6101 \ldots$

**The equation of regression line $x$ on $y$ is $x = 5.61 + 1.02y$.**
It is interesting to plot this on the scatter diagram, together with the regression line $y$ on $x$.

You know that the line must go through $(\bar{x}, \bar{y})$ so plot $(22.675, 16.75)$. Now choose two other $y$-coordinates, say $y = 16.4$ and $y = 17.0$ and calculate the value of $x$.

When $y = 16.4$. $x = 5.61 + 1.02 \times 16.4 = 22.3$

When $y = 17.0$, $x = 5.51 + 1.02 \times 17.0 = 22.95$

Plot $(22.3, 16.4)$ and $(22.95, 17.0)$ and join the three points with a straight line.

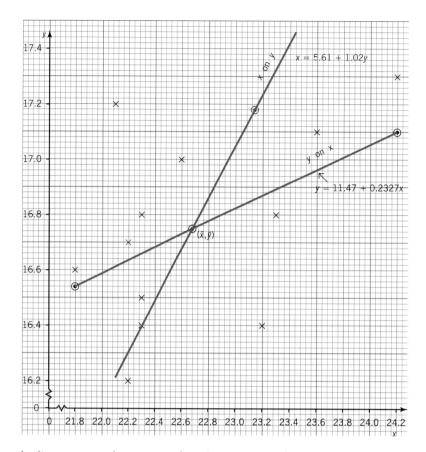

Notice that the lines are not the same; in fact they are quite far apart. You will see later that this indicates that the correlation is not very strong (page 139).

## Using a calculator to find the regression line x on y

The procedure, using linear regression mode, is similar to that described on page 126 for calculating the line $y$ on $x$. This time, however, input the data with the $y$-coordinate first. For the equation $x = c + dy$, the value of $c$ is given by $A$ on the calculator, and the value of $d$ by $B$.

The method is illustrated below, using the data for the cuckoos' eggs given in Example 2.2 on page 128.

| Casio 85W/85WA/570W | |
|---|---|
| Set LR mode | MODE 3 1 <br> or MODE MODE 2 1 |
| Clear memories | SHIFT Scl = |
| Input data | 16.5 , 22.3 DT <br> 17.1 , 23.6 DT <br> 17.3 , 24.2 DT <br> ⋮      ⋮ <br> 16.4 , 23.2 DT |

You now have access to

$A = 5.610 \dots (c)$    SHIFT 7 =

$B = 1.0187 \dots (d)$    SHIFT 8 =

You can check the following

$\Sigma y^2 = 3368.08$    RCL A

$\Sigma y = 201$    RCL B

$n = 5$    RCL C

$\Sigma x^2 = 6175.69$    RCL D

$\Sigma x = 272.1$    RCL E

$\Sigma xy = 4559.03$    RCL F

$\bar{y} = 16.75$    SHIFT 1 =

$s_{yy} = 0.1108 \dots$    SHIFT 2 = $x^2$ =

$\bar{x} = 22.675$    SHIFT 4 =

$s_{xx} = 0.4852 \dots$    SHIFT 5 = $x^2$ =

To clear LR mode    MODE 1

**Equation of regression line**

$x = c + dy$

ie $x = 5.61 + 1.02y$

Red letters on third row of calculator.

Note that if your calculator shows what is being found on the display, you should read $y$ for $x$ and $x$ for $y$ when checking these.

## Example 2.3

A student found the following data for the female life expectancy, $x$ years, and the Gross Domestic Production (GDP) per head, $\$y$, in six countries in South Asia in 1988.

| Country | $x$ | $y$ |
|---|---|---|
| Afghanistan | 42 | 143 |
| Bangladesh | 50 | 179 |
| Bhutan | 47 | 197 |
| India | 58 | 335 |
| Pakistan | 57 | 384 |
| Sri Lanka | 73 | 423 |

$[n = 6, \Sigma x = 327, \Sigma y = 1661, \Sigma x^2 = 18\ 415, \Sigma y^2 = 529\ 909, \Sigma xy = 96\ 412]$

(a) It is required to estimate the value of $x$ for Nepal, where the value of $y$ was 160.
  (i) Find the equation of a suitable line of regression. Simplify your answer as far as possible, giving the constants correct to three significant figures.
  (ii) Use your equation to obtain the required estimate.

(b) Use your equation to estimate the value of $x$ for North Korea, where the value of $y$ was 858.
  Comment on your answer. (C)

## Solution 2.3

(a) (i) Neither variable has been controlled in the given data and since you are required to estimate the life expectancy, $x$ years, when the Gross Domestic Product per head, $\$y$ is $\$160$, it is sensible to use the regression line of $x$ on $y$.

The least squares regression line of $x$ on $y$ has equation

$$x = c + dy \quad \text{where } c = \bar{x} - d\bar{y} \quad \text{and } d = \frac{s_{xy}}{s_{yy}} \quad \text{or } d = \frac{S_{xy}}{S_{yy}}$$

$$\bar{x} = \frac{\Sigma x}{n} = \frac{327}{6} \quad \text{and} \quad \bar{y} = \frac{\Sigma y}{n} = \frac{1661}{6}$$

Using small $s$ format to find $d$:

$$s_{xy} = \frac{1}{n}\Sigma xy - \bar{x}\bar{y} = \frac{1}{6} \times 96\,412 - \frac{327}{6} \times \frac{1661}{6} = 981.25$$

$$s_{yy} = \frac{1}{2}\Sigma y^2 - \bar{y}^2 = \frac{1}{6} \times 529\,909 - \left(\frac{1661}{6}\right)^2 = 11681.47\dots$$

$$d = \frac{s_{xy}}{s_{yy}} = \frac{981.25}{11681.47} = 0.08400\dots$$

Using big $S$ format to find $d$:

$$S_{xy} = \Sigma xy - \frac{\Sigma x \Sigma y}{n} = 96\,412 - \frac{327 \times 1661}{6} = 5887.5$$

$$S_{yy} = \Sigma y^2 - \frac{(\Sigma y)^2}{n} = 529\,909 - \frac{1661^2}{6} = 70\,088.83\dots$$

$$d = \frac{S_{xy}}{S_{yy}} = \frac{5887.5}{70\,088.83} = 0.08400\dots$$

Calculate $c$ using
$$c = \bar{x} - d\bar{y}$$
$$= \frac{327}{6} - 0.084\,00\dots \times \frac{1661}{6}$$
$$= 31.24\dots$$

Equation of regression line of $x$ on $y$ is $x = 31.2 + 0.0840\,y$ (3 s.f.).

(ii) When $y = 160$, $x = 31.2 + 0.0840 \times 160 = 45$ (2 s.f.)

**The estimated value of the life expectancy in Nepal is 45 years.**

(b) From the equation, when $y = 858$

$$x = 31.2 + 0.0840 \times 858 = 103 \text{ (3 s.f.)}$$

This would give the life expectancy in North Korea as 103 years, which is clearly not sensible. The value of $y = 858$ is a long way outside the range of the data, and should not be used to estimate a value of $x$.

## Note on using the calculator in LR mode

You should check whether the regulations of your examination board permit you to use the calculator in LR mode to find the equation of the regression lines without showing any supporting working. The equations are quick to find using the calculator, but a disadvantage is that if you make a slip when entering the data, your answer will be wrong, and this would result in the loss of *all* the marks. Supported by calculation, however, your answer, though wrong, would receive marks for *method*.

Sometimes data are presented in such a way that it is not possible to find the equations of the regression lines directly using the LR mode. This is the case when, for example you do not know the raw data, but just the values of the summary statistics, $\Sigma x$, $\Sigma x^2$, $\Sigma y$, $\Sigma y^2$, $\Sigma xy$ and $n$. If data are presented just in this form, then the appropriate formula must be used and the values calculated.

Consider also when data are given as in the following two examples:

## Example 2.4

For a given set of data it is known that $\bar{x} = 10$ and $\bar{y} = 4$. The gradient of the regression line $y$ on $x$ is 0.6.

Find the equation of this regression line and estimate $y$ when $x = 12$.

## Solution 2.4

The equation of the regression line is $y = a + bx$, where $b = 0.6$.

$$\therefore \quad y = a + 0.6x$$

The regression line goes through $(\bar{x}, \bar{y})$, so
$$\bar{y} = a + 0.6\bar{x}$$
$$4 = a + 0.6 \times 10$$
$$a = -2$$

Equation of regression line is $y = -2 + 0.6x$

When $x = 12$, $y = -2 + 0.6 \times 12 = 5.2$

## Example 2.5

Find the equation of the regression line of $x$ on $y$ if the line goes through $(1, 4)$ and has gradient 2.

**Solution 2.5**

Equation of regression line $x$ on $y$ is $x = c + dy$

Re-arranging

$$dy = x - c$$

$$y = \frac{1}{d}x - \frac{c}{d}$$

$$\text{Gradient} = \frac{1}{d}$$

$$\therefore \quad 2 = \frac{1}{d}$$

$$d = 0.5$$

So

$$x = c + 0.5y$$

You are given that $(1, 4)$ lies on the line

$$\therefore \quad 1 = c + 0.5 \times 4$$
$$c = -1$$

The equation of the regression line $x$ on $y$ is $x = -1 + 0.5y$.

---

# Exercise 2a   Equations of least squares regression lines

Use the method you prefer for calculating the equations of the regression lines. It is a good idea to be able to use the calculator in LR mode *and* to be competent at using the formula.

1. For each set of data, find
   (a) the equation of the regression line of $y$ on $x$,
   (b) the equation of the regression line of $x$ on $y$.
   Plot them both on a scatter diagram and comment.

   Data set 1

   | $x$ | 3 | 7 | 9 | 11 | 14 | 14 | 15 | 21 | 22 | 23 | 26 |
   |---|---|---|---|---|---|---|---|---|---|---|---|
   | $y$ | 5 | 12 | 5 | 12 | 10 | 17 | 23 | 16 | 10 | 20 | 25 |

   Data set 2

   | $x$ | $y$ |
   |---|---|
   | 1 | 85 |
   | 5 | 82 |
   | 5 | 85 |
   | 5 | 89 |
   | 6 | 78 |
   | 7.5 | 66 |
   | 7.5 | 77 |
   | 7.5 | 81 |
   | 10 | 70 |
   | 11 | 74 |
   | 12.5 | 65 |
   | 14 | 69 |
   | 14.5 | 63 |

2. The following data show, in convenient units, the yield ($y$) of a chemical reaction run at various different temperatures ($x$):

   | Temperature ($x$) | Yield ($y$) |
   |---|---|
   | 110 | 2.1 |
   | 120 | 4.3 |
   | 130 | 3.1 |
   | 140 | 3.4 |
   | 150 | 2.9 |
   | 160 | 5.5 |
   | 170 | 3.3 |

   (a) Plot the data. Comment on whether it appears that the usual simple linear regression model is appropriate.
   (b) Assuming that such a model is appropriate, estimate the regression line of yield on temperature.
   (c) Plot your estimated line on your graph, and indicate clearly on your graph the distances, the sum of whose squares is minimised by the linear regression procedure.     (*MEI*)

3. In a certain heathland region there is a large number of alder trees where the ground is marshy but very few where the ground is dry.

The number ($x$) of alder trees and the ground moisture content ($y$) are found in each of ten equal areas (which have been chosen to cover the range of $x$ in all such areas). The following is a summary of the results of the survey:

$\Sigma x = 500$, $\Sigma y = 300$, $\Sigma x^2 = 27\ 818$, $\Sigma xy = 16\ 837$, $\Sigma y^2 = 10\ 462$

Find the equation of the regression line of $y$ on $x$.

Estimate the ground moisture constant in an area equal to one of the chosen areas which contains 60 alder trees. (O & C)

4. To test the effect of a new drug twelve patients were examined before the drug was administered and given an initial score ($I$) depending on the severity of various symptoms. After taking the drug they were examined again and given a final score ($F$). A decrease in score represented an improvement. The scores for the twelve patients are given in the table below.

|  | Score | |
| --- | --- | --- |
| Patient | Initial ($I$) | Final ($F$) |
| 1 | 61 | 49 |
| 2 | 23 | 12 |
| 3 | 8 | 3 |
| 4 | 14 | 4 |
| 5 | 42 | 28 |
| 6 | 34 | 27 |
| 7 | 32 | 20 |
| 8 | 31 | 20 |
| 9 | 41 | 34 |
| 10 | 25 | 15 |
| 11 | 20 | 16 |
| 12 | 50 | 40 |

Calculate the equation of the line of regression of $F$ on $I$.

On the average what improvement would you expect for a patient whose initial score was 30? (MEI)

5. For a given set of data

$\Sigma x = 15$, $\Sigma x^2 = 55$, $\Sigma y = 43$, $\Sigma y^2 = 397$, $\Sigma xy = 145$, $n = 5$

Find the equations of the regression lines $y$ on $x$, and $x$ on $y$.

6. The following table shows the marks ($x$) obtained in a Christmas examination and the marks ($y$) obtained in the following summer examination by a group of nine students.

| Student | Christmas ($x$) | Summer ($y$) |
| --- | --- | --- |
| A | 57 | 66 |
| B | 35 | 51 |
| C | 56 | 63 |
| D | 57 | 34 |
| E | 66 | 47 |
| F | 79 | 70 |
| G | 81 | 84 |
| H | 84 | 84 |
| I | 52 | 53 |

It is given that $\Sigma x = 567$, $\Sigma y = 552$, $\Sigma xy = 36\ 261$, $\Sigma x^2 = 37\ 777$, $\Sigma y^2 = 36\ 112$.

(a) Find the equation of the estimated least squares regression line of $Y$ on $X$.
(b) A tenth student obtained a mark of 70 in the Christmas examination but was absent from the summer examination. Estimate the mark that this student would have obtained in the summer examination. (C)

7. For a period of three years a company monitors the number of units of output produced per quarter and the total cost of producing the units. The table below shows their results.

| Units of output ($x$) 1000's | Total cost ($y$) £1000 |
| --- | --- |
| 14 | 35 |
| 29 | 50 |
| 55 | 73 |
| 74 | 93 |
| 11 | 31 |
| 23 | 42 |
| 47 | 65 |
| 69 | 86 |
| 18 | 38 |
| 36 | 54 |
| 61 | 81 |
| 79 | 96 |

(Use $\Sigma x^2 = 28\ 740$; $\Sigma xy = 38\ 286$)
(a) Draw a scatter diagram of these data.
(b) Calculate the equation of the regression line of $y$ on $x$ and draw this line on your scatter diagram
The selling price of each unit of output is £1.60.
(c) Use your graph to estimate the level of output at which the total income and total costs are equal.
(d) Give a brief interpretation of this value. (AEB)

8. From a set of pairs of observations of the variables $x$ and $y$, it is found that the regression line of $y$ on $x$ passes through the point $(0, 1.8)$. If the means of the $x$ and $y$ values are 5.0 and 8.3 respectively, find the equation of the regression line of $y$ on $x$ in the form $y = a + bx$. (L)

9. For a set of 20 pairs of observations of the variables $x$ and $y$, it is known that $\Sigma x = 250$, $\Sigma y = 140$, and that the regression line of $y$ and $x$ passes through (15, 10). Find the equation of the regression line of $y$ on $x$ and use it to estimate $y$ when $x = 10$.

10. The gradient of the regression line $x$ on $y$ is $-0.2$ and the line passes through (0, 3). If the equation of the line is $x = c + dy$, find the value of $c$ and $d$ and sketch the line on a diagram.

11. A small firm negotiates an annual pay rise with each of its twelve employees. In an attempt to simplify the process it is proposed that each employee should be given a score ($x$) based on his/her level of responsibility. The annual salary ($y$) will be £$(a + bx)$ and the annual negotiations will only involve the values of $a$ and $b$. The following table gives last year's salaries (which were generally accepted as fair) and the proposed scores.

| Employee | $x$ | Annual salary (£), $y$ |
|---|---|---|
| A | 10 | 5 750 |
| B | 55 | 17 300 |
| C | 46 | 14 750 |
| D | 27 | 8 200 |
| E | 17 | 6 350 |
| F | 12 | 6 150 |
| G | 85 | 18 800 |
| H | 64 | 14 850 |
| I | 36 | 9 990 |
| J | 40 | 11 000 |
| K | 30 | 9 150 |
| L | 37 | 10 400 |

(You may assume that $\Sigma x = 459$, $\Sigma x^2 = 22\ 889$, $\Sigma y = 132\ 600$, $\Sigma xy = 6\ 094\ 750$)

(a) Plot the data on a scatter diagram.
(b) Estimate values that could have been used for $a$ and $b$ last year by fitting the regression line $y = a + bx$ to the data. Draw the line on the scatter diagram.
(c) Comment on whether the suggested method is likely to prove reasonably satisfactory in practice.
(d) Without recalculating the regression line find the appropriate values of $a$ and $b$ if every employee were to receive a rise of (i) £500 a year, (ii) 8%, (iii) 4% plus £300 per year.
(e) Two employees, B and C, had to work away from home for a large part of the year. In the light of this additional information, suggest an improvement to the model.
(AEB)

12. In a regression calculation for five pairs of observations one pair of values was lost when the data were filed. For the regression of $y$ on $x$ the equation was calculated as

$$y = 2x - 0.1$$

The four recorded pairs of values are

| $x$ | 0.1 | 0.2 | 0.4 | 0.3 |
|---|---|---|---|---|
| $y$ | 0.1 | 0.3 | 0.7 | 0.4 |

Find the missing pair of values, using the following data for the four pairs above:
$\Sigma x = 1$, $\Sigma x^2 = 0.3$, $\Sigma xy = 0.47$, $\Sigma y = 1.5$.
(MEI)

13. In an attempt to increase the yield (kg/h) of an industrial process a technician varies the percentage of a certain additive used, while keeping all other conditions as constant as possible. The results are shown below.

| Yield, $y$ | % additive, $x$ |
|---|---|
| 127.6 | 2.5 |
| 130.2 | 3.0 |
| 132.7 | 3.5 |
| 133.6 | 4.0 |
| 133.9 | 4.5 |
| 133.8 | 5.0 |
| 133.3 | 5.5 |
| 131.9 | 6.0 |

You may assume that $\Sigma x = 34$, $\Sigma y = 1057$, $\Sigma xy = 4504.55$, $\Sigma x^2 = 155$.

(a) Draw a scatter diagram of the data.
(b) Calculate the equation of the regression line of yield on percentage additive and draw it on the scatter diagram.

The technician now varies the temperature (°C) while keeping other conditions as constant as possible and obtains the following results

| Yield, $y$ | Temperature, $t$ |
|---|---|
| 127.6 | 70 |
| 128.7 | 75 |
| 130.4 | 80 |
| 131.2 | 85 |
| 133.6 | 90 |

He calculates (correctly) that the regression line is $y = 107.1 + 0.29t$.

(c) Draw a scatter diagram of these data together with the regression line.
(d) The technician reports as follows, 'The regression coefficient of yield on percentage additive is larger than that of yield on temperature, hence the most effective way of increasing the yield is to make the percentage additive as large as possible, within reason.'
Criticise the report and make your own recommendations on how to achieve the maximum yield.
(AEB)

# THE PRODUCT-MOMENT CORRELATION COEFFICIENT, *r*

The product-moment correlation coefficient, *r*, is a numerical value between −1 and 1 inclusive which indicates the linear degree of scatter.

$-1 \leqslant r \leqslant 1$

*r* = 1 indicates perfect positive linear correlation.
*r* = −1 indicates perfect negative linear correlation.
*r* = 0 indicates no correlation.

The nearer the value of *r* is to 1 or −1, the closer the points on the scatter diagram are to the regression line.

Here are some examples of the value of *r*:

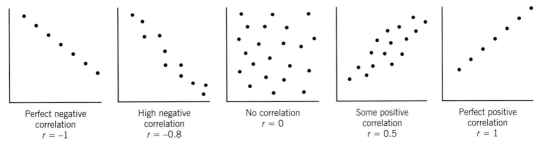

| Perfect negative correlation *r* = −1 | High negative correlation *r* = −0.8 | No correlation *r* = 0 | Some positive correlation *r* = 0.5 | Perfect positive correlation *r* = 1 |

Plotting the two regression lines, *y* on *x* and *x* on *y*, on a scatter diagram can also give a good idea of the value of *r*. The closer the two lines are together, the nearer *r* is to 1 or −1. This is illustrated in the following diagrams:

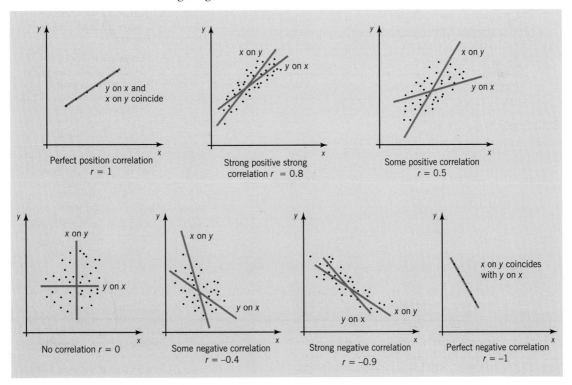

$r$ is a very useful measure because it is independent of the units of scale of the variables. It is calculated as follows.

Using small $s$ format:

$$r = \frac{s_{xy}}{s_x s_y} \quad \text{where} \quad s_{xy} = \frac{1}{n}\Sigma xy - \bar{x}\bar{y} = \frac{\Sigma xy}{n} - \bar{x}\bar{y}$$

$$s_x = \sqrt{s_{xx}} = \sqrt{\frac{1}{n}\Sigma x^2 - \bar{x}^2} = \sqrt{\frac{\Sigma x^2}{n} - \bar{x}^2}$$

$$s_y = \sqrt{s_{yy}} = \sqrt{\frac{1}{n}\Sigma y^2 - \bar{y}^2} = \sqrt{\frac{\Sigma y^2}{n} - \bar{y}^2}$$

Using big $S$ format:

$$r = \frac{S_{xy}}{S_x S_y} \quad \text{where} \quad S_{xy} = \Sigma xy - \frac{\Sigma x \Sigma y}{n}$$

$$S_x = \sqrt{S_{xx}} = \sqrt{\Sigma x^2 - \frac{(\Sigma x)^2}{n}}$$

$$S_y = \sqrt{S_{yy}} = \sqrt{\Sigma y^2 - \frac{(\Sigma y)^2}{n}}$$

## Example 2.6

The following table shows the marks of ten candidates in Physics and Mathematics. Find the product-moment correlation coefficient and comment on your value.

| Mark in Physics ($x$) | 18 | 20 | 30 | 40 | 46 | 54 | 60 | 80 | 88 | 92 |
|---|---|---|---|---|---|---|---|---|---|---|
| Mark in Mathematics ($y$) | 42 | 54 | 60 | 54 | 62 | 68 | 80 | 66 | 80 | 100 |

## Solution 2.6

| $x$ | $y$ | $x^2$ | $y^2$ | $xy$ |
|---|---|---|---|---|
| 18 | 42 | 324 | 1764 | 756 |
| 20 | 54 | 400 | 2916 | 1080 |
| 30 | 60 | 900 | 3600 | 1800 |
| 40 | 54 | 1600 | 2916 | 2160 |
| 46 | 62 | 2116 | 3844 | 2852 |
| 54 | 68 | 2916 | 4624 | 3672 |
| 60 | 80 | 3600 | 6400 | 4800 |
| 80 | 66 | 6400 | 4356 | 5280 |
| 88 | 80 | 7744 | 6400 | 7040 |
| 92 | 100 | 8464 | 10 000 | 9200 |
| $\Sigma x = 528$ | $\Sigma y = 666$ | $\Sigma x^2 = 34\ 464$ | $\Sigma y^2 = 46\ 820$ | $\Sigma xy = 38\ 640$ |

There are ten pairs of data, so $n = 10$.

$$\bar{x} = \frac{\Sigma x}{n} = \frac{528}{10} = 52.8 \quad \text{and} \quad \bar{y} = \frac{\Sigma y}{n} = \frac{666}{10} = 66.6$$

To find $r$ using small $s$ format:

$$s_{xy} = \frac{\Sigma xy}{n} - \bar{x}\bar{y} = \frac{38\,640}{10} - 52.8 \times 66.6 = 347.52$$

$$s_{xx} = \frac{\Sigma x^2}{n} - \bar{x}^2 = \frac{34\,464}{10} - 52.8^2 = 658.56$$

$$s_{yy} = \frac{\Sigma y^2}{n} - \bar{y}^2 = \frac{46\,820}{10} - 66.6^2 = 246.44$$

$$\therefore r = \frac{s_{xy}}{s_x s_y} = \frac{347.52}{\sqrt{658.56} \times \sqrt{246.44}} = 0.8626\ldots$$

To find $r$ using big $S$ format:

$$S_{xy} = \Sigma xy - \frac{\Sigma x\,\Sigma y}{n} = 38\,640 - \frac{528 \times 666}{10} = 3475.2$$

$$S_{xx} = \Sigma x^2 - \frac{(\Sigma x)^2}{n} = 34\,464 - \frac{528^2}{10} = 6585.6$$

$$S_{yy} = \Sigma y^2 - \frac{(\Sigma y)^2}{n} = 46\,820 - \frac{666^2}{10} = 2464.4$$

$$\therefore r = \frac{S_{xy}}{S_x S_y} = \frac{3475.2}{\sqrt{6585.6} \times \sqrt{2464.4}} = 0.8626\ldots$$

**The product moment correlation coefficient is 0.86 (2 s.f.), indicating good positive correlation.**

# Using the calculator in LR mode to find $r$

The value of $r$ can be found directly, for example:

| | Casio 85W/85WA/570W |
|---|---|
| Set LR mode | MODE 3 1 |
| | or MODE MODE 2 1 |
| Clear memories | SHIFT Scl = |
| Input data | 18 , 42 DT |
| | 20 , 54 DT |
| | ⋮       ⋮ |
| | 92 , 100 DT |
| Output | |
| $r = 0.826\ldots$ | SHIFT $r$ |
| Clear LR mode | MODE 1 |

*NOTE*: The value of $r$ should be considered in conjunction with a diagram.

By calculation, or using the data already in your calculator, you should find that the regression line $y$ on $x$ has equation $y = 38.7 + 0.527x$. See page 126 if you have forgotten how to obtain this.

Also, it can be shown that the regression line $x$ on $y$ has equation $x = -41.1 + 1.41y$. Check this yourself on your calculator.

The diagram shows the scatter diagram together with these two regression lines.

As expected, since $r$ is close to 1, the two regression lines are close together. The scatter diagram confirms good positive correlation.

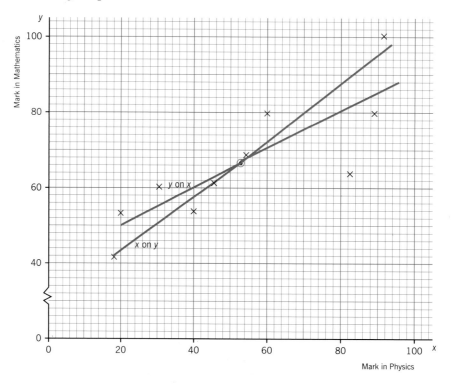

## Relationships between regression coefficients and *r*

The regression line $y$ on $x$ has equation $y = a + bx$ where $b = \dfrac{s_{xy}}{s_{xx}}$ or $b = \dfrac{S_{xy}}{S_{xx}}$.

The regression line $x$ on $y$ has equation $x = c + dy$ where $d = \dfrac{s_{xy}}{s_{yy}}$ or $d = \dfrac{S_{xy}}{S_{yy}}$.

Now $\quad b \times d = \dfrac{s_{xy}}{s_{xx}} \times \dfrac{s_{xy}}{s_{yy}}$ $\qquad$ or $\qquad$ $b \times d = \dfrac{S_{xy}}{S_{xx}} \times \dfrac{S_{xy}}{S_{yy}}$

$$= \frac{s_{xy}}{s_x s_x} \times \frac{s_{xy}}{s_y s_y} \qquad\qquad\qquad = \frac{S_{xy}}{S_x S_x} \times \frac{S_{xy}}{S_y S_y}$$

$$= \left(\frac{s_{xy}}{s_x s_y}\right)^2 \qquad\qquad\qquad\quad = \left(\frac{S_{xy}}{S_x S_y}\right)^2$$

$$= r^2 \qquad\qquad\qquad\qquad\qquad\quad\; = r^2$$

Since $r^2 > 0$, this implies that $b$ and $d$ are both positive or $b$ and $d$ are both negative.

If $b$ and $d$ are positive, then $r$ will be positive and $r = +\sqrt{bd}$.
If $b$ and $d$ are negative, then $r$ will be negative and $r = -\sqrt{bd}$.

In Example 2.6,

$b = 0.527$, $d = 1.41$ so $r^2 = b \times d = 0.743 \ldots$
$r = \sqrt{b \times d} = 0.86$ (2 d.p.)

## Example 2.7

Show that if $r = \pm 1$, the regression lines of $y$ on $x$ and $x$ on $y$ are identical. (This was illustrated in the diagrams on page 139.)

## Solution 2.7

The regression line $y$ on $x$, $y = a + bx$, has gradient $b$.

The regression line $x$ on $y$, $x = c + dy$, has gradient $\dfrac{1}{d}$.

Now if $\qquad\qquad r = \pm 1$
then $\qquad\qquad r^2 = 1$
Since $\quad r^2 = bd, \quad bd = 1$
so $\qquad\qquad b = \dfrac{1}{d}$

Therefore the two regression lines have the same gradient.

But you know that they both go through a common point $(\bar{x}, \bar{y})$, **so the regression lines must be identical.**

## Example 2.8

If $r = 0$, show that the two regression lines are at right angles.

## Solution 2.8

Since $\quad r = \dfrac{s_{xy}}{s_x s_y}$, $\quad$ if $r = 0$, $\qquad$ then $s_{xy} = 0$.

Now $\quad b = \dfrac{s_{xy}}{s_{xx}}$, so $b = 0$; $\qquad$ also $d = \dfrac{s_{xy}}{s_{yy}}$, so $d = 0$.

The equation of the regression line $y$ on $x$ is $y = a + bx$, but $b = 0$, therefore the equation is $\boldsymbol{y = a}$.

The equation of the regression line $x$ on $y$ is $x = c + dy$, but $d = 0$ therefore the equation is $\boldsymbol{x = c}$.

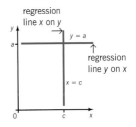

**Important note:**
The product-moment correlation coefficient $r$ is a measure of **linear correlation** only. It is important to consider the value of $r$ in conjunction with a scatter diagram. The following example illustrates this point.

## Example 2.9

For each set of bivariate data, find the product-moment correlation coefficient, draw a scatter diagram and then comment on your value of $r$.

(a)

| $x$ | -2 | -1 | 0 | 1 | 2 |
|---|---|---|---|---|---|
| $y$ | 4 | 1 | 0 | 1 | 4 |

(b)

| $x$ | 1 | 1 | 1 | 2 | 2 | 2 | 3 | 3 | 3 | 9 |
|---|---|---|---|---|---|---|---|---|---|---|
| $y$ | 1 | 2 | 3 | 1 | 2 | 3 | 1 | 2 | 3 | 8 |

## Solution 2.9

(a) Using a calculator for the first set of data, you should find that $r = 0$, indicating no linear correlation. But there could be some other relationship between the variables.

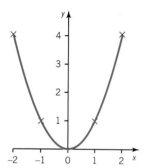

You may have noticed that the points all lie on the curve $y = x^2$.

There is a relationship between the variables – it is a quadratic one.

*NOTE*: $r = 0$ implies that either there is no correlation between the variables and they are independent, or the variables are related in a non-linear way.

(b) Using a calculator for the second set of data, you should find that $r = 0.86$ (2 d.p.), apparently indicating a strong degree of positive correlation.

**Scatter diagram to illustrate (b)**

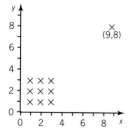

But you can see from the scatter diagram that there is not strong positive correlation.

The value of $r$ has been distorted by the point $(9, 8)$, known as an **outlier**.

So a value of $r$ close to 1, or $-1$, does not necessarily imply a strong degree of linear correlation. Always check by referring to a scatter diagram.

# Exercise 2b   Product-moment correlation coefficient

1. Calculate the value of the product-moment correlation coefficient for the following. Check using a calculator in LR mode if possible. Comment on your answers.

(a)

| $x$ | 5 | 10 | 15 | 20 | 25 |
|---|---|---|---|---|---|
| $y$ | 4.3 | 5.9 | 6.9 | 6.5 | 8.2 |

(b)

| $x$ | 12 | 14 | 16 | 18 | 20 | 22 |
|---|---|---|---|---|---|---|
| $y$ | 100 | 70 | 86 | 49 | 60 | 50 |

(c)

| $s$ | 1 | 2 | 3 | 4 | 5 | 6 | 7 | 8 |
|---|---|---|---|---|---|---|---|---|
| $t$ | 12.4 | 12.8 | 12.6 | 13.9 | 13.4 | 13.2 | 14 | 14.6 |

(d)

| $t$ | 27 | 43 | 62 | 89 | 72 |
|---|---|---|---|---|---|
| $z$ | 48 | 50 | 81 | 75 | 60 |

2. For a given set of data

   $\Sigma x = 680$   $\Sigma y = 996$   $\Sigma x^2 = 20\ 154$
   $\Sigma y^2 = 34\ 670$   $\Sigma xy = 24\ 844$   $n = 30$.

   Find the product-moment correlation coefficient.

3. The following data relate to the percentage unemployment and percentage change in wages over several years.

| % Unemployment ($x$) | % Change in wages ($y$) |
|---|---|
| 1.6 | 5.0 |
| 2.2 | 3.2 |
| 2.3 | 2.7 |
| 1.7 | 2.1 |
| 1.6 | 4.1 |
| 2.1 | 2.7 |
| 2.6 | 2.9 |
| 1.7 | 4.6 |
| 1.5 | 3.5 |
| 1.6 | 4.4 |

(a) Calculate the product-moment correlation coefficient between $x$ and $y$.

   (Use $\Sigma x = 18.9$, $\Sigma y = 35.2$, $\Sigma x^2 = 37.01$, $\Sigma y^2 = 132.22$, $\Sigma xy = 64.7$)

It has been suggested that low unemployment and a low rate of wage inflation cannot exist together.

(b) Without further calculation use your correlation coefficient to explain briefly whether or not you think the suggestion is justified.   (L)

4. Twelve students were given a prognostic test at the beginning of a course and their scores $X$ in the test were compared with their scores $Y$ obtained in an examination at the end of the course. The results were as follows:

| Student | $X$ | $Y$ |
|---|---|---|
| A | 1 | 3 |
| B | 2 | 4 |
| C | 2 | 5 |
| D | 4 | 5 |
| E | 5 | 4 |
| F | 5 | 8 |
| G | 6 | 6 |
| H | 7 | 6 |
| I | 8 | 6 |
| J | 8 | 7 |
| K | 9 | 8 |
| L | 9 | 10 |

Determine the product-moment correlation coefficient.

5. Ten boys compete in throwing a cricket ball, and the table shows the height of each boy ($x$ cm) measured to the nearest centimetre and the distance ($y$ m) to which he can throw the ball.

| Boy | $x$ | $y$ |
|---|---|---|
| A | 122 | 41 |
| B | 124 | 38 |
| C | 133 | 52 |
| D | 138 | 56 |
| E | 144 | 29 |
| F | 156 | 54 |
| G | 158 | 59 |
| H | 161 | 61 |
| I | 164 | 63 |
| J | 168 | 67 |

Calculate the product-moment correlation coefficient.
Calculate also the equations of the regression lines of $y$ on $x$ and $x$ on $y$.   (AEB)

NOTE: check your value of $r$ by using the regression coefficients obtained in the equations of the regression lines.

6. The heights $h$, in centimetres, and weight $W$, in kilograms, of ten people are measured. It is found that $\Sigma h = 1710$, $\Sigma W = 760$, $\Sigma h^2 = 293\ 162$, $\Sigma hW = 130\ 628$ and $\Sigma W^2 = 59\ 390$.
Calculate the correlation coefficient between the values of $h$ and $W$.
What is the equation of the regression line of $W$ on $h$?                                    (O & C)

7. For a set of data, the equations of the least squares regression lines are

$$y = 0.648x + 2.64 \quad (y \text{ on } x) \quad \text{and}$$
$$x = 0.917y - 1.91 \quad (x \text{ on } y)$$

find the product-moment correlation coefficient for the data.

8. For a given set of data the equations of the least squares regression lines are

$$y = -0.219x + 20.8 \quad (y \text{ on } x) \quad \text{and}$$
$$x = -0.785y + 16.2 \quad (x \text{ on } y)$$

Find the product-moment correlation coefficient for the data.

9. For a given set of data, the regression line $y$ on $x$ is $y = 0.4 + 1.3x$ and $x$ on $y$ is $x = -0.1 + 0.7y$.
Find (a) the product-moment correlation coefficient, (b) $\bar{x}$ and $\bar{y}$.

10. The body and heart masses of fourteen ten-month-old mice are tabulated below:

| Body mass ($x$ g) | Heart mass ($y$ mg) |
|---|---|
| 27 | 118 |
| 30 | 136 |
| 37 | 156 |
| 38 | 150 |
| 32 | 140 |
| 36 | 155 |
| 32 | 157 |
| 32 | 114 |
| 38 | 144 |
| 42 | 159 |
| 36 | 149 |
| 44 | 170 |
| 33 | 131 |
| 38 | 160 |

(a) Draw a scatter diagram of these data.
(b) Calculate the equation of the regression line of $y$ on $x$ and draw this line on the scatter diagram.
(c) Calculate the product–moment coefficient of correlation.                          (AEB)

# SPEARMAN'S COEFFICIENT OF RANK CORRELATION, $r_s$

You have used the product moment correlation coefficient, $r$, as a measure of the strength of the correlation between the paired data $(x_1, y_1)$, $(x_2, y_2)$, ..., $(x_n, y_n)$. This is reasonable provided that both $x$ and $y$ can be measured. Sometimes it is not possible to *measure* certain variables, but it is possible to *arrange them in order*.

For example, if two wine experts were asked to place six wines in order of preference, they would *rank* the six wines in order, using the numbers 1, 2, 3, 4, 5, 6.

The wine they liked best would be *ranked* 1.
The wine they liked least would be *ranked* 6.

It is possible to measure the strength of the correlation between the two rankings by using **Spearman's coefficient of rank correlation, $r_s$**.

In general, this is obtained as follows:

- Assign ranks 1, 2, 3, ..., $n$ to the values of each variable. This can be done by putting the values in descending order or in ascending order, but whichever you choose, you must use the same rule for both sets of data.
- For each pair of values, calculated $d$ where $d = \text{rank } x - \text{rank } y$.
- Calculate $r_s$ using the formula

$$r_s = 1 - \frac{6\Sigma d^2}{n(n^2 - 1)}$$

Consider this example: The five finalists in the County Dog Show were a Bulldog, a Poodle, a Red Setter, a Terrier and a Cocker Spaniel. Two judges ranked the dogs in order of preference. The dog they liked best was ranked 1 and the results are shown in the table:

| | | | Dog | | | |
|---|---|---|---|---|---|---|
| | | Bulldog | Poodle | Setter | Terrier | Spaniel |
| Judge | $X$ | 1 | 2 | 3 | 4 | 5 |
| | $Y$ | 3 | 2 | 4 | 1 | 5 |
| $d$ = rank $x$ – rank $y$ | $d$ | $-2$ | 0 | $-1$ | 3 | 0 |
| | $d^2$ | 4 | 0 | 1 | 9 | 0 | $\Sigma d^2 = 14$ |

To calculate Spearman's rank correlation coefficient, use

$$r_s = 1 - \frac{6\Sigma d^2}{n(n^2 - 1)} \quad \text{with } n = 5 \text{ and } \Sigma d^2 = 14$$

So $\qquad r_s = 1 - \frac{6 \times 14}{5 \times (25 - 1)} = 1 - \frac{7}{10} = 0.3$

But what does this value of $r_s$ tell you? In fact, Spearman's rank correlation coefficient is actually derived from the product-moment correlation coefficient, and is such that

$$-1 \leqslant r_s \leqslant 1$$

$r_s = 0.3$ indicates a weak positive correlation between the two rankings. To put it another way, it indicates a small degree of agreement between the two judges.

$r_s = +1$ means that the rankings are in **perfect agreement**.

$r_s = 0$ means that there is **no correlation** between the rankings.

$r_s = -1$ means that the rankings are in complete disagreement. In fact they are in **exact reverse order**.

To illustrate this, consider three different sets of judges at the Dog Show.

First pair of judges:

| | | Bulldog | Poodle | Setter | Terrier | Spaniel | |
|---|---|---|---|---|---|---|---|
| (Perfect | $A$ | 1 | 2 | 3 | 4 | 5 | |
| agreement) | $B$ | 1 | 2 | 3 | 4 | 5 | |
| | $d$ | 0 | 0 | 0 | 0 | 0 | |
| | $d^2$ | 0 | 0 | 0 | 0 | 0 | $\Sigma d^2 = 0$ |

$r_s = 1 - \frac{6\Sigma d^2}{n(n^2 - 1)} = 1 - 0 = 1$ and the **rankings are in perfect agreement.**

Second pair of judges:

| | | Bulldog | Poodle | Setter | Terrier | Spaniel | |
|---|---|---|---|---|---|---|---|
| (No | $C$ | 1 | 2 | 3 | 4 | 5 | |
| correlation) | $D$ | 4 | 1 | 3 | 5 | 2 | |
| | $d$ | $-3$ | 1 | 0 | $-1$ | 3 | |
| | $d^2$ | 9 | 1 | 0 | 1 | 9 | $\Sigma d^2 = 20$ |

$r_s = 1 - \frac{6\Sigma d^2}{n(n^2 - 1)} = 1 - \frac{6 \times 20}{5 \times 24} = 1 - 1 = 0$ and there **is no correlation between rankings.**

Third pair of judges:

|  |  | Bulldog | Poodle | Setter | Terrier | Spaniel |  |
|---|---|---|---|---|---|---|---|
| (Complete | E | 1 | 2 | 3 | 4 | 5 |  |
| disagreement) | F | 5 | 4 | 3 | 2 | 1 |  |
|  | d | −4 | −2 | 0 | 2 | 4 |  |
|  | $d^2$ | 16 | 4 | 0 | 4 | 16 | $\Sigma d^2 = 40$ |

$$r_s = 1 - \frac{6\Sigma d^2}{n(n^2 - 1)} = 1 - \frac{6 \times 40}{5 \times 24} = 1 - 2 = -1$$ and the **rankings are in exact reverse order**.

*NOTE*: the difference between the ranks, $d$, could be positive or negative. Since you are going to square this value to obtain $d^2$, you could just write the numerical value for the difference in the table. This is written $|d|$, so in the table above, for Bulldog

Rank $E$ − Rank $F = 1 - 5 = -4$ so $|d| = 4$ and $d^2 = 16$.

## Example 2.10

The marks of eight candidates in English and Mathematics are:

| Candidate | 1 | 2 | 3 | 4 | 5 | 6 | 7 | 8 |
|---|---|---|---|---|---|---|---|---|
| English ($x$) | 50 | 58 | 35 | 86 | 76 | 43 | 40 | 60 |
| Mathematics ($y$) | 65 | 72 | 54 | 82 | 32 | 74 | 40 | 53 |

Rank the results and hence find Spearman's rank correlation coefficient between the two sets of marks. Comment on the value obtained.

## Solution 2.10

There are eight pairs of data, so $n = 8$. Ranking the lowest mark 1 and the highest rank 8 gives the ranks as shown in the table.

| English ($x$) | 50 | 58 | 35 | 86 | 76 | 43 | 40 | 60 |  |
|---|---|---|---|---|---|---|---|---|---|
| Maths ($y$) | 65 | 72 | 54 | 82 | 32 | 74 | 40 | 53 |  |
| Rank $x$ | 4 | 5 | 1 | 8 | 7 | 3 | 2 | 6 |  |
| Rank $y$ | 5 | 6 | 4 | 8 | 1 | 7 | 2 | 3 |  |
| $|d|$ | 1 | 1 | 3 | 0 | 6 | 4 | 0 | 3 |  |
| $d^2$ | 1 | 1 | 9 | 0 | 36 | 16 | 0 | 9 | $\Sigma d^2 = 72$ |

$$r_s = 1 - \frac{6\Sigma d^2}{n(n^2 - 1)}$$

$$= 1 - \frac{6(72)}{8(64 - 1)}$$

$$= 0.14 \text{ (2 d.p.)}$$

Spearman's coefficient of rank correlation is **0.14 (2 d.p.)**.

**This appears to show a very weak positive correlation between the English and Mathematics rankings.**

It is interesting to compare the value of $r_s$ with the value of $r$, the product–moment correlation coefficient.

Using your calculator in linear regression mode, or using the formula, you should find that $r = 0.15$ (2 d.p.).

The two values of the correlation coefficient are very similar in this example.

Plotting a scatter diagram of the marks does not appear to indicate much correlation.

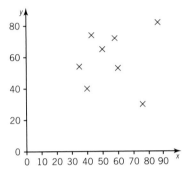

Spearman's coefficient of rank correlation can be found when data have already been ranked as in the following example.

## Example 2.11

Two judges rank the eight photographs in a competition as follows:

| Photograph | A | B | C | D | E | F | G | H |
|---|---|---|---|---|---|---|---|---|
| 1st judge | 2 | 5 | 3 | 6 | 1 | 4 | 7 | 8 |
| 2nd judge | 4 | 3 | 2 | 6 | 1 | 8 | 5 | 7 |

Calculate Spearman's coefficient or rank correlation for the data.

## Solution 2.11

In this example, the data have already been ranked.

| Rank $x$ | 2 | 5 | 3 | 6 | 1 | 4 | 7 | 8 | |
|---|---|---|---|---|---|---|---|---|---|
| Rank $y$ | 4 | 3 | 2 | 6 | 1 | 8 | 5 | 7 | |
| $|d|$ | 2 | 2 | 1 | 0 | 0 | 4 | 2 | 1 | |
| $d^2$ | 4 | 4 | 1 | 0 | 0 | 16 | 4 | 1 | $\Sigma d^2 = 30$ |

$$r_s = 1 - \frac{6\Sigma d^2}{n(n^2 - 1)} \quad \text{where } n = 8$$

$$= 1 - \frac{6(30)}{8(64 - 1)}$$

$$= 0.64 \text{ (2 d.p.)}$$

**Spearman's coefficient of rank correlation for the data is 0.64, indicating some agreement between the judges.**

*Note on ranking:*
The masses, in kilograms, of five men, 66, 68, 65, 69, 70, ranked in ascending order of magnitude gives

| $x$ | 66 | 68 | 65 | 69 | 70 |
|---|---|---|---|---|---|
| Rank $x$ | 2 | 3 | 1 | 4 | 5 |

If there are two equal values, as in 66, 68, 65, 68, 70, rank as follows:

| $x$ | 66 | 68 | 65 | 68 | 70 |
|---|---|---|---|---|---|
| Rank $x$ | 2 | 3.5 | 1 | 3.5 | 5 |

Since the 3rd and 4th places represent the same mass of 68 kg, assign the *average rank* of 3.5 to both places.

Note that if there are more than just a few equal values, this method is not appropriate.

Care must be taken when interpreting the value of the rank correlation coefficient as illustrated in the following example.

## Example 2.12

Find Spearman's rank correlation coefficient for the following data and interpret the value.

| $x$ | 1 | 2.5 | 6 | 7 | 4.5 | 3 | 6.5 |
|---|---|---|---|---|---|---|---|
| $y$ | 0.5 | 1 | 3.5 | 6.5 | 3 | 2.5 | 5.5 |

## Solution 2.12

| Rank $x$ | 1 | 2 | 5 | 7 | 4 | 3 | 6 |
|---|---|---|---|---|---|---|---|
| Rank $y$ | 1 | 2 | 5 | 7 | 4 | 3 | 6 |
| $|d|$ | 0 | 0 | 0 | 0 | 0 | 0 | 0 |

It is obvious that $\Sigma d^2 = 0$

$$r_s = 1 - \frac{6\Sigma d^2}{n(n^2 - 1)}$$
$$= 1 - 0$$
$$= 1$$

**Spearman's coefficient of rank correlation is 1.**

This indicates perfect agreement between the rankings, but if you draw a scatter diagram you will see that although there is good positive correlation between the data it is not perfect since the points do not lie on a straight line. In fact, if you calculate the product–moment correlation coefficient you will find that $r = 0.944$.

**Scatter diagram of data**

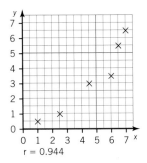

r = 0.944

**Diagram obtained by plotting the rankings**

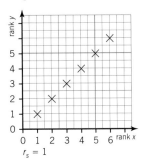

$r_s = 1$

Note that a value of $r_s = 1$ will be obtained for any set of values for which the values of $y$ increase as the values of $x$ increase.

Similarly $r_s = -1$ will be obtained for any set of values for which $y$ decreases as $x$ increases.

## Exercise 2c  Spearman's coefficient of rank correlation

1. The table shows the marks awarded to six children in a competition. Calculate a coefficient of rank correlation for the data:

| Child | A | B | C | D | E | F |
|-------|-----|-----|-----|-----|-----|-----|
| Judge 1 | 6.8 | 7.3 | 8.1 | 9.8 | 7.1 | 9.2 |
| Judge 2 | 7.8 | 9.4 | 7.9 | 9.6 | 8.9 | 6.9 |

2. At the end of a season a league of eight hockey clubs produced the following table showing the position of each club in the league and the average attendances (in hundreds) at home matches.

| Club | Position | Average Attendance |
|------|----------|--------------------|
| A | 1 | 27 |
| B | 2 | 29 |
| C | 3 | 9 |
| D | 4 | 16 |
| E | 5 | 24 |
| F | 6 | 15 |
| G | 7 | 12 |
| H | 8 | 22 |

(a) Calculate the Spearman rank correlation coefficient between position in the league and average attendance.

(b) Comment on your results. *(L)*

3. A record magazine asked critics and readers to vote for the 'Record of the Year' from a short list of nine. The numbers of votes cast were as follows.

| Recording | Critics | Readers (hundred) |
|-----------|---------|-------------------|
| A | 9 | 15 |
| B | 10 | 46 |
| C | 3 | 58 |
| D | 32 | 49 |
| E | 30 | 92 |
| F | 25 | 37 |
| G | 17 | 10 |
| H | 8 | 90 |
| I | 26 | 55 |

Calculate Spearman's rank correlation coefficient for the data. Explain what your result tells you about the opinions of these critics and readers. *(C)*

4. These are the marks obtained by eight pupils in Mathematics and Physics. Calculate Spearman's coefficient of rank correlation.

| Mathematics | Physics |
|-------------|---------|
| 67 | 70 |
| 42 | 59 |
| 85 | 71 |
| 51 | 38 |
| 39 | 55 |
| 97 | 62 |
| 81 | 80 |
| 70 | 76 |

Comment on your result.

5. In a skating competition the first judge awards the same mark to all four competitors. If the second judge awards a different mark to each competitor, show that the coefficient of rank correlation (Spearman's) is 0.5, irrespective of the marks awarded to the competitors by the second judge.

6. In a study of population density in eight suburbs of a town the statistics shown in the table were obtained. The population density is denoted by $p$, and the distance of the suburb from the centre of the town by $d$.

| Suburb | $p$ (persons/hectare) | $d$ (km) |
|--------|------------------------|----------|
| A | 55 | 0.7 |
| B | 11 | 3.8 |
| C | 68 | 1.7 |
| D | 38 | 2.6 |
| E | 46 | 1.5 |
| F | 43 | 2.6 |
| G | 21 | 3.4 |
| H | 25 | 1.9 |

(a) Plot $p$ against $d$ on a scatter diagram.
(b) Calculate and mark on the diagram the mean of the array.
(c) Calculate a coefficient of rank correlation between $p$ and $d$, stating the system of ranking adopted for both quantities.
(d) State what conclusions can be drawn from your answers to (a) and (c) concerning the general trend of the results.
(e) Giving a reason for your answer, state which suburb in your opinion fits the general trend least well. (L Additional)

7. Mr and Mrs Brown and their son John all drive the family car. Before ordering a new car they decide to list in order their preferences for five optional extras independently. The rank order of their choices is as shown:

| Optional extra | Mr Brown | Mrs Brown | John |
|----------------|----------|-----------|------|
| Heated rear window | 1st | 2nd | 3rd |
| Anti-rust treatment | 2nd | 4th | 2nd |
| Headrests | 3rd | 1st | 1st |
| Inertia-reel seat belts | 4th | 5th | 5th |
| Radio | 5th | 3rd | 4th |

(a) Calculate coefficients of rank correlation between each pair of members of the Brown family.
(b) A salesman offered to supply three of these extras free with the new car. The family agreed to choose those three which were ranked highest by the two members who agreed most. Which three did they choose, and in what order? (L Additional)

8. Seven army recruits ($A, B, ..., G$) were given two separate aptitude tests. Their orders of merit in each test were

| Order of merit | 1st | 2nd | 3rd | 4th | 5th | 6th | 7th |
|----------------|-----|-----|-----|-----|-----|-----|-----|
| 1st test | G | F | A | D | B | C | E |
| 2nd test | D | F | E | B | G | C | A |

Find Spearman's coefficient of rank correlation between the two orders and comment briefly on the correlation obtained. (O & C)

9. A doctor asked ten of his patients, who were smokers, how many years they had smoked. In addition, for each patient, he gave a grade between 0 and 100 indicating the extent of their lung damage. The following table shows the results:

| Patient | Number of years smoking | Lung damage grade |
|---------|--------------------------|-------------------|
| A | 15 | 30 |
| B | 22 | 50 |
| C | 25 | 55 |
| D | 28 | 30 |
| E | 31 | 57 |
| F | 33 | 35 |
| G | 36 | 60 |
| H | 39 | 72 |
| I | 42 | 70 |
| J | 48 | 75 |

Calculate Spearman's coefficient of rank correlation between the number of years of smoking and the extent of lung damage. Comment on the figure which you obtain. (C Additional)

10. Sketch scatter diagrams for which
(a) the product–moment correlation coefficient is $-1$,
(b) Spearman's correlation coefficient is $+1$, but the product moment correlation coefficient is *less than* 1.

Five independent observations of the random variables $X$ and $Y$ were:

| $X$ | 0 | 1 | 4 | 3 | 2 |
|-----|---|---|---|---|---|
| $Y$ | 11 | 8 | 5 | 4 | 7 |

Find
(c) the sample product–moment correlation coefficient,
(d) Spearman's correlation coefficient. (O & C)

11. Sketch two scatter diagrams illustrating the following situations:
    (a) two variables having a large, negative correlation;
    (b) two variables having a small, positive correlation.

The mean rainfall per day and the mean number of hours of sunshine per day observed at a weather station are given below.

| Month | Rainfall (mm) | Sunshine (hours) |
|---|---|---|
| January | 1.26 | 1.1 |
| February | 1.25 | 2.7 |
| March | 0.65 | 4.5 |
| April | 2.10 | 5.1 |
| May | 2.45 | 5.5 |
| June | 2.17 | 7.6 |
| July | 2.84 | 5.2 |
| August | 1.74 | 5.7 |
| September | 2.57 | 4.8 |
| October | 1.65 | 2.9 |
| November | 1.47 | 2.8 |
| December | 1.94 | 1.8 |

Calculate, correct to two decimal places, the rank correlation coefficient between rainfall and hours of sunshine.

What is the rank correlation coefficient between rainfall and minutes of sunshine?

12. (a) $X$ and $Y$ were judges at a beauty contest in which there were 10 competitors. Their rankings are shown below.

| Competitor | $X$ | $Y$ |
|---|---|---|
| A | 4 | 6 |
| B | 9 | 10 |
| C | 2 | 5 |
| D | 5= | 8 |
| E | 3 | 1 |
| F | 10 | 9 |
| G | 5= | 7 |
| H | 7 | 4 |
| I | 8 | 5 |
| J | 1 | 3 |

Calculate a coefficient of rank correlation between these two sets of ranks and comment briefly on your result.

(b) Illustrate by means of two scatter diagrams rank correlation coefficients of 0 and −1 between two variables $X$ and $Y$.

(C Additional)

13. A company is to replace its fleet of cars. Eight possible models are considered and the transport manager is asked to rank them, from 1 to 8, in order of preference. A saleswoman is asked to use each type of car for a week and grade them according to their suitability for the job ($A$ – very suitable to $E$ – unsuitable). The price is also recorded.

| Model | Transport manager's ranking | Saleswoman's grade | Price (£10s) |
|---|---|---|---|
| S | 5 | B | 611 |
| T | 1 | B+ | 811 |
| U | 7 | D− | 591 |
| V | 2 | C | 792 |
| W | 8 | B+ | 520 |
| X | 6 | D | 573 |
| Y | 4 | C+ | 683 |
| Z | 3 | A− | 716 |

(a) Calculate Spearman's rank correlation coefficient between
    (i) price and transport manager's rankings,
    (ii) price and saleswomenn's grades.
(b) Based on the results of (a) state, giving a reason, whether it would be necessary to use all three different methods of assessing the cars.
(c) A new employee is is asked to collect further data and to do some calculations. He produces the following results. The correlation coefficient between
    (i) price and boot capacity is 1.2,
    (ii) maximum speed and fuel consumption in miles per gallon is −0.7,
    (iii) price and engine capacity is −0.9
    For each of his results say, giving a reason, whether you think it is reasonable.
(d) Suggest two sets of circumstances where Spearman's rank correlation coefficient would be preferred to the product moment correlation coefficient as a measure of association.

(AEB)

14.

| Candidate | A | B | C | D | E | F |
|---|---|---|---|---|---|---|
| English | 38 | 62 | 56 | 42 | 59 | 48 |
| History | 64 | 84 | 84 | 60 | 73 | 69 |

The table shows the original marks of six candidates in two examinations. Calculate a coefficient of rank correlation and comment on the value of your results.

The History papers are re-marked and one of the six candidates is awarded five additional marks. Given that the other marks, and the coefficient of rank correlation, are unchanged, state, with reasons, which candidate received the extra marks.

(C Additional)

## Summary

- Least squares regression lines

  Regression line of $y$ on $x$      Regression line of $x$ on $y$

  $y = a + bx$                      $x = c + dy$

  $\Sigma m_i^2$ is a minimum          $\Sigma n_i^2$ is a minimum

- Useful formulae for regression and correlation work

  Small $s$ format                     Big $S$ format

  $$s_{xx} = \frac{1}{n}\Sigma x^2 - \bar{x}^2 = \frac{\Sigma x^2}{n} - \bar{x}^2 \qquad S_{xx} = \Sigma x^2 - \frac{(\Sigma x)^2}{n}$$

  $$s_{yy} = \frac{1}{n}\Sigma y^2 - \bar{y}^2 = \frac{\Sigma y^2}{n} - \bar{y}^2 \qquad S_{yy} = \Sigma y^2 - \frac{(\Sigma y)^2}{n}$$

  $$s_{xy} = \frac{1}{n}\Sigma xy - \bar{x}\bar{y} = \frac{\Sigma xy}{n} - \bar{x}\bar{y} \qquad S_{xy} = \Sigma xy - \frac{\Sigma x \, \Sigma y}{n}$$

- Least squares regression line $y$ on $x$ is

  $y = a + bx$   where   $a = \bar{y} - b\bar{x}$

  $$\text{and} \quad b = \frac{s_{xy}}{s_{xx}} = \frac{S_{xy}}{S_{xx}}$$

  Alternatively

  $$y - \bar{y} = b(x - \bar{x}) \quad \text{where} \quad b = \frac{x_{xy}}{s_{xx}} = \frac{S_{xy}}{S_{xx}}$$

- Least squares regression line $x$ on $y$ is

  $x = c + dy$   where   $c = \bar{x} - d\bar{y}$

  $$\text{and} \quad d = \frac{s_{xy}}{s_{yy}} = \frac{S_{xy}}{S_{yy}}$$

  Alternatively

  $$x - \bar{x} = d(y - \bar{y}) \quad \text{where} \quad d = \frac{s_{xy}}{s_{yy}} = \frac{S_{xy}}{S_{yy}}$$

- Linear correlation

  The product–moment correlation coefficient, $r$, is a measure of the strength of the linear correlation $-1 \leqslant r \leqslant 1$.

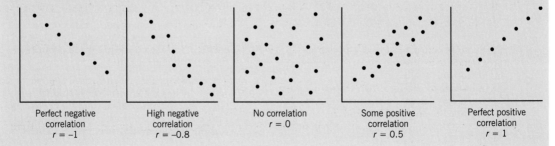

| Perfect negative correlation $r = -1$ | High negative correlation $r = -0.8$ | No correlation $r = 0$ | Some positive correlation $r = 0.5$ | Perfect positive correlation $r = 1$ |

- Formulae to calculate $r$

$$r = \frac{s_{xy}}{s_x s_y} \quad \text{where} \quad s_x = \sqrt{s_{xx}} \quad \text{and} \quad s_y = \sqrt{s_{yy}}$$

$$r = \frac{S_{xy}}{S_x S_y} \quad \text{where} \quad S_x = \sqrt{S_{xx}} \quad \text{and} \quad S_y = \sqrt{S_{yy}}$$

- Relationship between $r$ and the regression coefficients

  Regression coefficient of $y$ on $x$ is $b$ where $b = \dfrac{s_{xy}}{s_{xx}} = \dfrac{S_{xy}}{S_{xx}}$

  Regression coefficient of $x$ on $y$ is $d$ where $d = \dfrac{s_{xy}}{s_{yy}} = \dfrac{S_{xy}}{S_{yy}}$

  $r$ can be found using $r^2 = b \times d$

- Spearman's coefficient of rank correlation, $r_s$

$$r_s = 1 - \frac{6 \Sigma d^2}{n(n^2 - 1)} \quad \text{where } n \text{ is the number of pairs of values and } d = \text{rank } x - \text{rank } y.$$

$-1 \leqslant r_s \leqslant 1 \quad \text{where} \quad r_s = 1 \qquad$ means that the rankings are in perfect agreement.

$\qquad\qquad\qquad\qquad\qquad r_s = -1 \qquad$ means that the rankings are in exact reverse order.

# Miscellaneous worked examples

## Example 2.13

An old film is treated with a chemical in order to improve the contrast. Preliminary tests on nine samples drawn from a segment of the film produced the following results.

| Sample | A | B | C | D | E | F | G | H | I |
|---|---|---|---|---|---|---|---|---|---|
| $x$ | 1.0 | 1.5 | 2.0 | 2.5 | 3.0 | 3.5 | 4.0 | 4.5 | 5.0 |
| $y$ | 49 | 60 | 66 | 62 | 72 | 64 | 89 | 90 | 96 |

The quantity $x$ is a measure of the amount of chemical applied, and $y$ is the contrast index, which takes values between 0 (no contrast) and 100 (maximum contrast).

(a) Plot a scatter diagram to illustrate the data.
(b) It is subsequently discovered that one of the samples of film was damaged and produced an incorrect result. State which sample you think this was.

In all subsequent calculations this incorrect sample is ignored. The remaining data can be summarised as follows:

$$\Sigma x = 23.5, \quad \Sigma y = 584, \quad \Sigma x^2 = 83.75, \quad \Sigma y^2 = 44\,622, \quad \Sigma xy = 1883, \quad n = 8.$$

(c) Calculate the product moment correlation coefficient.
(d) State, with a reason, whether it is sensible to conclude from your answer to part (c) that $x$ and $y$ are linearly related.

(e) The line of regression of $y$ on $x$ has equation $y = a + bx$. Calculate the values of $a$ and $b$, each correct to three significant figures.

(f) Use your regression equation to estimate what the contrast index corresponding to the damaged piece of film would have been if the piece had been undamaged.

(g) State, with a reason, whether it would be sensible to use your regression equation to estimate the contrast index when the quantity of chemical applied to the film is zero.    (C)

## Solution 2.13

(a) Scatter diagram

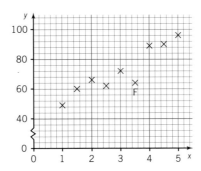

(b) **Sample $F$ was damaged.**

(c) $\bar{x} = \dfrac{\Sigma x}{n} = \dfrac{23.5}{8} = 2.9375$   and   $\bar{y} = \dfrac{\Sigma y}{n} = \dfrac{584}{8} = 73$

To calculate $r$:

Using small $s$ format

$$s_{xy} = \frac{1}{n}\Sigma xy - \bar{x}\bar{y} = \frac{1}{8} \times 1883 - 2.9375 \times 73 = 20.9375$$

$$s_{xx} = \frac{1}{n}\Sigma x^2 - \bar{x}^2 = \frac{1}{8} \times 83.75 - 2.9375^2 = 1.839 \ldots$$

$$s_{yy} = \frac{1}{n}\Sigma y^2 - \bar{y}^2 = \frac{1}{8} \times 44\,622 - 73^2 = 248.75$$

$$\therefore r = \frac{s_{xy}}{s_x s_y} = \frac{20.9375}{\sqrt{1.839\ldots}\,\sqrt{248.75}} = 0.9787 \ldots$$

Using big $S$ format

$$S_{xy} = \Sigma xy - \frac{\Sigma x\,\Sigma y}{n} = 1883 - \frac{23.5 \times 584}{8} = 167.5$$

$$S_{xx} = \Sigma x^2 - \frac{(\Sigma x)^2}{n} = 83.75 - \frac{23.5^2}{8} = 14.71$$

$$S_{yy} = \Sigma y^2 - \frac{(\Sigma y)^2}{n} = 44\,622 - \frac{584^2}{8} = 1990$$

$$\therefore r = \frac{S_{xy}}{S^x S_y} = \frac{167.5}{\sqrt{14.71}\,\sqrt{1990}} = 0.9787 \ldots$$

**So $r = 0.98$ (2 s.f.).**

(You should try this on your calculator, using LR mode.)

(d) Yes it is sensible to conclude that $x$ and $y$ are related. Since $r$ is very close to 1, it would appear to indicate a very strong position linear correlation.

(e) For the regression line $y = a + bx$, $a = \bar{y} - b\bar{x}$

and $\quad b = \dfrac{s_{xy}}{s_{xx}} = \dfrac{20.9375}{1.839\ldots} = 11.38\ldots$

or $\quad b = \dfrac{S_{xy}}{S_{xx}} = \dfrac{167.5}{14.38\ldots} = 11.38\ldots$

$a = \bar{y} - b\bar{x} = 73 - 11.38\ldots \times 2.9375 = 39.57\ldots$

$y = 39.6 + 11.4x$ **(3 s.f.)**

(f) When $x = 3.5$, $y = 38.57\ldots + 11.38\ldots \times 3.5 = 79$ (2 s.f.)
   The contrast index would have been 79.

(g) No it would not be sensible to use the regression equation when $x = 0$, since this is outside the range of data. Extrapolating outside the data is unreliable.

---

## Example 2.14

The rules for a flower competition at a village fate are as follows.

> Three judges each give a score out of 100 to each entry. The two judges whose rankings are in closest agreement are identified, and their scores for each entry are added. The three prize-winners are those whose total scores from these two judges are the highest. The scores of the third judge are ignored.

The judges awarded marks as shown in the table below.

| Contestant | A | B | C | D | E | F | G |
|---|---|---|---|---|---|---|---|
| Judge X | 89 | 83 | 80 | 72 | 69 | 54 | 41 |
| Judge Y | 77 | 84 | 85 | 65 | 79 | 72 | 69 |
| Judge Z | 73 | 83 | 89 | 80 | 67 | 75 | 69 |

The value of Spearman's rank correlation coefficient between X and Y is 0.5, and between X and Z is 0.46, correct to two decimal places. Calculate the value of Spearman's rank correlation coefficient between judges Y and Z, and hence establish which were the three prize-winners.

(C)

**Solution 2.14**

|                | A | B | C | D | E | F | G |   |
| -------------- | - | - | - | - | - | - | - | - |
| Rank Judge Y   | 4 | 6 | 7 | 1 | 5 | 3 | 2 |   |
| Rank Judge Z   | 3 | 6 | 7 | 5 | 1 | 4 | 2 |   |
| $\lvert d \rvert$ |   | 1 | 0 | 0 | 4 | 4 | 1 | 0 |
| $d^2$          |   | 1 | 0 | 0 | 16 | 16 | 1 | 0 | $\Sigma d^2 = 34$ |

$$r_s = 1 - \frac{6\Sigma d^2}{n(n^2 - 1)}$$
$$= 1 - \frac{6 \times 34}{7 \times 48}$$
$$= 0.39 \ (2 \ \text{d.p.})$$

Spearman's rank correlation coefficients are:

between X and Y: 0.5,    between X and Z: 0.46,    between Y and Z: 0.39.

The two judges whose rankings are in closest agreement are X and Y. So Judge Z is ignored.

Adding together the scores for X and Y, the final scores are:

| A | B | C | D | E | F | G |
| - | - | - | - | - | - | - |
| 166 | 167 | 165 | 137 | 148 | 126 | 110 |

**The three prize winners are A, B and C.**

---

**Example 2.15**

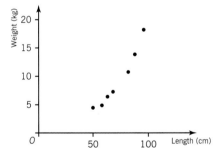

A mother monitored the growth of her baby and recorded the length $h$ cm and weight $y$ kg at various stages in the baby's development. The results were as follows.

| $h$ | 50 | 58 | 63 | 68 | 82 | 88 | 96 |
| --- | -- | -- | -- | -- | -- | -- | -- |
| $y$ | 4.43 | 4.88 | 6.31 | 7.18 | 10.63 | 13.60 | 17.95 |

The mother thought that a model of the form

$$y = p + qh$$

where $p$ and $q$ are constants, might be suitable to describe the relationship between $y$ and $h$.

The diagram shows a scatter diagram of these data.

(a) Comment on the suggested model.
(b) Suggest, giving reasons, a better model to represent the relationship between $y$ and $h$.

The new variable $x = \dfrac{h^3}{10\ 000}$ was calculated and the values of $x$ and $y$ are given in the table below.

| $x$ | 12.5 | 19.5 | 25.0 | 31.4 | 55.1 | 68.1 | 88.5 |
|-----|------|------|------|------|------|------|------|
| $y$ | 4.43 | 4.88 | 6.31 | 7.18 | 10.63 | 13.60 | 17.95 |

(c) On graph paper plot a scatter diagram of $y$ against $x$ and comment on whether a linear relationship between $y$ and $x$ is likely to provide a suitable model for the relationship between $y$ and $x$.
(d) Obtain the regression line of $y$ on $x$.

[You may use $\Sigma x^2 = 17\ 653.33$ and $\Sigma xy = 3634.185$]

(e) Estimate the weight of the baby when it was 75 cm long. (L)

## Solution 2.15

(a) A model of the form $y = p + qh$ suggests a linear relationship.
The graph, however, appears to suggest that the data could be modelled by a curve, though a straight line might be possible.
(b) Since the data suggest a curve, then a curve such as $y = kx^2$ or $y = ke^x$ might be a better model.
(c)

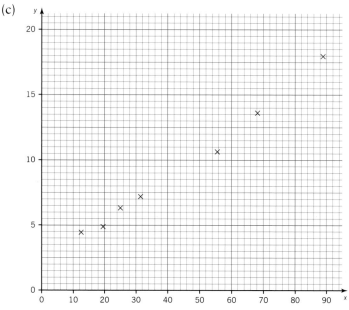

The scatter diagram of $y$ against $x$ suggests that a linear model would be a reasonable fit.
(d) $\Sigma x = 300.1$, $\Sigma y = 64.98$

Equation of regression line $y$ on $x$ is $y = a + bx$

To find $b$ using small $s$ format

$$s_{xy} = \frac{\Sigma xy}{n} - \bar{x}\bar{y} = \frac{3634.185}{7} - \frac{300.1}{7} \times \frac{64.98}{7} = 121.1999\ldots$$

$$s_{xx} = \frac{\Sigma x^2}{n} - \bar{x}^2 = \frac{17\,653.33}{7} - \left(\frac{300.1}{7}\right)^2 = 683.944\ldots$$

$$\therefore b = \frac{s_{xy}}{s_{xx}} = \frac{121.1999\ldots}{683.944\ldots} = 0.1772\ldots$$

To find $b$ using big $S$ format

$$S_{xy} = \Sigma xy - \frac{\Sigma x \Sigma y}{n} = 3634.185 - \frac{300.1 \times 64.98}{7} = 848.399\ldots$$

$$S_{xx} = \Sigma x^2 - \frac{(\Sigma x)^2}{n} = 17\,653.33 - \frac{(300.1)^2}{7} = 4787.614\ldots$$

$$\therefore b = \frac{S_{xy}}{S_{xx}} = \frac{848.399\ldots}{7487.614\ldots} = 0.1772\ldots$$

$$a = \bar{y} - b\bar{x} = \frac{64.98}{7} - 0.1772 \times \frac{300.1}{7} = 1.685\ldots$$

Giving values to three significant figures, the equation of the regression line is

$$y = 1.69 + 0.177x.$$

(e) When $h = 75$, $\quad x = \dfrac{h^3}{10\,000} = \dfrac{75^3}{10\,000} = 42.1875$

When $x = 42.1875$, $\quad y = 1.68 + 0.177 \times 42.1875 = 9.16$ (3 s.f.)

**When the baby is 75 cm long, an estimate of the weight is 9.16 kg.**

# Miscellaneous exercise 2d

1. A set of bivariate data can be summarised as follows:

$n = 6,$ $\quad \Sigma x = 21,$ $\quad \Sigma y = 43,$
$\Sigma x^2 = 91,$ $\quad \Sigma y^2 = 335,$ $\quad \Sigma xy = 171.$

(a) Calculate the equation of the regression line of $y$ on $x$. Give your answer in the form $y = a + bx$, where the values of $a$ and $b$ should be stated correct to three significant figures.

(b) It is required to estimate the value of $y$ for a given value of $x$. State circumstances under which the regression line of $x$ on $y$ should be used, rather than the regression line of $y$ on $x$. (C)

| $w$ | $t$ |
|-----|-----|
| 0 | 25 |
| 5 | 21 |
| 10 | 9 |
| 15 | 1 |
| 20 | −4 |
| 25 | −7 |
| 30 | −11 |
| 35 | −13 |
| 40 | −15 |
| 45 | −17 |
| 50 | −17 |

2. It is known that the wind causes a 'chill factor', so that the human body feels the temperature to be lower than the actual temperature. The following table gives the perceived temperature ($t\,°F$) for different wind speeds ($w$ miles per hour) when the actual temperature is 25 °F.

$[n = 11,$ $\quad \Sigma w = 275,$ $\quad \Sigma w^2 = 9625,$
$\Sigma t = -28,$ $\quad \Sigma t^2 = 2306,$ $\quad \Sigma wt = -3045.]$

(a) Calculate the equation of a suitable regression line from which a value of $t$ can be estimated for a given value of $w$. Simplify your answer as far as possible, giving the constants correct to three significant figures.

(b) Use your equation to estimate the perceived temperature when the wind speed is
(i) 38 miles per hour,
(ii) 55 miles per hour.

(c) Calculate the value of the product moment correlation coefficient for the data, and state what this indicates about the data.

(d) Comment on the reliability of the two estimates found in (b). (C)

3. The following data were collected during a study, under experimental conditions, of the effect of temperature, $x$ °C, on the pH, $y$, of skimmed milk.

| Temperature ($x$ °C) | pH ($y$) |
|---|---|
| 4 | 6.85 |
| 9 | 6.75 |
| 17 | 6.74 |
| 24 | 6.63 |
| 32 | 6.68 |
| 40 | 6.52 |
| 46 | 6.54 |
| 57 | 6.48 |
| 63 | 6.36 |
| 69 | 6.33 |
| 72 | 6.35 |
| 78 | 6.29 |

(a) Making reference to the following scatter diagram for these data, explain what it reveals about the relationship between $x$ and $y$.

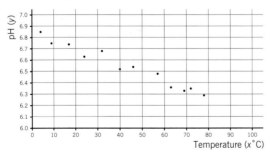

(b) Determine the equation of the least squares regression line of $y$ on $x$.
[You may make use the following information.
$$\Sigma x = 511, \quad \Sigma y = 78.52, \quad \Sigma x^2 = 28\,949,$$
$$\Sigma xy = 3291.88]$$

(c) Interpret your values for the gradient and intercept of the regression line found in (b).

(d) Estimate the pH of skimmed milk at 20 °C and at 95 °C. In **each** case indicate, **with a reason but without further calculation**, how reliable you think these estimates might be.

(e) Find the temperature at which you would expect skimmed milk to have a pH of 6.5.
(NEAB)

4. The price £$x$ of a certain cassette recorder is increased by £2 every six months. The number of recorders sold during the six months before the next increase is $y$ thousand. The values covering eight consecutive periods are shown in the table.

| $x$ | 40 | 42 | 44 | 46 | 48 | 50 | 52 | 54 |
|---|---|---|---|---|---|---|---|---|
| $y$ | 12.8 | 11.6 | 11.3 | 10.3 | 10.7 | 9.1 | 8.9 | 9.2 |

[$\Sigma x = 376, \quad \Sigma x^2 = 17\,840, \quad \Sigma y = 83.9,$
$\Sigma y^2 = 893.33, \quad \Sigma xy = 3898.4.$]

(a) Plot a scatter diagram for the data.

(b) Obtain, in the form $y = a + bx$, the equation of the regression line of $y$ on $x$, giving the values of $a$ and $b$ correct to three significant figures. Plot this line on your scatter diagram.

(c) Calculate an estimate of the number of recorders sold when the price is £58, and comment on the reliability of your estimate.

(d) Without further calculation, state whether the regression line of $x$ on $y$ will be the same as the line plotted in part (b). Give a reason for your answer. (C)

5. Explain, briefly, your understanding of the term 'correlation'.
Describe how you used, or could have used, correlation in a project or in classwork.
Twelve students sat two Biology tests, one theoretical and one practical. Their marks are shown in the table.

| Marks in theoretical test ($T$) | Marks in practical test ($P$) |
|---|---|
| 5 | 6 |
| 9 | 8 |
| 7 | 9 |
| 11 | 13 |
| 20 | 20 |
| 4 | 9 |
| 6 | 8 |
| 17 | 17 |
| 12 | 14 |
| 10 | 8 |
| 15 | 17 |
| 16 | 18 |

(a) Draw a scatter diagram to represent these data.

(b) Find, to three decimal places, the product–moment correlation coefficient.

(c) Using evidence from (a) and (b) explain why a straight line regression model is appropriate for these data.

Another student was absent from the practical test but scored 14 marks in the theoretical test.

(d) Find the equation of the appropriate regression line and use it to estimate a mark in the practical test for this student.    (L)

6. (a) State the quantity which is minimised when using the method of least squares. Use a sketch to illustrate your answer.

The heat output of wood is known to vary with the percentage moisture content. The table below shows, in suitable units, the data obtained from an experiment carried out to assess this variation.

| Moisture content ($x\%$) | Heat output ($y$) |
| --- | --- |
| 50 | 5.5 |
| 8 | 7.4 |
| 34 | 6.2 |
| 22 | 6.8 |
| 45 | 5.5 |
| 15 | 7.1 |
| 74 | 4.4 |
| 82 | 3.9 |
| 60 | 4.9 |
| 30 | 6.3 |

(b) Obtain the equation of the regression line for heat output on percentage moisture content, giving the values of the coefficients to two decimal places.

(c) Use your equation to estimate the heat output of wood with 40% moisture content. State any reservations you would have about making an estimate from the regression equation of the heat output for a 90% moisture content.

(d) Explain briefly the main implication of your analysis for a person wishing to use wood as a form of heating.    (L)

7. In the machine sewing section of a factory making high fashion clothes, a score is assigned to each finished item on the basis of its quality (the better the quality, the higher the score). Each seamstress's pay is, in part, dependent upon the number of items she finishes. The number of items finished by each of 12 seamstresses on a particular day and their mean quality score are shown.

| Seamstress | Number of items finished, $x$ | Mean quality score, $y$ |
| --- | --- | --- |
| 1 | 14 | 7.2 |
| 2 | 13 | 7.3 |
| 3 | 17 | 6.9 |
| 4 | 16 | 7.3 |
| 5 | 17 | 7.5 |
| 6 | 18 | 7.6 |
| 7 | 19 | 6.8 |
| 8 | 32 | 3.7 |
| 9 | 18 | 6.5 |
| 10 | 15 | 7.9 |
| 11 | 15 | 6.8 |
| 12 | 19 | 7.1 |

$\Sigma x = 213$,    $\Sigma y = 82.6$,    $\Sigma xy = 1414.1$,
$\Sigma x^2 = 4043$,    $\Sigma y^2 = 581.28$.

(a) Calculate the value of the product-moment correlation coefficient between $x$ and $y$, and interpret your value.

(b) Plot these data on a scatter diagram. Discuss, briefly, whether or not your interpretation in (a) should now be amended.

(c) When the results were presented at a Board meeting, the Personnel Manager explained that Seamstress 8 had been experiencing severe financial difficulties at home. Explain, briefly, the implications of this additional information on your conclusions.
    (NEAB)

8. A purchasing manager of a London-based company believes that the time in transit of goods sent by road depends upon the distance between the supplier and the company. In an attempt to measure this dependence, twelve packages, sent from different parts of the country, have their transit times ($y$ days) accurately recorded, together with the distance ($x$ miles) of the supplier from the company. The results are summarised as follows:

$\Sigma x = 1800$,    $\Sigma y = 36.0$,    $\Sigma xy = 6438.6$,
$\Sigma x^2 = 336\,296$,    $\Sigma y^2 = 126.34$.

Obtain the least squares straight line regression equation of $y$ on $x$.
Explain the significance of the regression coefficient.
Predict the transit time of a package sent from a supplier 200 miles away from the company.
Give two reasons why you would not use the equation to predict transit time for a package sent from a supplier 1500 miles away.
Calculate the product–moment correlation coefficient between $x$ and $y$.
Explain why the value you have obtained supports the purchasing manager's attempt to establish a regression equation of $y$ on $x$.    (AEB)

9. The government of a country considered making an investment to decrease the number of members of the population per doctor in order to try to reduce its infant mortality rate. (Infant mortality is measured as the number of infants per 1000 who die before reaching the age of five.) A study was made of several other similar countries and the variables $x$, population per doctor, and $y$, infant mortality, were examined. The data are summarised by the following statistics:

$$\bar{x} = 440.57, \bar{y} = 8.00, \quad S_{xy} = -1598.00,$$
$$S_{xx} = 174\,567.71.$$

(a) Calculate the equation of the regression line of $y$ on $x$.
(b) Given that the country at present has 380 people per doctor, estimate the infant mortality.
(c) Comment on the coefficient of $x$ in the light of the government's plans. (L)

10. Students on a French course were given an oral test, a listening test and a written test. The test results for the eight students on the course are given in the table. For the oral test, students were given a grade on a scale ranging from A, through A−, B+, B etc down to D−. For the listening test they were given a mark out of 25, and for the written test they were given a mark out of 100.

| Student | 1 | 2 | 3 | 4 | 5 | 6 | 7 | 8 |
|---|---|---|---|---|---|---|---|---|
| Oral test grade | C− | C+ | B− | A− | C | B | D+ | C |
| Listening test mark ($x$) | 10 | 21 | 22 | 19 | 17 | 14 | 13 | 16 |
| Written test mark ($y$) | 34 | 76 | 74 | 60 | 68 | 44 | 45 | 53 |

$\Sigma x = 132,$ $\Sigma x^2 = 2296,$ $\Sigma y = 454,$
$\Sigma y^2 = 27\,402,$ $\Sigma xy = 7909.$

(a) Calculate the value of the most appropriate measure of correlation between the results in the oral and listening tests, justifying your choice of measure. Interpret the value you obtain.
(b) Calculate the value of the most appropriate measure of correlation between the results in the listening and written tests, justifying your choice of measure. Interpret the value you obtain.
(c) The appropriate measure of correlation between the results in the oral and written tests has a value of 0.339. Comment on the indications given by the values of the three correlation coefficients about the performances of the students in the tests. (NEAB)

11. The table below shows the names of five toy construction kits which were bought from a catalogue, the numbers of pieces, $n$, found in each, and the corresponding prices paid, £$p$.

| Name | Set 1 | Set 3 | Set 4 | Set 5 | Set 6 |
|---|---|---|---|---|---|
| $n$ | 11 | 21 | 28 | 37 | 75 |
| $p$ | 11 | 26 | 34 | 41 | 88 |

[$\Sigma n = 172,$ $\Sigma p = 200,$ $\Sigma n^2 = 8340,$
$\Sigma p^2 = 11\,378,$ $\Sigma np = 9736.$]

(a) Plot a scatter diagram of the data, with $n$ on the horizontal axis and $p$ on the vertical axis.
(b) Calculate the equation of the regression line of $p$ on $n$, and plot this line on your scatter diagram. Use your equation to estimate the price of Set 2, which is not listed in the catalogue, but is thought to have 15 pieces. Give your answer correct to the nearest pound.
(c) Calculate the product moment correlation coefficient for the given data, giving your answer correct to three decimal places, and interpret the result in terms of your scatter diagram. (C)

12. The number of hours $x$ (correct to the nearest half-hour) spent studying for an examination by 12 students, together with the marks $y$ achieved in the examination, are given in the following table.

| $x$ | $y$ |
|---|---|
| 2 | 44 |
| 3 | 50 |
| 4 | 60 |
| 4.5 | 54 |
| 5 | 65 |
| 6 | 73 |
| 6.5 | 81 |
| 8 | 89 |
| 8.5 | 84 |
| 9 | 90 |
| 9.5 | 103 |
| 10 | 120 |

[$\Sigma x = 76,$ $\Sigma x^2 = 560,$ $\Sigma y = 913,$
$\Sigma y^2 = 75\,153,$ $\Sigma xy = 6425.$]

(a) Calculate the product moment correlation coefficient $r$ for the data.
(b) State what the value of $r$ indicates about the relation between $x$ and $y$.
(c) The value of Spearman's rank correlation coefficient for the above data is 0.986, correct to three decimal places. For the next examination the students each increased their study time by one hour and there was an increase of five marks in each of their examination scores. Without further calculation, state whether the new value of the rank correlation coefficient, correct to three decimal places, is less than, equal to or greater than 0.986. Give a reason for your answer. (C)

13. Over a period of ten years a survey was done on the number of cars owned per person in a particular county. The results are given in the table below.

| Year | $x$ = year – 1984 | $y$ = no. of cars per person |
|------|------|------|
| 1984 | 0 | 0.33 |
| 1985 | 1 | 0.35 |
| 1986 | 2 | 0.36 |
| 1987 | 3 | 0.37 |
| 1988 | 4 | 0.38 |
| 1989 | 5 | 0.39 |
| 1990 | 6 | 0.39 |
| 1991 | 7 | 0.40 |
| 1992 | 8 | 0.41 |
| 1993 | 9 | 0.41 |

You are given that $\Sigma xy = 17.76$, $\Sigma x = 45$ and $\Sigma y = 3.79$.
(a) Calculate the covariance of $x$ and $y$, giving your answer correct to three decimal places.
You are also given that the variance of the $x$-values is 8.25.
(b) Calculate the equation of the regression line of $y$ on $x$.
(c) State the value of $y$ which the regression equation found in part (b) predicted for the year 2000.
(d) Comment on the reliability of this prediction. (O)

14. The table is a summary of the maximum temperature recorded in Plymouth during each of the seven months from June to December 1986 inclusive.

| Month | $x$ | Maximum temp °C |
|------|------|------|
| Jun | 1 | 22.3 |
| Jul | 2 | 20.2 |
| Aug | 3 | 17.9 |
| Sep | 4 | 16.1 |
| Oct | 5 | 16.8 |
| Nov | 6 | 12.6 |
| Dec | 7 | 10.9 |

(a) Plot a scatter diagram of the data using as $x$ coordinates the coding shown in the table and the maximum temperature as the $y$ coordinate. Mark the mean point of the data on your graph.
(b) Given that $\Sigma xy = 416.7$, demonstrate that the gradient of the line of regression of $y$ on $x$ is $-1.80$ (to three significant figures). What is the physical meaning of this gradient?
(c) Calculate the full equation of regression of maximum temperature on month.

(d) Use your equation to predict the maximum temperature in May 1987. The actual maximum temperature was 15.3 °C. Why is your predicted value so different from reality? (O)

15. The following table gives the daily output of the substance creatinine from the body of each of ten nutrition students together with the student's body mass.

| Output of creatinine (grams) | Body mass (kilograms) |
|------|------|
| 1.32 | 55 |
| 1.54 | 48 |
| 1.45 | 55 |
| 1.06 | 53 |
| 2.13 | 74 |
| 1.00 | 44 |
| 0.90 | 49 |
| 2.00 | 68 |
| 2.70 | 78 |
| 0.75 | 51 |

Draw a scatter diagram for the data. Calculate, correct to two decimal places, the product–moment correlation coefficient. Comment on any relationship which is indicated by the scatter diagram and the correlation coefficient. (NEAB)

16. The yield of a particular crop on a farm is thought to depend principally on the amount of rainfall in the growing season. The values of the yield $y$, in tons per acre, and the rainfall $x$, in centimetres, for seven successive years are given in the table below.

| $x$ | 12.3 | 13.7 | 14.5 | 11.2 | 13.2 | 14.1 | 12.0 |
|------|------|------|------|------|------|------|------|
| $y$ | 6.25 | 8.02 | 8.42 | 5.27 | 7.21 | 8.71 | 5.68 |

[$\Sigma xy = 654.006$, $\Sigma x = 91$, $\Sigma x^2 = 1191.72$, $\Sigma y = 49.56$, $\Sigma y^2 = 362.1628$]
(a) Find the linear (product-moment) correlation coefficient between $x$ and $y$.
(b) Find the equation of the least squares regression line of $y$ on $x$ and also that of $x$ on $y$.
(c) Given that the rainfall in the growing season of a subsequent year was 14.0 cm, estimate the yield in that year.
(d) Given that the yield in a subsequent year was 8.08 tons per acre, estimate the rainfall in the growing season of that year. (C)

17. Following a leak of radioactivity from a nuclear power station an index of exposure to radioactivity was calculated for each of seven geographical areas close to the power station.

In the subsequent five years the incidence of death due to cancer (measured in deaths per 100 000 person-years) was recorded. The data were as follows:

| Area | Index ($x$) | Deaths ($y$) |
|---|---|---|
| 1 | 7.6 | 62 |
| 2 | 23.2 | 75 |
| 3 | 3.2 | 51 |
| 4 | 16.6 | 72 |
| 5 | 5.2 | 39 |
| 6 | 6.8 | 43 |
| 7 | 5.0 | 55 |

[$\Sigma x = 67.6$, $\Sigma x^2 = 980.08$, $\Sigma y = 397$, $\Sigma y^2 = 23\,649$, $\Sigma xy = 4339.8$.]

(a) Find the estimated regression line of $y$ on $x$.
(b) In another geographical area close to the power station the index of exposure was 6.0. Use the estimated regression line to predict the incidence, in this area, of death due to cancer (in deaths per 100 000 person-years).
(c) Estimate the incidence of death due to cancer (in deaths per 100 000 person-years) there would have been if there had been no leak from the power station (i.e. if the index of exposure to radioactivity were zero).  (C)

18. Suggest a value for the product–moment correlation coefficient between $x$ and $y$ in each of the following cases.

(a)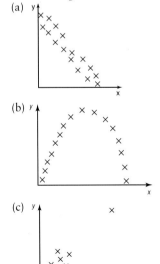

(b)

(c)

(d)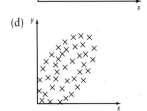

19. For twelve consecutive months a factory manager recorded the number of items produced by the factory and the total cost of their production. The following table summarises the manager's data.

| Number of items ($x$) thousands | Production cost ($y$) £1000 |
|---|---|
| 18 | 37 |
| 36 | 54 |
| 45 | 63 |
| 22 | 42 |
| 69 | 84 |
| 72 | 91 |
| 13 | 33 |
| 33 | 49 |
| 59 | 79 |
| 79 | 98 |
| 10 | 32 |
| 53 | 71 |

(a) Draw a scatter diagram for the data.
(b) Give a reason to support the use of the regression line
$$(y-\bar{y}) = b(x-\bar{x})$$
as a suitable model for the data.
(c) Giving the values of $\bar{x}$, $\bar{y}$ and $b$ to three decimal places, obtain the regression equation for $y$ on $x$ in the above form. (You may use $\Sigma x^2 = 27\,963$, $\Sigma xy = 37\,249$.)
(d) Rewrite the equation in the form
$$y = a + bx$$
giving $a$ to three significant figures.
(e) Give a practical interpretation of the values of $a$ and $b$.  (L)

20. An electric fire was switched on in a cold room and the temperature of the room was noted at five-minute intervals.

| Time, minutes, from switching on fire, $x$ | Temperature, °C, $y$ |
|---|---|
| 0 | 0.4 |
| 5 | 1.5 |
| 10 | 3.4 |
| 15 | 5.5 |
| 20 | 7.7 |
| 25 | 9.7 |
| 30 | 11.7 |
| 35 | 13.5 |
| 40 | 15.4 |

You may assume that $\Sigma x = 180$, $\Sigma y = 68.8$, $\Sigma xy = 1960$, $\Sigma x^2 = 5100$.

(a) Plot the data on a scatter diagram.
(b) Calculate the regression line $y = a + bx$ and draw it on your scatter diagram.
(c) Predict the temperature 60 minutes from switching on the fire. Why should this prediction be treated with caution?

# Mixed test 2A

1. The following table shows the amount of water, in centimetres, applied to seven similar plots on an experimental farm. It also shows the yield of hay in tonnes per acre.

| Amount of water (x) | Yield of hay (y) |
|---|---|
| 30 | 4.85 |
| 45 | 5.20 |
| 60 | 5.76 |
| 75 | 6.60 |
| 90 | 7.35 |
| 105 | 7.95 |
| 120 | 7.77 |

(Use $\Sigma x^2 = 45\ 675$; $\Sigma xy = 3648.75$)
(a) Find the equation of the regression line of $y$ on $x$ in the form $y = a + bx$.
(b) Interpret the coefficients of your regression line.
(c) What would you predict the yield to be for $x = 28$ and for $x = 150$? Comment on the reliability of each of your predicted yields.

(L)

2. In a physics experiment, a bottle of milk was brought from a cool room into a warm room. Its temperature, $y$ °C, was recorded at $t$ minutes after it was brought in, for 11 different values of $t$. The results are summarised as:

$\Sigma t = 44$, $\Sigma t^2 = 180.4$, $\Sigma ty = 824.5$, $\Sigma y = 205$.

(a) Calculate the equation of the line of regression of $y$ on $t$ in the form $y = a + bt$.
(b) Explain the practical significance of the value of $a$.
(c) Use your equation to estimate the values of $y$ at $t = 4.5$ and $t = 20.0$.
(d) State, with a reason, which of these estimates is likely to be the more reliable.

The experimenter plotted a graph of $y$ against $t$, but used only the data in the table below.

| Time (minutes), t | 3 | 3.4 | 3.8 | 4.2 | 4.6 | 5 |
|---|---|---|---|---|---|---|
| Temperature (°C), y | 17 | 18.3 | 18.6 | 18.9 | 19.3 | 19.4 |

(e) Plot this graph, and on it draw the line of regression.
(f) State why the linear model could not be valid for very large values of the time.
(g) Using your graph, comment on whether the model is a reasonable one, and state, giving a reason, whether you consider that a more refined model could be found.

(L)

3. Two people, $X$ and $Y$, were asked to give marks out of 20 for seven brands of fish finger. The results are recorded in the table.

| Brand | A | B | C | D | E | F | G |
|---|---|---|---|---|---|---|---|
| X's mark | 8 | 10 | 18 | 2 | 1 | 4 | 15 |
| Y's mark | 5 | 14 | 12 | 9 | 4 | 1 | 19 |

Construct a table of ranks and calculate Spearman's rank correlation coefficient.

(C)

4. Values of $x$ and $y$ for a set of bivariate data are given in the following table.

| x | y |
|---|---|
| 0.1 | 1.97 |
| 0.2 | 1.94 |
| 0.3 | 1.89 |
| 0.4 | 1.82 |
| 0.5 | 1.73 |
| 0.6 | 1.62 |
| 0.7 | 1.49 |
| 0.8 | 1.34 |
| 0.9 | 1.17 |

[$n = 9$, $\Sigma x = 4.5$, $\Sigma y = 14.97$, $\Sigma x^2 = 2.85$, $\Sigma y^2 = 25.5309$, $\Sigma xy = 6.885$.]

(a) Calculate the product moment correlation coefficient for this data and state what its value tells you about the relationship between $x$ and $y$.

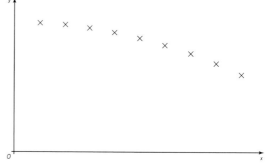

The scatter diagram representing this data is shown above.
(b) State the value of Spearman's rank correlation coefficient for this data, and state what further information its value gives about the relationship between $x$ and $y$.
(c) State which of the following best indicates the relationship between $x$ and $y$.
(i) The product moment correlation coefficient.
(ii) Spearman's rank correlation coefficient.
(iii) The scatter diagram.
Give a reason for your answer.

(C)

# Mixed test 2B

1. The average trade-in value of a particular make of used car depreciates with time according to the following table, in which the values of $x$ may be assumed to be exact.

| Age ($x$ years) | Value (£$y$ thousand) |
|---|---|
| 2.0 | 6.10 |
| 2.5 | 5.55 |
| 3.0 | 5.09 |
| 3.5 | 4.65 |
| 4.5 | 3.89 |
| 5.0 | 3.51 |
| 6.0 | 3.31 |
| 7.0 | 2.50 |

$[n = 8, \quad \Sigma x = 33.5, \quad \Sigma y = 34.6, \quad \Sigma x^2 = 161.75,$
$\Sigma y^2 = 160.2014, \quad \Sigma xy = 130.035.]$

(a) Calculate the product moment correlation coefficient between $x$ and $y$, and state what its value tells you about a scatter diagram illustrating the data.

(b) It is required to estimate the value of $y$ when $x$ is 4.0. Calculate the equation of a suitable line of regression, and use it to obtain the required estimate.

(c) Interpret the gradient of the line of regression in the context of this situation.

(d) State, with a reason in each case, whether you could use your equation to obtain a reliable estimate of
   (i) $y$ when $x = 10.0$,
   (ii) $x$ when $y = 3.00$. (C)

2. The following table gives $x$, the number of hours of sunshine, and $y$, the mid-day temperature in °C, at Springtown on the first seven days in May.

| Date | Hours of sunshine, $x$ | Mid-day temperature, $y$°C |
|---|---|---|
| May 1st | 10 | 17 |
| May 2nd | 11 | 21 |
| May 3rd | 2 | 12 |
| May 4th | 7 | 13 |
| May 5th | 5 | 18 |
| May 6th | 6 | 16 |
| May 7th | 12 | 15 |

$[\Sigma x = 53, \quad \Sigma y = 112, \quad \Sigma x^2 = 479,$
$\Sigma y^2 = 1848, \quad \Sigma xy = 882.]$

Plot the data on a scatter diagram.

Calculate the product moment correlation coefficient.
The regression line of $x$ on $y$ has equation $x = 0.607y - 2.14$, and the regression line of $y$ on $x$ has equation $y = 0.438x + 12.7$, where the coefficients are correct to three significant figures. Using the equation of the appropriate regression line, estimate the number of hours of sunshine expected on a day in May when the mid-day temperature is 18 °C.

Give a reason why this estimate differs from the actual number of hours of sunshine on May 5th.

Explain the concept of least squares by reference to your scatter diagram and the regression line of $y$ on $x$. (C)

3. A car manufacturer is testing the braking distance for a new model of car. The table shows the braking distance, $y$ metres, for different speeds, $x$ km/h, when the brakes were applied.

| Speed of car, $x$ km/h | 30 | 50 | 70 | 90 | 110 | 130 |
|---|---|---|---|---|---|---|
| Braking distance, $y$ metres (to the nearest 5 metres) | 25 | 50 | 85 | 155 | 235 | 350 |

$\Sigma x = 480, \quad \Sigma x^2 = 45\,400, \quad \Sigma y = 900,$
$\Sigma y^2 = 212\,100, \quad \Sigma xy = 94\,500.$

(a) Plot a scatter diagram.

(b) Calculate the equation of the regression line of $y$ on $x$ and draw the line on your scatter diagram.

(c) Use your regression equation to predict values of $y$ when $x = 100$ and $x = 150$. Comment, with reasons, on the likely accuracy of these predictions.

(d) Discuss briefly whether the regression line provides a good model or whether there is a better way of modelling the relationship between $y$ and $x$. (MEI)

4. In the two rounds of a show-jumping competition, seven riders recorded times, in seconds, given in the following table.

| Rider | A | B | C | D | E | F | G |
|---|---|---|---|---|---|---|---|
| Round 1 | 127 | 131 | 133 | 139 | 140 | 141 | 146 |
| Round 2 | 132 | 130 | 140 | 137 | 133 | 138 | 142 |

(a) Calculate Spearman's rank correlation coefficient between the times for the two rounds.

(b) It was subsequently discovered that rider $G$ had broken the rules of the competition and 10 seconds was added to his Round 2 time as a penalty. State, with a reason, what can be said about the value of Spearman's rank correlation coefficient calculated from the revised data.

(c) Later still it was discovered that, in Round 2, riders $A$ and $B$ had to have their times interchanged. State, with a reason but without further calculation, whether, as a result of this change, the value of Spearman's rank correlation coefficient would increase, decrease or stay the same. (C)

# 3

## Probability

*In this chapter you will learn*

- about different ways of estimating probabilities

- how to use probability notation

- about the probability laws including
  the rule for combined events
  the 'or' rule for mutually exclusive events,
  the 'and' rule for independent events

- about conditional probability

- how to use tree diagrams

- about arrangements, selections, permutations and combinations and their application to probability


The **probability** of an event is a measure of the likelihood that it will happen and it is given on a numerical scale from 0 to 1. The numbers representing probabilities can be written as percentages, fractions or decimals.

A probability of 0 indicates that the event is **impossible.**
A probability of 1 (i.e. 100%) indicates that the event is **certain** to happen.
All other events have a probability between 0 and 1.

For example

-   There is an evens chance of a coin coming down heads when tossed;
    the probability is $\frac{1}{2}$ or 0.5 or 50%.

-   There is a 1 in 4 chance of cutting a pack of cards at a diamond;
    the probability is $\frac{1}{4}$ or 0.25 or 25%.

-   The weather forecaster may say that there is a 70% chance of rain.

-   The likelihood of winning the lottery with one ticket can be shown to be approximately
    1 in 14 million so the probability is $\dfrac{1}{14\ 000\ 000} \approx 0.000\ 000\ 07$.

These probabilities can be shown on a probability scale:

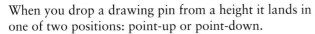

There are different ways of assigning numbers to the probabilities of events, depending on the situation being considered.

## EXPERIMENTAL PROBABILITY

When you drop a drawing pin from a height it lands in one of two positions: point-up or point-down.

Suppose you want to estimate the probability that a drawing pin will land with point-up. To do this

(a) take ten identical drawing pins and drop them from a height, say 30 cm, onto a flat surface,
(b) count the number out of the ten with points in the air,
(c) repeat the experiment so that it is carried out a total of 20 times, noting the cumulative number of 'points up' after each time,
(d) calculate the **relative frequency** of 'points-up' each time, where

$$\text{relative frequency} = \frac{\text{number of 'points-up'}}{\text{total number of pins thrown}}$$

Here is a table showing the results when this experiment was performed.

| Number of 'points-up' in 10 drawing pins | Cumulative number of 'points-up' | Cumulative number of pins thrown | Relative frequency of 'points-up' (2 d.p.) |
|---|---|---|---|
| 3 | 3 | 10 | $\frac{3}{10} = 0.30$ |
| 8 | 11 | 20 | $\frac{11}{20} = 0.55$ |
| 5 | 16 | 30 | $\frac{16}{30} = 0.53$ |
| 5 | 21 | 40 | $\frac{21}{40} = 0.53$ |
| 7 | 28 | 50 | $\frac{28}{50} = 0.56$ |
| 6 | 34 | 60 | $\frac{34}{60} = 0.57$ |
| 6 | 40 | 70 | $\frac{40}{70} = 0.57$ |
| 5 | 45 | 80 | $\frac{45}{80} = 0.56$ |
| 3 | 48 | 90 | $\frac{48}{90} = 0.53$ |
| 7 | 55 | 100 | $\frac{55}{100} = 0.55$ |
| 7 | 62 | 110 | $\frac{62}{110} = 0.56$ |
| 7 | 69 | 120 | $\frac{69}{120} = 0.58$ |
| 5 | 74 | 130 | $\frac{74}{130} = 0.57$ |
| 4 | 78 | 140 | $\frac{78}{140} = 0.56$ |
| 8 | 86 | 150 | $\frac{86}{150} = 0.57$ |
| 7 | 93 | 160 | $\frac{93}{160} = 0.58$ |
| 8 | 101 | 170 | $\frac{101}{170} = 0.59$ |
| 7 | 108 | 180 | $\frac{108}{180} = 0.60$ |
| 7 | 115 | 190 | $\frac{115}{190} = 0.61$ |
| 7 | 122 | 200 | $\frac{122}{200} = 0.61$ |

The results can be illustrated on a graph.

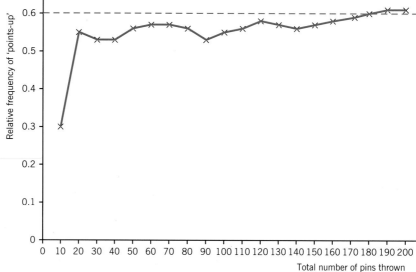

From the graph it appears that when the experiment is repeated a large number of times, the relative frequency approaches a limiting value which is around 0.6.

This limiting value is taken as an estimate of the probability, so

**P(drawing pin lands point-up) = 0.6**

In general, if an experiment is performed $n$ times under exactly similar conditions and a particular event occurs $r$ times, then the relative frequency $\dfrac{r}{n}$ is an estimate of the probability of this event. This is known as the experimental probability.

Note that the accuracy of the estimate increases as $n$ increases.

Writing $P(A)$ for the probability of event $A$,

$$\text{experimental probability } P(A) = \lim\left(\frac{r}{n}\right) \text{ as } n \to \infty.$$

where 'lim' means the limiting value to which $\dfrac{r}{n}$ settles as $n$ increases indefinitely.

### Experimental probability practicals

#### DOMINOES
Place a set of dominoes in a large bag. Use the relative frequency method to estimate the probability of drawing out of the bag at random two dominoes that have a number in common on one of their halves.

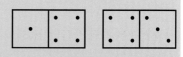

COUNTERS

You will need a supply of counters of two different colours. Ask someone to mix them up in a bag in a ratio known only to them.

Use relative frequency methods to estimate the proportion of each colour in the bag. Then check with the actual values to see how close your estimate was.

THREE COINS

Toss three coins a large number of times and use relative frequency methods to estimate the probability that on any given throw two tails and one head will be obtained.

## PROBABILITY WHEN OUTCOMES ARE EQUALLY LIKELY

When asked the probability of obtaining a head when a fair coin is tossed, you would probably give the answer $\frac{1}{2}$ (or 0.5 or 50%) without bothering to toss a coin a large number of times and working out the limiting value of the relative frequency of heads occurring. Intuitively you would have used the definition of probability that applies when the possible outcomes are **equally likely**.

For equally likely outcomes,
$$\text{probability} = \frac{\text{number of successful outcomes}}{\text{number of possible outcomes}}$$

When tossing a coin there are two possible outcomes, a head or a tail and if the coin is fair these are equally likely to occur. Only one of the outcomes is successful (obtaining a head) so $P(\text{head}) = \frac{1}{2}$.

## SUBJECTIVE PROBABILITIES

When you cannot estimate a probability using experimental methods or equally likely outcomes, you may need to employ a subjective method.

For example, you may wish to estimate the probability that it will snow on Christmas Day, or the likelihood that a particular make of car will be stolen. In these cases you have to form a **subjective probability** which you might base on past experience, such as weather records or crime figures, on expert opinion or on other factors. This method is, of course, open to error; two people faced with the same evidence may give different estimates of the probability. It is sometimes, however, the only method available.

## PROBABILITY NOTATION AND PROBABILITY LAWS

When deriving mathematical rules for probability it is useful to use the definition based on equally likely outcomes, but remember that the results hold for probability in general.

You need some preliminary definitions:

Any statistical **experiment** or **trial** has a number of possible **outcomes**.
The set of all possible outcomes is called the **possibility space S**.
An **event** *A* of the experiment is defined to be a **subset** of *S*.

Here are some examples:

- When a die is thrown, the outcomes are the numbers
  1 to 6.
  So *S* = (1, 2, 3, 4, 5, 6).
  Define *A* to be the event 'the score is less than 3'.
  Then *A* = (1, 2).

- When two dice are thrown, there are 36 possible
  outcomes, shown by dots on the possibility space
  diagram.
  Define *A* to be the event 'the sum of the two scores
  is 6'. These outcomes are shown ringed in the
  diagram.

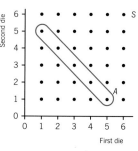

In general terms a **Venn diagram** is often used to show
*A* and *S*.

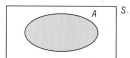

The number of outcomes in the possibility space is denoted by $n(S)$.
The number of outcomes in event *A* is denoted by $n(A)$.
Writing $P(A)$ for the probability of *A*,

$$P(A) = \frac{n(A)}{n(S)}$$

*A* is a subset of *S*, so $0 \leqslant n(A) \leqslant n(S)$.
Dividing throughout by $n(S)$ gives

$$0 \leqslant P(A) \leqslant 1$$

Remember that
$P(A) = 0$ means that event *A* is **impossible,**
$P(A) = 1$ means that event *A* is **certain to happen.**

## The complementary event *A'*

*A'* denotes the event *A* **does not occur.**

$$n(A') = n(S) - n(A)$$

so $\quad P(A') = \dfrac{n(S) - n(A)}{n(S)} = 1 - \dfrac{n(A)}{n(S)} = 1 - P(A)$

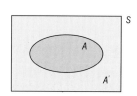

Therefore $\qquad P(A') = 1 - P(A)$

or $\qquad\qquad P(A) + P(A') = 1$

Note that sometimes $\overline{A}$ is written for the complementary event instead of *A'*.

## Example 3.1

A group of 20 university students contains eight who are in their first year of study. A student is picked at random to represent the group at a meeting. Find the probability that the student is not in the first year of study.

## Solution 3.1

Event $A$: student is in the first year of study.

$$P(A) = \frac{8}{20} = 0.4$$

so $P(A') = 1 - P(A) = 1 - 0.4 = 0.6$.

**The probability that the student is not in the first year of study is 0.6.**

## Example 3.2

Two fair coins are tossed. Show the possible outcomes on a possibility space diagram and find the probability that two heads are obtained.

## Solution 3.2

Each coin is equally likely to to show a head or a tail.

The possibility space for the outcomes is shown in the diagram, indicating that $n(S) = 4$.

Event A: Two heads are obtained.
There is just one outcome for this so $n(A) = 1$.

Therefore $P(A) = \dfrac{n(A)}{n(S)} = \dfrac{1}{4}$

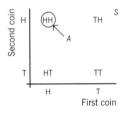

**The probability that two heads are obtained is $\dfrac{1}{4}$.**

# Exercise 3a   Elementary probability

1. An ordinary die is thrown. Find the probability that the number obtained is
   (a) a multiple of 3,
   (b) less than 7,
   (c) a factor of 6.

2. In a box of highlighters there are eight which have dried up and will not write. The box contains 10 red, 15 blue, 5 green and 10 yellow highlighters.

   A highlighter is picked at random from the box. Find the probability that
   (a) it is blue,
   (b) it is neither green nor yellow,
   (c) it is not yellow,
   (d) it is purple,
   (e) it will write.

3. The possibility space consists of the integers from 1 to 20 inclusive.

   $A$ is the event 'the number is a multiple of 3'.
   $B$ is the event 'the number is a multiple of 4'.
   An integer is picked at random.

   Find (a) $P(A)$,   (b) $P(B')$.

4. Dan carried out an experiment in which 16 coins were tossed together. The number of tails obtained from tossing the coins was counted.

   This procedure was carried out ten times in all and the results were

   Number of tails: 9, 7, 8, 6, 10, 7, 5, 5, 8, 9

   (a) Use Dan's data to calculate the probability of obtaining a tail.

The experiment was continued until the 16 coins were each tossed 100 times.

(b) Calculate the total number of tails that Dan would expect to obtain.

5. The probability of an event occurring is 0.27. What is the probability that it will not occur?

6. A card is drawn at random from an ordinary pack of 52 playing cards.

   (a) Find the probability that the card drawn is

      (i)   the four of spades,
      (ii)  the four of spades or any diamond,
      (iii) not a picture card (Jack or Queen or King) of any suit.

   (b) The card drawn is the three of diamonds. It is placed on the table and a second card is drawn. What is the probability that the second card drawn is not a diamond?

7. The pupils in a junior school class were asked how many brothers and sisters they had. Their answers are shown in the table.

| Number of brothers and sisters | 0 | 1 | 2 | 3 | 4 | 5 |
|---|---|---|---|---|---|---|
| Number of pupils | 4 | 12 | 8 | 3 | 2 | 1 |

   Find the probability that a child chosen at random from the class comes from a family with three children.

8. A cubical die, numbered 1 to 6, is weighted so that a six is twice as likely to occur as any other number. Find the probability of

   (a) a six occurring,
   (b) an odd number occurring.

9. A car manufacturer carried out a survey in which people were asked which factor from the following list influenced them most when buying a car:

   A – the colour range available,
   B – the servicing costs,
   C – driver air bag,
   D – fuel economy,
   E – range of optional extras.

   The pie chart shows the results from 90 people.

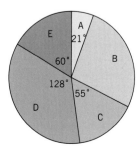

The names of those who took part were then placed in a prize draw.

Find the probability that someone who said 'servicing costs' will win the prize.

10. The durations of 60 telephone calls are summarised in the table below.

| Duration (minutes) | 0– | 9– | 18– | 27– | 36– | 45– |
|---|---|---|---|---|---|---|
| Number of calls | 6 | 10 | 21 | 20 | 3 | 0 |

   Use linear interpolation to estimate the probability that the duration of a call, selected at random from the 60 calls, exceeds 30 minutes. (C)

11. The table summarises the results of all the driving tests taken at a Test Centre during the first week of September.

|  | Male | Female |
|---|---|---|
| Pass | 32 | 43 |
| Fail | 8 | 15 |

   A person is chosen at random from those who took their test that week.

   (a) Find the probability that the person

      (i)   passed the driving test,
      (ii)  was a female who failed her driving test.

   (b) A male is chosen. What is the probability that he did not pass the test?

12. Wear tests on 100 components gave the following grouped frequency distribution of life length.

| Life length ($x$ hours) | Number of components |
|---|---|
| $500 \leqslant x < 530$ | 15 |
| $530 \leqslant x < 550$ | 24 |
| $550 \leqslant x < 570$ | 33 |
| $570 \leqslant x < 600$ | 21 |
| $600 \leqslant x < 650$ | 7 |

   Use linear interpolation to estimate the probability that a component drawn at random from the 100 has a life length between 540 and 580 hours. (C)

13. Two ordinary unbiased dice are thrown.

   Find the probability that

   (a) the sum on the two dice is 3,
   (b) the sum on the two dice exceeds 9,
   (c) the two dice show the same number,
   (d) the numbers on the two dice differ by more than 2.

14. Two fair cubical dice are thrown simultaneously and the scores multiplied. $P(n)$ denotes the probability that the number $n$ will be obtained.

(a) Calculate (i) $P(9)$ (ii) $P(4)$ (iii) $P(14)$.

(b) If $P(t) = \dfrac{1}{9}$, find the possible values of $t$.

## ILLUSTRATING TWO OR MORE EVENTS USING VENN DIAGRAMS

Suppose $A$ and $B$ are two events associated with the same experiment. Consider the outcomes described below

(a) $A \cup B$

In set language, the set that contains the outcomes that are in **A or B or both** is called the **union** of $A$ and $B$ and is written $A \cup B$.

To represent $A \cup B$ on the Venn diagram, shade the whole of the coloured 'figure-of-eight' shape.

Remember that although this outcome is written $A$ or $B$ it includes the events that are in both $A$ and $B$ as well.

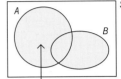

$A \cup B$ means $A$ or $B$ or both.

(b) $A \cap B$

In set language, the set that contains the outcomes that are in **both A and B** is called the **intersection** of $A$ and $B$ and is written $A \cap B$.

To represent $A \cap B$ on the Venn diagram, shade the overlap of $A$ and $B$. This outcome is often written $A$ and $B$.

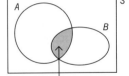

$A \cap B$ means $A$ and $B$.

## PROBABILITY RULE FOR COMBINED EVENTS

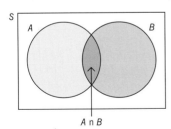

A ∩ B

If the number of outcomes in $A$ is $n(A)$ and the number of outcomes in $B$ is $n(B)$, then for two overlapping sets $A$ and $B$, if you add $n(A)$ and $n(B)$ together you will count the overlap twice.

So to find the number of outcomes in $A \cup B$ you have to take one overlap away like this:

$n(A \cap B) = n(A) + n(B) - n(A \cap B)$

Dividing by $n(S)$, this becomes

$P(A \cup B) = P(A) + P(B) - P(A \cap B)$

Alternatively

$P(A \text{ or } B) = P(A) + P(B) - P(A \text{ and } B)$

↑

Remember that the word **or** means **A or B or both**.

## Example 3.3

In a class of 20 children, 4 of the 9 boys and 3 of the 11 girls are in the athletics team. A person from the class is chosen to be in the 'egg and spoon' race on Sports Day. Find the probability that the person chosen is

(a) in the athletics team,
(b) female,
(c) a female member of the athletics team,
(d) a female or in the athletics team.

## Solution 3.3

Possibility space $S$: the class of 20 people

(a) Event $A$: member of the athletics team is chosen, $\quad P(A) = \dfrac{7}{20} = 0.35$

(b) Event $F$: a female is chosen, $\quad P(F) = \dfrac{11}{20} = 0.55$

(c) $P$(female and in the athletics team) = $P(A \text{ and } F)$
There are three girls in the athletics team, so
$$P(A \text{ and } F) = \frac{3}{20} = 0.15$$

(d) $P(A \text{ or } F) = P(A) + P(F) - P(A \text{ and } F)$
$$= 0.35 + 0.55 - 0.15$$
$$= 0.75$$

## Example 3.4

Events $C$ and $D$ are such that $P(C) = \dfrac{19}{30}$, $P(D) = \dfrac{2}{5}$ and $P(C \cup D) = \dfrac{4}{5}$.
Find $P(C \cap D)$.

## Solution 3.4

Using $\quad P(C \cup D) = P(C) + P(D) - P(C \cap D)$

$$\frac{4}{5} = \frac{19}{30} + \frac{2}{5} - P(C \cap D)$$

$$P(C \cap D) = \frac{19}{30} + \frac{2}{5} - \frac{4}{5}$$

$$= \frac{7}{30}$$

# Other useful results relating two events *A* and *B*

(a)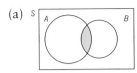

$$P(A \cap B) = P(B \cap A)$$

$\uparrow$        $\uparrow$

$P(A \text{ and } B)$   $P(B \text{ and } A)$

(b)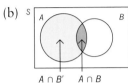

$$P(A) = P(A \cap B) + P(A \cap B')$$

$\uparrow$       $\uparrow$

$P(A \text{ and } B)$   $P(A \text{ but not } B)$

(c)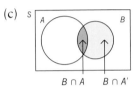

$$P(B) = P(B \cap A) + P(B \cap A')$$

$\uparrow$       $\uparrow$

$P(B \text{ and } A)$   $P(B \text{ but not } A)$

(d)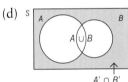

$P(\text{neither } A \text{ nor } B) = 1 - P(A \text{ or } B)$

i.e.     $P(A' \cap B') = 1 - P(A \cup B)$

## Example 3.5

In a survey, 15% of the participants said that they had never bought lottery tickets or a premium bonds, 73% had bought lottery tickets and 49% had bought premium bonds.

Find the probability that a person chosen at random from those taking part in the survey

(a) had bought lottery tickets or premium bonds,
(b) had bought lottery tickets and premium bonds,
(c) had bought lottery tickets only.

## Solution 3.5

*L*: person has bought lottery tickets,   $P(L) = 0.73$.
*B*: person has bought premium bonds,   $P(B) = 0.49$.
$P(\text{neither } L \text{ nor } B) = 0.15$

(a) $P(L \text{ or } B) = 1 - P(\text{neither } L \text{ nor } B)$
             $= 1 - 0.15$
             $= 0.85$

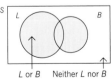

(b) Use    $P(L \text{ or } B) = P(L) + P(B) - P(L \text{ and } B)$
          $0.85 = 0.73 + 0.49 - P(L \text{ and } B)$
    $P(L \text{ and } B) = 0.73 + 0.49 - 0.85$
             $= 0.37$

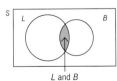

(c) $P(L \text{ only}) = P(L) - P(L \text{ and } B)$
           $= 0.73 - 0.37$
           $= 0.36$

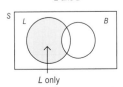

**Showing all the percentages on a Venn diagram:**

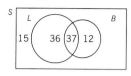

---

# Example 3.6

Events A and B are such that $P(A) = 0.3$, $P(B) = 0.4$, $P(A \cap B) = 0.1$.
Find (a) $P(A \cap B')$, (b) $P(A' \cap B')$

# Solution 3.6

(a)     $P(A) = P(A \cap B) + P(A \cap B')$
        $0.3 = 0.1 + P(A \cap B')$
    **$P(A \cap B') = 0.2$**

(b)     $P(A' \cap B') = 1 - P(A \cup B)$
        $P(A \cup B) = P(A) + P(B) - P(A \cap B)$
                    $= 0.3 + 0.4 - 0.1$
                    $= 0.6$

        $P(A' \cap B') = 1 - P(A \cup B)$
                    $= 1 - 0.6$
                    **$= 0.4$**

---

# Example 3.7

A group of 50 people was asked which of three newspapers, A, B or C they read. The results showed that 25 read A, 16 read B, 14 read C, 5 read both A and B, 4 read both B and C, 6 read both C and A and 2 read all 3.

(a)  Represent these data on a Venn diagram.

Find the probability that a person selected at random from this group reads

(b)  at least 1 of the newspapers,
(c)  only 1 of the newspapers,
(d)  only A.                                                                   (L)

# Solution 3.7

(a)  Draw 3 overlapping sets to represent A, B and C and fit in the numbers given.

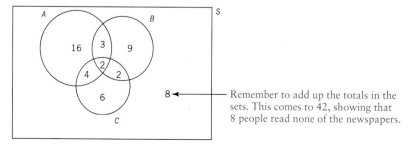

Remember to add up the totals in the sets. This comes to 42, showing that 8 people read none of the newspapers.

(b)  $P(\text{reads at least one}) = 1 - P(\text{reads none})$
$$= 1 - \tfrac{8}{50} = \tfrac{42}{50} = 0.84$$

(c)  $P(\text{reads only one}) = P(\text{reads only } A) + P(\text{reads only } B) + P(\text{reads only } C)$
$$= \tfrac{16}{50} + \tfrac{9}{50} + \tfrac{6}{50} = \tfrac{31}{50} = 0.62$$

(d)  $P(\text{reads only } A) = \tfrac{16}{50} = 0.32$

---

# EXCLUSIVE (OR MUTUALLY EXCLUSIVE) EVENTS

Consider events, $A$ and $B$, of the same experiment.
$A$ and $B$ are said to be **exclusive** (or **mutually exclusive**) if they cannot occur at the same time.
For example, with one throw of a die you cannot score a three and a five at the same time, so the events 'scoring a 3' and 'scoring a 5' are exclusive events.

If $A$ and $B$ are exclusive, then $P(A \cap B) = 0$ since $A \cap B$ is an impossible event. There is no overlap of $A$ and $B$.

For exclusive events, the rule for combined events becomes

$$P(A \cup B) = P(A) + P(B)$$

This is known as the **addition rule** for exclusive events.

It is also known as the '**or**' rule for exclusive events:

$$P(A \text{ or } B) = P(A) + P(B)$$

Extending this result to $n$ exclusive events,

$$P(A_1 \text{ or } A_2 \text{ or } A_3 \ldots \text{ or } A_n) = P(A_1) + P(A_2) + P(A_3) + \cdots + P(A_n)$$

## Example 3.8

In a race in which there are no dead heats, the probability that John wins is 0.3, the probability that Paul wins is 0.2 and the probability that Mark wins is 0.4.

Find the probability that

(a)  John or Mark wins,
(b)  John or Paul or Mark wins,
(c)  someone else wins.

## Solution 3.8

Since only one person wins, the events are mutually exclusive.

(a)  $P(\text{John or Mark wins}) = P(\text{John wins}) + P(\text{Paul wins})$
$$= 0.3 + 0.4 = 0.7$$

(b)  $P(\text{John or Paul or Mark wins}) = P(\text{John wins}) + P(\text{Paul wins}) + P(\text{Mark wins})$
$$= 0.3 + 0.4 + 0.2 = 0.9$$

(c)  $P(\text{someone else wins}) = 1 - 0.9 = 0.1$

## Example 3.9

A card is drawn from an ordinary pack of 52 playing cards. Find the probability that the card is

(a) a club or a diamond,
(b) a club or a King.

## Solution 3.9

Possibility space $S$: the pack of 52 cards, so $n(S) = 52$

$C$: a club is drawn, so $P(C) = \dfrac{n(C)}{n(S)} = \dfrac{13}{52} = \dfrac{1}{4}$.

$D$: a diamond is drawn, so $P(D) = \dfrac{n(D)}{n(S)} = \dfrac{13}{52} = \dfrac{1}{4}$.

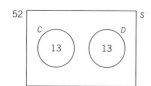

(a) Since a card cannot be both a club and a diamond, the events $C$ and $D$ are mutually exclusive.
Therefore $P(C \text{ or } D) = P(C) + P(D)$
$$= \frac{1}{4} + \frac{1}{4} = \frac{1}{2}$$

(b) Event $K$: a King is drawn, so $P(K) = \dfrac{n(K)}{n(S)} = \dfrac{4}{52} = \dfrac{1}{13}$.

The events $C$ and $K$ are **not** mutually exclusive since a card can be both a King and a club.
Therefore

$$P(C \text{ and } K) = P(\text{King of clubs}) = \frac{1}{52}.$$

$$P(C \text{ or } K) = P(C) + P(K) - P(C \text{ and } K)$$
$$= \frac{13}{52} + \frac{4}{52} - \frac{1}{52} = \frac{16}{52} = \frac{4}{13}.$$

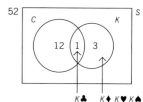

# EXHAUSTIVE EVENTS

If two events $A$ and $B$ are such that between them they make up the whole of the possibility space, then $A$ and $B$ are said to be exhaustive events and $P(A \cup B) = 1$.

For example, if
$S = $ (the integers from 1 to 10 inclusive),
$A = $ (the integers below 7) $= (1, 2, 3, 4, 5, 6)$,
$B = $ (the integers above 5) $= (6, 7, 8, 9, 10)$
then $A \cup B = (1, 2, 3, 4, 5, 6, 7, 8, 9, 10) = S$.

Special case:

Consider an event $A$ and its complementary event $A'$.

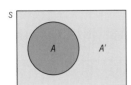

$P(A \cap A') = 0$

$P(A \cup A') = P(A) + P(A') = 1$

Any event $A$ and its complementary event $A'$ are both mutually exclusive and exhaustive.

Extending this to $n$ events:

If $A_1, A_2, A_3, ..., A_n$ are $n$ events which between them make up the whole possibility space without overlapping, then

$$P(A_1) + P(A_2) + P(A_3) + \cdots + P(A_n) = 1$$

and the $n$ events are both mutually exclusive and exhaustive.

# Exercise 3b   Probability – combined events

1. An ordinary die is thrown. Find the probability that the number obtained is
   (a) even,   (b) prime,   (c) even or prime.

2. In a group of 30 students all study at least one of the subjects Physics and Biology. 20 attend the Physics class and 21 attend the Biology class. Find the probability that a student chosen at random studies both Physics and Biology.

3. From an ordinary pack of 52 playing cards the seven of diamonds has been lost. A card is dealt from the well-shuffled pack. Find the probability that it is (a) a diamond, (b) a Queen, (c) a diamond or a Queen, (d) a diamond or a seven.

4. For events $A$ and $B$ it is known that $P(A) = \frac{2}{3}$, $P(A \cup B) = \frac{3}{4}$ and $P(A \cap B) = \frac{5}{12}$. Find $P(B)$.

5. For events $C$ and $D$,
   $P(C) = 0.7$,   $P(D \cup C) = 0.9$,   $P(C \cap D) = 0.3$.

   Find (a) $P(D)$,   (b) $P(D' \cap C)$,
   (c) $P(D \cap C')$,   (d) $P(D' \cap C')$.

6. Tests are carried out on three machines $A$, $B$ and $C$ to assess the likelihood that each machine will produce a faulty component. The results are summarised in the table.

|  | Faulty | Not faulty |
|---|---|---|
| Machine $A$ | 3 | 12 |
| Machine $B$ | 2 | 8 |
| Machine $C$ | 5 | 15 |

A component is chosen at random from those tested.

(a) Find the probability that the component chosen

   (i)   is from Machine $A$,
   (ii)  is a faulty component from Machine $C$,
   (iii) is not faulty or is from Machine $A$,

(b) It is known that the component chosen is faulty. Find the probability that it is from Machine $B$.

7. It is known that $P(X) = \frac{1}{2}$ and $P(Y) = \frac{1}{4}$. Given that $X$ and $Y$ are mutually exclusive, find
   (a) $P(X \cup Y)$,   (b) $P(Y \cap X)$,   (c) $P(Y \cap X')$.

8. For events $A$ and $B$ it is known that $P(A) = P(B)$, $P(A \cap B) = 0.1$ and $P(A \cup B) = 0.7$.

   Find $P(A')$.

9. The probability that a boy in Class 2 is in the football team is 0.4 and the probability that he is in the chess team is 0.5. If the probability that a boy in the class is in both teams is 0.2, find the probability that a boy chosen at random is in the football or the chess team.

10. Two ordinary dice are thrown. Find the probability that the sum of the scores obtained

    (a) is a multiple of 5,
    (b) is greater than 9,
    (c) is a multiple of 5 or is greater than 9,
    (d) is a multiple of 5 and is greater than 9.

11. Given that $P(A') = \frac{2}{3}$, $P(B) = \frac{1}{2}$ and $P(A \cap B) = \frac{1}{12}$, find $P(A \cup B)$.

12. Two ordinary dice are thrown. Find the probability that
    (a) at least one six is thrown,
    (b) at least one three is thrown,
    (c) at least one six or at least one three is thrown.

13. $A$ and $B$ are two events such that $P(A) = \frac{8}{15}$, $P(B) = \frac{2}{3}$ and $P(A \cap B) = \frac{1}{5}$. Are $A$ and $B$ exhaustive events?

14. Give two examples of events which are both mutually exclusive and exhaustive.

15. Two coins are tossed. $A$ is the event 'at least one head is obtained'. Describe an event $B$ such that $A$ and $B$ are exhaustive events.

16. In a large garden there are seven fruit trees and 13 other types of tree. Six of the trees have birds nesting in them but only two of these are fruit trees.

    (a) Copy and complete the table below to illustrate this information.

    |  | Fruit tree | Other tree | Total |
    |---|---|---|---|
    | Bird's nest | 2 |  | 6 |
    | No nest |  |  |  |
    | Total | 7 | 13 |  |

    The owner of the garden has given permission for Abdul to play in the garden but has instructed him not to climb any fruit trees or trees that have birds nesting in them. Abdul selects a tree at random to climb.

    (b) Find the probability that Abdul will obey the owner's instructions.

    Given that Abdul climbs a fruit tree,

    (c) find the probability that the tree has birds nesting in it. (L)

# CONDITIONAL PROBABILITY

If $A$ and $B$ are two events, not necessarily from the same experiment, then the **conditional probability** that $A$ occurs, **given that $B$ has already occurred**, is written $P(A$, given $B)$ or $P(A \mid B)$.

In the Venn diagram, the possibility space is reduced to just $B$, since $B$ has already occurred.

$A \cap B$

$$P(A, \text{given } B) = \frac{n(A \cap B)}{n(B)}$$

$$= \frac{\dfrac{n(A \cap B)}{n(S)}}{\dfrac{n(B)}{n(S)}} \longleftarrow \text{Dividing top and bottom by } n(S).$$

$$= \frac{P(A \cap B)}{P(B)}$$

So
$$P(A, \text{given } B) = \frac{P(A \text{ and } B)}{P(B)}$$

i.e.
$$P(A \mid B) = \frac{P(A \cap B)}{P(B)}$$

Rearranging:

$$P(A \cap B) = P(A \mid B) \times P(B) \longleftarrow \text{Remember } P(A \cap B) = P(B \cap A)$$

It is also true that

$$P(A \cap B) = P(B \mid A) \times P(A)$$

$$\therefore \quad P(A \mid B) \times P(B) = P(B \mid A) \times P(A)$$

## Example 3.10

When a die was thrown the score was an odd number. What is the probability that it was a prime number?

## Solution 3.10

$$P(\text{prime, given odd}) = \frac{P(\text{prime and odd})}{P(\text{odd})}$$

$$= \frac{\frac{2}{6}}{\frac{3}{6}} \quad \begin{array}{l} \longleftarrow \text{ There are two numbers, 3 and 5, that are prime and odd.} \\ \longleftarrow \text{ There are three odd numbers, 1, 3 and 5.} \end{array}$$

$$= \frac{2}{3}$$

$$P(\textbf{prime, given odd}) = \frac{2}{3}$$

It is possible to deduce this straightaway, since the possibility space has been reduced to the odd numbers 1, 3, 5 and two of these, 3 and 5, are prime.

## Example 3.11

In a certain college

65% of the students are full-time students,
55% of the students are female,
35% of the students are male full-time students.

Find the probability that

(a) a student chosen at random from all the students in the college is a part-time student,
(b) a student chosen at random from all the students in the college is female and a part-time student,
(c) a student chosen at random from all the **female** students in the college is a part-time student.

(*NEAB*)

## Solution 3.11

Define events as follows:
F: student is female,        $P(F) = 0.55$
M: student is male,        $P(M) = 1 - 0.55 = 0.45$
Full: student is full-time,    $P(\text{Full}) = 0.65$

(a) $P(\text{student is part-time}) = 1 - 0.65 = \textbf{0.35}$

(b) Given that 35% are male, full-time students

$$P(M \cap \text{Full}) = 0.35$$

Also $\quad P(\text{Full}) = P(M \cap \text{Full}) + P(F \cap \text{Full})$

$$0.65 = 0.35 + P(F \cap \text{Full})$$

$\therefore \quad P(F \cap \text{Full}) = 0.30$

$$P(F) = P(F \cap \text{Full}) + P(F \cap \text{Part})$$
$$0.55 = 0.30 + P(F \cap \text{Part})$$

$\therefore \quad$ **P(Female and part-time) = 0.25**

(c) $P(\text{Part, given } F) = \dfrac{P(\text{Part and } F)}{P(F)}$

$$= \frac{0.25}{0.55} = 0.45$$

**P(student chosen from female students is part-time) = 0.45**

---

## Example 3.12

$X$ and $Y$ are two events such that $P(X \mid Y) = 0.4$, $P(Y) = 0.25$ and $P(X) = 0.2$.

Find

(a) $P(Y \mid X)$ \qquad (b) $P(X \cap Y)$ \qquad (c) $P(X \cup Y)$

## Solution 3.12

(a) $P(Y \mid X) \times P(X) = P(X \mid Y) \times P(Y)$

$\qquad P(Y \mid X) \times 0.2 = 0.4 \times 0.25$

$\qquad\qquad\quad P(Y \mid X) = 0.5$

(b) $\qquad P(X \cap Y) = P(X \mid Y) \times P(Y)$

$\qquad\qquad\qquad\quad = 0.4 \times 0.25$

$\qquad\qquad\qquad\quad = 0.1$

$\qquad P(X \cap Y) = 0.1$

(c) $\qquad P(X \cup Y) = P(X) + P(Y) - P(X \cap Y)$

$\qquad\qquad\qquad\quad = 0.2 + 0.25 - 0.1$

$\qquad\qquad\qquad\quad = 0.35$

$\qquad P(X \cup Y) = 0.35$

---

## Example 3.13

A group of girls at a school is entered for Advanced Level Mathematics modules.
Each girl takes only module M1 or only module M2 or both M1 and M2.
The probability that a girl is taking M2 given that she is taking M1 is $\frac{1}{5}$.
The probability that a girl is taking M1 given that she is taking M2 is $\frac{1}{3}$.

Find the probability that

(a) a girl selected at random is taking both M1 and M2,
(b) a girl selected at random is taking only M1.

$(L)$

## Solution 3.13

**Events**
$M_1$: a girl takes module $M_1$
$M_2$: a girl takes module $M_2$

You are given that $P(M_2|M_1) = \frac{1}{5}$, $P(M_1|M_2) = \frac{1}{3}$

Since each girl takes one or both, $P(M_1 \cup M_2) = 1$

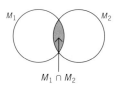

$M_1 \cap M_2$

(a)    Let $P(M_1 \cap M_2) = x$

$$P(M_2 \mid M_1) = \frac{P(M_2 \cap M_1)}{P(M_1)} \longleftarrow \quad P(M_2 \cap M_1) = P(M_1 \cap M_2)$$

$$\frac{1}{5} = \frac{x}{P(M_1)}$$

$$P(M_1) = 5x$$

Also    $P(M_1 \mid M_2) = \dfrac{P(M_1 \cap M_2)}{P(M_2)}$

$$\frac{1}{3} = \frac{x}{P(M_2)}$$

$$P(M_2) = 3x$$

$$P(M_1 \cup M_2) = P(M_1) + P(M_2) - P(M_1 \cap M_2)$$
But $M_1$ and $M_2$ are exhaustive events, so $P(M_1 \cup M_2) = 1$
$\therefore \qquad\qquad 1 = 5x + 3x - x$
$$1 = 7x$$
$$x = \tfrac{1}{7}$$
**$P$(a girl is taking $M_1$ and $M_2$) $= \frac{1}{7}$**

(b)  $P(M_1) = 5x = \frac{5}{7}$, $P(M_2) = 3x = \frac{3}{7}$
$P(\text{taking only } M_1) = P(M_1) - P(M_1 \cap M_2)$
$$= \tfrac{5}{7} - \tfrac{1}{7}$$
$$= \tfrac{4}{7}$$
**$P$(a girl is taking only $M_1$) $= \frac{4}{7}$**

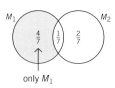

only $M_1$

---

# INDEPENDENT EVENTS

If either of the events $A$ and $B$ can occur without being affected by the other, then the two events are **independent**.

If $A$ and $B$ are independent, then $P(A$, given $B$ has occurred) is precisely the same as $P(A)$, since $A$ is not affected by $B$.
i.e.                     $P(A \mid B) = P(A)$.
It is also true that $P(B \mid A) = P(B)$

Now, since $P(A \cap B) = P(A \mid B) \times P(B)$, for independent events this becomes

$$P(A \cap B) = P(A) \times P(B)$$

This is the multiplication rule for independent events.

It is also known as the '**and**' rule for independent events.

$$P(A \text{ and } B) = P(A) \times P(B)$$

So there are three conditions for $A$ and $B$ to be independent and any one of them may be used as a test for independence

$$P(A \cap B) = P(A) \times P(B)$$
$$P(A \mid B) = P(A)$$
$$P(B \mid A) = P(B)$$

The multiplication law can be extended to any number of independent events

$$P(A_1 \text{ and } A_2 \text{ and } \dots A_n) = P(A_1) \times P(A_2) \times \cdots \times P(A_n)$$

## Example 3.14

A fair die is thrown twice. Find the probability that two fives are thrown.

## Solution 3.14

On one throw, $\quad P(5) = \frac{1}{6}$

On two throws, $\quad P(5_1 \text{ and } 5_2) = P(5_1) \times P(5_2)$ &larr;—— Independent events

$$= \frac{1}{6} \times \frac{1}{6}$$

$$= \frac{1}{36}$$

$P(\textbf{two fives are thrown}) = \frac{1}{36}$

## Example 3.15

In a group of 60 students, 20 study History, 24 study French and 8 study both History and French. Are the events 'a student studies History' and 'a student studies French' independent?

## Solution 3.15

From the information given:

$P(\text{History}) = \frac{20}{60} = \frac{1}{3}, \quad P(\text{French}) = \frac{24}{60} = \frac{2}{5} \quad P(\text{History and French}) = \frac{8}{60} = \frac{2}{15}$

Now $P(\text{History}) \times P(\text{French}) = \frac{1}{3} \times \frac{2}{5} = \frac{2}{15}$

So $\quad P(\text{History and French}) = P(\text{History}) \times P(\text{French})$

**The two events are independent.**

## Example 3.16

Events $A$ and $B$ are independent and $P(A) = \frac{1}{3}$, $P(A \cap B) = \frac{1}{12}$.
Find (a) $P(B)$   (b) $P(A \cup B)$.

## Solution 3.16

(a) Since $A$ and $B$ are independent
$$P(A \cap B) = P(A) \times P(B)$$
$$\frac{1}{12} = \frac{1}{3} \times P(B)$$
$$P(B) = \frac{1}{4}$$

(b) $P(A \cup B) = P(A) + P(B) - P(A \cap B)$
$$= \frac{1}{3} + \frac{1}{4} - \frac{1}{12}$$
$$= \frac{1}{2}$$
$$P(A \cup B) = \frac{1}{2}$$

## Example 3.17

The events $A$ and $B$ are such that $P(A \mid B) = 0.4$,   $P(B \mid A) = 0.25$,   $P(A \cap B) = 0.12$.

(a) Calculate the value of $P(B)$.
(b) Give a reason why $A$ and $B$ are not independent.
(c) Calculate the value of $P(A \cap B')$. $\hspace{4cm}$ (L)

## Solution 3.17

(a) $\quad P(A \mid B) = \dfrac{P(A \cap B)}{P(B)}$

$$0.4 = \frac{0.12}{P(B)}$$

$$\therefore \quad P(B) = \frac{0.12}{0.4} = 0.3$$

(b) $P(B \mid A) = 0.25$
$$\neq P(B)$$

**$A$ and $B$ are not independent.**

(c) $\hspace{2cm} P(A) = P(A \cap B) + P(A \cap B')$

Also $\quad P(B \mid A) = \dfrac{P(B \cap A)}{P(A)}$

$$0.25 = \frac{0.12}{P(A)}$$

$$P(A) = 0.48$$

So $\hspace{1.5cm} 0.48 = 0.12 + P(A \cap B')$

$$P(A \cap B') = 0.36$$

## Example 3.18

The events $A$ and $B$ are such that

$P(A) = 0.45$, $P(B) = 0.35$ and $P(A \cup B) = 0.7$.

(a) Find the value of $P(A \cap B)$.
(b) Explain why the events $A$ and $B$ are not independent.
(c) Find the value of $P(A \mid B)$. (L)

## Solution 3.18

(a) $\quad P(A \cup B) = P(A) + P(B) - P(A \cap B)$
$\therefore \quad\quad 0.7 = 0.45 + 0.35 - P(A \cap B)$
$\quad P(A \cap B) = 0.8 - 0.7 = \mathbf{0.1}$

(b) $\quad P(A) \times P(B) = 0.45 \times 0.35$
$= 0.1575$
$\neq P(A \cap B)$
$\therefore \quad$ **$A$ and $B$ are not independent.**

(c) $\quad P(A \mid B) = \dfrac{P(A \cap B)}{P(B)}$

$= \dfrac{0.1}{0.35}$

$= \mathbf{0.286} \text{ (3 d.p.)}$

---

It can be shown that if $A$ and $B$ are independent, then $A'$ and $B'$ are also independent.

For independent events $A$ and $B$

$\quad P(A' \text{ and } B') = P(A') \times P(B')$

and $\quad P(A' \mid B') = P(A')$
$\quad\quad P(B' \mid A') = P(B')$

## Example 3.19

The events $A$ and $B$ are independent and are such that $P(A) = x$, $P(B) = x + 0.2$, and $P(A \cap B) = 0.15$.

(a) Find the value of $x$.

For this value of $x$, find

(b) $P(A \cup B)$,

(c) $P(A' \mid B')$. (L)

## Solution 3.19

(a) Using the rule for independent events

$\quad P(A \cap B) = P(A) \times P(B)$
$\therefore \quad 0.15 = x(x + 0.2)$

By guesswork, $x = 0.3$, since $0.3 \times 0.5 = 0.15$.

Alternative algebraic method:

$$x^2 + 0.2x = 0.15$$
$$(x + 0.1)^2 - 0.01 = 0.15 \qquad \text{(completing the square)}$$
$$(x + 0.1)^2 = 0.16$$
$$x + 0.1 = \pm 0.4 \qquad \text{(taking the square root)}$$

Either $x = 0.3$ or $x = -0.5$

The negative value is impossible for a probability,

so $x = 0.3$

$P(A) = 0.3$ and $P(B) = 0.5$

(b) $P(A \cup B) = P(A) + P(B) - P(A \cap B)$
$$= 0.3 + 0.5 - 0.15$$
$$= 0.65$$

(c) Since $A$ and $B$ are independent, so are $A'$ and $B'$.
$$\therefore \quad P(A' \mid B') = P(A')$$
$$= 1 - P(A)$$
$$= 0.7$$

---

## Example 3.20

The probability that a certain type of machine will break down in the first month of operation is 0.1. If a firm has two such machines which are installed at the same time, find the probability that, at the end of the first month, just one has broken down.

Assume that the performances of the two machines are independent.

## Solution 3.20

$M_1$: machine 1 breaks down $\quad P(M_1) = 0.1, P(M_1') = 0.9$
$M_2$: machine 2 breaks down $\quad P(M_2) = 0.1, P(M_2') = 0.9$

If just one machine breaks down, then
either machine 1 has broken down and machine 2 is still working ($M_1 \cap M_2'$)
or machine 1 is still working and machine 2 has broken down ($M_1' \cap M_2$)

Now $M_1$ and $M_2'$ are independent, as are $M_1'$ and $M_2$
so $P(M_1 \cap M_2') + P(M_1' \cap M_2) = P(M_1) \times P(M_2') \quad + \quad P(M_1') \times P(M_2)$
$$= 0.1 \times 0.9 + 0.9 \times 0.1$$
$$= 0.18$$

**The probability that after one month just one machine has broken down is 0.18.**

---

## Example 3.21

Three people in an office decide to enter a marathon race. The respective probabilities that they will complete the marathon are 0.9, 0.7 and 0.6.

Assuming that their performances are independent, find the probability that

(a) they all complete the marathon,
(b) at least two complete the marathon.

## Solution 3.21

$A$: the first person completes the marathon,     $P(A) = 0.9, P(A') = 0.1$
$B$: the second person completes the marathon,    $P(B) = 0.7, P(B') = 0.3$
$C$: the third person completes the marathon      $P(C) = 0.6, P(C') = 0.4$

(a)   $P(\text{all three complete}) = P(A) \times P(B) \times P(C)$  ⟵ Independent events
$$= 0.9 \times 0.7 \times 0.6$$
$$= 0.378$$

(b)   If at least two complete the marathon then either two of them do, or all three do.
$P(\text{all three complete}) = 0.378$ from part (a)
$P(\text{two complete}) = P(A) \times P(B) \times P(C')$  +  $P(A) \times P(B') \times P(C)$  +  $P(A') \times P(B) \times P(C)$
$$= 0.9 \times 0.7 \times 0.4 \;+\; 0.9 \times 0.3 \times 0.4 \;+\; 0.1 \times 0.7 \times 0.6$$
$$= 0.456$$

$P(\text{at least two complete}) = 0.378 + 0.456$
$$= 0.834$$

---

It is important not to confuse the terms 'mutually exclusive' and 'independent'.

*Mutually exclusive events* are events that cannot happen together. They are usually the outcomes of one experiment.

*Independent events* are events that can happen simultaneously or can be seen to happen one after the other.

These three results are particularly useful. Learn them.

$P(A \text{ and } B) = P(A \text{ given } B) \times P(B)$
$P(A \cap B) = P(A \mid B) \times P(B)$

For mutually exclusive events

$P(A \text{ or } B) = P(A) + P(B)$  ⟵ The 'or' rule
$P(A \cup B) = P(A) + P(B)$

For independent events

$P(A \text{ and } B) = P(A) \times P(B)$  ⟵ The 'and' rule
$P(A \cap B) = P(A) \times P(B)$

## Example 3.22

The three events $E_1$, $E_2$ and $E_3$ are defined in the same sample space. The events $E_1$ and $E_3$ are mutually exclusive. The events $E_1$ and $E_2$ are independent.

Given that $P(E_1) = \frac{2}{5}$, $P(E_3) = \frac{1}{3}$ and $P(E_1 \cup E_2) = \frac{5}{8}$, find

(a)   $P(E_1 \cup E_3)$,
(b)   $P(E_2)$.                                                      (L)

## Solution 3.22

(a)   Since $E_1$ and $E_3$ are mutually exclusive,
$$P(E_1 \cup E_3) = P(E_1) + P(E_3)$$
$$= \frac{2}{5} + \frac{1}{3}$$
$$= \frac{11}{15}$$

(b) $P(E_1 \cup E_2) = P(E_1) + P(E_2) - P(E_1 \cap E_2)$

$$= \frac{2}{5} + P(E_2) - P(E_1) \times P(E_2) \longleftarrow E_1 \text{ and } E_2 \text{ are independent}$$

$\therefore \quad \dfrac{5}{8} = \dfrac{2}{5} + P(E_2) - \dfrac{2}{5} P(E_2)$

$\dfrac{9}{40} = \dfrac{3}{5} P(E_2)$

$P(E_2) = \dfrac{9}{40} \times \dfrac{5}{3}$

$\qquad = \dfrac{3}{8}$

## Example 3.23

Two ordinary fair dice, one red and one blue, are to be rolled once.

(a) Find the probabilities of the following events:

Event $A$: the number showing on the red die will be a 5 or a 6.
Event $B$: the total of the numbers showing on the two dice will be 7,
Event $C$: the total of the numbers showing on the two dice will be 8.

(b) State, with a reason, which two of the events $A$, $B$ and $C$ are mutually exclusive.

(c) Show that the events $A$ and $B$ are independent.　　　　　　　　　(NEAB)

## Solution 3.23

(a)

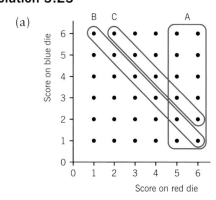

There are 36 equally likely outcomes, so $n(S) = 36$

$n(A) = 12 \qquad \therefore \quad P(A) = \dfrac{12}{36} = \dfrac{1}{3}$

$n(B) = 6 \qquad \therefore \quad P(B) = \dfrac{6}{36} = \dfrac{1}{6}$

$n(C) = 5 \qquad \therefore \quad P(C) = \dfrac{5}{36}$

(b) It is not possible to score 7 *and* 8 with one throw of the die, so events $B$ and $C$ do not overlap.

**Events $B$ and $C$ are mutually exclusive.**

(c) There are two ways to score 7 with the red die showing 5 or 6. These are $(5, 2)$ and $(6, 1)$.

So $n(A \text{ and } B) = 2$ and $P(A \text{ and } B) = \dfrac{2}{36} = \dfrac{1}{18}$

But $P(A) \times P(B) = \dfrac{1}{3} \times \dfrac{1}{6} = \dfrac{1}{18}$

So $P(A \text{ and } B) = P(A) \times P(B)$

**Events $A$ and $B$ are independent.**

## Exercise 3c    Combined events

1. A number is picked at random from the digits 1, 2, ..., 9. Given that the number is a multiple of 3, find the probability that the number is (a) even, (b) a multiple of 4.

2. In a large group of people it is known that 10% have a hot breakfast, 20% have a hot lunch and 25% have a hot breakfast or a hot lunch. Find the probability that a person chosen at random from this group
   (a) has a hot breakfast and a hot lunch,
   (b) has a hot lunch, given that the person chosen had a hot breakfast. (L)

3. If events $A$ and $B$ are such that they are independent and $P(A) = 0.3$, $P(B) = 0.5$, find (a) $P(A \cap B)$, (b) $P(A \cup B)$.

   Are events $A$ and $B$ mutually exclusive?

4. If $P(A \mid B) = \frac{2}{5}$, $P(B) = \frac{1}{4}$, $P(A) = \frac{1}{3}$, find (a) $P(B \mid A)$, (b) $P(A \cap B)$.

5. A die is thrown twice. Find the probability of obtaining a number less than three on both throws.

6. Events $A$ and $B$ are such that $P(A) = \frac{2}{3}$, $P(A \mid B) = \frac{2}{3}$, $P(B) = \frac{1}{4}$.
   Find (a) $P(B \mid A)$, (b) $P(A \cap B)$.

7. A card is picked at random from a pack of 20 cards numbered 1, 2, 3, ..., 20. Given that the card shows an even number, find the probability that it is a multiple of 4.

8. In a group of 100 people, 40 own a cat, 25 own a dog and 15 own a cat and a dog. Find the probability that a person chosen at random
   (a) owns a dog or a cat,
   (b) owns a dog or a cat, but not both,
   (c) owns a dog, given that he owns a cat,
   (d) does not own a cat, given that he owns a dog.

9. A card is picked from a pack containing 52 playing cards. It is then replaced and a second card is picked. Find the probability that
   (a) both cards are the seven of diamonds,
   (b) the first card is a heart and the second a spade,
   (c) one card is from a black suit and the other is from a red suit,
   (d) at least one card is a Queen.

10. A student investigating success in driving tests gathered information from 60 students in her school. Of these students, 25 were girls and 35 were boys. She found that 37 of the students had already taken a driving test, whilst 5, including 3 girls, were too young to take a driving test. Of the 37 who had taken a test, 16 boys and 8 girls had passed their test. The remainder, including 6 girls, had failed their test.

(a) Copy and complete the table.

|  | Boys | Girls |
| --- | --- | --- |
| Passed driving test | 16 | 8 |
| Taken driving test, but failed |  | 6 |
| Learning, but not yet taken a driving test |  |  |
| Too young to take a driving test |  |  |

Use your table to find the probability that
   (b) a student chosen at random has failed a driving test,
   (c) a girl chosen at random has taken a driving test,
   (d) a boy chosen at random has not yet taken a driving test,
   (e) 2 students, chosen at random, are both too young to take a driving test,
   (f) a boy and a girl, each chosen at random, have both passed their driving test. (C)

11. (a) Given that two events, $A$ and $B$, are such that $P(A \text{ and } B) = P(A) \times P(B)$, state what you can say about the events $A$ and $B$.

    If event $A$ is 'obtaining a 6 on a single throw of a die', suggest a possible description for event $B$.

    (b) Given that two events, $C$ and $D$, are such that $P(C \text{ or } D) = P(C) + P(D)$, state what you can say about the two events $C$ and $D$.

    Write down the value of $P(C \text{ and } D)$. (C)

12. The probability that a person in a particular evening class is left-handed is $\frac{1}{6}$. From a class of 15 women and 5 men a person is chosen at random. Assuming that 'left-handedness' is independent of the sex of a person, find the probability that the person chosen is a man or is left-handed.

13. $A$ and $B$ are exhaustive events and it is known that $P(A \mid B) = \frac{1}{4}$ and $P(B) = \frac{2}{3}$. Find $P(A)$.

14. A bag contains four red counters and six black counters. A counter is picked at random from the bag and not replaced. A second counter is then picked. Find the probability that
    (a) the second counter is red, given that the first counter is red,
    (b) both counters are red,
    (c) the counters are of different colours.

15. $A$ and $B$ are two independent events such that $P(A) = 0.2$ and $P(B) = 0.15$.
    Evaluate the following probabilities.
    (a) $P(A \mid B)$, (b) $P(A \cap B)$, (c) $P(A \cup B)$. (L)

16. Two events $A$ and $B$ are such that $P(A) = \frac{8}{15}$, $P(B) = \frac{1}{3}$, $P(A \mid B) = \frac{1}{5}$.

    Calculate the probabilities that
    (a) both events occur,
    (b) only one of the two events occurs,
    (c) neither event occurs. (NEAB)

17. All the answers to this question should be given either as fractions in their lowest terms or as decimals correct to three significant figures.

    (a) A man draws one card at random from a complete pack of 52 playing cards, replaces it and then draws another card at random from the pack.

    Calculate the probability that

    (i) both cards are clubs,
    (ii) exactly one of the cards is a Queen,
    (iii) the two cards are identical.

    (b) On another occasion the man draws **simultaneously** two cards at random from the pack of cards.

    Calculate the probability that

    (i) exactly one of the cards is a Queen,
    (ii) the two cards are identical. (C)

18. (a) The probability that an event $A$ occurs is $P(A) = 0.4$. $B$ is an event independent of $A$ and the probability of the union of $A$ and $B$ is $P(A \cup B) = 0.7$.

    Find $P(B)$.

    (b) $C$ and $D$ are two events such that $P(D \mid C) = \frac{1}{5}$ and $P(C \mid D) = \frac{1}{4}$.

    Given that $P(C \cap D) = p$, express in terms of $p$

    (i) $P(C)$, (ii) $P(D)$.

(c) Given also that $P(C \cup D) = \frac{1}{3}$, find the value of $p$. (O)

19. Events $A$ and $B$ are such that $P(A) = 0.4$ and $P(B) = 0.25$. If $A$ and $B$ are independent events, find
    (a) $P(A \cap B)$, (b) $P(A \cap B')$, (c) $P(A' \cap B')$.

20. Two tetrahedral dice, with faces labelled 1, 2, 3 and 4, are thrown and the number on which each lands is noted. The score is the sum of these two numbers. Find the probability that (a) the score is even, given that at least one die lands on a three, (b) at least one die lands on a three, given that the score is even.

21. Events $C$ and $D$ are such that $P(C) = \frac{4}{7}$, $P(C \cap D') = \frac{1}{3}$, $P(C \mid D) = \frac{5}{14}$.
    Find (a) $P(C \cap D)$, (b) $P(D)$, (c) $P(D \mid C)$.

22. Two athletes, $A$ and $B$, are attempting to qualify for an international competition in both the 5000 m and 10 000 m races. The probabilities of each qualifying are shown in the following table.

| Athlete | 5000 m | 10 000 m |
|---------|--------|----------|
| $A$ | $\frac{3}{5}$ | $\frac{1}{4}$ |
| $B$ | $\frac{2}{3}$ | $\frac{2}{5}$ |

Assuming that the probabilities are independent, calculate the probability that

(a) athlete $A$ will qualify for both races,
(b) exactly one of the athletes qualifies for the 5000 m race,
(c) both athletes qualify only for the 10 000 m race. (C)

# PROBABILITY TREES

A useful way of tackling many probability problems is to draw a **probability tree**. The method is illustrated in the following example.

## Example 3.24

In a certain selection of flower seeds $\frac{2}{3}$ have been treated to improve germination and $\frac{1}{3}$ have been left untreated. The seeds which have been treated have a probability of germination of 0.8, whereas the untreated seeds have a probability of germination of 0.5.

(a) Find the probability that a seed, selected at random, will germinate.

The seeds were sown and given time to germinate.

(b) Find the probability that a seed selected at random had been treated, given that it had germinated. (L)

## Solution 3.24

Events

$T$: seed is treated          $P(T) = \frac{2}{3}, P(T') = \frac{1}{3}$

$G$: seed germinates     $P(G \mid T) = 0.8, P(G \mid T') = 0.5$

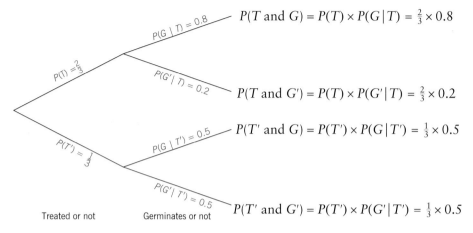

**How to use the tree:**

(i)   Multiply the probabilities along the branches to get the **end results**, so for the first outcome, use the fact that $P(T \text{ and } G) = P(T) \times P(G \text{ given } T)$

(ii)  On **any set** of branches that meet at a point, the probabilities must add up to 1

$$P(G \mid T) = 0.8$$
$$P(G' \mid T) = 0.2$$
$$0.8 + 0.2 = 1$$

(iii) Check that all the end results add up to 1.

(iv)  To answer any questions find the relevant end results. If more than one satisfy the requirements, *add* these end results together.

In practice you would usually label your tree more simply as follows.

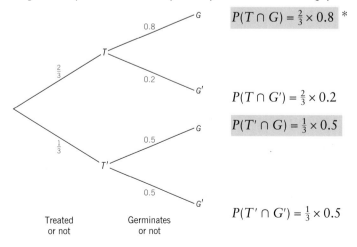

(a) $P(G) = P(T \cap G) + P(T' \cap G)$  ⟵ add the outcomes shaded above
$= \frac{2}{3} \times 0.8 + \frac{1}{3} \times 0.5$  ⟵ multiply along the branches
$= 0.7$

(b) $P(T, \text{given } G) = \dfrac{P(T \text{ and } G)}{P(G)}$  ⟵ marked * above

$= \dfrac{\frac{2}{3} \times 0.8}{0.7}$

$= 0.762$ (3 d.p.)

## Example 3.25

A manufacturer makes writing pens. The manufacturer employs an inspector to check the quality of his product. The inspector tested a random sample of the pens from a large batch and calculated the probability of any pen being defective as 0.025.

Carmel buys two of the pens made by the manufacturer.

(a) Calculate the probability that both pens are defective.
(b) Calculate the probability that exactly one of the pens is defective.  (C)

## Solution 3.25

$D$: a pen is defective,  $P(D) = 0.025$, $P(D') = 1 - 0.025 = 0.975$.

$P(D \cap D) = 0.025 \times 0.025^*$

$P(D \cap D') = 0.025 \times 0.975$

$P(D' \cap D) = 0.975 \times 0.025$

(a) $P(\text{both pens are defective}) = P(D \cap D)$  ⟵ * on diagram
$= 0.025 \times 0.025$
$= 0.000\ 625$

(b) $P(\text{exactly one pen is defective}) = P(D \cap D') + P(D' \cap D)$
$= 0.025 \times 0.975 + 0.975 \times 0.025$
$= 0.048\ 75$

## Example 3.26

Events $X$ and $Y$ are such that $P(X') = \frac{3}{5}$, $P(Y|X') = \frac{1}{3}$, $P(Y'|X) = \frac{1}{4}$.
By drawing a tree diagram, find
(a) $P(Y)$     (b) $P(X'|Y)$

## Solution 3.26

Draw a tree diagram, showing event $X$ followed by event $Y$, and write in all the given probabilities. Then work out the missing probabilities using the fact that probabilities on all the branches from a point add up to 1.

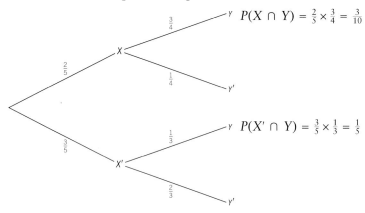

$$P(X \cap Y) = \tfrac{2}{5} \times \tfrac{3}{4} = \tfrac{3}{10}$$

$$P(X' \cap Y) = \tfrac{3}{5} \times \tfrac{1}{3} = \tfrac{1}{5}$$

(a) $P(Y) = P(X \cap Y) + P(X' \cap Y)$

$$= \frac{3}{10} + \frac{1}{5}$$

$$= \frac{1}{2}$$

(b) $P(X' \mid Y) \times P(Y) = P(Y \mid X') \times P(X')$

$$P(X' \mid Y) \times \frac{1}{2} = \frac{1}{3} \times \frac{3}{5}$$

$$P(X' \mid Y) = \frac{2}{5}$$

Alternatively

$$P(X' \mid Y) = \frac{P(X' \text{ and } Y)}{P(Y)}$$

$$= \frac{\dfrac{3}{5} \times \dfrac{1}{3}}{\dfrac{1}{2}}$$

$$= \frac{2}{5}$$

## Example 3.27

When a person needs a minicab, it is hired from one of three firms, $X$, $Y$ and $Z$. Of the hirings 40% are from $X$, 50% are from $Y$ and 10% are from $Z$. For cabs hired from $X$, 9% arrive late, the corresponding percentages for cabs hired from firms $Y$ and $Z$ being 6% and 20% respectively. Calculate the probability that the next cab hired

(a) will be from $X$ and will not arrive late,
(b) will arrive late.

Given that a call is made for a minicab and that it arrives late, find, to three decimal places, the probability that it came from $Y$.

(L)

## Solution 3.27

| *Events* | *Probabilities* |
|---|---|
| $X$: cab is from $X$ | $P(X) = 0.4$ |
| $Y$: cab is from $Y$ | $P(Y) = 0.5$ |
| $Z$: cab is from $Z$ | $P(Z) = 0.1$ |
| $L$: cab is late | $P(L \mid X) = 0.09, \quad P(L \mid Y) = 0.06, \quad P(L \mid Z) = 0.2$ |

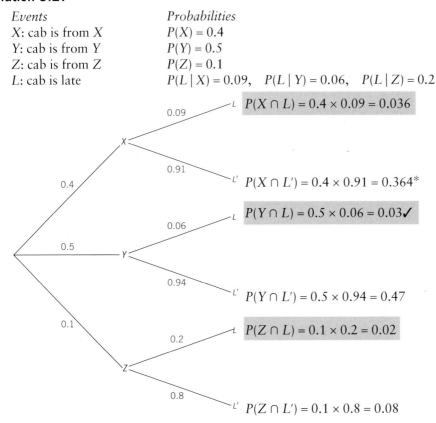

$P(X \cap L) = 0.4 \times 0.09 = 0.036$

$P(X \cap L') = 0.4 \times 0.91 = 0.364^*$

$P(Y \cap L) = 0.5 \times 0.06 = 0.03$✓

$P(Y \cap L') = 0.5 \times 0.94 = 0.47$

$P(Z \cap L) = 0.1 \times 0.2 = 0.02$

$P(Z \cap L') = 0.1 \times 0.8 = 0.08$

(a) $P(\text{from } X \text{ and not late}) = P(X \cap L') = \textbf{0.364}$    ⟵ $*$ on diagram

(b) $P(\text{arrives late}) = P(X \text{ and late}) + P(Y \text{ and late}) + P(Z \text{ and late})$

$$= P(X \cap L) \quad + P(Y \cap L) \quad + P(Z \cap L) \quad \text{⟵ shaded in diagram}$$
$$= 0.036 \quad + 0.03 \quad + 0.02$$
$$= \textbf{0.086}$$

The possibility space is now reduced to the outcomes when the cab arrives late, where $P(L) = 0.086$ (part b)

$$P(\text{from } Y \text{ given it was late}) = \frac{P(Y \text{ and late})}{P(\text{late})}$$

i.e.    $$P(Y \mid L) = \frac{P(Y \cap L)}{P(L)}$$

$$= \frac{0.03}{0.086} \quad \text{⟵ marked ✓ in diagram}$$

$$= \textbf{0.349 (3 d.p.)}$$

## BAYES' THEOREM

$P(Y \mid L)$ is easy to find from the tree diagram once you realise that the sample space has been reduced to the outcomes in which $L$ occurs. This is a useful method when you want to 'reverse the conditions', as in Example 3.27, when you know $P(L \mid Y)$ and you wanted $P(Y \mid L)$.

It is interesting to write out the full formulae used:

$$P(Y \text{ and } L) = P(L \mid Y) \times P(Y)$$

also $\qquad P(Y \text{ and } L) = P(Y \mid L) \times P(L)$

so $\qquad P(Y \mid L) \times P(L) = P(L \mid Y) \times P(Y)$

$$P(Y \mid L) = \frac{P(L \mid Y) \times P(Y)}{P(L)}$$

But $P(L) = P(X \cap L) + P(Y \cap L) + P(Z \cap L)$
$\qquad = P(L \mid X) \times P(X) \;+\; P(L \mid Y) \times P(Y) \;+\; P(L \mid Z) \times P(Z)$

so $\quad P(Y \mid L) = \dfrac{P(L \mid Y) \times P(Y)}{P(L \mid X) \times P(X) \;+\; P(L \mid Y) \times P(Y) \;+\; P(L \mid Z) \times P(Z)}$

This is an example of Bayes' Theorem, which can be written in general format as follows:

For $i = 1, 2, 3, \ldots, n$

$$P(A_i \mid B) = \frac{P(B \mid A_i) \times P(A_i)}{P(B \mid A_1) \times P(A_1) + P(B \mid A_2) \times P(A_2) + \cdots + P(B \mid A_n) \times P(A_n)}$$

The formula has been included here for reference. It is however easier to work from the format

$$P(A_i \mid B) = \frac{P(A_i \text{ and } B)}{P(B)}$$

especially when you have a tree diagram to illustrate the situation!

## Example 3.28

A computer program generates random questions in arithmetic that children have to answer within a fixed time. The probability of the first question being answered correctly is 0.8. Whenever a question is answered correctly, the next question generated is more difficult, and the probability of a correct answer being given is reduced by 0.1. Whenever a question is answered wrongly, the next question is of the same standard, and the probability of a correct answer being given remains unchanged. The following tree diagram shows this information for the first two questions generated.

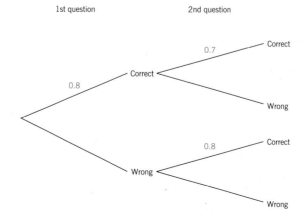

(a) Find the probability that the second question is answered correctly.

(b) By extending the tree diagram, or otherwise, find the probability that the second question is answered correctly given that the third question is answered correctly. (C)

### Solution 3.28

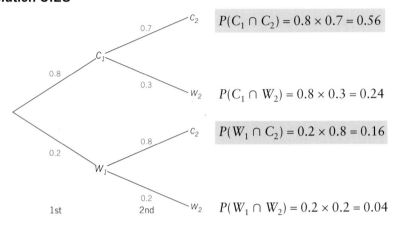

$$P(C_1 \cap C_2) = 0.8 \times 0.7 = 0.56$$

$$P(C_1 \cap W_2) = 0.8 \times 0.3 = 0.24$$

$$P(W_1 \cap C_2) = 0.2 \times 0.8 = 0.16$$

$$P(W_1 \cap W_2) = 0.2 \times 0.2 = 0.04$$

(a) $P$(2nd question answered correctly) $= 0.56 + 0.16 = \mathbf{0.72}$

(b)

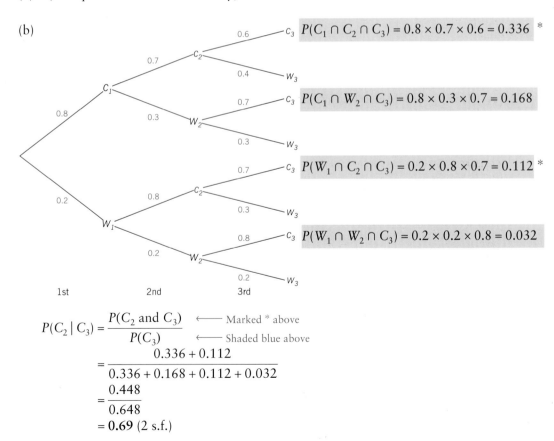

$$P(C_1 \cap C_2 \cap C_3) = 0.8 \times 0.7 \times 0.6 = 0.336 \quad *$$

$$P(C_1 \cap W_2 \cap C_3) = 0.8 \times 0.3 \times 0.7 = 0.168$$

$$P(W_1 \cap C_2 \cap C_3) = 0.2 \times 0.8 \times 0.7 = 0.112 \quad *$$

$$P(W_1 \cap W_2 \cap C_3) = 0.2 \times 0.2 \times 0.8 = 0.032$$

$$P(C_2 \mid C_3) = \frac{P(C_2 \text{ and } C_3)}{P(C_3)} \quad \longleftarrow \text{Marked * above}$$

$\longleftarrow$ Shaded blue above

$$= \frac{0.336 + 0.112}{0.336 + 0.168 + 0.112 + 0.032}$$

$$= \frac{0.448}{0.648}$$

$$= \mathbf{0.69} \text{ (2 s.f.)}$$

# Exercise 3d   Tree diagrams

### Section A

1. The probability that I am late for work is 0.05. Find the probability that, on two consecutive mornings, (a) I am late for work twice, (b) I am late for work once.

2. A mother and her daughter both enter the cake competition at a show. The probability that the mother wins a prize is $\frac{1}{6}$ and the probability that her daughter wins a prize is $\frac{2}{7}$.
   Assuming that the two events are independent, find the probability that
   (a) either the mother, or the daughter, but not both, wins a prize,
   (b) at least one of them wins a prize.

3. In a restaurant 40% of the customers choose steak for their main course. If a customer chooses steak, the probability that he will choose ice cream to follow is 0.6. If he does not have steak, the probability that he will choose ice cream is 0.3. Find the probability that a customer picked at random will choose
   (a) steak and ice cream,
   (b) ice cream.

4. A box contains six red pens and three blue pens.
   (a) A pen is selected at random, the colour is noted and the pen is returned to the box. This procedure is performed a second, then a third time. Find the probability of obtaining
      (i) three red pens,
      (ii) two red pens and one blue pen, in any order,
      (iii) more than one blue pen.
   (b) Repeat (a) but this time find the probabilities if, at each selection, the pen is not returned to the box.

5. Mass-produced glass bricks are inspected for defects. The probability that a brick has air bubbles is 0.002. If a brick has air bubbles the probability that it is also cracked is 0.5 while the probability that a brick free of air bubbles is cracked is 0.005. What is the probability that a brick chosen at random is cracked? The probability that a brick is discoloured is 0.006. Given that discolouration occurs independently of the other two defects, find the probability that a brick chosen at random has no defects. (O&C)

6. In each round of a certain game a player can score 1, 2 or 3 only. Copy and complete the table which shows the scores and two of the respective probabilities of these being scored in a single round.

| Score | 1 | 2 | 3 |
|---|---|---|---|
| Probability | $\frac{4}{7}$ | | $\frac{1}{7}$ |

Draw a tree diagram to show all the possible total scores and their respective probabilities after a player has completed two rounds.

Find the probability that a player has (a) a score of 4 after two rounds, (b) an odd number score after two rounds. (L Additional)

7. The probability that I have to wait at the traffic lights on my way to school is 0.25.
   Find the probability that, on two consecutive mornings, I have to wait on at least one morning.

8. A die is thrown three times. What is the probability of scoring a two on just one occasion?

9. A coin is tossed four times. Find the probability of obtaining less than two heads.

10. Two golfers, Smith and Jones, are attempting to qualify for a golf championship. It is estimated that the probability of Jones qualifying is 0.8, and that the probability of both Smith and Jones qualifying is 0.6. Given that the probability of Smith qualifying and the probability of Jones qualifying are independent, find the probability that only one of them will qualify. (C)

11. Whether or not Jonathan gets up in time for school depends on whether he remembers to set his alarm clock the evening before.
    For 85% of the time he remembers to set the clock; the other 15% of the time he forgets.
    If the clock is set, he gets up in time for school on 90% of the occasions.
    If the clock is not set, he does not get up in time for school on 60% of the occasions.
    On what proportion of the occasions does he get up in time for school? (NEAB)

12. In a game, a steel ball is dropped onto a set of nails arranged in three levels as shown.
    When a ball hits a nail, the probability of it moving right or left before reaching the next level is $\frac{1}{2}$.

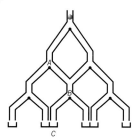

Calculate the probability of a ball
(a) reaching A
(b) reaching B,
(c) dropping into slot C. (NEAB)

13. A team needs to win at least two of its remaining three games to secure the championship. The probabilities that the team will win the games are assessed to be 0.6, 0.7 and 0.8, respectively. Calculate the probability, based on these assessed values, that the team will secure the championship. (C)

14. In the game of tennis a player has two serves.

    If the first serve is successful the game continues.

    If the first serve is not successful the player serves again. If this second service is successful the game continues.

    If both serves are unsuccessful the player has served a 'double fault' and loses the point.

    Gabriella plays tennis. She is successful with 60% of her first serves and 95% of her second serves.

    (a) Calculate the probability that Gabriella serves a double fault.

    If Gabriella is successful with her first serve she has a probability of 0.75 of winning the point.

    If she is successful with her second serve she has a probability of 0.5 of winning the point.

    (b) Calculate the probability that Gabriella wins the point. (MEG)

15. In a group of 12 international referees there are three from Africa, four from Asia and five from Europe. To officiate at a tournament, three referees are chosen at random from the group. Calculate the probability that

    (a) a referee is chosen from each continent,
    (b) exactly two referees are chosen from Asia,
    (c) the three referees are chosen from the same continent. (C)

16. A bag contains seven black and three white marbles. Three marbles are chosen at random and in succession, each marble being replaced after it has been taken out of the bag.

    Draw a tree diagram to show all possible selections.

    From your diagram, or otherwise, calculate, to two significant figures, the probability of choosing

    (a) three black marbles,
    (b) a white marble, a black marble and a white marble in that order,
    (c) two white marbles and a black marble in any order,
    (d) at least one black marble.

    State an event from this experiment which together with the event described in (d) would be both exhaustive and mutually exclusive. (L)

17. Alec and Bill frequently play each other in a series of games of table tennis. Records of the outcomes of these games indicate that whenever they play a series of games, Alec has the probability 0.6 of winning the first game and that in every

subsequent game in the series, Alec's probability of winning the game is 0.7 if he won the preceding game but only 0.5 if he lost the preceding game. A game cannot be drawn. Find the probability that Alec will win the third game in the next series he plays with Bill. (NEAB)

18. Three men, A, B and C agree to meet at the theatre. The man A cannot remember whether they agreed to meet at the Palace or the Queen's and tosses a coin to decide which theatre to go to. The man B also tosses a coin to decide between the Queen's and the Royalty. The man C tosses a coin to decide whether to go to the Palace or not and in this latter case he tosses again to decide between the Queen's and the Royalty. Find the probability that

    (a) A and B meet,
    (b) B and C meet,
    (c) A, B and C all meet,
    (d) A, B and C all go to different places,
    (e) at least two meet. (C)

## Section B

1. I travel to work by route A or route B. The probability that I choose route A is $\frac{1}{4}$. The probability that I am late for work if I go via route A is $\frac{2}{3}$ and the corresponding probability if I go via route B is $\frac{1}{3}$.

   (a) What is the probability that I am late for work on Monday?
   (b) Given that I am late for work, what is the probability that I went via route B?

2. A box contains 20 chocolates, of which 15 have soft centres and five have hard centres. Two chocolates are taken at random, one after the other. Calculate the probability that

   (a) both chocolates have soft centres,
   (b) one of each sort of chocolate is taken,
   (c) both chocolates have hard centres, given that the second chocolate has a hard centre. (C)

3. (a) Explain in words the meaning of the symbol $P(A \mid B)$ where A and B are two events. State the relationship between A and B when (i) $P(A \mid B) = 0$, (ii) $P(A \mid B) = P(A)$.

   (b) When a car owner needs her car serviced she phones one of three garages, A, B, or C. Of her phone calls to them, 30% are to garage A, 10% to B and 60% to C.

   The percentages of occasions when the garage phoned can take the car in on the day of phoning are 20% for A, 6% for B and 9% for C.

   Find the probability that the garage phoned will *not* be able to take the car in on the day of phoning.

   Given that the car owner phones a garage and the garage can take her car in on that day, find the probability that she phoned garage B. (L)

4. A shop stocks tinned cat food of two makes, *A* and *B*, and two sizes, large and small.

   Of the stock, 70% is of brand *A*, 30% is of brand *B*.

   Of the tins of brand *A*, 30% are small size whilst of the tins of brand *B*, 40% are small size.

   Using a tree diagram, or otherwise, find the probability that

   (a) a tin chosen at random from the stock will be of small size,
   (b) a small tin chosen at random from the stock will be of brand *A*. (L)

5. A die is known to be biased in such a way that, when it is thrown, the probability of a six showing is $\frac{1}{4}$. This biased die and an ordinary fair die are thrown. Find the probability that

   (a) the fair die shows a six and the biased die does not show a six,
   (b) at least one of the two dice shows a six,
   (c) exactly one of the two dice shows a six, given that at least one of them shows a six. (C)

6. A golfer observes that, when playing a particular hole at his local course, he hits a straight drive on 80% of the occasions when the weather is not windy but only on 30% of the occasions when the weather is windy. Local records suggest that the weather is windy on 55% of all days.

   (a) Show that the probability that, on a randomly chosen day, the golfer will hit a straight drive at the hole is 0.525.
   (b) Given that he fails to hit a straight drive at the hole, calculate the probability that the weather is windy. (NEAB)

7. In my bookcase there are four shelves and the number of books on each shelf is as shown in the table:

   | | Hardback | Paperback |
   |---|---|---|
   | Shelf 1 | 11 | 9 |
   | Shelf 2 | 8 | 12 |
   | Shelf 3 | 16 | 4 |
   | Shelf 4 | 9 | 3 |

   (a) If I choose a book at random, irrespective of its position in the bookcase, what is the probability that it is a paperback?
   (b) I am equally likely to choose any shelf. I choose a shelf at random and then choose a book. (i) What is the probability that it is a hardback? (ii) If the book chosen is a hardback, what is the probability that it is from shelf 3?

8. Of a group of pupils studying at A-level in schools in a certain area, 56% are boys and 44% are girls. The probability that a boy of this group is studying Chemistry is $\frac{1}{5}$ and the probability that a girl of this group is studying Chemistry is $\frac{1}{11}$.

   (a) Find the probability that a pupil selected at random from this group is a girl studying Chemistry.
   (b) Find the probability that a pupil selected at random from this group is not studying Chemistry.
   (c) Find the probability that a Chemistry pupil selected at random from this group is male.

   (You may leave your answers as fractions in their lowest terms.) (O&C)

9. Explain, by suitably defining events *A* and *B*, what is meant by 'the probability of *A* occurring given that *B* has occurred'.

   A local greengrocer sells conventionally grown and organically grown vegetables.

   Conventionally grown vegetables constitute 80% of his sales; carrots constitute 12% of the conventional sales and 30% of the organic sales.

   Display this information in an appropriately and accurately labelled tree diagram.

   One day a customer emerges from the shop and is questioned about her purchases. What is the probability that she bought

   (a) conventionally grown carrots,
   (b) carrots?

   Given that she did buy carrots, what is the probability that they were organically grown? What assumptions have you made in answering this question? (O)

10. In a simple model of the weather in October, each day is classified as either fine or rainy. The probability that a fine day is followed by a fine day is 0.8. The probability that a rainy day is followed by a fine day is 0.4. The probability that 1 October is fine is 0.75.

    (a) Find the probability that 2 October is fine and the probability that 3 October is fine.
    (b) Find the conditional probability that 3 October is rainy, given that 1 October is fine.
    (c) Find the conditional probability that 1 October is fine, given that 3 October is rainy. (C)

11. At the ninth hole on a certain golf course there is a pond. A golfer hits a grade B ball into the pond. Including the golfer's ball there are then six grade C, ten grade B and four grade A balls in the pond. The golfer uses a fishing net and 'catches' four balls. The events *X*, *Y* and *Z* are

defined as follows:

X: the catch consists of two grade A balls and two grade C balls

Y: the catch consists of two grade B balls and two other balls

Z: the catch includes the golfer's own ball

Assuming that the catch is a random selection from the balls in the pond, determine
(a) $P(X)$, (b) $P(Y)$, (c) $P(Z)$, (d) $P(Z \mid Y)$.
For each of the pairs X and Y, Y and Z, state, with a brief reason, whether the two events are (i) mutually exclusive, (ii) independent. (C)

12. [In this question, give your answers in decimal form, correct to three significant figures.]

A choir has seven sopranos, six altos, three tenors and four basses. The sopranos and altos are women and the tenors and basses are men. At a particular rehearsal, three members of the choir are chosen at random to make the tea.

(a) Find the probability that all three tenors are chosen.
(b) Find the probability that exactly one bass is chosen.
(c) Find the conditional probability that two women are chosen, given that exactly one bass is chosen.
(d) Find the probability that the chosen group contains exactly one tenor or exactly one bass (or both). (C)

13. Vehicles approaching a crossroads must go in one of three directions – left, right or straight on. Observations by traffic engineers showed that of vehicles approaching from the north, 45% turn left, 20% turn right and 35% go straight on. Assuming that the driver of each vehicle chooses direction independently, what is the probability that of the next three vehicles approaching from the north

(a) all go straight on,
(b) all go in the same direction,
(c) two turn left and one turns right,
(d) all go in different directions,
(e) exactly two turn left?

Given that three consecutive vehicles all go in the same direction, what is the probability that they all turned left? (AEB)

14. During an epidemic of a certain disease a doctor is consulted by 110 people suffering from symptoms commonly associated with the disease. Of the 110 people, 45 are female of whom 20 actually have the disease and 25 do not. Fifteen males have the disease and the rest do not.

(a) A person is selected at random. The event that this person is female is denoted by $A$ and the event that this person is suffering

from the disease is denoted by $B$.
Evaluate (i) $P(A)$, (ii) $P(A \cup B)$, (iii) $P(A \cap B)$, (iv) $P(A \mid B)$.

(b) If three different people are selected at random without replacement, what is the probability of (i) all three having the disease, (ii) exactly one of the three having the disease, (iii) one of the three being a female with the disease, one a male with the disease and one a female without the disease?

(c) Of people with the disease 96% react positively to a test for diagnosing the disease as do 8% of people without the disease. What is the probability of a person selected at random (i) reacting positively, (ii) having the disease given that he or she reacted positively? (AEB)

15. In an experiment two bags $A$ and $B$, containing red and green marbles are used. Bag $A$ contains four red marbles and one green marble and bag $B$ contains two red marbles and seven green marbles. An unbiased coin is tossed. If a head turns up, a marble is drawn at random from bag $A$ while if a tail turns up, a marble is drawn at random from bag $B$. Calculate the probability that a red marble is drawn in a single trial. Given that a red marble is selected, calculate the probability that when the coin was tossed a head was obtained. (L)

16. In a computer game played by a single player, the player has to find, within a fixed time, the path through a maze shown on the computer screen. On the first occasion that a particular player plays the game, the computer shows a simple maze, and the probability that the player succeeds in finding the path in the time allowed is $\frac{3}{4}$. On subsequent occasions, the maze shown depends on the result of the previous game. If the player succeeded on the previous occasion, the next maze is harder, and the probability that the player succeeds is one half of the probability of success on the previous occasion. If the player failed on the previous occasion, a simple maze is shown and the probability of the player succeeding is again $\frac{3}{4}$.

The player plays three games.

(a) Show that the probability that the player succeeds in all three games is $\frac{27}{512}$.
(b) Find the probability that the player succeeds in exactly one of the games.
(c) Find the probability that the player does not have two consecutive successes.
(d) Find the conditional probability that the player has two consecutive successes given that the player has exactly two successes. (C)

17. A sailing competition between two boats, $A$ and $B$, consists of a series of independent races, the competition being won by the first boat to win three races. Every race is won by either $A$ or $B$,

and their respective probabilities of winning are influenced by the weather. In rough weather the probability that $A$ will win is 0.9; in fine weather the probability that $A$ will win is 0.4. For each race the weather is either rough or fine, the probability of rough weather being 0.2. Show that the probability that $A$ will win the first race is 0.5.

Given that the first race was won by $A$, determine the conditional probability that

(a) the weather for the first race was rough,
(b) $A$ will win the competition. (C)

# SOME USEFUL METHODS

## (a) Problems involving an 'at least' situation

### Example 3.29

(a) Find the probability of obtaining at least one six when five dice are thrown.
(b) Find the probability of obtaining at least one six when $n$ dice are thrown.
(c) How many dice must be thrown so that the probability of obtaining at least one six is at least 0.99?

### Solution 3.29

(a) In one throw $P(6) = \frac{1}{6}$ and $P(\text{not } 6) = \frac{5}{6}$
When five dice are thrown,

$$P(\text{at least one six}) = 1 - P(\text{no sixes})$$
$$= 1 - \left(\frac{5}{6}\right)^5$$
$$= 0.598 \text{ (3 d.p.)}$$

(b) When $n$ dice are thrown,
$$P(\text{at least one six}) = 1 - \left(\frac{5}{6}\right)^n$$

(c) You need to find $n$ such that

$$1 - \left(\frac{5}{6}\right)^n \geqslant 0.99$$
i.e. $\qquad \left(\frac{5}{6}\right)^n \leqslant 0.01$

You could do this by trial and improvement:

$\left(\frac{5}{6}\right)^{20} = 0.026 \ldots \quad > 0.01$
$\left(\frac{5}{6}\right)^{25} = 0.0104 \ldots \quad > 0.01$
$\left(\frac{5}{6}\right)^{26} = 0.0087 \ldots \quad < 0.01$

So the least value of $n$ is 26.
$\therefore$ **26 dice must be thrown.**

*NOTE*: you could solve $\left(\frac{5}{6}\right)^n \leqslant 0.01$ using logarithms.

Take logs to the base 10 of both sides,

$n \log\left(\frac{5}{6}\right) \leqslant \log(0.01)$

Divide both sides by $\log(\frac{5}{6})$. Since $\log(\frac{5}{6})$ is negative, this will reverse the inequality sign.

$$n \geqslant \frac{\log(0.01)}{\log(\frac{5}{6})}$$

$$n \geqslant 25.3 \ldots$$

The least value of $n$ is 26

---

# (b) Problems involving the use of an infinite geometric progression (GP)

Many probability examples involve the use of GPs and the following formula is required.

If $S_\infty = a + ar + ar^2 + ar^3 + \cdots$ (to infinity),

then

$$S_\infty = \frac{a}{1 - r} \quad \text{for} \quad |r| < 1 \quad \text{where } a \text{ is the first term and } r \text{ is the common ratio}$$

### Example 3.30

Joe and Pete play a game in which they each throw a die in turn until someone throws a six. The person who throws the six wins the game. Joe starts the game. Find the probability that he wins.

### Solution 3.30

Joe will win the game if he wins on his first go, or on his second go, or on his third go, and so on.

$P(\text{Joe wins on his first go}) = \frac{1}{6}$

$P(\text{Joe wins on his second go}) = P(\text{Joe doesn't throw a six, Pete doesn't throw a six,}$
$$\qquad\qquad\qquad\qquad \text{then Joe throws a six})$$
$$= \tfrac{5}{6} \times \tfrac{5}{6} \times \tfrac{1}{6} = (\tfrac{5}{6})^2 \times \tfrac{1}{6}$$

$P(\text{Joe wins on his third go}) = \tfrac{5}{6} \times \tfrac{5}{6} \times \tfrac{5}{6} \times \tfrac{5}{6} \times \tfrac{1}{6} = (\tfrac{5}{6})^4 \times \tfrac{1}{6}$ and so on

$P(\text{Joe wins}) = \tfrac{1}{6} + (\tfrac{5}{6})^2 \times \tfrac{1}{6} + (\tfrac{5}{6})^4 \times (\tfrac{1}{6}) + \cdots$

$$= \tfrac{1}{6}(1 + (\tfrac{5}{6})^2 + (\tfrac{5}{6})^4 + \cdots)$$

Now $1 + (\tfrac{5}{6})^2 + (\tfrac{5}{6})^4 + \cdots$ is the sum of an infinite GP with $a = 1$, $r = (\tfrac{5}{6})^2 = \tfrac{25}{36}$.

$$S_\infty = \frac{a}{1 - r} = \frac{1}{1 - \frac{25}{36}} = \tfrac{36}{11} \quad \therefore \quad P(\text{Joe wins}) = \tfrac{1}{6} \times \tfrac{36}{11} = \tfrac{6}{11}$$

## Exercise 3e   Useful methods

1. A coin is biased so that the probability that it falls showing tails is 0.75.

   (a) Find the probability of obtaining at least one head when the coin is tossed five times.
   (b) How many times must the coin be tossed so that the probability of obtaining at least one head is greater than 0.98?

2. A missile is fired at a target and the probability that the target is hit is 0.7.

   (a) Find how many missiles should be fired so that the probability that the target is hit at least once is greater than 0.995.
   (b) Find how many missiles should be fired so that the probability that the target is not hit is less than 0.001.

3. A die is biased so that the probability of obtaining a three is $p$. When the die is thrown four times the probability that there is at least one three is 0.9375. Find the value of $p$.

   How many times should the die be thrown so that the probability that there are no threes is less than 0.03?

4. On a safe there are four alarms which are arranged so that any one will sound when someone tries to break into the safe. The probability that each alarm will function properly is 0.85, find the probability that at least one alarm will sound when someone tries to break into the safe.

5. For a certain strain of wallflower, the probability that, when sown, a seed produces a plant with yellow flowers is $\frac{1}{6}$. Find the minimum number of seeds that should be sown in order that the probability of obtaining at least one plant with yellow flowers is greater than 0.98.          (L)

6. Two people, $A$ and $B$, play a game. An ordinary die is thrown and the first person to throw a four wins. $A$ and $B$ take it in turns to throw the die, starting with $A$. Find the probability that $B$ wins.

7. $A$, $B$, $C$ and $D$ throw a coin, in turn, starting with $A$. The first to throw a head wins. The game can continue indefinitely until a head is thrown. However, $D$ objects because the others have their first turn before him.

   Compare the probability that $D$ wins with the probability that $A$ wins.

8. A box contains five black balls and one white ball. Alan and Bill take turns to draw a ball from the box, starting with Alan. The first boy to draw the white ball wins the game.

   Assuming that they do not replace the balls as they draw them out, find the probability that Bill wins the game.

   If the game is changed, so that, in the new game, they replace each ball after it has been drawn out, find the probabilities that:

   (a) Alan wins at his first attempt;
   (b) Alan wins at his second attempt;
   (c) Alan wins at his third attempt.

   Show that these answers are terms in a Geometric Progression. Hence find the probability that Alan wins the new game.

9. Two archers $A$ and $B$ shoot alternately at a target until one of them hits the centre of the target and is declared the winner.

   Independently, $A$ and $B$ have probabilities of $\frac{1}{3}$ and $\frac{1}{4}$, respectively, of hitting the centre of the target on each occasion they shoot.

   (a) Given that $A$ shoots first, find (i) the probability that $A$ wins on his second shot, (ii) the probability that $A$ wins on his third shot, (iii) the probability that $A$ wins.
   (b) Given that the archers toss a fair coin to determine who shoots first, find the probability that $A$ wins.          (NEAB)

## ARRANGEMENTS

In order to calculate the number of possible outcomes in a possibility space or an event, the following results are often used.

## Result 1

The number of ways of arranging $n$ unlike objects in a line is $n!$

NOTE: $n! = n \times (n-1) \times (n-2) \times \dots \times 3 \times 2 \times 1$.

For example, consider the letters A, B, C, D.

The first letter can be chosen in four ways (either A or B or C or D),
the second letter can be chosen in three ways,
the third letter can be chosen in two ways,
the fourth letter can be chosen in only one way.

Therefore the number of ways of arranging the four letters is $4 \times 3 \times 2 \times 1 = 4! = 24$.

On a calculator:  $\boxed{4}$  $\boxed{x!}$   (You may have to use $\boxed{\text{SHIFT}}$ key.)

The arrangements are

| ABCD | ABDC | ACBD | ACDB | ADCB | ADBC |
| BCDA | BCAD | BDAC | BDCA | BACD | BADC |
| CDBA | CDAB | CABD | CADB | CBAD | CBDA |
| DABC | DACB | DBCA | DBAC | DCAB | DCBA |

## Example 3.31

A witness reported that a car seen speeding away from the scene of the crime had a number plate that began with V or W, the digits were 4, 7 and 8 and the end letters were A, C, E. He could not however remember the order of the digits or the end letters. How many cars would need to be checked to be sure of including the suspect car?

## Solution 3.31

There are 3! ways of arranging the digits 4, 7, 8 and
    3! ways of arranging the letters A, C, E.
There are two choices for the initial letter.
The total number of different plates $= 2 \times 3! \times 3!$
$$= 72$$

**72 cars would need to be checked.**

# Result 2

The number of ways of arranging in a line $n$ objects, of which $p$ are alike, is $\dfrac{n!}{p!}$.

If instead of the letters A, B, C, D you have the letters A, A, A, D then the 24 arrangements listed previously reduce to the following:

AAAD    AADA    ADAA    DAAA

So the number of ways of arranging the four objects, of which three are alike

$$= \frac{4!}{3!} = \frac{4 \times 3 \times 2 \times 1}{3 \times 2 \times 1} = 4.$$    On a calculator: $\boxed{4}$ $\boxed{x!}$ $\boxed{\div}$ $\boxed{3}$ $\boxed{x!}$ $\boxed{=}$

The result can be extended as follows:

The number of ways of arranging in a line $n$ objects of which $p$ of one type are alike, $q$ of a second type are alike, $r$ of a third type are alike, and so on, is $\dfrac{n!}{p!\,q!\,r! \,...}$.

## Example 3.32

(a) In how many ways can the letters of the word STATISTICS be arranged?
(b) If the letters of the word MINIMUM are arranged in a line at random, what is the probability that the arrangement begins with MMM?

## Solution 3.32

(a) Consider the word STATISTICS.

There are ten letters and S occurs three times,
T occurs three times,
I occurs twice.

Therefore number of ways $= \dfrac{10!}{3!3!2!} = 50\ 400$

On a calculator: $\boxed{10}\ \boxed{x!}\ \boxed{\div}\ \boxed{3}\ \boxed{x!}\ \boxed{\div}\ \boxed{3}\ \boxed{x!}\ \boxed{\div}\ \boxed{2}\ \boxed{x!}\ \boxed{=}$

**There are 50 400 ways of arranging the letters in the word STATISTICS.**

(b) Consider the word MINIMUM.

The possibility space $S = $ (arrangements of MINIMUM).

$$n(S) = \frac{7!}{3!2!} = 420$$

$\uparrow\ \uparrow$
3 Ms  2 Is

Let $E$ be the event 'the arrangement begins with MMM'.

The letters must be arranged in the order MMMxxxx. There is only one way of arranging MMM; then the remaining four letters can be arranged in $\dfrac{4!}{2!} = 12$ ways.

$\therefore\quad n(E) = 12$

So $\quad P(E) = \dfrac{n(E)}{n(S)} = \dfrac{12}{420} = \dfrac{1}{35}$

**The probability that the arrangement begins MMM is $\frac{1}{35}$.**

---

## Example 3.33

Ten pupils are placed at random in a line. What is the probability that the two youngest pupils are separated?

## Solution 3.33

Let the possibility space be $S$, then $n(S) = 10!$

Let $E$ be the event 'the two youngest pupils are together'.

Treating these two together as one item, there are nine items to arrange.

Nine items can be arranged in 9! ways.

The two youngest can be arranged in two ways $(Y_1 Y_2$ or $Y_2 Y_1)$.

Therefore $n(E) = 2 \times 9!$

So $\quad P(E) = \dfrac{n(E)}{n(S)}$  $\qquad$ On a calculator: $\boxed{2}\ \boxed{\times}\ \boxed{9}\ \boxed{x!}\ \boxed{\div}\ \boxed{10}\ \boxed{x!}\ \boxed{=}$

$\qquad\qquad = \dfrac{2 \times 9!}{10!}$

$\qquad\qquad = 0.2$

$E'$ is the event 'the two youngest are not together'.

$\qquad P(E') = 1 - P(E)$

$\qquad\qquad = 1 - 0.2 = 0.8$

**The probability that the two youngest are separated is 0.8.**

## Example 3.34

If a four-digit number is formed from the digits 1, 2, 3 and 5 and repetitions are not allowed, find the probability that the number is divisible by 5.

## Solution 3.34

Let $S$ be the possibility space, then $n(S) = 4! = 24$.

Let $E$ be the event 'the number is divisible by 5'.

If the number is divisible by 5 then it must end with the digit 5.

$n(E)$ = number of ways of arranging the digits 1, 2, 3 = 3!

So $\quad P(E) = \dfrac{n(E)}{n(S)}$

$\qquad\qquad = \dfrac{3!}{24}$

$\qquad\qquad = \dfrac{1}{4}$

**The probability that the number is divisible by 5 is $\frac{1}{4}$.**

## Example 3.35

The letters of the word MATHEMATICS are written, one on each of 11 separate cards. The cards are laid out in a line.

(a) Calculate the number of different arrangements of these letters.

(b) Determine the probability that the vowels are all placed together. $\qquad\qquad (L)$

## Solution 3.35

(a) Number of different arrangements

$\qquad \dfrac{11!}{2! \times 2! \times 2!} = 4\ 989\ 600$

$\qquad\qquad \nearrow \qquad \uparrow \qquad \nwarrow$

$\qquad\quad$ 2Ms $\quad$ 2As $\quad$ 2Ts

(b) To find the number of ways with vowels together treat $\boxed{\text{AEAI}}$ as one item. So treat M, T, H, M, T, C, S and $\boxed{\text{AEAI}}$ as 8 items.

Then number of arrangements $= \dfrac{8!}{2!2!} = 10\,080$

$\nearrow$ 2Ms  $\nwarrow$ 2Ts

The vowels A, E, A, I, however, can be arranged in $\dfrac{4!}{2!} = 12$ ways,

so total number of arrangements $= 12 \times 10\,080 = 120\,960$

$P(\text{vowels together}) = \dfrac{120\,960}{4\,989\,600} = 0.024 \text{ (2 s.f.)}$

---

## Example 3.36

The six letters of the word LONDON are each written on a card and the six cards are then shuffled and placed in a line.

(a) Calculate the number of different arrangements.
(b) Find the probability that the middle two cards both have the letter N on them.
(c) Find the probability that the two cards with letter O are adjacent and the two cards with letter N are also adjacent.

The cards are shuffled again and placed in a line, face down. The first two cards in the line are turned over and reveal the letters L and O.

(d) Find the probability that when the other four cards are turned over the letters will spell LONDON.

(L)

## Solution 3.36

(a) Number of different arrangements of LONDON $= \dfrac{6!}{2! \times 2!} = 180$

$\nearrow$ 2 Os  $\nwarrow$ 2 Ns

(b) If the middle two letters are NN, then you need to find the number of different arrangements of LODO.

Number of arrangements $= \dfrac{4!}{2!} = 12$

$\uparrow$ 2 Os

$\therefore \quad P(\text{middle two letters are NN}) = \dfrac{12}{180} = \dfrac{1}{15}$

(c) If two Os and two Ns are adjacent then it is easier to think of each pair being glued together like this ⓄⓄ and ℕℕ

Number of different arrangements of L, ⓄⓄ, ℕℕ, D = 4! = 24

∴ $P$(two Os, two Ns are adjacent) $= \dfrac{24}{180} = \dfrac{2}{15}$

(d) If the first two cards are L, O then you need to find the number of different arrangements of N, D, O, N

Number of arrangements $= \dfrac{4!}{2!} = 12$

↑
2 Ns

It is quite easy to list these arrangements

| | |
|---|---|
| *NDON | DNNO |
| NDNO | DNON |
| NODN | DONN |
| NOND | ONND |
| NNDO | ONDN |
| NNOD | ODNN |

Of course, only one of these marked (*) will spell LONDON

So $P$(L, O and four remaining letters spell LONDON) $= \dfrac{1}{12}$

---

# Result 3

The number of ways of arranging $n$ unlike objects in a ring when clockwise and anticlockwise arrangements are different is $(n - 1)!$

For example, consider four people $A$, $B$, $C$ and $D$, who are to be seated at a round table. The following four arrangements are the same, as $A$ always has $D$ on his immediate right and $B$ on his immediate left.

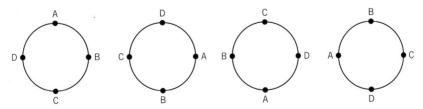

To find the number of different arrangements, fix $A$ and then consider the number of ways of arranging $B$, $C$ and $D$.

Therefore the number of different arrangements of four people around the table is 3!

# Result 4

The number of ways of arranging $n$ unlike objects in a ring, when clockwise and anticlockwise arrangements are the same, is $\dfrac{(n-1)!}{2}$

For example, if $A$, $B$, $C$ and $D$ are four different coloured beads which are threaded on a ring, then the following two arrangements are the same, since one is the other viewed from the other side.

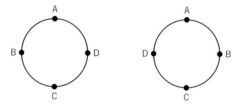

Therefore the number of arrangements of four beads on a ring is $\dfrac{3!}{2} = 3$.

## Example 3.37

Six bulbs are planted in a ring and two do not grow. What is the probability that the two that do not grow are next to each other?

## Solution 3.37

Let $S$ be the possibility space, then $n(S) = 5!$

Let $E$ be the event 'the bulbs that do not grow are next to each other'.

Consider the two bulbs that do not grow as one item. They can be arranged in 2! ways.

There are now five items to be arranged in a ring and this can be done in 4! ways.

Therefore $\qquad n(E) = 2!\,4!$

So $\qquad P(E) = \dfrac{n(E)}{n(S)}$

$\qquad\qquad = \dfrac{2!\,4!}{5!}$

$\qquad\qquad = \dfrac{2}{5}$

**The probability that the bulbs that do not grow are next to each other is $\frac{2}{5}$.**

## Example 3.38

One white, one blue, one red and two yellow beads are threaded on a ring to make a bracelet. Find the probability that the red and white beads are next to each other.

## Solution 3.38

Let $S$ be the possibility space.

If all the objects are unlike, the number of ways of arranging five beads on a ring is $\dfrac{4!}{2}$, but since there are two yellows, $n(S) = \dfrac{4!}{(2)(2!)} = 6$

Let $E$·be the event 'the red and the white beads are next to each other'.

Then $n(E) = \dfrac{2!3!}{2!2}$

┌── red and white can be arranged in 2! ways

←── number of ways of arranging four objects in a ring

←── anticlockwise and clockwise arrangements are the same

└── there are two yellows

So  $n(E) = 3$

and  $P(E) = \dfrac{n(E)}{n(S)} = \dfrac{3}{6} = \dfrac{1}{2}$

**The probability that the red and white beads are next to each other is $\tfrac{1}{2}$.**

This result can be shown diagrammatically:

### Ways of arranging the beads

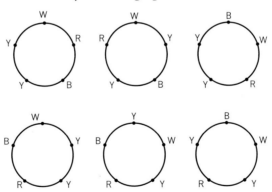

NOTE: as expected, in three of the six arrangements the red and white beads are next to each other.

## PERMUTATIONS OF $r$ OBJECTS FROM $n$ OBJECTS

Consider the number of ways of placing three of the letters A, B, C, D, E, F, G in three empty spaces.

The first space can be filled in seven ways. The second space can be filled in six ways. The third space can be filled in five ways. Therefore there are $7 \times 6 \times 5$ ways of arranging three letters taken from seven letters. This is the number of **permutations** of three objects taken from seven and it is written $^7P_3$.

So $\quad ^7P_3 = 7 \times 6 \times 5 = 210$

Now $7 \times 6 \times 5$ could be written $\dfrac{7 \times 6 \times 5 \times 4 \times 3 \times 2 \times 1}{4 \times 3 \times 2 \times 1}$

i.e. $\quad ^7P_3 = \dfrac{7!}{4!} = \dfrac{7!}{(7-3)!}$

On a calculator this can be obtained directly: $\boxed{7}$ $\boxed{^nP_r}$ $\boxed{3}$ $\boxed{=}$

(you may have to use the shift key.)

NOTE: the order in which the letters are arranged is important since ABC is a different permutation from ACB.

In general, the number of permutations, or ordered arrangements, of $r$ objects taken from $n$ unlike objects is written $^nP_r$ where

$$^nP_r = \frac{n!}{(n-r)!}$$

NOTE: Using the formula, $^nP_n = \dfrac{n!}{(n-n)!} = \dfrac{n!}{0!}$

But the number of ways of arranging $n$ unlike objects is $n!$

So 0! is defined to be 1, i.e.

$\quad 0! = 1$

Try it on your calculator.

## COMBINATIONS OF $r$ OBJECTS FROM $n$ OBJECTS

When considering the number of combinations of $r$ objects from $n$ objects, the order in which they are placed is not important.

For example, the one combination ABC gives rise to 3! permutations

$\quad$ ABC, ACB, BCA, BAC, CAB, CBA

Denoting the number of combinations of three letters from the seven letters A, B, C, D, E, F, G, by $^7C_3$ then

$\quad ^7C_3 \times 3! = {}^7P_3$

$$^7C_3 = \frac{^7P_3}{3!} = \frac{7!}{3!4!} = 35$$

On the calculator,

$^7C_3$ can be obtained directly: $\boxed{7}$ $\boxed{^nC_r}$ $\boxed{3}$ $\boxed{=}$ (You may have to use the shift key.)

In general, the number of combinations of $r$ objects from $n$ unlike objects is $^nC_r$ where

$$^nC_r = \frac{n!}{r!(n-r)!}$$

NOTE: $^nC_r$ is sometimes written $_nC_r$ or $\binom{n}{r}$.

## Example 3.39

In how many ways can a hand of four cards be dealt from an ordinary pack of 52 playing cards?

## Solution 3.39

You need to consider combinations, since the order in which the cards are dealt is not important.

$$^{52}C_4 = 270\ 725 \qquad \text{On a calculator:} \quad \boxed{52} \ \boxed{^nC_r} \ \boxed{4} \ \boxed{=}$$

**The number of ways of dealing the hand of four cards is 270 725.**

## Example 3.40

Four letters are chosen at random from the word RANDOMLY. Find the probability that all four letters chosen are consonants.

## Solution 3.40

Let $S$ be the possibility space, then $n(S) = {}^8C_4 = 70$

Let $E$ be the event 'four consonants are chosen'. Since there are six consonants
$n(E) = {}^6C_4 = 15$

$$P(E) = \frac{n(E)}{n(S)} = \frac{15}{70} = \frac{3}{14}$$

**The probability that the four letters chosen are consonants is $\frac{3}{14}$.**

## Example 3.41

A team of four is chosen at random from five girls and six boys.

(a) In how many ways can the team be chosen if

    (i) there are no restrictions;
    (ii) there must be more boys than girls?

(b) Find the probability that the team contains only one boy.

## Solution 3.41

(a) (i) There are 11 people, from whom four are chosen. The order in which they are chosen is not important.

Number of ways of choosing the team $= {}^{11}C_4 = 330$

**If there are no restrictions, the team can be chosen in 330 ways.**

(ii) If there are to be more boys than girls, then there must be three boys and one girl, or four boys.

Number of ways of choosing three boys and one girl $= {}^6C_3 \times {}^5C_1 = 100$

On a calculator: $\boxed{6}$ $\boxed{{}^nC_r}$ $\boxed{3}$ $\boxed{=}$ $\boxed{\times}$ $\boxed{5}$ $\boxed{{}^nC_r}$ $\boxed{1}$ $\boxed{=}$

Number of ways of choosing four boys $= {}^6C_4 = 15$

So number of ways of choosing three boys and a girl, or four boys $= 100 + 15 = 115$.

$\therefore$ Number of ways to choose the team with more boys than girls $= \mathbf{115}$.

(b) The possibility space $S =$ (all possible teams of four)

$\therefore$ $n(S) = 330$.

Let $E$ be the event 'only one boy is chosen'. If one boy is chosen, then three girls must be chosen,

so $n(E) = {}^6C_1 \times {}^5C_3 = 60$

$\therefore$ $P(E) = \dfrac{n(E)}{n(S)} = \dfrac{60}{330} = \dfrac{2}{11}$

**The probability that the team contains only one boy is $\frac{2}{11}$.**

---

## Example 3.42

If a diagonal of a polygon is defined to be a line joining any two non-adjacent vertices, how many diagonals are there in a polygon of (a) five sides, (b) six sides, (c) $n$ sides?

## Solution 3.42

(a) Number of ways to choose two points from five $= {}^5C_2 = 10$

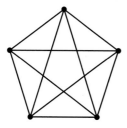

Note ${}^5C_2 = \dfrac{5!}{2!3!}$

$= \dfrac{5 \times 4}{2}$ $\longleftarrow$ 3! cancels on the top and on the bottom

So there are ten possible lines to draw, but as there are five sides, five of these are joining adjacent vertices.

$\therefore$ number of diagonals $= {}^5C_2 - 5 = \dfrac{5 \times 4}{2} - 5 = 5$.

**The number of diagonals for a polygon with five sides is 5.**

(b) Similarly, for the polygon of six sides,

$$\begin{aligned}
\text{the number of diagonals} &= {}^6C_2 - 6 \\
&= \frac{6 \times 5}{2} - 6 \\
&= 9
\end{aligned}$$

**The number of diagonals for a polygon with six sides is 9.**

(c) For a polygon with $n$ sides,

$$\begin{aligned}
\text{the number of diagonals} &= {}^nC_2, - n \\
&= \frac{n(n-1)}{2} - n \\
&= \frac{n^2 - n - 2n}{2} \\
&= \frac{n(n-3)}{2}
\end{aligned}$$

**The number of diagonals for a polygon with $n$ sides is $\dfrac{n(n-3)}{2}$.**

## Example 3.43

Three letters are selected at random from the word BIOLOGY. Find the probability that the selection (a) does not contain the letter O, (b) contains both of the letters O.

## Solution 3.43

You need to find the total number of selections and, because there are two letters O, find the number of selections with

no letters O
one letter O
two letters O.

Number of selections without the letter O

= number of ways to choose three letters from B, I, L, G, Y
= ${}^5C_3$
= 10

Number of selections with one letter O (e.g.  O, B, I ,  O, B, G , and so on)

= number of ways to choose two letters from B, I, L, G, Y
= ${}^5C_2$
= 10

Number of selections with two letters O (e.g.  O, O, B ,  O, O, G , and so on)

= number of ways to choose one letter from B, I, L, G, Y
= 5

Therefore, total number of selections = 10 + 10 + 5 = 25

(a) $P$(selection does not contain letter O) $= \frac{10}{25} = \frac{2}{5}$

(b) $P$(selection contains two letters O) $= \frac{5}{25} = \frac{1}{5}$

*NOTE*: it is easy to write them all out to check:

| | | |
|---|---|---|
| B, I, L | O, B, I | O, O, B |
| B, I, G | O, B, L | O, O, I |
| B, I, Y | O, B, G | O, O, L |
| B, L, G | O, B, Y | O, O, G |
| B, L, Y | O, I, L | O, O, Y |
| B, G, Y | O, I, G | 5 |
| I, L, G | O, I, Y | |
| I, L, Y | O, L, G | |
| I, G, Y | O, L, Y | |
| L, G, Y | O, G, Y | |
| 10 | 10 | |

## Example 3.44

Seven cards, labelled $A$, $B$, $C$, $D$, $E$, $F$, $G$, are thoroughly shuffled and then dealt out face upwards on a table.

Find the probabilities, giving each as a fraction in its simplest form, that

(a) the first three cards to appear are the cards labelled $A$, $B$, $C$, in that order,
(b) the first three cards to appear are the cards labelled $A$, $B$, $C$, but in any order,
(c) the seven cards appear in their original order: $A$, $B$, $C$, $D$, $E$, $F$, $G$.          (NEAB)

## Solution 3.44

(a) Number of ways to arrange three letters from seven

$$= {}^7P_3 = \frac{7!}{4!} = 7 \times 6 \times 5 = 210$$

∴   $P$(first three letters are $A$, $B$, $C$ in that order) $= \dfrac{1}{210}$

(b) Number of ways to *choose* three letters from seven

$$= {}^7C_3 = \frac{7!}{4!3!} = 35$$

∴   $P$(first three letters are $A$, $B$, $C$ in any order) $= \dfrac{1}{35}$

(c) Number of ways to arrange seven letters $= 7! = 5040$

∴   $P(A, B, C, D, E, F, G) = \dfrac{1}{5040}$

# Exercise 3f  Arrangements, permutations, combinations

1. In how many ways can the letters of the word FACETIOUS be arranged in a line?
   What is the probability that an arrangement begins with F and ends with S?

2. (a) In how many ways can seven people sit at a round table?
   (b) What is the probability that a husband and wife sit together?

3. On a shelf there are four mathematics books and eight English books.
   (a) If the books are to be arranged so that the mathematics books are together, in how many ways can this be done?
   (b) What is the probability that all the mathematics books will not be together?

4. The letters of the word PROBABILITY are arranged at random. Find the probability that the two Is are separated.

5. If the letters in the word ABSTEMIOUS are arranged at random, find the probability that the vowels and consonants appear alternately.

6. Nine children play a party game and hold hands in a circle.
   (a) In how many different ways can this be done?
   (b) What is the probability that Mary will be holding hands with her friends Natalie and Sarah?

7. (a) In how many different ways can the letters in the word ARRANGEMENTS be arranged?
   (b) Find the probability that an arrangement chosen at random begins with the letters EE.

8. From a group of ten boys and eight girls, two pupils are chosen at random. Find the probability that they are both girls.

9. From a group of six men and eight women, five people are chosen at random. Find the probability that there are more men chosen than women.

10. From a bag containing six white counters and eight blue counters, four counters are chosen at random. Find the probability that two white counters and two blue counters are chosen.

11. From a group of ten people, four are to be chosen to serve on a committee.
    (a) In how many different ways can the committee be chosen?
    (b) Among the ten people there is one married couple. Find the probability that both the husband and the wife will be chosen.
    (c) Find the probability that the three youngest people will be chosen.

12. In a group of six students, four are female and two are male. Determine how many committees of three members can be formed containing one male and two females.  (L)

13. Four persons are chosen at random from a group of ten persons consisting of four men and six women. Three of the women are sisters. Calculate the probabilities that the four persons chosen will:
    (a) consist of four women,
    (b) consist of two women and two men,
    (c) include the three sisters.  (NEAB)

14. A touring party of 20 cricketers consists of nine batsmen, eight bowlers and three wicket keepers. A team of 11 players must have at least five batsmen, four bowlers and one wicket keeper. How many different teams can be selected, (a) if all the players are available for selection, (b) if two batsmen and one bowler are injured and cannot play?

15. Find the number of ways in which ten different books can be shared between a boy and a girl if each is to receive an even number of books.

16. Four letters are picked from the word BREAKDOWN. What is the probability that there is at least one vowel among the letters?

17. Eight people sit in a minibus: four on the sunny side and four on the shady side. If two people want to sit on opposite sides to each other and another two people want to sit on the shady side, in how many ways can this be done?

18. Disco lights are arranged in a vertical line. How many different arrangements can be made from two green, three blue and four red lights (a) if all nine lights are used, (b) if at least eight lights are used?

19. A group consisting of 10 boys and 11 girls attends a course for special games coaching.
    (a) When they are introduced, each person hands a card containing his or her photograph and name and address to every other member of the group. State the total number of cards which are exchanged.
    (b) 5 boys are selected for basketball and 6 girls for netball. Find the number of different possible selections for each of these.
    (c) 5 particular boys and 5 particular girls are selected and placed in mixed pairs for tennis. Find the total number of different mixed pairs which can be made using these 10 children.
    (d) If 4 children are chosen at random from the whole group find the probability that there is a majority of girls in the 4 selected.
    (L Additional)

20. A competition has a first prize, a second prize, a third prize and a fourth prize. Ten competitors enter this competition and the prizes are awarded for the first, second, third and fourth competitors in order of merit.

    (a) Find the number of different ways in which these prizes could be won.

    Smith and Jones are two of the ten competitors. Find the number of different ways in which the prizes could be won if

    (b) neither Smith nor Jones wins a prize,
    (c) each of Smith and Jones wins a prize.  (C)

21. The number of applicants for a job is 15. Calculate the number of different ways in which six applicants can be selected for interview.

    The six selected applicants are interviewed on a particular day. Calculate the number of ways in which the order of the six interviews can be arranged.
    Of the six applicants interviewed, three have backgrounds in business, two have backgrounds in education and one has a background in recreation. Calculate the number of ways in which the order of the six interviews can be arranged, when applicants having the same background are interviewed successively.  (C)

22. Each of seven children, in turn, throws a ball once at a target. Calculate the number of ways the children can be arranged in order to take the throws.

    Given that three of the children are girls and four are boys, calculate the number of ways the children can be arranged in order that

    (a) successive throws are made by boys and girls alternately,
    (b) a girl takes the first throw and a boy takes the last throw.  (C)

23. To enter a cereal competition, competitors have to choose the eight most important features of a new car, from a possible 12 features, then list the eight in order of preference. Each cereal packet entry form contains space for five entries. A correct entry wins a new car.

    (a) What is the probability that a woman wins a new car if she completes the entry form from one packet?
    (b) How many entry forms would she need to complete, each entry showing different arrangements, if the probability that she wins a car is to be at least 0.8?

24. Three letters are selected at random from the word SCHOOL. Find the probability that the selection (a) does not contain the letter O, (b) contains both the letters O.

25. How many even numbers can be formed with the digits 3, 4, 5, 6, 7 by using some or all of the numbers (repetitions are not allowed)?

26.

Different coloured pegs, each of which is painted in one and only one of the six colours red, white, black, green, blue and yellow, are to be placed in four holes, as shown in the figure, with one peg in each hole. Pegs of the same colour are indistinguishable. Calculate how many different arrangements of pegs placed in the four holes so that they are all occupied can be made from

    (a) six pegs, all of different colours,
    (b) two red and two white pegs,
    (c) two red, one white and one black peg,
    (d) twelve pegs, two of each colour.
    (L Additional)

27. (a) Calculate how many different numbers altogether can be formed by taking one, two, three and four digits from the digits 9, 8, 3 and 2, repetitions not being allowed.
    (b) Calculate how many of the numbers in part (a) are odd and greater than 800.
    (c) If one of the numbers in part (a) is chosen at random, calculate the probability that it will be greater than 300.  (L Additional)

28. The positions of nine trees which are to be planted along the sides of a road, five on the north side and four on the south side, are shown in the figure.

    ○    ○    ○    ○    ○       N
    _____

    _____
        ○    ○    ○    ○       S

    (a) Find the number of ways in which this can be done if the trees are all of different species.
    (b) If the trees in (a) are planted at random, find the probability that two particular trees are next to each other on the same side of the road.
    (c) If there are three cupressus, four prunus and two magnolias, find the number of different ways in which these could be planted assuming that trees of the same species are identical.
    (d) If the trees in (c) are planted at random, find the probability that the two magnolias are on the opposite sides of the road.
    (L Additional)

29. A committee consisting of six persons is to be selected from five women and six men.

(a) Calculate the number of ways in which the chosen committee will contain exactly two men.

(b) Given that the committee is to contain at least two men, show that it can be selected in 456 ways.

(c) Given that these 456 ways are equally likely, calculate the probability that there will be more men than women on the committee.

(d) At a meeting the members of the chosen committee sit at a rectangular table in the fixed seats illustrated in the diagram:

(i) Given that each may sit in any of the six places, calculate the number of different ways they may be seated at the table.

(ii) Given that the committee consists of three men and three women and that the men and women must sit alternately round the table, calculate in how many different ways they may be seated.

(L Additional)

30. A committee of eight members consists of one married couple together with four other men and two other women. From the committee a working party of four persons is to be formed. Find the number of different working parties which can be formed.

Find also the number if the working party

(a) may not contain *both* the husband and his wife,

(b) must contain two men and two women,

(c) must contain at least one man and at least one woman.

The eight committee members sit round an octagonal table, their positions being decided by drawing lots. Find the probability of

(d) the man sitting next to his wife,

(e) the man sitting opposite to his wife,

(f) the three women sitting together. (AEB)

## Summary

- Experimental probability

$$P(A) = \lim_{n \to \infty} \left( \frac{r}{n} \right) \qquad \text{where } \frac{r}{n} \text{ is the relative frequency of } A.$$

- Equally likely outcomes

$$P(A) = \frac{n(A)}{n(S)} \qquad \text{where} \quad n(A) \text{ is the number of outcomes in } A$$

$$n(S) \text{ is the number of possible outcomes.}$$

$0 \leqslant P(A) \leqslant 1$      If $A$ is impossible, $P(A) = 0$

                  If $A$ is certain, $P(A) = 1$.

$P(A') = 1 - P(A)$ where $A'$ is the event 'A does not occur.'

- For events $A$ and $B$

$P(A \text{ or } B) = P(A) + P(B) - P(A \text{ and } B)$

$P(A \cup B) = P(A) + P(B) - P(A \cap B)$

- For mutually exclusive events $A$ and $B$, $P(A \cap B) = 0$

so $P(A \text{ or } B) = P(A) + P(B)$     'or' rule for exclusive events

i.e. $P(A \cup B) = P(A) + P(B)$

- For exhaustive events $A$ and $B$

$P(A \text{ or } B) = 1$, i.e. $P(A \cup B) = 1$

- Conditional probability

$$P(A \text{ given } B) = \frac{P(A \text{ and } B)}{P(B)}$$

  i.e.   $P(A \mid B) = \dfrac{P(A \cap B)}{P(B)}$

$$P(A \text{ and } B) = P(A \mid B)P(B) = P(B \mid A)P(A).$$

- For independent events $A$, $B$

$$P(A \mid B) = P(A)$$
$$P(B \mid A) = P(B)$$
$$P(A \text{ and } B) = P(A) \times P(B) \quad \textbf{'and' rule for independent events}$$

- Tree diagrams        (Multiply along the branches)

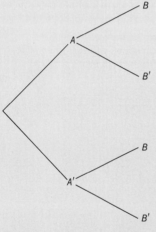

$$P(A \cap B) = P(A) \times P(B \mid A)$$

$$P(A \cap B') = P(A) \times P(B' \mid A)$$

$$P(A' \cap B) = P(A') \times P(B \mid A')$$

$$P(A' \cap B') = P(A') \times P(B' \mid A')$$

$$P(B) = P(A \cap B) + P(A' \cap B)$$

- Arrangements, permutations and combinations
  - The number of ways of arranging $n$ unlike objects in a line        $n!$
  - The number of ways of arranging in a line $n$ objects of which $p$ of one type are alike, $q$ of another type are alike, $r$ of a third type are alike, and so on        $\dfrac{n!}{p!q!r! \dots}$
  - The number of ways of arranging $n$ unlike objects in a ring when clockwise and anticlockwise arrangements are different        $(n-1)!$
  - The number of ways of arranging $n$ unlike objects in a ring when clockwise and anticlockwise arrangements are the same        $\dfrac{(n-1)!}{2}$
  - The number of permutations of $r$ objects taken from $n$ unlike objects        $^nP_r = \dfrac{n!}{(n-r)!}$
  - The number of combinations of $r$ objects taken from $n$ unlike objects        $^nC_r = \dfrac{n!}{r!(n-r)!}$

# Miscellaneous worked examples

### Example 3.45

A die is biased so that, when it is rolled, the probability of obtaining a score of 6 is $\frac{1}{4}$. The probabilities of obtaining each of the other five scores 1, 2, 3, 4, 5 are all equal. Calculate the probability of obtaining a score of five with this biased die.

(a) The biased die and an unbiased die are now rolled together. Calculate the probability that the total score is 11 or more.
(b) The two dice are rolled again. Given that the total score is 11 or more, calculate the probability that the score on the biased die is 6. (C)

### Solution 3.45

*Events*
$6_B$: score 6 on biased die    $6_U$: score 6 on unbiased die

$5_B$: score 5 on biased die    $5_U$: score 5 on unbiased die
For the biased die, $P(6_B) = \frac{1}{4}$

$\therefore\ P(\text{score is } 1, 2, 3, 4 \text{ or } 5) = \frac{3}{4}$

$$P(5_B) = \frac{1}{5} \times \frac{3}{4} = \frac{3}{20}$$

(a) $P(11 \text{ or more}) = P(6_B 6_U)^* + P(6_B 5_U)^* + P(5_B 6_U)$

$$= \frac{1}{4} \times \frac{1}{6} \quad + \frac{1}{4} \times \frac{1}{6} \quad + \frac{3}{20} \times \frac{1}{6}$$

$$= \frac{1}{6}\left[\frac{1}{4} + \frac{1}{4} + \frac{3}{20}\right]$$

$$= \frac{1}{6} \times \frac{13}{20}$$

$$= \frac{13}{120}$$

(b) $P(6_B \mid \text{score is 11 or more}) = \dfrac{P(6_B \text{ and score is 11 or more})}{P(\text{score is 11 or more})}$ ⟵ marked * above

$$= \frac{\dfrac{1}{4} \times \dfrac{1}{6} + \dfrac{1}{4} \times \dfrac{1}{6}}{\dfrac{13}{120}} = \frac{\dfrac{1}{12}}{\dfrac{13}{120}} = \frac{10}{13}$$

### Example 3.46

During 1996 a vet saw 125 dogs, each suspected of having a particular disease. Of the 125 dogs, 60 were female of whom 25 actually had the disease and 35 did not. Only 20 of the males had the disease, the rest did not. The case history of each dog was documented on a separate record card.

(a) A record card from 1996 is selected at random. Let $A$ represent the event that the dog referred to on the record card was female and $B$ represent the event that the dog referred to was suffering from the disease.

Find

(i) $P(A)$,
(ii) $P(A \cup B)$,
(iii) $P(A \cap B)$,
(iv) $P(A \mid B)$.

(b) if three different record cards are selected at random, *without replacement*, find the probability that

(i) all three record cards relate to dogs with the disease,
(ii) exactly one of the three record cards relates to a dog with the disease,
(iii) one record card relates to a female dog with the disease, one to a male dog with the disease and one to a female dog not suffering from the disease. (L)

## Solution 3.46

Summarising the information in a table:

|  | Diseased ($B$) | Not Diseased ($B'$) | Total |
|---|---|---|---|
| Female ($A$) | 25 | 35 | 60 |
| Male ($A'$) | 20 | 45 | 65 |
| Total | 45 | 80 | 125 |

(a) (i) $P(A) = \frac{60}{125} = 0.48$

(ii) $P(A \cup B) = \dfrac{25 + 35 + 20}{125} = \dfrac{80}{125} = 0.64$

(iii) $P(A \cap B) = \frac{25}{125} = 0.2$

(iv) $P(A \mid B) = \frac{25}{45} = \frac{5}{9}$

(b) (i) $P(BBB) = \frac{45}{125} \times \frac{44}{124} \times \frac{43}{123} = 0.045$ (2 s.f.)

(ii) Number of ways of arranging $B$, $B'$, $B'$ = 3
So $P(BB'B'$ in any order) $= 3 \times \frac{45}{125} \times \frac{80}{124} \times \frac{79}{123} = 0.44$ (2 s.f.)

(iii) Number of ways of arranging the cards = 3!
So $P$(female with disease, male with disease, female without disease)
$= 3! \times \frac{25}{125} \times \frac{20}{124} \times \frac{35}{123} = 0.055$ (2 s.f.)

## Example 3.47

A company needs to appoint three representatives, one to be based in Lancashire, one in Yorkshire and one in Cumbria. There are eight sales officers available for selection to the post of representative.

(a) Calculate the number of possible allocations of officers to representative posts.
(b) Calculate the number of different sets of three officers who could be appointed to represent the company.
(c) Two of the eight sales officers are members of the Brown family. Assuming that the three representatives are chosen at random from the eight officers, find the probability that both members of the Brown family will be chosen. Give your answer as a fraction in its simplest form. (NEAB)

## Solution 3.47

(a) The first post can be allocated in 8 possible ways.
The second post can be allocated in 7 possible ways.
The third post can be allocated in 6 possible ways.
Number of allocations = $8 \times 7 \times 6 = \mathbf{336}$

(b) Number of different sets of three officers = $^8C_3 = \mathbf{56}$

(c) If both the Browns are chosen,
number of ways to choose third representative = 6

So $P$(both Browns are chosen) $= \dfrac{6}{56} = \dfrac{3}{28}$

---

## Example 3.48

A factory has three machines $A$, $B$, $C$ producing large numbers of a certain item. Of the total daily production of the item, 50% are produced on $A$, 30% on $B$ and 20% on $C$. Records show that 2% of items produced on $A$ are defective, 3% of items produced on $B$ are defective and 4% of items produced on $C$ are defective. The occurrence of a defective item is independent of all other items.

One item is chosen at random from a day's total output.

(a) Show that the probability of its being defective is 0.027.

(b) Given that it is defective, find the probability that it was produced on machine $A$.　　(W)

## Solution 3.48

Events are defined as follows

$A$: Item produced on $A$　$P(A) = 0.5$　$P(D, \text{ given } A) = 0.02$
$B$: Item produced on $B$　$P(B) = 0.3$　$P(D, \text{ given } B) = 0.03$
$C$: Item produced on $C$　$P(C) = 0.2$　$P(D, \text{ given } C) = 0.04$
$D$: Item is defective

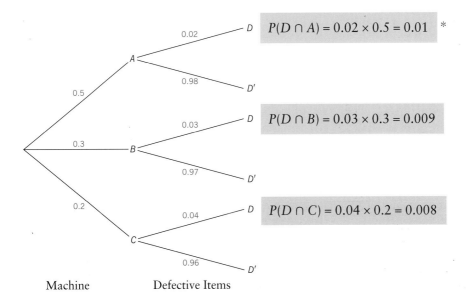

Machine　　　　　　Defective Items

(a) $P(D) = P(D \text{ and } A) + P(D \text{ and } B) + P(D \text{ and } C)$
$= 0.01 + 0.009 + 0.008$
$= 0.027$

(b) You already know that $P(D \text{ given } A) = 0.02$, but now you need to 'reverse the conditions' to find $P(A \text{ given } D)$

Use $P(A \text{ given } D) = \dfrac{P(A \text{ and } D)}{P(D)}$  ⟵ marked * on tree

$= \dfrac{0.01}{0.027}$  ⟵ found in part (a)

$= \mathbf{0.370}$ (3 d.p.).

## Example 3.49

A house is infested with mice and to combat this the householder acquired four cats, Albert, Belinda, Khalid and Poon. The householder observes that only half of the creatures caught are mice. A fifth are voles and the rest are birds.

20% of the catches are made by Albert, 45% by Belinda, 10% by Khalid and 25% by Poon.

(a) The probability of a catch being a mouse, a bird or a vole is independent of whether or not it is made by Albert. What is the probability of a randomly selected catch being a

  (i)  mouse caught by Albert,
  (ii) bird not caught by Albert?

(b) Belinda's catches are equally likely to be a mouse, a bird or a vole. What is the probability of a randomly selected catch being a mouse caught by Belinda?
(c) The probability of a randomly selected catch being a mouse caught by Khalid is 0.05. What is the probability that a catch made by Khalid is a mouse?
(d) Given that the probability that a randomly selected catch is a mouse caught by Poon is 0.2 verify that the probability of a randomly selected catch being a mouse is 0.5.
(e) What is the probability that a catch which is a mouse was made by Belinda?        (AEB)

## Solution 3.49

| Events | Probabilities |
|---|---|
| M: a mouse is caught | $P(M) = 0.5$ |
| V: a vole is caught | $P(V) = 0.2$ |
| B: a bird is caught | $P(B) = 1 - (0.5 + 0.2) = 0.3$ |
| | |
| A: Catch by Albert | $P(A) = 0.2$ |
| L: Catch by Belinda | $P(L) = 0.45$ |
| K: Catch by Khalid | $P(K) = 0.1$ |
| N: Catch by Poon | $P(N) = 0.25$ |

(a) (i)  $P(\text{Mouse caught by Albert}) = P(M \cap A)$
$= P(M) \times P(A)$  ⟵ M and A are independent
$= 0.5 \times 0.2$
$= \mathbf{0.1}$

(ii)  $P(\text{Bird not caught by Albert}) = P(B \cap A')$

$$= P(B) \times P(A') \longleftarrow B \text{ and } A' \text{ are independent}$$
$$= 0.3 \times 0.8$$
$$= \mathbf{0.24}$$

Before answering the next parts, it is useful to show all the given information on a tree diagram:

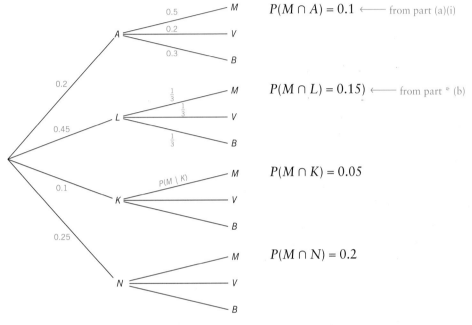

$P(M \cap A) = 0.1 \longleftarrow \text{from part (a)(i)}$

$P(M \cap L) = 0.15) \longleftarrow \text{from part * (b)}$

$P(M \cap K) = 0.05$

$P(M \cap N) = 0.2$

(b)  $P(\text{Mouse caught by Belinda}) = P(M \cap L)$
$$= 0.45 \times \tfrac{1}{3}$$
$$= \mathbf{0.15}$$

(c)  $P(\text{Catch is mouse caught by Khalid}) = P(M \cap K) = 0.05$

$P(\text{Catch by Khalid is a mouse}) = P(M \,|\, K)$
$$= \frac{P(M \cap K)}{P(K)}$$
$$= \frac{0.05}{0.1}$$
$$= \mathbf{0.5}$$

(d)  $P(\text{Catch is mouse caught by Poon}) = 0.2$

$P(\text{Catch is a mouse}) = P(M \cap A) + P(M \cap L) + P(M \cap K) + P(M \cap N)$
$$= \quad 0.1 \quad + \quad 0.15 \quad + \quad 0.05 \quad + \quad 0.5$$
$$= \mathbf{0.5}$$

(e)  $P(\text{Catch which is a mouse was caught by Belinda}) = P(L \,|\, M)$
$$= \frac{P(L \cap M)}{P(M)}$$
$$= \frac{0.15}{0.5} \longleftarrow \text{from part * (b)}$$
$$= \mathbf{0.3}$$

# Miscellaneous exercise 3g

1. Each time a table tennis player serves, the probability that she wins the point is 0.6, independently of the result of any preceding serves. At the start of a particular game, she serves for each of the first five points. Calculate the probability that, for the first two points of this game,

   (a) she wins both points,
   (b) she wins exactly one of these two points.

   Calculate the probability that, for the first five points of this game,

   (c) she loses all five points,
   (d) she wins at least one of these five points. (C)

2. A director of a company is selected at random.

   C denotes the event that the director's annual salary is more than £300 000.
   C' denotes the event that the director's annual salary is not more than £300 000.
   D denotes the event that the director's annual salary is less than £200 000.
   E denotes the event that the director's annual salary is less than £350 000.

   Write down two of the events C, C', D and E which are

   (a) complementary,
   (b) mutually exclusive but not exhaustive,
   (c) exhaustive but not mutually exclusive.

   (AEB)

3. Newborn babies are routinely screened for a serious disease which affects only two per 1000 babies. The result of screening can be positive or negative. A positive result suggests that the baby has the disease, but the test is not perfect. If a baby has the disease, the probability that the result will be negative is 0.01. If the baby does not have the disease, the probability that the result will be positive is 0.02.

   (a) Find the probability that a baby has the disease, given that the result of the test is positive.
   (b) Comment on the value you obtain. (L)

4. A penalty shoot-out in a game of hockey requires each of two players to take a penalty hit to try to score a goal. In a simple model, each player has a probability of 0.8 of scoring a goal, and independence is assumed. Calculate the probability that exactly one goal is scored from the two hits.

   In an alternative model, the probability of the second player scoring is reduced to 0.7 if the first player does not score. Calculate the probability that the second player has scored, given that only one goal is scored. (C)

5. Forty 17- and 18-year old students are the only people present at a party. The numbers of male and female students of each age are given in the following table.

   |        | 17-year old | 18-year old |
   |--------|-------------|-------------|
   | Male   | 9           | 13          |
   | Female | 7           | 11          |

   In the Grand Draw, each of the forty students has an equal chance of winning one of two prizes. The first prize is a gift token and the second prize is a box of chocolates. No student may win more than one prize. Find the probability that

   (a) the gift token will be won by an 18-year old male student,
   (b) both prizes will be won by female students,
   (c) the box of chocolates will be won by a 17-year old student, given that the gift token is won by a 17-year old male student. (C)

6. Each customer at a supermarket pays by one of cash, cheque or credit card. The probability of a randomly selected customer paying by cash is 0.54 and by cheque is 0.18.

   (a) Determine the probability of a randomly selected customer paying by credit card.

   Three customers are selected at random.

   (b) Find the probability of

   (i) all three paying by cash,
   (ii) exactly one paying by cheque,
   (iii) one paying by cash, one by cheque and one by credit card.

   The probability that the amount payable exceeds £30 is 0.26. If the amount payable does exceed £30, then the probability of it being paid by cheque is 0.28.

   (c) Find the probability that a randomly selected customer pays more than £30 and pays by cheque.
   (d) Hence find the probability that a randomly selected customer pays more than £30, given that the customer pays by cheque. (AEB)

7. A writer submits a poem for publication by a literary magazine. The poem will be accepted for publication if it is approved by at least two of the three members of the editorial staff who independently assess it. Given that the probabilities that the poem is approved by the three members are 0.9, 0.7 and 0.6 respectively, find the probability that the poem is not accepted.

The writer submits a different poem for each of three separate issues of the magazine. Given that the probabilities remain the same, calculate the probability that all three of her poems are accepted. (C)

8. At an art exhibition seven paintings are to be hung in a row along one wall. Find the number of possible arrangements.

Given that three paintings are by the same artist, find the number of arrangements in which

(a) these three paintings are hung side by side,
(b) any one of these three paintings is hung at the beginning of the row but neither of the other two is hung at the end of the row. (C)

9. A group of three pregnant women attend ante-natal classes together. Assuming that each woman is equally likely to give birth on each of the seven days in a week, find the probability that all three give birth

(a) on a Monday,
(b) on the same day of the week,
(c) on different days of the week,
(d) at a weekend (either a Saturday or Sunday).
(e) Find the probability of all three giving birth on the same day of the week given that they all give birth at a weekend.
(f) How large would the group need to be to make the probability of all the women in the group giving birth on different days of the week less than 0.05? (AEB)

10. The probability that for any married couple the husband has a degree is $\frac{6}{10}$ and the probability that the wife has a degree is $\frac{1}{2}$. The probability that the husband has a degree, given that the wife has a degree, is $\frac{11}{12}$.

A married couple is chosen at random.

Find the probability that

(a) both of them have degrees,
(b) only one of them has a degree,
(c) neither of them has a degree.

Two married couples are chosen at random.

(d) Find the probability that only one of the two husbands and only one of the two wives have a degree. (L)

11. A personal stereo system consists of a playing unit and a headphone unit. Each unit is tested for faults. If a unit is found to be faulty, an attempt is made to correct the fault and the unit is then retested. Any unit that is found to be faulty a second time is rejected.

(a) The probability of a randomly chosen playing unit being found to be faulty on the first test is 0.1. If a second test is needed, the probability of a playing unit being found to be faulty on the second test is 0.05.

(i) Calculate the probability that a randomly chosen playing unit is rejected.
(ii) Given that a playing unit is accepted, calculate the probability that a fault was found on the first test. Give your answer correct to three significant figures.

(b) The probability of a randomly chosen headphone unit being found to be faulty on the first test is 0.04. If a second test is needed, the probability of a headphone unit being found to be faulty on the second test is 0.02. Calculate the probability that a randomly chosen headphone unit is accepted. Give your answer correct to three significant figures.

(c) A randomly chosen playing unit that has been accepted and a randomly chosen headphone unit that has been accepted are combined to make a personal stereo system. Calculate the probability that at least one of the two units has been retested. Give your answer correct to three significant figures. (C)

12. A bag contains four red counters, three blue counters and three green counters. A counter is drawn at random from the bag and not replaced. A second counter is then drawn at random from the bag.

Assuming that at each stage each counter left in the bag has an equal chance of being drawn,

(a) find the probability, giving your answers as fractions in their lowest terms, that the second counter will be blue given that

(i) the first counter is red,
(ii) the first counter is blue,
(iii) the first counter is green.

(b) Find the probability, giving your answer as a fraction in its lowest terms, that the first counter will be red and the second counter will be blue.
(c) Find the probability, giving your answer as a fraction in its lowest terms, that the second counter will be blue regardless of the colour of the first counter. (C)

13. A particular firm has six vacancies to fill from 15 applicants. Calculate the number of ways in which these vacancies could be filled if there are no restrictions.

The firm decides that three of the six vacancies shall be filled by women and three by men. The applicants consist of seven women and eight men. Calculate the number of ways in which the six vacancies could be filled under these conditions.

One of the seven women is the wife of one of the eight men. Calculate the number of ways in which three women and three men could fill the six vacancies, given that both the wife and her husband are among those appointed.

Of all the possible selections of three women and three men, one is picked at random. Calculate the probability that this selection includes

(a) both the wife and her husband,
(b) either the wife or her husband, but not both.

(C)

14. Laura has 12 friends, seven girls and five boys, all of whom she wants to come to her birthday party. However, she is only allowed to invite five of them. Not wishing to show any favouritism, Laura chooses the five children to come to the party at random.

(a) How many different selections are possible?
(b) In how many selections are there exactly three boys?
(c) What is the probability that exactly three boys are invited to the party?

In fact, there are three girls at the party, including Laura, and three boys, including Liam and John. For the party tea they sit round a circular table, equally spaced, with Laura sitting in the position shown in the diagram.

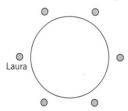

(d) In how many different ways can the other children fill the remaining seats?

With Laura sitting in her place, the other children take their seats at random.

(e) Find the probability that Laura sits next to Liam and John.
(f) Find the probability that boys and girls sit alternately.

(MEI)

15. A draw is being made for the quarter-finals of a knock-out table tennis tournament. Eight counters, alike in every respect except that they are numbered from one to eight inclusive, are placed in a bag and drawn one by one, without replacement. A typical draw might produce the numbers in the order 3, 5, 7, 2, 1, 8, 6, 4, resulting in the matches:

Match A     3 plays 5
Match B     7 plays 2
Match C     1 plays 8
Match D     6 plays 4

(a) In how many different orders can the counters be drawn from the bag?
(b) In how many ways can the counters be drawn such that
   (i) players 1 and 2 play each other in match A,
   (ii) players 1 and 2 play each other.

(c) Find the probability that
   (i) players 1 and 2 play each other,
   (ii) players 1 and 2 do not play each other.

In fact, players 1, 2, 3 and 4 are girls and the rest are boys.

(d) In how many ways can the counters be drawn such that the girls play each other in matches A and B and the boys play each other in matches C and D?
(e) What is the probability that no girl plays a boy in the quarter-finals? (MEI)

16. In a set of 28 dominoes each domino has from 0 to 6 spots at each end. Each domino is different from every other and the ends are indistinguishable so that, for example, the two diagrams in figure 1 represent the *same* domino.

Fig. 1

A domino which has the same number of spots at each end, or no spots at all, is called a 'double'. A domino is drawn at random from the set. Figure 2 shows a sample space diagram to represent the complete set of outcomes, each of which is equally likely.

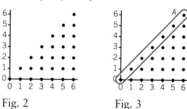

Fig. 2        Fig. 3

Let the event A be 'the domino is a double', event B 'the total number of spots on the domino is six' and event C be 'at least one end of the domino has five spots'.

Figure 3 shows the sample space with the event A marked.

(a) Write down the probability that event A occurs.
(b) Find the probability that either B or C or both occur.
(c) Determine whether or not events A and B are independent.
(d) Find the conditional probability $P(A \mid C)$. Explain why events A and C are *not* independent.

After the first domino has been drawn, a second domino is chosen at random from the remainder.

(e) Find the probability that at least one end of the first domino has the same number of spots as at least one end of the second domino.

[*HINT*: Consider separately the cases where the first domino is a double and where it is not.] (MEI)

# Mixed test 3A

1. A club social committee consists of eight people, two of whom are Nicky and Sam. Two of the eight committee members are to be chosen at random to organise the next club disco.

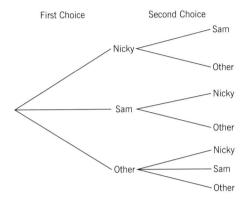

First Choice        Second Choice

By considering the above tree diagram, or otherwise,
   (a) find the probability that both Nicky and Sam are chosen,
   (b) find the probability that both Nicky and Sam are chosen, given that at least one of Nicky and Sam is chosen. (C)

2. A bag contains ten balls, of which four are red and six are blue. An experiment consists of drawing at random and without replacement three balls, one at a time, from the bag.

   (a) Draw a tree diagram to show all the possible outcomes of the experiment.

   Hence, or otherwise, find the probability that

   (b) the first two balls drawn will be of different colours,
   (c) the third ball will be red,
   (d) the third ball will be red, given that the first two balls drawn were both blue. (L)

3. Ann, Barry and Clare are three students taking a multiple choice examination paper. For each question a student has to select the correct answer from five that are offered. For Question 1, Ann has no idea of the correct answer, Barry correctly identifies one answer that is wrong and Clare correctly identifies two wrong answers. All three students decide to guess at random from the answers they think stand a chance of being correct. Calculate the probability that

   (a) none of the three students chooses the correct answer,
   (b) Clare is the only one to choose the correct answer,
   (c) exactly one of the three students chooses the correct answer. (NEAB)

4. Last year the employees of a firm either received no pay rise, a small pay rise or a large pay rise. The following table shows the number in each category, classified by whether they were weekly paid or monthly paid.

|  | No pay rise | Small pay rise | Large pay rise |
|---|---|---|---|
| Weekly paid | 25 | 85 | 5 |
| Monthly paid | 4 | 8 | 23 |

A tax inspector decides to investigate the tax affairs of an employee selected at random.

$D$ is the event that a weekly paid employee is selected.

$E$ is the event that an employee who received no pay rise is selected.

$D'$ and $E'$ are the events "not $D$" and "not $E$" respectively.

Find
(a) $P(D)$,
(b) $P(D \cup E)$,
(c) $P(D' \cap E')$.

$F$ is the event that an employee is female.

(d) Given that $P(F') = 0.8$, find the number of female employees.
(e) Interpret $P(D \mid F)$ in the context of this question.
(f) Given that $P(D \cap F) = 0.1$, find $P(D \mid F)$. (AEB)

5. The captain of a darts team is trying to arrange an evening match for next Monday, Tuesday, Wednesday or Thursday. He hopes that the leading players, $A$, $B$, $C$ and $D$, will all be free on one of these evenings. In fact each of the four players has arranged an engagement for exactly one of the four evenings.

   Assuming that each player is equally likely to have chosen any one of the four evenings, and that their choices are independent, find the probability that
   (a) $A$ and $B$ have both chosen Monday evening,
   (b) either $C$ or $D$ (or both) has chosen Monday evening,
   (c) the four players have chosen four different evenings,
   (d) there will be at least one evening when all four players are free. (NEAB)

# Mixed test 3B

1. A coin is biased so that, on each toss, the probability of obtaining a head is 0.4. The coin is tossed twice.
   (a) Calculate the probability that at least one head is obtained.
   (b) Calculate the conditional probability that exactly one head is obtained, given that at least one head is obtained. (C)

2. The probabilities of events A and B are $P(A)$ and $P(B)$ respectively.

   $P(A) = \frac{5}{12}$, $P(A \cap B) = \frac{1}{6}$, $P(A \cup B) = q$.

   Find, in terms of $q$,

   (a) $P(B)$,
   (b) $P(A \mid B)$.

   Given that A and B are independent events,

   (c) find the value of $q$. (L)

3. A questionnaire asks shareholders of a company to state whether they consider the chairman's salary to be too high, about right, or too low.

   Excluding shareholders who have no opinion, the probabilities of answers from a randomly selected shareholder are as follows:

   | | |
   |---|---|
   | Too high | 0.95 |
   | About right | 0.03 |
   | Too low | 0.02 |

   What is the probability that if three shareholders are selected at random,

   (a) they will all answer 'too high',
   (b) exactly two will answer 'too high',
   (c) exactly two will give the same answer,
   (d) exactly two will answer 'too high' given that exactly two give the same answer? (AEB)

4. A school has three photocopiers A, B and C. On any given day the independent probabilities of a breakdown are 0.1 for A, 0.05 for B and 0.04 for C.

   For a randomly chosen day, calculate the probability that

   (a) at least one of the copiers breaks down,
   (b) exactly one of the copiers breaks down,
   (c) given exactly one copier breaks down, then it is copier C. (NEAB)

5. Every year two teams, the Ramblers and the Strollers, meet each other for a quiz night. From past results it seems that in years when the Ramblers win, the probability of them winning the next year is 0.7 and in years when the Strollers win, the probability of them winning the next year is 0.5. It is not possible for the quiz to result in the scores being tied.

   The Ramblers won the quiz in 1996.

   (a) Draw a probability tree diagram for the three years up to 1999.
   (b) Find the probability that the Strollers will win in 1999.
   (c) If the Strollers win in 1999, what is the probability that it will be their first win for at least three years?
   (d) Assuming that the Strollers win in 1999, find the smallest value of $n$ such that the probability of the Ramblers winning the quiz for $n$ consecutive years after 1999 is less than 5%. (MEI)

# 4

# Probability distributions I – discrete variables

*In this chapter you will learn*

- about probability distributions for discrete random variables
- how to calculate and use $E(X)$, the expectation (mean)
- how to calculate and use $E(g(X))$, the expectation of a simple function of $X$
- how to calculate and use $\text{Var}(X)$, the variance of $X$
- about the cumulative distribution function $F(x)$
- about the results relating to expectation algebra for random variables $X$ and $Y$

This chapter is concerned with discrete variables. When a variable is **discrete**, it is possible to specify or describe all its possible numerical values, for example

- the number of females in a group of four students: the possible values are 0, 1, 2, 3, 4,
- the amount gained, in pence, in a game where the entry fee is 10p and the prizes are 50p and £1: the possible values are 10, 40, 90,
- the number of times you throw a die until a six appears: the possible values are 1, 2, 3, 4, 5, ..., to infinity.

## PROBABILITY DISTRIBUTIONS

A **probability distribution** gives the probability of each possible value of the variable.

Consider this situation:
By mistake, three faulty fuses are put into a box containing two good fuses. The faulty and good fuses become mixed up and are indistinguishable by sight. You take two fuses from the box. What is the probability that you take

(a) no faulty fuses,
(b) one faulty fuse,
(c) two faulty fuses.

It is possible to show the outcomes and probabilities on a tree diagram:

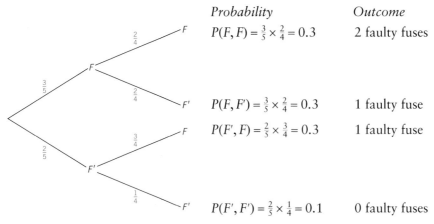

| | Probability | Outcome |
|---|---|---|
| | $P(F, F) = \frac{3}{5} \times \frac{2}{4} = 0.3$ | 2 faulty fuses |
| | $P(F, F') = \frac{3}{5} \times \frac{2}{4} = 0.3$ | 1 faulty fuse |
| | $P(F', F) = \frac{2}{5} \times \frac{3}{4} = 0.3$ | 1 faulty fuse |
| | $P(F', F') = \frac{2}{5} \times \frac{1}{4} = 0.1$ | 0 faulty fuses |

(a)  $P$(no faulty fuses) = 0.1
(b)  $P$(one faulty fuse) = 0.3 + 0.3 = 0.6
(c)  $P$(two faulty fuses) = 0.3

The variable being considered here is 'the number of faulty fuses' and it can be denoted by $X$. The values that $X$ can take are 0, 1 or 2.
The probability that there are no faulty fuses, i.e. the probability that the variable $X$ takes the value 0, can be written $P(X = 0)$, so $P(X = 0) = 0.1$.
Similarly $P(X = 1) = 0.6$ and $P(X = 2) = 0.3$.
Sometimes these are written $p_0 = 0.1$, $p_1 = 0.6$, $p_2 = 0.3$.

When defining variables, the variable is usually denoted by a capital letter ($X$, $Y$, $R$, etc) and a particular value that the variable takes by a small letter ($x$, $y$, $r$, etc), so that $P(X = x)$ means 'the probability that the variable $X$ takes the value $x$'.

The probability distribution for $X$ can be summarised in a table and illustrated in a vertical line graph.

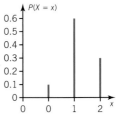

| $x$ | 0 | 1 | 2 |
|---|---|---|---|
| $P(X = x)$ | 0.1 | 0.6 | 0.3 |

If the sum of the probabilities is 1, the variable is said to be **random**.

In this example $P(X = 0) + P(X = 2) = 0.1 + 0.6 + 0.3 = 1$, so $X$ is a **discrete random variable**.

For a discrete random variable, the sum of the probabilities is 1,

i.e.   $\sum_{\text{all } x} P(X = x) = 1$    or    $\Sigma p_i = 1$   for $i = 1, 2, ..., n$

The function that is responsible for allocating probabilities, $P(X = x)$, is known as the **probability density function of X**, sometimes abbreviated to the **p.d.f. of X**. The probability density function can either list the probabilities individually or summarise them in a formula.

## Example 4.1

Two tetrahedral dice, each with faces labelled 1, 2, 3 and 4 are thrown and the score noted, where the score is the sum of the two numbers on which the dice land. Find the probability density function (p.d.f.) of $X$, where $X$ is the random variable 'the score when two dice are thrown'.

## Solution 4.1

The score for each possible outcome is shown in the possibility space:

| Second die | | | | |
|---|---|---|---|---|
| 4 | 5 | 6 | 7 | 8 |
| 3 | 4 | 5 | 6 | 7 |
| 2 | 3 | 4 | 5 | 6 |
| 1 | 2 | 3 | 4 | 5 |
| | 1 | 2 | 3 | 4 |

First die

From the diagram you can see that $X$ can take the values 2, 3, 4, 5, 6, 7, 8 only.

Since each outcome is equally likely, the probabilities can be found from the diagram.

For example, $P(X = 5) = \frac{4}{16}$ since 4 out of the possible 16 outcomes result in a score of 5.

The probability distribution is formed:

| $x$ | 2 | 3 | 4 | 5 | 6 | 7 | 8 |
|---|---|---|---|---|---|---|---|
| $P(X = x)$ | $\frac{1}{16}$ | $\frac{2}{16}$ | $\frac{3}{16}$ | $\frac{4}{16}$ | $\frac{3}{16}$ | $\frac{2}{16}$ | $\frac{1}{16}$ |

Notice the pattern for the probabilities relating to $x$ from 2 to 5.

$$P(X = x) = \frac{x - 1}{16} \quad \text{for } x = 2, 3, 4, 5$$

For $x$ from 6 to 8, there is a different pattern

$$P(X = x) = \frac{9 - x}{16} \quad \text{for } x = 6, 7, 8$$

These two formulae give the p.d.f. of $X$.

*NOTE:* $\sum_{\text{all } x} P(X = x) = \frac{1}{16}(1 + 2 + 3 + 4 + 3 + 2 + 1) = 1$, confirming that $X$ is a random variable.

## Example 4.2

The p.d.f. of a discrete random variable $Y$ is given by $P(Y = y) = cy^2$, for $y = 0, 1, 2, 3, 4$. Given that $c$ is a constant, find the value of $c$.

## Solution 4.2

It helps to write out the probability distribution of $Y$.

| $y$ | 0 | 1 | 2 | 3 | 4 |
|---|---|---|---|---|---|
| $P(Y = y)$ | 0 | $c$ | $4c$ | $9c$ | $16c$ |

Since $Y$ is a random variable, $\sum_{\text{all } y} P(Y = y) = 1$, i.e. the sum of all the probabilities is 1.

So $c + 4c + 9c + 16c = 1$

$$30c = 1$$

$$c = \frac{1}{30}$$

---

## Example 4.3

The discrete random variable $W$ has probability distribution as shown.

| $w$ | $-3$ | $-2$ | $-1$ | $0$ | $1$ |
|---|---|---|---|---|---|
| $P(W = w)$ | 0.1 | 0.25 | 0.3 | 0.15 | $d$ |

Find

(a) the value of $d$,  (b) $P(-3 \leqslant W < 0)$  (c) $P(W > -1)$,
(d) $P(-1 < W < 1)$,  (e) the mode.

## Solution 4.3

(a) Since $\displaystyle\sum_{\text{all } w} P(W = w) = 1$

$0.1 + 0.25 + 0.3 + 0.15 + d = 1$
$$0.8 + d = 1$$
$$d = \mathbf{0.2}$$

(b) $P(-3 \leqslant W < 0) = P(W = -3) + P(W = -2) + P(W = -1)$
$$= 0.1 + 0.25 + 0.3$$
$$= \mathbf{0.65}$$

(c) $P(W > -1) = P(W = 0) + P(W = 1)$
$$= 0.15 + 0.2$$
$$= \mathbf{0.35}$$

(d) $P(-1 < W < 1) = P(W = 0)$
$$= \mathbf{0.15}$$

(e) The value of $w$ with the highest probability is $-1$, so **the mode = -1.**

---

# Exercise 4a   Probability distributions

1. The discrete random variable $X$ has the given probability distribution.

| $x$ | 1 | 2 | 3 | 4 | 5 |
|---|---|---|---|---|---|
| $P(X = x)$ | 0.2 | 0.25 | 0.4 | $a$ | 0.05 |

(a) Find the value of $a$ and draw a vertical line graph to illustrate the distribution.
(b) Find (i) $P(1 \leqslant X \leqslant 3)$,  (ii) $P(X > 2)$,
(iii) $P(2 < X < 5)$, (iv) the mode.

2. The probability density function of a discrete random variable $X$ is given by $P(X = x) = kx$ for $x = 12, 13, 14$.
Write out the probability distribution and find the value of $k$.

3. The discrete random variable $X$ can take values 3, 5, 6, 8 and 10 only. Given that $p_3 = 0.1$, $p_5 = 0.05$, $p_6 = 0.45$ and $p_8 = 3p_{10}$, calculate $p_{10}$.

4. $X$ has probability distribution as shown in the table

| $x$ | 1 | 2 | 3 | 4 | 5 |
|---|---|---|---|---|---|
| $P(X = x)$ | $\dfrac{1}{10}$ | $\dfrac{3}{10}$ | $a$ | $\dfrac{1}{5}$ | $\dfrac{1}{20}$ |

   (a) Find the value of $a$.
   (b) Find $P(X \geqslant 4)$.
   (c) Find $P(X < 1)$.
   (d) Find $P(2 \leqslant X < 4)$.

5. Write out the probability distribution for each of these variables.

   (a) The number of heads, $X$, obtained when two fair coins are tossed.
   (b) The number of tails, $X$, obtained when three fair coins are tossed.

6. A drawer contains eight brown socks and four blue socks. A sock is taken from the drawer at random, its colour is noted and it is then replaced. This procedure is performed twice more. $X$ is the random variable the number of brown socks taken. Find the probability distribution for $X$.

7. The discrete random variable $R$ has p.d.f. $P(R = r) = c(3 - r)$ for $r = 0, 1, 2, 3$.

   (a) Find the value of the constant $c$.
   (b) Draw a vertical line graph to illustrate the distribution.
   (c) Find $P(1 \leqslant R < 3)$.

8 Write down the formula for the p.d.f. of $X$ where $X$ is the numerical value of a digit chosen from a set of random number tables.

9. A game consists of throwing tennis balls into a bucket from a given distance. The probability that William will get the tennis ball in the bucket is 0.4. A turn consists of three attempts.

   (a) Construct the probability distribution for $X$, the number of tennis balls that land in the bucket in a turn.

   (b) William wins a prize if, at the end of his turn, there are two or more tennis balls in the bucket. What is the probability that William does not win a prize?

10. Emma plays a game in which she throws two dice. If she gets two sixes, she wins 20p, if she gets one six she wins 10p, otherwise she wins nothing. She has to pay 5p to enter.

    Write out the probability distribution of $X$, the amount Emma gains in one turn.

11. A student has a fair coin and two six-sided dice, one of which is white and the other blue. The student tosses the coin and then rolls both dice. Let $X$ be a random variable such that if the coin falls heads, $X$ is the sum of the scores on the two dice, otherwise $X$ is the score on the white die only.

    Find the probability function of $X$ in the form of a table of possible values of $X$ and their associated probabilities.

    Find $P(3 \leqslant X \leqslant 7)$.

    State the assumption you made to enable you to evaluate the probability function.          (AEB)

12. $X$ can take values 5, 6, 7, 8 and 9. The vertical line graph to illustrate the distribution of $X$ is incomplete. Given that $P(X = 8) = 2P(X = 9)$, complete the line graph and describe the distribution.

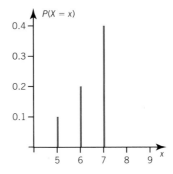

# EXPECTATION OF $X$, $E(X)$

$E(X)$ is read as '$E$ of $X$' and it gives an average or typical value of $X$, known as the expected value or **expectation** of $X$. This is comparable with the mean in descriptive statistics.

# Experimental approach

The **frequency distribution** shows the results when an unbiased die is thrown 120 times.

| Score, $x$ | 1 | 2 | 3 | 4 | 5 | 6 | |
|---|---|---|---|---|---|---|---|
| Frequency, $f$ | 15 | 22 | 23 | 19 | 23 | 18 | Total 120 |

The mean score, $\bar{x} = \dfrac{\Sigma fx}{\Sigma f} = \dfrac{1 \times 15 + 2 \times 22 + 3 \times 23 + 4 \times 19 + 5 \times 23 + 6 \times 18}{120} = 3.6(2 \text{ s.f.})$

You could write this out in a different way

$$\bar{x} = 1 \times \tfrac{15}{120} + 2 \times \tfrac{22}{120} + 3 \times \tfrac{23}{120} + 4 \times \tfrac{19}{120} + 5 \times \tfrac{23}{120} + 6 \times \tfrac{18}{120}$$

The fractions $\tfrac{15}{120}, \tfrac{22}{120}, \tfrac{23}{120}, \tfrac{19}{120}, \tfrac{23}{120}, \tfrac{18}{120}$ are the relative frequencies of the scores of 1, 2, 3, 4, 5, 6 respectively.

Notice that they are close to $\tfrac{20}{120} = \tfrac{1}{6}$.

If you throw the die a large number of times, you would expect each of these fractions to be closer to $\tfrac{1}{6}$, the limiting value of the relative frequency of a particular score on the die.

## Theoretical approach

When an unbiased die is thrown, the probability of obtaining a particular value is $\tfrac{1}{6}$. The probability distribution is $P(X = x) = \tfrac{1}{6}$ for $x = 1, 2, 3, 4, 5, 6$.

| Score, $x$ | 1 | 2 | 3 | 4 | 5 | 6 |
|---|---|---|---|---|---|---|
| $P(X = x)$ | $\tfrac{1}{6}$ | $\tfrac{1}{6}$ | $\tfrac{1}{6}$ | $\tfrac{1}{6}$ | $\tfrac{1}{6}$ | $\tfrac{1}{6}$ |

The expected mean, or expectation of $X$, is obtained by multiplying each score by its probability, then summing. It is written $E(X)$, so:

Expected mean, $E(X) = 1 \times \tfrac{1}{6} + 2 \times \tfrac{1}{6} + 3 \times \tfrac{1}{6} + 4 \times \tfrac{1}{6} + 5 \times \tfrac{1}{6} + 6 \times \tfrac{1}{6}$

$= 3.5$

The **expectation** or **expected mean** can be thought of as the **average value** when the number of experiments increases indefinitely.

In a statistical experiment

- a practical approach results in a frequency distribution and a mean value,
- a theoretical approach results in a probability distribution and an expected value, known as the **expectation**.

The expectation of $X$ (expected value or mean), written $E(X)$, is given by

$$E(X) = \sum_{\text{all } x} xP(X = x)$$

This can also be written

$$E(X) = \Sigma x_i p_i \qquad i = 1, 2, ..., n$$

The symbol $\mu$, pronounced 'mew' is often used for the expectation, where

$$\mu = E(X)$$

## Example 4.4

A random variable $X$ has probability distribution as shown. Find the expectation, $E(X)$.

| $x$ | $-2$ | $-1$ | 0 | 1 | 2 |
|---|---|---|---|---|---|
| $P(X = x)$ | 0.3 | 0.1 | 0.15 | 0.4 | 0.05 |

## Solution 4.4

$$E(X) = \sum_{\text{all } x} xP(X = x)$$
$$= (-2) \times 0.3 + (-1) \times 0.1 + 0 \times 0.15 + 1 \times 0.4 + 2 \times 0.05$$
$$= -0.2$$

## Example 4.5

Find the expected number of sixes when three fair dice are thrown.

## Solution 4.5

$X$ is the number of sixes and can take values 0, 1, 2, 3. Using the notation $\bar{6}$ to represent the event 'a six is not obtained',

$$P(X = 0) = P(\bar{6}, \bar{6}, \bar{6}) = (\tfrac{5}{6})^3 = \tfrac{125}{216}$$
$$P(X = 1) = P(\bar{6}, \bar{6}, 6) + P(\bar{6}, 6, \bar{6}) + P(6, \bar{6}, \bar{6})$$
$$= (\tfrac{5}{6})^2 \times (\tfrac{1}{6}) + (\tfrac{5}{6})^2 \times \tfrac{1}{6} + (\tfrac{5}{6})^2 \times \tfrac{1}{6}$$
$$= \tfrac{75}{216}$$
$$P(X = 2) = 3 \times P(\bar{6}, 6, 6) = 3 \times \tfrac{5}{6} \times (\tfrac{1}{6})^2 = \tfrac{15}{216}$$
$$P(X = 3) = P(6, 6, 6) = (\tfrac{1}{6})^3 = \tfrac{1}{216}$$

The probability distribution for $X$ is

| $x$ | 0 | 1 | 2 | 3 |
|---|---|---|---|---|
| $P(X = x)$ | $\tfrac{125}{216}$ | $\tfrac{75}{216}$ | $\tfrac{15}{216}$ | $\tfrac{1}{216}$ |

$$E(X) = \Sigma \, xP(X = x)$$
$$= 0 \times \tfrac{125}{215} + 1 \times \tfrac{75}{216} + 2 \times \tfrac{15}{216} + 3 \times \tfrac{1}{216}$$
$$= 0.5$$

**The expected number of sixes when three dice are thrown is 0.5.**

*NOTE:* in 50 throws you would expect 25 sixes. In practice you may not get 25 sixes. In 5000 throws though, you may get very close to 2500 sixes. The expected value gives you the long-term average value.

# Symmetrical probability distributions

An important property which some distributions possess is that of symmetry, for example

(a)

| $x$ | 1 | 2 | 3 | 4 | 5 |
|---|---|---|---|---|---|
| $P(X = x)$ | 0.1 | 0.2 | 0.4 | 0.2 | 0.1 |

It can be seen from the table or from the vertical line graph that the distribution is symmetrical about the central value $X = 3$, so $E(X) = 3$.

*Check:* $E(X) = \sum_{\text{all } x} xP(X = x)$

$= 1 \times 0.1 + 2 \times 0.2 + 3 \times 0.4 + 4 \times 0.2 + 5 \times 0.1 = 3$

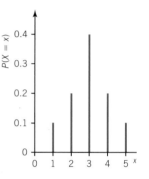

(b) If $X$ is the random variable 'the digit picked from a random number table', then the p.d.f. of $X$ is $P(X = x) = 0.1$ for $x = 0, 1, ..., 9$.

| $x$ | 0 | 1 | 2 | 3 | 4 | 5 | 6 | 7 | 8 | 9 |
|---|---|---|---|---|---|---|---|---|---|---|
| $P(X = x)$ | 0.1 | 0.1 | 0.1 | 0.1 | 0.1 | 0.1 | 0.1 | 0.1 | 0.1 | 0.1 |

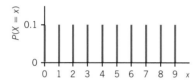

The distribution is symmetrical about the central value mid-way between 4 and 5 so $E(X) = 4.5$.

*NOTE:* the random variable $X$ with p.d.f. $P(X = x) = k$, for all possible values of $x$, where $k$ is a constant, is said to follow a **discrete uniform distribution**.

## Example 4.6

A fruit machine consists of three windows which operate independently. Each window shows pictures of fruits: lemons, apples, cherries or bananas. The probability that a window shows a particular fruit is as follows.

$P(\text{lemon}) = 0.4$       $P(\text{cherries}) = 0.2$

$P(\text{apple}) = 0.1$       $P(\text{banana}) = 0.3$

The rules for playing a game on the fruit machine are:

Only 10p to play!

You win £1

You win 50p

You win 40p

You win 80p

(In any order)

Find the expected gain or loss if you play a game.

## Solution 4.6

The variable $X$ is 'the amount **gained**, in pence in a game'.
Taking into account the cost of 10p to play, $X$ can take the values 90, 70, 40, 30, –10.

$P(X = 90)$ $= P(3 \text{ apples}) = 0.1 \times 0.1 \times 0.1 = 0.001$

$P(X = 70)$ $= P(2 \text{ apples and one with cherries, in any order})$
$= P(A, A, C) + P(A, C, A) + P(C, A, A)$
$= 3 \times 0.1^2 \times 0.2$
$= 0.006$

$P(X = 40)$ $= P(3 \text{ cherries}) = (0.2)^3 = 0.008$

$P(X = 30)$ $= P(3 \text{ lemons}) = (0.4)^3 = 0.064$

$P(X = -10) = P(\text{you win } none \text{ of these prizes})$
$= 1 - (0.001 + 0.006 + 0.008 + 0.064) = 0.921$

The probability distribution for $X$ is

| $x$ | 90 | 70 | 40 | 30 | –10 |
|---|---|---|---|---|---|
| $P(X = x)$ | 0.001 | 0.006 | 0.008 | 0.064 | 0.921 |

$E(X) = \sum_{\text{all } x} xP(X = x)$

$= 90 \times 0.001 + 70 \times 0.006 + 40 \times 0.008 + 30 \times 0.064 + (-10) \times 0.921$

$= -6.46$

The expected **loss** per turn is **6.46p.**

This means that, if you played the game, say 1000 times, on the average you could expect to lose £64.60!

## Example 4.7

A newsagent stocks 12 copies of a magazine each week. He has regular orders for nine copies, and the number of additional copies sold varies from week to week. The newsagent uses previous sales data to estimate the probability, for each possible total number of copies sold, as follows.

| Number of copies | 9 | 10 | 11 | 12 |
|---|---|---|---|---|
| Probability | 0.20 | 0.35 | 0.30 | 0.15 |

(a) Calculate an estimate of the mean number of copies that he sells in a week.
(b) The newsagent buys the magazines at 85p each and sells them at £1.45 each. Any copies not sold are destroyed.

    (i)   Find the profit on these magazines in a week when he sells 11 copies.
    (ii)  Construct a probability distribution table for the newsagent's weekly profit from the sale of these magazines. Hence, or otherwise, calculate an estimate of his mean weekly profit.

                                                                         (NEAB)

## Solution 4.7

(a) Let $X$ be the number of copies he sells in a week.
$$E(X) = \sum xP(X = x)$$
$$= 9 \times 0.20 + 10 \times 0.35 + 11 \times 0.30 + 12 \times 0.15$$
$$= 10.4$$
**An estimate of the mean number sold in a week is 10.4.**

(b) (i)   When he sells 11 copies, profit = $11 \times £1.45 - 12 \times £0.85 = £5.75$
    (ii)  When he sells  9 copies, profit =  $9 \times £1.45 - 12 \times £0.85 = £2.85$
           When he sells 10 copies, profit = $10 \times £1.45 - 12 \times £0.85 = £4.30$
           When he sells 12 copies, profit = $12 \times £1.45 - 12 \times £0.85 = £7.20$
           Let $£Y$ be the weekly profit. The probability distribution of $Y$ is

| $y$ | 2.85 | 4.30 | 5.75 | 7.20 |
|---|---|---|---|---|
| $P(Y = y)$ | 0.20 | 0.35 | 0.30 | 0.15 |

$$E(Y) = 2.85 \times 0.2 + 4.30 \times 0.35 + 5.75 \times 0.30 + 7.20 \times 0.15$$
$$= 4.88$$

**An estimate of his mean weekly profit is £4.88.**

## Example 4.8

In a game, three dice are thrown. If you get a one or a six on any of the dice, you win £1, otherwise you have to pay £5. How much would you expect to win or lose when you play nine games?

You are now given the opportunity to change the rule for the amount you win when a one or a six appears.

To make the game worthwhile to yourself, what is the minimum amount that you would suggest?

## Solution 4.8

When a die is thrown,

$$P(1 \text{ or } 6) = \tfrac{2}{6} = \tfrac{1}{3} \quad \text{so} \quad P(\text{neither 1 nor 6}) = 1 - \tfrac{1}{3} = \tfrac{2}{3}$$

When three dice are thrown,

$$P(\text{neither 1 nor 6 on all three}) = \left(\tfrac{2}{3}\right)^3 = \tfrac{8}{27}$$

so $P(1 \text{ or } 6 \text{ turns up}) = 1 - \tfrac{8}{27} = \tfrac{19}{27}$

Let £$X$ be the amount won in a game.

If a 1 or a 6 turns up, you **win** £1. Then $X = 1$ and $P(X = 1) = \tfrac{19}{27}$.
If neither a 1 nor a 6 turns up, you **pay** £5. Then $X = -5$ and $P(X = -5) = \tfrac{8}{27}$.

The probability distribution of $X$ is

| $x$ | $-5$ | 1 |
|---|---|---|
| $P(X = x)$ | $\tfrac{8}{27}$ | $\tfrac{19}{27}$ |

So
$$E(X) = \sum_{\text{all } x} xP(X = x)$$
$$= (-5) \times \frac{8}{27} + 1 \times \frac{19}{27}$$
$$= -\frac{7}{9}$$

The negative amount indicates that you would expect to make a **loss** of £$\tfrac{7}{9}$ per game.

After nine games, **expected loss** = £$\tfrac{7}{9}$ × 9 = **£7**.

In making the game worthwhile, you would obviously want to ensure that you didn't make a loss.

Change the rule for payment to £$y$ when you get 1 or 6 on any of the dice. The probability distribution is

| $x$ | $-5$ | $y$ |
|---|---|---|
| $P(X = x)$ | $\tfrac{8}{27}$ | $\tfrac{19}{27}$ |

$$E(X) = (-5) \times \frac{8}{27} + y \times \frac{19}{27}$$
$$= \frac{-40 + 19y}{27}$$

You want $E(X) > 0$

i.e. $\dfrac{-40 + 19y}{27} > 0$

$-40 + 19y > 0$

$19y > 40$

$y > 2.105\ldots$

The minimum amount you should suggest that you win if you get a one or a six is **£2.11**. To make the game worthwhile, perhaps suggest **£2.50**.

## Exercise 4b   Expectation

1. The probability distribution for the random variable $X$ is shown in the table:

| $x$ | 0 | 1 | 2 | 3 | 4 |
|-----|---|---|---|---|---|
| $P(X = x)$ | $\frac{1}{6}$ | $\frac{1}{12}$ | $\frac{1}{4}$ | $\frac{1}{3}$ | $\frac{1}{6}$ |

Find $E(X)$.

2. The random variable $X$ has p.d.f. $P(X = x)$ for $x = 5, 6, 7, 8, 9$ as defined in the table:

| $x$ | 5 | 6 | 7 | 8 | 9 |
|-----|---|---|---|---|---|
| $P(X = x)$ | $\frac{3}{11}$ | $\frac{2}{11}$ | $\frac{1}{11}$ | $\frac{2}{11}$ | $\frac{3}{11}$ |

Find $\mu$.

3. The probability distribution of a random variable $X$ is as shown in the table:

| $x$ | 1 | 2 | 3 | 4 | 5 |
|-----|---|---|---|---|---|
| $P(X = x)$ | 0.1 | 0.3 | $y$ | 0.2 | 0.1 |

Find (a) the value of $y$, (b) $E(X)$.

4. Find the expected number of heads when two fair coins are tossed.

5. A bag contains five black counters and six red counters. Two counters are drawn, one at a time, and not replaced. Let $X$ be 'the number of red counters drawn'. Find $E(X)$.

6. An unbiased tetrahedral die has faces marked 1, 2, 3, 4. If the die lands on the face marked 1, the player has to pay 10p.

   If it lands lands on a face marked with a 2 or a 4, the players wins 5p and if it lands on a 3, the player wins 3p. Find the expected gain in one throw.

7. A discrete random variable $X$ can take values 10 and 20 only. If $E(X) = 16$, write out the probability distribution of $X$.

8. The discrete random variable $X$ can take values 0, 1, 2 and 3 only. Given $P(X \leqslant 2) = 0.9$, $P(X \leqslant 1) = 0.5$ and $E(X) = 1.4$, find (a) $P(X = 1)$, (b) $P(X = 0)$.

9. 
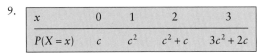

| $x$ | 0 | 1 | 2 | 3 |
|-----|---|---|---|---|
| $P(X = x)$ | $c$ | $c^2$ | $c^2 + c$ | $3c^2 + 2c$ |

The above table shows the probability distribution for a random variable $X$.

Calculate (a) $c$, (b) $E(X)$.   (L Additional)

10. A bag contains three red balls and one blue ball. A second bag contains one red ball and one blue ball. A ball is picked out of each bag and is then placed in the other bag. What is the expected number of red balls in the first bag?

11. In a game, a player rolls two balls down an inclined plane so that each ball finally settles in one of five slots and scores the number of points allotted to that slot as shown in the diagram below:

It is possible for both balls to settle in one slot and it may be assumed that each slot is equally likely to accept either ball.
The player's score is the sum of the points scored by each ball.
Draw up a table showing all the possible scores and the probability of each.
If the player pays 10p for each game and receives back a number of pence equal to his score, calculate the player's expected gain or loss per 50 games.   (C Additional)

12. In a game a player tosses three fair coins. He wins £10 if three heads occur, £$x$ if two heads occur, £3 if one head occurs and £2 if no heads occur. Express in terms of $x$ his expected gain from each game.
Given that he pays £4.50 to play each game, calculate

   (a) the value of $x$ for which the game is fair,
   (b) his expected gain or loss over 100 games if $x = 4.90$.   (C Additional)

13. In an examination a candidate is given the four answers to four questions but is not told which answer applies to which question. He is asked to write down each of the four answers next to its appropriate question.

   (a) Calculate in how many different ways he could write down the four answers.
   (b) Explain why it is impossible for him to have just three answers in the correct places and show that there are six ways of getting just two answers in the correct places.
   (c) If a candidate guesses at random where the four answers are to go and $X$ is the number of correct guesses he makes, draw up the probability distribution for $X$ in tabular form.
   (d) Calculate $E(X)$.   (L Additional)

14. The discrete random variable $X$ has p.d.f. $P(X = x) = kx$ for $x = 1, 2, 3, 4, 5$ where $k$ is constant. Find $E(X)$.

15. A woman has three keys on a ring, just one of which opens the front door. As she approaches the front door she selects one key after another at random without replacement. Draw a tree diagram to illustrate the various selections before she finds the correct key. Use this diagram to calculate the expected number of keys that she will use before opening the door.   (L Additional)

## THE EXPECTATION OF ANY FUNCTION OF $X$, $E(g(X))$

The definition of expectation can be extended to any function of $X$,
such as $10X$, $X^2$, $\dfrac{1}{X}$, $X - 4$, etc.

In general, if $g(X)$ is any function of the discrete random variable $X$, then

$$E(g(X)) = \sum_{\text{all } x} g(x) \times P(X = x)$$

**For example,**

$$E(10X) = \Sigma\, 10x P(X = x)$$
$$E(X^2) = \Sigma\, x^2 P(X = x)$$
$$E\left(\frac{1}{X}\right) = \Sigma\frac{1}{x} P(X = x)$$
$$E(X - 4) = \Sigma(x - 4)P(X = x)$$

### Example 4.9

The random variable $X$ has p.d.f. $P(X = x)$ for $x = 1, 2, 3$ as shown.

| $x$ | 1 | 2 | 3 |
|---|---|---|---|
| $P(X = x)$ | 0.1 | 0.6 | 0.3 |

Calculate

(a)  $E(X)$,          (b)  $E(3)$,          (c)  $E(5X)$,          (d)  $E(5X + 3)$.

### Solution 4.9

(a)  $E(X) = \Sigma\, x P(X = x)$
$\qquad\quad = 1 \times 0.1 + 2 \times 0.6 + 3 \times 0.3$
$\qquad\quad = \mathbf{2.2}$

(b)  $E(3)\ = \Sigma\, 3 P(X = x)$
$\qquad\quad = 3 \times 0.1 + 3 \times 0.6 + 3 \times 0.3$
$\qquad\quad = \mathbf{3}$

Notice that the expected value of a constant is equal to the constant.

(c)  $E(5X) = \Sigma\, 5x P(X = x)$
$\qquad\qquad = 5 \times 0.1 + 10 \times 0.6 + 15 \times 0.3$
$\qquad\qquad = \mathbf{11}$

Notice that $5E(X) = 5 \times 2.2 = 11$
so $E(5X) = 5E(X)$.

(d) $E(5X + 3) = \Sigma (5x + 3)P(X = x)$
$\qquad\qquad = 8 \times 0.1 + 13 \times 0.6 + 18 \times 0.3$
$\qquad\qquad = \mathbf{14}$

Notice that $E(5X) + E(3) = 11 + 3 = 14$

so $\quad E(5X + 3) = E(5X) + E(3)$

i.e. $\mathbf{E(5X + 3) = 5E(X) + 3}$

---

In general, for constants $a$ and $b$,

$E(a) = a$
$E(aX) = aE(X)$
$E(aX + b) = aE(X) + b$

## Example 4.10

A six-sided die has faces marked with the numbers 1, 3, 5, 7, 9 and 11. It is biased so that the probability of obtaining the number $R$ in a single roll of the die is proportional to $R$.

(a) Show that the probability distribution of $R$ is given by

$$P(R = r) = \frac{r}{36}, \qquad r = 1, 3, 5, 7, 9, 11.$$

(b) The die is to be rolled and a rectangle drawn with sides of lengths 6 cm and $R$ cm. Calculate the expected value of the area of the rectangle.

(c) The die is to be rolled again and a square drawn with sides of length $24R^{-1}$ cm. Calculate the expected value of the perimeter of the square. (*NEAB*)

## Solution 4.10

(a)

| $r$ | 1 | 3 | 5 | 7 | 9 | 11 |
|---|---|---|---|---|---|---|
| $P(R = r)$ | $k$ | $3k$ | $5k$ | $7k$ | $9k$ | $11k$ |

$\Sigma P(R = r) = 1 \qquad \therefore \quad k + 3k + 5k + 7k + 9k + 11k = 1$
$$\qquad\qquad\qquad\qquad\qquad\qquad 36k = 1$$
$$\qquad\qquad\qquad\qquad\qquad\qquad k = \frac{1}{36}$$

The distribution is

| $r$ | 1 | 3 | 5 | 7 | 9 | 11 |
|---|---|---|---|---|---|---|
| $P(R = r)$ | $\frac{1}{36}$ | $\frac{3}{36}$ | $\frac{5}{36}$ | $\frac{7}{36}$ | $\frac{9}{36}$ | $\frac{11}{36}$ |

$\therefore \quad P(R = r) = \dfrac{r}{36} \qquad$ for $r = 1, 3, 5, 7, 9, 11$

(b) $\qquad A = 6R$

$\therefore E(A) = E(6R)$

$\qquad\qquad = 6E(R)$

$E(R) = \Sigma rP(R = r)$

$\quad = 1 \times \frac{1}{36} + 3 \times \frac{3}{36} + 5 \times \frac{5}{36} + 7 \times \frac{7}{36} + 9 \times \frac{9}{36} + 11 \times \frac{11}{36}$

$\quad = 7\frac{17}{18}$

$E(A) = 6 \times 7\frac{17}{18} = 47\frac{2}{3}$

**The expected value of the area is $47\frac{2}{3}$ cm$^2$.**

(c) $\qquad P = 4 \times 24R^{-1} = \dfrac{96}{R}$

$24\ R^{-1}$

$\therefore E(P) = E\left(\dfrac{96}{R}\right)$

$\qquad = 96E\left(\dfrac{1}{R}\right)$

$E\left(\dfrac{1}{R}\right) = \Sigma \dfrac{1}{r} P(R = r)$

$\qquad = \dfrac{1}{1} \times \dfrac{1}{36} \;+\; \dfrac{1}{3} \times \dfrac{3}{36} \;+\; \dfrac{1}{5} \times \dfrac{5}{36} \;+\; \dfrac{1}{7} \times \dfrac{7}{36} \;+\; \dfrac{1}{9} \times \dfrac{9}{36} \;+\; \dfrac{1}{11} \times \dfrac{11}{36}$

$\qquad = \dfrac{1}{6}$

$E(P) = 96 \times \dfrac{1}{6} = 16$

**The expected value of the perimeter is 16 cm.**

---

## Example 4.11

$X$ is the number of heads obtained when two coins are tossed. Find

(a) the expected number of heads,
(b) $E(X^2)$,
(c) $E(X^2 - X)$.

## Solution 4.11

$P(X = 0) = P(T, T) = \frac{1}{2} \times \frac{1}{2} = \frac{1}{4}$

$P(X = 1) = P(T, H) + P(H, T) = \frac{1}{2} \times \frac{1}{2} + \frac{1}{2} \times \frac{1}{2} = \frac{1}{2}$

$P(X = 2) = P(H, H) = \frac{1}{2} \times \frac{1}{2} = \frac{1}{4}$

The probability distribution for $X$ is

| $x$ | 0 | 1 | 2 |
|---|---|---|---|
| $P(X = x)$ | $\frac{1}{4}$ | $\frac{1}{2}$ | $\frac{1}{4}$ |

(a) $E(X) = 1$ (by symmetry)

(b) $E(X^2) = \Sigma x^2 P(X = x)$

$\qquad = 0^2 \times \frac{1}{4} + 1^2 \times \frac{1}{2} + 2^2 \times \frac{1}{4}$

$\qquad = 1\frac{1}{2}$

(c) $E(X^2 - X) = \Sigma(x^2 - x)P(X = x)$

$\qquad = 0 \times \frac{1}{4} + 0 \times \frac{1}{2} + 2 \times \frac{1}{4}$

$\qquad = \frac{1}{2}$

Notice that $E(X^2) - E(X) = 1\frac{1}{2} - 1 = \frac{1}{2}$ and $E(X^2 - X) = \frac{1}{2}$,
so $E(X^2 - X) = E(X^2) - E(X)$

---

In general, for two functions of X, $g(x)$ and $h(x)$

$\qquad E(g(X) + h(X)) = E(g(X)) + E(h(X))$

For example

$$E\left(X^2 + \frac{1}{X}\right) = E(X^2) + E\left(\frac{1}{X}\right), \qquad E(3X - 4X^2) = 3E(X) - 4E(X^2)$$

## Example 4.12

The discrete random variable X has the following probability distribution.

| $x$ | 0 | 1 | 2 | 3 | 4 |
|---|---|---|---|---|---|
| $P(X = x)$ | 0.20 | 0.20 | 0.20 | 0.20 | 0.20 |

(a) Write down the name of the distribution of X.
(b) Find $P(0 \leqslant X < 2)$.
(c) Find $E(X)$.
(d) Find $E(X^2 + 3X)$. $\hfill$ (L)

## Solution 4.12

(a) It is a **discrete uniform distribution**.
(b) $P(0 \leqslant X < 2) = P(X = 0) + P(X = 1) = 0.2 + 0.2 = \mathbf{0.4}$
(c) By symmetry, $E(X) = \mathbf{2}$
(d) $E(X^2 + 3X) = E(X^2) + 3E(X)$
$\qquad E(X^2) = \Sigma x^2 P(X = x)$
$\qquad\qquad = 0.20 \times (0^2 + 1^2 + 2^2 + 3^2 + 4^2)$
$\qquad\qquad = 6$
so $E(X^2 + 3X) = 6 + 3E(X) = 6 + 6 = \mathbf{12}$

---

# VARIANCE OF X, VAR(X) OR V(X)

Remember that variance = (standard deviation)$^2$.

# Experimental approach

For a frequency distribution with mean $\bar{x}$, the variance $s^2$ is given by

$$s^2 = \frac{\Sigma f(x - \bar{x})^2}{\Sigma f}.$$

This can also be written

$$s^2 = \frac{\Sigma fx^2}{\Sigma f} - \bar{x}^2.$$

# Theoretical approach

For a discrete random variable $X$, with $E(X) = \mu$, the variance is defined as follows:

The variance of $X$ written $\text{Var}(X)$ is given by
$$\text{Var}(X) = E(X - \mu)^2$$

Alternatively, 
$$\begin{aligned}
\text{Var}(X) &= E(X - \mu)^2 \\
&= E(X^2 - 2\mu X + \mu^2) \\
&= E(X^2) - 2\mu E(X) + E(\mu^2) \\
&= E(X^2) - 2\mu^2 + \mu^2 \\
&= E(X^2) - \mu^2
\end{aligned}$$

$$\text{Var}(X) = E(X^2) - \mu^2$$

This format is usually easier to work with.

NOTE:   $\mu = E(X)$   so $\mu^2 = (E(X))^2$

This is very cumbersome to write, so it is often written $E^2(X)$. This is similar to the notation used in trigonometry where $(\cos A)^2$ is written $\cos^2 A$.

You could write   $\text{Var}(X) = E(X^2) - E^2(X)$

$\text{Var}(X)$ is sometimes written as $\sigma^2$ ($\sigma$ is pronounced 'sigma').

$$\sigma = \sqrt{\text{Var}(X)} = \text{standard deviation of } X$$

## Example 4.13

The random variable $X$ has probability distribution as shown in the table:

| $x$ | 1 | 2 | 3 | 4 | 5 |
|---|---|---|---|---|---|
| $P(X = x)$ | 0.1 | 0.3 | 0.2 | 0.3 | 0.1 |

Find
(a) $\mu = E(X)$,
(b) $E(X^2)$,
(c) $\text{Var}(X)$,
(d) $\sigma$, the standard deviation of $X$.

## Solution 4.13

(a) By symmetry, $\mu = E(X) = 3$

(b) $\begin{aligned}[t]
E(X^2) &= \Sigma x^2 P(X = x) \\
&= 1 \times 0.1 + 4 \times 0.3 + 9 \times 0.2 + 16 \times 0.3 + 25 \times 0.1 \\
&= 10.4
\end{aligned}$

(c) $\begin{aligned}[t]
\text{Var}(X) &= E(X^2) - \mu^2 \\
&= 10.4 - 9 \\
&= 1.4
\end{aligned}$

(d) $\begin{aligned}[t]
\sigma &= \sqrt{1.4} \\
&= 1.18 (2 \text{ d.p.})
\end{aligned}$

## Example 4.14

Two boxes each contain three cards. The first box contains cards labelled 1, 3 and 5; the second box contains cards labelled 2, 6 and 8. In a game, a player draws one card at random from each box and his score, $X$, is the sum of the numbers on the two cards.

(a) Obtain the six possible values of $X$ and find the corresponding probabilities.
(b) Calculate $E(X)$, $E(X^2)$ and the variance of $X$. *(C Additional)*

## Solution 4.14

(a)            *Possibility space*              *Probability distribution*

|  | | Second box | | |
|---|---|---|---|---|
| | | 2 | 6 | 8 |
| | 1 | 3 | 7 | 9 |
| First box | 3 | 5 | 9 | 11 |
| | 5 | 7 | 11 | 13 |

| $x$ | 3 | 5 | 7 | 9 | 11 | 13 |
|---|---|---|---|---|---|---|
| $P(X = x)$ | $\frac{1}{9}$ | $\frac{1}{9}$ | $\frac{2}{9}$ | $\frac{2}{9}$ | $\frac{2}{9}$ | $\frac{1}{9}$ |

(b) $E(X) = \Sigma\, x P(X = x)$

$\qquad = 3 \times \frac{1}{9} + 5 \times \frac{1}{9} + 7 \times \frac{2}{9} + 9 \times \frac{2}{9} + 11 \times \frac{2}{9} + 13 \times \frac{1}{9}$

$\qquad = 8\frac{1}{3}$

$E(X^2) = \Sigma\, x^2 P(X = x)$

$\qquad = 9 \times \frac{1}{9} + 25 \times \frac{1}{9} + 49 \times \frac{2}{9} + 81 \times \frac{2}{9} + 121 \times \frac{2}{9} + 169 \times \frac{1}{9}$

$\qquad = 78\frac{1}{3}$

$Var(X) = E(X^2) - E^2(X)$

$\qquad = 78\frac{1}{3} - 8\frac{1}{3}^2$

$\qquad = 8\frac{8}{9}$

---

The following results relating to variance are useful.

If $a$ and $b$ are any constants,

$\qquad Var(a) = 0$
$\qquad Var(aX) = a^2\, Var(X)$
$\qquad Var(aX + b) = a^2\, Var(X)$

For example

$\qquad Var(2X) = 2^2\, Var(X)$
$\qquad\qquad = 4\, Var(X)$
$\qquad Var(2X + 3) = 2^2\, Var(X)$
$\qquad\qquad = 4\, Var(X)$
$\qquad Var(5 - X) = (-1)^2\, Var(X)$
$\qquad\qquad = Var(X)$

# Exercise 4c   Expectation and variance

1. The discrete random variable $X$ has p.d.f.
   $P(X = x)$ for $x = 1, 2, 3$.

   | $x$ | 1 | 2 | 3 |
   |---|---|---|---|
   | $P(X = x)$ | 0.2 | 0.3 | 0.5 |

   Find (a) $E(X)$, (b) $E(X^2)$ (c) $Var(X)$.

2. The discrete random variable $X$ has the
   probability distribution specified in the following
   table.

   | $x$ | −1 | 0 | 1 | 2 |
   |---|---|---|---|---|
   | $P(X = x)$ | 0.25 | 0.10 | 0.45 | 0.20 |

   (a)  Find $P(-1 \leqslant X < 1)$.
   (b)  Find $E(2X + 3)$.

3. The discrete random variable $X$ has p.d.f.
   $P(X = 0) = 0.05$, $P(X = 1) = 0.45$
   $P(X = 2) = 0.5$. Find
   (a) $\mu = E(X)$,   (b) $E(X^2)$,   (c) $E(5X^2 + 2X - 3)$.

4. The discrete random variable $X$ has p.d.f.
   $P(X = x) = k$ for $x = 1, 2, 3, 4, 5, 6$. Find
   (a) $E(X)$,   (b) $E(X^2)$,
   (c) $E(3X + 4)$,   (d) $Var(X)$.

5. The random variable $X$ takes values 2, 4, 6, 8,
   and its probability distribution is represented in
   the vertical line graph.

   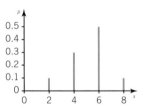

   Find $Var(X)$.

6. A roulette wheel is divided into six sectors of
   unequal area, marked with the numbers 1, 2, 3,
   4, 5, and 6. The wheel is spun and $X$ is the
   random variable 'the number on which the wheel
   stops'. The probability distribution of $X$ is as
   follows:

   | $x$ | 1 | 2 | 3 | 4 | 5 | 6 |
   |---|---|---|---|---|---|---|
   | $P(X = x)$ | $\frac{1}{16}$ | $\frac{3}{16}$ | $\frac{1}{4}$ | $\frac{1}{4}$ | $\frac{3}{16}$ | $\frac{1}{16}$ |

   Calculate (a) $E(X)$,   (b) $E(X^2)$,   (c) $E(3X - 5)$,
   (d) $E(6X^2)$,   (e) $Var(X)$.

7. The random variable $X$ has p.d.f. $P(X = x)$ as
   shown in the table:

   | $x$ | −2 | −1 | 0 | 1 | $c$ |
   |---|---|---|---|---|---|
   | $P(X = x)$ | 0.1 | 0.1 | 0.3 | 0.4 | 0.1 |

   Find the value of $c$ (a) if $E(X) = 0.3$, (b) if
   $E(X^2) = 1.8$.

8. The discrete random variable $X$ has probability
   function given by
   $$p(x) = \begin{cases} (\frac{1}{2})^x & x = 1, 2, 3, 4, 5, \\ c & x = 6, \\ 0 & \text{otherwise,} \end{cases}$$
   where $c$ is a constant.
   Determine the value of $c$ and hence the mode and
   mean of $X$.                                    (L)

9. A game consists of tossing four unbiased coins
   simultaneously. The total score is calculated by
   giving three points for each head and one point
   for each tail. The random variable $X$ represents
   the total score.

   (a)  Show that $P(X = 8) = \frac{3}{8}$.
   (b)  Copy and complete the table, given below,
        for the symmetrical probability distribution
        of $X$.

   | $x$ | 4 | 6 | 8 | 10 | 12 |
   |---|---|---|---|---|---|
   | $P(X = x)$ | | | $\frac{3}{8}$ | | |

   (c)  Calculate the variance of $X$.        (NEAB)

10. Find $Var(X)$ for each of the following probability
    distributions:

    (a)

    | $x$ | −3 | −2 | 0 | 2 | 3 |
    |---|---|---|---|---|---|
    | $P(X = x)$ | 0.3 | 0.3 | 0.2 | 0.1 | 0.1 |

    (b)

    | $x$ | 1 | 3 | 5 | 7 | 9 |
    |---|---|---|---|---|---|
    | $P(X = x)$ | $\frac{1}{6}$ | $\frac{1}{4}$ | $\frac{1}{6}$ | $\frac{1}{4}$ | $\frac{1}{6}$ |

    (c)

    | $x$ | 0 | 2 | 5 | 6 |
    |---|---|---|---|---|
    | $P(X = x)$ | 0.11 | 0.35 | 0.46 | 0.08 |

11. $X$ is the random variable 'the number on a
    biased die', and the p.d.f. of $X$ is as shown,

    | $x$ | 1 | 2 | 3 | 4 | 5 | 6 |
    |---|---|---|---|---|---|---|
    | $P(X = x)$ | $\frac{1}{6}$ | $\frac{1}{6}$ | $\frac{1}{5}$ | $y$ | $\frac{1}{5}$ | $\frac{1}{6}$ |

    Find (a) the value of $y$,   (b) $E(X)$,   (c) $E(X^2)$,
    (d) $Var(X)$,   (e) $Var(4X)$.

12. A team of three is to be chosen from four boys and five girls. If $X$ is the random variable 'the number of girls in the team', find (a) $E(X)$, (b) $E(X^2)$, (c) $Var(X)$.

13. Two discs are drawn without replacement from a box containing three red and four white discs. If $X$ is the random variable 'the number of white discs drawn', construct a probability distribution table. Find (a) $E(X)$, (b) $E(X^2)$, (c) $Var(X)$, (d) $Var(3X-4)$.

14. If $X$ is the random variable 'the sum of the scores on two tetrahedral dice', where the score is the number on which the die lands, find (a) $E(X)$, (b) $Var(X)$, (c) $Var(2X)$, (d) $Var(2X+3)$.

15. The discrete random variable $X$ has probability distribution as shown in the table. Find $Var(2X+3)$.

| $x$ | 10 | 20 | 30 |
|---|---|---|---|
| $P(X=x)$ | 0.1 | 0.6 | 0.3 |

16. Two discs are drawn, without replacement, from a box containing three red discs and four white discs. The discs are drawn at random. If $X$ is the random variable 'the number of red discs drawn', find (a) the expected number of red discs, (b) the standard deviation of $X$.

17. Ten identically shaped discs are in a bag; two of them are black, the rest white. Discs are drawn at random from the bag in turn and not replaced. Let $X$ be the number of discs drawn up to and including the first black one.

List the values of $X$ and the associated theoretical probabilities.
Calculate the mean value of $X$ and its standard deviation. What is the most likely value of $X$?
If, instead, each disc is replaced before the next is drawn, construct a similar list of values and point out the chief differences between the two lists.

18. The discrete random variable $X$ has p.d.f.
$$P(X=x) = k|x|$$
where $x$ takes the values $-3, -2, -1, 0, 1, 2, 3$.
Find (a) the value of the constant $k$,
(b) $E(X)$,
(c) $E(X^2)$
(d) the standard deviation of $X$.

19. The random variable $X$ takes integer values only and has p.d.f.
$$P(X=x) = kx \qquad x = 1, 2, 3, 4, 5$$
$$P(X=x) = k(10-x) \qquad x = 6, 7, 8, 9$$
Find
(a) the value of the constant $k$, (b) $E(X)$,
(c) $Var(X)$, (d) $E(2X-3)$, (e) $Var(2X-3)$.

20. (a) In a game a player pays £5 to toss three fair coins. Depending on the number of tails he obtains he receives a sum of money as shown in the table below.

| Number of tails | 3 | 2 | 1 | 0 |
|---|---|---|---|---|
| Sum received | £10 | £6 | £3 | £1 |

Calculate the player's expected gain or loss over 12 games.

(b) A variable $X$ has a probability distribution shown in the table below.

| Value of $X$ | 10 | 20 | 50 | 100 |
|---|---|---|---|---|
| Probability | 0.5 | 0.3 | $p$ | $q$ |

Given that $X$ can only take the values 10, 20, 50 or 100, and that $E(X) = 25$, calculate
(i) the value of $p$ and of $q$.
(ii) the variance of $X$.

In a fairground game, a player rolls discs on to a board containing squares, each of which bears one of the numbers, 10, 20, 50 or 100. If a disc falls entirely within a square, the player receives the same number of pence as the number in the square; if it does not, the player does not receive anything. The probability that a player will receive money from any given roll is $\frac{1}{4}$. If a player does receive money, the probabilities of receiving 10p, 20p, 50p or £1 are the same as those connected with the values of $X$ above. How many discs should a player be allowed to roll for 50p, if the game is to be fair? (C Additional)

21. (a) A man takes part in a game in which he throws two fair dice and scores the sum of two numbers shown. The rewards for the scores are given in the following table.

| Score | 12 | 10 | 7 | 5 | other |
|---|---|---|---|---|---|
| Reward (£) | 16 | 6 | 3 | 5 | 0 |

Calculate the expected reward for a throw of the two dice.

(b) A bag contains five identical discs, two of which are marked with the letter $A$ and three with the letter $B$. The discs are randomly drawn, one at a time without replacement, until both discs marked $A$ are obtained. Show that the probability that three draws are required is $\frac{2}{10}$.
Given that $X$ denotes the number of draws required to obtain both discs marked $A$, copy and complete the following table.

| Value of $X$ | 2 | 3 | 4 | 5 |
|---|---|---|---|---|
| Probability of $X$ | | $\frac{2}{10}$ | | |

Evaluate (i) $E(X)$, (ii) $E(X^2)$
(iii) the variance of $X$. (C Additional)

# THE CUMULATIVE DISTRIBUTION FUNCTION, $F(x)$

In a frequency distribution, the cumulative frequencies are obtained by summing all the frequencies up to a particular value.

In the same way, in a probability distribution, the probabilities up to a certain value are summed to give a cumulative probability. The cumulative probability function is written $F(x)$.

Consider the following probability distribution

| $x$ | 1 | 2 | 3 | 4 | 5 |
|---|---|---|---|---|---|
| $P(X = x)$ | 0.05 | 0.4 | 0.3 | 0.15 | 0.1 |

$F(1) = P(X \leqslant 1) = 0.05$
$F(2) = P(X \leqslant 2) = P(X = 1) + P(X = 2) = 0.05 + 0.4 = 0.45$
$F(3) = P(X \leqslant 3) = 0.75$
$F(4) = P(X \leqslant 4) = 0.9$
$F(5) = P(X \leqslant 5) = 1$

Notice that $F(5)$ give the *total* probability.

The **cumulative distribution function** is

| $x$ | 1 | 2 | 3 | 4 | 5 |
|---|---|---|---|---|---|
| $F(x)$ | 0.05 | 0.45 | 0.75 | 0.9 | 1 |

In general, for the discrete random variable $X$,
the cumulative distribution function is given by $F(x)$ where

$$F(x) = P(X \leqslant x)$$

Sometimes $F(x)$ can be given by a formula as in the following example.

## Example 4.15

The discrete random variable $X$ has cumulative distribution function $F(x) = \dfrac{x}{6}$ for $x = 1, 2, ..., 6$. Write out the probability distribution and suggest what $X$ represents.

## Solution 4.15

The cumulative distribution function is

| $x$ | 1 | 2 | 3 | 4 | 5 | 6 |
|---|---|---|---|---|---|---|
| $F(x)$ | $\frac{1}{6}$ | $\frac{2}{6}$ | $\frac{3}{6}$ | $\frac{4}{6}$ | $\frac{5}{6}$ | 1 |

You can find the probability distribution from the table.

$P(X = 1) = \frac{1}{6}$
$P(X = 2) = F(2) - F(1) = \frac{2}{6} - \frac{1}{6} = \frac{1}{6}$
$P(X = 3) = F(3) - F(2) = \frac{3}{6} - \frac{2}{6} = \frac{1}{6}$   and so on
The probability distribution is

| $x$ | 1 | 2 | 3 | 4 | 5 | 6 |
|---|---|---|---|---|---|---|
| $P(X = x)$ | $\frac{1}{6}$ | $\frac{1}{6}$ | $\frac{1}{6}$ | $\frac{1}{6}$ | $\frac{1}{6}$ | $\frac{1}{6}$ |

This is the uniform distribution, $P(X = x) = \frac{1}{6}$, $x = 1, 2, ..., 6$.
$X$ could be the score when a die is thrown.

## Example 4.16

For a discrete random variable $X$ the cumulative distribution function $F(x)$ is as shown:

| $x$ | 1 | 2 | 3 | 4 | 5 |
|---|---|---|---|---|---|
| $F(x)$ | 0.2 | 0.32 | 0.67 | 0.9 | 1 |

Find (a) $P(X = 3)$,    (b) $P(X > 2)$.

## Solution 4.16

(a)  From the table,

$$F(3) = P(X \leqslant 3) = P(X = 1) + P(X = 2) + P(X = 3) = 0.67$$
$$F(2) = P(X \leqslant 2) = P(X = 1) + P(X = 2) = 0.32$$

$$P(X = 3) = F(3) - F(2)$$
$$= 0.67 - 0.32 = \mathbf{0.35}$$

(b)  $P(X > 2) = 1 - P(X \leqslant 2)$
$$= 1 - F(2)$$
$$= 1 - 0.32$$
$$= \mathbf{0.68}$$

## Example 4.17

The cumulative probabilities for a random variable $X$ are given in the following table, where $X$ takes the values 0, 1, 2, ..., 10.

| $x$ | $F(x)$ |
|---|---|
| 0 | 0.0388 |
| 1 | 0.1756 |
| 2 | 0.4049 |
| 3 | 0.6477 |
| 4 | 0.8298 |
| 5 | 0.9327 |
| 6 | 0.9781 |
| 7 | 0.9941 |
| 8 | 0.9987 |
| 9 | 0.9998 |
| 10 | 1 |

Use the table to find
(a) $P(X \leqslant 5)$,
(b) $P(X > 3)$,
(c) $P(3 \leqslant X \leqslant 7)$,
(d) $P(X = 7)$,
(e) $P(X \geqslant 8)$.

## Solution 4.17

(a)  $P(X \leqslant 5) = F(5) = \mathbf{0.9327}$

(b)  $P(X > 3) = 1 - P(X \leqslant 3) = 1 - 0.6477 = \mathbf{0.3523}$

(c)  $P(3 \leqslant X \leqslant 7) = F(7) - F(2) = 0.9941 - 0.4049 = \mathbf{0.5892}$

(d)  $P(X = 7) = P(X \leqslant 7) - P(X \leqslant 6) = 0.9941 - 0.9781 = \mathbf{0.016}$

(e)  $P(X \geqslant 8) = 1 - P(X \leqslant 7) = 1 - F(7) = 1 - 0.9941 = \mathbf{0.0059}$

# Exercise 4d   Cumulative distribution function

1. The probability distribution for the random variable $Y$ is shown in the table:

| $y$ | 0.1 | 0.2 | 0.3 | 0.4 | 0.5 |
|---|---|---|---|---|---|
| $P(Y=y)$ | 0.05 | 0.25 | 0.3 | 0.15 | 0.25 |

   Construct the cumulative distribution table.

2. For a discrete random variable $R$ the cumulative distribution function $F(r)$ is as shown in the table:

| $r$ | 1 | 2 | 3 | 4 |
|---|---|---|---|---|
| $F(r)$ | 0.13 | 0.54 | 0.75 | 1 |

   Find (a) $P(R = 2)$,   (b) $P(R > 1)$,   (c) $P(R \geqslant 3)$,
   (d) $P(R < 2)$,   (e) $E(R)$.

3. Construct the cumulative distribution tables for the following discrete random variables:
   (a) the number of sixes obtained when two ordinary dice are thrown,
   (b) the smaller number when two ordinary dice are thrown,
   (c) the number of heads when three fair coins are tossed.

4. For the discrete random variable $X$ the cumulative distribution function $F(x)$ is as shown:

| $x$ | 3 | 4 | 5 | 6 | 7 |
|---|---|---|---|---|---|
| $F(x)$ | 0.01 | 0.23 | 0.64 | 0.86 | 1 |

   Construct the probability distribution of $X$, and find $Var(X)$.

5. For a discrete random variable $X$ the cumulative distribution function is given by

   $$F(x) = \frac{x^2}{9} \text{ for } x = 1, 2, 3.$$

   (a) Find $F(2)$.
   (b) Find $P(X = 2)$.
   (c) Write out the probability distribution of $X$.
   (d) Find $E(2X - 3)$.

6. For a discrete random variable $X$ the cumulative distribution function is given by $F(x) = kx$, $x = 1, 2, 3$. Find (a) the value of the constant $k$, (b) $P(X < 3)$, (c) the probability distribution of $X$, (d) the standard deviation of $X$.

7. The discrete random variable $X$ has distribution function $F(x)$ where

   $$F(x) = 1 - (1 - \tfrac{1}{4}x)^x \quad \text{for } x = 1, 2, 3, 4$$

   (a) Show that $F(3) = \frac{63}{64}$ and $F(2) = \frac{3}{4}$.
   (b) Obtain the probability distribution of $X$.
   (c) Find $E(X)$ and $Var(X)$.
   (d) Find $P(X > E(X))$.

8. The cumulative probabilities for $X$ are given in the following table, where $X$ takes the values $0, 1, 2, \ldots 12$.

| $x$ | $F(x)$ |
|---|---|
| 0 | 0.0115 |
| 1 | 0.0692 |
| 2 | 0.2061 |
| 3 | 0.4114 |
| 4 | 0.6296 |
| 5 | 0.8042 |
| 6 | 0.9133 |
| 7 | 0.9679 |
| 8 | 0.9900 |
| 9 | 0.9974 |
| 10 | 0.9994 |
| 11 | 0.9999 |
| 12 | 1 |

   Use the table to find

   (a) $P(X \leqslant 8)$,
   (b) $P(X = 5)$,
   (c) $P(X \geqslant 4)$,
   (d) $P(3 < X \leqslant 7)$,
   (e) $P(1 \leqslant X < 9)$.

## TWO INDEPENDENT RANDOM VARIABLES

If $X$ and $Y$ are any two random variables, then

$$E(X + Y) = E(X) + E(Y)$$

If $X$ and $Y$ are *independent* random variables, then

$$\text{Var}(X + Y) = \text{Var}(X) + \text{Var}(Y)$$

To illustrate this, consider two independent random variables $X$ and $Y$.

| $x$ | 0 | 1 | 2 |
|---|---|---|---|
| $P(X = x)$ | 0.1 | 0.5 | 0.4 |

| $y$ | 1 | 2 | 3 |
|---|---|---|---|
| $P(Y = y)$ | 0.3 | 0.2 | 0.5 |

$$
\begin{aligned}
E(X) &= \Sigma\, xP(X = x) \\
&= 0 \times 0.1 + 1 \times 0.5 + 2 \times 0.4 \\
&= 1.3 \\
E(X^2) &= \Sigma\, x^2 P(X = x) \\
&= 0^2 \times 0.1 + 1^2 \times 0.5 + 2^2 \times 0.4 \\
&= 2.1 \\
\text{Var}(X) &= E(X^2) - E^2(X) \\
&= 2.1 - 1.3^2 \\
&= 0.41
\end{aligned}
$$

$$
\begin{aligned}
E(Y) &= \Sigma\, yP(Y = y) \\
&= 1 \times 0.3 + 2 \times 0.2 + 3 \times 0.5 \\
&= 2.2 \\
E(Y^2) &= \Sigma\, y^2 P(Y = y) \\
&= 1^2 \times 0.3 + 2^2 \times 0.2 + 3^2 \times 0.5 \\
&= 5.6 \\
\text{Var}(Y) &= E(Y^2) - E^2(Y) \\
&= 5.6 - 2.2^2 \\
&= 0.76
\end{aligned}
$$

Notice that

$$E(X) + E(Y) = 1.3 + 2.2 = \mathbf{3.5} \qquad \text{...①}$$
$$\text{Var}(X) + \text{Var}(Y) = 0.41 + 0.76 = \mathbf{1.17} \qquad \text{...②}$$

Now consider the distribution $X + Y$ where $X + Y$ can take the values 1, 2, 3, 4, 5. For example,

$$
\begin{aligned}
P(X + Y = 4) &= P(X = 1 \text{ and } Y = 3) + P(X = 2 \text{ and } Y = 2) \\
&= 0.5 \times 0.5 \qquad\quad + 0.4 \times 0.2 \\
&= 0.33
\end{aligned}
$$

A tree diagram shows all the outcomes:

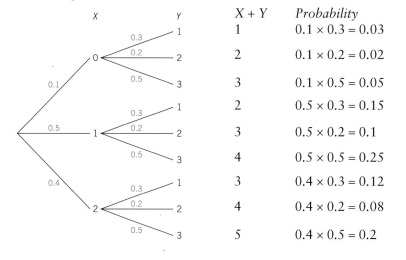

| | | | $X + Y$ | *Probability* |
|---|---|---|---|---|
| | | 1 | 1 | $0.1 \times 0.3 = 0.03$ |
| | | 2 | 2 | $0.1 \times 0.2 = 0.02$ |
| | | 3 | 3 | $0.1 \times 0.5 = 0.05$ |
| | | 1 | 2 | $0.5 \times 0.3 = 0.15$ |
| | | 2 | 3 | $0.5 \times 0.2 = 0.1$ |
| | | 3 | 4 | $0.5 \times 0.5 = 0.25$ |
| | | 1 | 3 | $0.4 \times 0.3 = 0.12$ |
| | | 2 | 4 | $0.4 \times 0.2 = 0.08$ |
| | | 3 | 5 | $0.4 \times 0.5 = 0.2$ |

The probability distribution for $X + Y$ is

| $x + y$ | 1 | 2 | 3 | 4 | 5 |
|---|---|---|---|---|---|
| Probability | 0.03 | 0.17 | 0.27 | 0.33 | 0.2 |

$$E(X + Y) = 1 \times 0.03 + 2 \times 0.17 + 3 \times 0.27 + 4 \times 0.33 + 5 \times 0.2$$
$$= 3.5$$
$$= E(X) + E(Y) \text{ from } ① \quad \text{So } E(X + Y) = E(X) + E(Y)$$

To find the variance, consider first $E((X + Y)^2)$

$$E((X + Y)^2) = 1 \times 0.03 + 4 \times 0.17 + 9 \times 0.27 + 16 \times 0.33 + 25 \times 0.2$$
$$= 13.42$$
$$\text{Var}(X + Y) = 13.42 - 3.5^2$$
$$= 1.17$$
$$= \text{Var}(X) + \text{Var}(Y) \text{ from } ② \quad \text{So } \mathbf{Var}(X + Y) = \mathbf{Var}(X) + \mathbf{Var}(Y)$$

If you perform similar calculations to find the expectation and variance of $X - Y$, you will find that

$$E(X - Y) = E(X) - E(Y) \text{ but Var}(X - Y) = \text{Var}(X) + \text{Var}(Y)$$

↑<br>Notice the<br>+ sign here.

The following general results are useful:

In general, for random variables $X$ and $Y$ and constants $a$ and $b$

$$E(aX + bY) = aE(X) + bE(Y)$$
$$E(aX - bY) = aE(X) - bE(Y)$$

If $X$ and $Y$ are *independent*, then

$$\text{Var}(aX + bY) = a^2 \text{Var}(X) + b^2 \text{Var}(Y)$$
$$\text{Var}(aX - bY) = a^2 \text{Var}(X) + b^2 \text{Var}(Y)$$

↑<br>Notice the + sign here.

## Example 4.18

$X$ and $Y$ are independent random variables such that

$$E(X) = 10, \quad \text{Var}(X) = 2, \quad E(Y) = 8, \quad \text{Var}(Y) = 3.$$

Find (a) $E(5X + 4Y)$, (b) $\text{Var}(5X + 4Y)$, (c) $\text{Var}(\frac{1}{2}X - Y)$, (d) $\text{Var}(\frac{1}{2}X + Y)$.

## Solution 4.18

(a) $E(5X + 4Y) = 5E(X) + 4E(Y) = 5 \times 10 + 4 \times 8 = \mathbf{82}$

(b) $\text{Var}(5X + 4Y) = 5^2\text{Var}(X) + 4^2\text{Var}(Y) = 25 \times 2 + 16 \times 3 = \mathbf{98}$

(c) $\text{Var}(\frac{1}{2}X - Y) = (\frac{1}{2})^2 \text{Var}(X) + \text{Var}(Y) = \frac{1}{4} \times 2 + 3 = \mathbf{3.5}$

(d) $\text{Var}(\frac{1}{2}X + Y) = (\frac{1}{2})^2 \text{Var}(X) + \text{Var}(Y) = \frac{1}{4} \times 2 + 3 = \mathbf{3.5}$

# DISTRIBUTION OF $X_1 + X_2 + \cdots + X_n$

Consider the random variable $X$ where $E(X) = \mu$ and $Var(X) = \sigma^2$.
Take two observations $X_1$ and $X_2$ from $X$.

$$E(X_1) = \mu, \quad E(X_2) = \mu, \quad Var(X_1) = \sigma^2, \quad Var(X_2) = \sigma^2.$$

$$E(X_1 + X_2) = E(X_1) + E(X_2) = \mu + \mu = 2\mu = 2E(X)$$

If the observations are independent

$$Var(X_1 + X_2) = Var(X_1) + Var(X_2) = \sigma^2 + \sigma^2 = 2\sigma^2 = 2\ Var(X)$$

This result can be extended to $n$ observations.

$$E(X_1 + X_2) = 2E(X)$$
$$E(X_1 + X_2 + \cdots + X_n) = nE(X)$$

If the observations are independent

$$Var(X_1 + X_2) = 2\ Var(X)$$
$$Var(X_1 + X_2 + \cdots + X_n) = n\ Var(X)$$

## Example 4.19

Find the expectation and variance of the number of heads obtained when six coins are tossed.

## Solution 4.19

Let $X$ be the number of heads when a coin is tossed, where $X$ can take the values 0, 1.
First find the expectation and variance of $X$. The probability distribution is

| $x$ | 0 | 1 |
|---|---|---|
| $P(X = x)$ | 0.5 | 0.5 |

$$E(X) = 0.5 \quad \text{(by symmetry)}$$
$$E(X^2) = 1 \times 0.5 = 0.5$$
so $\quad Var(X) = E(X^2) - E^2(X) = 0.5 - 0.5^2 = 0.25$

Now consider $Y = X_1 + X_2 + \cdots + X_6$ where $Y$ is the number of heads when six heads are tossed.

$$E(Y) = 6E(X) \qquad\qquad Var(Y) = 6\ Var(X)$$
$$= 6(0.5) \qquad\qquad\qquad = 6(0.25)$$
$$= 3 \qquad\qquad\qquad\qquad = 1.5$$

**The expected number of heads is 3 and the variance is 1.5.**

## COMPARING THE DISTRIBUTIONS OF $X_1 + X_2$ and $2X$

Confusion sometimes arises between the random variable $X_1 + X_2$, where $X_1$, $X_2$ are two independent observations of $X$, and the random variable $2X$.

You will see from the following example that the distributions of the two random variables, $X_1 + X_2$ and $2X$ are very different.

### Example 4.20

When a tetrahedral die is thrown, the number on the face on which it lands, $X$, has probability distribution as shown, with $E(X) = 2.5$ and $\text{Var}(X) = 1.25$.

| $x$ | 1 | 2 | 3 | 4 |
|---|---|---|---|---|
| $P(X = x)$ | 0.25 | 0.25 | 0.25 | 0.25 |

(a) Find the probability distribution of $S$, the sum of the two numbers obtained when the die is thrown twice, where $S = X_1 + X_2$ and illustrate it by drawing a vertical line graph. Find $E(S)$ and $\text{Var}(S)$.

(b) Find the probability distribution of $D$, where $D$ is double the number on which the die lands when it is thrown once. Illustrate by drawing a vertical line graph. Find $E(D)$ and $\text{Var}(D)$.

### Solution 4.20

(a) Consider the sum when the die is thrown twice and illustrate the outcomes on a possibility space diagram.

$S$ can take the values 2, 3, 4, 5, 6, 7, 8 and the outcomes (all equally likely) are shown in the diagram:

Second throw, $X_2$

| 4 | 5 | 6 | 7 | 8 |
| 3 | 4 | 5 | 6 | 7 |
| 2 | 3 | 4 | 5 | 6 |
| 1 | 2 | 3 | 4 | 5 |
|   | 1 | 2 | 3 | 4 |

First throw, $X_1$

$S = X_1 + X_2$

The probability distribution of $S$ is:

| $s$ | 2 | 3 | 4 | 5 | 6 | 7 | 8 |
|---|---|---|---|---|---|---|---|
| $P(S = s)$ | $\frac{1}{16}$ | $\frac{2}{16}$ | $\frac{3}{16}$ | $\frac{4}{16}$ | $\frac{3}{16}$ | $\frac{2}{16}$ | $\frac{1}{16}$ |

$E(S) = 5.$      (by symmetry)

$E(S^2) = \Sigma s^2 P(S = s)$

$= \frac{1}{16}(4 + 18 + 48 + 100 + 108 + 98 + 64)$

$= 27.5$

$$Var(S) = E(S^2) - E^2(S)$$
$$= 27.5 - 25$$
$$= \mathbf{2.5}$$

As expected,

$$E(S) = E(X_1 + X_2) = 2E(X) = 5$$
$$Var(S) = Var(X_1 + X_2) = 2\,Var(X) = 2.5$$

(b) $D$ is double the number on which the die lands, so $D = 2X$. The probability distribution for $D$ is

| $d$ | 2 | 4 | 6 | 8 |
|---|---|---|---|---|
| $P(D = d)$ | 0.25 | 0.25 | 0.25 | 0.25 |

$$E(D) = 5 \qquad \text{(by symmetry)}$$
$$E(D^2) = \Sigma\, d^2 P(D = d)$$
$$= 0.25(4 + 16 + 36 + 64)$$
$$= 30$$
$$Var(D) = E(D^2) - E^2(D)$$
$$= 30 - 25$$
$$= 5$$

As expected,

$$E(D) = E(2X) = 2E(X) = 5$$
$$Var(D) = Var(2X) = 4\,Var(X) = 5$$

Although the means of the two distributions are the same, the variances are not. The variable 'double the number' has the greater variance.

Summarising these results:

| | Multiples | Sums |
|---|---|---|
| For two observations | $E(2X) = 2E(X)$<br>$Var(2X) = 4\,Var(X)$ | $E(X_1 + X_2) = 2E(X)$<br>$Var(X_1 + X_2) = 2\,Var(X)$ |
| For $n$ observations | $E(nX) = nE(X)$<br>$Var(nX) = n^2\,Var(X)$ | $E(X_1 + X_2 + \cdots + X_n) = nE(X)$<br>$Var(X_1 + X_2 + \cdots + X_n) = n\,Var(X)$ |

**It is important that you understand whether multiples or sums are being considered. Think carefully about this point.**

# Exercise 4e    Combinations of random variables

1. Independent random variables $X$ and $Y$ are such that $E(X) = 4$, $E(Y) = 5$, $\text{Var}(X) = 1$, $\text{Var}(Y) = 2$. Find
   (a)  $E(4X + 2Y)$,
   (b)  $E(5X - Y)$,
   (c)  $\text{Var}(3X + 2Y)$,
   (d)  $\text{Var}(5Y - 3X)$,
   (e)  $\text{Var}(3X - 5Y)$.

2. Independent random variables $X$ and $Y$ are such that $E(X^2) = 14$, $E(Y^2) = 20$, $\text{Var}(X) = 10$, $\text{Var}(Y) = 11$. Find
   (a) $E(3X - 2Y)$, (b) $\text{Var}(5X - 2Y)$.

3. Independent random variables $X$ and $Y$ are such that $E(X) = 3$, $E(X^2) = 12$, $E(Y) = 4$, $E(Y^2) = 18$. Find the value of
   (a)  $E(3X - 2Y)$,
   (b)  $E(2Y - 3X)$,
   (c)  $E(6X + 4Y)$,
   (d)  $\text{Var}(2X - Y)$,
   (e)  $\text{Var}(2X + Y)$,
   (f)  $\text{Var}(3Y + 2X)$.

4. Independent random variables $X$ and $Y$ have probability distributions as shown in the tables:

| $x$ | 0 | 1 | 2 | 3 |
|---|---|---|---|---|
| $P(X = x)$ | 0.3 | 0.2 | 0.4 | 0.1 |

| $y$ | 0 | 1 | 2 |
|---|---|---|---|
| $P(Y = y)$ | 0.4 | 0.2 | 0.4 |

   (a)  Find $E(X)$, $E(Y)$, $\text{Var}(X)$, $\text{Var}(Y)$.
   (b)  Construct the probability distribution for the random variable $X + Y$.
   (c)  Verify that $E(X + Y) = E(X) + E(Y)$.
   (d)  Verify that $\text{Var}(X + Y) = \text{Var}(X) + \text{Var}(Y)$.

   (e)  Construct the probability distribution for the random variable $X - Y$.
   (f)  Verify that $E(X - Y) = E(X) - E(Y)$.
   (g)  Verify that $\text{Var}(X - Y) = \text{Var}(X) + \text{Var}(Y)$.

5. Rods of length 2 m or 3 m are selected at random with probabilities 0.4 and 0.6 respectively.

   (a)  Find the expectation and variance of the length of a rod.
   (b)  Two lengths are now selected at random. Find the expectation and variance of the sum of the two lengths.
   (c)  Three lengths are now selected at random. Show that the probability distribution of $Y$, the sum of the three lengths, is.

| $y$ | 6 | 7 | 8 | 9 |
|---|---|---|---|---|
| $P(Y = y)$ | 0.064 | 0.288 | 0.432 | 0.216 |

   and find $E(Y)$ and $\text{Var}(Y)$. Comment on your results

6. Find the variance of the sum of the scores when an ordinary die is thrown ten times.

7. $X$ has a p.d.f. given by $P(X = x) = kx$, $x = 1, 2, 3, 4$. Find
   (a)  $k$,
   (b)  $E(X)$,
   (c)  $\text{Var}(X)$,
   (d)  $P(X_1 + X_2 = 5)$,
   (e)  $E(4X)$,
   (f)  $\text{Var}(X_1 + X_2 + X_3)$.

## Summary

- For the discrete random variable $X$ with probability density function $P(X = x)$

  $\Sigma P(X = x) = 1$

  Cumulative distribution function $F(x) = P(X \leqslant x)$

  $\mu = E(X) = \Sigma\, xP(X = x)$   where $\mu$ is the expectation of $X$

  $E(g(X)) = \Sigma\, g(x)P(X = x)$

  $E(g(x) + h(X)) = E(g(X)) + E(h(X))$

  $E(X^2) = \Sigma\, x^2 P(X = x)$

  $\sigma^2 = \mathrm{Var}(X) = E(X - \mu)^2$   where $\mathrm{Var}(X)$ is the variance of $X$

  $\qquad\qquad = E(X^2) - \mu^2$   (or $\mathrm{Var}(X) = E(X^2) - E^2(X)$)

  $\qquad\qquad = \Sigma\, x^2 P(X = x) - \mu^2$

  $\sigma = $ standard deviation of $X = \sqrt{\mathrm{Var}(X)}$

- For the random variable $X$ and constants $a$ and $b$,

  | | |
  |---|---|
  | $E(a) = a$ | $\mathrm{Var}(a) = 0$ |
  | $E(aX) = aE(X)$ | $\mathrm{Var}(aX) = a^2\,\mathrm{Var}(X)$ |
  | $E(aX + b) = aE(X) + b$ | $\mathrm{Var}(aX + b) = a^2\,\mathrm{Var}(X)$ |

- For any two random variables $X$ and $Y$ and constants $a$ and $b$,

  $E(X + Y) = E(X) + E(Y)$

  $E(X - Y) = E(X) - E(Y)$

  $E(aX + bY) = aE(X) + bE(Y)$

  $E(aX - bY) = aE(X) - bE(Y)$

- For independent random variables $X$ and $Y$ and constants $a$ and $b$,

  $\mathrm{Var}(X + Y) = \mathrm{Var}(X) + \mathrm{Var}(Y)$

  $\mathrm{Var}(X - Y) = \mathrm{Var}(X) + \mathrm{Var}(Y)$

  $\mathrm{Var}(aX + bY) = a^2\,\mathrm{Var}(X) + b^2\,\mathrm{Var}(Y)$

  $\mathrm{Var}(aX - bY) = a^2\,\mathrm{Var}(X) + b^2\,\mathrm{Var}(Y)$

- For $n$ independent observations of $X$,

  $E(X_1 + X_2 + \cdots + X_n) = nE(X)$

  $\mathrm{Var}(X_1 + X_2 + \cdots + X_n) = n\,\mathrm{Var}(X)$

- For multiples of $X$,

  $E(nX) = nE(X)$

  $\mathrm{Var}(nX) = n^2\mathrm{Var}(X)$

## Miscellaneous worked examples

### Example 4.21

The discrete random variable $X$ has probability function

$$P(X = x) = \begin{cases} \dfrac{kx}{(x^2 + 1)}, & x = 2, 3 \\[3mm] \dfrac{2kx}{(x^2 - 1)}, & x = 4, 5 \\[3mm] 0, & \text{otherwise} \end{cases}$$

(a) Show that the value of $k$ is $\frac{20}{33}$.
(b) Find the probability that $X$ is less than 3 or greater than 4.
(c) Find $F(3.2)$.
(d) Find (i) $E(X)$, (ii) $\text{Var}(X)$.

(L)

### Solution 4.21

(a) When $x = 2$, $\quad P(X = 2) = \dfrac{2k}{2^2 + 1} = \dfrac{2k}{5}$

When $x = 3$, $\quad P(X = 3) = \dfrac{3k}{10}$

When $x = 4$, $\quad P(X = 4) = \dfrac{8k}{15}$

When $x = 5$, $\quad P(X = 5) = \dfrac{10k}{24}$

Now $\Sigma P(X = x) = 1$

$\therefore \quad \dfrac{2k}{5} + \dfrac{3k}{10} + \dfrac{8k}{15} + \dfrac{10k}{24} = 1$

$\dfrac{33k}{20} = 1$

$k = \dfrac{20}{33}$

Substituting this value for $k$, the probability distribution for $X$ is

| $x$ | 2 | 3 | 4 | 5 |
|---|---|---|---|---|
| $P(X = x)$ | $\frac{8}{33}$ | $\frac{2}{11}$ | $\frac{32}{99}$ | $\frac{25}{99}$ |

(b) $P(X < 3 \text{ or } X > 4) = P(X = 2) + P(X = 5)$

$= \dfrac{8}{33} + \dfrac{25}{99}$

$= \dfrac{49}{99}$

(c) $F(3.2) = P(X \leqslant 3.2) = P(X = 2) + P(X = 3)$

$= \dfrac{8}{33} + \dfrac{2}{11}$

$= \dfrac{14}{33}$

(d) (i) $E(X) = \Sigma\, xP(X = x)$

$\qquad = 2 \times \frac{8}{33} + 3 \times \frac{2}{11} + 4 \times \frac{32}{99} + 5 \times \frac{25}{99}$

$\qquad = 3\frac{58}{99}$

(ii) $E(X^2) = \Sigma\, x^2 P(X = x)$

$\qquad = 4 \times \frac{8}{33} + 9 \times \frac{2}{11} + 16 \times \frac{32}{99} + 25 \times \frac{25}{99}$

$\qquad = 14\frac{1}{11}$

$\quad \text{Var}(X) = E(X^2) - E^2(X)$

$\qquad = 14\frac{1}{11} - (3\frac{58}{99})^2$

$\qquad = 1.23 \ (3 \ \text{s.f.})$

---

## Example 4.22

Anne plays a game in which a fair six-sided die is thrown once. If the score is 1, 2 or 3, Anne loses £10. If the score is 4 or 5, Anne wins £$x$. If the score is 6, Anne wins £$2x$.

(a) Show that the expectation of Anne's profit is £$(\frac{2}{3}x - 5)$ in a single game.
(b) Calculate the value of $x$ for which, on average, Anne's profit is zero.
(c) Given that $x = 12$, calculate the variance of Anne's profit in a single game. $\qquad$ (C)

## Solution 4.22

Let £$X$ be Anne's profit.

$P(\text{score 1, 2 or 3}) = \frac{3}{6} = \frac{1}{2}$, therefore $P(X = -10) = \frac{1}{2}$

$P(\text{score 4 or 5}) = \frac{2}{6} = \frac{1}{3}$, therefore $P(X = x) = \frac{1}{3}$

$P(\text{score 6}) = \frac{1}{6}$, therefore $P(X = 2x) = \frac{1}{6}$

The probability distribution for $X$ is

| $x$ | $-10$ | $x$ | $2x$ |
|---|---|---|---|
| $P(X = x)$ | $\frac{1}{2}$ | $\frac{1}{3}$ | $\frac{1}{6}$ |

(a) $E(X) = \Sigma\, xP(X = x)$

$\qquad = -10 \times (\frac{1}{2}) + x \times (\frac{1}{3}) + 2x \times (\frac{1}{6})$

$\qquad = -5 + \dfrac{1}{3}x + \dfrac{2x}{6}$

$\qquad = \dfrac{2}{3}x - 5$

So the expectation of Anne's profit in a single game is £$(\frac{2}{3}x - 5)$.

(b) If $E(X) = 0$ then $\dfrac{2}{3}x - 5 = 0$

$\qquad\qquad\qquad\qquad x = 7.5$

(c) When $x = 12$, the probability distribution becomes

| $x$ | $-10$ | $12$ | $24$ |
|---|---|---|---|
| $P(X = x)$ | $\frac{1}{2}$ | $\frac{1}{3}$ | $\frac{1}{6}$ |

$E(X) = \frac{2}{3}x - 5 = \frac{2}{3} \times 12 - 5 = 3$   from (a)

$E(X^2) = \Sigma x^2 P(X = x)$

$\qquad = 100 \times \frac{1}{2} + 144 \times \frac{1}{3} + 576 \times \frac{1}{6}$

$\qquad = 194$

$\text{Var}(X) = E(X^2) - E^2(X)$

$\qquad\quad = 194 - 9$

$\qquad\quad = 185$

**The variance of Anne's profit in a single game is $185(\pounds^2)$.**

---

## Example 4.23

Any integer may be reduced to a single digit by the method illustrated below.

$51 \rightarrow 5 + 1 \rightarrow 6$
$58 \rightarrow 5 + 8 \rightarrow 13 \rightarrow 1 + 3 \rightarrow 4$

The random variable $D$ denotes the digit that results from the reduction of an integer, selected at random, from the twenty integers $50, 51, 52, \ldots, 69$.

(a) Show that $P(D = 5) = 0.15$
(b) Determine the probability for each of the other possible values of $D$.
(c) Calculate the expected value of $D$.
(d) Calculate, to two decimal places, the variance of $D$.                                    (NEAB)

## Solution 4.23

To calculate $D$, consider the following

$50 \rightarrow 5 + 0 \rightarrow ⑤$          $60 \rightarrow 6 + 0 \rightarrow 6$
$51 \rightarrow 5 + 1 \rightarrow 6$          $61 \rightarrow 6 + 1 \rightarrow 7$
$52 \rightarrow 5 + 2 \rightarrow 7$          $62 \rightarrow 6 + 2 \rightarrow 8$
$53 \rightarrow 5 + 3 \rightarrow 8$          $63 \rightarrow 6 + 3 \rightarrow 9$
$54 \rightarrow 5 + 4 \rightarrow 9$          $64 \rightarrow 10 \rightarrow 1 + 0 \rightarrow 1$
$55 \rightarrow 10 \rightarrow 1 + 0 \rightarrow 1$          $65 \rightarrow 11 \rightarrow 1 + 1 \rightarrow 2$
$56 \rightarrow 11 \rightarrow 1 + 1 \rightarrow 2$          $66 \rightarrow 12 \rightarrow 1 + 2 \rightarrow 3$
$57 \rightarrow 12 \rightarrow 1 + 2 \rightarrow 3$          $67 \rightarrow 13 \rightarrow 1 + 3 \rightarrow 4$
$58 \rightarrow 13 \rightarrow 1 + 3 \rightarrow 4$          $68 \rightarrow 14 \rightarrow 1 + 4 \rightarrow ⑤$
$59 \rightarrow 14 \rightarrow 1 + 4 \rightarrow ⑤$          $69 \rightarrow 15 \rightarrow 1 + 5 \rightarrow 6$

Three integers out of the twenty reduce to 5. These have been ringed in the list above.

(a) $P(D = 5) = \frac{3}{20} = 0.15$

(b)

| $d$ | 1 | 2 | 3 | 4 | 5 | 6 | 7 | 8 | 9 |
|---|---|---|---|---|---|---|---|---|---|
| $P(D = d)$ | $\frac{2}{20}$ | $\frac{2}{20}$ | $\frac{2}{20}$ | $\frac{2}{20}$ | $\frac{3}{20}$ | $\frac{3}{20}$ | $\frac{2}{20}$ | $\frac{2}{20}$ | $\frac{2}{20}$ |

(c) $E(D) = \Sigma \, dP(D = d)$

$= 1 \times \frac{2}{20} + 2 \times \frac{2}{20} + 3 \times \frac{2}{20} + 4 \times \frac{2}{20} + 5 \times \frac{3}{20} + 6 \times \frac{3}{20} + 7 \times \frac{2}{20} + 8 \times \frac{2}{20} + 9 \times \frac{2}{20}$

$= \frac{1}{20}(2 + 4 + 6 + 8 + 15 + 18 + 14 + 16 + 18)$

$= \mathbf{5.05}$

(d) $E(D^2) = 1 \times \frac{2}{20} + 4 \times \frac{2}{20} + 9 \times \frac{2}{20} + 16 \times \frac{2}{20} + 25 \times \frac{3}{20} + 36 \times \frac{3}{20} + 49 \times \frac{2}{20} + 64 \times \frac{2}{20} + 81 \times \frac{2}{20}$

$= \frac{1}{20}(2 + 8 + 18 + 32 + 75 + 108 + 98 + 128 + 162)$

$= \mathbf{31.55}$

$Var(D) = E(D^2) - E^2(D) = 31.55 - 5.05^2$

$= \mathbf{6.05}[2 \text{ d.p.}]$

## Miscellaneous exercise 4f

1. Fertiliser is sold in 25 kg sacks. The probability that a sack is underweight by $Y$ kg, measured to the nearest 0.1 kg, is given in the table below.

| $Y$ | 0.1 | 0.2 | 0.3 | 0.4 | 0.5 | more than 0.5 |
|---|---|---|---|---|---|---|
| Probability | 0.5 | 0.3 | 0.1 | 0.075 | 0.025 | 0 |

Find the expected loss in weight per sack.

The price quoted by the manufacturers to a farmer for 1000 kg of fertiliser, packed in 25 kg sacks, is £240. Estimate by how much this price exceeds the value of the fertiliser that would actually be supplied to the farmer. (C)

2. A discrete random variable $X$ can take only the values 0, 1, 2 or 3, and its probability distribution is given by $P(X = 0) = k$, $P(X = 1) = 3k$, $P(X = 2) = 4k$, $P(X = 3) = 5k$, where $k$ is a constant. Find
   (a) the value of $k$,
   (b) the mean and variances of $X$. (NEAB)

3. A random variable $R$ takes the integer value $r$ with probability $P(r)$ where
   $P(r) = kr^3$,    $r = 1, 2, 3, 4$,
   $P(r) = 0$,    otherwise
   Find
   (a) the value of $k$, and display the distribution on graph paper,
   (b) the mean and the variance of the distribution,
   (c) the mean and the variance of $5R - 3$. (L)

4. A curiously shaped six-faced die produces a score, $X$, for which the probability distribution is given in the following table.

| $r$ | 1 | 2 | 3 | 4 | 5 | 6 |
|---|---|---|---|---|---|---|
| $P(X = x)$ | $k$ | $k/2$ | $k/3$ | $k/4$ | $k/5$ | $k/6$ |

Show that the constant $k$ is $\frac{20}{49}$. Find the mean and variance of $X$.

The die is thrown twice. Show that the probability of obtaining equal scores is approximately $\frac{1}{4}$. (MEI)

5. A random variable $R$ takes the integer value $r$ with probability $P(r)$ defined by
   $P(r) = kr^2$,    $r = 1, 2, 3$,
   $P(r) = k(7 - r)^2$,    $r = 4, 5, 6$,
   $P(r) = 0$,    otherwise.

   Find the value of $k$ and the mean and variance of the probability distribution. Exhibit this distribution by a suitable diagram. Determine the mean and the variance of the variable $Y$ where $Y \equiv 4R - 2$. (L)

6. A discrete random variable $X$ has the distribution function

| $x$ | 1 | 2 | 4 | 5 |
|---|---|---|---|---|
| $F(x)$ | $\frac{1}{12}$ | $\frac{1}{2}$ | $\frac{5}{6}$ | 1 |

   (a) Write down the probability distribution of $X$.
   (b) Find the probability distribution of the sum of two independent observations from $X$ and find the mean and variance of the distribution of this sum.

7. The probability of there being $X$ unusable matches in a full box of Surelite matches is given by $P(X = 0) = 8k$, $P(X = 1) = 5k$, $P(X = 2) = P(X = 3) = k$, $P(X \geqslant 4) = 0$. Determine the constant $k$ and the expectation and variance of $X$.

   Two full boxes of Surelite matches are chosen at random and the total number $Y$ of unusable matches is determined. Calculate $P(Y > 4)$, and state the values of the expectation and variance of $Y$. (C)

8. Two unbiased four-sided dice, having the numbers 1, 2, 3 and 4 on their faces, are thrown together. The random variable $D$ represents the modulus of the difference between the numbers on the two hidden faces.

   (a) Show that $P(D = 1) = \frac{3}{8}$.
   (b) Calculate the probability for each of the other possible values of $D$.
   (c) Calculate the expected value of $D$. (NEAB)

9. $A$ and $B$ play a series of tennis matches. The probability that $A$ wins any single match in the series is 0.6. The winner of the series is the first player to win either two matches in succession or a total of three matches. Show that the probability

   (a) that the series lasts exactly two matches is 0.52,
   (b) that the series lasts exactly three matches is 0.24.

   Calculate the probability that the series lasts exactly four matches.
   Hence, or otherwise, show that the probability the series last five matches is 0.1152.
   Calculate the expectation of $n$, the number of matches in the series.

   The prize-money involved depends on $n$ and is shown in the table below.

   | $n$ | 2 | 3 | 4 or 5 |
   |---|---|---|---|
   | Prize-money | £1000 | £1240 | £1510 |

   Tickets are sold, each of which entitles the purchaser to see the whole series of matches. Given that each ticket costs £5, calculate the number of tickets which must be sold to cover the expected value of the prize-money. (C)

10. A fair cubical die has two yellow faces and four blue faces. The die is rolled repeatedly until a yellow face appears uppermost or the die has been rolled four times. The random variable $B$ represents the number of times a blue face appears uppermost and the random variable $R$ represents the number of times the die is rolled.

    (a) Show that $P(B = 3) = \frac{8}{81}$.
    (b) Find the probability distribution of $B$.
    (c) Find $E(B)$.
    (d) Show that $P(R = 4) = \frac{8}{27}$.
    (e) Find $P(R = B)$. (L)

11. The discrete random variable $X$ has the probability distribution given in the following table.

    | $x$ | 1 | 2 | 3 | 4 |
    |---|---|---|---|---|
    | $P(X = x)$ | 0.4 | 0.3 | 0.1 | 0.2 |

    Two independent observations of $X$ are made. The value of the random variable $Y$ is found by subtracting the smaller of the two values of $X$ from the larger. If the two values of $X$ are equal, $Y$ is zero. Show that $P(Y = 1) = 0.34$ and tabulate the complete probability distribution of $Y$.
    Find
    (a) $E(Y)$,
    (b) $\text{Var}(Y)$,
    (c) $P(Y \geqslant E(Y))$. (C)

12. A box contains five discs, labelled 1, 2, 4, 5 and 6. In a game a player draws a disc at random, replaces it and then draws again. The player's score is the sum of the numbers on the two discs drawn.
    Construct a table showing the 11 possible scores and their probabilities. Find the expected score.
    In a social club this game is played and the prize is £1 for each point scored. The players pay £7.50 each time they play. Find the expected profit to the club after 250 games have been played. (C)

13. On a long train journey, a statistician is invited by a gambler to play a dice game. The game uses two ordinary dice which the statistician is to throw. If the total score is 12, the statistician is paid £6 by the gambler. If the total score is 8, the statistician is paid £3 by the gambler. However if both or either dice show a 1, the statistician pays the gambler £2. Let £$X$ be the amount paid to the statistician by the gambler after the dice are thrown once.
    Determine, the probability that (a) $X = 6$, (b) $X = 3$, (c) $X = -2$.
    Find the expected value of $X$ and show that, if the statistician played the game 100 times, his expected loss would be £2.78, to the nearest penny.
    Find the amount, £$a$, that the £6 would have to be changed to in order to make the game unbiased.

14. The discrete random variable $X$ can take only the values 0, 1, 2, 3, 4, 5. The probability distribution of $X$ is given by the following, where $a$ and $b$ are constants.

$$P(X = 0) = P(X = 1) = P(X = 2) = a$$
$$P(X = 3) = P(X = 4) = P(X = 5) = b$$
$$P(X \geqslant 2) = 3P(X < 2)$$

(a) Determine the values of $a$ and $b$.
(b) Show that the expectation of $X$ is $\frac{23}{8}$ and determine the variance of $X$.
(c) Determine the probability that the sum of two independent observations from this distribution exceeds 7. (C)

15. A gambling machine works in the following way. The player inserts a penny into one of five slots, which are coloured Blue, Red, Orange, Yellow and Green corresponding to five coloured light bulbs. The player can choose whichever coloured slot he likes. After the penny has been inserted one of the five bulbs lights up. If the bulb lit up is the same colour as the slot selected by the player, then the player wins and receives from the machine $R$ pennies, where

$$P(R = 2) = \tfrac{1}{2}, \qquad P(R = 4) = \tfrac{1}{4}$$
$$P(R = 6) = \tfrac{3}{20}, \qquad \text{and}$$
$$P(R = 8) = P(R = 10) = \tfrac{1}{20}$$

If the colour of the bulb lit up and the slot selected are not the same, the player receives nothing from the machine. In either case the player does not get back the penny that he inserted. Assuming that each of the colours is equally likely to light up, and that the machine selects the bulbs at random, determine
(a) the probability that the player receives nothing from the machine,
(b) the expected value of the amount gained by the player from a single try,
(c) the variance of the amount gained by the player from a single try. (C)

16. (a) A regular customer at a small clothes shop observes that the number of customers, $X$, in the shop when she enters has the following probability distribution.

| Number of customers, $x$ | 0 | 1 | 2 | 3 | 4 |
|---|---|---|---|---|---|
| Probability $p(x)$ | 0.15 | 0.34 | 0.27 | 0.14 | 0.10 |

Find the mean and standard deviation of $X$.

(b) She also observes that the average waiting time, $Y$, before being served, is as follows.

| Number of customers, $x$ | 0 | 1 | 2 | 3 | 4 |
|---|---|---|---|---|---|
| Average waiting time, $v$ minutes | 0 | 2 | 6 | 9 | 12 |

Find her mean waiting time. (AEB)

17. During winter a family requests four bottles of milk every day, and these are left on the door-step. Three of the bottles have silver tops and the fourth has a gold top. A thirsty blue-tit attempts to remove the tops from these bottles. The probability distribution of $X$, the number of silver tops removed by the blue-tit, is the same each day and is given by

$$P(X = 0) = \tfrac{5}{15}, \qquad P(X = 1) = \tfrac{6}{15}.$$
$$P(X = 2) = \tfrac{3}{15}, \qquad P(X = 3) = \tfrac{1}{15}.$$

The blue-tit finds the gold top particularly attractive, and the probability that this top is removed is $\frac{3}{5}$, independent of the number of silver tops removed. Determine the expectation and variance of
(a) the number of silver tops removed in a day,
(b) the number of gold tops removed in a day,
(c) the total number of tops (silver and gold) removed in seven days.
Find also the probability distribution of the total number of tops (silver and gold) removed in a day. (C)

18. A player throws a die whose faces are numbered 1 to 6 inclusive. If the player obtains a six he throws the die a second time, and in this case his score is the sum of 6 and the second number; otherwise his score is the number obtained. The player has no more than two throws.
Let $X$ be the random variable denoting the player's score. Write down the probability distribution of $X$, and determine the mean of $X$. Show that the probability that the sum of two successive scores is 8 or more is $\frac{17}{36}$.
Determine the probability that the first of two successive scores is 7 or more, given that their sum is 8 or more. (C)

# Mixed test 4A

1. A discrete random variable $X$ has the following probability distribution and can only take the values tabulated.

| $X$ | 1 | 3 | 6 | $n$ | 12 |
|---|---|---|---|---|---|
| Probability | 0.1 | 0.3 | $k$ | 0.25 | 0.15 |

   (a) Find the value of $k$.
   Given that $E(X) = 6.0$, find
   (b) the value of $n$,
   (c) the variance of $X$.        (C)

2. When a certain type of cell is subjected to radiation, the cell may die, survive as a single cell or divide into two cells with probabilities $\frac{1}{2}, \frac{1}{3}, \frac{1}{6}$ respectively.
   Two cells are independently subjected to radiation. The random variable $X$ represents the total number of cells in existence after this experiment.

   (a) Show that $P(X = 2) = \frac{5}{18}$.
   (b) Find the probability distribution for $X$.
   (c) Evaluate $E(X)$.
   (d) Show that $\text{Var}(X) = \frac{10}{9}$.
   Another two cells are submitted to radiation in a similar experiment and the random variable $Y$ represents the total number of cells in existence after this experiment. The random variable $Z$ is defined as $Z = X - Y$.
   (e) Find $E(Z)$ and $\text{Var}(Z)$.     (L)

3. In a game two fair, cubical dice with faces numbered 1 to 6 are thrown. The score in the game is the positive difference between the numbers showing uppermost on the two dice.
   (a) Tabulate the probability distribution for the score.
   (b) Calculate the expected value of the score.
   (c) State the probability that the score is less than the expected value.    (NEAB)

# Mixed test 4B

1. The discrete random variable $X$ has the probability function shown in the table below.

| $x$ | 1 | 2 | 3 | 4 | 5 |
|---|---|---|---|---|---|
| $P(X = x)$ | 0.2 | 0.3 | 0.3 | 0.1 | 0.1 |

   Find
   (a) $P(2 < X \leqslant 4)$,
   (b) $F(3.7)$,
   (c) $E(X)$,
   (d) $\text{Var}(X)$,
   (e) $E(X^2 + 4X - 3)$.    (L)

2. A box contains six discs, of which two are labelled 2, three are labelled 3 and one is labelled 6. A game consists of a player drawing two discs simultaneously from the box. The sum of the numbers on the two discs is denoted by $X$.
   (a) Find the probability distribution of $X$.
   (b) Calculate $E(X)$, $E(X^2)$ and the variance of $X$.

   A player pays £20 for 30 games and is paid £$kX$ for each value of $X$ he obtains.
   (c) Calculate the expected profit or loss for 30 games if $k = 0.1$.
   (d) Calculate the value of $k$ for which the game would be fair.    (C)

3. An unbiased four-sided die has faces numbered 1, 2, 3 and 6. The die and a fair coin are tossed together. The random variable $R$ denotes the number on the hidden face of the die. If the coin shows heads, the score recorded, $S$, is equal to $2R$, otherwise $S = R$.
   (a) Tabulate the probability distribution for $S$.
   (b) Calculate the expected value of $S$.
   (c) Calculate the variance of $S$.    (NEAB)

# 5

## Special discrete probability distributions

*In this chapter you will learn*

- about the conditions needed to model a situation for a discrete variable using
  - a uniform distribution
  - a geometric distribution, $X \sim \text{Geo}(p)$
  - a binomial distribution, $X \sim B(n, p)$
  - a Poisson distribution, $X \sim \text{Po}(\lambda)$
- how to calculate probabilities for these distributions and also the mean and variance
- about the use of the Poisson distribution as an approximation to the binomial distribution
- about the distribution of the sum of two or more independent Poisson variables

## THE UNIFORM DISTRIBUTION

Throw an ordinary die. The probability distribution of $X$, the number on the die, is shown in the table and illustrated by the vertical line graph.

| $x$ | 1 | 2 | 3 | 4 | 5 | 6 |
|-----|---|---|---|---|---|---|
| $P(X = x)$ | $\frac{1}{6}$ | $\frac{1}{6}$ | $\frac{1}{6}$ | $\frac{1}{6}$ | $\frac{1}{6}$ | $\frac{1}{6}$ |

$P(X = x) = \frac{1}{6}$ for $x = 1, 2, 3, 4, 5, 6$

This is an example of a **discrete uniform distribution**.

## Conditions for a uniform model

For a situation to be described using a **discrete uniform model**,

- the discrete random variable $X$ is defined over the set of $n$ distinct values $x_1, x_2, ..., x_n$
- each value is equally likely to occur and

$$P(X = x_r) = \frac{1}{n} \quad \text{for} \quad r = 1, 2, ..., n$$

## Example 5.1

The discrete variable $X$ is such that $P(X = x) = c$ for $x = 20, 30, 45, 50$. Find

(a) the probability distribution of $X$,
(b) $\mu$, the expectation of $X$,
(c) $P(X < \mu)$,
(d) $\sigma$, the standard deviation of $X$.

## Solution 5.1

(a)

| $x$ | 20 | 30 | 45 | 50 |
|---|---|---|---|---|
| $P(X = x)$ | $c$ | $c$ | $c$ | $c$ |

$\sum P(X = x) = 1$
$\therefore \qquad 4c = 1$
$\qquad c = 0.25$
$P(X = x) = 0.25 \quad$ for $\quad x = 20, 30, 45, 50$

NOTE: There are four values, each of which is equally likely to occur and $P(X = x_r) = \frac{1}{4} = 0.25$ for $r = 1, 2, 3, 4$. The distribution is **uniform**.

(b) $\mu = E(X) = \sum x P(X = x)$
$\qquad = 20 \times 0.25 + 30 \times 0.25 + 45 \times 0.25 + 50 \times 0.25$
$\qquad = \mathbf{36.25}$

(c) $P(X < \mu) = P(X < 36.25)$
$\qquad = P(X = 20) + P(X = 30)$
$\qquad = 0.25 + 0.25$
$\qquad = \mathbf{0.5}$

(d) $\qquad E(X^2) = \sum x^2 P(X = x)$
$\qquad \qquad = 0.25(20^2 + 30^2 + 45^2 + 50^2)$
$\qquad \qquad = 1456.25$
$\qquad \mathrm{Var}(X) = E(X^2) - \mu^2$
$\qquad \qquad = 1456.25 - 36.25^2$
$\qquad \qquad = 142.1875$
$\qquad \qquad \sigma = \sqrt{142.1875} = \mathbf{11.9 (3\ s.f.)}.$

# THE GEOMETRIC DISTRIBUTION

Plastic models of animals are given away in packets of breakfast cereal. The probability that a packet contains a model of a rabbit is 0.1. Consider the probability distribution of $X$, the number of packets you open until you get a rabbit.

$P(X = 1) = P(\text{first packet contains a rabbit}) = 0.1$
$P(X = 2) = P(\text{first doesn't, second packet does}) = 0.9 \times 0.1 = 0.09$
$P(X = 3) = P(\text{first doesn't, second doesn't, third packet does}) = 0.9 \times 0.9 \times 0.1 = 0.081$

Similarly

$P(X = 4) = 0.9 \times 0.9 \times 0.9 \times 0.1 = (0.9)^3 \times 0.1$
$P(X = 5) = (0.9)^4 \times 0.1$
$P(X = 6) = (0.9)^5 \times 0.1$

and so on. A geometric model is being used in this example.

## Conditions for a geometric model

For a situation to be described using a **geometric model,**

- independent trials are carried out,
- the outcome of each trial is deemed either a success or a failure,
- the probability, $p$, of a successful outcome is the same for each trial.

The discrete random variable, $X$, is **the number of trials needed to obtain the first successful outcome.**
If the above conditions are satisfied, $X$ is said to follow a **geometric distribution.** This is written

$$X \sim \text{Geo}(p)$$

The probability of success, $p$, is all that is needed to describe the distribution completely. It is known as the parameter of the distribution.

Writing $P(\text{failure})$ as $q$, where $q = 1 - p$:

If $X \sim \text{Geo}(p)$, the probability that the first success is obtained at the $r$th attempt is $P(X = r)$ where

$$P(X = r) = q^{r-1} \times p \text{ for } r = 1, 2, 3, 4, \ldots$$

so that $P(X = 1) = q^0 p = p$
$\qquad P(X = 2) = q^1 p = qp$
$\qquad P(X = 3) = q^2 p \quad$ and so on.

*NOTE:*

- $X$ cannot take the value 0,
- the number of trials could be infinite, although this is unlikely in practice!

Here are some diagrammatic illustrations of geometric distributions:

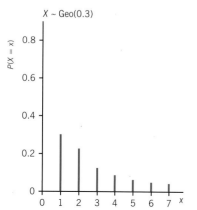

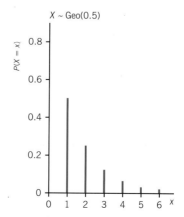

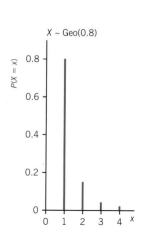

# The mode of the geometric distribution

From the diagrams, you can see that the **mode** of any geometric distribution is 1. This means that for any value of $p$, one attempt is the most likely number of attempts to obtain the first success. This is quite a surprising result.

$P(X = 1) = p$
$P(X = 2) = qp$

Since $0 < q < 1$, $qp < p$.

Also $P(X = 3) = q^2 p < qp < p$ and so on.

For example, if $X \sim \text{Geo}(0.3)$

$P(X = 1) = 0.3$
$P(X = 2) = 0.7 \times 0.3 = 0.21 < 0.3$
$P(X = 3) = 0.7^2 \times 0.3 = 0.147 < 0.21 < 0.3$

## Example 5.2

Jack is playing a board game in which he needs to throws a six with an ordinary die in order to start the game. Find the probability that

(a) exactly four attempts are needed to obtain a six,
(b) at least two attempts are needed,
(c) he is successful in throwing a six in three or fewer attempts,
(d) he needs more than three attempts to obtain a six.

## Solution 5.2

$X$ is the number of attempts up to and including the first occurrence of a six.
$P(\text{six}) = \frac{1}{6}$, so using a geometric model with $p = \frac{1}{6}$, $q = \frac{5}{6}$, $X \sim \text{Geo}(\frac{1}{6})$.

(a) $P(X = 4) = q^3 p$
$\qquad\qquad = (\frac{5}{6})^3 \times (\frac{1}{6})$
$\qquad\qquad = 0.096$ (2 s.f.)

(b) $P(X \geqslant 2) = 1 - P(X = 1)$
$\qquad\qquad = 1 - p$
$\qquad\qquad = 1 - \frac{1}{6}$
$\qquad\qquad = \frac{5}{6}$

(c) $P(X \leqslant 3) = P(X = 1) + P(X = 2) + P(X = 3)$
$\qquad\qquad = p + qp + q^2 p$
$\qquad\qquad = \frac{1}{6} + \frac{5}{6} \times \frac{1}{6} + (\frac{5}{6})^2 \times \frac{1}{6}$
$\qquad\qquad = 0.42$ (2 s.f.)

Alternatively,

$P(X \leqslant 3) = P(\text{success at some trial in the first three trials})$
$\qquad\qquad = 1 - P(\text{no success in first three trials})$
$\qquad\qquad = 1 - q^3$
$\qquad\qquad = 1 - (\frac{5}{6})^3$
$\qquad\qquad = 0.42$ (2 s.f.)

(d) $(X > 3) = 1 - P(X \leqslant 3)$
$$= 1 - (1 - q^3)$$
$$= q^3$$
$$= (\tfrac{5}{6})^3$$
$$= 0.58 \textbf{ (2 s.f.)}$$

---

These two results, illustrated above, are very useful:

If $X \sim \text{Geo}(p)$ and $q = 1 - p$:

$$P(X \leqslant x) = 1 - q^x$$
$$P(X > x) = q^x$$

## Example 5.3

On a particular production line the probability that an item is faulty is 0.08. In a quality control test, items are selected at random from the production line. It is assumed that quality of an item is independent of that of other items.

(a)  Find the probability that the first faulty item

  (i)   does not occur in the first six selected,
  (ii)  occurs in fewer than five selections.

(b)  There is to be at least a 90% chance of picking a faulty item on or before the $n$th attempt. What is the smallest number $n$?

## Solution 5.3

$X$ is the number of items picked until a faulty one is selected.
Using a geometric model with $p = 0.08$, $q = 0.92$, $X \sim \text{Geo}(0.08)$.

(a) (i)   $P(X > 6) = q^6 = 0.92^6 = \textbf{0.61 (2 s.f.)}$
    (ii)  $P(X < 5) = P(X \leqslant 4) = 1 - q^4 = 1 - 0.92^4 = \textbf{0.28 (2 s.f.)}$

(b)  You need to find $n$ such that $P(X \leqslant n) \geqslant 0.9$
    But $P(X \leqslant n) = 1 - q^n$
    $$= 1 - 0.92^n$$

    So   $1 - 0.92^n \geqslant 0.9$
    $$0.1 \geqslant 0.92^n$$
    $$0.92^n \leqslant 0.1$$

By trial and improvement,

$$0.92^{25} = 0.124 \ldots > 0.1$$
$$0.92^{26} = 0.114 \ldots > 0.1$$
$$0.92^{27} = 0.105 \ldots > 0.1$$
$$0.92^{28} = 0.096 \ldots < 0.1$$

**The smallest value of $n$ is 28.**

If you have studied logarithms in Pure Mathematics, you could use them to find $n$:

$0.92^n \leqslant 0.1$

$n \log 0.92 \leqslant \log 0.1$ ←—— Taking logs to base 10 of both sides

$n \geqslant \dfrac{\log 0.1}{\log 0.92}$ ←—— log 0.92 is negative, and dividing by a negative quantity reverses the inequality.

i.e. $n \geqslant 27.6 \ldots$

The smallest value of $n$ is 28, as before.

## EXPECTATION AND VARIANCE OF THE GEOMETRIC DISTRIBUTION

If $X \sim \text{Geo}(p)$,

$$E(X) = \frac{1}{p}, \qquad \text{Var}(X) = \frac{q}{p^2}$$

### Example 5.4

When I make a telephone call to an office, the probability of not getting through is 0.45. If I do not get through, then I try again later. Let $X$ denote the number of attempts I have to make in order to get through. Stating any necessary assumptions, identify the probability distribution of $X$. Hence, calculate

(a) $P(X \geqslant 4)$,

(b) $E(X)$ and $\text{Var}(X)$. (C)

### Solution 5.4

$X$ is the number of attempts I have to make in order to get through. Assuming that the attempts are independent and the probability of getting through is the same for each attempt, then $X$ follows a geometric distribution with $p = 0.55$, $q = 0.45$.

(a) $P(X \geqslant 4) = P(X > 3) = q^3 = (0.45)^3 = \mathbf{0.091}$ **(2 s.f.)**

(b) $E(X) = \dfrac{1}{p} = \dfrac{1}{0.55} = \mathbf{1.8}$ **(2 s.f.)**

$\text{Var}(X) = \dfrac{q}{p^2} = \dfrac{0.45}{0.55^2} = \mathbf{1.5}$ **(2 s.f.)**

### Example 5.5

Identical independent trials of an experiment are carried out. The probability of a successful outcome is $p$. On average, five trials are required until a successful outcome occurs.

(a) Find the value of $p$.

(b) Find the probability that the first successful outcome occurs on the fifth trial.

### Solution 5.5

$X$ is the number of trials up to and including the first success.

$X \sim \text{Geo}(p)$ and $E(X) = 5$.

(a) $E(X) = \dfrac{1}{p}$

$\therefore \quad 5 = \dfrac{1}{p}$

$p = \dfrac{1}{5} = 0.2$

(b) $X \sim \text{Geo}(0.2)$, i.e. $p = 0.2$, $q = 0.8$
$$P(X = 5) = q^4 p$$
$$= 0.8^4 \times 0.2$$
$$= 0.08192$$

## Example 5.6

$X \sim \text{Geo}(p)$ and it is known that $P(X = 2) = 0.21$ and $p < 0.5$. Find $P(X = 1)$.

## Solution 5.6

$$P(X = 2) = qp \quad \text{where} \quad q = 1 - p$$

so
$$0.21 = (1 - p) \times p$$
$$0.21 = p - p^2$$
$$p^2 - p + 0.21 = 0$$
$$(p - 0.3)(p - 0.7) = 0$$
$$p = 0.3 \quad \text{or} \quad p = 0.7$$
$$\text{Since } p < 0.5, \, p = 0.3$$

$$P(X = 1) = p = 0.3$$

## Exercise 5a    The uniform and geometric distributions

1. The probability distribution for the random variable $X$ is shown in the table.

| $x$ | 6 | 7 | 8 | 9 | 10 |
|---|---|---|---|---|---|
| $P(X = x)$ | $a$ | $a$ | $a$ | $a$ | $a$ |

   Find

   (a) the value of $a$,
   (b) the mean of $X$,
   (c) the probability that $X$ is the smaller than the mean.

2. The random variable $X$ is Geo(0.35). Calculate

   (a) $P(X = 4)$,    (b) $P(X > 4)$,
   (c) $P(X \leqslant 3)$,    (d) $E(X)$.

3. A coin is biased so that the probability of obtaining a head is 0.6.
   The random variable $X$ is the number of tosses up to and including the first head. Find

   (a) $P(X \leqslant 4)$,
   (b) $P(X > 5)$,
   (c) the most likely number of tosses until a head is obtained,

   (d) the expected number of tosses until a head is obtained,
   (e) the expected number of tosses until a tail is obtained.

4. A sixth former is waiting for a bus to take him to town. He passes the time by counting the number of buses, up to and including the one that he wants, that come along his side of the road.
   If 30% of the buses travelling on that side of the road go to town, what is

   (a) the most likely count he makes to the arrival of one that will take him into town,
   (b) the probability that he will count, at most, four buses?    (O)

5. During January the probability that it will rain on any given day is 0.55.
   Stating a necessary assumption, find the probability that

   (a) the first rainy day in January is on 5 January,
   (b) it does not rain before 8 January.

6. A random number machine generates random digits between 0 and 9. Each of the ten digits is equally likely to be generated.

   (a) $X$ is the value of the digit generated. Find
       (i) $P(X < 6)$,
       (ii) $P(X \geq 7)$,
       (iii) $E(X)$,
       (iv) the standard deviation of $X$.

   (b) $X$ is the number of digits generated to the first occurrence of a 5. Find
       (i) the probability that the first occurrence of the digit 5 is at the seventh number generated,
       (ii) the most likely number of digits generated to obtain a 5,
       (iii) the mean number of digits generated to obtain a 5.

7. $X \sim \text{Geo}(0.5)$. Find

   (a) the mode,
   (b) the mean of $X$,
   (c) the standard deviation of $X$.

8. A darts player practises throwing a dart at the bull's eye on a dart board. Independently for each throw, her probability of hitting the bull's eye is 0.2. Let $X$ be the number of throws she makes, up to and including her first success.

   (a) Find the probability that she is successful for the first time on the third throw.
   (b) Write down the distribution of $X$ and give the name of the distribution.
   (c) Write down the probability that she will have at least three failures before her first success. (L)

9. The random variable $X$ follows the geometric distribution with probability $p = 0.3$.

   (a) Write down the probability $P(X = 4)$.
   (b) Carefully explain why $P(X = n)$ is $0.7^{n-1} \, 0.3$.
   (c) Describe in words a situation that has probability $0.7^{n-1}$. (O)

10. $X \sim \text{Geo}(p)$ and the probability that the first success is obtained on the second attempt is 0.1275. If $p > 0.5$, find $P(X > 2)$.

11. The probability that a telephone box is occupied is 0.2. Find, to two significant figures, the probability that a person wishing to make a telephone call will find a telephone box which is not occupied only at the sixth box tried. (L)

12. An unbiased coin is tossed repeatedly until a tail appears. Find the expected number of tosses.

13. In a computer game, the probability that the player hits the target is 0.4 for each attempt and the result of each attempt is independent of all others. Find

   (a) the probability that he hits the target for the first time on the fourth attempt,
   (b) the mean number of attempts needed to hit the target,
   (c) the standard deviation of the number of attempts,
   (d) the most likely number of attempts to hit the target,
   (e) the probability that he takes more than seven attempts to hit the target.

14. Alice runs a stall at a fete in which each player is guaranteed to win £10. Players pay a certain amount each time they throw a die and must keep throwing the die until a four occurs. When a four is obtained, Alice gives the player £10.

   On average Alice expects to make a profit of 50p per game. How much does she charge per throw?

15. During the winter in Glen Shee, the probability that snow will fall on any given day is 0.1. Taking 1 November as the first day of winter and assuming independence from day to day, find to two significant figures, the probability that the first snow of winter will fall in Glen Shee on the last day of November (30th).

   Given that no snow has fallen at Glen Shee during the whole of November, a teacher decides not to wait any longer to book a skiing holiday. The teacher decides to book for the earliest date for which the probability that snow will have fallen on or before that date is at least 0.9. Find the date of the booking. (L)

16. In many board games it is necessary to 'throw a six with an ordinary die' before a player can start the game. Write down, as a fraction, the probability of a player

   (a) starting on his first attempt,
   (b) not starting until his third attempt,
   (c) requiring more than three attempts before starting.

   What is
   (d) the most common number of throws required to obtain a six,
   (e) the mean number of throws required to obtain a six?

   Prove that the probability of a player requiring more than $n$ attempts before starting is $(\frac{5}{6})^n$.

   (f) What is the smallest value of $n$ if there is to be at least a 95% chance of starting on or before the $n$th attempt? (O)

## THE BINOMIAL DISTRIBUTION

In a particular population, 10% of people have blood type $B$. If **three people** are selected at random from the population, what is the probability that exactly two of them have blood type $B$?

Since the people are selected at random, assume that the blood type of one person is independent of that of another so $P(\text{type } B) = P(B) = 0.1$, $P(\text{not type } B) = P(B') = 0.9$.

To calculate the probability you could use a tree diagram.

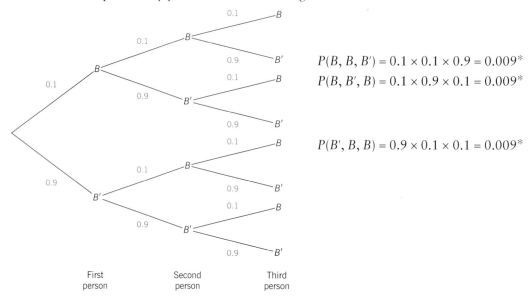

$$P(B, B, B') = 0.1 \times 0.1 \times 0.9 = 0.009^*$$
$$P(B, B', B) = 0.1 \times 0.9 \times 0.1 = 0.009^*$$

$$P(B', B, B) = 0.9 \times 0.1 \times 0.1 = 0.009^*$$

| First person | Second person | Third person |
|---|---|---|

$$P(\text{exactly two type } B) = P(B, B, B') + P(B, B', B) + P(B', B, B) \quad \longleftarrow \text{ marked * on diagram}$$
$$= 3 \times 0.9 \times 0.1^2$$
$$= \mathbf{0.027}$$

Now consider the situation when **eight people** are selected. What is the probability that exactly two of the eight people will have blood type $B$?

You could extend your tree, but it would become very complicated. It is possible to find the probability as follows since you want two with type $B$ and six who do not have blood type $B$.

$$P(\text{choose 6 } B' \text{ then 2 } B) = P(B', B', B', B', B', B', B, B)$$
$$= 0.9^6 \times 0.1^2$$

But there are several arrangements of this outcome,
for example $B', B', B, B', B', B, B', B'$ or $B, B', B', B', B, B', B', B',$
each with a probability of occurring of $0.9^6 \times 0.1^2$.

The number of different arrangements is given by $^8C_2$, sometimes written $_8C_2$ or $\binom{8}{2}$ (see page 215).

This can be found directly on your calculator using the $_nC_r$ key:

(You may have to press $\boxed{\text{SHIFT}}$ ) $\boxed{8}$ $\boxed{_nC_r}$ $\boxed{2}$ $\boxed{=}$

otherwise use the formula $^nC_r = \dfrac{n!}{r!(n-r)!}$ so $^8C_2 = \dfrac{8!}{2!\,6!}$

On calculator:

$\boxed{8}$ $\boxed{!}$ $\boxed{\div}$ $\boxed{2}$ $\boxed{!}$ $\boxed{\div}$ $\boxed{6}$ $\boxed{!}$ $\boxed{=}$

You should find that $^8C_2 = 28$. So there are 28 different arrangements of two who have blood type $B$ and six who do not have blood type $B$.

Therefore $P(\text{exactly 2 have type } B) = 28 \times 0.9^6 \times 0.1^2 = \mathbf{0.15}$ (**2 s.f.**)

Using a similar argument, you could find the probability that exactly two have blood type $B$ in a randomly selected group of **12 people**. In this case, ten will not have type $B$ and

$P(\text{exactly 2 have type } B) = {}^{12}C_2 \times 0.9^{10} \times 0.1^2 = \mathbf{0.23}$ (**2 s.f.**)

The above three situations have been described using a binomial model.

## Conditions for a binomial model

For a situation to be described using a **binomial model**,

- a finite number, $n$, trials are carried out,
- the trials are independent,
- the outcome of each trial is deemed either a success or a failure,
- the probability, $p$, of a successful outcome is the same for each trial.

The discrete random variable, $X$, is **the number of successful outcomes in $n$ trials**.

If the above conditions are satisfied, $X$ is said to follow a **binomial distribution**. This is written

$$X \sim B(n, p) \quad \text{or} \quad X \sim \text{Bin}(n, p)$$

NOTE: The number of trials, $n$, and the probability of success, $p$, are both needed to describe the distribution completely. They are known as the parameters of the binomial distribution.

Writing $P(\text{failure})$ as $q$ where $q = 1 - p$:

If $X \sim B(n, p)$, the probability of obtaining $r$ successes in $n$ trials is $P(X = r)$ where

$$P(X = r) = {}^nC_r q^{n-r} p^r \quad \text{for} \quad r = 0, 1, 2, 3, \ldots, n$$

For the three situations described above:

When 3 people are selected, $n = 3$, $p = 0.1$, $q = 0.9$.
$X$ is the number of successful outcomes in 3 trials, so $X \sim B(3, 0.1)$.

$$\begin{aligned} P(X = 2) &= {}^3C_2 q^1 p^2 \qquad\qquad \text{Note that } {}^3C_2 = 3 \\ &= 3 \times 0.9 \times 0.1^2 \\ &= \mathbf{0.072} \end{aligned}$$

When 8 people are selected, $n = 8$, $p = 0.1$, $q = 0.9$.
$X$ is the number of successful outcomes in 8 trials, so $X \sim B(8, 0.1)$.

$$\begin{aligned} P(X = 2) &= {}^8C_2 q^6 p^2 \qquad\qquad \text{Note that } {}^8C_2 = 28 \\ &= 28 \times 0.9^6 \times 0.1^2 \\ &= \mathbf{0.15} \text{ (2 s.f.)} \end{aligned}$$

When 12 people are selected, $n = 12$, $p = 0.1$, $q = 0.9$.
$X$ is the number of successful outcomes in 12 trials, so $X \sim B(12, 0.1)$.

$$P(X = 2) = {}^{12}C_2 q^{10} p^2 \qquad \text{Note that } {}^{12}C_2 = 66$$
$$= 66 \times 0.9^{10} \times 0.1^2$$
$$= 0.23 \ (2 \text{ s.f.})$$

## Example 5.7

At Sellitall Supermarket, 60% of customers pay by credit card. Find the probability that in a randomly selected sample of ten customers,

(a) exactly two pay by credit card,
(b) more than seven pay by credit card.

## Solution 5.7

$X$ is the number of customers in a sample of ten who pay by credit card.

Consider 'paying by credit card' as success, $p = 0.6$, $q = 1 - p = 0.4$.
Assuming independence, a binomial model can be used, with $n = 10$,
so $X \sim B(10, 0.6)$.

(a) $P(X = 2) = {}^{10}C_2 q^8 p^2 \qquad \qquad$ Notice that the index numbers add up to 10.
$\qquad \qquad = 45 \times 0.4^8 \times 0.6^2$
$\qquad \qquad = 0.011 \ (2 \text{ s.f.})$

(b) $P(X > 7) = P(X = 8) + P(X = 9) + P(X = 10)$
$\qquad \quad = {}^{10}C_8 q^2 p^8 + {}^{10}C_9 q^1 p^9 + {}^{10}C_{10} q^0 p^{10}$
$\qquad \quad = 45 \times 0.4^2 \times 0.6^8 + 10 \times 0.4^1 \times 0.6^9 + 0.6^{10}$
$\qquad \quad = 0.17 \ (2 \text{ s.f.})$

It is useful to note that, for any binomial distribution,

$$P(X = 0) = {}^nC_0 q^n p^0 \quad \text{but} \quad p^0 = 1 \quad \text{and} \quad {}^nC_0 = \frac{n!}{0! \, n!} = 1, \quad \text{since} \quad 0! = 1$$

so $\quad P(X = 0) = q^n$

Also $P(X = n) = {}^nC_n q^0 p^n$ but $q^0 = 1$ and ${}^nC_n = \dfrac{n!}{n! \, 0!} = 1$

so $\quad P(X = n) = p^n$

There is a link between the probabilities in the binomial distribution and the terms in the binomial expansion of $(q + p)^n$ which you may have studied in Pure Mathematics. This is illustrated in the following example.

## Example 5.8

Five independent trials of an experiment are carried out. The probability of a successful outcome is $p$ and the probability of failure is $1 - p = q$.

Write out the probability distribution of $X$, where $X$ is the number of successful outcomes in five trials. Comment on your answer.

## Solution 5.8

$X \sim B(5, p)$ and $X$ takes the values 0, 1, 2, 3, 4, 5.

$P(X = 0) = {}^5C_0 q^5 p^0 = q^5$
$P(X = 1) = {}^5C_1 q^4 p^1 = 5q^4 p^1$
$P(X = 2) = {}^5C_2 q^3 p^2 = 10q^3 p^2$
$P(X = 3) = {}^5C_3 q^2 p^3 = 10q^2 p^3$
$P(X = 4) = {}^5C_4 q^1 p^4 = 5q^1 p^4$
$P(X = 5) = {}^5C_5 q^0 p^5 = p^5$

*Notice that the powers of p and q add up to 5 each time.*

The terms $q^5, 5q^4 p^1, ..., p^5$ are the terms in the binomial expansion of $(q + p)^5$.

So $\quad q^5 \quad + \quad 5q^4 p \quad + \quad 10q^3 p^2 \quad + \quad 10q^2 p^3 \quad + \quad 5qp^4 \quad + \quad p^5 \quad = (q + p)^5$

$\qquad \underset{P(X=0)}{\uparrow} \qquad \underset{P(X=1)}{\uparrow} \qquad \underset{P(X=2)}{\uparrow} \qquad \underset{P(X=3)}{\uparrow} \qquad \underset{P(X=4)}{\uparrow} \qquad \underset{P(X=5)}{\uparrow}$

But $(q + p)^5 = 1$, since $q + p = 1$,

$\therefore P(X = 0) + P(X = 1) + ... + P(X = 5) = 1$.

This confirms that the total sum of the probabilities is 1.

*NOTE*: Some vertical line graphs illustrating the binomial distribution are given on page 289.

---

## Example 5.9

The random variable $X$ is distributed $B(7, 0.2)$. Find, correct to three decimal places,

(a) $P(X = 3)$,
(b) $P(1 < X \leqslant 4)$,
(c) $P(X > 1)$.

## Solution 5.9

$p = 0.2, q = 1 - p = 0.8, n = 7$

(a) $\quad P(X = 3) = {}^7C_3 q^4 p^3$
$\qquad \qquad \qquad = 35 \times 0.8^4 \times 0.2^3$
$\qquad \qquad \qquad = 0.115 \text{ (3 d.p.)}$

(b) $P(1 < X \leqslant 4) = P(X = 2) + P(X = 3) + P(X = 4)$
$\qquad \qquad \qquad = {}^7C_2 q^5 p^2 + {}^7C_3 q^4 p^3 + {}^7C_4 q^3 p^4$
$\qquad \qquad \qquad = 21 \times 0.8^5 \times 0.2^2 + 35 \times 0.8^4 \times 0.2^3 + 35 \times 0.8^3 \times 0.2^4$
$\qquad \qquad \qquad = 0.419 \text{ (3 d.p.)}$

(c)     $P(X > 1) = P(X = 2) + P(X = 3) + \cdots + P(X = 7)$

Rather than calculate all these terms, it is much quicker to find

$$P(X > 1) = 1 - P(X \leqslant 1)$$
$$= 1 - (P(X = 0) + P(X = 1))$$
$$= 1 - (q^7 + {}^7C_1 q^6 p)$$
$$= 1 - (0.8^7 + 7 \times (0.8)^6 \times 0.7)$$
$$= \mathbf{0.423} \ (3 \ \mathbf{d.p.})$$

## Example 5.10

A box contains a large number of pens. The probability that a pen is faulty is 0.1.
How many pens would you need to select to be more than 95% certain of picking at least one
faulty one?

## Solution 5.10

Let $n$ be the number of pens you need to select. $X$ is the number of faulty pens in $n$.
Assuming independence and using a binomial model, $X \sim B(n, 0.1)$, with $p = 0.1$, $q = 0.9$.

You want     $P(X \geqslant 1) > 0.95$
But          $P(X \geqslant 1) = 1 - P(X = 0)$
             $\qquad\qquad = 1 - 0.9^n$
So           $1 - 0.9^n > 0.95$
             $1 - 0.95 > 0.9^n$
             $0.05 > 0.9^n$
i.e.         $0.9^n < 0.05$

By trial and improvement

$0.9^{25} = 0.071 \ldots$      (greater than 0.05)
$0.9^{30} = 0.042 \ldots$      (less than 0.05)

So the value of $n$ lies between 25 and 30.

$0.9^{26} = 0.0646 \ldots$      (greater than 0.05)
$0.9^{27} = 0.058 \ldots$      (greater than 0.05)

NOTE: On the calculator $\boxed{0.9}$ $\boxed{x^y}$ $\boxed{26}$ $\boxed{=}$ (0.0646 ...).

To get $0.9^{27}$ all you have to do is press $\boxed{\times}$ $\boxed{0.9}$ $\boxed{=}$ (0.0581 ...) and so on. The answers
are of course getting smaller because you are multiplying by a number between 0 and 1.

$0.9^{28} = 0.0523 \ldots$      (greater than 0.05)
$0.9^{29} = 0.0471 \ldots$      (less than 0.05)

**You need to select at least 29 pens.**

Alternatively, using logarithms:

$\quad 0.9^n \qquad < 0.05$
Taking logs to base 10 of both sides,

$\quad n \log 0.9 < \log 0.05$

From the calculator, you find that log 0.9 = −0.045 ..., so divide both sides by log 0.9 and reverse the inequality (as you are dividing by a negative quantity).

$$n > \frac{\log 0.05}{\log 0.9}$$

$$n > 28.4 \ldots$$

The least value of $n$ is 29, as before.

---

# Using cumulative binomial probability tables

If you have access to these tables, you may wish to use them to calculate probabilities.

The tables are printed on page 645. They give $P(X \leqslant r)$ for various values of $n$ and $p$. Here is an extract for $B(7, 0.2)$, the distribution used in Example 5.9.

| $n = 7$ | $p = 0.2$ | | |
|---|---|---|---|
| $r = 0$ | 0.2097 | ← | $P(X \leqslant 0)$ |
| 1 | 0.5767 | ← | $P(X \leqslant 1)$ |
| 2 | 0.8520 | ← | $P(X \leqslant 2)$ |
| 3 | 0.9667 | ← | $P(X \leqslant 3)$ |
| 4 | 0.9953 | ← | $P(X \leqslant 4)$ |
| 5 | 0.9996 | ← | $P(X \leqslant 5)$ |
| 6 | 1.0000 | ← | $P(X \leqslant 6)$ |
| 7 | 1.0000 | ← | $P(X \leqslant 7)$ |

Using the tables to work out the probabilities required in Example 5.9:

(a)  $P(X = 3) = P(X \leqslant 3) - P(X \leqslant 2)$
$= 0.9667 - 0.8520$
$= \textbf{0.115 (3 d.p.)}$

(b)  $P(1 < X \leqslant 4) = P(X = 2) + P(X = 3) + P(X = 4)$
$= P(X \leqslant 4) - P(X \leqslant 1)$
$= 0.9953 - 0.5767$
$= \textbf{0.419 (3 d.p.)}$

(c)  $P(X > 1) = 1 - P(X \leqslant 1)$
$= 1 - 0.5767$
$= \textbf{0.423 (3 d.p.)}$

# Example 5.11

The random variable $X$ is distributed $B(5, 0.3)$. Giving your answers to three decimal places, use the extract from the cumulative binomial probability tables to find

(a)  $P(X \leqslant 4)$
(b)  $P(X = 2)$
(c)  $P(X < 3)$
(d)  $P(X > 1)$
(e)  $P(X \geqslant 3)$

| $n = 5$ | $p = 0.3$ |
|---|---|
| $r = 0$ | 0.1681 |
| $r = 1$ | 0.5282 |
| $r = 2$ | 0.8369 |
| $r = 3$ | 0.9692 |
| $r = 4$ | 0.9976 |
| $r = 5$ | 1.0000 |

## Solution 5.11

(a) $P(X \leqslant 4) = 0.9976 = \mathbf{0.998}$ (**3 d.p.**)

(b) $P(X = 2) = P(X \leqslant 2) - P(X \leqslant 1) = 0.8369 - 0.5282 = 0.3087 = \mathbf{0.309}$ (**3 d.p.**)

(c) $P(X < 3) = P(X \leqslant 2) = 0.8369 = \mathbf{0.837}$ (**3 d.p.**)

(d) $P(X > 1) = 1 - P(X \leqslant 1) = 1 - 0.5282 = 0.4718 = \mathbf{0.472}$ (**3 d.p.**)

(e) $P(X \geqslant 3) = 1 - P(X \leqslant 2) = 1 - 0.8369 = 0.1631 = \mathbf{0.163}$ (**3 d.p.**)

---

# Using symmetry properties to read binomial tables

In some versions of the cumulative binomial tables, values of $p$ are given only up to $p = 0.5$.
To use the tables for values of $p > 0.5$, you need to use the symmetry properties of the binomial distribution.

This is illustrated in the sketches of the probability distributions of $B(5, 0.3)$ and $B(5, 0.7)$.
In both these distributions $n = 5$ and note that $0.7 = 1 - 0.3$.

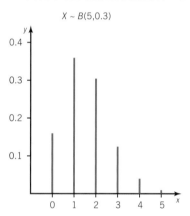

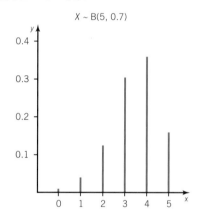

You can see that

$P(X = 0 \mid p = 0.3) = 0.17 = P(X = 5 \mid p = 0.7)$
$P(X = 1 \mid p = 0.3) = 0.36 = P(X = 4 \mid p = 0.7)$
$P(X = 2 \mid p = 0.3) = 0.31 = P(X = 3 \mid p = 0.7)$
and so on.

Also $P(X \leqslant 2 \mid p = 0.3) = 0.84 = P(X \geqslant 3 \mid p = 0.7)$

In general

$P(X = r \mid X \sim B(n, p)) = P(X = n - r \mid X \sim B(n, 1 - p))$
$P(X \leqslant r \mid X \sim B(n, p)) = P(X \geqslant n - r \mid X \sim B(n, 1 - p))$
$P(X \geqslant r \mid X \sim B(n, p)) = P(X \leqslant n - r \mid X \sim B(n, 1 - p))$

## Example 5.12

The random variable $X$ is $B(8, 0.6)$.

Use the extract of the cumulative binomial tables for $X \sim B(8, 0.4)$ to find

(a) $P(X \geqslant 3)$

(b) $P(X \leqslant 2)$

(c) $P(X = 5)$

| $n = 8$ | $p = 0.4$ |
|---|---|
| $r = 0$ | 0.0168 |
| 1 | 0.1064 |
| 2 | 0.3154 |
| 3 | 0.5941 |
| 4 | 0.8263 |
| 5 | 0.9502 |
| 6 | 0.9915 |
| 7 | 0.9993 |
| 8 | 1.0000 |

## Solution 5.12

Using $n = 8$,

These two values add up to 1

(a) $P(X \geqslant 3 \mid p = 0.6) = P(X \leqslant 5 \mid p = 0.4) = \mathbf{0.9502}$

These two values add up to $n$

(b) $P(X \leqslant 2 \mid p = 0.6) = P(X \geqslant 6 \mid p = 0.4)$
$$= 1 - P(X \leqslant 5 \mid p = 0.4)$$
$$= 1 - 0.9502$$
$$= \mathbf{0.0498}$$

(c) $P(X = 5 \mid p = 0.6) = P(X = 3 \mid p = 0.4)$
$$= P(X \leqslant 3 \mid p = 0.4) - P(X \leqslant 2 \mid p = 0.4)$$
$$= 0.5941 - 0.3154$$
$$= \mathbf{0.2787}$$

---

*NOTE:*

- It is sometimes quicker to use the cumulative tables, but you should make sure that you know how to calculate the probabilities directly.
- The tables are not available for all possible values of $p$.
- The values given in the tables should agree with the calculated values to three decimal places.

## Exercise 5b   The binomial distribution

Give answers to three significant figures.

1. 30% of pupils in a school travel to school by bus.
   From a sample of ten pupils chosen at random, find the probability that
   (a) only three travel by bus,
   (b) less than half travel by bus.

2. In a survey on washing powder, it is found that the probability that a shopper chooses Soapysuds is 0.25. Find the probability that in a random sample of nine shoppers
   (a) exactly three choose Soapysuds,
   (b) more than seven choose Soapysuds.

3. A bag contains counters of which 40% are red and the rest yellow. A counter is taken from the bag, its colour noted and then replaced. This is performed eight times in all.
   Calculate the probability that
   (a) exactly three will be red,
   (b) at least one will be red,
   (c) more than four will be yellow.

4. The random variable $X$ is $B(6, 0.42)$. Find
   (a) $P(X = 6)$,   (b) $P(X = 4)$,   (c) $P(X \leqslant 2)$.

5. An unbiased die is thrown seven times. Find the probability of throwing at least 5 sixes.

6. The probability that it will rain on any given day in September is 0.3. Stating any assumption made, calculate the probability that in a given week in September, it will rain on
   (a) exactly two days,
   (b) at least two days,
   (c) at most two days,
   (d) exactly three days that are consecutive.

7. A fair coin is tossed six times. Find the probability of throwing at least four heads.

8. Assuming that a couple are equally likely to produce a boy or a girl, find the probability that in a family of five children there are more boys than girls.

9. $X$ is $B(4, p)$ and $P(X = 4) = 0.0256$.
   Find $P(X = 2)$.

10. Charlie finds that when she takes a cutting from a particular plant, the probability that it roots successfully is $\frac{1}{3}$.
    (a) She takes nine cuttings. Find the probability that
        (i) more than five cuttings root successfully,
        (ii) at least three cuttings root successfully.
    (b) Find the number of cuttings that she should take in order to be 99% certain that at least one cutting roots successfully.

11. An experiment consists of taking seven shots at a target and counting the number of hits.
The probability of hitting the target with a single shot is 0.6. Using a binomial model, find the probability that in seven attempts the target is hit at most twice.
Give a reason why the binomial model may not be a good one to use in this situation.

12. In the mass production of bolts it is found that 5% are defective. Bolts are selected at random and put into packets of ten.
A packet is selected at random. Find the probability that it contains

    (a) three defective bolts,
    (b) less than three defective bolts.

    Two packets are selected at random.

    (c) Find the probability that there are no defective bolts in either packet.

13. A coin is biased so that it is twice as likely to show heads as tails. The coin is tossed five times. Calculate the probability that

    (a) exactly three heads are obtained,
    (b) more than three are obtained.

14. The random variable $X$ can be modelled by a binomial distribution with $n = 6$ and $p = 0.5$. Construct the probability distribution and illustrate it graphically. Comment on the distribution.

15. The probability that a target is hit is 0.3. Find the least number of shots which should be fired if the probability that the target is hit at least once is greater than 0.95.
State any assumptions that you have made.

16. 1% of light bulbs in a box are faulty. Using a binomial model, find the largest sample size which can be taken if it is required that the probability that there are no faulty bulbs in the sample is greater than 0.5.
Comment on the use of the binomial model in this situation.

17. In a test there are ten multiple choice questions. For each question there is a choice of four answers, only one of which is correct. A student guesses each of the answers.

    (a) Find the probability that he gets more than seven correct.

    He needs to obtain over half marks to pass and each question carries equal weight.

    (b) Find the probability that he passes the test.

18. $X \sim B(n, 0.3)$. Find the least possible value of $n$ such that $P(X \geqslant 1) = 0.8$.

19. Given that $X \sim B(7, 0.85)$ use the cumulative binomial probability tables on page 646 to write out the probability distribution of $X$.

20. The random variable $X$ is $B(n, 0.6)$ and $P(X < 1) = 0.0256$. Find the value of $n$.

21. For each of the experiments described below, state, giving a reason, whether a binomial distribution is appropriate.
Experiment 1: A bag contains black, white and red marbles that are selected one at a time, with replacement. The colour of each marble is noted.
Experiment 2: This experiment is a repeat of experiment 1 except that the bag contains black and white marbles only.
Experiment 3: This experiment is a repeat of experiment 2 except that the marbles are not replaced after each selection. (L)

# EXPECTATION AND VARIANCE OF THE BINOMIAL DISTRIBUTION

It can be shown that

If $X \sim B(n, p)$

$E(X) = np$ and $\mathrm{Var}(X) = npq$ where $q = 1 - p$

These results can be quoted and should be learnt. They are illustrated in the following example.

## Example 5.13

The random variable $X$ is $B(4, 0.8)$. Construct the probability distribution for $X$ and find the expectation and variance. Verify that $E(X) = np$ and $\mathrm{Var}(X) = npq$.

## Solution 5.13

$X$ is $B(4, 0.8)$ so $n = 4$ and $p = 0.8$.

$$P(X = 0) = 0.2^4 = 0.0016$$
$$P(X = 1) = 4 \times 0.2^3 \times 0.8 = 0.0256$$
$$P(X = 2) = {}^4C_2 \times 0.2^2 \times 0.8^2 = 0.1536$$
$$P(X = 3) = {}^4C_3 \times 0.2 \times 0.8^3 = 0.4096$$
$$P(X = 4) = 0.8^4 = 0.4096$$

The probability distribution for $X$ is

| $x$ | 0 | 1 | 2 | 3 | 4 |
|---|---|---|---|---|---|
| $P(X = x)$ | 0.0016 | 0.0256 | 0.1536 | 0.4096 | 0.4096 |

$$
\begin{aligned}
E(X) &= \Sigma x P(X = x) \\
&= 0 \times 0.0016 + 1 \times 0.0256 + 2 \times 0.1536 + 3 \times 0.4096 + 4 \times 0.4096 \\
&= \mathbf{3.2}
\end{aligned}
$$

$$
\begin{aligned}
E(X^2) &= \Sigma x^2 P(X = x) \\
&= 1 \times 0.0256 + 4 \times 0.1536 + 9 \times 0.4096 + 16 \times 0.4096 \\
&= 10.88
\end{aligned}
$$

$$
\begin{aligned}
\text{Var}(X) &= E(X^2) - E^2(X) \\
&= 10.88 - 3.2^2 \\
&= \mathbf{0.64}
\end{aligned}
$$

Now    $np = 8 \times 0.4 = 3.2$            $\therefore$   $E(X) = np$

$npq = 8 \times 0.4 \times 0.6 = 0.64$      $\therefore$   $\text{Var}(X) = npq$

## Example 5.14

The probability that it will be a fine day is 0.4. Find the expected number of fine days in a week and also the standard deviation.

## Solution 5.14

Let $X$ be the number of fine days in a week. Assuming that the weather on any particular day is independent of the weather on other days,

$X \sim B(n, p)$ with $n = 7$ and $p = 0.4$

$$
\begin{aligned}
\text{The expected number of fine days} &= E(X) \\
&= np \\
&= 7 \times 0.4 \\
&= \mathbf{2.8}
\end{aligned}
$$

$$
\begin{aligned}
\text{Standard deviation of} &= \sqrt{\text{Var}(X)} \\
&= \sqrt{npq} \\
&= \sqrt{7 \times 0.4 \times 0.6} \\
&= \mathbf{1.3 \text{ days (2 s.f.)}}
\end{aligned}
$$

**Example 5.15**

$X$ is $B(n, p)$ with mean 5 and standard deviation 2. Find the values of $n$ and $p$.

**Solution 5.15**

$E(X) = np$,        therefore $np = 5$      ... ①

$\text{Var}(X) = npq$,      therefore $npq = 2^2 = 4$    ... ②

Substituting for $np$ in equation ②      $5q = 4$

$$q = 0.8$$

So $$p = 1 - q$$

$$p = 0.2$$

Substituting for $p$ in equation ①      $n \times 0.2 = 5$

$$n = 25$$

# Fitting a theoretical distribution to practical data

It is sometimes useful to compare experimental results with theoretical data as illustrated in the following example.

**Example 5.16**

A biased coin is tossed four times and the number of heads noted. The experiment is performed 500 times in all and the results are summarised in the table:

| Number of heads | 0 | 1 | 2 | 3 | 4 |
|---|---|---|---|---|---|
| Frequency | 12 | 50 | 151 | 200 | 87 |

(a) From the experimental data, estimate the probability of obtaining a head when the coin is tossed.

(b) Using a binomial distribution with the same mean, calculate the theoretical probabilities of obtaining 0, 1, 2, 3 and 4 heads.

**Solution 5.16**

(a) For the frequency distribution,

$$\text{mean} = \bar{x} = \frac{\Sigma fx}{\Sigma f} = \frac{1300}{500} = 2.6$$

Let $X$ be the number of heads in four tosses. Then $X \sim B(4, p)$.

For a distribution with the same mean,    $4p = 2.6$

$$p = 0.65$$

**An estimate of the probability that the coin shows heads is 0.65.**

(b) Using $X \sim B(4, 0.65)$ calculate the probabilities of 0, 1, 2, 3 and 4 heads and multiply these by 500 to obtain the theoretical frequencies.

| $x$ | $P(X = x)$ | | Frequency (nearest integer) |
|---|---|---|---|
| 0 | $0.35^4$ | $= 0.015 \ldots$ | 8 |
| 1 | $4 \times 0.35^3 \times 0.65$ | $= 0.111 \ldots$ | 56 |
| 2 | $6 \times 0.35^2 \times 0.65^2$ | $= 0.310 \ldots$ | 155 |
| 3 | $4 \times 0.35 \times 0.65^3$ | $= 0.384$ | 192 |
| 4 | $0.65^4$ | $= 0.178 \ldots$ | 89 |
| | | | Total 500 |

These compare reasonably well with the original distribution.
A statistical test to compare the two sets of data, the $\chi^2$ test, is illustrated on page 571.

## Diagrammatic representation of the binomial distribution

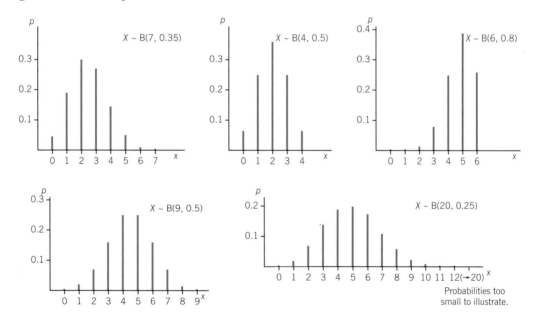

## The mode of the binomial distribution

The mode is the value of $X$ that is most likely to occur.
From the probability distribution sketches above, it can be seen that

- when $p = 0.5$ and $n$ is odd, there are two modes,
- otherwise the distribution has one mode.

The mode can be found by calculating all the probabilities and finding value of $X$ with the highest probability. This is however very tedious; it is usually only necessary to consider the probabilities of values of $X$ close to the mean $np$.

## Example 5.17

The probability that a student is awarded a distinction in the Mathematics examination is 0.05.

In a randomly selected group of 50 students, what is the most likely number of students awarded a distinction?

## Solution 5.17

$X$ is the number of students who are awarded a distinction in 50, so $X \sim B(50, 0.05)$.

$E(X) = np = 50 \times 0.05 = 2.5$, so calculate the probabilities for values of $X$ near 2.5.

$P(X = 1) = 50 \times 0.95^{49} \times 0.05 = 0.202 \ldots$

$P(X = 2) = {}^{50}C_2 \times 0.95^{48} \times 0.05^2 = \boxed{0.261} \ldots$

$P(X = 3) = {}^{50}C_3 \times 0.95^{47} \times 0.05^3 = 0.219 \ldots$

From the list, you can see that the value of $X$ with the highest probability is 2.

**The most likely number of students awarded a distinction in a group of 50 is two.**

---

## Exercise 5c    Expectation, variance and mode of the binomial distribution

1. 10% of the articles from a certain production line are defective. A sample of 25 articles is taken. Find the expected number of defective items and the standard deviation.

2. The probability that an apple picked at random from a sack is bad is 0.15.

    (a) Find the standard deviation of the number of bad apples in a sample of 15 apples.
    (b) What is the most likely number of bad apples in a sample of 30 apples?

3. The random variable $X$ is $B(n, 0.3)$ and $E(X) = 2.4$. Find $n$ and the standard deviation of $X$.

4. In a group of people the expected number who wear glasses is two and the variance is 1.6. Find the probability that

    (a) a person chosen at random from the group wears glasses,
    (b) six people in the group wear glasses.

5. The random variable $X$ is $B(10, p)$ where $p < 0.5$. The variance of $X$ is 1.875. Find

    (a) the value of $p$,
    (b) $E(X)$,
    (c) $P(X = 2)$.

6. A die is biased and the probability, $p$, of throwing a six is known to be less than $\frac{1}{6}$. An experiment consists of recording the number of sixes in 25 throws of the die.

In a large number of experiments the standard deviation of the number of sixes is 1.5. Calculate the value of $p$ and hence determine, to two places of decimals, the probability that exactly three sixes are recorded during a particular experiment.                                    (C)

7. In a certain African village, 80% of the villagers are known to have a particular eye disorder. Twelve people are waiting to see the nurse.

    (a) What is the most likely number to have the eye disorder?
    (b) Find the probability that fewer than half have the eye disorder.

8. In a bag there are six red counters, eight yellow counters and six green counters. An experiment consists of taking a counter at random from the bag, noting its colour and then replacing it in the bag. This procedure is carried out ten times in all. Find

    (a) the expected number of red counters drawn,
    (b) the most likely number of green counters drawn,
    (c) the probability that no more than four yellow counters are drawn.

9. The random variable $X$ is distributed binomially with mean 2 and variance 1.6. Find

    (a) the probability that $X$ is less than 6,
    (b) the most likely value of $X$.

10. Seeds are planted in rows of six and after 14 days the number of seeds which have germinated in each of the 100 rows is noted.
The results are shown in the table:

| Number of seeds germinating | 0 | 1 | 2 | 3 | 4 | 5 | 6 |
|---|---|---|---|---|---|---|---|
| Number of rows | 2 | 1 | 2 | 10 | 30 | 35 | 20 |

Find the theoretical frequencies of 0, 1, ..., 6 seeds germinating in a row, using the associated theoretical binomial distribution.

11. Each day a bakery delivers the same number of loaves to a certain shop which sells, on average, 98% of them. Assuming that the number of loaves sold per day has a binomial distribution with a standard deviation of 7, find the number of loaves the shop would expect to sell per day.
   (C Additional)

12. In a large batch of items from a production line the probability that an item is faulty is $p$.
400 samples, each of size 5, are taken and the number of faulty items in each batch is noted.
From the frequency distribution below estimate $p$ and work out the expected frequencies of 0, 1, 2, 3, 4, 5 faulty items per batch for a theoretical binomial distribution having the same mean.

| Number of faulty items | 0 | 1 | 2 | 3 | 4 | 5 |
|---|---|---|---|---|---|---|
| Frequency | 297 | 90 | 10 | 2 | 1 | 0 |

13. On average 20% of the bolts produced by a machine in a factory are faulty. Samples of ten bolts are to be selected at random each day.
Each bolt will be selected and replaced in the set of bolts which have been produced on that day.

(a) Calculate, to two significant figures, the probability that, in any one sample, two bolts or less will be faulty.
(b) Find the expected value and the variance of the number of bolts in a sample which will not be faulty. (L Additional)

14. An experiment consists of taking 12 shots at a target and counting the number of hits.
When this experiment was repeated a large number of times the mean number of hits was found to be 3. Calculate

(a) the probability of hitting the target with a single shot,
(b) the standard deviation of the number of hits in an experiment. (C Additional)

15. In an experiment a certain number of dice are thrown and the number of sixes obtained is recorded. The dice are all biased and the probability of obtaining a six with each individual die is $p$. In all there were 60 experiments and the results are shown in the table.

| Number of sixes obtained in an experiment | 0 | 1 | 2 | 3 | 4 | >4 |
|---|---|---|---|---|---|---|
| Frequency | 19 | 26 | 12 | 2 | 1 | 0 |

Calculate the mean and the standard deviation of these data.
By comparing these answers with those expected for a binomial distribution, estimate

(a) the number of dice thrown in each experiment,
(b) the value of $p$. (C Additional)

# THE POISSON DISTRIBUTION

Consider these random variables

- the number of emergency calls received by an ambulance control in an hour,
- the number of vehicles approaching a motorway toll bridge in a five-minute interval,
- the number of flaws in a metre length of material,
- the number of white corpuscles on a slide.

Assuming that each occurs randomly, they are all examples of variables that can be modelled using a **Poisson distribution**.

## Conditions for a Poisson model

- Events occur singly and at random in a given interval of time or space,
- $\lambda$, the mean number of occurrences in the given interval, is known and is finite.

The variable $X$ is **the number of occurrences in the given interval**.

If the above conditions are satisfied, $X$ is said to follow a Poisson distribution, written

$X \sim \text{Po}(\lambda)$, where

$$P(X = x) = e^{-\lambda}\frac{\lambda^x}{x!} \quad \text{for } x = 0, 1, 2, 3, \dots \text{ to infinity}$$

### Example 5.18

A student finds that the average number of amoebas in 10 ml of pond water from a particular pond is four. Assuming that the number of amoebas follows a Poisson distribution,

find the probability that in a 10 ml sample

(a) there are exactly five amoebas,
(b) there are no amoebas,
(c) there are fewer than three amoebas.

### Solution 5.18

$X$ is the number of amoebas in 10 ml of pond water, where $X \sim \text{Po}(4)$.

Using $P(X = x) = e^{-\lambda}\dfrac{\lambda^x}{x!}$ with $\lambda = 4$,

(a) $P(X = 5) = e^{-4}\dfrac{4^5}{5!}$

$= 0.156$ (3 s.f.)

(b) $P(X = 0) = e^{-4}\dfrac{4^0}{0!}$

$= 0.183$ (3 s.f.)

(c) $P(X < 3) = P(X = 0) + P(X = 1) + P(X = 2)$

$= e^{-4}\dfrac{4^0}{0!} + e^{-4}\dfrac{4^1}{1!} + e^{-4}\dfrac{4^2}{2!}$

$= e^{-4}(1 + 4 + 8)$

$= 13e^{-4}$

$= 0.238$ (3 s.f.)

*NOTE:*

- $P(X = 0) = e^{-4}\dfrac{4^0}{0!}$ but $4^0 = 1$ and $0! = 1$, so $P(X = 0) = e^{-4}$

- $P(X = 1) = e^{-4}\dfrac{4^1}{1!}$ but $4^1 = 4$ and $1! = 1$, so $P(X = 1) = 4e^{-4}$

These two results are useful in general

If $X \sim \text{Po}(\lambda)$,

then $P(X = 0) = e^{-\lambda}$ and $P(X = 1) = \lambda e^{-\lambda}$

# Unit interval

Care must be taken to specify the interval being considered.

In Example 5.18 the mean number of amoebas in 10 ml of pond water from a particular pond is four so the number in 10 ml is distributed Po(4).

Now suppose you want to find a probability relating to the number of amoebas in 5 ml of water from the same pond. The mean number of amoebas in 5 ml is two, so the number in 5 ml is distributed Po(2).

Similarly, the number of amoebas in 1 ml of pond water is distributed Po(0.4).

## Example 5.19

On average the school photocopier breaks down eight times during the school week (Monday to Friday). Assuming that the number of breakdowns can be modelled by a Poisson distribution, find the probability that it breaks down

(a) five times in a given week,
(b) once on Monday,
(c) eight times in a fortnight.

## Solution 5.19

(a) $X$ is the number of breakdowns in a week, where $X \sim$ Po(8).

$$P(X = 5) = e^{-8}\frac{8^5}{5!}$$
$$= 0.0916 \text{ (3 s.f.)}$$

(b) Let $Y$ be the number of breakdowns in a day.
The mean number of breakdowns in a day is $\frac{8}{5} = 1.6$, so $Y \sim$ Po(1.6).

$$P(Y = 1) = 1.6e^{-1.6}$$
$$= 0.323 \text{ (3 s.f.)}$$

(c) Let $F$ be the number of breakdowns in a fortnight.
The mean number of breakdowns in a fortnight is $2 \times 8 = 16$, so $F \sim$ Po(16).

$$P(F = 8) = e^{-16}\frac{16^8}{8!}$$
$$= 0.0120 \text{ (3 s.f.)}$$

# Mean and variance of the Poisson distribution

The mean number of occurrences in the interval, $\lambda$, is all that is needed to define the distribution completely; $\lambda$ is the only parameter of the distribution.

In a Poisson distribution, it is obvious that the mean, $E(X) = \lambda$, but it is also the case that $\text{Var}(X) = \lambda$. The following should be learnt:

If $X \sim$ Po($\lambda$)

$E(X) = \lambda$ and $\text{Var}(X) = \lambda$

## Example 5.20

$X$ follows a Poisson distribution with standard deviation 1.5. Find $P(X \geqslant 3)$.

## Solution 5.20

If $X \sim \text{Po}(\lambda)$ then $\text{Var}(X) = \lambda$.

But $\text{Var}(X) = (\text{standard deviation})^2 = 1.5^2 = 2.25$,

so $\lambda = 2.25$ and $X \sim \text{Po}(2.25)$.

$$
\begin{aligned}
P(X \geqslant 3) &= 1 - P(X < 3) \\
&= 1 - (P(X = 0) + P(X = 1) + P(X = 2) \\
&= 1 - e^{-2.25}\left(1 + 2.25 + \frac{2.25^2}{2!}\right) \\
&= 1 - 0.6093 \ldots \\
&= \mathbf{0.391} \text{ (3 s.f.)}
\end{aligned}
$$

# Using cumulative Poisson probability tables

If you have access to these tables you may wish to use them to calculate probabilities. The tables are printed on page 647. As with the cumulative binomial tables, they give $P(X \leqslant r)$ for various values $\lambda$, where $X \sim \text{Po}(\lambda)$.

Here is an extract for $\text{Po}(1.6)$.

|  | $\lambda = 1.6$ |  |  |
|---|---|---|---|
| $r = 0$ | 0.2019 | $\leftarrow$ | $P(X \leqslant 0)$ |
| 1 | 0.5249 | $\leftarrow$ | $P(X \leqslant 1)$ |
| 2 | 0.7834 | $\leftarrow$ | $P(X \leqslant 2)$ |
| 3 | 0.9212 | $\leftarrow$ | $P(X \leqslant 3)$ |
| 4 | 0.9763 | $\leftarrow$ | $P(X \leqslant 4)$ |
| 5 | 0.9940 | $\leftarrow$ | $P(X \leqslant 5)$ |
| 6 | 0.9987 | $\leftarrow$ | $P(X \leqslant 6)$ |
| 7 | 0.9997 | $\leftarrow$ | $P(X \leqslant 7)$ |
| 8 | 1.0000 | $\leftarrow$ | $P(X \leqslant 8)$ |

## Example 5.21

Given that $X \sim \text{Po}(1.6)$, use cumulative Poisson probability tables to find, to three decimal places,

(a) $P(X \leqslant 6)$,
(b) $P(X = 5)$,
(c) $P(X \geqslant 3)$,
(d) $P(X = 10)$.

Find also the smallest integer $n$ such that $P(X > n) < 0.01$.

## Solution 5.21

Using the table printed above,

(a) $P(X \leqslant 6) = 0.9987 = \mathbf{0.999}$ (3 d.p.)

(b) $\begin{aligned}[t] P(X = 5) &= P(X \leqslant 5) - P(X \leqslant 4) \\ &= 0.9940 - 0.9763 \\ &= \mathbf{0.018} \text{ (3 d.p.)} \end{aligned}$

(c) $P(X \geqslant 3) = 1 - P(X \leqslant 2)$
$$= 1 - 0.7834$$
$$= 0.217 \text{ (3 d.p.)}$$

(d) $X$ takes the values 0, 1, 2, ..., to infinity, but from the tables, $P(X \leqslant 8) = 1.0000$ to four decimal places. This implies that for values of $X$ greater than 8, the probabilities are very small, so to three decimal places, $P(X = 10) = 0.000$.

In fact, using the formula, $P(X = 10) = e^{-1.6} \times \dfrac{1.6^{10}}{10!} = 0.000\,006\,117\ldots$

If $\quad\quad P(X > n) < 0.01$
$$1 - P(X \leqslant n) < 0.01$$
$$P(X \leqslant n) > 0.99$$

From tables $\quad P(X \leqslant 4) = 0.9763 < 0.99$
$$P(X \leqslant 5) = 0.9940 > 0.99$$

The smallest integer $n$ is 5.

---

# Diagrammatic representation of the Poisson distribution

Notice that for small values of $\lambda$, the distribution is very skew, but it becomes more symmetrical as $\lambda$ increases.

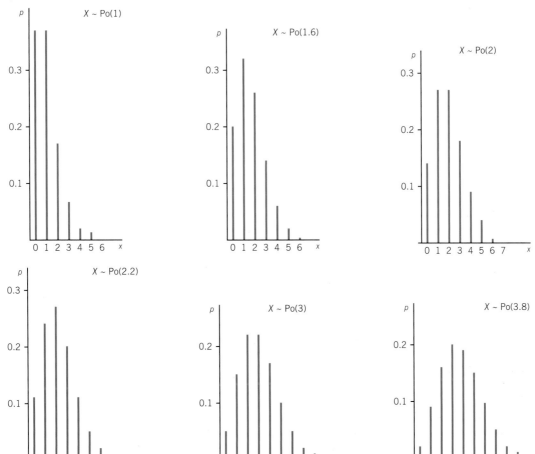

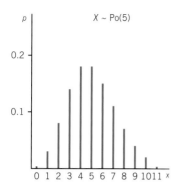

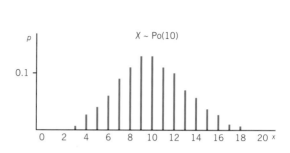

## The mode of the Poisson distribution

The mode is the value of $X$ that is most likely to occur, i.e. the one with the greatest probability.

From the diagrams, you can see that

    when $\lambda = 1$, there are two modes, 0 and 1,

    when $\lambda = 2$, there are two modes, 1 and 2,

    when $\lambda = 3$, there are two modes, 2 and 3.

In general, if $\lambda$ is an integer, there are two modes, $\lambda - 1$ and $\lambda$.

For example, if $X \sim Po(8)$, the modes are 7 and 8.

Notice also that

    when $\lambda = 1.6$, the mode is 1,

    when $\lambda = 2.2$, the mode is 2,

    when $\lambda = 3.8$, the mode is 3.

In general, if $\lambda$ is not an integer, the mode is the integer below $\lambda$.

For example, if $X \sim Po(4.9)$, the mode is 4.

## Fitting a theoretical distribution to practical data

As with the binomial distribution it is possible to fit a theoretical Poisson distribution to experimental data.

### Example 5.22

I recorded the number of e-mails I received over a period of 150 days with the following results:

| Number of e-mails | 0 | 1 | 2 | 3 | 4 |
|---|---|---|---|---|---|
| Number of days | 51 | 54 | 36 | 6 | 3 |

(a) Find the mean number of e-mails per day.

(b) Calculate the frequencies of the Poisson distribution having the same mean.

**Solution 5.22**

(a) $\bar{x} = \dfrac{\Sigma fx}{\Sigma f} = \dfrac{156}{150} = \mathbf{1.04}$

(b) Let $X$ be the number of e-mails received in a day. For a Poisson distribution with the same mean, use $X \sim Po(1.04)$ and calculate the probabilities of 0, 1, 2, 3, 4, ... e-mails. Multiply these by 150 to obtain the theoretical frequencies.

| $x$ | $P(X = x)$ | | Frequency (nearest integer) |
|---|---|---|---|
| 0 | $e^{-1.04}$ | $= 0.3534 \ldots$ | 53 |
| 1 | $1.04e^{-1.04}$ | $= 0.3675 \ldots$ | 55 |
| 2 | | $= 0.1911 \ldots$ | 29 |
| 3 | $e^{-1.04} \times \dfrac{1.04^3}{3!}$ | $= 0.0662 \ldots$ | 10 |
| 4 | $e^{-1.04} \times \dfrac{1.04^4}{4!}$ | $= 0.0172 \ldots$ | 3 |
| >4 | $1 - P(X \leqslant 4)$ | $= 0.000\,431 \ldots$ | 0 |
| | | | Total 150 |

These compare reasonably well with the original distribution.

A statistical test to compare the two sets of data, the $\chi^2$ test, is illustrated on page 573.

## Exercise 5d   The Poisson distribution

1. An insurance company receives on average two claims per week from a particular factory. Assuming that the number of claims can be modelled by a Poisson distribution, find the probability that it receives

   (a) three claims in a given week,
   (b) more than four claims in a given week,
   (c) four claims in a given fortnight,
   (d) no claims on a given day, assuming that the factory operates on a five-day week.

2. A sales manger receives six telephone calls on average between 9.30 a.m. and 10.30 a.m. on a weekday. Find the probability that

   (a) she will receive two or more calls between 9.30 a.m. and 10.30 a.m. on Tuesday,
   (b) she will receive exactly two calls between 9.30 a.m. and 9.40 a.m. on Wednesday,
   (c) during a five-day working week, there will be exactly three days on which she receives no calls between 10.00 a.m. and 10.10 a.m.

3. The number of bacterial colonies on a petri dish can be modelled by a Poisson distribution with average number 2.5 per cm$^2$. Find the probability that

   (a) in 1 cm$^2$ there are no bacterial colonies,
   (b) in 2 cm$^2$ there are more than four bacterial colonies,
   (c) in 4 cm$^2$ there are six bacterial colonies.

4. On a particular motorway bridge, breakdowns occur at a rate of 3.2 a week. Assuming that the number of breakdowns can be modelled by a Poisson distribution, find the probability that

   (a) fewer than the mean number of breakdowns occur in a particular week,
   (b) more than five breakdowns occur in a given fortnight,
   (c) exactly three breakdowns occur in each of four successive weeks.

5. Cars arrive at a petrol station at an average rate of 30 per hour. Assuming that the cars arrive at random, find the probability that

   (a) no cars arrive during a particular five-minute interval,
   (b) more than three cars arrive during a five-minute interval,
   (c) more than five cars arrive in a 15-minute interval,
   (d) in a half hour period, ten cars arrive,
   (e) fewer than three cars arrive in a ten-minute interval.

6. Flaws occur randomly in a roll of fabric at an average rate of 1.5 per metre length.

   (a) Find the probability that in a randomly chosen one-metre length there are more than two flaws.
   (b) Find the probability that in a randomly chosen two-metre length there are no flaws.
   (c) What is the standard deviation of the number of flaws in a four-metre length?

7. The number of calls made to a Health Centre can be modelled by a Poisson distribution with standard deviation 2 per five-minute interval. Find the probability that in a given five-minute interval, the number of calls is more than the average for a five-minute interval.

8. The average number of misprints on each page in the first draft of a novel is four. Find the probability that on a randomly selected double page

   (a) there are three misprints on each page
   (b) there are six misprints in total.

9. The number of goals scored in a match by Random Rovers can be modelled using a Poisson distribution. The probability, to three decimal places, that the team scores no goals is 0.135. Given that the mean number of goals scored in a match is an integer, find the probability that the team scores fewer than three goals in a match.

10. The number of accidents occurring in a week in a certain factory follows a Poisson distribution with variance 3.2. Find

    (a) the most likely number of accidents in a given week,
    (b) the probability that exactly seven accidents happen in a given fortnight.

11. For each of the following sets of data, fit a theoretical Poisson distribution with the same mean.

    (a)

    | $x$ | 0 | 1 | 2 | 3 | 4 | 5 |
    |-----|-----|-----|-----|-----|-----|-----|
    | $f$ | 110 | 50 | 20 | 12 | 7 | 1 |

    (b)

    | $x$ | 0 | 1 | 2 | 3 | 4 |
    |-----|-----|-----|-----|-----|-----|
    | $f$ | 45 | 44 | 20 | 8 | 3 |

12. A firm investigated the number of employees suffering injuries whilst at work. The results recorded below were obtained for a 52-week period:

    | Number of employees injured in a week | Number of weeks |
    |-----|-----|
    | 0 | 31 |
    | 1 | 17 |
    | 2 | 3 |
    | 3 | 1 |
    | 4 or more | 0 |

    Give reasons why one might expect this distribution to approximate to a Poisson distribution. Evaluate the mean and variance of the data and explain why this gives further evidence in favour of a Poisson distribution. Using the calculated value of the mean, find the theoretical frequences of a Poisson distribution for the number of weeks in which 0, 1, 2, 3, 4 or more, employees were injured. (C)

13. Along a stretch of motorway, breakdowns require the summoning of the breakdown services occur with a frequency of 2.4 per day, on average. Assuming that the breakdowns occur randomly and that they follow a Poisson distribution, find

    (a) the probability that there will be exactly two breakdowns on a given day,
    (b) the smallest integer $n$ such that the probability of more than $n$ breakdowns in a day is less 0.03.

# USING THE POISSON DISTRIBUTION AS AN APPROXIMATION TO THE BINOMIAL DISTRIBUTION

When $n$ is large ($n > 50$) and $p$ is small ($p < 0.1$), the binomial distribution

$$X \sim B(n, p)$$

can be approximated using a Poisson distribution with the same mean, i.e. $X \sim Po(np)$.
The larger the value of $n$ and the smaller the value of $p$ the better the approximation.

### Example 5.23

Eggs are packed into boxes of 500. On average 0.7% of the eggs are found to be broken when the eggs are unpacked. Find, correct to two significant figures, the probability that in a box of 500 eggs,

(a) exactly three are broken,
(b) at least two are broken.

### Solution 5.23

Let $X$ be the number of broken eggs in a box of 500.
$P$(egg is broken) = 0.007, so $X \sim B(500, 0.007)$.

$E(X) = np = 500 \times 0.007 = 3.5$
Since $n > 50$ and $p < 0.1$, use a Poisson approximation, $X \sim Po(3.5)$.

(a)  $P(X = 3) = e^{-3.5} \dfrac{3.5^3}{3!}$
    $= 0.22$ (2 s.f.)

(b)  $P(X \geqslant 2) = 1 - (P(X = 0) + P(X = 1))$
    $= 1 - (e^{-3.5} + 3.5e^{-3.5})$
    $= 0.86$ (2 s.f.)

---

### Example 5.24

A Christmas draw aims to sell 5000 tickets, 50 of which will win a prize.

(a) A syndicate buys 200 tickets. Let $X$ represent the number of these tickets that win a prize.
    (i)   Justify the use of the Poisson approximation for the distribution of $X$.
    (ii)  Calculate $P(X \leqslant 3)$.

(b) Calculate how many tickets should be bought in order for there to be a 90% probability of winning at least one prize.                                              (C)

### Solution 5.24

$P$(a ticket wins a prize) = $\frac{50}{5000} = 0.01$

(a) Let $X$ be the number of these tickets that win a prize.
    Strictly speaking you do not have independent trials, but since $n$ is very large, $X$ can be considered to be modelled by a binomial distribution where $X \sim B(200, 0.01)$

$E(X) = np = 200 \times 0.01 = 2$.

(i) Since $n > 50$ and $p < 0.1$, use a Poisson approximation, $X \sim \text{Po}(2)$.

(ii) $P(X \leqslant 3) = P(X = 0) + P(X = 1) + P(X = 2) + P(X = 3)$

$$= e^{-2} + 2e^{-2} + \frac{2^2}{2!}e^{-2} + \frac{2^3}{3!}e^{-2}$$

$$= 0.86 \text{ (2 s.f.)}$$

(b) Let $X$ be the number of these tickets that win a prize in $n$ tickets, so $X \sim \text{B}(n, 0.01)$ and
$E(X) = np = 0.01n$
Assuming $n > 50$ and $p < 0.1$, use $X \sim \text{Po}(0.01n)$.

You want $P(X \geqslant 1) = 0.9$
But $\qquad P(X \geqslant 1) = 1 - P(X = 0)$
$$\qquad\qquad\qquad = 1 - e^{-0.01n}$$
So $\qquad 0.9 = 1 - e^{-0.01n}$
$$e^{-0.01n} = 0.1$$

Taking logs to base e

$$-0.01n = \ln 0.1$$
$$n = -\frac{\ln 0.1}{0.01}$$
$$n = 230.25 \ldots$$

So the least integer value of $n$ must be 231.

Check: If $n = 230$, $np = 230 \times 0.01 = 2.3$ and $1 - e^{-2.3} = 0.8997 \ldots < 0.9$
If $n = 231$, $np = 231 \times 0.01 = 2.31$ and $1 - e^{-2.31} = 0.9007 \ldots > 0.9$

**231 tickets should be bought.**

Note that $n$ can be found by trial and improvement methods if logarithms are not used.

---

## Exercise 5e    The Poisson approximation to the binomial

1. The random variable $X$ is $B(100, 0.03)$.
   Find the following probabilities using
   (i)   the binomial distribution
   (ii)  a Poisson approximation

   (a) $P(X = 0)$,   (b) $P(X = 2)$,   (c) $P(X = 4)$.

2. The probability that a bolt is defective is 0.2%.
   Bolts are packed in boxes of 500.

   (a) Find the probability that in a randomly
       chosen box,
       (i)   there are two defective bolts,
       (ii)  there are more than three defective
             bolts.
   (b) Two boxes are picked at random from the
       production line. Find the probability that
       one has two defective bolts and the other
       has no defective bolts.
   (c) Three boxes are selected at random. Find
       the probability that they contain no
       defective bolts.

3. On average one in 200 cars breaks down on a
   certain stretch of road per day. Find the
   probability that, on a randomly chosen day,

   (a) none of a sample of 250 cars break down,
   (b) more than two of a sample of 300 cars
       break down.

4. Two dice are thrown

   (a) What is the probability of throwing a
       double six?

   Two dice are thrown a total of 90 times.

   (b) What is the probability that at least two
       double sixes are thrown?

5. An aircraft has 116 seats. The airline has found, from long experience, that on average 2.5% of people who have bought tickets for a flight do not arrive for that flight. The airline sells 120 tickets for a particular flight.

   (a) Calculate, using a suitable approximation, the probability that more than 116 people arrive for the flight.
   (b) Calculate also the probability that there are empty seats on the flight. (C)

6. In a large town one person in 80, on average, has blood of type X. If 200 blood donors are sampled at random, find an approximation to the probability that they include at least five people with blood type X.
   How many donors must be sampled in order that the probability of including at least one donor of type X is 90% or more? (AEB)

7. A lottery has a very large number of tickets, one in every 500 of which entitles the purchaser to prize. An agent sells 1000 tickets for the lottery. Using a Poisson approximation, find, to three decimal places, the probability that the number of prize-winning tickets sold by the agent is

   (a) less than three
   (b) more than five.

   Calculate the minimum number of tickets the agent must sell to have a 95% chance of selling at least one prize-winning ticket. (NEAB)

8. A manufacturer has found that 3% of seeds produced do not germinate. Using a Poisson approximation, find, to two significant figures, the probability that in a pack containing 150 seeds,

   (a) more than four fail to germinate,
   (b) at least 145 germinate

9. X is $B(250, p)$. The value of $p$ is such that it is valid to apply a Poisson approximation. When this is done, it is found that $P(X = 0) = 0.0235$. Find the value of $p$.

10. The probability that I dial a wrong number when making a telephone call is 0.015. In a typical week I will make 50 telephone calls. Using a Poisson approximation to a binomial model find, correct to two decimal places, the probability that in such a week,

    (a) I dial no wrong numbers,
    (b) I dial more than two wrong numbers.

    Comment on the suitability of the binomial model and of the Poisson approximation. (C)

11. A newspaper reports that 8.6% of adults in the U.K. painted the outside of their houses. A sample of 55 adults in the U.K. was selected. Stating any necessary assumptions, show that the number in the sample that painted the outsides of their own houses can be approximated by a Poisson distribution.
    Using this approximation, find the probability that fewer than four people in the sample painted the outsides of their own houses. (C)

# THE SUM OF INDEPENDENT POISSON VARIABLES

For independent variables, $X$ and $Y$, if $X \sim \text{Po}(m)$ and $Y \sim \text{Po}(n)$,

then   $X + Y \sim \text{Po}(m + n)$

## Example 5.25

Two identical racing cars are being tested on a circuit. For each car, the number of mechanical breakdowns can be modelled by a Poisson distribution with a mean of one breakdown in 100 laps. If a car breaks down it is attended and continues on the circuit. The first car is tested for 20 laps and the second car for 40 laps.

Find the probability that the service team is called out to attend to breakdowns

(a) once,
(b) more than twice.

## Solution 5.25

Since the average number of breakdowns in 100 laps is one, the average number in 20 laps is 0.2 and the average number in 40 laps is 0.4.

Let $X$ be the number of breakdowns of the first car, then $X \sim \text{Po}(0.2)$
Let $Y$ be the number of breakdowns of the second car, then $Y \sim \text{Po}(0.4)$
Let $T$ be the total number of breakdowns,
Then $\quad T = X + Y$ and $T \sim \text{Po}(0.2 + 0.4)$, i.e. $T \sim \text{Po}(0.6)$

(a) $P(T = 1) = 0.6e^{-0.6}$
$\qquad\qquad = \mathbf{0.329}$ (3 d.p.)

(b) $P(T > 2) = 1 - (P(T = 0) + P(T = 1) + P(T = 2))$
$\qquad\quad = 1 - e^{-0.6}\left(1 + 0.6 + \dfrac{0.6^2}{2!}\right)$
$\qquad\quad = \mathbf{0.023}$ (3 d.p)

---

## Example 5.26

The centre pages of the Weekly Sentinel consist of a page of film and theatre reviews and a page of classified advertisement. The number of misprints in the reviews can be modelled using a Poisson distribution with mean 2.3 and the number of misprints in the classified section can be modelled by a Poisson distribution with mean 1.7.

Using cumulative Poisson probability tables, find
(a) the probability that on the centre pages there will be more than five misprints,
(b) the smallest integer $n$ such that the probability that there are more than $n$ misprints on the centre pages is less than 5%.

## Solution 5.26

Let $X$ be the number of misprints in the reviews, then $X \sim \text{Po}(2.3)$
Let $Y$ be the number of misprints in the classified advertisements, then $Y \sim \text{Po}(1.7)$
Let $T$ be the total number of misprints on the centre pages, then $T = X + Y$ and
$T \sim \text{Po}(2.3 + 1.7)$, i.e. $T \sim \text{Po}(4)$.

The cumulative tables are printed on page 647 and the relevant extract is shown here:

| $\lambda = 4.0$ | $P(X \leqslant r)$ |
|---|---|
| $r = 0$ | 0.0183 |
| 1 | 0.0916 |
| 2 | 0.2381 |
| 3 | 0.4335 |
| 4 | 0.6288 |
| 5 | 0.7851 |
| 6 | 0.8893 |
| 7 | 0.9489 |
| 8 | 0.9786 |
| 9 | 0.9919 |
| 10 | 0.9972 |
| 11 | 0.9991 |
| 12 | 0.9997 |
| 13 | 0.9999 |
| 14 | 1.0000 |
| 15 | |

(a) $P(T > 5) = 1 - P(T \leqslant 5)$
$\qquad\qquad = 1 - 0.7851$
$\qquad\qquad = \mathbf{0.215}$ (3 d.p.)

(b) You need the smallest value of $n$ such that

$$P(T > n) < 0.05$$
i.e. $\qquad\qquad 1 - P(T \leqslant n) < 0.05$
so $\qquad\qquad P(T \leqslant n) > 0.95$

From the tables, $\qquad P(T \leqslant 7) = 0.9489 < 0.95$
$\qquad\qquad\qquad\quad P(T \leqslant 8) = 0.9786 > 0.95$

The **smallest value of $n$ is 8.**

# Exercise 5f   Sums of Poisson variables

1. Telephone calls reach a secretary independently and at random, internal ones at a mean rate of two in any five-minute period, and external ones at a mean rate of one in any five-minute period. Calculate the probability that there will be more than two calls in any period of two minutes.

   (O & C)

2. During a weekday, heavy lorries pass a census point $P$ on a village high street independently and at random times. The mean rate for westward travelling lorries is two in any 30-minutes period, and for eastward travelling lorries is three in any 30-minute period.

   Find the probability

   (a) that there will be no lorries passing $P$ in a given ten-minute period,
   (b) that at least one lorry from each direction will pass $P$ in a given ten-minute period,
   (c) that there will be exactly four lorries passing $P$ in a given 20-minute period.     (O & C)

3. A large number of screwdrivers from a trial production run is inspected. It is found that the cellulose acetate handles are defective on 1% and that the chrome steel blades are defective on $1\frac{1}{2}$% of the screwdrivers, the defects occurring independently.

   (a) What is the probability that a sample of 80 contains more than two defective screwdrivers?
   (b) What is the probability that a sample of 80 contains at least one screwdriver with both a defective handle and defective blade?

   (O & C)

4. A restaurant kitchen has two food mixers, $A$ and $B$. The number of times per week that $A$ breaks down has a Poisson distribution with mean 0.4, while independently the number of times that $B$ breaks down in a week has a Poisson distribution with mean 0.1. Find, to three decimal places, the probability that in the next three weeks

   (a) $A$ will not break down at all,
   (b) each mixer will break down exactly once,
   (c) there will be a total of two breakdowns.  (L)

## Summary

- Uniform distribution

$$P(X = x_r) = \frac{1}{n} \text{ for } r = x_1, x_2, \ldots, x_n$$

- Geometric distribution $X \sim \text{Geo}(p)$

  $p$ is the probability of a successful outcome

  $X$ is the number of independent trials needed to obtain the first successful outcome.

  $P(X = x) = q^{x-1} \times p$ for $x = 1, 2, 3, \ldots$, to infinity, where $q = 1 - p$

  Note that $X$ cannot be zero.

  $$E(X) = \frac{1}{p}, \quad \text{Var}(X) = \frac{q}{p^2}, \quad \text{mode} = 1.$$

- Binomial distribution $X \sim B(n, p)$

  $p$ is the probability of a successful outcome.

  $X$ is the number of successful outcomes in $n$ independent trials

  $P(X = x) = {}^nC_x\, q^{n-x} \times p^x$ for $x = 0, 1, 2, \ldots, n$, where $q = 1 - p$

  $E(X) = np$ and $\text{Var}(X) = npq$.

- Poisson distribution $X \sim \text{Po}(\lambda)$

  $X$ is the number of occurrences of an event in a given interval of time or space, when the mean number of occurrences in the given interval is $\lambda$.

  $$P(X = x) = e^{-\lambda}\frac{\lambda^x}{x!} \text{ for } x = 0, 1, 2, 3, \ldots, \text{ to infinity}$$

  $E(X) = \lambda$ and $\text{Var}(X) = \lambda$.

- Poisson approximation to the binomial distribution

  If $X \sim B(n, p)$ with $n > 50$ and $p < 0.1$, then $X \sim \text{Po}(np)$ approximately.

- Sum of independent Poisson variables

  If $X \sim \text{Po}(m)$ and $X \sim \text{Po}(n)$ then $X + Y \sim \text{Po}(m + n)$.

# Miscellaneous worked examples

## Example 5.27

Every working day Mr Driver pulls out from his drive on to a main road in such a way that there is a very small probability $p$ that his car will be involved in a collision.

(a) Show that in a five-day week the probability that there will be no collision is $(1 - p)^5$.
(b) State one assumption that is made in this calculation.
(c) Using a binomial expansion, show that the probability that there will be at least one collision in a five-day week is approximately $5p$.
(d) Given that $p = 0.001$, use a calculator to find the probability that Mr Driver will avoid a collision in 500 working days. (NEAB)

## Solution 5.27

(a) $X$ is the number of collisions in five days, $X \sim B(5, p)$.

$$P(X = 0) = q^5$$
$$= (1 - p)^5 \quad \text{where } q = 1 - p.$$

(b) The circumstances remain the same for the five days; the events are independent.

(c) $\quad P(X \geqslant 1) = 1 - P(X = 0)$
$$= 1 - (1 - p)^5$$
$$= 1 - (1 - 5p + 10p^2 + \cdots)$$
$$\approx 1 - 1 + 5p \quad \text{(ignoring higher powers of } p \text{ since } p \text{ is small)}$$
$$\therefore \quad P(X \geqslant 1) \approx 5p$$

(d) $X$ is the number of collisions in 500 days, $X \sim B(500, 0.001)$
Using the binomial distribution

$$P(X = 0) = 0.999^{500} = \textbf{0.606 (3 d.p.)}$$

Alternatively, since $n > 50$ and $p < 0.1$, using a Poisson approximation with mean $= np = 0.5$, $X \sim Po(0.5)$, so

$$P(X = 0) = e^{-0.5} = \textbf{0.606 (3 d.p.)}$$

## Example 5.28

A salesman sells goods by telephone. The probability that any particular call achieves a sale is $\frac{1}{12}$, independently of all other calls. The salesman continues to make calls until one call achieves a sale.

(a) Name an appropriate distribution with which to model this situation.

(b) Calculate the probability that the call that achieves a sale
  (i) is the fifth call made,
  (ii) does not occur in the first five calls.

(c) Obtain the mean and variance of the number of calls the salesman makes. (C)

## Solution 5.28

(a)  $X$ is the number of calls until a call achieves a sale.
$X$ can be modelled by a geometric distribution, $X \sim \text{Geo}(\frac{1}{12})$.

(b)  (i)   $P(X = 5) = q^4 p$
$$= (\tfrac{11}{12})^4 \times \tfrac{1}{12}$$
$$= 0.059 \ (2 \text{ s.f.})$$

 (ii)   $P(X > 5) = q^5$
$$= (\tfrac{11}{12})^5$$
$$= 0.65 \ (2 \text{ s.f.})$$

(c)  $E(X) = \dfrac{1}{p} = \dfrac{1}{\frac{1}{12}} = 12$

$\text{Var}(X) = \dfrac{q}{p^2} = \dfrac{\frac{11}{12}}{(\frac{1}{12})^2} = 132$

So $E(X) = 12$ and $\text{Var}(X) = 132$.

---

## Example 5.29

The number of births announced in the personal column of a local weekly newspaper may be modelled by a Poisson distribution with mean 2.4.

Find the probability that, in a particular week,

(a)  three or fewer births will be announced,
(b)  exactly four births will be announced.                                           (*AEB*)

## Solution 5.29

$X$ is the number of birth announcements in a week, $X \sim \text{Po}(2.4)$.

(a)  $P(X \leqslant 3) = P(X = 0) + P(X = 1) + P(X = 2) + P(X = 3)$
$$= e^{-2.4} + 2.4 e^{-2.4} + \frac{2.4^2}{2!} e^{-2.4} + \frac{2.4^3}{3!} e^{-2.4}$$
$$= e^{-2.4}\left(1 + 2.4 + \frac{2.4^2}{2!} + \frac{2.4^3}{3!}\right)$$
$$= 0.779 \ (3 \text{ s.f.})$$

(b)  $P(X = 4) = e^{-2.4} \dfrac{2.4^4}{4!}$
$$= 0.125 \ (3 \text{ s.f.})$$

---

## Example 5.30

Weak spots occur at random in the manufacture of a certain cable at an average rate of one per 100 metres.
If $X$ represents the number of weak spots in 100 metres of cable, write down the distribution of $X$.

Lengths of this cable are wound on to drums. Each drum carries 50 metres of cable.
Find the probability that a drum will have three or more weak spots.
A contractor buys five such drums. Find the probability that two have just one weak spot each
and the other three have none. *(AEB)*

## Solution 5.30

$X$ is the number of weak spots in 100 m of cable, $X \sim Po(1)$.
Let $Y$ be the number of weak spots in 50 m of cable, $Y \sim Po(0.5)$.

$$
\begin{aligned}
P(Y \geqslant 3) &= 1 - P(Y < 3) \\
&= 1 - (P(Y = 0) + P(Y = 1) + P(Y = 2)) \\
&= 1 - \left( e^{-0.5} + 0.5\, e^{-0.5} + \frac{0.5^2}{2!}\, e^{-0.5} \right) \\
&= 1 - e^{-0.5}\left( 1 + 0.5 + \frac{0.5^2}{2!} \right) \\
&= 0.01438 \ldots \\
&= \mathbf{0.014} \textbf{ (2 s.f.)}
\end{aligned}
$$

$P(\text{a drum has one weak spot}) = P(Y = 1) = 0.5 e^{-0.5}$

$P(\text{a drum has no weak spot}) = P(Y = 0) = e^{-0.5}$

In five drums,

$P(\text{2 have one weak spot, 3 have none})$

$$
\begin{aligned}
&= {}^5C_3 \times (P(Y = 1))^2 \times (P(Y = 0))^3 \\
&= 10 \times (0.5\, e^{-0.5})^2 \times (e^{-0.5})^3 \\
&= 10 \times 0.25\, e^{-1} \times e^{-1.5} \\
&= 2.5 \times e^{-2.5} \\
&= \mathbf{0.21} \textbf{ (2 s.f.)}
\end{aligned}
$$

# Miscellaneous exercise 5g

1. The random variable $X$ has the binomial
   distribution $B(10, 0.35)$. Find $P(X \leqslant 4)$.

   The random variable $Y$ has the Poisson
   distribution with mean 3.5. Find $P(2 < Y \leqslant 5)$. *(L)*

2. The number of white corpuscles on a slide has a
   Poisson distribution with mean 3.2.

   (a) Find the most likely number of white
       corpuscles on a slide.
   (b) Calculate correct to three decimal places the
       probability of obtaining this number.
   (c) If two such slides are prepared, what is the
       probability, correct to three decimal places,
       of obtaining at least two white corpuscles in
       total on the two slides?

3. Copies of an advertisement for a course in
   practical statistics are sent to Mathematics
   teachers in a large city. For each teacher who
   receives a copy, the probability of subsequently
   attending the course is 0.09.
   Twenty teachers receive a copy of the advertisement. What is the probability that the number
   who subsequently attend the course will be

   (a) two or fewer,
   (b) exactly four. *(AEB)*

4.  (a)  Experience shows that a charity receives
         replies to letters at the rate of eight per 100.
         Calculate, giving each answer to two
         decimal places, the probability that the
         number of replies from ten letters is

         (i) 0,   (ii) 1,   (iii) 2,   (iv) more than 2.

    (b)  On average, out of every 1000 items made
         by a certain factory worker, one item is
         defective. The items are inspected in batches.
         A large number of batches, each of $n$ items,
         made by the worker is inspected. Evaluate $n$
         in each of the following cases.

         (i)   The mean number of defective items per
               batch is 0.045.
         (ii)  The standard deviation of the number
               of defective items per batch 0.333.

         The probability of at least one defective item
         in a batch of $N$ items is greater than 0.02.
         Use this information to write down an
         inequality which is satisfied by $N$.   (C)

5.  A store sells word processors. The proportion
    which are returned as faulty has been found to
    be 0.035. During the Christmas period of 1995,
    the store sold 104 word processors. The number
    of these which will be returned as faulty is $X$.
    Assuming independence, state the exact
    distribution of $X$.
    Give reasons why this distribution can be
    approximated by a Poisson distribution.
    Calculate the probability that at most three of
    the word processors will be returned as faulty.
                                                    (C)

6.  (a)  The probability that a seed of a particular
         variety of bean will germinate when sown is
         0.96.
         Seeds are sold in packets of 50. If a packet is
         selected at random, calculate the probability
         that the number of seeds which will
         germinate when sown is exactly

         (i) 50,   (ii) 49,   (iii) 48.

         If 200 packets of seeds are selected, estimate
         the number of packets from each of which
         fewer than 48 seeds will germinate.
         If three packets of seeds are selected,
         calculate, to three decimal places, the
         probability that at least 149 of the 150 seeds
         will germinate.

    (b)  A self-employed worker contacts an agency
         every morning in an attempt to obtain work
         for the day. The probability that work is
         available on any given day is 0.9. Calculate,
         for a period of 100 working days, the mean
         and the standard deviation of the number of
         days on which work is available.   (C)

7.  A car hire firm has three cars, which it hires out
    on daily basis. The number of cars demanded
    per day follows a Poisson distribution with
    mean 2.1.

    (a)  Find the probability that exactly two cars
         are hired out on any one day.
    (b)  Find the probability that all cars are in use
         on any one day.
    (c)  Find the probability that all cars are in use
         on exactly three days of a five-day week.
    (d)  Find the probability that exactly ten cars are
         demanded in a five-day week. Explain
         whether or not such a demand could always
         be met.
    (e)  It costs the firm £20 a day to run each car,
         whether it is hired out or not. The daily hire
         charge per car is £50. Find the expected
         daily profit.   (MEI)

8.  A factory produces a particular type of electronic
    component. The probability of a component
    being acceptable is 0.95. The components are
    packed in boxes of 24.

    (a)  Calculate the probability that a box, chosen
         at random, contains exactly 22 acceptable
         components.

    All boxes are inspected and a box is rejected if it
    contains fewer than 22 acceptable components.

    (b)  Calculate the probability that a box, chosen
         at random, is rejected.

    The factory produces 80 boxes per day over a
    long period of time.

    (c)  Estimate the mean and the standard
         deviation of the number of boxes rejected
         per day.

    It is proposed to introduce an alternative policy
    with regard to packing and inspection, as
    follows:

    The daily production of components is to be
    packed in 160 boxes, each containing 12
    components, and boxes containing fewer than 11
    acceptable components are to be rejected.

    (d)  Estimate the mean number of boxes rejected
         per day under this alternative policy.
    (e)  Explain whether or not this alternative
         policy would lead to a decrease in the
         expected number of components rejected per
         day.   (C)

9.  A large bin contains 5250 used golf balls, 1260
    of which are unusable. The random variable $R$
    denotes the number of unusable balls in a
    random sample of ten balls, selected without
    replacement, from the bin.

    (a)  Explain why $R$ may be approximated as a
         binomial random variable with parameters
         10 and 0.24.
    (b)  Hence calculate the probability that the
         sample contains
         (i)   exactly three unusable balls,
         (ii)  at most three unusable balls.   (NEAB)

10. The number of night calls to a fire station in a small town can be modelled by a Poisson distribution with mean 4.2 per night. Find the probability that on a particular night there will be three or more calls to the fire station.
State what needs to be assumed about the calls to the fire station in order to justify a Poisson model. (C)

11. A television repair company uses a particular spare part at a rate of four per week.
Assuming that requests for this spare part occur at random, find the probability that
(a) exactly six are used in a particular week,
(b) at least ten are used in a two-week period,
(c) exactly six are used in each of three consecutive weeks.

The manager decides to replenish the stock of this spare part to a constant level $n$ at the start of each week.

(d) Find the value of $n$ such that, on average, the stock will be insufficient no more than once in a 52-week year. (L)

12. In the Growmore Market Garden plants are inspected for the presence of the deadly red angus leaf bug. The number of bugs per leaf is known to follow a Poisson distribution with mean one. What is the probability that any one leaf on a given plant will have been attacked (at least one bug is found on it)?

A random sample of 12 plants is taken. For each plant ten leaves are selected at random and inspected for these bugs. If more than eight leaves on any particular plant have been attacked then the plant is destroyed. What is the probability that exactly two of these 12 plants are destroyed? (AEB)

13. In Blackbury it is known that 0.4% of people have blood group $AB^-$.
Blackbury High School has 1000 pupils, with 28 pupils in class 4T.
(a) (i) Write down a distribution that could be used to model the number of pupils in class 4T with blood group $AB^-$.
    (ii) Hence calculate the probability that there are exactly two pupils in class 4T with blood group $AB^-$.
(b) Using an appropriate distributional approximation, calculate the probability that there are fewer than six pupils at the school with blood group $AB^-$.
(c) State an assumption that you have made in answering this question. (NEAB)

14. The probability that a fisherman has a successful day's fishing is 0.6. Given that he fishes for six days every week, find the probability that in any week he has
(a) exactly four successful days,
(b) at least two successful days.

The fisherman fishes for six days every week for many weeks. Estimate the mean and the standard deviation of the number of successful days per week over this period (C)

15. A large number of groups, each consisting of 12 adults, are selected at random from the population of a particular town. Given that 30% of the adults in this town are car owners, calculate
(a) the probability that a group contains not more than two car owners,
(b) the mean and the standard deviation of the number of car owners in the groups. (C)

16. In a large city one person in five is left-handed.
(a) Find the probability that in a random sample of ten people
    (i) exactly three will be left-handed,
    (ii) more than half will be left-handed.
(b) Find the most likely number of left-handed people in a random sample of 12 people.
(c) Find the mean and the standard deviation of the number of left-handed people in a random sample of 25 people.
(d) How large must a random sample be if the probability that it contains at least one left-handed person is to be greater than 0.95?

17. Batches of 400 shells in the First World War were classified as 'accepted' or 'rejected' by testing a small number of shells from the batch. Tested shells are either 'good' or 'bad'; the probability that a randomly selected shell is good is $p$.
(a) In one testing method, eight shells from a batch (of 400) are selected at random and tested. The batch is accepted if at least three of these eight shells are good. Use a binomial distribution, with $p = 0.2$, to find the probability that the batch is accepted.
(b) In a second testing method, each batch of 400 is subdivided into four sub-batches of 100 shells each. Two shells from each sub-batch are tested, and the sub-batch is accepted if at least one of the two shells is good. Use a binomial distribution, with $p = 0.2$,
    (i) to show that the probability that one particular sub-batch is accepted is 0.36,
    (ii) to find the probability that, out of four sub-batches, at least three are accepted.
(c) In a third testing method, four shells are selected and the batch (of 400) is accepted if all four of the shells are good. The probability that the batch is accepted is 0.01. Assuming a binomial distribution, find the value of $p$.
(d) State one condition which must be satisfied by the shells if a binomial model is to be valid, and give a reason why it may not be satisfied in this context. (C)

18. Define the Poisson distribution and state its mean and variance.

The number of telephone calls received at a switchboard in any time interval of length $T$ minutes has a Poisson distribution with mean $\frac{1}{2}T$. The operator leaves the switchboard unattended for five minutes.

Calculate to three decimal places the probabilities that there are (a) no calls, (b) four or more calls in her absence.

Find to three significant figures the maximum length of time in seconds for which the operator could be absent with a 95% probability of not missing a call. (NEAB)

19. A shop sells a particular make of radio at a rate of four per week on average. The number sold in a week has a Poisson distribution.

(a) Find the probability that the shop sells at least two in a week.

(b) Find the smallest number that can be in stock at the beginning of a week in order to have at least a 99% chance of being able to meet all demands during that week. (L)

20. The independent Poisson random variables $X$ and $Y$ have means 2 and 5, respectively. Obtain the mean and variance of the random variables

(a) $Y - X$,

(b) $2Y + 10$.

For **each** of these random variables give **one** reason why the distribution is **not** Poisson. (NEAB)

21. Fanfold paper for computer printers is made by putting perforations every 30 cm in a continuous roll of paper. A box of fanfold paper contains 2000 sheets. State the length of the continuous roll from which the box of paper is produced. The manufacturers claim that faults occur at random and at an average rate of one per 240 metres of paper. State an appropriate distribution for the number of faults per box of paper. Find the probability that a box of paper has no faults and also the probability that it has more than four faults.

Two copies of a report which runs to 100 sheets per copy are printed on this sort of paper. Find the probability that there are no faults in either copy of the report and also the probability that just one copy is faulty. (MEI)

22. A randomly chosen doctor in general practice sees, on average, one case of a broken nose per year and each case is independent of other similar cases.

(a) Regarding a month as a twelfth part of a year,

(i) show that the probability that, between them, three such doctors see no cases of a broken nose in a period of one month is 0.779, correct to three significant figures,

(ii) find the variance of the number of cases seen by three such doctors in a period of six months.

(b) Find the probability that, between them, three such doctors see at least three cases in one year

(c) Find the probability that, of three such doctors, one sees three cases and the other two see no cases in one year. (C)

23. Lemons are packed in boxes, each box containing 200. It is found that, on average, 0.45% of the lemons are bad when the boxes are opened. Use the Poisson distribution to find the probabilities of 0, 1, 2, and more than two bad lemons in a box.

A buyer who is considering buying a consignment of several hundred boxes checks the quality of the consignment by having a box opened. If the box opened contains no bad lemons he buys the consignment. If it contains more than two bad lemons he refuses to buy, and if it contains one or two bad lemons he has another box opened and buys the consignment if the second box contains fewer than two bad lemons. What is the probability that he buys the consignment?

Another buyer checks consignments on a different basis. He has one box opened; if that box contains more than one bad lemon he asks for another to be opened and does not buy if the second also contains more than one bad lemon. What is the probability that he refuses to buy the consignment?

24. A hire company has two electric lawnmowers which it hires out by the day. The number of demands per day for a lawnmower has the form of a Poisson distribution with mean 1.50. In a period of 100 working days, how many times do you expect

(a) neither of the lawnmowers to be used,

(b) some requests for the lawnmowers to have to be refused?

If each lawnmower is to be used an equal amount, on how many days in a period of 100 working days would you expect a particular lawnmower not to be in use? (MEI)

25. The number of oil tankers arriving at a port between successive high tides has a Poisson distribution with mean 2. The depth of the water is such that loaded vessels can enter the dock area only on the high tide. The port has dock space for only three tankers, which are discharged and leave the dock area before the next tide. Only the first three loaded tankers waiting at any high tide go into the dock area; any others must await another high tide. Starting from an evening high tide after which no ships remain waiting their turn, find (to three decimal places) the probabilities that after the next morning's high tide

   (a) the three dock berths remain empty,
   (b) the three berths are all filled.

   Find (to two decimal places) the probability that no tankers are left waiting outside the dock area after the following evening's high tide. (NEAB)

26. In the manufacture of commercial carpet, small faults occur at random in the carpet at an average rate of 0.95 per 20 m$^2$. Find the probability that in a randomly selected 20 m$^2$ area of this carpet

   (a) there are no faults,
   (b) there are at most two faults.

The ground floor of a new office block has 10 rooms. Each room has an area of 80 m$^2$ and has been carpeted using the same commercial carpet described above. For any one of these rooms, determine the probability that the carpet in the room

   (c) contains at least two faults,
   (d) contains exactly three faults,
   (e) contains at most five faults.

Find the probability that in exactly half of these ten rooms the carpets will contain exactly three faults. (AEB)

27. During each working day in a certain factory a number of accidents occur independently according to a Poisson distribution with mean 0.5.
Calculate the probability that

   (a) during any one day there are two or more accidents,
   (b) during two consecutive days there are exactly three accidents altogether.

Out of 50 consecutive five-day weeks how many would you expect to be accident-free?

# Mixed test 5A

1. A series of $n$ experiments is carried out and in each experiment the only possible outcomes are 'success' and 'failure'. The total number of successes is denoted by $X$. State two conditions which must be satisfied for the distribution of $X$ to be modelled by a binomial distribution. Gromit invites 11 friends to a party. For each friend, the probability that he or she will accept the invitation may be taken to be $\frac{2}{3}$. Use a binomial distribution to calculate the probability that

   (a) exactly nine,
   (b) fewer than nine,

   of the friends will accept the invitation.
   Give a reason why a binomial distribution might not be a good model in this situation. *(C)*

2. The weekly number of detached dwellings sold by an estate agent may be modelled by a Poisson distribution with mean 2.75 and, independently, the weekly number of other dwellings sold may be modelled by a Poisson distribution with mean 3.25.
   Determine the probability that the estate agent sells

   (a) exactly four detached dwellings in a week,
   (b) between ten and 15, inclusive, detached dwellings over a four-week period,
   (c) fewer than five dwellings in a week. *(NEAB)*

3. In one part of the country, one person in 80 has blood of Type P. A random sample of 150 blood donors is chosen from that part of the country. Let $X$ represent the number of donors in the sample having blood of Type P.

   (a) State the distribution of $X$. Find the parameter of the Poisson distribution which can be used as an approximation. Give a reason why a Poisson approximation is appropriate.
   (b) Using the Poisson distribution, calculate the probability that in the sample of 150 donors at least two have blood of Type P.
   (c) A hospital urgently requires blood of Type P. How large a random sample of donors must be taken in order that the probability of finding at least one donor of Type P should be 0.99 or more. *(MEI)*

4. A geography student is studying the distribution of telephone boxes in a large rural area where there is an average of 300 boxes per 500 km$^2$. A map of part of the area is divided into 50 squares, each of area 1 km$^2$, and the student wishes to model the number of telephone boxes per square.

   (a) Suggest a suitable simple model the student could use and specify any parameters required.

   One of the squares is picked at random.

   (b) Find the probability that this square does not contain any telephone boxes.
   (c) Find the probability that this square contains at least three telephone boxes.

   The student suggests using this model on another map of a large city and surrounding villages.

   (d) Comment, giving your reason briefly, on the suitability of the model in this situation. *(L)*

5. A crossword puzzle is published in *The Times* each day of the week, except Sunday. A woman is able to complete, on average, eight out of ten of the crossword puzzles.

   (a) Find the expected value and the standard deviation of the number of completed crosswords in a given week.
   (b) Show that the probability that she will complete at least five in a given week is 0.655 (to three significant figures).
   (c) Given that she completes the puzzle on Monday, find, to three significant figures, the probability that she will complete at least four in the rest of the week.
   (d) Find, to three significant figures, the probability that, in a period of four weeks, she completes four or less in only one of the four weeks. *(C)*

# Mixed test 5B

1. In practising the high jump a certain athlete has five attempts at a particular height. The probability that she succeeds at any one attempt is $p$. Find an expression, in terms of $p$, for the probability that she succeeds

   (a) exactly four times,
   (b) exactly two times.

   The probability that she succeeds exactly four times is twice the probability that she succeeds exactly two times. Find the value of $p$.     (C)

2. Before starting to play the game 'Snakes and Ladders' each player throws an ordinary unbiased die until a six is obtained. The number of throws before a player starts is the random variable $Y$, where $Y$ takes the values 1, 2, 3, ... .

   (a) Name the probability distribution of $Y$, stating a necessary assumption.
   (b) Find Var$(Y)$.
   (c) Two people play Snakes and Ladders. Calculate the probability that they will each need at least five throws before starting.     (C)

3. State, giving your reasons, the distribution which you would expect to be appropriate in describing

   (a) the number of heads in ten throws of a penny,
   (b) the number of blemishes per square metre of sheet metal.

   A building has an automatic telephone exchange. The number $X$ of wrong connections in any one day is a Poisson variable with parameter $\lambda$. Find, in terms of $\lambda$, the probability that in any one day there will be

   (c) exactly three wrong connections,
   (d) three or more wrong connections.

   Evaluate, to three decimal places, these probabilities when $\lambda = 0.5$. Find, to three decimal places, the largest value of $\lambda$ for the probability of one or more wrong connections in any day to be at most $\frac{1}{6}$.     (L)

4. The number of customers entering a certain branch of a bank on a Monday lunchtime may be modelled by a Poisson distribution with mean 2.4 per minute.

   (a) Find the probability that, during a particular minute, four or more customers enter the branch.

   The probability that a customer, who enters the branch, intends to open a new account is 0.002 and is independent of the intentions of other customers. During a particular morning 450 customers enter the bank.

   (b) Use a suitable approximation to find the probability that three or fewer of these 450 customers intend to open new accounts.
   
   (AEB)

5. A process for making plate glass produces small bubbles (imperfections) scattered at random in the glass, at an average rate of four small bubbles per 10 m$^2$.
   Assuming a Poisson model for the number of small bubbles, determine, to three decimal places, the probability that a piece of glass 2.2 m $\times$ 3.0 m will contain

   (a) exactly two small bubbles,
   (b) at least one small bubble,
   (c) at most two small bubbles.

   Show that the probability that five pieces of glass, each 2.5 m by 2.0 m, will all be free of small bubbles is e$^{-10}$.
   Find, to three decimal places, the probability that five pieces of glass, each 2.5 m by 2.0 m, will contain a total of at least ten small bubbles.     (L)

# Probability distributions II – continuous variables

*In this chapter you will learn*

- about probability density functions for continuous random variables
- how to find probabilities by calculating areas under curves
- how to find
    - the expectation, $E(X)$ of the continuous random variable, $X$
    - the expectation of any function of $X$
    - the variance of $X$
    - the mode
- about the cumulative distribution function, $F(x)$
- how to find the median, quartiles and other percentiles,
- how to obtain the probability density function $f(x)$ from the cumulative function $F(x)$
- about the rectangular (uniform) distribution

## CONTINUOUS RANDOM VARIABLES

The following are examples of **continuous random variables:**

- the mass, in grams, of a bag of sugar packaged by a particular machine
- the time taken, in minutes, to perform a task,
- the height, in centimetres, of a five-year-old girl,
- the lifetime, in hours, of a 100-watt light bulb.

## PROBABILITY DENSITY FUNCTION (P.D.F.)

A continuous random variable $X$ is given by its **probability density function (p.d.f.),** which is specified for the range of values for which $x$ is valid. The function can be illustrated by a curve, $y = f(x)$. Note that this function cannot be negative throughout the specified range.

**Probabilities** are given by the **area under the curve.** It is sometimes possible to find an area by geometry, for example by using formulae for the area of a triangle or a trapezium. Often, however, areas need to be calculated using integration.

## Example 6.1

$X$ is the delay, in hours, of a flight from Chicago, where

$$f(x) = 0.2 - 0.2x, \qquad 0 \leqslant x \leqslant 10$$

Find

(a) the probability that the delay will be less than four hours,
(b) the probability that the delay will be between two and six hours.

## Solution 6.1

It is useful to draw a sketch of $f(x)$.

Note that since $f(x)$ is valid for $0 \leqslant x \leqslant 10$, the delay can be between 0 and 10 hours.

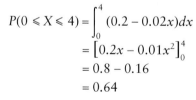

(a) The probability that the delay will be less than four hours is given by the area under the curve between 0 and 4.

*Method 1 – using geometry*
In this example it is easy to calculate the area using $A = \frac{1}{2}(a+b)h$, the formula for the area of a trapezium.

$a = 0.2$, $h = 4$
$b = f(4) = 0.2 - 0.02 \times 4 = 0.12$
$A = \frac{1}{2}(a+b)h$
$\quad = \frac{1}{2}(0.2 + 0.12) \times 4$
$\quad = 0.64$

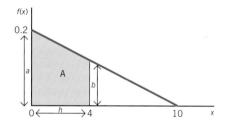

*Method 2 – using integration*

$$P(0 \leqslant X \leqslant 4) = \int_0^4 (0.2 - 0.02x)dx$$
$$= \left[ 0.2x - 0.01x^2 \right]_0^4$$
$$= 0.8 - 0.16$$
$$= 0.64$$

**The probability that the delay will be less than four hours is 0.64.**

(b) The probability that the delay will be between two and six hours is given by the area under the curve between 2 and 6.

*Method 1 – using geometry:*

$f(2) = 0.2 - 0.02 \times 2 = 0.16$
$f(6) = 0.2 - 0.02 \times 6 = 0.08$
$\quad A = \frac{1}{2}(a+b)h$
$\qquad = \frac{1}{2}(0.16 + 0.08) \times 4$
$\qquad = 0.48$
**P$(2 \leqslant X \leqslant 6) = 0.48$**

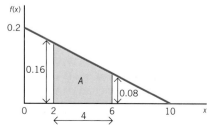

*Method 2* – using integration:

$$P(2 \leqslant X \leqslant 6) = \int_2^6 (0.2 - 0.02x)dx$$
$$= \left[0.2x - 0.01x^2\right]_2^6$$
$$= 1.2 - 0.36 - (0.4 - 0.04)$$
$$= 0.48$$

**The probability that the delay will be between two and six hours is 0.48.**

---

Notice that the total area under the curve gives the total probability.
In the above example it is easy to check by finding the area of the triangle.

Area of triangle $= \frac{1}{2}$ base $\times$ height
$= \frac{1}{2} \times 10 \times 0.2$
$= 1$

Alternatively, $\int_0^{10} f(x)dx = \int_0^{10} (0.2 - 0.02x)dx$
$$= \left[0.2x - 0.01x^2\right]_0^{10}$$
$$= 1$$

Note that it is not possible to find the probability that the delay is, say, exactly three hours.
If you try to integrate, you get

$$P(X = 3) = \int_3^3 f(x)dx = 0$$

You can only find the probability that $X$ lies within a particular range.
It is also not possible to distinguish between

$P(2 \leqslant X \leqslant 6),$
$P(2 < X \leqslant 6),$
$P(2 \leqslant X < 6),$
$P(2 < X < 6),$

so there is no need to worry about whether the inequality is strict or not.

## Example 6.2

$X$ is the continuous variable, the mass, in kilograms, of a substance produced per minute in an industrial process, where

$$f(x) = \begin{cases} \frac{1}{36}x(6-x) & (0 \leqslant x \leqslant 6) \\ 0 & \text{otherwise} \end{cases}$$

Find the probability that the mass is more than 5 kg.

## Solution 6.2

Note that $f(x)$ is a quadratic function and use this to help to draw the sketch of $f(x)$, noting that $f(x) = 0$ when $x = 0$ and $x = 6$.

Since you want the probability that $x$ is more than 5, shade the area between 5 and 6.

You will need to find this by integrating:

$$P(X > 5) = \int_5^6 \frac{1}{36} x(6 - x)dx$$

$$= \frac{1}{36} \int_5^6 (6x - x^2)dx$$

$$= \frac{1}{36} \left[ 3x^2 - \frac{x^3}{3} \right]_5^6$$

$$= 0.074 \quad (3 \text{ d.p.})$$

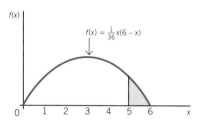

**The probability that the mass is more than 5 kg is 0.074 (3 d.p.).**

In general, for a continuous random variable $X$, with p.d.f. $f(x)$ valid over the range $a \leqslant x \leqslant b$

(a) $\displaystyle\int_a^b f(x)dx = 1$

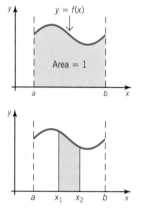

(b) for $a \leqslant x_1 \leqslant x_2 \leqslant b$

$$P(x_1 \leqslant X \leqslant x_2) = \int_{x_1}^{x_2} f(x)dx$$

Remember that in an experimental approach, the area under the histogram represents frequency. In a theoretical approach, the area under the curve $y = f(x)$ represents probability.

## Example 6.3

A continuous random variable has p.d.f. $f(x) = kx^2$ for $0 \leqslant x \leqslant 4$.

(a) Find the value of the constant $k$.
(b) Find $P(1 \leqslant X \leqslant 3)$.

## Solution 6.3

(a) $\displaystyle\int_{\text{all } x} f(x)dx = 1$

$$\int_0^4 kx^2 dx = 1$$

$$\left[ \frac{kx^3}{3} \right]_0^4 = 1$$

$$64\frac{k}{3} = 1$$

$$k = \frac{3}{64}$$

$\therefore \quad f(x) = \dfrac{3}{64}x^2, \, 0 \leqslant x \leqslant 4$

(b) $P(1 \leqslant X \leqslant 3) = \int_1^3 \frac{3}{64} x^2 \, dx$

$\qquad = \frac{3}{64} \left[ \frac{x^3}{3} \right]_1^3$

$\qquad = 0.406\ 25$

$\qquad = 0.41$ (2 s.f.)

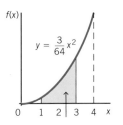

$P(1 \leqslant x \leqslant 3)$ is given by this area

---

## Example 6.4

The continuous random variable $X$ has p.d.f. $f(x)$ where

$$f(x) = \begin{cases} k(x+2)^2 & -2 \leqslant x < 0 \\ 4k & 0 \leqslant x \leqslant 1\frac{1}{3} \\ 0 & \text{otherwise} \end{cases}$$

(a) Find the value of the constant $k$.
(b) Sketch $y = f(x)$.
(c) Find $P(-1 \leqslant X \leqslant 1)$.
(d) Find $P(X > 1)$.

## Solution 6.4

(a) To find $k$, you need to use the result $\int_{\text{all } x} f(x) \, dx = 1$

$f(x)$ has been given in two parts, so you will need to calculate two separate integrals, as follows:

$$\int_{-2}^0 k(x+2)^2 \, dx + \int_0^{1\frac{1}{3}} 4k \, dx = 1$$

$$\frac{k}{3} \left[ (x+2)^3 \right]_{-2}^0 + 4k \left[ x \right]_0^{1\frac{1}{3}} = 1$$

$$\frac{k}{3}(8) + 4k \left( \frac{4}{3} \right) = 1$$

$$8k = 1$$

$$k = \frac{1}{8}$$

(b) The p.d.f. of $f(x)$ is

$$f(x) = \begin{cases} \frac{1}{8}(x+2)^2 & -2 \leqslant x < 0 \\ \frac{1}{2} & 0 \leqslant x \leqslant 1\frac{1}{3} \\ 0 & \text{otherwise} \end{cases}$$

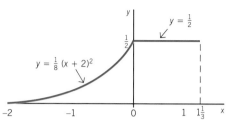

(c) $P(-1 \leqslant X < 1)$ is given by the shaded area.

It must be found in two stages:

$$P(-1 \leqslant X \leqslant 0) = \int_{-1}^{0} \frac{1}{8}(x+2)^2 dx$$

$$= \frac{1}{24}\left[(x+2)^3\right]_{-1}^{0}$$

$$= \frac{1}{24}(8-1)$$

$$= \frac{7}{24}$$

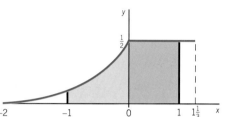

and $P(0 \leqslant X \leqslant 1) = $ area of rectangle

$$= \frac{1}{2}$$

$$\therefore \quad P(-1 \leqslant X \leqslant 1) = \frac{7}{24} + \frac{1}{2} = \frac{19}{24}$$

(d) From the diagram,

$$P(X > 1) = \text{area of shaded rectangle}$$

$$= \frac{1}{3} \times \frac{1}{2}$$

$$\therefore \quad P(X > 1) = \frac{1}{6}$$

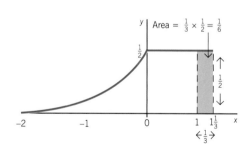

---

# Exercise 6a  Calculating probabilities

1. The continuous random variable $X$ has a p.d.f. $f(x)$ where $f(x) = kx^2$, $0 \leqslant x \leqslant 2$.

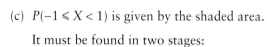

   (a) Find the value of the constant $k$.
   (b) Find $P(X \geqslant 1)$.
   (c) Find $P(0.5 \leqslant X \leqslant 1.5)$.

2. The continuous random variable $X$ has p.d.f. $f(x)$ where $f(x) = k$, $-2 \leqslant x \leqslant 3$.

   (a) Sketch $y = f(x)$.
   (b) Find the value of the constant $k$.
   (c) Find $P(-1.6 \leqslant X \leqslant 2.1)$.

3. The continuous random variable $X$ has p.d.f. $f(x)$ where $f(x) = k(4-x)$, $1 \leqslant x \leqslant 3$.

   (a) Find the value of the constant $k$.
   (b) Sketch $y = f(x)$.
   (c) Find $P(1.2 \leqslant X \leqslant 2.4)$.

4. The continuous random variable $X$ has p.d.f. $f(x)$ where $f(x) = k(x+2)^2$, $0 \leqslant x \leqslant 2$.

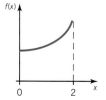

   (a) Find the value of the constant $k$.
   (b) Find $P(0 \leqslant X \leqslant 1)$ and hence find $P(X > 1)$.

5.  The continuous random variable $X$ has p.d.f. $f(x)$ where $f(x) = kx^3$, $0 \leqslant x \leqslant c$ and $P(X \leqslant \frac{1}{2}) = \frac{1}{16}$.

    Find the values of the constants $c$ and $k$.

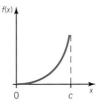

6.  A continuous random variable has p.d.f. $f(x)$ where $f(x) = kx$, $0 \leqslant x \leqslant 4$.

    (a)  Find the value of the constant $k$.
    (b)  Sketch $y = f(x)$.
    (c)  Find $P(1 \leqslant X \leqslant 2.5)$.

7.  The continuous random variable $X$ has p.d.f. $f(x)$ where

$$f(x) = \begin{cases} k & 0 \leqslant x < 2 \\ k(2x-3) & 2 \leqslant x \leqslant 3 \\ 0 & \text{otherwise} \end{cases}$$

    (a)  Find the value of the constant $k$.
    (b)  Sketch $y = f(x)$.
    (c)  Find $P(X \leqslant 1)$.
    (d)  Find $P(X > 2.5)$.
    (e)  Find $P(1 \leqslant X \leqslant 2.3)$.

# EXPECTATION OF $X$, $E(X)$

For a continuous random variable with p.d.f. $f(x)$,

$$E(X) = \int_{\text{all } x} xf(x)dx$$

$E(X)$ is referred to as the mean or expectation of $X$ and is often denoted by $\mu$.

## Example 6.5

The sketch shows the p.d.f. of $X$ where $f(x) = \frac{1}{9}x^2$, $0 \leqslant x \leqslant 3$.

(a)  Find $\mu$, the mean of $X$.
(b)  Find $P(X < \mu)$.

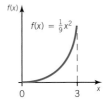

## Solution 6.5

(a)  $\mu = E(X)$

$$= \int_{\text{all } x} xf(x)dx$$

$$= \int_0^3 x \times \frac{1}{9}x^2 dx$$

$$= \frac{1}{9}\int_0^3 x^3 dx$$

$$= \frac{1}{9}\left[\frac{x^4}{4}\right]_0^3$$

$$= 2.25$$

(b)  $P(X < \mu) = P(X < 2.25)$

$$= \int_0^{2.25} \frac{1}{9}x^2 dx$$

$$= \frac{1}{9}\left[\frac{x^3}{3}\right]_0^{2.25}$$

$$= 0.42 \text{ (2 s.f.)}$$

If $f(x)$ has a line of symmetry in the specified range, then $E(X)$ can be found directly as in the following example.

## Example 6.6

A continuous random variable $X$ has p.d.f. $f(x)$ where

$$f(x) = \begin{cases} 0.25x & 0 \leqslant x < 2 \\ 1 - 0.25x & 2 \leqslant x \leqslant 4 \\ 0 & \text{otherwise} \end{cases}$$

Sketch $y = f(x)$ and find $E(X)$.

## Solution 6.6

Sketch of $y = f(x)$

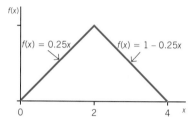

From the sketch, you can see that there is symmetry about $x = 2$.

$$\therefore \quad E(X) = 2$$

Check by integration:

$$E(X) = \int_{\text{all } x} xf(x)dx$$

$$= \int_0^2 x \times 0.25x\,dx + \int_2^4 x \times (1 - 0.25x)dx$$

$$= 0.25 \int_0^2 x^2\,dx + \int_2^4 (x - 0.25x^2)dx$$

$$= 0.25 \left[ \frac{x^3}{3} \right]_0^2 + \left[ \frac{x^2}{2} - 0.25\frac{x^3}{3} \right]_2^4$$

$$= \frac{2}{3} + \left( 8 - \frac{16}{3} - \left( 2 - \frac{2}{3} \right) \right)$$

$$= 2$$

## Example 6.7

A teacher of young children is thinking of asking her class to guess her height in metres. The teacher considers that the height guessed by a randomly selected child can be modelled by the random variable $H$ with probability density function

$$f(h) = \begin{cases} \frac{3}{16}(4h - h^2) & 0 \leqslant h \leqslant 2 \\ 0 & \text{otherwise} \end{cases}$$

Using this model,

(a) find $P(H < 1)$,
(b) show that $E(H) = 1.25$.

A friend of the teacher suggests that the random variable $X$ with probability density function

$$g(x) = \begin{cases} kx^3 & 0 \leqslant x \leqslant 2 \\ 0 & \text{otherwise} \end{cases}$$

where $k$ is a constant, might be a more suitable model.

(c) Show that $k = \frac{1}{4}$.
(d) Find $P(X < 1)$.
(e) Find $E(X)$.
(f) Using your calculations in (a), (b), (d) and (e), state, giving reasons, which of the random variables $H$ or $X$ is likely to be the more appropriate model in this instance.    (L)

## Solution 6.7

(a) $P(H < 1) = \displaystyle\int_0^1 f(h)\,dh$

$$= \int_0^1 \frac{3}{16}(4h - h^2)\,dh$$

$$= \frac{3}{16}\left[2h^2 - \frac{h^3}{3}\right]_0^1$$

$$= \frac{5}{16} = 0.3125$$

(b) $E(H) = \displaystyle\int_0^2 hf(h)\,dh$

$$= \frac{3}{16}\int_0^2 (4h^2 - h^3)\,dh$$

$$= \frac{3}{16}\left[\frac{4h^3}{3} - \frac{h^4}{4}\right]_0^2$$

$$= 1.25$$

(c) $\displaystyle\int_{\text{all } x} g(x)\,dx = 1$

$$\therefore \quad \int_0^2 kx^3\,dx = 1$$

$$k\left[\frac{x^4}{4}\right]_0^2 = 1$$

$$4k = 1$$

$$k = \frac{1}{4}$$

(d) $P(X < 1) = \displaystyle\int_0^1 \frac{1}{4}x^3\,dx$

$$= \frac{1}{4}\left[\frac{x^4}{4}\right]_0^1$$

$$= \frac{1}{16} = 0.0625$$

f(h)    sketch of f(h)

g(x)    sketch of g(x)

(e) $E(X) = \int_0^2 xg(x)dx$

$= \dfrac{1}{4} \int_0^2 x^4 dx$

$= \dfrac{1}{4} \left[ \dfrac{x^5}{5} \right]_0^2$

$= \mathbf{1.6}$

(f) For $H$, $P(H < 1) = 0.3125$, so 31% of children guess the teacher's height to be less than 1 m (i.e. 3 ft 3 in).
$E(H) = 1.25$, so the average guess for height of the teacher is 1.25 m (i.e. 4 ft 1 in).

For $X$,
$P(X < 1) = 0.062\,55$, so only 6% of children guess the height to be less than 3 ft 3 in.
$E(X) = 1.6$, so the average guess for the height of the teacher is 1.6 m (i.e. 5 ft 2 in).

**$X$ is the more appropriate model.**

---

## Exercise 6b   Expectation $E(X)$

1. Find $E(X)$ for each of the following continuous random variables.

   (a) $f(x) = \frac{3}{4}(x^2 + 1)$, $0 \leqslant x \leqslant 1$.

   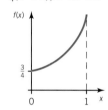

   (b) $f(x) = \frac{3}{4}x(2 - x)$, $0 \leqslant x \leqslant 2$.

   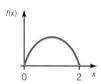

   (c) $f(x) = \frac{1}{18}(6 - x)$, $0 \leqslant x \leqslant 6$.

   (d) $f(x) = kx^3$, $0 \leqslant x \leqslant 2$.

   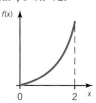

   (e) $f(x) = \begin{cases} \frac{3}{8} & \frac{2}{3} \leqslant x < 2 \\ \frac{3}{32}x(4 - x) & 2 \leqslant x \leqslant 4 \\ 0 & \text{otherwise} \end{cases}$

   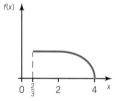

2. The continuous random variable $X$ has p.d.f. $f(x)$ where

   $f(x) = \begin{cases} kx & 0 \leqslant x < 1 \\ k & 1 \leqslant x < 3 \\ k(4 - x) & 3 \leqslant x \leqslant 4 \\ 0 & \text{otherwise} \end{cases}$

   (a) Draw a sketch of $y = f(x)$.
   (b) Find $k$.
   (c) Find $E(X)$.

3. $X$ is a continuous random variable with p.d.f. $f(x) = kx^2$, $0 \leqslant x \leqslant 4$.
   Find $E(X)$.

4. In a game a wooden block is propelled with a stick across a flat deck. On each attempt the distance, $x$ metres, reached by the block lies between 0 and 10 m, and the variation is modelled by the probability density function

   $$f(x) = 0.0012x^2(10 - x).$$

   Calculate the mean distance reached by the block. *(SMP)*

5. The continuous random variable $X$ has the probability density function $f$ given by $f(x) = kx$, $5 < x < 10$, $f(x) = 0$ otherwise.

(a) Find the value of $k$.
(b) Find the expected value of $X$.
(c) Find the probability that $X > 8$.

The annual income from money invested in a Unit Trust Fund is $X$ per cent of the amount invested, where $X$ has the above distribution. Suppose that you have a sum of money to invest and that you are prepared to leave the money invested over a period of several years. State, with your reasons, whether you would invest in the Unit Trust Fund or in a Money Bond offering a guaranteed annual income of 8% on the money invested. (NEAB)

6. The lifetime $X$ in *tens of hours* of a torch battery is a random variable with probability density function

$$f(x) = \begin{cases} \frac{3}{4}(1 - (x - 2)^2) & 1 \leq x \leq 3, \\ 0 & \text{otherwise} \end{cases}$$

Calculate the mean of $X$.
A torch runs on two batteries, both of which have to be working for the torch to function. If two new batteries are put in the torch, what is the probability that the torch will function for at least 22 hours, on the assumption that the lifetimes of the batteries are independent? (O & C)

7. A random variable $X$ has a probability density function $f$ given by

$$f(x) = \begin{cases} cx(5 - x) & 0 \leq x \leq 5 \\ 0 & \text{otherwise} \end{cases}$$

Show that $c = \dfrac{6}{125}$ and find the mean of $X$.

The lifetime $X$ in years of an electric light bulb has this distribution. Given that a lamp standard is fitted with two such new bulbs and that their failures are independent, find the probability that neither bulb fails in the first year and the probability that exactly one bulb fails within two years. (MEI)

8. The mass, $X$ kg, of a particular substance produced per hour in a chemical process is a continuous random variable whose probability density function is given by

$$f(x) = \frac{3x^2}{32} \qquad 0 \leq x < 2$$

$$f(x) = \frac{3(6 - x)}{32} \qquad 2 \leq x \leq 6$$

$$f(x) = 0 \qquad \text{otherwise}$$

(a) Find the mean mass produced per hour.
(b) The substance produced is sold at £2 per kilogram and the total running cost of the process is £1 per hour. Find the expected profit per hour and the probability that in an hour the profit will exceed £7. (NEAB)

9. A continuous random variable $X$ has the probability density function $f$ defined by

$$f(x) = \frac{c}{3}x \qquad 0 \leq x < 3$$

$$f(x) = c \qquad 3 \leq x \leq 4$$

$$f(x) = 0 \qquad \text{otherwise}$$

where $c$ is a positive constant. Find

(a) the value of $c$,
(b) the mean of $X$,
(c) the value, $a$, for there to be a probability of 0.85 that a randomly observed value of $X$ will exceed $a$. (NEAB)

# THE EXPECTATION OF ANY FUNCTION OF $X$

If $g(x)$ is any function of the continuous random variable, $X$, having p.d.f. $f(x)$, then

$$E(g(x)) = \int_{\text{all } x} g(x)f(x)dx$$

In particular $\quad E(X^2) = \displaystyle\int_{\text{all } x} x^2\, f(x)dx$

As in the case of the discrete random variable (see pages 246 and 248), the following results also hold when $X$ is continuous; $a$ and $b$ are constants:

1. $E(a) = a$
2. $E(aX) = aE(X)$
3. $E(aX + b) = aE(X) + b$
4. $E(g(X) + h(X)) = E(g(X)) + E(h(X))$

## Example 6.8

The continuous random variable $X$ has p.d.f. $f(x)$ where $f(x) = \frac{1}{20}(x + 3)$, $0 \leqslant x \leqslant 4$.

(a) Find $E(X)$.
(b) Find $E(2X + 5)$.
(c) Find $E(X^2)$.
(d) Find $E(X^2 + 2X - 3)$.

## Solution 6.8

Sketch of $y = f(x)$.

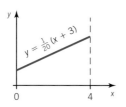

Note from the sketch that there is no symmetry.

(a) $E(X) = \displaystyle\int_{\text{all } x} x f(x) dx$

$\displaystyle = \int_0^4 \frac{1}{20} x(x + 3) dx$

$\displaystyle = \frac{1}{20} \int_0^4 (x^2 + 3x) dx$

$\displaystyle = \frac{1}{20} \left[ \frac{x^3}{3} + \frac{3x^2}{2} \right]_0^4$

$= 2.266\ldots$

$= \mathbf{2.3}$ (**2 s.f.**)

(b) $E(2X + 5) = E(2X) + 5$      (Result 3)

$\qquad\qquad = 2E(X) + 5$      (Result 2)

$\qquad\qquad = 2(2.266\ldots) + 5$

$\qquad\qquad = 9.533\ldots$

$\qquad\qquad = \mathbf{9.5}$ (**2 s.f.**)

(c) $E(X^2) = \displaystyle\int_{\text{all } x} x^2 f(x) dx$

$\displaystyle = \frac{1}{20} \int_0^4 x^2(x + 3) dx$

$\displaystyle = \frac{1}{20} \int_0^4 (x^3 + 3x^2) dx$

$\displaystyle = \frac{1}{20} \left[ \frac{x^4}{4} + x^3 \right]_0^4$

$= \mathbf{6.4}$

(d) $E(X^2 + 2X - 3) = E(X^2) + E(2X) - E(3)$      (Result 4)

$\qquad\qquad\qquad = E(X^2) + 2E(X) - 3$      (Results 1, 2)

$\qquad\qquad\qquad = 6.4 + 2(2.266\ldots) - 3$

$\qquad\qquad\qquad = 7.933\ldots$

$\qquad\qquad\qquad = \mathbf{7.9}$ (**2 s.f.**)

## Example 6.9

The mass, $X$ kg, of a particular substance produced in one hour in a chemical process is modelled by a continuous random variable with probability density function given by

$$f(x) = \tfrac{3}{32}x^2, \qquad 0 \leqslant x < 2,$$
$$f(x) = \tfrac{3}{32}(6-x), \qquad 2 \leqslant x \leqslant 6,$$
$$f(x) = 0, \qquad \text{otherwise}$$

(a) Sketch the graph of $f$.
(b) Find $P(X < 4)$.
(c) Find the mean mass produced per hour.
(d) The substance is sold at £100 per kilogram and the running cost of the process is £20 per hour. Taking £$Y$ as the profit made in each hour, express $Y$ in terms of $X$.
(e) Find the expected value of $Y$. (NEAB)

## Solution 6.9

(a)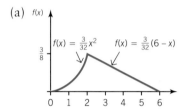

(b) $P(X < 4) = \displaystyle\int_0^2 \frac{3}{32} x^2 \, dx + \int_2^4 \frac{3}{32}(6-x)\,dx$

$\qquad = \dfrac{1}{32}\left[x^3\right]_0^2 + \dfrac{3}{32}\left[6x - \dfrac{x^2}{2}\right]_2^4$

$\qquad = \dfrac{8}{32} + \dfrac{3}{32}(24 - 8 - (12 - 2))$

$\qquad = \dfrac{13}{16}$

(c) $E(X) = \displaystyle\int_0^2 \frac{3}{32} x^3 \, dx + \int_2^6 \frac{3}{32}(6x - x^2)\,dx$

$\qquad = \dfrac{3}{32}\left[\dfrac{x^4}{4}\right]_0^2 + \dfrac{3}{32}\left[3x^2 - \dfrac{x^3}{3}\right]_2^6$

$\qquad = \dfrac{3}{8} + \dfrac{3}{32}\left(108 - 72 - \left(12 - \dfrac{8}{3}\right)\right)$

$\qquad = 2\dfrac{7}{8}$

(d) $Y = 100X - 20$

(e) $E(Y) = E(100X - 20)$
$\qquad = 100E(X) - 20$
$\qquad = 267\tfrac{1}{2}$

So the expected profit is £267.50.

## Example 6.10

The continuous random variable $X$ has p.d.f. $f(x)$ where

$$f(x) = \begin{cases} \frac{6}{7}x & 0 \leqslant x \leqslant 1 \\ \frac{6}{7}x(2-x) & 1 \leqslant x \leqslant 2 \\ 0 & \text{otherwise} \end{cases}$$

Find $E(X^2)$.

## Solution 6.10

$$\begin{aligned}
E(X^2) &= \int_{\text{all } x} x^2 f(x)dx \\
&= \int_0^1 \frac{6}{7}x^3 dx + \int_1^2 \frac{6}{7}x^3(2-x)dx \\
&= \frac{6}{7}\int_0^1 x^3 dx + \frac{6}{7}\int_1^2 (2x^3 - x^4)dx \\
&= \frac{6}{7}\left[\frac{x^4}{4}\right]_0^1 + \frac{6}{7}\left[\frac{x^4}{2} - \frac{x^5}{5}\right]_1^2 \\
&= 1.328\ldots \\
&= \textbf{1.3 (2 s.f.)}
\end{aligned}$$

---

NOTE: $E(X^2)$ is an important value which is needed when calculating the variance of $X$.

# VARIANCE OF X, Var(X)

For a random variable $X$,

$$\text{Var}(X) = E(X - \mu)^2 \qquad \text{where } \mu = E(X).$$

As in the discrete case (see page 249) the formula can be written:

$$\begin{aligned}
\text{Var}(X) &= E(X^2) - E^2(X) \\
&= E(X^2) - \mu^2
\end{aligned}$$

If $X$ is a continuous random variable with p.d.f. $f(x)$, then

$$\text{Var}(X) = \int_{\text{all } x} x^2 f(x)dx - \mu^2$$

where $\qquad \mu = E(X) = \int_{\text{all } x} xf(x)dx$

The standard deviation of $X$ is often written as $\sigma$, so $\sigma = \sqrt{\text{Var}(X)}$.

As in the case of the discrete random variable (see page 250), the following results also hold when $X$ is continuous; where $a$ and $b$ are constants

1. $\text{Var}(a) = 0$
2. $\text{Var}(aX) = a^2 \text{Var}(X)$
3. $\text{Var}(aX + b) = a^2 \text{Var}(X)$

## Example 6.11

The continuous random variable $X$ has p.d.f. $f(x)$ where $f(x) = \frac{1}{8}x$, $0 \leqslant x \leqslant 4$. Find

(a) $E(X)$,
(b) $E(X^2)$,
(c) $\text{Var}(X)$,
(d) $\sigma$, the standard deviation of $X$,
(e) $\text{Var}(3X + 2)$.

## Solution 6.11

(a) $E(X) = \displaystyle\int_{\text{all } x} xf(x)dx$

$\qquad = \displaystyle\int_0^4 \frac{1}{8}x^2 dx$

$\qquad = \dfrac{1}{8}\left[\dfrac{x^3}{3}\right]_0^4$

$\qquad = 2.666\ldots$

$\qquad = \mathbf{2.7}$ **(2 s.f.)**

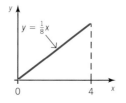

Note that there is no symmetry.

(b) $E(X^2) = \displaystyle\int_{\text{all } x} x^2 f(x)dx$

$\qquad = \displaystyle\int_0^4 \frac{1}{8}x^3 dx$

$\qquad = \dfrac{1}{8}\left[\dfrac{x^4}{4}\right]_0^4$

$\qquad = \dfrac{1}{8}(64)$

$\qquad = 8$

(c) $\text{Var}(X) = E(X^2) - E^2(X)$
$\qquad\qquad = 8 - (2.666\ldots)^2$
$\qquad\qquad = 0.888\ldots$
$\qquad\qquad = \mathbf{0.89}$ **(2 s.f.)**

(d) $\qquad \sigma = \sqrt{\text{Var}(X)}$
$\qquad\qquad = \sqrt{0.888\ldots}$
$\qquad\qquad = 0.9428\ldots$
$\qquad\qquad = \mathbf{0.94}$ **(2 s.f.)**

(e) $\text{Var}(3X + 2) = 9\,\text{Var}(X)$ \qquad (using variance result 3)
$\qquad\qquad\qquad = 9(0.888\ldots)$
$\qquad\qquad\qquad = 8$

**Example 6.12**

As an experiment a temporary roundabout is installed at the crossroads. The time, $X$ minutes, which vehicles have to wait before entering the roundabout has probability density function

$$f(x) = \begin{cases} 0.8 - 0.32x & 0 \leqslant x \leqslant 2.5 \\ 0 & \text{otherwise} \end{cases}$$

Find the mean and the standard deviation of $X$. (AEB)

**Solution 6.12**

$$E(X) = \int_{\text{all } x} xf(x)dx$$

$$= \int_0^{2.5} (0.8x - 0.32x^2)dx$$

$$= \left[ 0.8\frac{x^2}{2} - 0.32\frac{x^3}{3} \right]_0^{2.5}$$

$$= 0.833 \ldots \text{ minutes}$$

$$= 50 \text{ seconds}$$

**The mean time is 50 seconds**

$$E(X^2) = \int_{\text{all } x} x^2 f(x)dx$$

$$= \int_0^{2.5} (0.8x^2 - 0.32x^3)dx$$

$$= \left[ 0.8\frac{x^3}{3} - 0.32\frac{x^4}{4} \right]_0^{2.5}$$

$$= 1.041 \ldots$$

$$\text{Var}(X) = E(X^2) - E^2(X)$$

$$= 1.041 \ldots (-0.833 \ldots)^2$$

$$= 0.347 \ldots$$

$$\text{s.d. of } X = \sqrt{0.347 \ldots}$$

$$= 0.589 \ldots \text{ minutes}$$

$$= \textbf{35 seconds (2 s.f.)}$$

# THE MODE

The mode is the value of $X$ for which $f(x)$ is greatest in the given range of $X$.
To locate the mode it is a good idea to draw a sketch. Sometimes the mode can be deduced immediately.

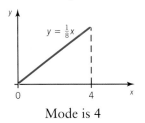

Mode is 4

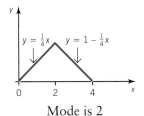

Mode is 2

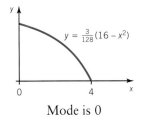

Mode is 0

For some probability density functions you will need to determine the maximum point on the curve $y = f(x)$ using the fact that, at a maximum point, $f'(x) = 0$, where $f'(x) = \dfrac{d}{dx} f(x)$.

Note that a maximum point is confirmed if $f''(x) < 0$, where $f''(x) = \dfrac{d}{dx} f'(x)$.

### Example 6.13

$X$ has p.d.f. defined by $f(x) = \frac{3}{80}(2 + x)(4 - x)$, for $0 \leqslant x \leqslant 4$ and is illustrated in the diagram. Find the mode.

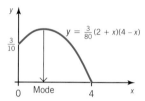

### Solution 6.13

$f(x) = \frac{3}{80}(2 + x)(4 - x) = \frac{3}{80}(8 + 2x - x^2)$

The mode is the value of $x$ at the maximum point.

Differentiate to find $f'(x)$.

$f'(x) = \frac{3}{80}(2 - 2x)$

$\therefore \quad f'(x) = 0 \qquad$ when $x = 1$

Differentiate again to find $f''(x)$

$f''(x) = \frac{3}{80} \times (-2) = -\frac{3}{40}$,

so $f''(x) < 0$ for all values of $x$, indicating that there is a maximum point when $x = 1$.

**The mode = 1**

### Example 6.14

A random variable $X$ has a probability density function

$f(x) = Ax(6 - x)^2 \qquad 0 \leqslant x \leqslant 6$

$\phantom{f(x)} = 0 \qquad\qquad$ elsewhere.

(a) Find the value of the constant $A$.
(b) Calculate
    (i) the mean,     (ii) the mode,
    (iii) the variance,   (iv) the standard deviation of $X$.

(AEB)

## Solution 6.14

(a) Since $X$ is a random variable, $\displaystyle\int_{\text{all } x} f(x)dx = 1$

$$\therefore \quad 1 = \int_0^6 Ax(6-x)^2 dx$$

$$= A \int_0^6 (36x - 12x^2 + x^3)dx$$

$$= A \left[ 18x^2 - 4x^3 + \frac{1}{4}x^4 \right]_0^6$$

$$= 108A$$

$$A = \frac{1}{108}$$

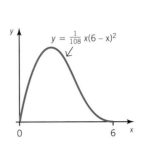

$y = \frac{1}{108} x(6-x)^2$

(b) $f(x) = \frac{1}{108}x(6-x)^2 \qquad 0 \leqslant x \leqslant 6$

(i) The mean is $E(X)$ where $\displaystyle E(X) = \int_{\text{all } x} xf(x)dx$

$$E(X) = \frac{1}{108} \int_0^6 x^2(6-x)^2 dx$$

$$= \frac{1}{108} \int_0^6 (36x^2 - 12x^3 + x^4)dx$$

$$= \frac{1}{108} \left[ 12x^3 - 3x^4 + \frac{x^5}{5} \right]_0^6$$

$$= 2.4$$

(ii) To find the mode, find the value of $x$ for which $f(x)$ is a maximum, $0 \leqslant x \leqslant 6$.

$f(x) = \frac{1}{108}(36x - 12x^2 + x^3)$

Differentiating

$f'(x) = \frac{1}{108}(36 - 24x + 3x^2)$

$\qquad = \frac{3}{108}(6-x)(2-x)$

$\therefore \quad f'(x) = 0 \quad$ when $x = 2$ and when $x = 6$

$f''(x) = \frac{3}{108}(6x - 24)$

To check maximum or minimum, consider $f''(x)$.

When $x = 2$, $f''(x) < 0$ and when $x = 6$, $f''(x) > 0$.

$f(x)$ is maximum when $x = 2$, so the mode is **2**.

(iii) To find the variance of $X$, first find $E(X^2)$.

$$E(X^2) = \int_{\text{all } x} x^2 f(x)dx$$

$$= \frac{1}{108} \int_0^6 (36x^3 - 12x^4 + x^5)dx$$

$$= \frac{1}{108} \left[ 9x^4 - \frac{12x^5}{5} + \frac{x^6}{6} \right]_0^6$$

$$= 7.2$$

$$\text{Var}(X) = E(X^2) - E^2(X) = 7.2 - (2.4)^2$$
$$= \mathbf{1.44}$$

(iv) Standard deviation of $X = \sqrt{\text{Var}(X)}$
$$= \sqrt{1.44}$$
$$= \mathbf{1.2}$$

---

## Example 6.15

The time taken to perform a particular task, $t$ hours, has the probability density function

$$f(t) = \begin{cases} 10ct^2 & 0 \leqslant t < 0.6 \\ 9c(1-t) & 0.6 \leqslant t \leqslant 1.0 \\ 0 & \text{otherwise,} \end{cases}$$

where $c$ is a constant.

(a) Find the value of $c$ and sketch the graph of this distribution.
(b) Write down the most likely time.
(c) Find the expected time.
(d) Determine the probability that the time will be
    (i)   more than 48 minutes,
    (ii)  between 24 and 48 minutes.

## Solution 6.15

(a)
$$1 = \int_{\text{all } t} f(t)dt$$
$$= 10c \int_0^{0.6} t^2 \, dt + 9c \int_{0.6}^{1.0} (1-t)dt$$
$$= \frac{10c}{3} \left[ t^3 \right]_0^{0.6} + 9c \left[ t - \frac{t^2}{2} \right]_{0.6}^{1.0}$$
$$= 0.72c + 0.72c$$
$$= 1.44c$$

$\therefore$
$$c = \frac{1}{1.44} = \frac{100}{144} = \frac{25}{36}$$

The probability density function is

$$f(t) = \begin{cases} \frac{125}{18}t^2 & 0 \leqslant t < 0.6 \\ \frac{25}{4}(1-t) & 0.6 \leqslant t \leqslant 1.0 \\ 0 & \text{otherwise} \end{cases}$$

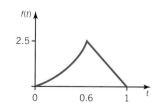

(b) From the sketch, $t = 0.6$ gives the maximum value of $f(t)$.
Therefore the mode is 0.6 hours = 36 minutes.
**The most likely time is 36 minutes.**

(c) $E(t) = \displaystyle\int_{\text{all } t} tf(t)\,dt$

$$= 10c \int_0^{0.6} t^3\,dt + 9c \int_{0.6}^{1.0} (t - t^2)\,dt$$

$$= \frac{10c}{4} \left[ t^4 \right]_{0.6}^{1.0} + 9c \left[ \frac{t^2}{2} - \frac{t^3}{3} \right]_{0.6}^{1.0}$$

$$= 0.225 + 0.366\ldots$$
$$= 0.591\ldots \text{ hours}$$
$$= 35.5 \text{ minutes}$$

**The expected time is 35.5 minutes.**

(d) (i)  48 minutes = 0.8 hours

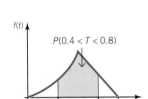

$$P(T > 0.8) = 9c \int_{0.8}^{1.0} (1 - t)\,dt$$

$$= 9c \left[ t - \frac{t^2}{2} \right]_{0.8}^{1.0}$$

$$= 0.125$$

**The probability that the time will be more than 48 minutes is 0.125.**

(ii)  24 minutes = 0.4 hours

$$P(0.4 < T < 0.8) = 1 - P(T > 0.8) - P(T < 0.4)$$

$$P(T < 0.4) = 10c \int_0^{0.4} t^2\,dt$$

$$= \frac{10c}{3} \left[ t^3 \right]_0^{0.4}$$

$$= 0.1481\ldots$$

$$\therefore \quad P(0.4 < T < 0.8) = 1 - 0.125 - 0.1481\ldots$$
$$= 0.727 \text{ (3 s.f.)}$$

**The probability that the time will be between 24 and 48 minutes is 0.727.**

## Exercise 6c   Standard deviation and variance

In Questions 1–7 find

(a) $E(X)$,   (b) $E(X^2)$,   (c) $\text{Var}(X)$,   (d) the standard deviation of $X$.

It is assumed that the value of the function is zero outside the range(s) stated. Do not forget to look for symmetry when considering $E(X)$.

*NOTE*: some of these functions were given in Exercise 6a and you may wish to refer to your previous sketches.

1. $f(x) = \frac{3}{8}x^2$      $0 \leqslant x \leqslant 2$
2. $f(x) = \frac{1}{5}$      $-2 \leqslant x \leqslant 3$
3. $f(x) = \frac{1}{4}(4 - x)$      $1 \leqslant x \leqslant 3$
4. $f(x) = \frac{3}{56}(x + 2)^2$      $0 \leqslant x \leqslant 2$
5. $f(x) = 4x^3$      $0 \leqslant x \leqslant 1$

6. $f(x) = \begin{cases} \frac{1}{4} & 0 \leqslant x \leqslant 2 \\ \frac{1}{4}(2x - 3) & 2 \leqslant x \leqslant 3 \end{cases}$

7. $f(x) = \begin{cases} \frac{1}{8}(x + 2)^2 & -2 \leqslant x \leqslant 0 \\ \frac{1}{2} & 0 \leqslant x \leqslant 1\frac{1}{3} \end{cases}$

8. A continuous random variable $X$ has p.d.f.
$f(x) = kx^2, 0 \leqslant x \leqslant 4$.

(a) Find the value of $k$, and sketch $y = f(x)$.
(b) Find $E(X)$ and $\mathrm{Var}(X)$.
(c) Find $P(1 < X < 2)$.

9. A continuous random variable $X$ has p.d.f. $f(x)$
where

$$f(x) = \begin{cases} kx & 0 \leqslant x < 1 \\ k(2-x) & 1 \leqslant x \leqslant 2 \\ 0 & \text{otherwise} \end{cases}$$

Find
(a) the value of the constant $k$,
(b) $E(X)$,  (c) $\mathrm{Var}(X)$,
(d) $P(\frac{3}{4} \leqslant X \leqslant 1\frac{1}{2})$,  (e) the mode.

10. The continuous random variable $X$ has p.d.f.
given by $f(x)$ where

$$f(x) = \begin{cases} \frac{1}{27}x^2 & 0 \leqslant x < 3 \\ \frac{1}{3} & 3 \leqslant x \leqslant 5 \\ 0 & \text{otherwise} \end{cases}$$

(a) Sketch $y = f(x)$.
(b) Find $E(X)$.
(c) Find $E(X^2)$.
(d) Find the standard deviation $\sigma$ of $X$.

11. A continuous random variable $X$ has a
probability density function $f$ given by

$$f(x) = \frac{k}{x(4-x)} \quad 1 \leqslant x \leqslant 3$$

$$f(x) = 0 \qquad \text{otherwise}$$

(a) Show that $k = \dfrac{2}{\ln 3}$.

(b) Calculate the mean and the variance of $X$.
(NEAB)

12. The probability density function of $X$ is given by

$$f(x) = \begin{cases} k(ax - x^2) & 0 \leqslant x \leqslant 2 \\ 0 & x < 0, \quad x > 2 \end{cases}$$

where $k$ and $a$ are positive constants.

Show that $a \geqslant 2$ and that $k = \dfrac{3}{6a-8}$.

Given that the mean value of $X$ is 1, calculate the
values of $a$ and $k$.

For these values of $a$ and $k$ sketch the graph of
the probability density function and find the
variance of $X$.
(NEAB)

13. A continuous random variable $X$ has probability
density function $f(x)$ defined by

$$f(x) = \begin{cases} 12(x^2 - x^3) & 0 \leqslant \ \leqslant 1 \\ 0 & \text{otherwise} \end{cases}$$

Find the mean and standard deviation of $X$.
(O&C)

# THE CUMULATIVE DISTRIBUTION FUNCTION, $F(x)$

In Chapter 4 (page 253) you met the idea of a cumulative distribution function, $F(x)$, for a
discrete random variable and in Chapter 5 (pages 283 and 294) you used cumulative
probability tables giving $F(r) = P(X \leqslant r)$ for binomial and Poisson distributions.

In the same way, if $X$ is a continuous random variable with p.d.f. $f(x)$, you can find the
**cumulative distribution function** $F(x)$.

For a particular value, $t$, in the range of the function,

$$F(t) = P(X \leqslant t) = \int_{-\infty}^{t} f(x)dx.$$

The lower limit is given as $-\infty$, but in practice it is the smallest possible value of $x$ in the range
for which $x$ is valid.

So if $f(x)$ is valid in the range $a \leqslant x \leqslant b$,

then $\quad F(t) = \int_{a}^{t} f(x)dx$

$\uparrow$

lower limit

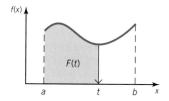

Remember that $F(t)$ gives the area under the curve $f(x)$ *up to* a particular value $t$.

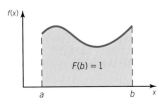

Notice that $\quad F(b) = P(X \leqslant b)$

$$= \int_a^b f(x)dx$$

$$= 1$$

This is as expected, since the **total area under the curve is 1.**

# Using $F(x)$ to find $P(x_1 \leqslant X \leqslant x_2)$

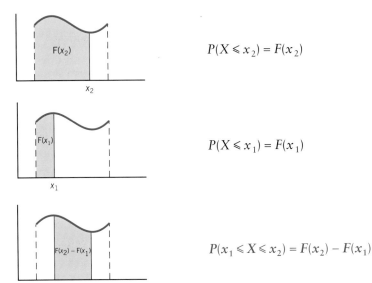

$P(X \leqslant x_2) = F(x_2)$

$P(X \leqslant x_1) = F(x_1)$

$P(x_1 \leqslant X \leqslant x_2) = F(x_2) - F(x_1)$

## Finding the median, quartiles and other percentiles

The median is the value 50% of the way through the distribution. It splits the area under the curve $y = f(x)$ into two halves. If $m$ is the median, then for $f(x)$ defined for $a \leqslant x \leqslant b$,

$$\int_a^m f(x)dx = 0.5$$

i.e. $\quad F(m) = 0.5$

For example:

(a)

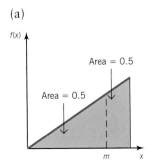

(b)

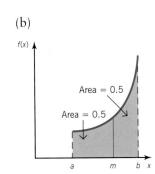

Note that if $f(x)$ is symmetrical in the given range, the mean and median will coincide.

The **lower quartile**, $q_1$, is the value 25% of the way through the distribution, so

$$\int_a^{q_1} f(x) = dx = 0.25$$

i.e.   $F(q_1) = 0.25$

The **upper quartile**, $q_3$, is the value 75% of the way through the distribution, so

$$\int_a^{q_3} f(x)dx = 0.75$$

i.e.   $F(q_3) = 0.75$

Similarly for other percentiles, for example

$F(\text{10th percentile}) = 0.1$ and $F(\text{35th percentile}) = 0.35$.

In general

$$F(n\text{th percentile}) = \frac{n}{100}$$

## Example 6.16

$X$ is a continuous random variable with p.d.f. as shown.

$f(x) = \frac{1}{8}x, \quad 0 \leqslant x \leqslant 4$

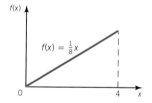

Find

(a) the cumulative distribution function $F(x)$ and sketch $y = F(x)$,
(b) $P(0.3 \leqslant X \leqslant 1.8)$,
(c) the median, $m$,
(d) the interquartile range.

## Solution 6.16

(a) For values of $t$ between 0 and 4,

$$F(t) = \int_0^t \frac{1}{8} x\, dx$$

$$= \left[ \frac{x^2}{16} \right]_0^t$$

$$= \frac{t^2}{16}$$

Check $F(4)$

$F(4) = \dfrac{4^2}{16} = 1$, as expected.

The cumulative distribution function can now be written in terms of $x$ as follows:

$$F(x) = \begin{cases} 0 & x \leqslant 0 \\ \dfrac{x^2}{16} & 0 \leqslant x \leqslant 4 \\ 1 & x \geqslant 4 \end{cases}$$

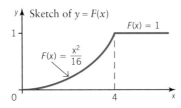

(b) $P(0.3 \leqslant X \leqslant 1.8) = F(1.8) - F(0.3)$

$$F(1.8) = \frac{1.8^2}{16} = 0.2025$$

$$F(0.3) = \frac{0.3^2}{16} = 0.005\,625$$

$P(0.3 \leqslant X \leqslant 1.8) = 0.025 - 0.005\,625 = \mathbf{0.197}$ **(3 d.p.)**

(c) For the median $m$,

$$F(m) = 0.5$$

i.e. $$\frac{m^2}{16} = 0.5$$

$$m^2 = 8$$

$$m = 2.828\ldots$$

The median $m = \mathbf{2.83}$ **(2 d.p.)**.

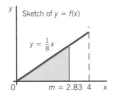

*NOTE:* take the positive square root, since $0 \leqslant m \leqslant 4$.

(d) $F(q_1) = 0.25 \qquad \therefore \quad \dfrac{q_1^{\,2}}{16} = 0.25$

$$q_1^{\,2} = 4$$

$$q_1 = 2$$

$F(q_3) = 0.75 \qquad \therefore \quad \dfrac{q_3^{\,2}}{16} = 0.75$

$$q_3^{\,2} = 12$$

$$q_3 = \sqrt{12} = 3.464\ldots$$

Interquartile range $= q_3 - q_1 = 3.464\ldots - 2 = \mathbf{1.5}$ **(2 s.f.)**.

---

## Example 6.17

$X$ is a continuous random variable with p.d.f. $f(x)$ where

$$f(x) = \begin{cases} \dfrac{x}{3} & 0 \leqslant x \leqslant 2 \\ -\dfrac{2x}{3} + 2 & 2 \leqslant x \leqslant 3 \\ 0 & \text{otherwise} \end{cases}$$

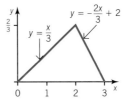

(a) Find the cumulative distribution function $F(x)$ and sketch it.
(b) Find $P(1 \leqslant X \leqslant 2.5)$.
(c) Find the median, $m$.

## Solution 6.17

(a) $F(t) = \displaystyle\int_0^t f(x)dx$

Since $f(x)$ is given in two parts, $F(x)$ must be found in two stages.
First consider $t$ where $0 \leqslant t \leqslant 2$.

$F(t) = \displaystyle\int_0^t \frac{x}{3}\, dx$

$\quad = \left[\dfrac{x^2}{6}\right]_0^t$

$\quad = \dfrac{t^2}{6}$

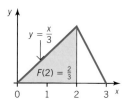

So, for $0 \leqslant x \leqslant 2$, $F(x) = \dfrac{x^2}{6}$

NOTE: $F(2) = \frac{4}{6} = \frac{2}{3}$

Now consider $t$ such that $2 \leqslant t \leqslant 3$.

$F(t) = F(2) + \displaystyle\int_2^t \left(-\frac{2x}{3} + 2\right)dx$

$\quad = F(2) + \left[-\dfrac{x^2}{3} + 2x\right]_2^t$

$\quad = \dfrac{2}{3} + \left\{-\dfrac{t^2}{3} + 2t - \left(-\dfrac{4}{3} + 4\right)\right\}$

$\quad = -\dfrac{t^2}{3} + 2t - 2$

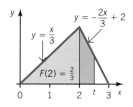

Check F(3)

$F(3) = -\dfrac{3^2}{3} + 6 - 2 = 1$

Writing the answer as a general formula in terms of $x$,

$$F(x) = \begin{cases} \dfrac{x^2}{6} & 0 \leqslant x \leqslant 2 \\[2mm] -\dfrac{x^2}{3} + 2x - 2 & 2 \leqslant x \leqslant 3 \\[2mm] 1 & x \geqslant 3 \end{cases}$$

Sketch of $y = F(x)$

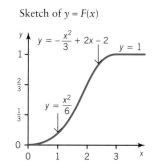

(b) $P(1 \leqslant X \leqslant 2.5) = F(2.5) - F(1)$

To find $F(2.5)$, use $F(x)$ in the interval $2 \leqslant x \leqslant 3$.

$$F(x) = -\frac{x^2}{3} + 2x - 2$$

$$\therefore \quad F(2.5) = -\frac{(2.5)^2}{3} + 2(2.5) - 2$$

$$= \frac{11}{12}$$

To find $F(1)$, use $F(x)$ in the interval $0 \leqslant x \leqslant 2$.

$$F(x) = \frac{x^2}{6}$$

$$\therefore \quad F(1) = \frac{1}{6}$$

$$P(1 \leqslant X \leqslant 2.5) = F(2.5) - F(1)$$

$$= \frac{11}{12} - \frac{1}{6}$$

$$= 0.75$$

(c) $F(m) = 0.5$, where $m$ is the median.

Since $F(2) = \frac{2}{3}$, the median must be less than 2, so consider $F(x)$ in the range $0 \leqslant x \leqslant 2$.

Therefore $\quad F(m) = \frac{m^2}{6}$

So $\qquad \frac{m^2}{6} = 0.5$

$$m^2 = 3$$

$$m = 1.73 \text{ (2 d.p.)}$$

---

## Exercise 6d   Cumulative distribution function

1. The random variable $X$ has probability density function

$$f(x) = \tfrac{3}{8}x^2, \quad 0 \leqslant x \leqslant 2$$

Find

(a) the cumulative distribution function, $F(x)$ and draw a sketch of $y = F(x)$,
(b) the median, $m$.

2. The random variable $X$ has probability density function

$$f(x) = \tfrac{1}{4}(4 - x), \quad 1 \leqslant x \leqslant 3$$

Find

(a) the cumulative distribution function $F(x)$,
(b) $P(1.5 < X < 2)$.

3. The random variable $X$ has probability density function

$$f(x) = k, \quad 1 \leqslant x \leqslant 6$$

(a) Find $k$.
(b) Find the cumulative distribution function $F(x)$.
(c) Find the 20th percentile.
(d) Find the interquartile range.

4. The random variable $X$ has probability density function

$$f(x) = \begin{cases} \tfrac{1}{4} & 0 \leqslant x \leqslant 2 \\ \tfrac{1}{4}(2x - 3) & 2 \leqslant x \leqslant 3 \end{cases}$$

Find

(a) the cumulative distribution function $F(x)$,
(b) the median $m$.

5. The random variable $X$ has cumulative distribution function

$$F(x) = \begin{cases} 0 & x < 0 \\ x^4 & 0 \leqslant x \leqslant 1 \\ 1 & x \geqslant 1 \end{cases}$$

Find

(a) $P(0.3 < X < 0.6)$,
(b) the median $m$,
(c) the value of $a$ such that $P(X > a) = 0.4$.

6. The continuous random variable $X$ has p.d.f. $f(x) = \frac{1}{3}$, $0 \leqslant x \leqslant 3$. Find

(a) $E(X)$,           (b) $Var(X)$,
(c) $F(x)$ and sketch $y = F(x)$, (d) $P(X \geqslant 1.8)$,
(e) $P(1.1 \leqslant X \leqslant 1.7)$.

7. $X$ is the continuous random variable with p.d.f. $f(x) = kx^2$, $1 \leqslant x \leqslant 2$. Find (a) the constant $k$ and sketch $y = f(x)$, (b) the standard deviation $\sigma$, (c) the cumulative distribution function $F(x)$, (d) the median, $m$.

8. The continuous random variable $X$ has probability density function $f$ given by

$$f(x) = \begin{cases} k(4 - x^2) & \text{for } 0 \leqslant x \leqslant 2 \\ 0 & \text{otherwise} \end{cases}$$

where $k$ is a constant. Show that $k = \frac{3}{16}$ and find the values of $E(X)$ and $Var(X)$.
Find the cumulative distribution function of $X$, and verify by calculation that the median value of $X$ is between 0.69 and 0.70
Find also $P(0.69 < X < 0.70)$, giving your answer correct to one significant figure. (C)

9. The continuous random variable $X$ has continuous p.d.f. $f(x)$ where

$$f(x) = \begin{cases} \dfrac{x}{3} - \dfrac{2}{3} & 2 \leqslant x \leqslant 3 \\ \alpha & 3 \leqslant x \leqslant 5 \\ 2 - \beta x & 5 \leqslant x \leqslant 6 \\ 0 & \text{otherwise} \end{cases}$$

Find (a) $\alpha$ and $\beta$, (b) $F(x)$ and sketch $y = F(x)$, (c) $P(2 \leqslant X \leqslant 3.5)$, (d) $P(X \geqslant 5.5)$.

10. The continuous random variable $X$ has probability density function

$$f(x) = \begin{cases} \dfrac{1 + x}{6} & 1 \leqslant x \leqslant 3 \\ 0 & \text{otherwise} \end{cases}$$

(a) Sketch the probability density function of $X$.
(b) Calculate the mean of $X$.
(c) Specify fully the cumulative distribution function of $X$.
(d) Find $m$ such that $P(X \leqslant m) = \frac{1}{2}$. (L)

11. A factory is supplied with flour at the beginning of each week. The weekly demand, $X$ thousand tonnes, for flour from this factory is a continuous random variable having the probability density function

$$f(x) = k(1 - x)^4, \qquad 0 \leqslant x \leqslant 1$$
$$f(x) = 0, \qquad\qquad \text{elsewhere}$$

Find

(a) the value of $k$,
(b) the mean value of $X$,
(c) the variance of $X$, to three decimal places.

Sketch the probability density function.
Find, to the nearest tonne, the quantity of flour that the factory should have in stock at the beginning of a week in order that there is a probability of 0.98 that the demand in that week will be met. (L)

12. A continuous random variable $X$ has probability density function, $f$, defined by

$$f(x) = \frac{1}{4}, \qquad 0 \leqslant x \leqslant 1$$
$$f(x) = \frac{x^3}{5}, \qquad 1 \leqslant x \leqslant 2$$
$$f(x) = 0, \qquad \text{otherwise}$$

Obtain the distribution function and hence, or otherwise, find, to three decimal places, the median and the interquartile range of the distribution (L)

13. The continuous random variable $X$ has probability density function $f$ given by

$$f(x) = \begin{cases} k(x + 3), & -3 \leqslant x \leqslant 3 \\ 0, & \text{otherwise} \end{cases}$$

where $k$ is a constant.

(a) Show that $k = \frac{1}{18}$.
(b) Find $E(X)$ and $Var(X)$.
(c) Find the lower quartile of $X$, i.e. the value $q$ such that $P(X \leqslant q) = \frac{1}{4}$.
(d) Let $Y = aX + b$, where $a$ and $b$ are constants with $a > 0$. Find the values of $a$ and $b$ for which $E(Y) = 0$ and $Var(Y) = 1$. (C)

14. The continuous random variable, $X$, has probability density function defined by

$$f(x) = \begin{cases} kx, & 0 \leqslant x \leqslant 8 \\ 8k, & 8 < x \leqslant 9 \\ 0 & \text{otherwise} \end{cases}$$

where $k$ is a constant.

(a) Sketch the graph of $f(x)$.
(b) Show that $k = 0.025$.
(c) Determine, for all $x$, the distribution function $F(x)$.
(d) Calculate the probability that an observed value of $X$ exceeds 6. (NEAB)

15. A continuous random variable, $X$, has probability density function given by

$$f(x) = ax - bx^2 \quad \text{for} \quad 0 \leqslant x \leqslant 2$$
$$= 0 \qquad \text{elsewhere}$$

Observations on $X$ indicate that the mean is 1.
(a) Obtain two simultaneous equations for $a$ and $b$, show that $a = 1.5$ and find the value of $b$.
(b) Find the variance of $X$.
(c) If $F(x)$ is the probability that $X \leqslant x$ find $F(x)$ and verify that $F(2) = 1$.
(d) If two independent observations are made on $X$ what is the probability that at least one of them is less than $\frac{1}{2}$?

16. The continuous random variable X has probability density function given by

$$f(x) = \begin{cases} \dfrac{k}{x} & \text{for } 1 \leqslant x \leqslant 9, \\ 0 & \text{otherwise} \end{cases}$$

where $k$ is a constant. Giving your answers correct to three significant figures where appropriate, find
(a) the value of $k$, and also the median value of $X$,
(b) the mean and variance of $X$,
(c) the cumulative distribution function, $F$, of $X$, and sketch the graph of $y = F(x)$. (C)

# OBTAINING THE P.D.F., $f(x)$, FROM THE CUMULATIVE DISTRIBUTION FUNCTION $F(x)$

Since $F$ can be obtained by *integrating* $f$, it follows that $f$ can be obtained by *differentiating* $F$.

$$f(x) = \frac{d}{dx} F(x)$$
$$= F'(x)$$

NOTE: the gradient of the $F(x)$ curve gives the value of $f(x)$.

## Example 6.18

The continuous random variable $X$ has cumulative distribution function $F(x)$ where

$$F(x) = \begin{cases} 0 & x \leqslant 0 \\ \dfrac{x^3}{27} & 0 \leqslant x \leqslant 3 \\ 1 & x \geqslant 3 \end{cases}$$

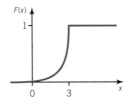

(a) State the range of values for which the probability density function $f(x)$ is valid.
(b) Find $f(x)$ and illustrate it in a sketch.

## Solution 6.18

(a) Since $F(x)$ is *unchanging* in the regions $x \leqslant 0$ and $x \geqslant 3$ it follows that $f(x)$ must be zero for $x \leqslant 0$ and $x \geqslant 3$.
So $f(x)$ is valid for $0 \leqslant x \leqslant 3$ and $f(x) = 0$ otherwise.

(b) $f(x) = \dfrac{d}{dx} F(x)$

$$= \frac{d}{dx} \left( \frac{x^3}{27} \right)$$
$$= \frac{3x^2}{27}$$
$$= \frac{x^2}{9}$$

The p.d.f. for $X$ is $f(x)$ where

$$f(x) = \begin{cases} \dfrac{x^2}{9} & 0 \leqslant x \leqslant 3 \\ 0 & \text{otherwise} \end{cases}$$

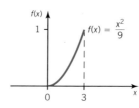

---

## Example 6.19

The continuous random variable $X$ has cumulative distribution function $F(x)$ as shown in the sketch.

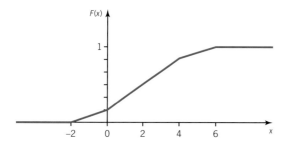

$$F(x) = \begin{cases} 0 & x < -2 \\ \frac{1}{12}(2 + x) & -2 \leqslant x < 0 \\ \frac{1}{6}(1 + x) & 0 \leqslant x < 4 \\ \frac{1}{12}(6 + x) & 4 \leqslant x < 6 \\ 1 & x \geqslant 6 \end{cases}$$

(a) Find the p.d.f. of $X$, $f(x)$, and sketch $y = f(x)$.
(b) Find $E(X)$.

## Solution 6.19

(a) Since $F(x)$ is unchanging for $x < -2$ and $x \geqslant 6$, it follows that $f(x)$ must be zero for $x < -2$ and $x \geqslant 6$.

Since $f(x) = \dfrac{d}{dx} F(x)$,

for $\quad -2 \leqslant x < 0, \quad f(x) = \dfrac{d}{dx} \dfrac{1}{12}(2 + x) = \dfrac{1}{12}$

for $\quad 0 \leqslant x < 4, \quad f(x) = \dfrac{d}{dx} \dfrac{1}{6}(1 + x) = \dfrac{1}{6}$

for $\quad 4 \leqslant x < 6, \quad f(x) = \dfrac{d}{dx} \dfrac{1}{12}(6 + x) = \dfrac{1}{12}$

The sketch of $y = f(x)$ is shown:

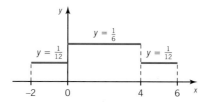

(b) Since $f(x)$ is symmetrical, $E(X) = 2$.

---

### Example 6.20

The continuous random variable $X$ has cumulative distribution function given by

$$F(x) = \begin{cases} 0 & x < 0 \\ 2x - x^2 & 0 \leqslant x \leqslant 1 \\ 1 & x > 1 \end{cases}$$

(a) Show that $P(X < \frac{1}{2}) = \frac{3}{4}$.

(b) Find the interquartile range of $X$. (C)

### Solution 6.20

(a) $P(X < \frac{1}{2}) = F(\frac{1}{2}) = 2 \times \frac{1}{2} - (\frac{1}{2})^2 = \mathbf{0.75}$

(b) To find the interquartile range, you need to find the upper quartile and lower quartile.

Upper quartile $q_3$ is such that $F(q_3) = 0.75$.
From (a)  $F(\frac{1}{2}) = 0.75$
$$\therefore \quad q_3 = \tfrac{1}{2}$$

Lower quartile $q_1$ is such that $F(q_1) = 0.25$

$$F(q_1) = 2q_1 - q_1^2$$
$$\therefore \quad 2q_1 - q_1^2 = 0.25$$
$$q_1^2 - 2q_1 + 0.25 = 0$$
$$(q_1 - 1)^2 - 1 + 0.25 = 0$$
$$(q_1 - 1)^2 = 0.75$$
$$q_1 - 1 = \pm\sqrt{0.75}$$

So $q_1 = 1 + \sqrt{0.75}$ or $q_1 = 1 - \sqrt{0.75}$

Since $F(x)$ is unchanging for $x > 1$, $f(x) = 0$ for $x > 1$.
So $1 + \sqrt{0.75}$ is outside the range of $f(x)$.

$$\therefore \quad q_1 = 1 - \sqrt{0.75} = 0.1339\ldots$$

Interquartile range $= q_3 - q_1$

$$= 0.5 - 0.1339\ldots$$
$$= \mathbf{0.37 \ (2 \ s.f.)}$$

## Exercise 6e   Obtaining $f(x)$ from $F(x)$

1. The cumulative distribution function of $X$ is given by

$$F(x) = \begin{cases} 0 & x \leqslant 2 \\ 0.25x - 0.5 & 2 \leqslant x \leqslant 6 \\ 1 & x \geqslant 6 \end{cases}$$

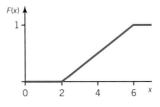

(a) Find the probability density function $f(x)$.
(b) Sketch $y = f(x)$.
(c) Find $E(X)$.
(d) Find the interquartile range.

2. The cumulative distribution function of $X$ is given by

$$F(x) = \begin{cases} 0 & x \leqslant 0 \\ x^3 & 0 \leqslant x \leqslant 1 \\ 1 & x \geqslant 1 \end{cases}$$

Find

(a) the median,
(b) the mean.

3. The cumulative distribution function of $X$ is given by

$$F(x) = \begin{cases} 0 & x \leqslant 0 \\ x - kx^2 & 0 \leqslant x \leqslant 2 \\ 1 & x \geqslant 2 \end{cases}$$

Find

(a) the value of $k$,
(b) the probability density function $f(x)$,
(c) the median of $X$,
(d) the variance of $X$.

4. The continuous random variable $X$ has cumulative distribution function $F(x)$ where

$$F(x) = \begin{cases} 0 & x \leqslant 0 \\ \dfrac{2x}{3} & 0 \leqslant x \leqslant 1 \\ \dfrac{x}{3} + k & 1 \leqslant x \leqslant 2 \\ 1 & x \geqslant 2 \end{cases}$$

Find

(a) the value of $k$,
(b) the p.d.f. $f(x)$ and sketch it,
(c) the mean $\mu$,
(d) the standard deviation $\sigma$.

5. The continuous random variable $X$ has cumulative distribution function $F(x)$ where

$$F(x) = \begin{cases} 0 & x \leqslant 1 \\ \dfrac{(x-1)^2}{12} & 1 \leqslant x \leqslant 3 \\ \dfrac{(14x - x^2 - 25)}{24} & 3 \leqslant x \leqslant 7 \\ 1 & x \geqslant 7 \end{cases}$$

Find

(a) the p.d.f. $f(x)$ and sketch it,
(b) $E(X)$
(c) $\text{Var}(X)$,
(d) the median of $X$,
(e) $P(2.8 \leqslant X \leqslant 5.2)$.

6. The continuous random variable $X$ has (cumulative) distribution function given by

$$F(x) = \begin{cases} \dfrac{1+x}{8} & -1 \leqslant x \leqslant 0 \\ \dfrac{1+3x}{8} & 0 \leqslant x \leqslant 2 \\ \dfrac{5+x}{8} & 2 \leqslant x \leqslant 3 \end{cases}$$

where $F(x) = 0$ for $x < -1$, and $F(x) = 1$ for $x > 3$.

(a) Sketch the graph of the probability density function $f(x)$.
(b) Determine the expectation of $X$ and the variance of $X$.
(c) Determine $P(3 \leqslant 2X \leqslant 5)$. (C)

7. A continuous random variable $X$ takes values in the interval 0 to 3. It is given that $P(X > x) = a + bx^3$, $0 \leqslant x \leqslant 3$.

(a) Find the values of the constants $a$ and $b$.
(b) Find the cumulative distribution function $F(x)$.
(c) Find the probability density function $f(x)$.
(d) Show that $E(X) = 2.25$.

8. The length $X$ of an offcut of wooden planking is a random variable which can take any value up to 0.5 m. It is known that the probability of the length being not more than $x$ metres ($0 \leqslant x \leqslant 0.5$) is equal to $kx$. Determine

(a) the value of $k$,
(b) the probability density function of $X$,
(c) the expected value of $X$,
(d) the standard deviation of $X$ (correct to three significant figures). (C)

# THE CONTINUOUS UNIFORM (OR RECTANGULAR) DISTRIBUTION

Consider the continuous random variable $X$ with probability density function
$f(x) = k$ for $1 \leqslant x \leqslant 6$.

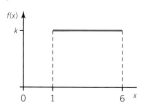

Since the total area under the curve is 1,

$$5k = 1$$
$$k = 0.2$$
$$\therefore \quad f(x) = 0.2, \qquad 1 \leqslant x \leqslant 6$$

$X$ is said to follow a continuous uniform, or rectangular, distribution between 1 and 6.
This can be written $X \sim R(1, 6)$.

In general:

The probability density function for a continuous random variable, distributed uniformly in
the range $a \leqslant x \leqslant b$ is

$$f(x) = \frac{1}{b - a}$$

This is written $X \sim R(a, b)$

$a$ and $b$ are known as the parameters of the distribution.

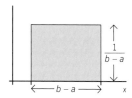

*NOTE:* It is easy to see from the diagram that the total area is 1.

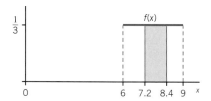

$$\text{Area} = (b - a) \times \frac{1}{(b - a)}$$
$$= 1$$

## Example 6.21

$X$ is distributed uniformly, where $6 \leqslant x \leqslant 9$.

Find $P(7.2 \leqslant X \leqslant 8.4)$.

## Solution 6.21

$$f(x) = \frac{1}{b - a} = \frac{1}{9 - 6} = \frac{1}{3}$$
$$P(7.2 \leqslant X \leqslant 8.4) = \tfrac{1}{3}(8.4 - 7.2)$$
$$= 0.4$$

## Example 6.22

The lengths of metal rods are measured to the nearest 5 mm. What is the distribution of the random variable $E$, the rounding error made when measuring? Give its probability density function $f(e)$.

## Solution 6.22

The error is the difference between the true length and the recorded length after rounding to the nearest 5 mm.

Suppose you have recorded a length to be 75 mm, to the nearest 5 mm. The true length could have been any length in the interval

$$72.5 \text{ mm} \leqslant l < 77.5 \text{ mm}$$

So the error, $E$, could be anywhere in the interval $-2.5 \leqslant E < 2.5$.

All points in this interval are equally likely 'stopping places' for $E$, so $E$ is *uniformly distributed* in the interval, i.e.

$E \sim R(-2.5, 2.5)$

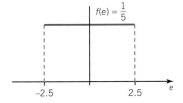

$$f(e) = \frac{1}{2.5 - (-2.5)}$$
$$= \frac{1}{5}, \quad -2.5 < e \leqslant 2.5$$

## Example 6.23

Rosie spins a 'Spinning Jenny' at a fair. When the wheel stops, the shorter distance of an arrow measured along the circumference from Rosie is denoted by $C$. What is the distribution of $C$?

## Solution 6.23

All the points on the circumference are equally likely stopping places for the arrow, so $C$ is uniformly distributed between 0 (when the arrow is next to Rosie) and $\pi r$ (when the arrow is diametrically opposite Rosie).

So   $C \sim R(0, \pi r)$

$$f(c) = \frac{1}{\pi r - 0}$$
$$= \frac{1}{\pi r}, \quad 0 \leqslant c \leqslant \pi r.$$

### Example 6.24

The error, in grams, made by a greengrocer's scales may be modelled by the random variable, $X$, with probability density function

$$f(x) = \begin{cases} 0.1 & -3 \leqslant x \leqslant 7 \\ 0 & \text{otherwise.} \end{cases}$$

Find the probability that

(a) an error is positive,
(b) the magnitude of an error exceeds 2 grams (i.e. $|X| > 2$),
(c) the magnitude of an error is less than 4 grams (i.e. $|X| < 4$). (AEB)

### Solution 6.24

(a)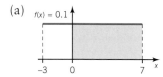

$$P(X > 0) = 7 \times 0.1 = 0.7$$

(b)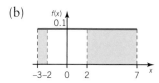

$$P(|X| > 2) = 1 - P(|X| < 2)$$
$$= 1 - P(-2 < X < 2)$$
$$= 1 - 4 \times 0.1$$
$$= 0.6$$

(c)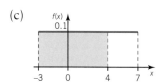

$$P(|X| < 4) = P(-4 < X < 4)$$
Since $f(x) = 0$ when $x < -3$, find $P(-3 < X < 4)$.
$$P(-3 < X < 4) = 7 \times 0.1$$
$$= 0.7$$
So $P(|X| < 4) = 0.7$

## EXPECTATION AND VARIANCE OF THE UNIFORM DISTRIBUTION

### Example 6.25

The continuous random variable $Y$ has a rectangular distribution

$$f(y) = \begin{cases} \dfrac{1}{\pi} & -\dfrac{\pi}{2} \leqslant y \leqslant \dfrac{\pi}{2} \\ 0 & \text{otherwise} \end{cases}$$

(a) Find the mean of $Y$.
(b) Find the variance of $Y$. (L)

**Solution 6.25**

Sketch $f(y)$.

(a)

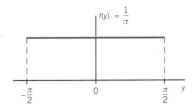

By symmetry

$E(Y) = 0$

**The mean of $Y$ is 0.**

(b) To find Var$(Y)$, find $E(Y^2)$ first

$$E(Y^2) = \int_{all\ y} y^2\, f(y)dy$$

$$= \int_{-\frac{\pi}{2}}^{\frac{\pi}{2}} y^2 \frac{1}{\pi}\, dy$$

$$= \frac{1}{\pi} \left[ \frac{y^3}{3} \right]_{-\frac{\pi}{2}}^{\frac{\pi}{2}}$$

$$= \frac{1}{3\pi} \left( \frac{\pi^3}{8} - \left( -\frac{\pi^3}{8} \right) \right)$$

$$= \frac{1}{3\pi} \left( \frac{\pi^3}{4} \right)$$

$$= \frac{\pi^2}{12}$$

$$Var(Y) = E(Y^2) - E^2(Y)$$

$$= \frac{\pi^2}{12} - 0$$

$$= \frac{\pi^2}{12}$$

The variance of $Y$ is $\dfrac{\pi^2}{12}$.

---

It is possible to write the mean and the variance of a uniform distribution in general formulae.

If the continuous variable $X$ is uniformly distributed over the interval $(a, b)$, then

$X \sim R(a, b)$

By symmetry

$E(X) = \frac{1}{2}(a + b)$

It can also be shown that

$Var(X) = \frac{1}{12}(b - a)^2$

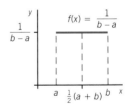

# THE CUMULATIVE DISTRIBUTION FUNCTION, *F(x)*, FOR A UNIFORM DISTRIBUTION

## Example 6.26

$X$ has probability density function $f(x) = \frac{1}{4}$, $5 \leqslant x \leqslant 9$. Find $F(x)$.

## Solution 6.26

By integration:

If $5 \leqslant t \leqslant 9$

$$F(t) = \int_5^t f(x)dx$$

$$= \int_5^t \frac{1}{4}\,dx$$

$$= \left[\frac{1}{4}x\right]_5^t$$

$$= \frac{t}{4} - \frac{5}{4}$$

$$= \frac{t-5}{4}$$

So $F(x) = \begin{cases} 0 & x < 5 \\ \dfrac{x-5}{4} & 5 \leqslant x \leqslant 9 \\ 1 & x \geqslant 9 \end{cases}$

Diagrammatically:

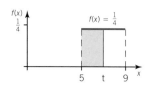

$$F(t) = (t-5) \times \frac{1}{4}$$

$$= \frac{t-5}{4}$$

In general, for $X$ distributed uniformly, $a \leqslant x \leqslant b$, $F(t) = \dfrac{t-a}{b-a}$

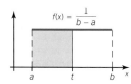

$$\therefore \quad F(x) = \begin{cases} 0 & x < a \\ \dfrac{x-a}{b-a} & a \leqslant x \leqslant b \\ 1 & x \geqslant b \end{cases}$$

$F(x)$ can be illustrated diagrammatically.

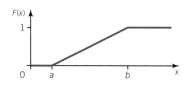

## Exercise 6f   Uniform distribution

1. $X$ follows a uniform distribution with probability density function

   $$f(x) = k, \qquad 3 \leqslant x \leqslant 6.$$

   Find

   (a) $k$,
   (b) $E(X)$,
   (c) $Var(X)$,
   (d) $P(X > 5)$.

2. $X$ is distributed uniformly, $-5 \leqslant x \leqslant -2$.

   Find

   (a) $P(-4.3 < X < -2.8)$,
   (b) $E(X)$,
   (c) the standard deviation of $X$.

3. The continuous random variable $X$ has p.d.f. $f(x)$ as shown in the diagram:

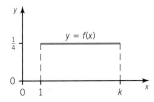

Find

(a) the value of $k$,
(b) $P(2.1 < X < 3.4)$,
(c) $E(X)$,
(d) $Var(X)$.

4. The random variable $X$ has p.d.f. $f(x)$ as shown in the diagram.

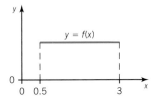

If two independent observations of $X$ are made, find the probability that one is less than 1.5 and the other is greater than the mean.

5. The random variable $Y$ has probability density function given by

$$f(y) = \begin{cases} 0.2 & 32 \leqslant y \leqslant 37 \\ 0 & \text{otherwise} \end{cases}$$

Find the probability that $Y$ lies within one standard deviation of the mean.

6. $X$ has cumulative distribution function

$$F(x) = \frac{x-2}{5}, \qquad 2 \leqslant x \leqslant 7$$

Find

(a) $E(X)$,
(b) $Var(X)$.

7. The continuous random variable $X$ is uniformly distributed in the interval $a < x < b$.
The lower quartile is 5 and the upper quartile is 9.

Find

(a) the values of $a$ and $b$,
(b) $P(6 < X < 7)$,
(c) the cumulative distribution function $F(x)$.

8. $X$ has cumulative distribution function $F(x)$ illustrated as follows

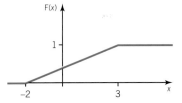

(a) Find the probability density function $f(x)$.
(b) Find the standard deviation of $X$.
(c) Find the interquartile range.
(d) Find the 20th percentile

## Summary

- For a continuous random variable $X$, with p.d.f. $f(x)$ for $a \leqslant x \leqslant b$

$$\int_{\text{all } x} f(x)dx = 1$$

- $P(c \leqslant X \leqslant d) = \displaystyle\int_c^d f(x)dx$   where $a \leqslant c < d \leqslant b$.

- Expectation $E(X) = \displaystyle\int_{\text{all } x} xf(x)dx$

- Variance, $\text{Var}(X) = \displaystyle\int_{\text{all } x} x^2 f(x)dx - E^2(X)$

- The cumulative distribution function $F(x)$

$$F(t) = \int_a^t f(x)dx \qquad \text{for } a \leqslant t \leqslant b.$$

- To obtain $f(x)$ from $F(x)$, differentiate $F(x)$

$$f(x) = \frac{d}{dx}F(x) = F'(x).$$

- Median, quartiles and other percentiles

  Median $m$: $\qquad\qquad F(m) = 0.5$

  Lower quartile $q_1$: $\qquad F(q_1) = 0.25$

  Upper quartile $q_3$: $\qquad F(q_3) = 0.75$

  $n$th percentile $\qquad\qquad F(n\text{th percentile}) = \dfrac{n}{100}$

  Interquartile range $= q_3 - q_1$

- The continuous uniform (rectangular) distribution

  If $f(x) = \dfrac{1}{b-a}$   $a \leqslant x \leqslant b$, then $X \sim R(a, b)$

  $E(X) = \frac{1}{2}(a + b)$

  $\text{Var}(X) = \frac{1}{12}(b - a)^2$

  $$F(x) = \begin{cases} 0 & x \leqslant a \\ \dfrac{x-a}{b-a} & a \leqslant x \leqslant b \\ 1 & x \geqslant 1 \end{cases}$$

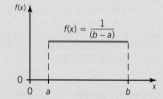

## Miscellaneous worked examples

### Example 6.27

The random variable $X$ has probability density function

$$f(x) = \begin{cases} 3x^k & 0 \leqslant x \leqslant 1, \\ 0 & \text{otherwise,} \end{cases}$$

where $k$ is a positive integer.

Find

(a) the value of $k$,
(b) the mean of $X$,
(c) the value, $x$, such that $P(X \leqslant x) = 0.5$.

### Solution 6.27

(a) Since $X$ is a random variable, $\displaystyle\int_{\text{all } x} f(x)dx = 1$

Therefore $\displaystyle\int_0^1 3x^k dx = 1$

$$3\left[\frac{x^{k+1}}{(k+1)}\right]_0^1 = 1$$

$$\frac{3}{k+1} = 1$$

$$k+1 = 3$$

$$k = 2$$

$$f(x) = \begin{cases} 3x^2 & 0 \leqslant x \leqslant 1 \\ 0 & \text{otherwise} \end{cases}$$

(b) $\displaystyle E(X) = \int_0^1 xf(x)dx$

$$= \int_0^1 3x^3 dx$$

$$= \frac{3}{4}\left[x^4\right]_0^1$$

$$= 0.75$$

**The mean of $X$ is 0.75.**

(c) Let $P(X \leqslant x_1) = 0.5$

Therefore $\displaystyle\int_0^{x_1} 3x^2 dx = 0.5$

$$\left[x^3\right]_0^{x_1} = 0.5$$

$$x_1^3 = 0.5$$

$$x_1 = (0.5)^{\frac{1}{3}}$$

$$= 0.794 \text{ (3 d.p.)}$$

**So $x = 0.794$ (3 d.p.)**

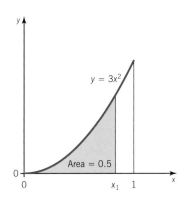

## Example 6.28

The length of blades of grass mown from a lawn are modelled by a uniform distribution between 1 cm and 5 cm.

(a) Find the standard deviation of this distribution.
(b) Find the percentage of blades of grass whose lengths lie within one standard deviation of the mean length.
(c) A better model may be a triangular distribution as shown.

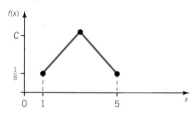

Find the value of $C$.                                                      (NEAB)

## Solution 6.28

(a) $X$ is the length, in centimetres, of a blade of grass.

$f(x) = \frac{1}{4}, \quad 1 \leqslant x \leqslant 5.$               (see page 345)

$\text{Var}(X) = \frac{1}{12}(5-1)^2 = \frac{16}{12} = \frac{4}{3}$         (see page 348)

**Standard deviation of $X = \sqrt{\frac{4}{3}} = 1.15$ (2 d.p.)**

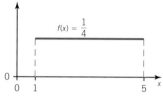

(b) $E(X) = \dfrac{1+5}{2} = 3$

$P(3 - \sqrt{\frac{4}{3}} \leqslant X \leqslant 3 + \sqrt{\frac{4}{3}}) = 2\sqrt{\frac{4}{3}} \times \frac{1}{4}$

$\qquad\qquad\qquad\qquad\quad = 0.577\ldots$

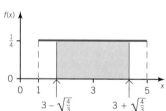

**So approximately 58% of blades of grass have length within one standard deviation of the mean.**

(c) Total area = 1

Area of rectangle = $4 \times \frac{1}{8} = \frac{1}{2}$

Area of triangle = $\frac{1}{2} \times 4 \times h = 2h$

$\therefore \quad \frac{1}{2} + 2h = 1$

$\qquad\quad h = \frac{1}{4}$

$C = h + \frac{1}{8} = \frac{1}{4} + \frac{1}{8} = \frac{3}{8}$

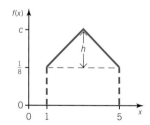

## Example 6.29

On any day, the amount of time, measured in hours, that Mr Goggle spends watching television is a continuous random variable $T$, with cumulative distribution function given by

$$F(t) = \begin{cases} 0 & t \leqslant 0 \\ 1 - k(15 - t)^2 & 0 \leqslant t \leqslant 15 \\ 1 & t \geqslant 15 \end{cases}$$

where $k$ is a constant.

(a) Show that $k = \frac{1}{225}$ and find $P(5 \leqslant T \leqslant 10)$.

(b) Show that, for $0 \leqslant t \leqslant 15$, the probability density function of $T$ is given by

$$f(t) = \frac{2}{15} - \frac{2}{225} t.$$

(c) Find the median of $T$. (C)

## Solution 6.29

(a) When $t = 0$, $F(t) = 0$

Using $F(t) = 1 - k(15 - t)^2$,

when $\quad t = 0$

$$0 = 1 - k \times 15^2$$

$$\therefore \quad k = \frac{1}{225}$$

$$P(5 \leqslant T \leqslant 10) = F(10) - F(5)$$
$$= 1 - \tfrac{1}{225}(15 - 10)^2 - (1 - \tfrac{1}{225}(15 - 5)^2)$$
$$= 1 - \tfrac{25}{225} - (1 - \tfrac{100}{225})$$
$$= \tfrac{1}{3}$$

(b) For $t \leqslant 0$ or $t \geqslant 15$, $f(t) = 0$.

For $0 \leqslant t \leqslant 15$, $f(t) = F'(t)$

$$\therefore \quad f(t) = -\tfrac{2}{225}(15 - t) \times (-1)$$
$$= \tfrac{2}{225}(15 - t)$$
$$= \tfrac{2}{15} - \tfrac{2}{225}t$$

(c) Let the median of $T$ be $m$ so $F(m) = 0.5$

$$\therefore \quad 1 - \tfrac{1}{225}(15 - m)^2 = 0.5$$
$$(15 - m)^2 = 112.5$$
$$15 - m = \pm\sqrt{112.5}$$
$$m = 15 - \sqrt{112.5} \quad \text{or} \quad m = 15 + \sqrt{112.5}$$

Since $f(t)$ is valid only for $0 \leqslant t \leqslant 15$,

$$m = 15 - \sqrt{112.5}$$
$$= 4.393\ldots$$

**Median = 4.4 (2 s.f.)**

# Miscellaneous exercise 6g

1. A continuous random variable $X$ has a probability density function, $f$, defined by

   $f(x) = \frac{1}{2}x \quad 0 \leqslant x \leqslant 2,$

   $f(x) = 0 \quad$ otherwise.

   Find the expected value of

   (a) $X$,
   (b) $2X + 4$              (NEAB)

2. (a) A continuous variable $X$ is distributed at random between the values 2 and 3 and has a probability density function of $\dfrac{6}{x^2}$.

   Find the median value of $X$.

   (b) A continuous random variable $X$ takes values between 0 and 1, with a probability density function of $Ax(1 - x)^3$. Find the value of $A$, and the mean and standard deviation of $X$.

3. A continuous random variable $X$ has probability density function $f(x)$ given by $f(x) = 0$ for $x < 0$ and $x > 3$ and between $x = 0$ and $x = 3$ its form is as shown in the graph.

   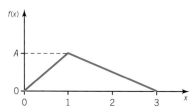

   (a) Find the value of $A$.
   (b) Express $f(x)$ algebraically and obtain the mean and variance of $X$.
   (c) Find the median value of $X$.

   A sample $X_1$, $X_2$ and $X_3$ is obtained. What is the probability that at least one is greater than the median value?

4. The number of kilograms of metal extracted from 10 kg of ore from a certain mine is a continuous random variable $X$ with probability density function $f(x)$, where $f(x) = cx(2 - x)^2$ if $0 \leqslant x \leqslant 2$ and $f(x) = 0$ otherwise, where $c$ is a constant.
   Show that $c = 0.75$, and find the mean and variance of $X$.
   The cost of extracting the metal from 10 kg of ore is £$10x$. Find the expected cost of extracting the metal from 10 kg of ore.    (MEI)

5. The continuous random variable $X$ has probability density function $f(x)$ defined by

   $$f(x) = \begin{cases} \dfrac{c}{x^4} & (x < -1) \\ c(2 - x^2) & (-1 \leqslant x \leqslant 1) \\ \dfrac{c}{x^4} & (x > 1) \end{cases}$$

   (a) Show that $c = \frac{1}{4}$.
   (b) Sketch the graph of $f(x)$.
   (c) Determine the cumulative distribution function $F(x)$.
   (d) Determine the expected value of $X$ and the variance of $X$.      (C)

6. A continuous variable $X$ is distributed at random between the values $x = 0$ and $x = 2$, and has a probability density function of $ax^2 + bx$. The mean is 1.25.

   (a) Show that $b = \frac{3}{4}$, and find the value of $a$.
   (b) Find the variance of $X$.
   (c) Verify that the median value of $X$ is approximately 1.3.
   (d) Find the mode.

7. The continuous random variable $X$ has probability density function given by

   $$f(x) = \begin{cases} cx^2 & 0 \leqslant x \leqslant 2 \\ 2c(4 - x) & 2 \leqslant x \leqslant 4 \\ 0 & \text{otherwise} \end{cases}$$

   where $c$ is a constant.

   (a) Show that $c = 0.15$.
   (b) Find the mean of $X$.
   (c) Find the lower quartile of $X$.
   (d) Find the probability that a single observation of $X$ lies between the lower quartile and the mean.
   (e) Three independent observations of $X$ are taken. Find the probability that one of the observations is greater than the mean and the other two are less than the median value of $X$.      (C)

8. The total number of radio taxi calls received at a control centre in a month is modelled by a random variable $X$ (in tens of thousands of calls) having the probability density function

   $$f(x) = \begin{cases} cx, & 0 < x < 1 \\ c(2 - x) & 1 \leqslant x < 2 \\ 0, & \text{otherwise} \end{cases}$$

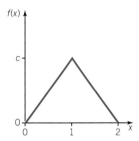

(a)  Show that the value of $c$ is 1.
(b)  Write down the probability that $X \leqslant 1$.
(c)  Show that the cumulative distribution function of $X$ is

$$F(x) = \begin{cases} 0 & x < 0 \\ \frac{1}{2}x^2 & 0 \leqslant x < 1 \\ 2x - \frac{1}{2}x^2 - 1 & 1 \leqslant x < 2 \\ 1 & x \geqslant 2 \end{cases}$$

(d)  Find the probability that the control centre receives between 8000 and 12 000 calls in a month.

A colleague criticises the model on the grounds that the number of radio calls must be discrete, while the model used for $X$ is continuous.

(e)  State briefly whether you consider that it was reasonable to use this model for $X$.
(f)  Give two reasons why the probability density function in the diagram might be unsuitable as a model.
(g)  Sketch the shape of a more suitable probability density function.       (L)

9.  The lifetime, in tens of hours, of a certain delicate electrical component is modelled by the random variable $X$ with probability density function

$$f(x) = \begin{cases} k(9 - x) & 0 \leqslant x \leqslant 9 \\ 0 & \text{otherwise} \end{cases}$$

where $k$ is a positive constant.

(a)  Show that $k = \dfrac{2}{81}$.

(b)  Find the mean lifetime of a component.
(c)  Show that the standard deviation of lifetimes is 21.2 hours.
(d)  Find the probability that a component lasts at most 50 hours.

A particular device requires two of these components and it will not operate if one or more of the components fail. The device has just been fitted with two new components. The lifetimes of components are independent.

(e)  Find the probability that the device will work for more than 50 hours.
(f)  Give a reason why the above distribution may not be realistic as a model for the distribution of lifetimes of these electrical components.       (L)

10.  The times, in excess of two hours, taken to complete a marathon road race are modelled by the continuous random variable $T$ hours, where $T$ has the probability density function

$$f(t) = \frac{4}{27}t^2(3 - t) \qquad 0 \leqslant t \leqslant 3$$
$$f(t) = 0 \qquad \text{otherwise}$$

The diagram shows a sketch of the probability density function.

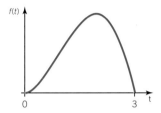

(a)  Find the mean and variance of the times taken to complete the race.
(b)  Find the modal time taken to complete the race.
(c)  What proportion of competitors complete the race in less than the modal time?
(d)  Show that the median time to complete the race lies between the mean and the mode.       (MEI)

11.  A tennis player hits a ball against a wall, aiming at a fixed horizontal line on the wall. The vertical distance from the horizontal line to the point where the ball strikes the wall is recorded as positive for points above the line and negative for points below the line.
It is assumed that the distribution of this vertical distance, $X$ metres, may be modelled by the probability density function

$$f(x) = \begin{cases} 1.5(1 - 4x^2) & -0.5 \leqslant x \leqslant 0.5 \\ 0 & \text{otherwise} \end{cases}$$

(a)  State the probability that the ball strikes the wall precisely 0.25 m above the line.
(b)  Determine the probability that the ball strikes the wall more than 0.25 m above the line.
(c)  (i)  Give a reason why the mean value of $X$ is zero.
     (ii)  Calculate the variance of $X$.
(d)  Give one reason why the above probability model may not be appropriate.
(e)  Suggest one likely effect of repeated practice on the above probability model.       (NEAB)

12. A horticultural firm is studying the number of hours that daffodils will last in a vase of water with a new additive. The random variable $X$ (in hundreds of hours) with probability density function

$$f(x) = \begin{cases} k(4 - x^2) & 0 \leqslant x \leqslant 2 \\ 0 & \text{otherwise} \end{cases}$$

is proposed as a model.

(a) Show that the value of the constant $k$ is $\frac{3}{16}$.
(b) Find the mean number of hours that a daffodil will last, according to this model.
(c) Use this model to find the probability that a daffodil will last for more than 100 hours.

The new additive is tested on carnations and it is found that several of these last for more than 250 hours.

(d) Explain why the random variable $X$, with probability density function $f(x)$ as defined above, would not be a suitable model in this case.
(e) Suggest how the probability density function could be changed to model the time carnations will last. (L)

13. Each batch of a chemical used in drug manufacture is tested for impurities. The percentage of impurity is $X$, where $X$ is a random variable with probability density function given by

$$f(x) = \begin{cases} kx & 0 < x \leqslant 1 \\ \frac{1}{3}k(4 - x) & 1 < x \leqslant 4 \\ 0 & \text{otherwise} \end{cases}$$

where $k$ is a constant.

(a) Sketch the graph of $f(x)$.
(b) Show that $k = \frac{1}{2}$.
(c) Determine, for all $x$, the distribution function $F(x)$.

In order to purify the chemical it is subjected to one of four possible purification processes, the percentage impurity in the batch determining the actual process used. The process used and its cost, for each level of percentage impurity, is shown in the table.

| Percentage Impurity $x$ | Process used | Batch cost (£) |
|---|---|---|
| $0 < x \leqslant 1$ | A | 200 |
| $1 < x \leqslant 2$ | B | 250 |
| $2 < x \leqslant 3$ | C | 350 |
| $3 < x \leqslant 4$ | D | 500 |

(d) Determine the expected cost per batch of removing the impurities.
(e) Determine the probability that the cost of purifying a batch exceeds the expected cost. (NEAB)

14. An ironmonger is supplied with paraffin once a week. The weekly demand, $X$ hundred litres, has the probability density function $f$, where

$$f(x) = c(1 - x)^7 \qquad 0 \leqslant x \leqslant 1$$
$$f(x) = 0 \qquad \text{otherwise}$$

where $c$ is a constant. Find the value of $c$. Find the mean value of $X$, and, to the nearest litre, the minimum capacity of his paraffin tank if the probability that it will be exhausted in a given week is not to exceed 0.02. (L)

15. The probability that a randomly chosen flight from Stanston Airport is delayed by more than $x$ hours is $\frac{1}{100}(x - 10)^2$, for $x \in R$, $0 \leqslant x \leqslant 10$. No flights leave early, and none is delayed for more than ten hours. The delay, in hours, for a randomly chosen flight is denoted by $X$.

(a) Find the median, $m$, of $X$, correct to three significant figures.
(b) Find the cumulative distribution function, $F$, of $X$ and sketch the graph of $F$.
(c) Find the probability density function, $f$, of $X$ and sketch the graph of $f$.
(d) Show that $E(X) = \frac{10}{3}$.

A random sample of two flights is taken. Find the probability that both flights are delayed by more than $m$ hours, where $m$ is the median of $X$. (C)

16. The continuous random variable $X$ has probability density function given by

$$f(x) = \begin{cases} kx & 0 \leqslant x \leqslant 1, \\ kx^2 & 1 \leqslant x \leqslant 2, \\ 0 & \text{otherwise} \end{cases}$$

(a) Show that $k = \frac{6}{17}$.
(b) Find the cumulative distribution function of $X$.
(c) Find, correct to two decimal places, the median, $m$, of $X$.
(d) Find, correct to two decimal places, $P(|X - m| < 0.75)$. (C)

17. Determine $\lambda$ such that

$$f(x) = \begin{cases} 0 & x < 0 \\ \lambda/2 & 0 \leqslant x \leqslant 1 \\ 0 & 1 < x < 2 \\ \dfrac{3\lambda}{2} - \dfrac{3\lambda(x - 3)^2}{4} & 2 \leqslant x \leqslant 4 \\ 0 & x > 4 \end{cases}$$

is a probability density function of the distribution of a random variable $X$. Sketch the density function and find $E(X)$ and $P(X \leqslant 3.5)$. (MEI)

18. A random variable $X$ has cumulative (distribution) function $F(x)$ where

$$F(x) = \begin{cases} 0 & x < -1 \\ ax + a & -1 \leqslant x < 0 \\ 2ax + a & 0 \leqslant x < 1 \\ 3a & 1 \leqslant x \end{cases}$$

Determine

(a) the value of $a$,
(b) the frequency function $f(x)$ of $X$,
(c) the expected value $\mu$ of $X$,
(d) the standard deviation $\sigma$ of $X$,
(e) the probability that $|X - \mu|$ exceeds $\frac{1}{3}$.     (C)

19. (a) A discrete random variable $R$ takes integer values between 0 and 4 inclusive with probabilities given by

$$P(R = r) = \begin{cases} \dfrac{r + 1}{10} & (r = 0, 1, 2) \\ \dfrac{9 - 2r}{10} & (r = 3, 4) \end{cases}$$

Find the expectation and variance of $R$.

(b) A continuous random variable $X$ takes values in the interval $x \geqslant 0$. The probability density function of $X$ is defined by

$$f(x) = \begin{cases} kx & \text{if } 0 \leqslant x \leqslant 1 \\ \dfrac{k}{x^4} & \text{if } x > 1 \end{cases}$$

Prove that $k = \frac{6}{5}$ and find the expectation and variance of $X$.     (C)

# Mixed test 6A

1. A survey of 491 households, in part of the Midlands, gave the following results for gross weekly income, £$y$.

| Income ($y$) | No. of households |
|---|---|
| $0 \leqslant y < 80$ | 68 |
| $80 \leqslant y < 130$ | 38 |
| $130 \leqslant y < 170$ | 46 |
| $170 \leqslant y < 220$ | 40 |
| $220 \leqslant y < 270$ | 50 |
| $270 \leqslant y < 320$ | 45 |
| $320 \leqslant y < 400$ | 60 |
| $400 \leqslant y < 800$ | 144 |

(a) Draw a histogram on graph paper to illustrate these data. Label your scales and axes clearly.

A statistician suggests that a suitable model for the gross weekly income in £100 units is the continuous random variable $X$ with probability density function

$$f(x) = \begin{cases} 3k & 0 \leqslant x < 4 \\ k & 4 \leqslant x \leqslant 8 \\ 0 & \text{otherwise} \end{cases}$$

where $k$ is a constant.
(b) Find the value of $k$.
(c) Use this model to estimate how many of these 491 households have a gross weekly income in the range £0–£130.
(d) Comment on your findings.     (L)

2. The random variable $X$ has a probability density function given by

$$f(x) = \begin{cases} kx(1 - x^2) & 0 \leqslant x \leqslant 1 \\ 0 & \text{elsewhere} \end{cases}$$

$k$ being a constant. Find the value of $k$ and find also the mean and variance of this distribution. Find the median of the distribution.     (O & C)

3. The amount of vegetables eaten by a family in a week is a random variable $W$ kg. The probability density function is given by

$$f(w) = \begin{cases} \dfrac{20}{5^5} w^3(5 - w) & 0 \leqslant w \leqslant 5 \\ 0 & \text{otherwise} \end{cases}$$

(a) Find the cumulative distribution function of $W$.
(b) Find, to three decimal places, the probability that the family eats between 2 kg and 4 kg of vegetables in one week.
(c) Given that the mean of the distribution is $3\frac{1}{3}$, find, to three decimal places, the variance of $W$.
(d) Find the mode of the distribution.
(e) Verify that the amount, $m$, of vegetables such that the family is equally likely to eat more or less than $m$ in any week is about 3.431 kg.
(f) Use the information above to comment on the skewness of the distribution.     (L)

# Mixed test 6B

1. The continuous random variable $X$ has probability density function given by

$$f(x) = \begin{cases} \frac{1}{4}x^3 & 0 \leqslant x \leqslant 2 \\ 0 & \text{otherwise.} \end{cases}$$

   (a) Sketch the graph of $f$.
   (b) Calculate the mean of $X$.
   (c) Calculate the standard deviation of $X$.
   (d) Show why the median value of $X$ must be greater than the mean.  (NEAB)

2. The random variable $X$ has probability density function

$$f(x) = \begin{cases} a(x - x^3) & 0 \leqslant x \leqslant 1 \\ 0 & \text{otherwise} \end{cases}$$

   (a) Show that $a = 4$.
   (b) Find $E(X)$
   (c) Find the mode of the distribution of $X$.  (L)

3. A firm has a large number of employees. The distance in miles they have to travel each day from home to work can be modelled by a continuous random variable $X$ whose cumulative distribution function is given by

$$F(1) = 0$$

$$F(x) = k\left(1 - \frac{1}{x}\right) \qquad 1 \leqslant x \leqslant b$$

$$F(b) = 1$$

   where $b$ represents the farthest distance anyone lives from work.
   The diagram shows a sketch of this cumulative distribution function.

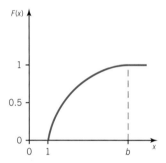

   A survey suggests that $b = 5$. Use this parameter for parts (a) to (d).

   (a) Show that $k = 1.25$.
   (b) Write down and solve an equation to find the median distance travelled to work.
   (c) Find the probability that an employee lives within half a mile of the median.
   (d) Derive the probability density function for $X$ and illustrate it with a sketch.
   (e) Show that, for any value of $b$ greater than 1, the median distance travelled does not exceed 2.  (MEI)

# 7

# The normal distribution

*In this chapter you will learn how to*

- standardise a normal variable and use standard normal tables
- use the normal distribution as a model to solve problems
- use the normal distribution as an approximation to the binomial distribution and to the Poisson distribution

The **normal distribution** is one of the most important distributions in statistics. Many measured quantities in the natural sciences follow a normal distribution and under certain circumstances it is also a useful approximation to the binomial distribution and to the Poisson distribution.

The normal variable $X$ is continuous. Its probability density function $f(x)$ depends on its

mean $\mu$ and standard deviation $\sigma$, where $f(x) = \dfrac{1}{\sigma\sqrt{2\pi}}\, e^{-(x-\mu)^2/2\sigma^2}$ , $-\infty < x < \infty$.

This is very complicated and has been included just for reference. You would not be expected to remember it!

To describe the distribution, write

$$X \sim N(\mu, \sigma^2)$$
$$\uparrow \quad \uparrow$$

      mean   variance

Notice that the description gives the variance $\sigma^2$, rather than the standard deviation, $\sigma$.

The normal distribution curve has the following features:

- It is bell-shaped
- It is symmetrical about $\mu$
- It extends from $-\infty$ to $+\infty$

- The maximum value of $f(x)$ is $\dfrac{1}{\sigma\sqrt{2\pi}}$

- The total area under the curve is 1

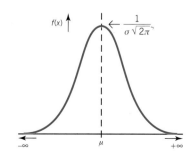

Notice also that

- approximately 95% of the distribution lies within two standard deviations of the mean

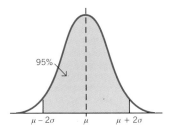

- approximately 99.9% (very nearly all) of the distribution lies within three standard deviations of the mean

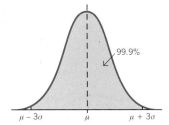

The spread of the distribution depends on $\sigma$. Here are some normal curves, each drawn to the same scale:

$X \sim N(0, 1)$

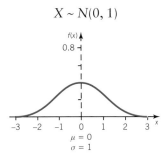

$X \sim N(4, \frac{1}{4})$

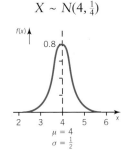

$X \sim N(50, 4)$

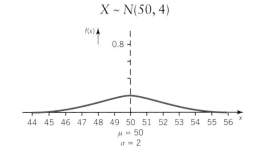

# FINDING PROBABILITIES

The probability that $X$ lies between $a$ and $b$ is written $P(a < X < b)$. To find this probability, you need to find the **area under the normal curve** between $a$ and $b$.

One way of finding areas is to integrate, but since the normal function is complicated and very difficult to integrate, tables are used instead.

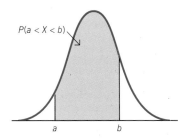

# THE STANDARD NORMAL VARIABLE, Z

In order to use the same set of tables for **all possible values** of $\mu$ and $\sigma^2$, the variable $X$ is **standardised** so that the mean is 0 and the standard deviation is 1. Notice that since the variance is the square of the standard deviation, the variance is also 1. This **standardised normal variable** is called $Z$ and $Z \sim N(0, 1)$.

To illustrate how the variable $X$ is standardised to the variable $Z$, consider $X$ distributed normally with a mean 50 and a variance 4,

i.e. $X \sim N(50, 4)$.     $\mu = 50$ and $\sigma^2 = 4$, so $\sigma = 2$.

The maximum value of $f(x)$ is $\dfrac{1}{2\sqrt{2\pi}} \approx 0.2$ and the curve is shown in the right-hand section of the diagram below.

Now translate the curve 50 units to the left so that the mean is 0. This is shown on the left hand section of the diagram. The standard deviation $\sigma$ is still 2, so the maximum value is again approximately 0.2.

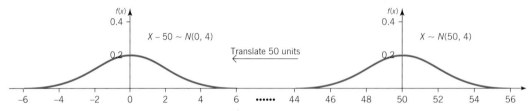

Now 'squash' the curve towards the vertical axis so that the standard deviation is 1. This is done by dividing by the standard deviation ($\sigma = 2$).

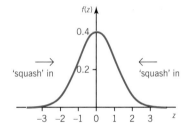

$$\frac{X - 50}{2} \sim N(0, 1)$$

You write $\quad Z = \dfrac{X - 50}{2}$

so that $\quad Z \sim N(0, 1)$

In general

To standardise $X$, where $X \sim N(\mu, \sigma^2)$

- subtract the mean $\mu$
- then divide by the standard deviation $\sigma$

to obtain

$$Z = \frac{X - \mu}{\sigma} \qquad \text{where } Z \sim N(0, 1)$$

## USING STANDARD NORMAL TABLES

The standard normal tables give the area under the curve as far as a particular value $z$. This is written $\Phi(z)$.

This area gives the probability that $Z$ is less than a particular value $z$, so $P(Z < z) = \Phi(z)$.

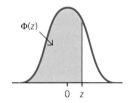

Note that $\Phi$ is a Greek letter, pronounced phi.

The tables are printed on page 649. On the following page is an extract from the first section. The highlighted values are referred to in the following text. Notice that the values of $\Phi(z)$ are given to four decimal places in the tables.

| $z$ | 0 | 1 | 2 | 3 | 4 | 5 | 6 | 7 | 8 | 9 | 1 | 2 | 3 | 4 | 5 | 6 | 7 | 8 | 9 |
|---|---|---|---|---|---|---|---|---|---|---|---|---|---|---|---|---|---|---|---|
| | | | | | | | | | | | | | | ADD | | | | | |
| 0.0 | 0.5000 | 0.5040 | 0.5080 | 0.5120 | 0.5160 | 0.5199 | 0.5239 | 0.5279 | 0.5319 | 0.5359 | 4 | 8 | 12 | 16 | 20 | 24 | 28 | 32 | 36 |
| (a) 0.1 | 0.5398 | 0.5438 | 0.5478 | 0.5517 | 0.5557 | 0.5596 | 0.5636 | 0.5675 | 0.5714 | 0.5753 | 4 | 8 | 12 | 16 | 20 | 24 | 28 | 32 | 36 |
| 0.2 | 0.5793 | 0.5832 | 0.5871 | 0.5910 | 0.5948 | 0.5987 | 0.6026 | 0.6064 | 0.6103 | 0.6141 | 4 | 8 | 12 | 15 | 19 | 23 | 27 | 31 | 35 |
| (b) 0.3 | 0.6179 | 0.6217 | 0.6255 | 0.6293 | 0.6331 | 0.6368 | 0.6404 | 0.6443 | 0.6480 | 0.6517 | 4 | 7 | 11 | 15 | 19 | 22 | 26 | 30 | 34 |
| (c) 0.4 | 0.6554 | 0.6591 | 0.6628 | 0.6664 | 0.6700 | 0.6736 | 0.6772 | 0.6808 | 0.6884 | 0.6879 | 4 | 7 | 11 | 14 | 18 | 22 | 25 | 29 | 32 |
| 0.5 | 0.6915 | 0.6950 | 0.6985 | 0.7019 | 0.7054 | 0.7088 | 0.7123 | 0.7157 | 0.7190 | 0.7224 | 3 | 7 | 10 | 14 | 17 | 20 | 24 | 27 | 31 |
| 0.6 | 0.7257 | 0.7291 | 0.7324 | 0.7357 | 0.7389 | 0.7422 | 0.7454 | 0.7486 | 0.7517 | 0.7549 | 3 | 7 | 10 | 13 | 16 | 19 | 23 | 26 | 29 |
| 0.7 | 0.7580 | 0.7611 | 0.7642 | 0.7673 | 0.7704 | 0.7734 | 0.7764 | 0.7794 | 0.7823 | 0.7852 | 3 | 6 | 9 | 12 | 15 | 18 | 21 | 24 | 27 |
| 0.8 | 0.7881 | 0.7910 | 0.7939 | 0.7967 | 0.7995 | 0.8023 | 0.8051 | 0.8078 | 0.8106 | 0.8133 | 3 | 5 | 8 | 11 | 14 | 16 | 19 | 22 | 25 |
| 0.9 | 0.8159 | 0.8186 | 0.8212 | 0.8238 | 0.8264 | 0.8289 | 0.8315 | 0.8340 | 0.8365 | 0.8389 | 3 | 5 | 8 | 10 | 13 | 15 | 18 | 20 | 23 |

(a)  To find $P(Z < 0.16)$, read off the value of $\Phi(0.16)$:

- find row 0.1 and go across to column 6. This gives 0.5636.

   $P(Z < 0.16) = 0.5636$

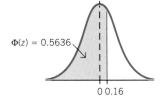

$\Phi(z) = 0.5636$

0  0.16

(b)  To find $P(Z < 0.345)$, read off the value of $\Phi(0.345)$:

- Find the value when $z = 0.34$ from row 0.3, column 4. This is 0.6331.
- Now go to the right-hand section and read the number along that row in column 5. This is 19.
- Note the instruction to ADD. This means that 19 is added to the **digits** 6331.

$$\begin{array}{r} 6331 \\ + \quad 19 \\ \hline 6350 \end{array} \qquad \therefore \quad P(Z < 0.345) = 0.6350$$

(c)  To find $P(Z < 0.429)$, read off $\Phi(0.429)$:

- Find row 0.4, column 2, right-hand section 9.

$P(Z < 0.429) = 0.6660$

$$\begin{array}{r} 6228 \\ + \quad 32 \\ \hline 6660 \end{array}$$

When calculating probabilities, remember that the total area under the standardised normal curve is 1.

## Example 7.1

Using the standard normal tables printed on page 649, find

(a)  $P(Z < 0.85)$      (b)  $P(Z > 0.85)$

## Solution 7.1

(a)

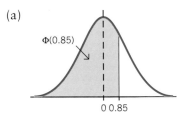

$\Phi(0.85)$

0  0.85

$P(Z < 0.85) = \Phi(0.85)$
$= 0.8023$

(b)

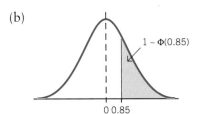

$1 - \Phi(0.85)$

0  0.85

$P(Z > 0.85) = 1 - \Phi(0.85)$
$= 1 - 0.8023$
$= 0.1977$

In general

$$P(Z > a) = 1 - \Phi(a)$$

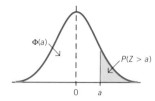

# Finding probabilities involving negative values of *z*

The standard normal tables start at $z = 0$. You can however find probabilities relating to negative values of $z$ by using the symmetrical properties of the curve. Look at these diagrams:

To find $P(Z < -a)$, where $a > 0$

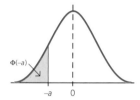
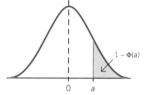

$$P(Z < -a) = \Phi(-a)$$
$$= 1 - \Phi(a)$$

To find $P(Z > -a)$, where $a > 0$

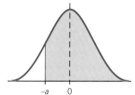
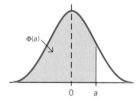

$$P(Z > -a) = \Phi(a)$$

From the diagrams, it is obvious that

$$P(Z < -a) = P(Z > a)$$
$$P(Z > -a) = P(Z < a)$$

## Example 7.2

$Z \sim N(0, 1)$

(a) Using the standard normal tables on page 649, find $P(Z < 1.377)$,
(b) Drawing sketches to illustrate your answers, find
    (i)   $P(Z > -1.377)$
    (ii)  $P(Z > 1.377)$
    (iii) $P(Z < -1.377)$

(Give your answers correct to two significant figures.)

## Solution 7.2

(a)

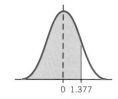

$$P(Z < 1.377) = \Phi(1.377)$$
$$= 0.9158$$
$$= \mathbf{0.92 \ (2 \ s.f.)}$$

(b) (i)

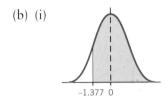

Using $P(Z > -a) = P(Z < a) = \Phi(a)$
$$\begin{aligned} P(Z > -1.377) &= \Phi(1.377) \\ &= 0.9158 \\ &= \mathbf{0.92} \text{ (2 s.f.)} \end{aligned}$$

(ii)

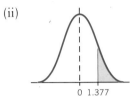

Using $P(Z > a) = 1 - \Phi(a)$
$$\begin{aligned} P(Z > 1.377) &= 1 - \Phi(1.377) \\ &= 1 - 0.9158 \\ &= 0.0842 \\ &= \mathbf{0.084} \text{ (2 s.f.)} \end{aligned}$$

(iii)

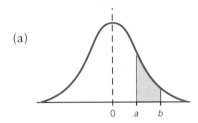

Using $P(Z < -a) = P(Z > a) = 1 - \Phi(a)$
$$\begin{aligned} P(Z < -1.377) &= 1 - \Phi(1.377) \\ &= 1 - 0.9158 \\ &= 0.0842 \\ &= \mathbf{0.084} \text{ (2 s.f.)}. \end{aligned}$$

## Important results – these are worth learning.

In the following, $a > 0$, $b > 0$ and $a < b$.

Examples:

(a)

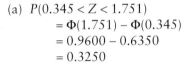

(a) $$\begin{aligned} P(0.345 < Z < 1.751) &= \Phi(1.751) - \Phi(0.345) \\ &= 0.9600 - 0.6350 \\ &= 0.3250 \end{aligned}$$

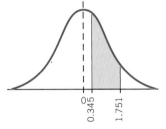

$$P(a < Z < b) = \Phi(b) - \Phi(a)$$

(b)

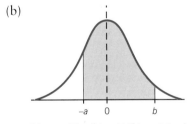

$$\begin{aligned} P(-a < Z < b) &= \Phi(b) - \Phi(-a) \\ &= \Phi(b) - (1 - \Phi(a)) \\ &= \Phi(b) - 1 + \Phi(a) \\ &= \Phi(a) + \Phi(b) - 1 \end{aligned}$$

$$P(-a < Z < b) = \Phi(a) + \Phi(b) - 1$$

(b) $$\begin{aligned} P(-2.696 < Z < 1.865) &= \Phi(1.865) - \Phi(-2.696) \\ &= \Phi(1.865) - (1 - \Phi(2.696)) \\ &= \Phi(1.865) + \Phi(2.696) - 1 \\ &= 0.9690 + 0.9965 - 1 \\ &= 0.9655 \end{aligned}$$

In practice, you can just write

$$\begin{aligned} P(-2.696 < Z < 1.865) \\ = \Phi(2.696) + \Phi(1.865) - 1 \end{aligned}$$

and go on from there.

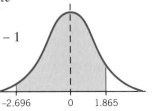

(c)

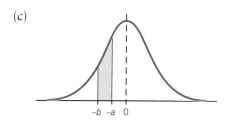

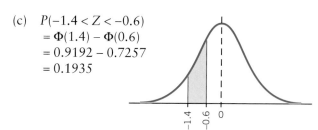

(c) $P(-1.4 < Z < -0.6)$
$= \Phi(1.4) - \Phi(0.6)$
$= 0.9192 - 0.7257$
$= 0.1935$

$$P(-b < Z < -a) = \Phi(-a) - \Phi(-b)$$
$$= 1 - \Phi(a) - (1 - \Phi(b))$$
$$= \Phi(b) - \Phi(a)$$

$$P(-b < Z < -a) = \Phi(b) - \Phi(a)$$

(d)

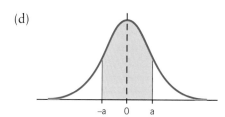

(d) $P(|Z| < 1.433)$
$= P(-1.433 < Z < 1.433)$
$= 2\Phi(1.433) - 1$
$= 2 \times 0.9240 - 1$
$= 0.8480$

$$P(|Z| < a) = P(-a < Z < a)$$
$$= \Phi(a) + \Phi(a) - 1 \quad \text{Result (b)}$$
$$= 2\Phi(a) - 1$$

$$P(|Z| < a) = 2\Phi(a) - 1$$

(e)

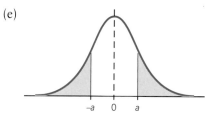

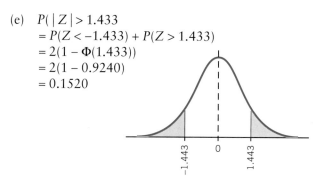

(e) $P(|Z| > 1.433$
$= P(Z < -1.433) + P(Z > 1.433)$
$= 2(1 - \Phi(1.433))$
$= 2(1 - 0.9240)$
$= 0.1520$

$$P(|Z| > a) = P(Z < -a) + P(Z > a)$$
$$= 1 - \Phi(a) + 1 - \Phi(a)$$
$$= 2(1 - \Phi(a))$$

$$P(|Z| > a) = 2(1 - \Phi(a))$$

It is also useful to remember that

$$P(|Z| > a) = 1 - P(|Z| < a)$$

## Example 7.3

$Z \sim N(0, 1)$. Show that

(a) $P(-1.96 < Z < 1.96) = 0.95$
(b) $P(-2.575 < Z < 2.575) = 0.99$

## Solution 7.3

(a) $P(-1.96 < Z < 1.96) = 2\Phi(1.96) - 1$
$$= 2(0.975) - 1$$
$$= 0.95$$

$\therefore \quad P(-1.96 < Z < 1.96) = 0.95.$

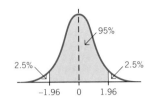

The central 95% of the distribution lies between ±1.96.

(b) $P(-2.575 < Z < 2.575) = 2\Phi(2.575) - 1$
$$= 2(0.995) - 1$$
$$= 0.99$$

$\therefore \quad P(-2.575 < Z < 2.575) = 0.99.$

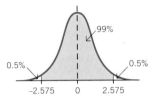

The central 99% of the distribution lies between ±2.575.

*NOTE*: These are important results which will be used in later work.

## Exercise 7a   Finding probabilities, where $Z \sim N(0, 1)$

Draw sketches to illustrate your answer and consider whether your answer is sensible.

1. If $Z \sim N(0, 1)$, find
   (a) $P(Z < 0.874)$,  (b) $P(Z > -0.874)$,
   (c) $P(Z > 0.874)$,  (d) $P(Z < -0.874)$.

2. If $Z \sim N(0, 1)$, find
   (a) $P(Z > 1.8)$,     (b) $P(Z < -0.65)$,
   (c) $P(Z > -2.46)$,   (d) $P(Z < 1.36)$,
   (e) $P(Z > 2.58)$,    (f) $P(Z > -2.37)$,
   (g) $P(Z < 1.86)$,    (h) $P(Z < -0.725)$,
   (i) $P(Z > 1.863)$,   (j) $P(Z < 1.63)$,
   (k) $P(Z > -2.061)$,  (l) $P(Z < -2.875)$.

3. If $Z \sim N(0, 1)$, find
   (a) $P(Z > 1.645)$,   (b) $P(Z < -1.645)$,
   (c) $P(Z > 1.282)$,   (d) $P(Z > 1.96)$,
   (e) $P(Z > 2.575)$,   (f) $P(Z > 2.326)$,
   (g) $P(Z > 2.808)$,   (h) $P(Z < 1.96)$.

4. $Z \sim N(0, 1)$
   Find the probabilities represented by the shaded areas in the diagrams.

   (a)        (b)

   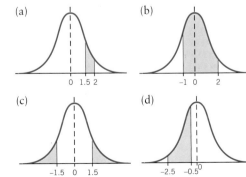

   (c)        (d)

5. If $Z \sim N(0, 1)$, find
   (a) $P(0.829 < Z < 1.834)$,
   (b) $P(-2.56 < Z < 0.134)$,
   (c) $P(-1.762 < Z < -0.246)$,
   (d) $P(0 < Z < 1.73)$,
   (e) $P(-2.05 < Z < 0)$,
   (f) $P(-2.08 < Z < 2.08)$,
   (g) $P(1.764 < Z < 2.567)$,
   (h) $P(-1.65 < Z < 1.725)$,
   (i) $P(-0.98 < Z < -0.16)$,
   (j) $P(Z < -1.97 \text{ or } Z > 2.5)$,
   (k) $P(|Z| < 1.78)$,
   (l) $P(|Z| > 0.754)$,
   (m) $P(-1.645 < Z < 1.645)$,
   (n) $P(|Z| > 2.326)$.

6. $Z \sim N(0, 1)$

   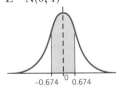

   Complete this statement:

   The central ...% of the distribution lies between ±0.674.

7. $Z \sim N(0, 1)$ and $P(Z < a) = 0.3$,
   $P(a < Z < b) = 0.6$.
   Find
   (a) $P(Z < b)$,
   (b) $P(Z > a)$.

8. $Z \sim N(0, 1)$ and $P(Z < a) = 0.7$, $P(Z > b) = 0.45$.
   Find
   (a) $\Phi(b)$,
   (b) $P(b < Z < a)$.

9. $Z \sim N(0, 1)$ and $P(|Z| < a) = 0.8$.
   Find
   (a) $P(Z < a)$,
   (b) $P(Z > a)$.

# USING STANDARD NORMAL TABLES FOR ANY NORMAL VARIABLE X

Remember that to standardise $X$, where $X \sim N(\mu, \sigma^2)$,

- subtract the mean $\mu$
- then divide by the standard deviation $\sigma$

$$\text{to give } Z = \frac{X - \mu}{\sigma} \quad \text{where } Z \sim N(0, 1)$$

The procedure is illustrated in the following example:

## Example 7.4

Lengths of metal strips produced by a machine are normally distributed with mean length of 150 cm and a standard deviation of 10 cm.
Find the probability that the length of a randomly selected strip is

(a) shorter than 165 cm,
(b) within 5 cm of the mean.

## Solution 7.4

$X$ is the length, in centimetres, of a metal strip.

Since $\mu = 150$ and $\sigma = 10$, $X \sim N(150, 10^2)$

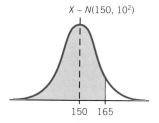

(a) You need to find the probability that the length is shorter that 165 cm, i.e. $P(X < 165)$.

To be able to use the standard normal tables, standardise the $X$ variable by subtracting the mean, 150, then dividing by the standard deviation, 10. Apply this to both sides of the inequality $X < 165$.

$$X \text{ becomes } \frac{X - 150}{10} = Z,$$

$$165 \text{ becomes } \frac{165 - 150}{10} = 1.5,$$

so $P(X < 165)$ becomes $P(Z < 1.5)$

$$\begin{aligned}
P(X < 165) &= P(Z < 1.5) \\
&= \Phi(1.5) \\
&= 0.9332 \\
&= 0.93 \text{ (2 s.f.)}
\end{aligned}$$

**The probability that the length is shorter than 165 cm is 0.93.**

*NOTE:* Although the $X$ and $Z$ distributions have different spreads, in practice it is convenient to show the values for both distributions on one sketch.

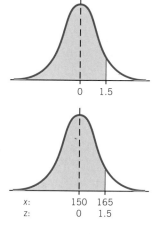

(b) To find the probability that the length is within 5 cm of the mean, you need to find $P(|X - 150| < 5)$.

Dividing by the standard deviation gives $P\left(\dfrac{|X - 150|}{10} < \dfrac{5}{10}\right)$ i.e. $P(|Z| < 0.5)$

$\begin{aligned} P(|Z| < 0.5) &= 2\Phi(0.5) - 1 \qquad \text{\textcolor{gray}{result (d) page 366}} \\ &= 2 \times 0.6915 - 1 \\ &= 0.383 \\ &= 0.38 \text{ (2 s.f.)} \end{aligned}$

**The probability that the length is within 5 cm of the mean is 0.38.**

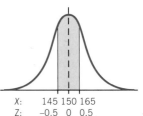

X:    145 150 165
Z:    −0.5 0 0.5

---

## Example 7.5

The time taken by the milkman to deliver to the High Street is normally distributed with a mean of 12 minutes and a standard deviation of 2 minutes. He delivers milk every day. Estimate the number of days during the year when he takes

(a) longer than 17 minutes,
(b) less than ten minutes,
(c) between nine and 13 minutes.

## Solution 7.5

$X$ is the time, in minutes, taken to deliver milk to the High Street.

$X \sim N(12, 2^2)$

Standardise $X$ using $Z = \dfrac{X - \mu}{\sigma}$, i.e. $Z = \dfrac{X - 12}{2}$.

(a) $\begin{aligned} P(X > 17) &= P\left(Z > \dfrac{17 - 12}{2}\right) \\ &= P(Z > 2.5) \\ &= 1 - \Phi(2.5) \\ &= 1 - 0.9938 \\ &= 0.0062 \end{aligned}$

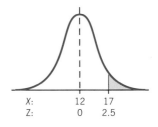

X:    12  17
Z:    0  2.5

To find the number of days, multiply by 365.

$365 \times 0.0062 = 2.263 \approx 2$

**On two days in the year he takes longer than 17 minutes.**

(b) $\begin{aligned} P(X < 10) &= P\left(Z < \dfrac{10 - 12}{2}\right) \\ &= P(Z < -1) \\ &= 1 - \Phi(1) \\ &= 1 - 0.8413 \\ &= 0.1587 \end{aligned}$

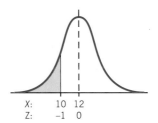

X:    10 12
Z:    −1 0

Now $365 \times 0.1587 = 57.92 \approx 58$

**On 58 days in the year he takes less than ten minutes.**

(c) $P(9 < X < 13) = P\left(\dfrac{9 - 12}{2} < Z < \dfrac{13 - 12}{2}\right)$

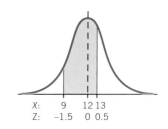

$= P(-1.5 < Z < 0.5)$
$= \Phi(0.5) + \Phi(1.5) - 1$
$= 0.6915 + 0.9332 - 1$
$= 0.6247$

| X: | 9 | 12 13 |
| Z: | −1.5 | 0 0.5 |

Now $365 \times 0.6247 = 228.01 \approx 228$

**On 228 days in the year he takes between nine and 13 minutes.**

NOTE: Since $X$ is a continuous variable, the following are indistinguishable:

$9 < X < 13,$
$9 \leqslant X < 13,$
$9 < X \leqslant 13,$
$9 \leqslant X \leqslant 13.$

## Exercise 7b   Finding probabilities using $X \sim N(\mu, \sigma^2)$

1. The masses of packages from a particular machine are normally distributed with a mean of 200 g and a standard deviation of 2 g. Find the probability that a randomly selected package from the machine weighs
   (a) less than 197 g,
   (b) more than 200.5 g,
   (c) between 198.5 g and 199.5 g.

2. The heights of boys at a particular age follow a normal distribution with mean 150.3 cm and variance 25 cm.
   Find the probability that a boy picked at random from this age group has height
   (a) less than 153 cm,
   (b) more than 158 cm,
   (c) between 150 cm and 158 cm,
   (d) more than 10 cm difference from the mean height.

3. $X \sim N(300, 25)$
   Find the probabilities represented by the shaded areas in the diagrams:
   (a)

   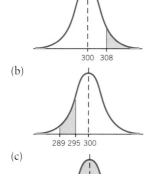

   300  308

   (b)

   289 295 300

   (c)

   290   300   310

4. The random variable $X$ is distributed normally such that $X \sim N(50, 20)$. Find
   (a) $P(X > 60.3)$,
   (b) $P(X < 59.8)$.

5. $X \sim N(-8, 12)$. Find
   (a) $P(X < -9.8)$,
   (b) $P(X > -8.2)$,
   (c) $P(-7 < X < 0.5)$.

6. The masses of a certain type of cabbage are normally distributed with a mean of 1000 g and a standard deviation of 0.15 kg.
   In a batch of 800 cabbages, estimate how many have a mass between 750 g and 1290 g.

7. The number of hours of life of a torch battery is normally distributed with a mean of 150 hours and standard deviation of 12 hours. In a quality control test, two batteries are chosen at random from a batch. If both batteries have a life less than 120 hours, the batch is rejected.
   Find the probability that the batch is rejected.

8. Cartons of milk from a particular supermarket are advertised as containing 1 litre, but in fact the volume of the contents is normally distributed with a mean of 1012 ml and a standard deviation of 5 ml.
   (a) Find the probability that a randomly chosen carton contains more than 1010 ml.
   (b) In a batch of 1000 cartons, estimate the number of cartons that contain less than the advertised volume of milk.

9. A random variable $X$ is such that $X \sim N(-5, 9)$.
   (a) Find the probability that a randomly chosen item from the population will have a positive value.
   (b) Find the probability that out of ten items chosen at random, exactly four will have a positive value.

10. $X \sim N(100, 81)$. Find
    (a) $P(|X - 100| < 18)$,
    (b) $P(|X - 100| > 5)$,
    (c) $P(12 < X - 100 < 15)$.

11. The life of a certain make of electric light bulb is known to be normally distributed with a mean life of 2000 hours and a standard deviation of 120 hours. Estimate the probability that the life of such a bulb will be
    (a) greater than 2150 hours,
    (b) greater than 1910 hours,
    (c) within the range 1850 hours to 2090 hours. *(C)*

12. The weights of vegetable marrows supplied to retailers by a wholesaler have a normal distribution with mean 1.5 kg and standard deviation 0.6 kg. The wholesaler supplies three sizes of marrow:
    Size 1, under 0.9 kg,
    Size 2, from 0.9 kg to 2.4 kg,
    Size 3, over 2.4 kg.

    Find, three decimal places, the proportions of marrows in the three sizes.
    The prices of the marrows are 16p for Size 1, 40p for Size 2 and 60p for Size 3. Calculate the expected total cost of 100 marrows chosen at random from those supplied. *(L)*

13. The random variable $Y$ is such that $Y \sim N(8, 25)$. Show that, correct to three decimal places, $P(|Y - 8| < 6.2) = 0.785$.
    Three random observations of $Y$ are made. Find the probability that exactly two observations will lie in the interval defined by $|Y - 8| < 6.2$. *(C)*

14. The manufacturers of a new model of car state that, when travelling at 56 miles per hour, the petrol consumption has a mean value of 32.4 miles per gallon with standard deviation 1.4 miles per gallon.
    Assuming a normal distribution, calculate the probability that a randomly chosen car of that model will have a petrol consumption greater than 30 miles per gallon when travelling at 56 miles per hour. *(C)*

# USING THE STANDARD NORMAL TABLES IN REVERSE TO FIND z WHEN Φ(z) IS KNOWN

The procedure is illustrated below using the extract taken from the standard normal tables. The highlighted values are referred to in the examples.

|   | $z$ | 0 | 1 | 2 | 3 | 4 | 5 | 6 | 7 | 8 | 9 | 1 | 2 | 3 | 4 | 5 | 6 | 7 | 8 | 9 |
|---|-----|---|---|---|---|---|---|---|---|---|---|---|---|---|---|---|---|---|---|---|
| | | | | | | | | | | | | | | | | ADD | | | | |
| (a) | 1.5 | 0.9332 | 0.9345 | 0.9357 | 0.9370 | 0.9382 | 0.9394 | 0.9406 | 0.9418 | 0.9429 | 0.9441 | 1 | 2 | 4 | 5 | 6 | 7 | 8 | 10 | 11 |
| | 1.6 | 0.9452 | 0.9463 | 0.9474 | 0.9484 | 0.9495 | 0.9505 | 0.9515 | 0.9525 | 0.9535 | 0.9545 | 1 | 2 | 3 | 4 | 5 | 6 | 7 | 8 | 9 |
| (b) | 1.7 | 0.9554 | 0.9564 | 0.9573 | 0.9582 | 0.9591 | 0.9599 | 0.9608 | 0.9616 | 0.9625 | 0.9633 | 1 | 2 | 3 | 4 | 4 | 5 | 6 | 7 | 8 |
| | 1.8 | 0.9641 | 0.9649 | 0.9656 | 0.9664 | 0.9671 | 0.9678 | 0.9686 | 0.9693 | 0.9699 | 0.9706 | 1 | 1 | 2 | 3 | 4 | 4 | 5 | 6 | 6 |
| | 1.9 | 0.9713 | 0.9719 | 0.9726 | 0.9732 | 0.9738 | 0.9744 | 0.9750 | 0.9756 | 0.9761 | 0.9767 | 1 | 1 | 2 | 2 | 3 | 4 | 4 | 5 | 5 |
| | 2.0 | 0.9772 | 0.9778 | 0.9783 | 0.9788 | 0.9793 | 0.9798 | 0.9803 | 0.9808 | 0.9812 | 0.9817 | 0 | 1 | 1 | 2 | 2 | 3 | 3 | 4 | 4 |
| (c) | 2.1 | 0.9821 | 0.9826 | 0.9830 | 0.9834 | 0.9838 | 0.9842 | 0.9846 | 0.9850 | 0.9854 | 0.9857 | 0 | 1 | 1 | 2 | 2 | 2 | 3 | 3 | 4 |
| | 2.2 | 0.9861 | 0.9864 | 0.9868 | 0.9871 | 0.9875 | 0.9878 | 0.9881 | 0.9884 | 0.9887 | 0.9890 | 0 | 1 | 1 | 1 | 2 | 2 | 2 | 3 | 3 |

(a) If you are given that $\Phi(z) = 0.9406$,
to find $z$, look for 0.9406 in the main body of the table.
This occurs when $z = 1.56$,
so if $\Phi(z) = 0.9406$, then $z = 1.56$.
Using notation similar to the used in trigonometry where, for example, if $\sin \theta = 0.82$, then $\theta = \sin^{-1} 0.82$, you could write

$\Phi^{-1}(0.9406) = 1.56$

This means that the value of $z$ such that $\Phi(z) = 0.9046$ is 1.56.

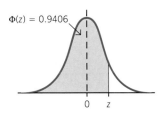

$\Phi(z) = 0.9406$

(b) Find $z$ if $P(Z < z) = 0.9579$

$$\Phi(z) = 0.9579$$

so $\qquad z = \Phi^{-1}(0.9579)$

Look for 0.9579 in the main body of the table. It does not appear, so look for the number *below* it. This is 0.9573 and it occurs when $z = 1.72$.

To get the digits 9579 you need to add 6 to 9573. Look at the far right-hand section and find 6. It is in column 7. This means that the $z$ value required is 1.727.

So $\quad z = \Phi^{-1}(0.9579) = 1.727$.

(c) Find $z$ if $P(Z < z) = 0.9832$

$$\Phi(z) = 0.9832$$

so $\qquad z = \Phi^{-1}(0.9832)$

Look for 0.9832 in the main body of the table; note that $\phi(2.12) = 0.9830$.
Refer to the end column and you find that

$\Phi(2.124) = 0.9832$
$\Phi(2.125) = 0.9832$
$\Phi(2.126) = 0.9832$

The probabilities have been given to four decimal places and it is not possible to distinguish between the $z$ values, so just decide on one of them, say 2.124.

So $\quad z = \Phi^{-1}(0.9832) = 2.124$.

NOTE: If you cannot find the value for the probability in the table, choose the value that is closest to the required probability.

Often final answers are given to two or three significant figures, so these discrepancies are not important.

## Example 7.6

If $Z \sim N(0, 1)$, find the value of $a$ if

(a) $P(Z < a) = 0.9693$
(b) $P(Z > a) = 0.3802$
(c) $P(Z > a) = 0.7367$
(d) $P(Z < a) = 0.0793$

## Solution 7.6

(a) $\quad P(Z < a) = 0.9693$

i.e. $\quad \Phi(a) = 0.9693$

$a = \Phi^{-1}(0.9693)$

$= 1.87$

(b) $\qquad P(Z > a) = 0.3802$

i.e. $\quad 1 - \Phi(a) = 0.3802$

so $\qquad \Phi(a) = 1 - 0.3802$

$= 0.6198$

$a = \Phi^{-1}(0.6198)$

$= 0.305$

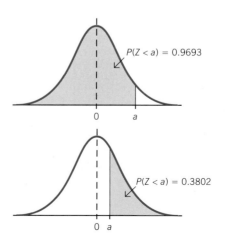

(c)  $P(Z > a) = 0.7367$
Since the probability is greater than 0.5, $a$ must be negative, and therefore $-a$ is positive.
Using symmetry, $\Phi(-a) = 0.7367$

$-a = \Phi^{-1}(0.7367)$
$\quad = 0.633$
$\quad a = -0.633$

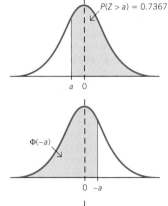

(d)  $P(Z < a) = 0.0793$
Since the probability is less than 0.5, $a$ must be negative.
Using symmetry,

$\Phi(-a) = 1 - 0.0793$
$\quad\quad = 0.9207$
$\quad -a = \Phi^{-1}(0.9207)$
$\quad\quad = 1.41$
$\quad\quad a = -1.41$

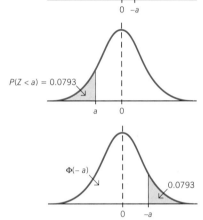

## Example 7.7

If $Z \sim N(0, 1)$ find $a$ such that $P(\,|\,Z\,|\, < a) = 0.9$.

## Solution 7.7

$P(\,|\,Z\,|\, < a) = 0.9$,
i.e.  $P(-a < Z < a) = 0.9$.

From symmetry, using result (d) on page 366,

$2\Phi(a) - 1 = 0.9$
$\quad 2\Phi(a) = 1.9$
$\quad\quad \Phi(a) = 0.95$
$\quad\quad\quad a = \Phi^{-1}(0.95)$
$\quad\quad\quad\quad = 1.645$

This means that the central 90% of the standard normal distribution lies between $\pm 1.645$.

Alternatively,

If $P(-a < Z < a) = 0.9$

then the value of $a$ corresponds to an upper tail probability of 0.05, and a lower tail probability of 0.95.

$\quad\therefore \quad \Phi(a) = 0.95$
$\quad\quad\quad a = \Phi^{-1}(0.95)$
$\quad\quad\quad\quad = 1.645$

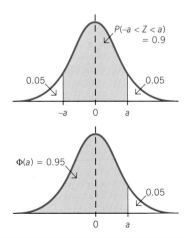

## USING THE TABLES IN REVERSE FOR *ANY* NORMAL VARIABLE *X*

### Example 7.8

The heights of female students at a particular college are normally distributed with a mean of 169 cm and a standard deviation of 9 cm.

(a) Given that 80% of these female students have a height less than $h$ cm, find the value of $h$.
(b) Given that 60% of these female students have a height greater than $s$ cm, find the value of $s$.

### Solution 7.8

$X$ is the height, in centimetres, of a female student.
$X \sim N(169, 9^2)$

(a) Given $P(X < h) = 0.8$
Standardising

$$P\left(Z < \frac{h - 169}{9}\right) = 0.8$$

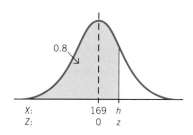

Let $z = \dfrac{h - 169}{9}$

$$P(Z < z) = 0.8$$
$$z = \Phi^{-1}(0.8)$$
$$= 0.842$$

$$\therefore \quad \frac{h - 169}{9} = 0.842$$

$$h = 169 + 9 \times 0.842$$
$$= 176.38$$
$$= 176.4 \text{ (1 d.p.)}$$

(b) Given $P(X > s) = 0.6$

Standardising

$$P\left(Z > \frac{s - 196}{9}\right) = 0.6$$

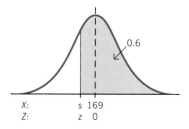

Let $z = \dfrac{s - 196}{9}$

$$P(Z > z) = 0.6$$

$z$ must be negative
and $\quad \Phi(-z) = 0.6$
$$-z = \Phi^{-1}(0.6)$$
$$= 0.253$$
$$z = -0.253$$

$$\therefore \quad \frac{s - 169}{9} = -0.253$$

$$s = 169 - 9 \times 0.253$$
$$= 166.723$$
$$= 166.7 \text{ (1 d.p.)}$$

## Example 7.9

The marks of 500 candidates in an examination are normally distributed with a mean of 45 marks and a standard deviation of 20 marks.

(a) Given that the pass mark is 41, estimate the number of candidates who passed the examination.

(b) If 5% of the candidates obtain a distinction by scoring $x$ marks or more, estimate the value of $x$.

(c) Estimate the interquartile range of the distribution. *(L Additional)*

## Solution 7.9

$X$ is the examination mark.

$X \sim N(45, 20^2)$

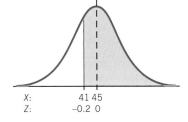

(a) $P(X > 41) = P\left(Z > \dfrac{41 - 45}{20}\right)$

$= P(Z > -0.2)$

$= \Phi(0.2)$

$= 0.5793$

$\therefore \quad P(\text{Pass}) = 0.5793$

Since there are 500 candidates, to find the number of candidates who pass, multiply the probability by 500.

$500 \times 0.5793 = 289.65$

$\therefore \quad$ **290 candidates passed.**

(b) $P(X > x) = 0.05$

Writing $z$ for the standardised value of $x$,

$P(Z > z) = 0.05 \quad$ where $z = \dfrac{x - 45}{20}$

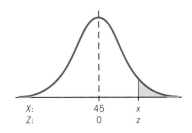

$\Phi(z) = 1 - 0.05$

$= 0.95$

$z = \Phi^{-1}(0.95)$

$= 1.645$

$\therefore \quad \dfrac{x - 45}{20} = 1.645$

$x = 45 + 1.645 \times 20$

$= 78 \ (2 \text{ s.f.})$

**A distinction is awarded for a mark of 78 or more.**

(c) The interquartile range encloses the central 50% of the distribution between the lower quartile $q_1$, and upper quartile, $q_3$.

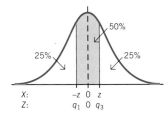

If $P(-z < Z < z) = 50\%$ then $z$ corresponds to an upper tail probability of 25%.

So $\quad \Phi(z) = 0.75$

$z = \Phi^{-1}(0.75)$

$= 0.674$

Now $z$ is the standardised value of the upper quartile, $q_3$,

so $\dfrac{q_3 - 45}{20} = 0.674$

$$q_3 = 45 + 20 \times 0.674$$
$$= 58.48$$

Lower quartile $q_1$ is such that $\dfrac{q_1 - 45}{20} = -0.674$

$$q_1 = 45 - 20 \times 0.674$$
$$= 31.52$$

Interquartile range $= q_3 - q_1$
$$= 58.48 - 31.52$$
$$= 27 \ (2 \text{ s.f.})$$

# Exercise 7c  Using the standard normal tables in reverse

1. In the following, find the value of $z$; where $Z \sim N(0, 1)$.

(a)

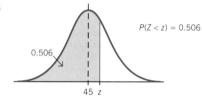

$P(Z < z) = 0.506$

(b)

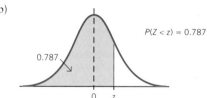

$P(Z < z) = 0.787$

(c)

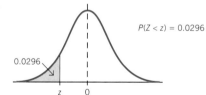

$P(Z < z) = 0.0296$

(d)

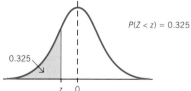

$P(Z < z) = 0.325$

(e)

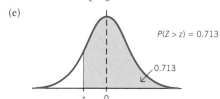

$P(Z > z) = 0.713$

(f)

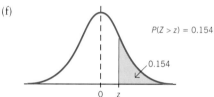

$P(Z > z) = 0.154$

(g)

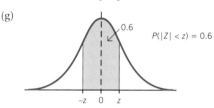

$P(|Z| < z) = 0.6$

2. $Z \sim N(0, 1)$.

Find the value of a, where

(a) $P(Z < a) = 0.9738$
(b) $P(Z < a) = 0.2435$
(c) $P(Z > a) = 0.82$
(d) $P(Z > a) = 0.2351$

3. $Z \sim N(0, 1)$.

Find $a$ if

(a) $P(|Z| < a) = 0.6372$,
(b) $P(|Z| > a) = 0.097$,
(c) $P(|Z| < a) = 0.5$,
(d) $P(|Z| > a) = 0.0404$.

4. If $Z \sim N(0, 1)$, find the upper quartile and the lower quartile of the distribution. Find also the 70th percentile.

5. Find $x$ in each of the following.

(a) $X \sim N(60, 25)$

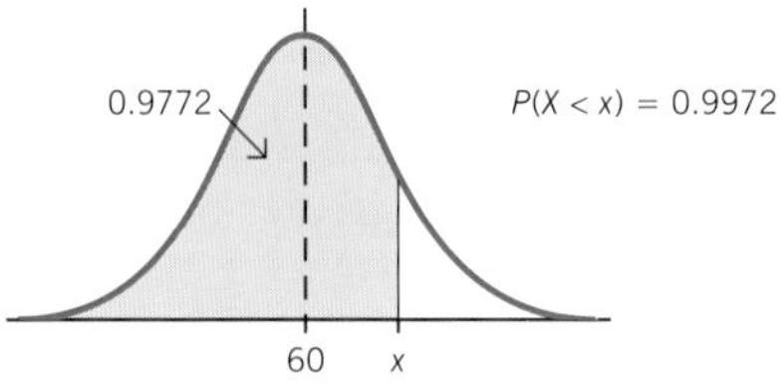


(b) $X \sim N(5, \frac{4}{9})$

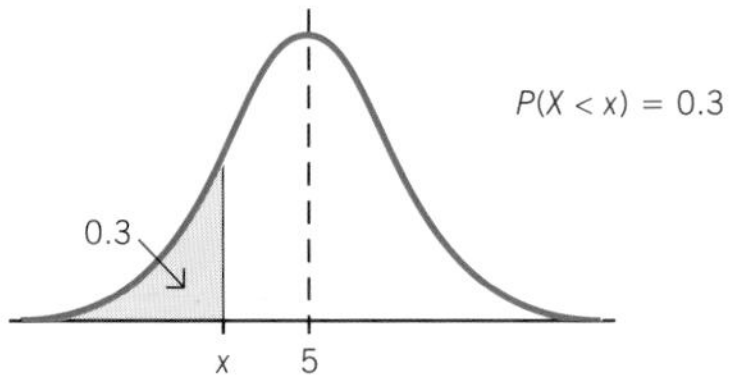


(c) $X \sim N(200, 36)$

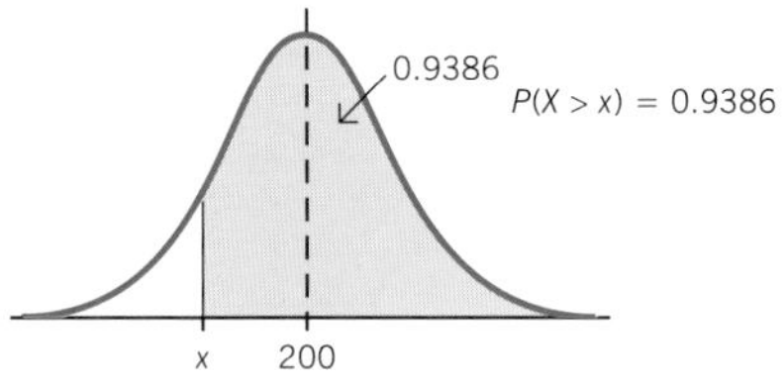


(d) $X \sim N(0, 4)$

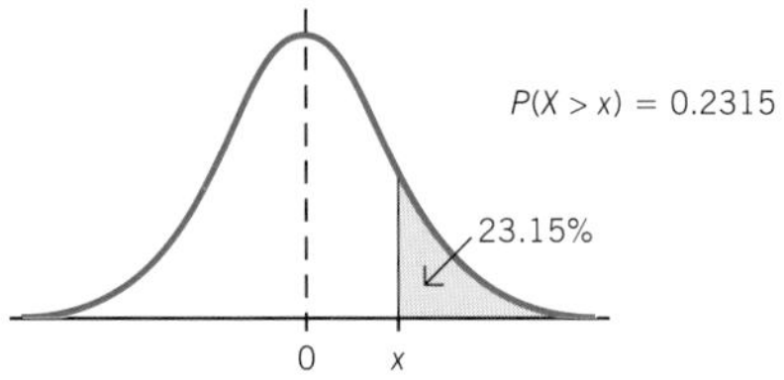


6. Bags of flour packed by a particular machine have masses which are normally distributed with mean 500 g and standard deviation 20 g. 2% of the bags are rejected for being underweight and 1% of the bags are rejected for being overweight. Between what range of values should the mass of a bag of flour lie if it is to be accepted?

7. The masses of cos lettuces sold at a market are normally distributed with mean mass 600 g and standard deviation 20 g.

(a) If a lettuce is chosen at random, find the probability that its mass lies between 570 g and 610 g.
(b) Find the mass exceeded by 7% of the lettuces.
(c) In one day, 1000 lettuces are sold.
Estimate how many weigh less than 545 g.

8. A sample of 100 apples is taken from a load. The apples have the following distribution of sizes

| Diameter to nearest cm | 6 | 7 | 8 | 9 | 10 |
|---|---|---|---|---|---|
| Frequency | 11 | 21 | 38 | 17 | 13 |

Determine the mean and standard deviation of these diameters.
Assuming that the distribution is approximately normal with this mean and this standard deviation find the range of size of apples for packing, if 5% are to be rejected as too small and 5% are to be rejected as too large. (*O & C*)

9. $X \sim N(400, 64)$.

(a) Find the limits within which the central 95% of the distribution lies.
(b) Find the interquartile range of the distribution.

10. The lengths of metal strips are normally distributed with a mean of 120 cm and a standard deviation of 10 cm. Find the probability that a strip selected at random has a length

(a) greater than 105 cm,
(b) within 5 cm of the mean.

Strips that are shorter than $L$ cm are rejected. Estimate the value of $L$, correct to one decimal place, if 9% or all strips are rejected.
In a sample of 500 strips, estimate the number having a length over 126 cm. (*C*)

11. The numbers of shirts sold in a week by the world's largest menswear store are normally distributed with a mean of 2080 and a standard deviation of 50. Estimate

(a) the probability that in a given week fewer than 2000 shirts are sold,
(b) the number of weeks in a year that between 2060 and 2130 shirts are sold,
(c) the interquartile range of the distribution,
(d) the least number $n$ of shirts such that the probability that more than $n$ are sold in a given week is less than 0.02. (*C*)

12. Batteries for a transistor radio have a mean life under normal usage of 160 hours, with a standard deviation of 30 hours. Assuming the battery life follows a normal distribution,

(a) calculate the percentage of batteries which have a life between 150 hours and 180 hours,
(b) calculate the range, symmetrical about the mean, within which 75% of the battery lives lie.

If a radio takes four of these batteries and requires all of them to be working, calculate

(c) the probability that the radio will run for at least 135 hours. (*O & C*)

# FINDING THE VALUE OF $\mu$ OR $\sigma$ OR BOTH

## Example 7.10

The lengths of certain items follow a normal distribution with mean $\mu$ cm and standard deviation 6 cm. It is known that 4.78% of the items have length greater than 82 cm. Find the value of the mean $\mu$.

## Solution 7.10

$X$ is the length, in centimetres, of an item.
$X \sim N(\mu, 6^2)$ and $P(X > 82) = 0.0478$

Since $P(X > 82)$ is less than 0.5, 82 must be greater than $\mu$.

$$P(X > 82) = P(Z > z) \text{ where}$$
$$z = \frac{82 - \mu}{6}$$
so $P(Z > z) = 0.0478$
$$\Phi(z) = 1 - 0.0478$$
$$= 0.9522$$
$$z = \Phi^{-1}(0.9522)$$
$$= 1.667$$
$$\therefore \quad \frac{82 - \mu}{6} = 1.667$$
$$82 - \mu = 1.667 \times 6$$
$$\mu = 82 - 1.667 \times 6 = 71.998$$

**The mean, $\mu = 72$ cm (2 s.f.)**

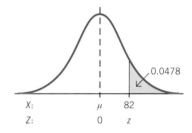

## Example 7.11

$X \sim N(100, \sigma^2)$ and $P(X < 106) = 0.8849$.
Find the value of the standard deviation $\sigma$.

## Solution 7.11

$$P(X < 106) = 0.8849$$
so $P(Z < z) = 0.8849$
where $z = \frac{106 - 100}{\sigma} = \frac{6}{\sigma}$
$$\Phi(z) = 0.8849$$
$$z = \Phi^{-1}(0.8849)$$
$$= 1.2$$
$$\therefore \quad \frac{6}{\sigma} = 1.2$$
$$\sigma = \frac{6}{1.2} = 5$$

**The standard deviation, $\sigma = 5$**

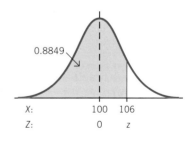

## Example 7.12

The masses of boxes of oranges are normally distributed such that 30% of them are greater than 4.00 kg and 20% are greater than 4.53 kg. Estimate the mean and standard deviation of the masses. (C)

## Solution 7.12

$X$ is the mass, in kilograms, of a box of oranges.
$X \sim N(\mu, \sigma^2)$ where $\mu$ and $\sigma$ are to be found.

$$P(X > 4.00) = 0.3$$
$$P(Z > z) = 0.3$$

where
$$z = \frac{4.00 - \mu}{\sigma}$$

$$\phi(z) = 1 - 0.3$$
$$= 0.7$$
$$z = \phi^{-1}(0.7)$$
$$= 0.524$$

$\therefore \quad \dfrac{4.00 - \mu}{\sigma} = 0.524$

$$4.00 - \mu = 0.524\sigma$$
$$4.00 = \mu + 0.524\sigma \quad......①$$

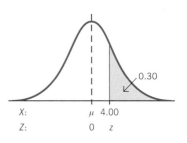

$$P(X > 4.53) = 0.2$$
$$P(Z > z) = 0.2$$

where
$$z = \frac{4.53 - \mu}{\sigma}$$

$$\phi(z) = 1 - 0.2$$
$$= 0.8$$
$$z = \phi^{-1}(0.8)$$
$$= 0.842$$

$\therefore \quad \dfrac{4.53 - \mu}{\sigma} = 0.842$

$$4.53 - \mu = 0.842\sigma$$
$$4.53 = \mu + 0.842\sigma \quad......②$$

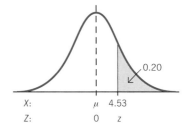

Equation ② – equation ① gives

$$0.53 = 0.318\sigma$$
$$\sigma = 1.666 \ldots = 1.67 \text{ (3 s.f.)}$$

Substituting in equation ①

$$4.00 = \mu + 0.524 \times 1.666 \ldots$$
$$\mu = 3.126 \ldots = 3.13 \text{ (3 s.f.)}.$$

$\therefore \quad \mu = 3.13 \text{ kg and } \sigma = 1.67 \text{ kg}$

## Example 7.13

The speeds of cars passing a certain point on a motorway can be taken to be normally distributed. Observations show that of cars passing the point, 95% are travelling at less than 85 m.p.h. and 10% are travelling at less than 55 m.p.h.

(a) Find the average speed of the cars passing the point.
(b) Find the proportion of cars that travel at more than 70 m.p.h. $\hspace{3cm}$ (L)

## Solution 7.13

$X$ is the speed, in m.p.h., of a car passing a certain point.
$X \sim N(\mu, \sigma^2)$.

(a) $\quad P(X < 85) = 0.95$

$\quad$ i.e. $P(Z < z_1) = 0.95 \quad$ where $z_1 = \dfrac{85 - \mu}{\sigma}$

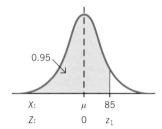

$$z_1 = \phi^{-1}(0.95)$$
$$= 1.645$$

$\therefore \dfrac{85 - \mu}{\sigma} = 1.645 \quad \Rightarrow \quad 85 = \mu + 1.645\sigma \dots \text{①}$

$\quad P(X < 55) = 0.1$

$\quad$ i.e. $P(Z < z_2) = 0.1 \quad$ where $z_2 = \dfrac{55 - \mu}{\sigma}$

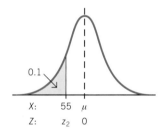

$$\phi(-z_2) = 0.9$$
$$-z_2 = \phi^{-1}(0.9)$$
$$= 1.282$$
$$\therefore \quad z_2 = -1.282$$

$\therefore \dfrac{55 - \mu}{\sigma} = -1.282 \quad \Rightarrow \quad 55 = \mu - 1.282\sigma \dots \text{②}$

① − ② gives $\quad\quad 30 = 2.927\sigma$
$\quad\quad\quad\quad\quad\quad\quad \sigma = 10.249 \dots$

Substituting in ① $\quad 85 = \mu + 1.645 \times 10.249 \dots$
$\quad\quad\quad\quad\quad\quad\quad\quad \mu = 68.139 \dots$

**The average speed is 68 m.p.h. (2 s.f.).**

(b) $P(Z > 70) = P\left(Z > \dfrac{70 - 68.13 \dots}{10.24 \dots}\right)$

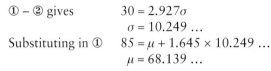

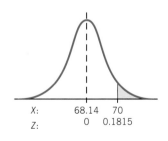

$$= P(Z > 0.1815)$$
$$= 1 - \phi(0.1815)$$
$$= 1 - 0.5718$$
$$= 0.4282$$

**The proportion of cars travelling at more than 70 m.p.h. is 0.43 (2 s.f.).**

## Exercise 7d  Finding $\mu$ or $\sigma$ or both, where $X \sim N(\mu, \sigma^2)$

You are advised to draw sketches to illustrate your answers.

1. The random variable $X$ is distributed $N(\mu, \sigma^2)$, with $\sigma = 25$.
If $P(X < 27.5) = 0.3085$, find the value of $\mu$.

2. The random variable $X$ is normally distributed with a mean of 45. The probability that $X$ is greater than 51 is 0.288. Find the standard deviation of the distribution.

3. The volumes of drinks in cans are normally distributed with a mean of 333 ml.
Given that 20% of the cans contain more than 340 ml, find the standard deviation of the volume of drink in a can. Find also the percentage of cans that contain less than than 330 ml.

4. The random variable $X$ is distributed $N(\mu, 12)$ and it is known that $P(X > 32) = 0.8438$. Find the value of $\mu$.

5. The heights, measured in metres, of 500 people are normally distributed with a standard deviation of 0.080 m. Given that the heights of 129 of these people are greater than the mean height, but less than 1.806 m, estimate the mean height. (C)

6. The random variable $X$ is distributed $N(\mu, \sigma^2)$.
$P(X > 80) = 0.0113$ and $P(X < 30) = 0.0287$.
Find $\mu$ and $\sigma$.

7. The masses of boxes of apples are normally distributed such that 20% of them are greater than 5.08 kg and 15% are greater than 5.62 kg. Estimate the mean and standard deviation of the masses.

8. Metal rods produced by a machine have lengths that are normally distributed.
2% of the rods are rejected as being too short and 5% are rejected as being too long.

   (a) Given that the least and greatest acceptable lengths of the rods are 6.32 cm and 7.52 cm, calculate the mean and variance of the lengths of the rods.
   (b) If ten rods are chosen at random from a batch produced by the machine, find the probability that exactly three of them are rejected as being too long.

9. The random variable $X$ is distributed $N(\mu, \sigma^2)$.
$P(X < 35) = 0.2$ and $P(35 < X < 45) = 0.65$.
Find $\mu$ and $\sigma$.

10. The marks in an examination were found to be normally distributed.
10% of the candidates were awarded a distinction for obtaining over 75.
20% of the candidates failed the examination with a mark of under 40. Find the mean and standard deviation of the distribution of marks.

11. A farmer cuts hazel twigs to make into bean poles to sell at the market. He says that a stick is 240 cm long. In fact the lengths of the sticks are normally distributed and 55% are over 240 cm long. 10% are over 250 cm long.
Find the probability that a randomly selected stick is shorter than 235 cm.

12. The diameters of bolts produced by a particular machine follow a normal distribution with mean 1.34 cm and standard deviation 0.04 cm. A bolt is rejected if its diameter is less than 1.24 cm or more than 1.40 cm. Find the percentage of bolts which are accepted.
The setting of the machine is altered so that the mean diameter changes but the standard deviation remains the same. With the new setting, 3% of the bolts are rejected because they are too large in diameter. Find the new mean diameter of bolts produced by the machine. Find also the percentage of bolts that are now rejected because they are too small in diameter.

13. Tea is sold in packages marked 750 g. The masses of the packages are normally distributed with a mean of 760 g. It is known that less than 1% of packages are underweight. What is the maximum value of the standard deviation of the distribution?

14. The random variable $X$ is normally distributed. The probability that $X$ is less than 53 is 0.04 and the probability that $X$ is less than 65 is 0.97. Find the interquartile range of the distribution.

15. A certain make of car tyre can be safely used for 25 000 km on average before it is replaced. The makers guarantee to pay compensation to anyone whose tyre does not last for 22 000 km. They expect 7.5% of all tyres sold to qualify for compensation. Assuming that the distance $X$ travelled before a tyre is replaced has a normal probability distribution, draw a diagram illustrating the facts given above.
Calculate, to three significant figures, the standard deviation of $X$.
Estimate the number of tyres per 1000 which will not have been replaced when they have covered 26 500 km. (L Additional)

16. Two firms, Goodline and Megadelay, produce delay lines for use in communications. The delay time for a delay line is measured in nanoseconds (ns).

    (a) The delay times for the output of Goodline may be modelled by a normal distribution with mean 283 ns and standard deviation 8 ns. What is the probability that the delay time of one line selected at random from Goodline's output is between 275 ns and 286 ns?

    (b) It is found that, in the output of Megadelay, 10% of the delay times are less than 274.6 ns and 7.5% are more than 288.2 ns. Again assuming a normal distribution, calculate the mean and standard deviation of the delay times for Megadelay. Give your answers correct to three significant figures.

    (C)

17. Machine components are mass-produced at a factory. A customer requires that the components should be 5.2 cm long but they will be acceptable if they are within limits 5.195 cm to 5.205 cm. The customer tests the components and finds that 10.75% of those supplied are over-size and 4.95% are under-size. Find the mean and standard deviation of the lengths of the components supplied, assuming that they are normally distributed.
    If three of the components are selected at random what is the probability that one is under-size, one is over-size and one satisfactory?

18. A machine dispenses peanuts into bags so that the weight of peanuts in a bag is normally distributed.

    (a) Initially the mean weight of peanuts in a bag is 128.5 g and the standard deviation is 1.5 g. Find the probability that the weight of peanuts in a randomly chosen bag exceeds 130 g.

    (b) The machine is given a minor overhaul that changes the mean weight, $\mu$, of peanuts in a bag without affecting the standard deviation. Following the overhaul, 14% of bags contain more than 130 g of peanuts. Find, to four significant figures, the new value for $\mu$.

    (c) Later the machine requires a major repair, following which the mean weight of peanuts in a bag is 128.3 g, and 4% of bags contain less than 126 g. Find, to three significant figures, the standard deviation of the weight of peanuts in a bag after this major repair.

    (NEAB)

19. A machine is used to fill cans of soup with a nominal volume of 0.450 litres. Suppose that the machine delivers a quantity of soup which is normally distributed with mean $\mu$ litres and standard deviation $\sigma$ litres. Given that $\mu = 0.457$ and $\sigma = 0.004$, find the probability that a randomly chosen can will contain less than the nominal volume.
    It is required by law that no more than 1% of cans contain less than the nominal volume. Find

    (a) the least value of $\mu$ which will comply with the law if $\sigma = 0.004$,

    (b) the greatest value of $\sigma$ which will comply with the law if $\mu = 0.457$. (MEI)

20. The masses of packets of sugar are normally distributed. In a large consignment of packets of sugar, it is found that 5% of them have a mass greater than 510 g and 2% have a mass greater than 515 g. Estimate the mean and the standard deviation of this distribution. (C)

21. On a particular day, 50% of the employees in a large company had arrived at work by 8.30 a.m., and 10% had not arrived by 8.55 a.m.

    (a) Assuming a normal model, find the standard deviation of the arrival times, in minutes.

    (b) It is given that only 5% of the employees had arrived by 8.05 a.m. Without further calculation, explain why this might suggest that a normal model is not appropriate.

    (c) Eighty employees are selected at random. Find the expectation of the number of these employees that arrived between 8.30 and 8.55 a.m. (C)

# THE NORMAL APPROXIMATION TO THE BINOMIAL DISTRIBUTION

Under certain circumstances the normal distribution can be used as an approximation to the binomial distribution. One practical advantage is that the calculations for finding probabilities are much less tedious to perform.

The diagrams opposite illustrate the distribution $B(n, p)$ for $p = 0.2$ and $p = 0.5$, for various values of $n$. In each case a vertical line graph has been drawn, and to make comparison easier, a curve has been superimposed on each.

(a) $n = 5$

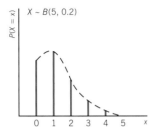

(b) $n = 12$

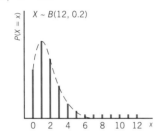

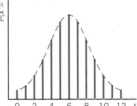

(c) $n = 20$

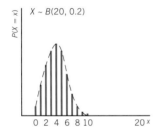

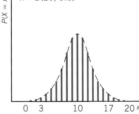

Notice

- when $p = 0.5$, the distributions are symmetric, and for larger values of $n$, the distribution takes on the characteristic normal shape,
- when $p = 0.2$, the distribution is positively skewed for small values of $n$, but when $n = 20$, the distribution is almost symmetrical and bell-shaped.

For the discrete random variable $X$, distributed binomially where $X \sim B(n, p)$, the mean $\mu = E(X) = np$ and the variance $\sigma^2 = \text{Var}(X) = npq$ (see page 286).

When $n$ is large and $p$ is not too far from 0.5, a normal distribution with mean $np$ and variance $npq$ can be used as an approximation to the binomial distribution.

A rule that can be used is as follows:

If $X \sim B(n, p)$ and $n$ and $p$ are such that $np > 5$ and $nq > 5$, where $q = 1 - p$, then $X \sim N(np, npq)$ approximately.

## CONTINUITY CORRECTIONS

The following example compares probabilities obtained using a binomial distribution and a normal approximation. It also illustrates the use of a **continuity correction,** needed when using a continuous distribution (the normal) as an approximation for a discrete distribution (the binomial).

## Example 7.14

Find the probability of obtaining 4, 5, 6 or 7 heads when a fair coin is tossed 12 times

(a) using the binomial distribution,
(b) using a normal approximation to the binomial distribution.

## Solution 7.14

$X$ is the number of heads in 12 tosses.
Since the coin is fair, $P(\text{head}) = 0.5$, so $X \sim B(12, 0.5)$.

(a) Using the binomial distribution,

$$P(X = 4) = {}^{12}C_4(0.5)^8 \times (0.5)^4 = {}^{12}C_4(0.5)^{12} = 0.1208 \ldots$$
$$P(X = 5) = {}^{12}C_5(0.5)^{12} = 0.1933 \ldots$$
$$P(X = 6) = {}^{12}C_6(0.5)^{12} = 0.2255 \ldots$$
$$P(X = 7) = {}^{12}C_7(0.5)^{12} = 0.1933 \ldots$$

$$P(4 \leqslant X \leqslant 7) = 0.733 \text{ (3 d.p.)}$$

(b) The diagram below shows the probability distribution for $X \sim B(12, 0.5)$. Note that the vertical lines have been replaced by rectangles to help illustrate the intention to use a continuous distribution as an approximation for a discrete one. The required binomial probability is represented by the sum of the areas of the shaded rectangles.

First check the conditions for a normal approximation:

$$np = 12 \times 0.5 = 6, \text{ so } np > 5$$
$$nq = 12 \times 0.5 = 6, \text{ so } nq > 5$$

Since $np > 5$ and $nq > 5$, use the normal approximation

$$X \sim N(np, npq) \text{ with } np = 6, npq = 12 \times 0.5 \times 0.5 = 3$$
$$\text{So } X \sim N(6, 3).$$

Superimposing the curve which is approximately $N(6, 3)$, the probability of obtaining 4, 5, 6 or 7 heads is found by considering the area under this normal curve from $x = 3.5$ to $x = 7.5$.

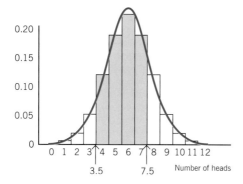

$P(4 \leqslant X \leqslant 7)$ transforms to $P(3.5 < X < 7.5)$ using a continuity correction.

Writing the symbol $\rightarrow$ to represent 'transforms to',

$$P(4 \leqslant X \leqslant 7) \rightarrow P(3.5 < X < 7.5)$$
$$= P\left(\frac{3.5 - 6}{\sqrt{3}} < Z < \frac{7.5 - 6}{\sqrt{3}}\right)$$
$$= P(-1.433 < Z < 0.866)$$
$$= \Phi(1.433) + \Phi(0.866) - 1$$
$$= 0.732 \text{ (3 d.p.)}$$

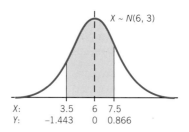

| X: | 3.5 | 6 | 7.5 |
|---|---|---|---|
| Y: | −1.443 | 0 | 0.866 |

Note that the probabilities found by the two different methods compare well and the working for part (b) is quicker to perform. The approximation is good because, although $n$ is not very large, $p = 0.5$.

# More about continuity corrections

Continuity corrections sometimes cause difficulties, so these are considered in more detail, using the diagram for the distribution of the number of heads when a coin is tossed 12 times.

If you want the probability that there are three heads or fewer, i.e. $P(X \leqslant 3)$, then consider $P(X < 3.5)$.

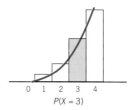

$P(X \leqslant 3)$ rectangle for 3 included

If you want the probability that there are fewer than three heads, i.e. $P(X < 3)$, then consider $P(X < 2.5)$.

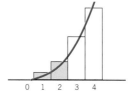

$P(X < 3)$ rectangle for 3 not included

If you want the probability that there are exactly three heads, i.e. $P(X = 3)$, then consider $P(2.5 < X < 3.5)$.

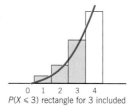

$P(X = 3)$

Consider these further examples.

| | |
|---|---|
| $P(5 \leqslant X \leqslant 8) \rightarrow P(4.5 < X < 8.5)$ | (5, 6, 7 or 8 heads) |
| $P(5 < X \leqslant 8) \rightarrow P(5.5 < X < 8.5)$ | (6, 7 or 8 heads) |
| $P(5 \leqslant X < 8) \rightarrow P(4.5 < X < 7.5)$ | (5, 6 or 7 heads) |
| $P(5 < X < 8) \rightarrow P(5.5 < X < 7.5)$ | (6 or 7 heads) |
| $P(X < 4) \rightarrow P(X < 3.5)$ | (0, 1, 2, 3 heads) |
| $P(X \leqslant 4) \rightarrow P(X < 4.5)$ | (0, 1, 2, 3 or 4 heads) |
| $P(X \geqslant 4) \rightarrow P(X > 3.5)$ | (4, 5, 6, … , 12 heads) |
| $P(X > 4) \rightarrow P(X > 4.5)$ | (5, 6, 7, … , 12 heads) |
| $P(X = 9) \rightarrow P(8.5 < X < 9.5)$ | (9 heads) |
| $P(X = 7) \rightarrow P(6.5 < X < 7.5)$ | (7 heads) |
| $P(X \geqslant 0) \rightarrow P(X > -0.5)$ | (0, 1, 2, … , 12 heads) |
| $P(X > 0) \rightarrow P(X > 0.5)$ | (1, 2, 3, … , 12 heads) |
| $P(X = 0) \rightarrow P(-0.5 < X < 0.5)$ | (0 heads) |

## Exercise 7e  Continuity corrections

Write down the transformations for each of the following, when a normal distribution is to be used as an approximation for a binomial distribution.

1. $P(3 \leqslant X \leqslant 9)$
2. $P(3 < X < 9)$
3. $P(10 < X \leqslant 24)$
4. $P(2 \leqslant X < 8)$
5. $P(X > 54)$
6. $P(X \geqslant 76)$
7. $P(45 < X < 67)$
8. $P(X < 109)$
9. $P(X \leqslant 45)$
10. $P(X = 56)$

11. $P(400 < X \leqslant 560)$
12. $P(X = 67)$
13. $P(X > 59)$
14. $P(X = 100)$
15. $P(34 \leqslant X < 43)$
16. $P(X = 7)$
17. $P(X \geqslant 509)$
18. $P(X < 7)$
19. $P(27 \leqslant X < 29)$
20. $P(X = 53)$

## Example 7.15

In a sack of mixed grass seeds, the probability that a seed is ryegrass is 0.35. Find the probability that in a random sample of 400 seeds from the sack,

(a) less than 120 are ryegrass seeds,
(b) between 120 and 150 (inclusive) are ryegrass,
(c) more than 160 are ryegrass seeds.

## Solution 7.15

$X$ is the number of ryegrass seeds in a sample of 400 seeds.
$n = 400$, $p = 0.35$, $q = 0.65$, so $X \sim B(400, 0.35)$

To see whether a normal approximation is suitable, check the value of $np$ and $nq$:

$np = 400 \times 0.35 = 140$ and $nq = 400 \times 0.65 = 260$.

Since $np > 5$ and $nq < 5$, use the normal approximation

$X \sim N(np, npq)$ with $np = 140$, $npq = 400 \times 0.35 \times 0.65 = 91$

So $X \sim N(140, 91)$

(a) $P(X < 120) \rightarrow P(X < 119.5)$  (continuity correction)

$$P(X < 119.5) = P\left(Z < \frac{119.5 - 140}{\sqrt{91}}\right)$$
$$= P(Z < -2.149)$$
$$= 1 - \Phi(2.149)$$
$$= 0.0158$$

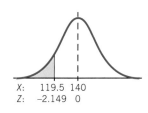

X:  119.5 140
Z:  −2.149  0

The probability that there are less than 120 ryegrass seeds is 0.016 (2 s.f.).

(b) $P(120 \leqslant X \leqslant 150) \rightarrow P(119.5 < X < 150.5)$  (continuity correction)

$$P(119.5 < X < 150.5) = P\left(\frac{119.5 - 140}{\sqrt{91}} < Z < \frac{150.5 - 140}{\sqrt{91}}\right)$$
$$= P(-2.149 < Z < 1.101)$$
$$= \Phi(2.149) + \Phi(1.101) - 1$$
$$= 0.8487$$

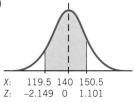

X:  119.5 140 150.5
Z:  −2.149  0  1.101

The probability that there are between 120 and 150 ryegrass seeds is 0.85 (2 s.f.).

(c) $P(X > 160) \rightarrow P(X > 160.5)$ (continuity correction)

$$P(X > 160.5) = P\left(Z > \frac{160.5 - 140}{\sqrt{91}}\right)$$
$$= P(Z > 2.149)$$
$$= 1 - \Phi(2.149)$$
$$= 0.0158$$

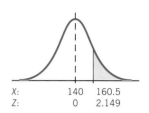

$X$: 140  160.5
$Z$: 0  2.149

**The probability that there are more than 160 ryegrass seeds is 0.016 (2 s.f.).**

*NOTE*: You should define $X$ as binomial, then check that the conditions are suitable before defining the approximate normal distribution.

---

## Example 7.16

It is given that 40% of the population support the Gamboge Party. One hundred and fifty members of the population are selected at random. Use a suitable approximation to find the probability that more than 55 out of the 150 support the Gamboge Party. (C)

## Solution 7.16

$X$ is the number in 150 who support the Gamboge Party.
$n = 150$, $p = 0.4$, $q = 0.6$
so $X \sim B(n, p)$ with $n = 150$, $p = 0.4$, $q = 0.6$

Check $np$ and $nq$:

$np = 150 \times 0.4 = 60$, $nq = 150 \times 0.6 = 90$

Since $np > 5$ and $nq > 5$, use the normal approximation
$X \sim N(np, npq)$ with $np = 60$, $npq = 150 \times 0.4 \times 0.6 = 36$
So $X \sim N(60, 36)$

$P(X > 55) \rightarrow P(X > 55.5)$ (continuity correction)

$$= P\left(Z > \frac{55.5 - 60}{6}\right)$$
$$= P(Z > -0.75)$$
$$= \Phi(0.75)$$
$$= 0.7734$$
$$= 0.77 \text{ (2 s.f.)}$$

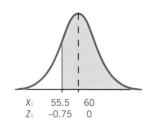

$X$: 55.5  60
$Z$: −0.75  0

---

# DECIDING WHEN TO USE A NORMAL APPROXIMATION AND WHEN TO USE A POISSON APPROXIMATION FOR A BINOMIAL DISTRIBUTION

For $X \sim B(n, p)$

- a Poisson approximation can be used when $n$ is large ($n > 50$) and $p$ is small ($p < 0.1$).
    Then $X \sim Po(np)$ approximately.
- a normal approximation can be used when $n$ and $p$ are such that $np > 5$ and $nq > 5$.
    Then $X \sim N(np, npq)$ approximately.

## Example 7.17

A number of different types of fungi are distributed at random in a field. Eighty % of these fungi are mushrooms, and the remainder are toadstools. Five % of the toadstools are poisonous. A man, who cannot distinguish between mushrooms and toadstools, wanders across the field and picks a total of 100 fungi. Determine, correct to two significant figures, using appropriate approximations, the probability that the man has picked

(a) at least 20 toadstools,
(b) exactly two poisonous toadstools. (C)

## Solution 7.17

$P(\text{mushroom}) = 0.8$, $P(\text{toadstool}) = 0.2$, $P(\text{poisonous toadstool}) = 0.05 \times 0.2 = 0.01$

(a) $X$ is the number of toadstools picked in a sample of 100.
$X \sim B(100, 0.2)$.
$np = 100 \times 0.2 = 20$, $nq = 100 \times 0.8 = 80$
Since $np > 5$, $nq > 5$, use a normal approximation
with mean $= np = 20$,
variance $= npq = 100 \times 0.2 \times 0.8 = 16$.
$X \sim N(20, 16)$.

$$P(X \geqslant 20) \rightarrow P(X > 19.5)$$
$$= P\left(Z > \frac{19.5 - 20}{4}\right)$$
$$= P(Z > -0.125)$$
$$= \Phi(0.125)$$
$$= 0.5498$$
$$= \mathbf{0.55 \ (2 \ s.f.)}$$

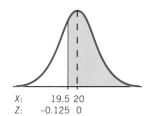

X:    19.5 20
Z:    −0.125 0

(b) $X$ is the number of poisonous toadstools in 100 fungi.
$X \sim B(100, 0.01)$

$np = 100 \times 0.01 = 1$

Use a Poisson distribution, since $n > 50$, $p < 0.1$.
$X \sim Po(1)$

$$P(X = 2) = e^{-1} \frac{1^2}{2!}$$
$$= 0.1839 \ldots$$
$$= \mathbf{0.18 \ (2 \ s.f.)}$$

# Exercise 7f   The normal approximation to the binomial

1. An ordinary unbiased die is thrown 120 times. Using a suitable approximation, find the probability of obtaining at least 24 sixes.

2. State conditions under which the distribution $B(n, p)$ can be approximated by a normal distribution.
   The random variable $X$ has the distribution $B(25, 0.38)$.
   (a) Verify that the distribution can be approximated by a normal distribution.
   (b) Use the normal approximation to calculate the probability that $X$ takes the values 15, 16, 17, 18 or 19.
   (c) Use the normal approximation to calculate the probability that $X$ takes the value 12. (C)

3. 10% of the chocolates produced in a factory are mis-shapes. A random sample of 1000 chocolates is taken. Find the probability that
   (a) less than 80 are mis-shapes,
   (b) between 90 and 115 (inclusive) are mis-shapes,
   (c) 120 or more are mis-shapes.

4. When I try to send a fax, the probability that I can successfully send it is 0.85.
   (a) I try to send eight faxes. Use a binomial model to find the probability that I can successfully send at least seven of the faxes.
   (b) I try to send 50 faxes. Use a normal approximation to the binomial model to find the probability that I can successfully send at least 45 faxes. (C)

5. At a particular hospital, records show that each day, on average, only 80% of people keep their appointment at the outpatients' clinic.
   Find the probability that on a day when 200 appointments have been booked,
   (a) more than 170 patients keep their appointments,
   (b) at least 155 patients keep their appointments.

6. The random variable $X$ is distributed $B(200, 0.7)$.
   Use the normal approximation to the binomial distribution to find
   (a) $P(X \geqslant 130)$,
   (b) $P(136 \leqslant X < 148)$,
   (c) $P(X < 142)$,
   (d) $P(X = 152)$.

7. One-fifth of a given population has a minor eye defect. Use the normal distribution as an approximation to the binomial distribution to estimate the probability that the number of people with the defect is
   (a) more than 20 in a random sample of 100 people,
   (b) exactly 20 in a random sample of 100 people,
   (c) more than 200 in a random sample of 1000 people.

8. It is estimated that one-fifth of the population of England watched last year's Cup Final on television. If random samples of 100 people are interviewed, calculate the mean and variance of the number of people from these samples who watched the Cup Final on television.
   Estimate, to two significant figures, the probability that in a random sample of 100 people, more than 30 watched the Cup Final on television. (L Additional)

9. In a series of $n$ independent trials, the probability of a success at each trial is $p$. If $R$ is the random variable denoting the total number of successes, state the probability that $R = r$. State also the mean and variance of $R$.
   A certain variety of flower seed is sold in packets containing 1000 seeds. It is claimed on the packet that 40% will bloom white and 60% will bloom red. This may be assumed to be accurate. Five seeds are planted. Find the probability that
   (a) exactly three will bloom white,
   (b) at least one will bloom white.

   100 seeds are planted. Use the normal approximation to estimate the probability of obtaining between 30 and 45 (inclusive) white flowers.

10. A certain tribe is distinguished by the fact that 45% of the males have six toes on their right foot. Find the probability that, in a group of 200 males from the tribe, more than 97 have six toes on their right foot.

11. A lorry load of potatoes has, on average, one rotten potato in six. A greengrocer decides to refuse the consignment if she finds more than 18 rotten potatoes in a random sample of 100. Find the probability that she accepts the consignment.

12. State conditions under which a binomial probability model can be well approximated by a normal model.
    $X$ is a random variable with the distribution $B(12, 0.42)$.
    (a) Anne uses the binomial distribution to calculate the probability that $X < 4$ and gives four significant figures in her answer. What answer should she get?
    (b) Ben uses a normal distribution to calculate an approximation for the probability that $X < 4$ and gives four significant figures in his answer. What answer should he get?
    (c) Given that Ben's working is correct, calculate the percentage error in his answer.
    (C)

## THE NORMAL APPROXIMATION TO THE POISSON DISTRIBUTION

If $X$ follows a Poisson distribution with parameter $\lambda$, i.e. $X \sim Po(\lambda)$,

then $E(X) = \lambda$ and $Var(X) = \lambda$

When $\lambda$ is large (say $\lambda > 15$), the normal distribution can be used as an approximation, where

$X \sim N(\lambda, \lambda)$.

As with the normal approximation to the binomial distribution, a continuity correction is needed, since you are using a continuous distribution as an approximation to a discrete one.

### Example 7.18

A radioactive disintegration gives counts that follow a Poisson distribution with a mean count of 25 per second.
Find the probability that in a one-second interval the count is between 23 and 27 inclusive.

### Solution 7.18

$X$ is the radioactive count in a one-second interval.

$X \sim Po(25)$

$E(X) = 25$, $Var(X) = 25$

Using a normal approximation,

$X \sim N(25, 25)$

$P(23 \leqslant X \leqslant 27) \rightarrow P(22.5 < X < 27.5)$    (continuity correction)

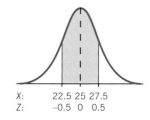

$$= P\left(\frac{22.5 - 25}{5} < Z < \frac{27.5 - 25}{5}\right)$$

$$= P(-0.5 < Z < 0.5)$$

$$= 2\Phi(0.5) - 1$$

$$= 0.383 \ (3 \text{ d.p.})$$

| $X$: | 22.5 25 27.5 |
| $Z$: | −0.5 0 0.5 |

*NOTE*: this compares very well with the value given using the Poisson distribution. Check if for yourself.

## Exercise 7g  The normal approximation to the Poisson distribution

1. If $X \sim Po(24)$, use the normal approximation to find
   (a) $P(X \leqslant 25)$,
   (b) $P(22 \leqslant X \leqslant 26)$,
   (c) $P(X > 23)$.

2. If $X \sim Po(35)$, use the normal approximation to find
   (a) $P(X \leqslant 33)$,
   (b) $P(33 < X < 37)$,
   (c) $P(X > 37)$,
   (d) $P(X = 37)$.

3. If $X \sim Po(60)$, use the normal approximation to find
   (a) $P(50 < X \leqslant 58)$,
   (b) $P(57 \leqslant X < 68)$,
   (c) $P(X > 52)$,
   (d) $P(X \geqslant 70)$.

4. The number of calls per hour received by an office switchboard follows a Poisson distribution with parameter 30. Using the normal approximation to the Poisson distribution, find the probability that, in one hour,
   (a) there are more than 33 calls,
   (b) there are between 25 and 28 calls (inclusive),
   (c) there are 34 calls.

5. In a certain factory the number of accidents occurring in a month follows a Poisson distribution with mean 4. Find the probability that there will be at least 40 accidents during one year.

6. The number of bacteria on a plate viewed under a microscope follows a Poisson distribution with a parameter 60. Find the probability that there are between 55 and 75 bacteria on a plate.

   A plate is rejected if less than 38 bacteria are found. If 2000 such plates are viewed, how many will be rejected?

7. In an experiment with a radioactive substance the number of particles reaching a counter over a given period of time follows a Poisson distribution with mean 22. Find the probability that the number of particles reaching the counter over the given period of time is

   (a) less than 22,
   (b) between 25 and 30,
   (c) 18 or more.

8. The number of accidents on a certain railway line occur at an average rate of one every two months. Find the probability that

   (a) there are 25 or more accidents in four years,
   (b) there are 30 or fewer accidents in five years.

9. The number of eggs laid by an insect follows a Poisson distribution with parameter 200.

   (a) Find the probability that
       (i)   more than 170 eggs are laid,
       (ii)  more than 205 eggs are laid,
       (iii) between 180 and 240 eggs (inclusive) are laid.
   (b) If the probability that an egg develops is 0.1, show that the number of survivors follows a Poisson distribution with parameter 20 and find the probability that there are more than 30 survivors.

10. When a trainee typist types a document the number of mistakes made on any one page is a Poisson variable with mean 3, independently of the number of mistakes made on any other page. Use tables, or otherwise, to find, to three significant figures,

    (a) the probability that the number of mistakes on the first page is less than two,
    (b) the probability that the number of mistakes on the first page is more than four.

    When the typist types a 48-page document the total number of mistakes made by the typist is a Poisson variable with mean 144.
    Use a suitable approximate method to find, to three decimal places, the probability that this total number of mistakes is greater than 130.   (NEAB)

11. Tomatoes from a particular nursery are packed in boxes and sent to a market.
    Assuming that the number of bad tomatoes in a box has a Poisson distribution with mean 0.44, find, to three significant figures, the probability of there being
    (a) fewer than two,
    (b) more than two bad tomatoes in a box when it is opened.

    Use a normal approximation to find, to three decimal places, the probability that in 50 randomly chosen boxes there will be fewer than 20 bad tomatoes in total.   (L)

12. A large silo is filled with grain harvested by a farmer. The grain is contaminated with insect pests called weevils. The farmer finds that there are on average three weevils per litre of grain.

    (a) State two conditions which are necessary for the Poisson distribution to be a suitable model for the number of weevils which would be found in a given volume of grain.

    Assume that the Poisson distribution can be used in this case.

    (b) Find, to three decimal places,
        (i)  the probability that 1 litre of grain contains at least one weevil,
        (ii) the probability that 4 litres of grain contain exactly ten weevils.
    (c) Use an appropriate distributional approximation to estimate the probability that 10 litres of grain contain fewer than 25 weevils, giving your answer to three decimal places.   (NEAB)

13. A biologist gathers leaves of a certain plant in order to collect insects of a particular type. From past experience she knows that the distribution of the number of insects on $n$ leaves may be modelled by a Poisson distribution with mean $0.8n$.

    (a) Calculate, to three decimal places, the probability that the number of insects on the next leaf to be examined will be fewer than three.
    (b) Determine, to three decimal places, the probability that the total number of insects on the next ten leaves to be examined will lie between six and 12 (both inclusive).
    (c) Use a distributional approximation to find, to two decimal places, the probability that the total number of insects on the next 50 leaves to be examined will exceed 45.   (NEAB)

14. (a) Give two conditions which must apply when modelling a random variable by a Poisson distribution.

    A particular make of kettle is sold by a shop at an average rate of five per week. The random variable $X$ represents the number of kettles sold in any one week and $X$ is modelled by a Poisson distribution.
    The shop manager notices that at the beginning of a particular week there are seven kettles in stock.

    (b) Find the probability that the shop will not be able to meet all the demands for kettles that week, assuming that it is not possible to restock during the week.

    In order to increase sales performance, the manager decides to have in stock at the beginning of each week sufficient kettles to have at least a 99% chance of being able to meet all demands during that week.

    (c) Find the smallest number of kettles that should be in stock at the beginning of each week.
    (d) Using a suitable approximation find the probability that the shop sells at least 18 kettles in a four-week period, subject to stock always being available to meet demand.   (L)

## Summary

- Normal variable $X$

  $X \sim N(\mu, \sigma^2)$   $E(X) = \mu$, $\text{Var}(X) = \sigma^2$.

- Standard normal variable $Z$

  $Z \sim N(0, 1)$   $E(Z) = 0$, $\text{Var}(Z) = 1$.

  To standardise $X$, use $Z = \dfrac{X - \mu}{\sigma}$.

- Using the standard normal tables:

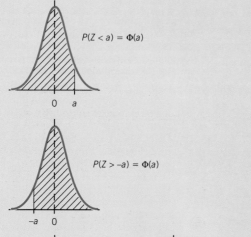

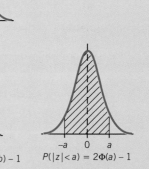

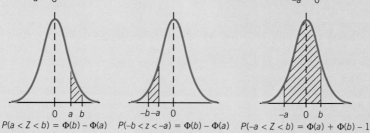

- Using the tables in reverse:

  If $\Phi(a) = k$, i.e. $P(Z < a) = k$, then $a = \Phi^{-1}(k)$

- The normal approximation to the binomial distribution.

  If $X \sim B(n, p)$ and $np > 5$, $nq > 5$     then $X \sim N(np, npq)$.

- The normal approximation to the Poisson distribution

  If $X \sim Po(\lambda)$ and $\lambda > 15$     then $X \sim N(\lambda, \lambda)$.

- A Poisson approximation to the binomial distribution

  If $X \sim B(n, p)$ and $n$ is large ($n > 50$) and $p$ is small ($p < 0.1$) then $X \sim Po(np)$.

- Continuity corrections

  These must be used when using a continuous distribution (e.g. normal) as an approximation to a discrete distribution (e.g. binomial, Poisson)

## Miscellaneous worked examples

### Example 7.19

A product is sold in packets whose masses are normally distributed with a mean of 1.42 kg and a standard deviation of 0.025 kg.

(a) Find the probability that the mass of a packet, selected at random, lies between 1.37 kg and 1.45 kg.
(b) Estimate the number of packets, in an output of 5000, whose mass is less than 1.35 kg. (C)

### Solution 7.19

$X$ is the mass, in kilograms, of a packet.
$X \sim N(1.42, 0.025^2)$

(a) $P(1.37 < X < 1.45)$

Standardising, using $Z = \dfrac{X - 1.42}{0.025}$

$P\left(\dfrac{1.37 - 1.42}{0.025} < Z < \dfrac{1.45 - 1.42}{0.025}\right)$

$= P(-2 < Z < 1.2)$
$= \Phi(1.2) + \Phi(2) - 1$
$= 0.8849 + 0.9772 - 1$
$= 0.8621$

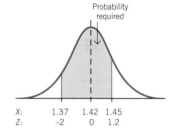

The probability that the mass lies between 1.37 kg and 1.45 kg is 0.86 (2 s.f.).

(b) $P(X < 1.35) = P\left(Z < \dfrac{1.35 - 1.42}{0.025}\right)$

$= P(Z < -2.8)$
$= 1 - \Phi(2.8)$
$= 1 - 0.9974$
$= 0.0026$

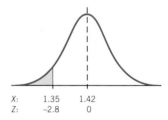

Since there are 5000 packets, multiply the probability by 5000.

$5000 \times 0.0026 = 13$

**13 packets have a mass less than 1.35 kg.**

### Example 7.20

In a certain cross country running competition the times that each of the 136 runners took to complete the course were recorded to the nearest minute. The winner completed the course in 23 minutes and the final runner came in with a time of 78 minutes. The full results are summarised in the table below.

| Recorded time | 20–29 | 30–39 | 40–49 | 50–59 | 60–69 | 70–79 |
|---|---|---|---|---|---|---|
| Frequency | 7 | 21 | 42 | 37 | 20 | 9 |

(a) Use linear interpolation to estimate the median time.

The upper and lower quartiles of the time taken are 58.1 and 40.9 respectively.

(b) On graph paper, draw a box and whisker plot for the results from this competition. You should mark the end points, the median and the quartiles clearly on your diagram.

(c) Comment on the skewness.

Assume that the time taken by the runners to complete the course follows a normal distribution with the values for the quartiles as given above.

(d) Calculate the mean of this normal distribution.

(e) Calculate the standard deviation of this normal distribution.  (L)

## Solution 7.20

| Recorded time | <29.5 | <39.5 | <49.5 | <59.5 | <69.5 | <79.5 |
|---|---|---|---|---|---|---|
| Cumulative frequency | 7 | 28 | 70 | 107 | 127 | 136 |

(a) For grouped data, the median, $Q_2$, is the the $\frac{1}{2}n$th value, i.e. $Q_2$ is the 68th value. This lies in the interval $39.5 - 49.5$.
This interval has 42 items in it and is of width 10.

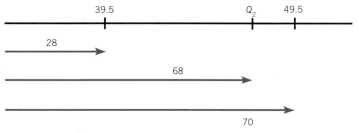

$$Q_2 \approx 39.5 + \tfrac{40}{42} \times 10 = 49.0 \ (1 \ d.p.)$$

**The median time is 49.0 minutes.**

(b)

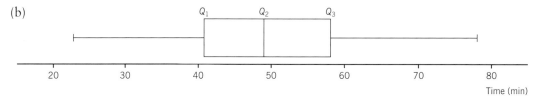

(c) The distribution appears to be symmetrical

$$Q_3 - Q_2 = 58.1 - 49.0 = 9.1$$
$$Q_2 - Q_1 = 49.0 - 40.9 = 8.1$$

So $\quad Q_3 - Q_2 > Q_2 - Q_1$, but only just.

**The distribution has a slight positive skew.**

(d)  Assuming $Q_1 = 40.9$ and $Q_3 = 58.1$
$$\mu = \tfrac{1}{2}(40.9 + 58.1)$$
$$= 49.5$$

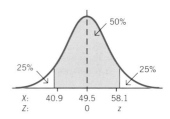

If $X$ is the recorded time, in minutes

$X \sim N(49.5, \sigma^2)$ and $P(X < 58.1) = 0.75$

The central 50% of the distribution is enclosed between $Q_1$ and $Q_3$, so the $z$ value for $Q_3$ corresponds to an upper tail probability of 0.25, i.e. a lower tail probability of 0.75.

$$\therefore\ P(Z < z) = 0.75 \ \text{ where } \ z = \frac{58.1 - 49.5}{\sigma} = \frac{8.6}{\sigma}$$

$$\Phi(z) = 0.75$$
$$z = \Phi^{-1}(0.75)$$
$$= 0.674$$

$$\therefore\quad \frac{8.6}{\sigma} = 0.674$$

$$\sigma = \frac{8.6}{0.674}$$
$$= 12.75 \dots$$

**The standard deviation is 12.8 minutes (1 d.p.).**

---

## Example 7.21

Machine $A$, used for filling bags with ground coffee, can be set to dispense any required mean weight of coffee per bag. At any setting the weight of coffee in a bag can be modelled by a normal distribution with a standard deviation of 1.95 g.

(a)  If the machine is set to dispense a mean weight of 128 g of coffee per bag, calculate the percentage of bags that contain less than 125 g.

(b)  To meet an official regulation the setting on a machine must be adjusted so that no more than 1% of bags contain less than 125 g.
   (i)   Calculate the smallest mean weight to which machine $A$ should be set to meet the regulation.
   (ii)  Machine $B$ will only just meet the regulation when it is set to dispense a mean weight of 128.5 g. Assuming that the weight of coffee is a bag filled by Machine $B$ can be modelled by a normal distribution, calculate the standard deviation of this distribution.
                                                                           *(NEAB)*

## Solution 7.21

$X$ is the weight, in grams of coffee in a bag from Machine $A$.
$X \sim N(\mu, 1.95^2)$.

(a)  $\mu = 128$, so $X \sim N(128, 1.95^2)$.

$$P(X < 125) = P\left(Z < \frac{125 - 128}{1.95}\right)$$
$$= P(Z < -1.538)$$
$$= 1 - \Phi(1.538)$$
$$= 1 - 0.9380$$
$$= 0.062$$
$$= 6.2\%$$

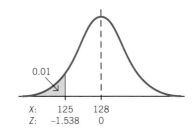

**6.2% of bags contain less than 125 g.**

(b) $X \sim N(\mu, 1.95^2)$ and $P(X < 125) \leqslant 0.01$

   (i)  Standardising, you need to find $z$ such that

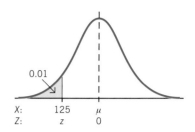

$$P(Z < z) = 0.01$$
i.e. $\qquad \Phi(z) = 0.01$
so $\qquad \Phi(-z) = 0.99$
$$-z = \Phi^{-1}(0.99)$$
$$= 2.326$$
$$z = -2.326$$
$$\therefore \quad \frac{125 - \mu}{1.95} \leqslant -2.326$$
$$125 - \mu \leqslant -2.326 \times 1.95$$
$$\mu \geqslant 125 + 2.326 \times 1.95$$
$$\mu \geqslant 129.53 \ldots$$

**The smallest mean weight is 129.5 g (1 d.p.).**

  (ii)  $Y$ is the weight, in grams of coffee in a bag from machine $B$.

$Y \sim N(128.5, \sigma^2)$ and $P(Y < 125) = 0.01$

Standardising:

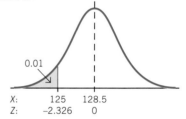

$$P(Z < z) = 0.01 \text{ where } Z = \frac{125 - 128.5}{\sigma}$$
$$= \frac{-3.5}{\sigma}$$

From part (i) $z = -2.326$
$$\therefore \quad -2.326 = \frac{-3.5}{\sigma}$$
$$\sigma = \frac{3.5}{2.326}$$
$$= 1.504 \ldots$$

**The standard deviation is 1.5 g (1 d.p.).**

---

## Example 7.22

It is estimated that, on average, one match in five in the Football League is drawn, and that one match in two is a home win.

(a) Twelve matches are selected at random. Calculate the probability that the number of drawn matches is
   (i)   exactly three,
  (ii)  at least four.

(b) Ninety matches are selected at random. Use a suitable approximation to calculate the probability that between 13 and 20 (inclusive) of the matches are drawn.

(c) Twenty matches are selected at random. The random variables $D$ and $H$ are the numbers of drawn matches and home wins, respectively, in these matches. State, with a reason, which of $D$ and $H$ can be better approximated by a normal variable.     (C)

## Solution 7.22

$X$ is the number of drawn matches in 12.

$X \sim B(12, 0.2)$ since $P(\text{draw}) = \frac{1}{5} = 0.2$

(a) (i) $P(X = 3) = {}^{12}C_3(0.8)^9(0.2)^3$

$\qquad\qquad = 0.24$ (2 s.f.)

(ii) $P(X \geqslant 4) = 1 - P(X < 4)$

$\qquad\qquad = 1 - ((0.8)^{12} + 12(0.8)^{11}(0.2) + {}^{12}C_2(0.8)^{10}(0.2)^2 + {}^{12}C_3(0.8)^9(0.2)^3)$

$\qquad\qquad = 1 - 0.794 \ldots$

$\qquad\qquad = 0.21$ (2 s.f.).

(b) $X$ is the number of drawn matches in 90.

Then $X \sim B(n, p)$ with $n = 90$, $p = 0.2$, $q = 0.8$

Now $np = 90 \times 0.2 = 18$, $nq = 90 \times 0.8 = 72$

Since $np > 5$, $nq > 5$, use a normal approximation

$X \sim N(np, npq)$ with $np = 18$, and $npq = 90 \times 0.2 \times 0.8 = 14.4$,

so $X \sim N(18, 14.4)$.

$P(13 \leqslant X \leqslant 20) \rightarrow P(12.5 < X < 20.5)$ $\quad$ (continuity correction)

$$= P\left(\frac{12.5 - 18}{\sqrt{14.4}} < Z < \frac{20.5 - 18}{\sqrt{14.4}}\right)$$

$$= P(-1.449 < Z < 0.659)$$

$$= \Phi(1.449) + \Phi(0.659) - 1$$

$$= 0.9264 + 0.7451 - 1$$

$$= 0.67 \text{ (2 s.f.)}$$

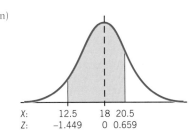

| $X$: | 12.5 | 18 | 20.5 |
| $Z$: | −1.449 | 0 | 0.659 |

(c) $D$ is the number of drawn matches.

$D \sim B(20, 0.2)$ $\qquad np = 20 \times 0.2 = 4$, so $np < 5$.

$H$ is the number of home wins.

$H \sim B(20, 0.5)$ $\qquad np = 20 \times 0.5 = 10 > 5$, $nq = 10 > 5$.

For $H$, $np > 5$ and $nq > 5$, so $H$ can be better approximated by a normal variable.

# Miscellaneous exercise 7h

1. Squash balls, dropped onto a concrete floor from a given point, rebound to heights which can be modelled by a normal distribution with mean 0.8 m and standard deviation 0.2 m. The balls are classified by height of rebound, in order of decreasing height, into these categories: Fast, Medium, Slow, Super-Slow and Rejected.

   (a) Balls which rebound to heights between 0.65 m and 0.9 m are classified as Slow. Calculate the percentage of balls classified as Slow.
   (b) Given that 9% of balls are classified as Rejected, calculate the maximum height of rebound of these balls.
   (c) The percentage of balls classified as Fast and as Medium are equal. Calculate the minimum height of rebound of a ball classified as Fast, giving your answer correct to two decimal places. (C)

2. The mass of grapes sold per day in a supermarket can be modelled by a normal distribution. It is found that, over a long period, the mean mass sold per day is 35.0 kg, and that, on average, less than 15.0 kg are sold on one day in twenty.

   (a) Show that the standard deviation of the mass of grapes sold per day is 12.2 kg, correct to three significant figures.
   (b) Calculate the probability that, on a day chosen at random, more than 53.0 kg are sold.
   (c) Ten days are chosen at random. Assuming independence, find the probability that less than 15.0 kg will be sold on exactly two of these days. (C)

3. (a) Give **two** reasons why the normal distribution is important in statistics.
   (b) An airline has a regular flight from one airport to another. The airline models the duration of a flight as a normally distributed random variable with a mean of 246 minutes and a standard deviation of five minutes. Use this model to calculate, to one decimal place, the percentage of these flights that are completed in less than four hours. (NEAB)

4. The random variable $X$ is normally distributed with mean $\mu$ and variance $\sigma^2$.

   Given that $\qquad P(X > 58.37) = 0.02$

   and $\qquad\qquad P(X < 40.85) = 0.01$

   find $\mu$ and $\sigma$. (L)

5. Alan is a member of an athletics club. In long jump competitions, his jumps are normally distributed with a mean of 7.6 m and a standard deviation of 0.16 m.

   (a) Calculate the probability of him jumping
   (i) more than 8.0 m,
   (ii) between 7.50 m and 7.75 m.
   (b) Determine the distance exceeded by 75% of his jumps.

   Brian also belongs to the athletics club. In long jump competitions, his jumps are normally distributed with a mean of 7.45 m and 95.2% of them exceed 7.0 m.

   (c) Calculate, correct to two decimal places, the standard deviation of Brian's jumps.

   The athletics club has to select either Alan or Brian to be its long jump competitor at a major athletics meeting. In order to qualify for the final rounds of jumps at the meeting, it is necessary to achieve a jump of at least 8.0 m in the preliminary rounds.

   (d) State, with justification, which of the two athletes should be selected. (NEAB)

6. The time required to complete a certain car journey has been found from experience to have mean 2 hours 20 minutes and standard deviation 15 minutes.

   (a) Use a normal model to calculate the probability that, on one day chosen at random, the journey requires between 1 hour 50 minutes and 2 hours 40 minutes.
   (b) It is known that delays occur rarely on this journey, but that when they do occur they are lengthy. Give a reason why this information suggests that a normal distribution might not be a good model. (C)

7. A machine is producing a type of circular gasket. The specifications for the use of these gaskets in the manufacture of a certain make of engine are that the thickness should lie between 5.45 mm and 5.55 mm, and the diameter should lie between 8.45 mm and 8.54 mm. The machine is producing the gaskets so that their thicknesses are $N(5.5, 0.0004)$, that is, normally distributed with mean 5.5 mm and variance 0.0004 mm$^2$, and their diameters are independently distributed $N(8.54, 0.0025)$.
   Calculate, to one decimal place, the percentage of gaskets produced which will not meet

   (a) the specified thickness limits,
   (b) the specified diameter limits,
   (c) the specifications.

   Find, to three decimal places, the probability that, if six gaskets made by the machine are chosen at random, exactly five of them will meet the specifications. (L)

8. A machine is used to fill tubes, of nominal content 100 ml, with toothpaste. The amount of toothpaste delivered by the machine is normally distributed and may be set to any required mean value. Immediately after the machine has been overhauled, the standard deviation of the amount delivered is 2 ml. As time passes, this standard deviation increases until the machine is again overhauled. The following three conditions are necessary for a batch of tubes of toothpaste to comply with current legislation:

I     the average content of the tubes must be at least 100 ml,

II    not more than 2.5% of the tubes may contain less than 95.5 ml,

III   not more than 0.1% of the tubes may contain less than 91 ml.

(a) For a batch of tubes with mean content 98.8 ml and standard deviation 2 ml, find the proportion of tubes which contain
(i)   less than 95.5 ml,
(ii)  less than 91 ml.

Hence state which, if any, of the three conditions above are **not** satisfied.

(b) If the standard deviation is 5 ml, find the mean in **each** of the following cases:
(i)   exactly 2.5% of tubes contain less than 95.5 ml,
(ii)  exactly 0.1% of tubes contain less than 91 ml.

Hence state the smallest value of the mean which would enable all three conditions to be met when the standard deviation is 5 ml.

(c) Currently exactly 0.1% of tubes contain less than 91 ml and exactly 2.5% contain less than 95.5 ml.
(i)   Find the current values of the mean and the standard deviation.
(ii)  State, giving a reason, whether you would recommend that the machine is overhauled immediately.   (*AEB*)

$(\Phi^{-1}(0.999) = 3.09)$

9. A wholesaler buys cauliflowers from a farmer for distribution to retail greengrocers.
The wholesaler classifies the lightest 15% of cauliflowers as small, the heaviest 25% as large, and the rest as medium.

(a) Given that the wholesaler makes a profit of 2 pence on each small cauliflower, 12 pence on each medium one and 27 pence on each large one, calculate the wholesaler's mean profit per cauliflower.

The weights of the cauliflowers can be modelled by a normal distribution with a mean of 628 g and a standard deviation of 160 g.

(b) Calculate the weight that a cauliflower must exceed to be classified as large.

(c) Calculate the weight that a cauliflower must fall below to be classified as small.   (*NEAB*)

10. In 1994 an insurance company received claims from 20% of the motorists it had insured.

(a) For a random sample of 14 motorists insured with the company in 1994, find the probability that
(i)   exactly three claimed on their insurance,
(ii)  between two and five inclusive claimed on their insurance,
(iii)  a majority claimed on their insurance.

(b) For a random sample of 90 motorists insured with the company, use an appropriate approximating distribution to determine the probability that at least 25 claimed on their insurance in 1994.  (*NEAB*)

11. A horticulturalist knows from experience that when taking cuttings from bay trees only 15 in every 100 successfully take root.

(a) In a batch of ten randomly selected cuttings, find the probability that
(i)   none of the cuttings take root,
(ii)  fewer than three of the cuttings take root.

(b) Let $n$ be the smallest number of cuttings which need to be examined before there is at least a 95% chance that one or more of them will have taken root.
(i)   Show that $n$ satisfies $(0.85)^n \leqslant 0.05$.
(ii)  Given that $(0.85)^{17} = 0.0631$, find the value of $n$.

(c) Using a suitable approximation, estimate the probability that fewer than six in a batch of 50 cuttings take root.   (*L*)

12. A large bag of seeds contains three varieties in the ratios $4 : 2 : 1$ and their germination rates are 50%, 60% and 80% respectively.
Show that the probability that a seed chosen at random from the bag will germinate is $\frac{4}{7}$.
Find, to three decimal places, the probability that of four seeds chosen at random from the bag, exactly two of them will germinate.
Given that 150 seeds are chosen at random from the bag, estimate, to three decimal places, the probability that fewer than 90 of them will germinate.   (*L*)

13. A building society announces its intention to convert to a bank. During the first day following the announcement, the number of calls per minute answered by the society's hotline may be modelled satisfactorily by a Poisson distribution with mean 12.

(a) Calculate the probability that the hotline answers more than ten calls in a one-minute period.

(b) Estimate the probability that the hotline answers fewer than 700 calls in one hour.   (*NEAB*)

14. (a) A trade union asked 300 of its members whether they were full-time workers or part-time workers, and the number of hours they worked in a particular week. The table below shows an analysis of this survey.

| | Number workers | Mean number of hours worked | Standard deviation of hours worked |
|---|---|---|---|
| Full-time | 100 | 40 | 4.5 |
| Part-time | 200 | 20 | 6.9 |

The hours, both for the full-time workers and for the part-time workers, are normally distributed.
   (i)   Calculate the total number of workers who worked more than 32 hours.
   (ii)  Given that only 6% of the full-time workers worked for less than $T_1$ hours, calculate $T_1$.
   (iii) Given that only 3% of the part-time workers worked for more than $T_2$ hours, calculate $T_2$.

   (b) A set of numbers is normally distributed; 1.5% of the numbers exceed 1434 and 16.6% of the numbers exceed 1194. Calculate the mean and the standard deviation of the distribution. (C)

15. During an advertising campaign, the manufacturer of Wolfitt (a dog food) claimed that 60% of dog owners preferred to buy Wolfitt. Assuming that the manufacturer's claim is correct for the population of dog owners, calculate

   (a) using the binomial distribution, and
   (b) using a normal approximation to the binomial;

   the probability that at least six of a random sample of eight dog owners prefer to buy Wolfitt.
   Comment on the agreement, or disagreement, between your two values. Would the agreement be better or worse if the proportion had been 80% instead of 60%?
   Continuing to assume that the manufacturer's figure of 60% is correct, use the normal approximation to the binomial to estimate the probability that, of a random sample of 100 dog owners, the number preferring Wolfitt is between 60 and 70 inclusive. (MEI)

16. Six hundred rounds are fired from a gun at a horizontal target 50 m long which extends from 950 m to 1000 m in range from the gun. The trajectories of the rounds all lie in the vertical plane through the gun and the target. It is found that 27 rounds fall short of the target and 69 rounds fall beyond it.

Assuming that the range of rounds is normally distributed, find the mean and standard deviation of the range.
Estimate the number of rounds falling within 5 m of the centre of the target. (C)

17. A traffic survey is being undertaken on a main road to determine whether or not a pedestrian crossing should be installed. On five successive days, from Monday to Friday, the hour between 8 a.m. and 9 a.m. was split up into 30-second intervals, and the number of vehicles passing a certain point in each of these intervals was recorded.
The random variable $X$ represents the number of cars travelling *from* the town centre per 30-second interval. For the 600 observations the mean and variance were 3.1 and 3.27 respectively.

   (a) Explain why $X$ might be modelled by a Poisson distribution.
   (b) Using the sample mean as an estimate for the Poisson parameter, calculate the probability of recording exactly three vehicles travelling *from* the town centre in a 30-second interval.
   (c) Calculate the probability of recording at least six vehicles travelling from the town centre in a 60-second interval.

   The mean number of vehicles per 30-second interval passing the survey point travelling *towards* the town centre during the same survey period was 7.9.

   (d) Show that there is roughly a 12% chance that the *total* number of vehicles passing per 30-second interval is ten.
   (e) Using a suitable approximation, estimate the probability of between 16 and 24 vehicles (inclusive) passing the survey point in a 60-second interval. (MEI)

18. [In this question give three places of decimals in each answer.]
When a telephone call is made in the country of Japonica, the probability of getting the intended number is 0.95.

   (a) Ten independent calls are made. Find the probability of getting eight or more of the intended numbers. Find also the conditional probability of getting all ten intended numbers given that at least eight of the intended numbers are obtained.
   (b) Three hundred independent calls are made. Find the probability of failing to get the intended number on a least ten but not more than twenty of the calls.
   (c) Four hundred independent calls are made. For each call the probability of getting 'number unobtainable' is 0.004. Find the probability of getting 'number unobtainable' fewer than three times. (C)

19. An old car is never garaged at night. On the morning following a wet night, the probability that the car does not start is $\frac{1}{3}$.
On the morning following a dry night, this probability is $\frac{1}{25}$. The starting performance of the car each morning is independent of its performance on previous mornings.

(a) There are six consecutive wet nights. Determine the probability that the car does not start on at least two of the six mornings.

(b) During a wet autumn there are 32 wet nights. Using a suitable approximation, determine the probability that the car does not start on fewer than 16 of the 32 mornings.

(c) During a long summer drought there are 100 dry nights. Using a Poisson approximation, determine the probability that the car does not start on five or more of the 100 mornings.

(Give three decimal places in your answers.) (C)

20. The life, in years, of a randomly chosen Flashpan car battery is normally distributed with mean 2 and standard deviation 0.4.
Show that the probability that a randomly chosen Flashpan battery has a life less than one year is 0.006 21, correct to five places of decimals.

(a) A farmer buys two randomly chosen Flashpan batteries. Find the probability that the batteries each have a life more than one year.

(b) A wholesaler buys 500 randomly chosen Flashpan batteries. Using a suitable approximation, find the probability that at most three have lives each less than one year.

(c) A retailer buys ten randomly chosen Flashpan batteries. Find the probability that at least four have lives each exceeding two years. (C)

21. Describe, briefly, the conditions under which the binomial distribution Bin$(n, p)$ may be approximated by

(a) a normal distribution,
(b) a Poisson distribution,

giving the parameters of each of the approximate distributions.
Among the blood cells of a certain animal species, the proportion of cells which are of type A is 0.37 and the proportion of cells which are of type B is 0.004. Find, to three decimal places, the probability that in a random sample of eight blood cells at least two will be of type A.
Find, to three decimal places, an approximate value for the probability that

(c) in a random sample of 200 blood cells the combined number of type A and type B cells is 81 or more,

(d) there will be four or more cells of type B in a random sample of 300 blood cells. (L)

# Mixed test 7A

1. A smoker's blood nicotine level, measured in ng/ml, may be modelled by a normal random variable with mean 310 and standard deviation 110.

(a) What proportion of smokers have blood nicotine levels lower than 250?
(b) What blood nicotine level is exceeded by 20% of smokers? (AEB)

2. The number of hours of sunshine at a resort has been recorded for each month for many years. One year is selected at random and $H$ is the number of hours of sunshine in August of that year. $H$ can be modelled by a normal variable with mean 130.

(a) Given that $P(H < 179) = 0.975$, calculate the standard deviation of $H$.
(b) Calculate $P(100 < H < 150)$. (C)

3. In a large university 90% of the students are right-handed.

(a) Show that the probability that in a random sample of eight students exactly six will be right-handed is approximately 0.149.

(b) Find the probability that in a random sample of 20 students fewer than 15 will be right-handed.
(c) Determine, to two decimal places, an approximate value for the probability that in a random sample of 200 students at most 184 will be right-handed. (NEAB)

4. The random variable $X$ represents the weight, in grams, of chocolate chips in packets sold by a supermarket. It is suggested that $X$ can be modelled by a normal distribution with $X \sim N(100, 25)$.

(a) Find $P(X > 108)$.
(b) Show that $P(|X - 100| < 6.8) = 0.8262$.

Three packets are selected at random from the packets of chocolate chips on the supermarket shelf.

(c) Find the probability that exactly two of them will have weights in the range $|X - 100| < 6.8$.
(d) Comment on the suitability of the normal distribution as a model for $X$. (L)

# Mixed test 7B

1. The area that can be painted using one litre of Luxibrite paint is normally distributed with mean 13.2 m$^2$ and standard deviation 0.197 m$^2$. The corresponding figures for one of Maxigloss paint are 13.4 m$^2$ and 0.343 m$^2$. It is required to paint an area of 12.9 m$^2$. Find which paint gives the greater probability that one litre will be sufficient, and obtain this probability. (C)

2. Soup is sold in tins which are filled by a machine. The actual weight of soup delivered to a tin by the filling machine is always normally distributed about the mean weight with a standard deviation of 8 g. The machine is set originally to deliver a mean weight of 810 g.

   (a) Determine the probability that the weight of soup in a tin, selected at random, is less than 800 g.
   (b) Determine the probability that the weight of soup in a tin, selected at random, is between 795 g and 820 g.

   Proposed legislation requires that not more than 2.5% of tins may contain less than the nominal net weight of 800 g.

   (c) Assuming that the value of the standard deviation remains unchanged, determine the minimum mean weight that the machine should be set to deliver in order to comply with this requirement. (NEAB)

3. Consultants employed by a large library reported that the time spent in the library by a user could be modelled by a normal distribution with mean 65 minutes and standard deviation 20 minutes.

   (a) Assuming that this model is adequate, what is the probability that a user spends
   (i) less than 90 minutes in the library,
   (ii) between 60 and 90 minutes in the library?

   The library closes at 9.00 p.m.

   (b) Explain why the model above could not apply to a user who entered the library at 8.00 p.m.
   (c) Estimate an approximate latest time of entry for which the model above could still be plausible. (AEB)

4. Frugal Bakeries claim that packs of ten of their buns contain on average 75 raisins. A Poisson distribution is used to model the number of raisins in a randomly selected bun.

   (a) Specify the value of the parameter.
   (b) State any assumption required about the distribution of raisins in the production process for this model to be valid.
   (c) Show that the probability that a randomly selected bun contains more than eight raisins is 0.338.
   (d) Find the probability that in a pack of ten buns at least two buns contain more than eight raisins.
   (e) Using a suitable approximation, find the probability that in a pack of ten buns there are more than 80 raisins. (L)

5. An engineering firm sets an aptitude test when applicants first apply for training. The times taken to complete the test are normally distributed with mean 40.5 minutes and standard deviation 7.5 minutes. Applicants who complete the test in less than 30 minutes are immediately accepted for training. Those who take between 30 and 36 minutes are required to take a further test. All other applicants are rejected.

   (a) For a randomly chosen applicant calculate the probability of
   (i) immediate acceptance for training,
   (ii) requirement to take a further test.
   (b) Given that a randomly chosen applicant was not rejected after this first test, calculate, to three decimal places, the probability that the applicant was immediately accepted for training.
   (c) On a certain occasion there were 100 applicants. Use a suitable distributional approximation to calculate the probability that more than 25 applicants were required to take a further test. (NEAB)

# 8

# Linear combinations of normal variables

*In this chapter you will learn about the distributions for*

- the sum of independent normal variables
- the difference of independent normal variables
- multiples of independent normal variables

You will need the following results, first introduced on pages 256 and 257.

If $X$ and $Y$ are any two random variables, discrete or continuous, and $a$ and $b$ are any two constants,

**Sums**                                          **Differences**

$E(X + Y) = E(X) + E(Y$      ... ①     $E(X - Y) = E(X) - E(Y)$      ... ②

$E(aX + bY) = aE(X) + bE(Y)$      ... ③     $E(aX - bY) = aE(X) - bE(Y)$      ... ④

Also, if $X$ and $Y$ are independent, then

$\text{Var}(X + Y) = \text{Var}(X) + \text{Var}(Y)$      ... ⑤     $\text{Var}(X - Y) = \text{Var}(X) + \text{Var}(Y)$      ... ⑥

$\text{Var}(aX + bY) = a^2 \text{Var}(X) + b^2 \text{Var}(Y)$ ... ⑦     $\text{Var}(aX - bY) = a^2 \text{Var}(X) + b^2 \text{Var}(Y)$ ... ⑧

$$\uparrow$$

(Remember the + sign here)

## THE SUM OF INDEPENDENT NORMAL VARIABLES

Consider this example which involves the sum of independent normal variables.

### Example 8.1

A coffee machine is installed in a students' common room. It dispenses white coffee by first releasing a quantity of black coffee, normally distributed with mean 122.5 ml and standard deviation 7.5 ml, and then adding a quantity of milk, normally distributed with mean 30 ml and standard deviation 5 ml.

Each cup is marked to a level of 137.5 ml and if this level is not attained the customer receives the drink free of charge.

What percentage of cups of white coffee will be given free of charge?

## Solution 8.1

$B$ is the amount, in millilitres, of black coffee, where $B \sim N(122.5, 7.5^2)$.
$M$ is the amount, in millilitres, of milk, where $M \sim N(30, 5^2)$.
$B$ and $M$ are independent normal variables.

Consider $W$, the amount, in millilitres, of white coffee, made by combining the black coffee and milk, so $W = B + M$ and

$$E(W) = E(B) + E(M) = 122.5 + 30 = 152.5 \qquad \text{(using Result 1 above)}$$
$$\text{Var}(W) = \text{Var}(B) + \text{Var}(M) = 7.5^2 + 5^2 = 81.25 \qquad \text{(using Result 5 above)}$$

So $W = B + M$ has a mean of 152.5 and a variance of 81.25.

For **independent normal variables**, it is true that the **sum** of these variables is also **normally distributed**, so

$$B + M \sim N(152.5, 81.25)$$
i.e. $\qquad W \sim N(152.5, 81.25)$

The drink is free of charge if $W < 137.5$

$$
\begin{aligned}
P(W < 137.5) &= P\left(Z < \frac{137.5 - 152.5}{\sqrt{81.25}}\right) \\
&= P(Z < -1.664) \\
&= 1 - \Phi(1.664) \\
&= 1 - 0.9519 \\
&= 0.0481 \\
&= 4.81\%
\end{aligned}
$$

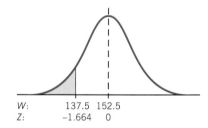

$W$: 137.5  152.5
$Z$: −1.664   0

**So approximately 5% of the cups of white coffee will be given free of charge.**

In general

If $X \sim N(\mu_1, \sigma_1^2)$ and $Y \sim N(\mu_2, \sigma_2^2)$
then $X + Y \sim N(\mu_1 + \mu_2, \sigma_1^2 + \sigma_2^2)$

This result can be extended to any set of independent normal variables $X_1, X_2, \dots, X_n$ where, with obvious notation

$$X_1 + X_2 + \dots + X_n \sim N(\mu_1 + \mu_2 + \dots + \mu_n, \sigma_1^2 + \sigma_2^2 + \dots + \sigma_n^2)$$

## Example 8.2

Four runners, Andy, Bob, Chris and Dai, train to take part in a 1600 m relay race in which Andy is to run 100 m, Bob 200 m, Chris 500 m and Dai 800 m.
During training their individual times, recorded in seconds, follow normal distributions. With obvious notation, these are:

$$A \sim N(10.8, 0.2^2), \ B \sim N(23.7, 0.3^2), \ C \sim N(62.8, 0.9^2) \text{ and } D \sim N(121.2, 2.1^2).$$

Find the probability that they run the relay race in less than 3 minutes 35 seconds.

## Solution 8.2

Let $T$ be the total time, in seconds, for the relay race.
Then $T = A + B + C + D$

$$E(T) = E(A) + E(B) + E(C) + E(D)$$
$$= 10.8 + 23.7 + 62.8 + 121.2$$
$$= 218.5$$

$$\mathrm{Var}(T) = \mathrm{Var}(A) + \mathrm{Var}(B) + \mathrm{Var}(C) + \mathrm{Var}(D)$$
$$= 0.2^2 + 0.3^2 + 0.9^2 + 2.1^2$$
$$= 5.35$$

$\therefore \quad T \sim N(218.5, 5.35)$

To find the probability that the total time is less than 3 minutes 35 seconds, i.e. 215 seconds,

find $P(T < 215) = P\left(Z < \dfrac{215 - 218.5}{\sqrt{5.35}}\right)$

$$= P(Z < -1.513)$$
$$= 1 - \Phi(1.513)$$
$$= 1 - 0.9349$$
$$= 0.0651$$

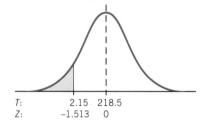

| $T$: | 2.15 | 218.5 |
| $Z$: | −1.513 | 0 |

**The probability that the runners take less than 3 minutes 35 seconds is 0.065 (2 s.f.).**

---

Consider now the special case when $X_1, X_2, \ldots, X_n$ are $n$ independent observations from the *same* normal distribution

so $X_1 \sim N(\mu, \sigma^2), \quad X_2 \sim N(\mu, \sigma^2), \quad \ldots, \quad X_n \sim N(\mu, \sigma^2)$

then $E(X_1 + X_2 + \cdots + X_n) = E(X_1) + E(X_2) + \cdots + E(X_n)$
$$= \mu + \mu + \cdots + \mu$$
$$= n\mu$$

$\mathrm{Var}(X_1 + X_2 + \cdots + X_n) = \mathrm{Var}(X_1) + \mathrm{Var}(X_2) + \cdots + \mathrm{Var}(X_n)$
$$= \sigma^2 + \sigma^2 + \cdots + \sigma^2$$
$$= n\sigma^2$$

So $\quad X_1 + X_2 + \cdots + X_n \sim N(n\mu, n\sigma^2)$

## Example 8.3

Masses of a particular article are normally distributed with mean 20 g and standard deviation 2 g. A random sample of 12 such articles is chosen. Find the probability that the total mass is greater than 230 g.

## Solution 8.3

$X$ is the mass, in grams, of an article.
$X \sim N(20, 2^2)$.

So $\quad X_1 \sim N(20, 4) \qquad E(X_1) = 20, \quad \mathrm{Var}(X_1) = 4$
$\qquad X_2 \sim N(20, 4) \qquad E(X_2) = 20, \quad \mathrm{Var}(X_2) = 4$
$\qquad \vdots \qquad\qquad\qquad\qquad \vdots$
$\qquad X_{12} \sim N(20, 4) \qquad E(X_{12}) = 20, \quad \mathrm{Var}(X_{12}) = 4$

Let $T = X_1 + X_2 + \cdots + X_{12}$
then $E(T) = E(X_1) + E(X_2) + \cdots + E(X_{12})$
$\qquad = 12E(X)$
$\qquad = 240$
$\quad \text{Var}(T) = \text{Var}(X_1) + \text{Var}(X_2) + \cdots + \text{Var}(X_{12})$
$\qquad\quad = 12\text{Var}(X)$
$\qquad\quad = 48$
So $T \sim N(240, 48)$.

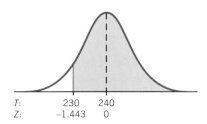

$P(T > 230) = P\left(Z > \dfrac{230 - 240}{\sqrt{48}}\right)$
$\qquad\qquad = P(Z > -1.443)$
$\qquad\qquad = \Phi(1.443)$
$\qquad\qquad = 0.9255$

T: 230  240
Z: −1.443  0

**The probability that the total mass is greater than 230 g is 0.93 (2 s.f.).**

## Example 8.4

The maximum load a lift can carry is 450 kg. The weights of men are normally distributed with mean 60 kg and standard deviation 10 kg. The weights of women are normally distributed with mean 55 kg and standard deviation 5 kg. Find the probability that the lift will be overloaded by five men and two women, if their weights are independent. (L)

## Solution 8.4

Let $M$ be the weight, in kilograms, of a man. Then $M \sim N(60, 10^2)$.
Let $W$ be the weight, in kilograms, of a woman. Then $W \sim N(55, 5^2)$.

The lift will be overloaded if

$M_1 + M_2 + M_3 + M_4 + M_5 + W_1 + W_2 > 450$.

Let $T = M_1 + M_2 + \cdots + M_5 + W_1 + W_2$
$E(T) = 5E(M) + 2E(W)$
$\qquad = 300 + 110$
$\qquad = 410$
$\text{Var}(T) = 5\,\text{Var}(M) + 2\,\text{Var}(W)$
$\qquad\quad = 500 + 50$
$\qquad\quad = 550$

Since $M$ and $W$ are normally distributed, $T$ is also normal.

So $T \sim N(410, 550)$.

$P(\text{lift is overloaded}) = P(T > 450)$
$\qquad\qquad\qquad\quad = P\left(Z > \dfrac{450 - 410}{\sqrt{550}}\right)$
$\qquad\qquad\qquad\quad = P(Z > 1.706)$
$\qquad\qquad\qquad\quad = 1 - \Phi(1.706)$
$\qquad\qquad\qquad\quad = 0.0441$

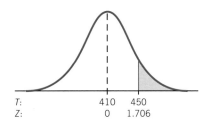

T: 410  450
Z: 0  1.706

**The probability that the lift will be overloaded by five men and two women is 0.044 (2 s.f.)**

## THE DIFFERENCE OF INDEPENDENT NORMAL VARIABLES

For two independent variables $X$ and $Y$, where $X \sim N(\mu_1, \sigma_1^2)$ and $Y \sim N(\mu_2, \sigma_2^2)$

$E(X - Y) = E(X) - E(Y) = \mu_1 - \mu_2$           Result 2, page 403

$\mathrm{Var}(X - Y) = \mathrm{Var}(X) + \mathrm{Var}(Y) = \sigma_1^2 + \sigma_2^2$       Result 6, page 403

$X - Y$ is *normally distributed*, so

$$X - Y \sim N(\mu_1 - \mu_2, \sigma_1^2 + \sigma_2^2)$$
$$\uparrow$$

(Remember the + sign here.)

## Example 8.5

A machine produces rubber balls whose diameters are normally distributed with mean 5.50 cm and standard deviation 0.08 cm.
The balls are packed in cylindrical tubes whose internal diameters are normally distributed with mean 5.70 cm and standard deviation 0.12 cm.
If a ball, selected at random, is placed in a tube, selected at random, what is the distribution of the clearance? (The clearance is the internal diameter of the tube minus the diameter of the ball.)
What is the probability that the clearance is between 0.05 cm and 0.25 cm?

## Solution 8.5

Let $B$ be the diameter, in centimetres, of a rubber ball. Then $B \sim N(5.50, 0.08^2)$
Let $T$ be the internal diameter, in centimetres, of a cylindrical tube. Then $T \sim N(5.70, 0.12^2)$

Let $C$ be the clearance, in centimetres, so $C = T - B$

$E(C) = E(T) - E(B) = 5.70 - 5.50 = 0.2$
$\mathrm{Var}(C) = \mathrm{Var}(T) + \mathrm{Var}(B) = 0.08^2 + 0.12^2 = 0.0208$

so $C \sim N(0.2, 0.0208)$

To find the probability that the clearance is between 0.05 cm and 0.25 cm, find

$$P(0.05 < C < 0.25) = P\left(\frac{0.05 - 0.2}{\sqrt{0.0208}} < Z < \frac{0.25 - 0.2}{\sqrt{0.0208}}\right)$$

$$= P(-1.040 < Z < 0.347)$$
$$= \Phi(1.040) + \Phi(0.347) - 1$$
$$= 0.8508 + 0.6357 - 1$$
$$= 0.4865$$

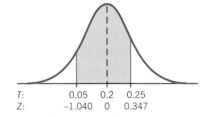

| $T$: | 0.05 | 0.2 | 0.25 |
|------|------|-----|------|
| $Z$: | −1.040 | 0 | 0.347 |

The probability that the clearance is between 0.05 cm and 0.25 cm is 0.49 (2 s.f.).

## Example 8.6

A certain liquid drug is marketed in bottles containing a nominal 20 ml of drug. Tests on a large number of bottles indicate that the volume of liquid in each bottle is distributed normally with mean 20.42 ml and standard deviation 0.429 ml.
If the capacity of the bottles is normally distributed with mean 21.77 ml and standard deviation 0.210 ml, estimate what percentage of bottles will overflow during filling.

## Solution 8.6

$X$ is the volume, in millilitres, of liquid and $X \sim N(20.42, 0.429^2)$.
$Y$ is the capacity, in millilitres, of a bottle and $Y \sim N(21.77, 0.210^2)$.

The bottle will overflow if the quantity of liquid is greater than the capacity of the bottle, i.e. if $X > Y$ so $X - Y > 0$

Let $D = X - Y$

$E(D) = E(X) - E(Y) = 20.42 - 21.77 = -1.35$

$\text{Var}(D) = \text{Var}(X) + \text{Var}(Y) = 0.429^2 + 0.210^2 = 0.2281$

$D \sim N(-1.35, 0.2281)$

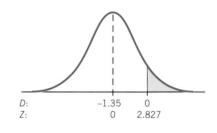

$$P(D > 0) = P\left(Z > \frac{0 - (-1.35)}{\sqrt{0.2281}}\right)$$
$$= P(Z > 2.827)$$
$$= 1 - \Phi(2.827)$$
$$= 1 - 0.9976$$
$$= 0.0024$$

$\therefore$   **0.24% of bottles will overflow during filling.**

## Example 8.7

In a cafeteria, baked beans are served either in ordinary portions or in children's portions. The quantity given for an ordinary portion is a normal variable with mean 90 g and standard deviation 3 g and the quantity given for a child's portion is a normal variable with mean 43 g and standard deviation 2 g.

What is the probability that Tom, who has two children's portions, is given more than his father, who has an ordinary portion?

## Solution 8.7

$C$ is the quantity, in grams, in a child's portion. Then $C \sim N(43, 4)$
$A$ is the quantity, in grams, in an ordinary portion. Then $A \sim N(90, 9)$

You need to find $P(C_1 + C_2 > A)$, i.e. $P(C_1 + C_2 - A > 0)$

Let $W = C_1 + C_2 - A$

$E(W) = E(C_1) + E(C_2) - E(A)$
$\quad\quad = 2E(C) - E(A)$
$\quad\quad = 86 - 90$
$\quad\quad = -4$

$\text{Var}(W) = \text{Var}(C_1) + \text{Var}(C_2) + \text{Var}(A)$
$\quad\quad\quad = 2\,\text{Var}(C) + \text{Var}(A)$
$\quad\quad\quad = 8 + 9$
$\quad\quad\quad = 17$

So $W \sim N(-4, 17)$

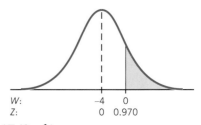

$$P(W > 0) = P\left(Z > \frac{0 - (-4)}{\sqrt{17}}\right)$$
$$= P(Z > 0.970)$$
$$= 1 - \Phi(0.970)$$
$$= 0.166$$

**The probability that Tom is given more than his father is 0.17 (2 s.f.).**

# Exercise 8a   Sums and differences of normal variables

1. $X$ and $Y$ are independent normal variables with $X \sim N(100, 49)$ and $Y \sim N(110, 576)$.

   (a) Find the mean and the standard deviation of the distribution $X + Y$.
   (b) Describe the distribution of $X + Y$.
   (c) Find $P(X + Y > 200)$.
   (d) Find $P(180 < X + Y < 240)$.

2. Each weekday Mr Harper goes to the local library to read the newspapers. The time he spends travelling is a normal variable with mean 15 minutes and standard deviation 2 minutes. The time he spends in the library is normally distributed with mean 25 minutes and standard deviation 4 minutes.
   Find the probability that, on a particular day, Mr Harper

   (a) is away from the house for more than 45 minutes,
   (b) spends more time travelling than in the library.

3. Bolts are manufactured which are to fit in holes in steel plates.
   The diameter of the bolts is normally distributed with mean 2.60 cm and standard deviation 0.03 cm. The diameter of the holes is normally distributed with mean of 2.71 cm and standard deviation of 0.04 cm.

   (a) Verify that, if a bolt and a hole are selected at random, the probability that the bolt is too large to enter the hole is 0.0139.
   (b) The random selection of a bolt and hole is carried out five times. Find the probability that in every case the bolt will be able to enter the hole. (C)

4. The mass of a particular article follows a normal distribution with mean 20 g and variance 4 g². A random sample of 12 items is tested. Find the probability that the total mass is less than 230 g.

5. Fiona, Carly, Jenny and Vicky swim in the $4 \times 100$ m freestyle relay team, with each one swimming 100 m. The times in seconds taken by each of the girls to swim 100 m are independent normal variables, distributed as follows:

   $F \sim N(52.5, 0.3^2)$,   $C \sim N(52.0, 0.6^2)$,
   $J \sim N(53.5, 1.2^2)$,   $V \sim N(51.5, 0.6^2)$.

   Calculate the probability that in a particular race,

   (a) Fiona will swim her leg in less than 52.5 seconds,
   (b) the relay team will take longer than 3 minutes 31.3 seconds to swim the race,
   (c) Carly will swim her leg faster than Vicky.

6. The mass, in grams, of a Chocolate Delight cake is normally distributed with mean 20 g and standard deviation 2 g. The cakes are sold in packets of six and the mass of the packing material is normally distributed with a mean of 30 g and a standard deviation of 4 g.

   (a) Find the probability that the mass of six cakes is less than 110 g.
   (b) Find the probability that the total mass of a packet containing six cakes is
   (i) more than 162 g,
   (ii) less that 137 g,
   (iii) between 140 g and 153 g.

7. In a certain village, the heights of the women are normally distributed with a mean of 164 cm and a standard deviation of 5 cm. The heights of the men are normally distributed with a mean of 173 cm and a standard deviation of 6 cm.
   A man and a woman are picked at random from the people in the village.
   Find the probability that

   (a) the woman is taller than the man,
   (b) the man is more than 5 cm taller than the woman.

8. The mass of a certain grade of apple is normally distributed with mean mass 120 g and standard deviation 10 g.

   (a) An apple of this grade is selected at random. Find the probability that its mass lies between 100.5 g and 124 g.
   (b) Four apples of this grade are selected at random. Find the probability that their total mass exceeds 505 g.

9. Rods are produced in two lengths, called 'short' and 'long'.
   $S$ is the length, in centimetres, of a short rod, where $S \sim N(5, 0.25)$.
   $L$ is the length, in centimetres, of a long rod, where $L \sim N(10, 1)$.
   Rods are joined to give longer lengths. Find the probability that a length consisting of

   (a) two short rods and four long rods is longer than 52 cm,
   (b) three short rods and two long rods is between 33 cm and 36 cm long,
   (c) six short rods is longer than a length consisting of three long rods.

10. Mr Smith has five dogs, two of which are male and three are female. The masses of food they eat in any given week are normally distributed as follows:

| | Mean (kg) | Standard deviation (kg) |
|---|---|---|
| Male | 3.5 | '0.4 |
| Female | 2.5 | 0.3 |

Find the probability that the two males eat more than the three females in a particular week.

11. The time taken to carry out a standard service on a car of type A is known, to a good approximation, to be a normal variable with mean 1 hour and standard deviation 10 minutes. Assuming that only one car is serviced at a time, find the probability that it will take more than 6.5 hours to service six cars.
The time taken to carry out a standard service on a car of type B is a normal variable with mean 1.5 hours and standard deviation 15 minutes. Find the probability that five cars of type B can be serviced more quickly than eight cars of type A. (C)

12. The process of painting the body-work of a mass-produced lorry consists of giving it one coat of paint A, three coats of paint B and two coats of paint C. A record of the quantity of each type of paint used for each coat is kept for each lorry produced over a long period. The following table gives the means and standard deviations of these quantities measured in litres:

| | Mean | Standard deviation |
|---|---|---|
| The coat of paint A | 3.7 | 0.42 |
| Each coat of paint B | 1.3 | 0.15 |
| Each coat of paint C | 1.0 | 0.12 |

Assuming independence of the distribution for each coat, calculate the mean and standard deviation for the total quantity of paint used on each lorry.
Assuming that the quantities of paint used for each coat are normally distributed, calculate

(a) the percentage of lorries receiving less than 8.5 litres of paint,
(b) the percentage of lorries receiving more than 10.0 litres of paint. (C)

13. The values of two types of resistors are normally distributed as follows:

Type A: mean: 100 ohms; standard deviation: 2 ohms
Type B: mean: 50 ohms; standard deviation: 1.3 ohms

(a) What tolerances would be permitted for type A if only 0.5% were rejected?
(b) 300-ohm resistors are made by connecting together three of the type A resistors, drawn from the total production. What percentage of the 300-ohm resistors may be expected to have resistances greater than 295 ohms?
(c) Pairs of resistors, one of 100 ohms and one of 50 ohms, drawn from the total production for types A and B respectively, are connected together to make 150-ohm resistors. What percentage of the resulting resistors may be expected to have resistances in the range 150 ohms to 151.4 ohms? (AEB)

14. The time of departure of my train from Temple Meads Station is distributed normally about the scheduled time of 08:25 with a standard deviation of 1 minute. I arrive at Temple Meads Station on another train whose time of arrival is normally distributed about the scheduled time of 08:20 with standard deviation of 1 minute. It takes me three minutes to change platforms. If I miss the train from Temple Meads, I am late for work.

(a) Find the probability that I am late for work.
(b) Find the probability that I miss the train from Temple Meads Station every day from Monday to Friday in a given week.

# MULTIPLES OF INDEPENDENT NORMAL VARIABLES

Remember that, for any constant $a$,

$E(aX) = aE(X)$ (page 246) and $\text{Var}(aX) = a^2 \text{Var}(X)$ (page 250)

If $X$ is a *normal* variable such that $X \sim N(\mu, \sigma^2)$

then $E(aX) = aE(X) = a\mu$
$\text{Var}(aX) = a^2 \text{Var}(X) = a^2\sigma^2$

It can be shown that $aX$ is also normally distributed

so $aX \sim N(a\mu, a^2\sigma^2)$

Now consider two independent normal variables $X$ and $Y$ where $X \sim N(\mu_1, \sigma_1^2)$, $Y \sim N(\mu_2, \sigma_2^2)$

For any constants $a$, $b$, using the results on page 403

$$E(aX + bY) = aE(X) + bE(Y) = a\mu_1 + b\mu_2 \quad \longleftarrow \text{Result ③}$$
$$E(aX - bY) = aE(X) - bE(Y) = a\mu_1 - b\mu_2 \quad \longleftarrow \text{Result ④}$$

$$\text{Var}(aX + bY) = a^2\, \text{Var}(X) + b^2\, \text{Var}(Y) = a^2\sigma_1^2 + b^2\sigma_2^2 \quad \longleftarrow \text{Result ⑦}$$
$$\text{Var}(aX - bY) = a^2\, \text{Var}(X) + b^2\, \text{Var}(Y) = a^2\sigma_1^2 + b^2\sigma_2^2 \quad \longleftarrow \text{Result ⑧}$$

↑<br>(Remember the + sign here.)

$aX + bY$ and $aX - bY$ are also **normally distributed**, so

$$aX + bY \sim N(a\mu_1 + b\mu_2,\ a^2\sigma_1^2 + b^2\sigma_2^2)$$
$$aX - bY \sim N(a\mu_1 - b\mu_2,\ a^2\sigma_1^2 + b^2\sigma_2^2)$$

## Example 8.8

$X$ and $Y$ are independent random variables and $X \sim N(100, 8)$, $Y \sim N(55, 10)$. Find the probability that an observation from the population of $X$ is more than twice the value of an observation from the population of $Y$.

## Solution 8.8

You need to find $P(X > 2Y)$, i.e. $P(X - 2Y > 0)$.
Let $D = X - 2Y$
$E(D) = E(X) - 2E(Y) = 100 - 110 = -10$
$\text{Var}(D) = \text{Var}(X) + 2^2\, \text{Var}(Y) = 8 + 4 \times 10 = 48$

So $D \sim N(-10, 48)$

$$P(D > 0) = P\left(Z > \frac{0 - (-10)}{\sqrt{48}}\right)$$
$$= P(Z > 1.443)$$
$$= 1 - \Phi(1.443)$$
$$= 1 - 0.9255$$
$$= 0.0745$$

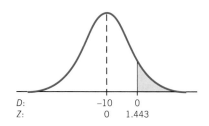

D:     −10   0
Z:      0   1.443

**The probability that an observation from the population of $X$ is more than twice the value of an observation from the population of $Y$ is 0.075 (2 s.f.).**

---

Great care must be taken in distinguishing between a sum of random variables and a multiple of a random variable.
For example, if $X$ is the weight of a small loaf, then the **sum** $X_1 + X_2 + X_3$ is the **total weight of three loaves**.

If $X \sim N(\mu, \sigma^2)$ then $X_1 + X_2 + X_3 \sim N(3\mu, 3\sigma^2)$.

But if there is a large economy-size loaf which is *three times the weight of a small loaf*, then the weight of an economy loaf is $3X$ (a **multiple**)

and $3X \sim N(3\mu, 9\sigma^2)$.

In general, for $X \sim N(\mu, \sigma^2)$

Sum: $\qquad X_1 + X_2 + \cdots + X_n \sim N(n\mu, n\sigma^2)$
Multiple: $\qquad\qquad\qquad nX \sim N(n\mu, n^2\sigma^2)$

Notice that the means are the same but the variances are not.
The distribution for the multiple is more spread out.

Look carefully at the following example.

## Example 8.9

A soft drinks manufacturer sells bottles of drinks in two sizes. The amount in each bottle, in

| | Mean (ml) | Variance (ml²) |
|---|---|---|
| Small | 252 | 4 |
| Large | 1012 | 25 |

millilitres, is normally distributed as shown in the table:
(a) A bottle of each size is selected at random. Find the probability that the large bottle contains less than four times the amount in the small bottle.
(b) One large and four small bottles are selected at random. Find the probability that the amount in the large bottle is less than the total amount in the four small bottles.

## Solution 8.9

Let $S$ be the amount, in millilitres, in a small bottle. Then $S \sim N(252, 4)$.
Let $L$ be the amount, in millilitres, in a large bottle. Then $L \sim N(1012, 25)$.

(a) To find the probability that the large bottle contains less than four times the amount in a small bottle, you need $P(L < 4S)$
   i.e. $P(L - 4S < 0)$.
   Now $\quad E(L - 4S) = E(L) - E(4S) \qquad$ (Multiple of $S$)
   $\qquad\qquad\qquad = E(L) - 4E(S)$
   $\qquad\qquad\qquad = 1012 - 1008$
   $\qquad\qquad\qquad = 4$

   $\qquad \mathrm{Var}(L - 4S) = \mathrm{Var}(L) + \mathrm{Var}(4S) \qquad$ (Remember the + sign.)
   $\qquad\qquad\qquad = \mathrm{Var}(L) + 16\,\mathrm{Var}(S)$
   $\qquad\qquad\qquad = 25 + 64$
   $\qquad\qquad\qquad = 89$

So $\quad L - 4S \sim N(4, 89)$

$P(L - 4S < 0) = P\left(X < \dfrac{0 - 4}{\sqrt{89}}\right)$
$\qquad\qquad\qquad = P(Z < -0.424)$
$\qquad\qquad\qquad = 1 - \Phi(0.424)$
$\qquad\qquad\qquad = 1 - 0.6642$
$\qquad\qquad\qquad = 0.3358$

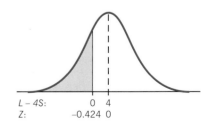

$L - 4S$: $\qquad$ 0 $\quad$ 4
$Z$: $\qquad$ −0.424 0

The probability that a large bottle contains less than four times the amount in a small bottle is 0.34 (2 s.f.).

(b) To find the probability that the amount in a large bottle is less than the total amount in four small bottles you need $P(L < S_1 + S_2 + S_3 + S_4) = P(L - (S_1 + S_2 + S_3 + S_4) < 0)$

$$E(L - (S_1 + \cdots + S_4)) = E(L) - E(S_1 + \cdots + S_4) \qquad \text{sum of normal variables}$$
$$= E(L) - 4E(S)$$
$$= 1012 - 1008$$
$$= 4$$
$$\mathrm{Var}(L - (S_1 + \cdots + S_4)) = \mathrm{Var}(L) + \mathrm{Var}(S_1 + \cdots + S_4) \qquad \text{Remember the + sign}$$
$$= \mathrm{Var}(L) + 4\,\mathrm{Var}(S)$$
$$= 25 + 16$$
$$= 41$$

Therefore $L - (S_1 + \cdots + S_4) \sim N(4, 41)$

$$P(L - (S_1 + \cdots + S_4) < 0) = P\left(Z < \frac{0 - 4}{\sqrt{41}}\right)$$
$$= P(Z < -0.625)$$
$$= 1 - \Phi(0.625)$$
$$= 0.266$$

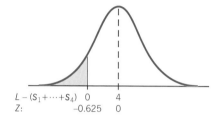

**The probability that a large bottle contains less than four small bottles is 0.27 (2 s.f.).**

It is very important to distinguish between

the multiple of $S$ in part (a) and
the sum of $S_1, S_2, S_3, S_4$ in part (b).

Note that $\left.\begin{array}{l} E(L - 4S) = 4 \\ E(L - (S_1 + S_2 + S_3 + S_4)) = 4 \end{array}\right\}$ The means are the same.

$\left.\begin{array}{l} \mathrm{Var}(L - 4S) = 89 \\ \mathrm{Var}(L - (S_1 + S_2 + S_3 + S_4)) = 41 \end{array}\right\}$ The variances are different.

## Exercise 8b  Multiples of normal variables

1. $X$ and $Y$ are independent normal variables such that $X \sim N(40, 12)$ and $Y \sim N(60, 15)$. Find

   (a) $P(2X + Y > 130)$
   (b) $P(3X - 2Y < 20)$

2. The time taken by Simon to do his Mathematics homework can be modelled by a normal distribution with mean 50 minutes and standard deviation 10 minutes. The time taken by Belinda is $N(30, 25)$.

   (a) Find the probability that, for a particular homework, Simon takes more than twice as long as Belinda.
   (b) Find the probability that Belinda spends less time in total on Monday's homework and Thursday's homework than Simon spends on Monday's homework.

3. The thickness, $P$ cm, of a randomly chosen paperback book may be regarded as an observation from a normal distribution with mean 2.0 and variance 0.730.
   The thickness, $H$ cm, of a randomly chosen hardback book may be regarded as an observation from a normal distribution with mean 4.9 and variance 1.920.

   (a) Determine the probability that the combined thickness of four randomly chosen paperbacks is greater than the combined thickness of two randomly chosen hardbacks.
   (b) By considering $X = 2P - H$, or otherwise, determine the probability that a randomly chosen paperback is less than half as thick as a randomly chosen hardback.
   (c) Determine the probability that a randomly chosen collection of 16 paperbacks and 8 hardbacks will have a combined thickness of less than 70 cm.

   (Give three decimal places in your answers.)  (C)

4. The random variable $X$ is distributed normally with mean $\mu$ and variance 6, and the random variable $Y$ is normally distributed with mean 8 and variance $\sigma^2$.

   $2X - 3Y$ is distributed normally with mean $-12$ and variance 42. Find

   (a)  the value of $\mu$ and the value of $\sigma$,
   (b)  $P(X > 8)$,
   (c)  $P(Y < 9)$,
   (d)  $P(-4 < 3X - 2Y < 7)$.

5. A single observation is taken from each of the distributions
   $A \sim N(82, 1.5^2)$, $B \sim N(42, 0.3^2)$ and
   $C \sim N(85, 0.7^2)$
   Find the probability that the mean of these observations, $\frac{1}{3}(A + B + C)$, is greater than 70.

6. Next May, an ornithologist intends to trap one male cuckoo and one female cuckoo. The mass $M$ of the male cuckoo may be regarded as being a normal random variable with mean 116 g and standard deviation 16 g. The mass $F$ of the female cuckoo may be regarded as being independent of $M$ and as being a normal random variable with mean 106 g and standard deviation 12 g. Determine

   (a)  the probability that the mass of the two birds together will be more than 230 g,
   (b)  the probability that the mass of the male will be more than the mass of the female.

   By considering $X = 9M - 16F$, or otherwise, determine the probability that the mass of the female will be less than nine-sixteenths of that of the male.
   Suppose that one of the two trapped birds escapes. Assuming that the remaining bird will be equally likely to be the male or the female, determine the probability that its mass will be more than 118 g.                                    (C)

## Summary

- For two independent normal variables such that
  $X \sim N(\mu_1, \sigma_1^2)$ and $Y \sim N(\mu_2, \sigma_2^2)$
  $$X + Y \sim N(\mu_1 + \mu_2, \sigma_1^2 + \sigma_2^2)$$
  $$X - Y \sim N(\mu_1 - \mu_2, \sigma_1^2 + \sigma_2^2)$$

- For $n$ independent normal variables such that $X_i \sim N(\mu_i, \sigma_i^2)$
  $$X_1 + X_2 + \cdots + X_n \sim N(\mu_1 + \mu_2 + \cdots + \mu_n, \sigma_1^2 + \sigma_2^2 + \cdots + \sigma_n^2)$$

- For $n$ independent observations of the random variable $X$ where $X \sim N(\mu, \sigma^2)$,
  $$X_1 + X_2 + \cdots + X_n \sim N(n\mu, n\sigma^2)$$

- For the normal variable such that $X \sim N(\mu, \sigma^2)$, and for any constant $a$
  $$aX \sim N(a\mu, a^2\sigma^2)$$

- For two independent normal variables such that
  $X \sim N(\mu_1, \sigma_1^2)$ and $Y \sim N(\mu_2, \sigma_2^2)$ and for any constants $a$ and $b$
  $$aX + bY \sim N(a\mu_1 + b\mu_2, a^2\sigma_1^2 + b^2\sigma_2^2)$$
  $$aX - bY \sim N(a\mu_1 - b\mu_2, a^2\sigma_1^2 + b^2\sigma_2^2)$$

## Miscellaneous worked examples

### Example 8.10

The distribution of the masses of adult husky dogs may be modelled by the normal distribution with mean 37 kg and standard deviation 5 kg.

(a) Calculate the probability that an adult husky has a mass greater than 30 kg.
(b) Calculate the probability that a randomly chosen team of six huskies has a total mass lying between 198 kg and 240 kg, giving your answer to three decimal places.     (NEAB)

### Solution 8.10

$H$ is the mass, in kilograms, of a husky dog. Then $H \sim N(37, 5^2)$.

(a) $P(H > 30) = P\left(Z > \dfrac{30 - 37}{5}\right)$

$\qquad = P(Z > -1.4)$
$\qquad = \Phi(1.4)$
$\qquad = 0.9192$

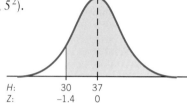

The probability that an adult husky dog has a mass greater than 30 kg is 0.919 (3 d.p.).

(b) Let $T = H_1 + H_2 + \cdots + H_6$
$E(T) = 6E(H) = 6 \times 37 = 222$ and $\text{Var}(T) = 6\,\text{Var}(H) = 6 \times 25 = 150$
$\therefore \quad T \sim N(222, 150)$

$P(198 < T < 240) = P\left(\dfrac{198 - 222}{\sqrt{150}} < Z < \dfrac{240 - 222}{\sqrt{150}}\right)$

$\qquad = P(-1.960 < Z < 1.470)$
$\qquad = \Phi(1.960) + \Phi(1.470) - 1$
$\qquad = 0.9750 + 0.9292 - 1$
$\qquad = 0.9042$

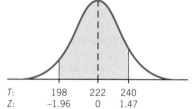

The probability that six huskies have a total mass lying between 198 g and 240 kg is 0.904 (3 d.p.).

## Example 8.11

The lifetimes of Econ light bulbs are normally distributed with mean 1000 h and standard deviation 25 h.

(a) Find, to three decimal places, the probability that an Econ light bulb will have a lifetime between 975 h and 1020 h.

(b) Calculate, to three decimal places, the probability that the sum of the lifetimes of eight Econ light bulbs will exceed 7930 h. Indicate clearly the stage in your calculation when an assumption concerning independence is essential.

The lifetimes of Enersaver light bulbs are normally distributed with mean 7900 h and standard deviation 50 h.

(c) Calculate, to three decimal places, the probability that an Enersaver light bulb will last at least eight times as long as an Econ light bulb. (*NEAB*)

## Solution 8.11

$X$ is the lifetime, in hours, of an Econ light bulb. Then $X \sim N(1000, 25^2)$.

(a) $P(975 < X < 1020)$

$$= P\left(\frac{975 - 1000}{25} < Z < \frac{1020 - 1000}{25}\right)$$
$$= P(-1 < Z < 0.8)$$
$$= \Phi(1) + \Phi(0.8) - 1$$
$$= 0.8413 + 0.7881 - 1$$
$$= 0.6294$$

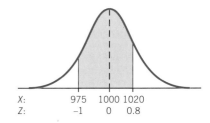

X: 975 1000 1020
Z: −1 0 0.8

**The probability that an Econ light bulb has a lifetime between 975 h and 1020 h is 0.629 (3 d.p.).**

(b) $S$ is the sum of the lifetimes of eight Econ light bulbs, so $S = X_1 + X_2 + \cdots + X_8$
$E(S) = 8E(X) = 8000$
$Var(S) = 8\,Var(X) = 8 \times 25^2 = 5000$ (assuming the lifetimes are independent)
$\therefore \quad S \sim N(8000, 5000)$

$$P(S > 7930) = P\left(Z > \frac{7930 - 8000}{\sqrt{5000}}\right)$$
$$= P(Z > -0.990)$$
$$= \Phi(0.990)$$
$$= 0.8389$$

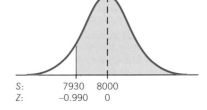

S: 7930 8000
Z: −0.990 0

**The probability that the sum of the lifetimes of eight Econ light bulbs exceeds 7930 h is 0.839 (3 d.p.).**

(c) $Y$ is the lifetime of an Enersaver light bulb and $Y \sim N(7900, 50^2)$.

$P(Y \geqslant 8X)$ is needed, i.e. $P(Y - 8X \geqslant 0)$.
$E(Y - 8X) = E(Y) - 8E(X) = 7900 - 8000 = -100$
$Var(Y - 8X) = Var(Y) + 8^2 Var(X) = 50^2 + 64 \times 25^2 = 42\ 500$ (assuming independence).
$Y - 8X \sim N(-100, 42\ 500)$

$$P(Y - 8X \geqslant 0) = P\left(Z \geqslant \frac{0 - (-100)}{\sqrt{42\ 500}}\right)$$

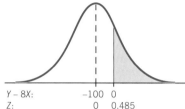

$$= P(Z > 0.485)$$
$$= 1 - \Phi(0.485)$$
$$= 1 - 0.6862$$
$$= 0.3138$$

**The probability that an Enersaver light bulb lasts at least eight times as long as an Econ light bulb is 0.314 (3 d.p.).**

# Miscellaneous exercise 8c

1. The weights of grade $A$ oranges are normally distributed with mean 200 g and standard deviation 12 g. Determine, correct to two significant figures, the probability that

   (a) a grade $A$ orange weighs more than 190 g but less than 210 g,
   (b) a sample of 4 grade $A$ oranges weighs more than 820 g.

   The weights of grade $B$ oranges are normally distributed with mean 175 g and standard deviation 9 g. Determine, correct to two significant figures, the probability that

   (c) a grade $B$ orange weighs less than a grade $A$ orange,
   (d) a sample of 8 grade $B$ oranges weighs more than a sample of seven grade $A$ oranges.  (C)

2. Prints from two types of film $C$ and $D$ have developing times which can be modelled by normal variables, $C$ with mean 16.18 s and standard deviation 0.11 s and $D$ with mean 15.88 s and standard deviation 0.10 s.

   (a) What is the probability that a type $C$ print will take less than 16 s to develop?
   (b) A type $C$ print is developed and immediately afterwards a type $D$ print is developed. What is the probability that the total time is greater than 32.5 s?
   (c) What is the probability of a type $C$ print taking longer to develop than a type $D$ print?

3. In testing the length of life of electric light bulbs of a particular type, it is found that 12.3% of the bulbs tested fail within 800 hours and that 28.1% are still operating 1100 hours after the start of the test.
   Assuming that the distribution of the length of life is normal, calculate, to the nearest hour in each case, the mean, $\mu$, and the standard deviation, $\sigma$, of the distribution.
   A light fitting takes a single bulb of this type. A packet of three bulbs is bought, to be used one after the other in this fitting. State the mean and variance of the total life of the 3 bulbs in the packet in terms of $\mu$ and $\sigma$ and calculate, to two decimal places, the probability that the total life is more than 3300 hours.
   Calculate the probability that all 3 bulbs have lives in excess of 1100 hours, so that again the total life is more than 3300 hours. Explain why this answer should be different from the previous one.  (NEAB)

4. The weight of a large loaf of bread is a normal variable with mean 420 g and standard deviation 30 g. The weight of a small loaf of bread is a normal variable with mean 220 g and standard deviation 10 g.

   (a) Find the probability that 5 large loaves weigh more than 10 small loaves.
   (b) Find the probability that the total weight of 5 large loaves and 10 small loaves lies between 4.25 kg and 4.4 kg.  (C)

5. The tensile strengths, measured in newtons (N), of a large number of ropes of equal length are independently and normally distributed such that 5% are under 706 N and 5% over 1294 N. Four such ropes are randomly selected and joined end-to-end to form a single rope; the strength of the combined rope is equal to the strength of the weakest of the 4 selected ropes. Derive the probabilities that this combined rope will not break under tensions of 1000 N and 900 N, respectively.
A further 4 ropes are randomly selected and attached between two rings, the strength of the arrangement being the sum of the strengths of the 4 separate ropes. Derive the probabilities that this arrangement will break under tensions of 4000 N and 4200 N, respectively. (NEAB)

6. $X$ and $Y$ are independent normally distributed random variables such that $X$ has mean 32 and variance 25, and $Y$ has mean 43 and variance 96. Find

(a) $P(X > 43)$,
(b) $P(X - Y > 0)$,
(c) $P(2X - Y > 0)$. (NEAB)

7. The times taken by two runners $A$ and $B$ to run 400 m races are independent and normally distributed with means 45.0 s and 45.2 s, and standard deviations 0.5 s and 0.8 s respectively. The two runners are to complete in a 400 m race for which there is a track record of 44.5 s.

(a) Calculate, to three decimal places, the probability of runner $A$ breaking the track record.
(b) Show that the probability of runner $B$ breaking the track record is greater than that of runner $A$.
(c) Calculate, to three decimal places, the probability of runner $A$ beating runner $B$. (NEAB)

8. In a packaging factory, the empty containers for a certain product have a mean weight of 400 g with a standard deviation of 10 g. The mean weight of the contents of a full container is 800 g with a standard deviation of 15 g. Find the expected total weight of 10 full containers and the standard deviation of this weight, assuming that the weights of containers and contents are independent.
Assuming further that these weights are normally distributed random variables, find the proportion of batches of 10 full containers which weigh more than 12.1 kg. (O&C)

9. Jam is packed into tins of advertised weight 1 kg. The weight of a randomly selected tin of jam is normally distributed about a target weight with a standard deviation of 12 g.

(a) If the target weight is 1 kg, find the probability that a randomly chosen tin weighs
(i) less than 985 g,
(ii) between 970 g and 1015 g.
(b) If not more than one tin in 100 is to weigh less than the advertised weight, find the minimum target weight required to meet this condition.
(c) The target weight is fixed at 1 kg. The resulting tins are packed in boxes of six and the weight of the box is normally distributed with mean weight 250 g and standard deviation 10 g. Find the probability that a randomly chosen box of 6 tins will weigh less than 6.2 kg. (L)

10 (a) The lifetime in hours of an electrical component has a normal distribution with mean 150 hours and standard deviation 8 hours.
Find the probability that
(i) a new component lasts at least 160 hours,
(ii) a component which has already operated for 145 hours will last at least another 15 hours.
(b) The weight of these components is normally distributed with mean 250 g and standard deviation 10 g. Each component is in its own box, the weight of which is also normally distributed with mean 50 g and standard deviation 5 g. There are 10 boxed components to a carton and the weight of the carton is normally distributed with mean 75 g and standard deviation 7 g. Find the probability that a carton of 10 boxed components weighs less than 3 kg. (L)

11. Jim Longlegs is an athlete whose specialist event is the triple jump. This is made up of a *hop*, a *step* and a *jump*. Over a season the lengths of the *hop*, *step* and *jump* sections, denoted by $H$, $S$ and $J$ respectively, are measured, from which the following models are proposed:

$H \sim N(5.5, 0.5^2)$, $S \sim N(5.1, 0.6^2)$, $J \sim N(6.2, 0.8^2)$

where all distances are in metres. Assume that $H$, $S$ and $J$ are independent.

(a) In what proportion of his triple jumps will Jim's total distance exceed 18 m?
(b) In 6 successive independent attempts, what is the probability that at least one total distance will exceed 18 m?
(c) What total distance will Jim exceed 95% of the time?
(d) Find the probability that, in Jim's next triple jump, his step will be greater than his hop. (MEI)

12. [In this question give three places of decimals in each answer.]

The mass of tea in 'Supacuppa' teabags has a normal distribution with mean 4.1 g and standard deviation 0.12 g. The mass of tea in 'Bumpacuppa' teabags has a normal distribution with mean 5.2 g and standard deviation 0.15 g.

(a) Find the probability that a randomly chosen Supacuppa teabag contains more than 4.0 g of tea.

(b) Find the probability that, of 2 randomly chosen Supacuppa teabags, one contains more than 4.0 g of tea and one contains less than 4.0 g of tea.

(c) Find the probability that 5 randomly chosen Supacuppa teabags contain a total of more than 20.8 g of tea.

(d) Find the probability that the total mass of tea in 5 randomly chosen Supacuppa teabags is more than the total mass of tea in 4 randomly chosen Bumpacuppa teabags.

(C)

13. A small bank has two cashiers dealing with customers wanting to withdraw or deposit cash. For each cashier, the time taken to deal with a customer is a random variable having a normal distribution with mean 150 s and standard deviation 45 s.

(a) Find the probability that the time taken for a randomly chosen customer to be dealt with by a cashier is more than 180 s.

(b) One of the cashiers deals with two customers, one straight after the other. Assuming that the times for the customers are independent of each other, find the probability that the total time taken by the cashier is less than 200 s.

(c) At a certain time, one cashier has a queue of 4 customers and the other cashier has a queue of 3 customers, and the cashiers begin to deal with the customers at the front of their queues. Assuming that the cashiers work independently, find the probability that the 4 customers in the first queue will all be dealt with before the 3 customers in the second queue are all dealt with.

(C)

# Mixed test 8A

1. A country baker makes biscuits whose masses are normally distributed with mean 30 g and standard deviation 2.3 g. She packs them by hand into either a small carton (containing 20 biscuits) or a large carton (containing 30 biscuits).

(a) State the distribution of the total mass, $S$, of biscuits in a small carton and find the probability that $S$ is greater than 615 g.

(b) Six small and four large cartons are placed in a box. Find the probability that the total mass of biscuits in the 10 cartons lies between 7150 g and 7250 g.

(c) Find the probability that 3 small cartons contain at least 25 g more than 2 large ones.

The label on a large carton of biscuits reads 'Net mass 900 g'. A trading standards officer insists that 90% of such cartons should contain biscuits with a total mass of at least 900 g.

(d) Assuming the standard deviation remains unchanged, find the least value of the mean mass of a biscuit consistent with this requirement.

(MEI)

2. Foster's Fancy Cakes are sold in packets of six. The mass of each cake is a normally distributed random variable having mean 25 g and standard deviation 0.4 g. The mass of the packaging is a normally distributed random variable having mean 20 g and standard deviation 1 g. Find, to three decimal places, the probabilities that

(a) the mass of a randomly chosen cake is between 24.7 g and 25.7 g,

(b) the total mass of a randomly chosen packet is less than 173 g.

State one assumption that you have made in answering (b).

(NEAB)

3. Monto sherry is sold in bottles of two sizes: standard and large. For each size, the content, in litres, of a randomly chosen bottle is normally distributed with mean and standard deviation as given in the table.

|  | Mean | Standard deviation |
|---|---|---|
| Standard bottle | 0.760 | 0.008 |
| Large bottle | 1.010 | 0.009 |

(a) Show that the probability that a randomly chosen standard bottle contains less than 0.750 litres is 0.1056, correct to four places of decimals.

(b) Find the probability that a box of 10 randomly chosen standard bottles contains at least 3 bottles whose contents are each less than 0.750 litres. Give three significant figures in your answer.

(c) Find the probability that there is more sherry in 4 randomly chosen standard bottles than in 3 randomly chosen large bottles.

(C)

# Mixed test 8B

1. The continuous random variables $X$ and $Y$ represent the masses of male and female students who attend my local College.
   Both $X$ and $Y$ are normally distributed such that $X \sim N(75, 6^2)$ and $Y \sim N(65, 5^2)$, where all masses are given in kilograms.

   (a) Find the probability that, if a male student and a female student are chosen at random, they both have a mass exceeding 70 kg.
   (b) State carefully the distribution of the combined mass of a random sample of $m$ male and $f$ female students.
   A lift in the college has a notice

   > **MAXIMUM 8 PEOPLE** *or* 650 kg

   Find the probability that the combined mass of a random sample of 8 students will exceed the mass restriction if it consists of
   (i)   8 males,
   (ii)  5 males and 3 females.
   (c) What is the probability that a randomly selected female student has a greater mass than a randomly selected male student?

   (*MEI*)

2. The mass of a cheese biscuit has a normal distribution with mean 6 g and standard deviation 0.2 g. Determine the probability that

   (a) a collection of twenty-five cheese biscuits has a mass of more than 149 g,
   (b) a collection of 30 cheese biscuits has a mass of less than 180 g,
   (c) twenty-five times the mass of a cheese biscuit is less than 149 g.

The mass of a ginger biscuit has a normal distribution with mean 10 g and standard deviation 0.3 g. Determine the probability that a collection of 7 cheese biscuits has a mass greater than a collection of 4 ginger biscuits.
(It may be assumed that all the biscuits were sampled at random from their respective populations.)

(*C*)

3. Certain components for a revolutionary new sewing machine are assembled by inserting a part of one type (sprotsil) into a part of another type (weavil). Sprotsils have external dimensions which are normally distributed with mean 2.50 cm and standard deviation 0.018 cm. Weavils have internal dimensions which are normally distributed with mean 2.54 cm and standard deviation 0.024 cm. Under suitable pressure, the two types fit together satisfactorily if the dimensions differ by not more than $\pm 0.035$ cm. Show that, if pairs of parts are chosen at random, the difference

   $D =$ internal dimension of a weavil
           $-$ external dimension of a sprotsil

   is distributed with mean 0.04 cm and standard deviation 0.030 cm. Hence show that approximately 42.8% of randomly selected pairs will fit together satisfactorily. Now, if it is known that the internal dimension of a given weavil is 2.517 cm, what is the probability that a randomly chosen sprotsil will fit this weavil satisfactorily?

   (*AEB*)

# 9

# Sampling and estimation

*In this chapter you will learn about*

- sampling methods including random and non-random sampling
- how to simulate a random sample from a given distribution
- the expectation and variance of the sample mean
- the distribution of the sample mean
- the use of the central limit theorem
- the distribution of the sample proportion
- estimates of population parameters:
  - mean
  - variance
  - proportion
- confidence intervals for:
  - a population mean, involving the *z*-distribution
  - a population mean, involving the *t*-distribution
  - a population proportion

## SAMPLING

## Population

In a statistical enquiry you often need information about a particular group. This group is known as the **population** or the **target population,** and it could be small, large or even infinite. Note that the word 'population' does not necessarily mean 'people'.
Here are some examples of populations:

- pupils in a class,
- people in England in full time employment,
- hospitals in Wales,
- cans of soft drink produced in a factory,
- ferns in a wood,
- rational numbers between 0 and 10.

# SURVEYS

Information is collected by means of a survey. There are two types:

(a)  a census,
(b)  a sample survey.

## (a)  Census

In a census **every** member of the population is surveyed.

When the population is small, this could be a straightforward exercise. For example, it would be easy to find out how each pupil in a class travelled to school on a particular morning. When populations are large, taking a census can be very time consuming and difficult to do with accuracy. Each year the government carries out a census in schools on the third Thursday in January. This requests the number of boys and girls in each age group on the roll of every school in the country. Its accuracy, though, relates only to that day. Even more difficult to carry out accurately is the population census taken every ten years. This attempts to provide details of different age groups for every area in Britain. When populations are very large, or infinite, it is not possible to survey every member.

On occasions it would not be sensible to survey every member. For example, if you performed a census to establish the length of life of a particular brand of light bulb, you would test each bulb until it failed and so you would destroy the population!

## (b)  Sample survey

When a survey covers less than 100% of the population, it is known as a **sample survey**. In many circumstances, taking a sample is preferable to carrying out a census. Sample data can be obtained relatively cheaply and quickly and, if the sample is representative of the population, a sample survey can give an accurate indication of the population characteristic being studied.

The size of the sample does not depend on the size of the population. It often depends on the time and money available to collect information. Note that large samples are more likely to give more reliable information than small ones. The next time that you read the results of a public opinion poll in the newspaper, look at the size of the sample – it is usually over 1000.

## Sample design

Once the purpose of a survey has been stated precisely, the **target population** must be defined, for example

- all the primary schools in England,
- all the oak trees in Hampshire,
- all the people admitted to the General Hospital in January suffering from a heart attack.

The **sampling units** must be defined clearly. These are the people or items to be sampled, for example

- the primary school,
- the oak tree,
- the person suffering from a heart attack.

Once the sampling units within a population are individually named or numbered to form a list, then this list of sampling units is called a **sampling frame**. It could take various forms (e.g. a list, a map, a set of maps), and should be as accurate as possible.

Ideally the sampling frame should be the same as the target population. For example, if the target population is all the first year students in a college, then the sampling frame and the target population should be the same, provided that the register is up-to-date and accurate. A sampling frame for people in Britain eligible to vote, however, is more difficult to form. The electoral register attempts to list all those who are eligible to vote throughout all the areas in the country, but it is never completely accurate, since many changes occur during the time that the information is being processed. Some people do not return the forms, people move in and out of the area, people die etc.

In some instances it is not possible to enumerate all the population, for example, the fish in a lake.

## Example 9.1

(a) Explain briefly what you understand by
  (i) a population,
  (ii) a sampling frame.

(b) A market research organisation wants to take a sample of
  (i) owners of diesel motor cars in the UK,
  (ii) persons living in Oxford who suffered from injuries to the back during July 1996.

Suggest a suitable sampling frame in each case. *(L)*

## Solution 9.1

(a) (i) A population is a particular group of individuals or items.
  (ii) Once the individual members of a population have been numbered to form a list, this list is called a sampling frame.

(b) (i) The list of registered owners as kept by DVLA in Swansea.
  (ii) A list made from information supplied by Health Clinics in Oxford during July 1996.

# Bias

The purpose of sampling is to gain information about the whole population by selecting a sample from that population. You want the sample to be representative of the population so you must give every member of the population an equal chance of being included in the sample. This should eliminate any **bias** in the selection of the sample.

Sources of bias include

(a) the lack of a good sampling frame:
- using the telephone directory misses all those who do not have a telephone or whose number is ex-directory,
- using the electoral register in a city area misses the more mobile section of the population.
(b) the wrong choice of sampling unit:
- choosing an individual rather than a particular group such as 'household'.
(c) non-response by some of the chosen units:
- it might be difficult to locate the particular unit,
- the cooperation of the respondent might not have been obtained,
- the enquiry might not have been understood, for example, a questionnaire might have been badly designed. Questionnaires should be clear, specific, unambiguous and easily understood. Questions should be worded neutrally, especially in opinion surveys, to avoid bias caused by pointing towards a particular response.
(d) bias introduced by the person conducting the survey:
- the interviewer might not question someone who appears uncooperative,
- the style of questioning may influence the response.

It should be noted that a sample can only be representative of the population from which it is selected. If you select a sample of teachers from one school, the sample is representative of the teachers in that school, not of all teachers in all schools.

## SAMPLING METHODS

Once a sampling frame has been established, you can choose a method of sampling. These fall into two categories:

- random sampling e.g. simple, systematic, stratified;
- non-random sampling e.g. quota, cluster

## Simple random sampling

Suppose a population consists of $N$ sampling units and you require a sample of $n$ of these units. A sample of size $n$ is called a **simple random sample** if all possible samples of size $n$ are equally likely to be selected. Some form of random processes must be used to make the selection.

If the unit selected at each draw is **replaced** into the population before the next draw, then it can appear more than once in the sample. This is known as **sampling with replacement**.

If the unit selected at each draw is **not replaced** into the population before the next draw, this is known as **sampling without replacement**.

The second method of sampling without replacement is known as **simple random sampling**.

Two methods of simple random sampling are commonly used

- drawing lots,
- random number sampling.

For each, make a list of all $N$ members of the population and give each member a different number.

# Drawing lots

For each member, place a coloured ball into a container and then draw $n$ balls out of the container at random and without replacement. If you wanted a sample of size 20, you would draw out 20 balls. This is suitable for a small population. Note, however, that the sample must be large enough to provide sufficiently accurate information about the population.

The sample should be selected at random. Any hint of possible bias should be avoided.

If the population is large then the method of drawing lots, sometimes described as 'drawing out of a hat' is not practical. You could instead make the choice by referring to random number tables. For your reference, a set is printed on page 653.

# Using random number tables

Random number tables consist of lists of digits 0, 1, 2, 3, ..., 9, such that each digit has an equal chance of occurring, so for example, the probability that a 3 occurs is 0.1. In random number tables the digits may appear singly or be grouped in some way. This is solely for convenience of printing.

## Example 9.2

Here is an extract from a set of random number tables

```
6  8  7  2  5  3  8  1  5  9
2  5  3  4  7  0  5  4  9  5
3  2  6  8  7  4  4  7  0  5
```

Use it to select a random sample of

(a)  eight people from a group of 100 people,
(b)  eight people from a group of 60.

## Solution 9.2

(a)  To select a group of eight people from a target population of 100 people, allocate a two-digit number to each person, for example allocate 01 to the first on the list, 02 to the second, ... up to 98, 99, 00, calling the hundredth person 00 for convenience.

Using the list, starting at the beginning of the first row and reading along the rows, you would select people corresponding to the following numbers:

68   72   53   81   59   25   34   70

Alternatively, you could decide to read the digits backwards, from bottom right, in which case your sample would consist of people corresponding to the numbers

50   74   47   86   23   59   45   07

(b)  To select a group of eight from a target population of 60 people, allocate each person a number from 01 to 60.

Using the tables, disregard any two-digit number outside the range.

Starting at the beginning of the first row and grouping in pairs gives

~~68~~  ~~72~~  53  ~~81~~  59  25  34  ~~70~~  54  ~~95~~  32  ~~68~~  ~~74~~  47  05

**So you would choose people corresponding to the numbers**

**53,  59,  25,  34,  54,  32,  47,  05.**

## Example 9.3

Use the following extract from random number tables to select a random sample of 12 numbers, each to two decimal places, from the continuous range $0 \leqslant x < 10$.

| 52 | 74 | 54 | 80 | 68 | 72 | 51 | 96 | 08 | 00 |
|----|----|----|----|----|----|----|----|----|----|
| 02 | 52 | 09 | 93 | 60 | 43 | 57 | 42 | 13 | 44 |

## Solution 9.3

Since the sample values are required to two decimal place accuracy, consider groups of three digits, inserting the decimal point between the first and second digit.

In this case your sample would consist of the values

5.27,   4.54,   8.06,   8.72,   5.19,   6.08,   0.00,   2.52,   0.99,   3.60,   4.35,   7.42

## Example 9.4

Here is a set of random numbers

848051   386103   153842   242330   580007   479971

Use it to select a random sample of four numbers, each to three decimal places, from the continuous range $0 \leqslant x < 5$.

## Solution 9.4

Consider groups of four digits, inserting the decimal point between the first and second digit. Disregard any values that are out of range. This gives

~~8.480~~  ~~5.138~~  ~~6.103~~  1.538  4.224  2.330  ~~5.800~~  0.747

**So the numbers chosen are 1.538, 4.224, 2.330, 0.747.**

# Calculator random number generator

You probably have a **random number generator** key $\boxed{\text{Ran\#}}$ on your calculator, which produces a number, for example 0.398, every time you press it. The numbers generated are in fact obtained using a mathematical formula and are really *pseudo* random numbers, but they suit the purpose very well indeed.

Suppose you want to use your calculator to select a random sample of six numbers between 1 and 49 for your entry in the National Lottery.

To do this, you probably need to press $\boxed{\text{Shift}}$ then $\boxed{\text{Ran\#}}$ $\boxed{=}$.

Suppose the numbers you get are

0.730, 0.798, 0.369, 0.499, 0.491, 0.310, 0.135, 0.112, 0.593, 0.652, 0.015, 0.346

You can interpret them in various ways, for example:

- If you decide to use the first two digits to the right of the decimal point each time, you would obtain the numbers ~~73~~, ~~79~~, 36, 49, 49, 31, 13, 11, ~~59~~, ~~65~~, 01, 34.
  Ignoring repeats and numbers bigger than 49, the six numbers would be
  **36, 49, 31, 13, 11, 1.**
- Suppose instead you decide to choose the second and third digits to the right of the decimal point and ignore repeats and numbers bigger than 49. In this case your numbers would be
  **30, 10, 35, 12, 15, 46.**
- If you decide to use all the digits after the decimal point, you would be choosing from the digits 730798369499491310135112593652015346. Grouping these as two-digit numbers gives ~~73~~, 07, ~~98~~, 36, ~~94~~, ~~99~~, 49, 13, 10, ~~13~~, ~~51~~, 12, 59, 36, 52, 01, 53, 46.
  Ignoring repeats and numbers bigger than 49 gives the six numbers as
  **7, 36, 49, 13, 10, 12.**

The lists are endless!

# Systematic sampling

Random sampling from a *very* large population is very cumbersome.
An alternative procedure is to list the population in some order, for example alphabetically or in order of completion on a production line, and then choose every $k$th member from the list after obtaining a random starting point. If you choose every tenth member from the list, for example every tenth vehicle passing a checkpoint, you would form a 10% sample. If you choose every twentieth item, for example every twentieth card in an index file, you would form a 5% sample.

## Example 9.5

Describe how to choose a systematic sample of eight members from a list of 300.

## Solution 9.5

Since you are going to choose every $k$th member, you need to find a suitable value for $k$. To do this, choose a convenient value close to $\dfrac{N}{n}$.

In this case, $\dfrac{N}{n} = \dfrac{300}{8} = 37.5$, so $k = 40$ will do.

Now choose a random starting point, for example if $\boxed{\text{Ran\#}}$ on your calculator gives 0.870 take the first member of the sample as 87 and then add 40 each time . The other members are 127, 167, 207, 247, 287, 27 and 67. Note that when you reach the end of the list, go back to the beginning.

So the sample consists of 27, 67, 87, 127, 167 207, 247, 287.

---

The advantages of systematic sampling are that it is quick to carry out and it is easy to check for errors. For large scale sampling, systematic selection is usually used in preference to taking simple random samples.

The disadvantage of this system is that there may be a periodic cycle within the frame itself. For example a machine may operate in such a manner that every tenth item is faulty. Systematic sampling of every fifth item, starting at 5, would result in half the items in the sample being faulty, whereas starting at 2 would produce a sample with no faulty items. Of course, if the periodic cycle is recognised then different samples could be taken by varying the starting points and the length of the interval between the chosen items.

# Stratified sampling

Stratified sampling is used when the population is split into distinguishable layers or strata that are quite different from each other and which together cover the whole population, for example

- age groups,
- occupational groups,
- topographical regions.

Separate random samples are then taken from each stratum and put together to form the sample from the population.

It is usual to represent the population proportionately in the strata, as in the following example.

## Example 9.6

Competent Carriers employs 320 drivers, 80 administrative staff and 40 mechanics. A committee to represent all the employees is to be formed. The committee is to have 11 members and the selection is to be made so that there is as close a representation as possible without bias towards any individuals or groups. Explain how this could be done.

## Solution 9.6

If you were to take a simple random sample of all 440 employees this would mean that every employee would have an equal chance of being selected. There is a high probability that the committee would consist of 11 drivers and therefore would not be representative of all employees.

A stratified random sample would provide a more accurate representation of the population and could be formed as follows:

Taking into account that drivers make up $\frac{320}{440}$ of the work force,

$$\text{number of drivers} = \frac{320}{440} \times 11 = 8$$

Similarly

$$\text{number of administrative staff} = \frac{80}{440} \times 11 = 2$$
$$\text{number of mechanics} = \frac{40}{440} \times 11 = 1$$

The required representation on the committee is eight drivers, two from the administrative staff and one mechanic. The people to be included can then be selected from each stratum by using simple random sampling or systematic sampling.

# Non-random sampling

### (a) Cluster sampling

Sometimes there is a natural sub-grouping of the population and these subgroups are called **clusters**. For example, in a population consisting of all children in the country attending state primary schools, the local education authorities form natural clusters. When a sample survey is carried out on a population that can be broken into clusters it is often more convenient to first choose a random sample of clusters and then to sample within each cluster chosen.

Unlike *stratified sampling* where the strata are as *different* from each other as possible, each *cluster* should be as *similar* to other clusters as possible.

One advantage of cluster sampling is that there is no need to have a complete sampling frame of the whole population. For the primary school children, you would need only a list of the pupils in the chosen local authority. Another advantage is that it is usually far less costly than random sampling. Consider the fees and travelling expenses paid to interviewers. Far less travelling and time is involved in an interviewer visiting individuals in a cluster than visiting individuals in the whole population.

The disadvantage of cluster sampling is that it is non-random. Suppose that a town has 7500 primary school children in 250 classes, each with an average class size of 30. If you want to select a sample of 90 children then you could use simple random sampling. It would however be quicker to use the classes as clusters and to take a sample consisting of three classes. This would give a sample of 90 children. The problem is that within each class there will be a certain amount of similarity between the children in say age, ability, home background. In selecting one whole class or cluster you are in fact selecting 30 similar children instead of 30 randomly chosen children from throughout the town. Therefore three clusters will not give as precise a picture of the whole population as 90 children chosen at random from throughout the town.

### (b) Quota sampling

Quota sampling is widely used in market research where the population is divided into groups in terms of age, sex, income level and so on. Then the interviewer is told how many people to interview within each specified group, but is given no specific instructions about how to locate them and fulfil the quota. This is the method generally used in street interview surveys commonly carried out in shopping centres. It is quick to use, complications are kept to a minimum and, unlike random sampling, any member of the sample may be replaced by another member with the same characteristics.

If no sampling frame exists, then quota sampling may be the only practical method of obtaining a sample. The disadvantage of quota sampling, however, is that it is non-random. There is a possibility of bias in the selection process if, for example, the interviewer selects those easiest to question or those who look cooperative. The location of such surveys in shopping centres excludes a substantial part of the population in that area. It is difficult to find out about those who refuse to cooperate and they are simply replaced. One of the reasons put forward to explain the inaccuracy of the opinion polls before the British general election in 1992 was the high refusal rates of Conservative voters to take part in surveys.

# Exercise 9a   Sampling methods

1. Explain briefly the difference between a census and a sample survey.
   Give an example to illustrate the practical use of each method.
   A school held an evening disco which was attended by 500 pupils. The disco organisers were keen to assess the success of the evening. Having decided to obtain information from those attending the disco, they were undecided whether to use a census or a sample survey.
   Which method would you recommend them to use?
   Give one advantage and one disadvantage associated with your recommendation. (L)

2. A school of 1000 pupils is divided into year groups as follows

   | Year | Number of pupils |
   | --- | --- |
   | 7 | 150 |
   | 8 | 150 |
   | 9 | 150 |
   | 10 | 150 |
   | 11 | 150 |
   | 12 | 125 |
   | 13 | 125 |

   A survey is to be carried out and a committee representative of the school is to be formed consisting of 40 pupils.
   It is decided that stratified sampling should be used.

   (a) Calculate the number of pupils chosen from each year group.
   (b) Explain how to choose the pupils from Year 7.

3. (a) Explain briefly
       (i) why it is often desirable to take samples,
       (ii) what you understand by a sampling frame.

   (b) State two circumstances when you would consider using
       (i) clustering,
       (ii) stratification,

       when sampling from a population.

   (c) Give two advantages and two disadvantages associated with quota sampling (L)

4. (a) A television company wishes to estimate the popularity of a particular television series by street interviews. Describe how the method of *quota sampling* might be used for this investigation.

   (b) A meat canning factory supplies a supermarket with cans of meat in three sizes: large, medium and small.
   The regular consignment is of 300 large cans, 500 medium cans and 400 small cans. Describe how the supermarket could apply the method of *stratified random sampling* to a sample of 60 cans to test the quality of these goods.

5. Write brief notes on

   (a) simple random sampling,
   (b) quota sampling.

   Your notes should include a description of each method, and an advantage and a disadvantage associated with it. (L)

6. In a school year group of 140 pupils there are 60 girls and 80 boys. A survey is to be taken to find methods to improve the school's meal services. A sample of 14 members of this group is needed for the survey.
   The school decides to use one of the following methods to obtain the names of pupils for the sample:

   A: Every tenth name on the year group register is selected for the sample.
   B: Each of the 140 names is allocated a different number from 1 to 140 inclusive; the school's computer then picks 14 different random numbers between 1 and 140 inclusive

   (a) State *briefly* one advantage and one disadvantage of each method.
   (b) Explain what is meant by a stratified random sample and describe how method B could be changed to give a stratified random sample.

7. Explain *briefly* the difference between a *census* and *sample*, and give **two** reasons why a sample may be preferred to a census.
   Explain the meaning and purpose of a *sampling frame* in random sampling.
   It is required to obtain the views of the pupils of a school about the school magazine. It is decided to do this by means of a small panel of pupils.
   Describe **briefly** how you would select such a panel using

   (a) simple random sampling,
   (b) stratified random sampling.

   State, with reason, which of these two sampling methods you consider to be the more appropriate for this situation. (AEB)

8. A research study into the use of hormone replacement therapy for women in the United Kingdom involved in a survey of women in three general medical practices in Greater London. The designer of the survey describes his method of obtaining his sample as follows.
'I obtained the names and addresses of 5025 women aged between 45 and 65 from the practices' age–sex registers. The women were sent a questionnaire that asked whether they had received hormone replacement therapy'
Source: *British Medical Journal*
December 1989

(a) Suggest *one* advantage and *one* disadvantage of this sampling method.
(b) Of the 5025 women contacted, 3238 returned a completed questionnaire, and 330 of these had received hormone replacement therapy. Given that there are about 703 000 women in the 45–65 age group living in Greater London, obtain an estimate for the number of 45–65 year-old women in Greater London who have received hormone replacement therapy. With reference to the sampling method used, comment on the reliability of this estimate.
(c) Suggest an alternative method of obtaining such an estimate. (NEAB)

## SIMULATING RANDOM SAMPLES FROM GIVEN DISTRIBUTIONS

A good way to simulate a random sample from a given distribution is to use cumulative proportional frequencies or cumulative probabilities, as illustrated in the following examples.

## (a) From a frequency distribution

### Example 9.7

Use the sequence of random digits 364294 588330 923918 400300 to generate five simulated observations from the following frequency distribution.

| $x$ | 1 | 2 | 3 | 4 | |
|---|---|---|---|---|---|
| $f$ | 8 | 12 | 14 | 6 | Total 40 |

### Solution 9.7

Consider first the cumulative frequencies and then transfer them to cumulative proportional frequencies with a total proportion of 1. Then allocate the random numbers in a convenient way in accordance with the cumulative proportional frequencies.

| $x$ | $f$ | Cumulative frequency | Cumulative proportional frequency | Corresponding random numbers |
|---|---|---|---|---|
| 1 | 8 | 8 | $\frac{8}{40} = 0.20$ | 01 to 20 |
| 2 | 12 | 20 | $\frac{20}{40} = 0.50$ | 21 to 50 |
| 3 | 14 | 34 | $\frac{34}{40} = 0.85$ | 51 to 85 |
| 4 | 6 | 40 | $\frac{40}{40} = 1$ | 86 to 99 and 00 |

Since the cumulative proportional frequencies contain two decimal places, it is convenient to use two-digit random numbers. Note that 00 has been allocated to the $x$-value of 4 for convenience.

Take 5 two-digit random numbers from the list:  36,  42,  94,  58,  83

Match these up with the corresponding sample values:  2,  2,  4,  3,  3

So a random sample of size 5 from the given distribution is **2, 2, 3, 3, 4.**

## (b) From a probability distribution

### Example 9.8

Generate a random sample size 10 from the given probability distribution, using the random numbers 3  7  4  7  6  5  3  3  9  0.

| $x$ | 0 | 1 | 2 | 3 |
|---|---|---|---|---|
| $P(X = x)$ | 0.1 | 0.2 | 0.4 | 0.3 |

### Solution 9.8

Form the cumulative distribution function $F(x)$ and then allocate random numbers in a convenient way.

| $x$ | $P(X = x)$ | $F(x)$ | Corresponding random numbers |
|---|---|---|---|
| 0 | 0.1 | 0.1 | 1 |
| 1 | 0.2 | 0.3 | 2, 3 |
| 2 | 0.4 | 0.7 | 4, 5, 6, 7 |
| 3 | 0.3 | 1 | 8, 9, 0 |

Take the 10 random numbers given and convert them to sample values:

Random number    3  7  4  7  6  5  3  3  9  0
Sample values    1  2  2  2  2  2  1  1  3  3

So the sample values are **1, 1, 1, 2, 2, 2, 2, 2, 3, 3.**

### Example 9.9

Generate a random sample of size 4 from the binomial distribution $X \sim B(4, 0.2)$, using the random numbers 2811 5747 6157 8988.

## Solution 9.9

Calculate the cumulative probabilities, either by calculating probabilities first or using cumulative probability tables directly (see page 682).

Remember that $P(X = x) = {}^4C_x 0.8^{4-x} 0.2^x$ for $x = 0,1,2,3,4$.

| $x$ | $P(X = x)$ | | $F(x)$ | Corresponding random numbers |
|---|---|---|---|---|
| 0 | $0.8^4$ | $= 0.4096$ | 0.4096 | 0001 to 4096 |
| 1 | $4 \times 0.8^3 \times 0.2$ | $= 0.4096$ | 0.8192 | 4097 to 8192 |
| 2 | $6 \times 0.8^2 \times 0.2^2$ | $= 0.1536$ | 0.9728 | 8193 to 9728 |
| 3 | $4 \times 0.8 \times 0.2^3$ | $= 0.0256$ | 0.9984 | 9729 to 9984 |
| 4 | $0.2^4$ | $= 0.0016$ | 1 | 9985 to 9999 and 0000 |

The random number 2811 is in the range 0001 to 4096 and so corresponds to $x = 0$.

Similarly   5747 corresponds to $x = 1$
6157 corresponds to $x = 1$
8988 corresponds to $x = 2$

So the random sample of four observations from the binomial distribution consists of the values **0, 1 1, 2.**

## Example 9.10

Using the random number 8135 take a single random observation from a Poisson distribution with parameter 3.

## Solution 9.10

$X \sim Po(3)$.

Using cumulative Poisson probability tables (see page 648) and arranging the results in a table together with a convenient corresponding random number allocation gives:

| $x$ | $F(x)$ | Corresponding random numbers |
|---|---|---|
| 0 | 0.0498 | 0001 to 0498 |
| 1 | 0.1991 | 0499 to 1991 |
| 2 | 0.4232 | 1992 to 4232 |
| 3 | 0.6472 | 4233 to 6472 |
| 4 | 0.8153 | 6473 to 8153 |
| 5 | 0.9161 | 8154 to 9161 |
| 6 | 0.9665 | 9162 to 9665 |
| 7 | 0.9881 | 9666 to 9881 |
| 8 or over | 1 | 9882 to 9999 and 0000 |

The given random number 8135 is in the range 6473 to 8153, so the random observation corresponds to $x = 4$.

## Example 9.11

Using the random digits 723 850 take a random sample of size 2 from the continuous distribution with probability density function

$$f(x) = \frac{3}{8}x^2 \quad \text{for } 0 \leqslant x \leqslant 2$$

## Solution 9.11

The cumulative distribution function is given by

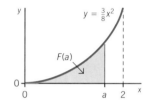

$$F(x) = \int_0^x \frac{3}{8}x^2 dx$$

$$= \frac{x^3}{8}$$

Taking the first three random numbers:

if $F(x) = 0.723$, then

$$\frac{x^3}{8} = 0.723$$

and $\quad x = \sqrt[3]{8 \times 0.723} = 1.80$ (2 d.p.)

Taking the next three random numbers:

if $F(x) = 0.850$, then

$$\frac{x^3}{8} = 0.850$$

and $\quad x = \sqrt[3]{8 \times 0.850} = 1.89$ (2 d.p.)

**So the two random observations are $x = 1.80$ and $x = 1.89$.**

## Example 9.12

Use the random numbers 382 824 to take a random sample of 2 from the normal distribution $N(30, 4)$.

## Solution 9.12

$X \sim N(30, 4)$.
Cumulative probabilities $\Phi(z)$ are given in the standard normal tables (see page 649).
Taking the first three digits of the random number list

$$\Phi(z) = 0.382$$

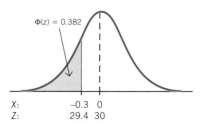

$$z = \Phi^{-1}(0.382)$$

$$= -0.3$$

$$\therefore \quad \frac{x - 30}{2} = -0.3$$

$$x = 30 - 0.6$$

$$= 29.4$$

Now take the second three digits

$$\Phi = 0.824$$
$$z = \Phi^{-1}(0.824)$$
$$= 0.931$$
$$\therefore \quad \frac{x - 30}{2} = 0.931$$
$$x = 30 + 1.862$$
$$= 31.9 \quad (1 \text{ d.p.})$$

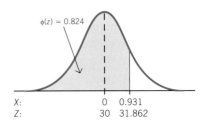

**So the two random observations are 29.4 and 31.9.**

## Exercise 9b   Simulating random samples from given distributions

In the following, use the random number tables on page 653 if random numbers have not been given in the question.

1.  Select a random sample of size 10 (to 3 d.p.) from the continuous range $3 \leqslant x < 9$.

2.  Draw up a random sample of 100 numbers from the discrete integer range 0 to 9. Find the mean and variance of the sample values and compare them with the theoretical mean and variance.

3.  The discrete random variable $X$ has probability distribution

    | $x$ | 5 | 6 | 7 | 8 | 9 |
    |---|---|---|---|---|---|
    | $P(X = x)$ | 0.15 | 0.2 | 0.33 | 0.21 | 0.11 |

    Simulate a sample of size 12 from the distribution of $X$. Compare the mean and variance of this sample with $E(X)$ and $Var(X)$.

4.  The discrete random variable $X$ has distribution function $F(x) = \frac{1}{4}(x - 2)$, $x = 3, 4, 5, 6$. Using random number tables, generate 10 observations of $X$, showing your working clearly.
    Describe how you would select a random sample of 30 pupils from a school containing 850 pupils.

5.  You wish to select a person at random from a group of 58 people. The following procedure is suggested:
    Allocate the numbers 1 to 58 to the people. Choose a line in a table of random numbers and call the first two digits $x$ and $y$. Let $z = 10x + y$. If $1 \leqslant z \leqslant 58$ then the person who was allocated the number is selected. Otherwise, the person allocated the number $z - 58$ is selected.
    Comment on this method of selection.

6.  Take a random sample of size 6 from the distribution:

    | $x$ | 15 | 16 | 17 | 18 | 19 |
    |---|---|---|---|---|---|
    | $f$ | 13 | 15 | 12 | 6 | 4 |

7.  Take a random sample of size 3 from the distribution:

    | $x$ | 2.3 | 2.4 | 2.5 | 2.6 | 2.7 |
    |---|---|---|---|---|---|
    | $f$ | 40 | 60 | 90 | 50 | 60 |

8.  Take a random sample of size 10 from each of the following probability distributions. In each case, find the sample mean and variance and compare with $E(X)$ and $Var(X)$.

    (a)

    | $x$ | 1 | 2 | 3 | 4 |
    |---|---|---|---|---|
    | $P(X = x)$ | 0.11 | 0.2 | 0.45 | 0.24 |

    (b)  $P(X = x) = kx$, $x = 0, 1, 2, 3$.

9.  Take a random sample of size 5 from the distribution of $X$ where $F(x) = \frac{1}{5}x$, $x = 2, 3, 4, 5$.

10. (a)  The discrete random variable $X$ is such that $X \sim B(3, 0.4)$. Take a random sample of size 5 from this distribution, using the random numbers

    407   315   401   203   972

    (b)  Using the random number 6143 take a single random observation from the Poisson distribution with parameter 4.

11. Using the random numbers 267 394 018 take a random sample of size 3 from the normal distribution with mean 35 and variance 9.

12. Using the random numbers 2654 9342, make two random observations from each of the following distributions:

    (a) The number of seeds that germinate in a group of 5 selected at random, given that 75% are expected to germinate.
    (b) The number of goals in a football match, where the number of goals follows a Poisson distribution with variance 2.4.
    (c) The mass of a bag of sugar, where the mass is normally distributed with mean 1010 g and standard deviation 4.5 g.

13. Using the random number 256 construct a random observation of the continuous random variable $X$ where

    $F(x) = \frac{1}{9}x^2, \quad 0 \le x \le 3$.

14. Take 20 samples, each of size 2, from the following distribution:

    | $x$ | 1 | 2 | 3 | 4 | 5 |
    |---|---|---|---|---|---|
    | $f$ | 10 | 15 | 25 | 35 | 15 |

    Calculate the mean of each sample and find the mean and variance of the sample means. Find the mean and variance of the original distribution. Comment.

15. You are given the random number 431. Use this number to obtain a sample observation from

    (a) a binomial distribution with $n = 12$ and $p = 0.4$.
    (b) a normal distribution with mean 6.2 and standard deviation 0.1.

    You are expected to explain clearly how you obtain the sample observations. (O)

16 The digits 8453276 are obtained from a table of random digits. Use them to obtain a random observation from each of the following distributions:

    (a) the number of the winning ticket in a lottery in which there are 500 ticket numbers from 1 to 500 and every ticket has the same chance of being selected.
    (b) the number of babies born in a cottage hospital in a week, assuming that on average one baby is born every three days and that births are independent (and ignoring the possibility of multiple births). (O)

## SAMPLE STATISTICS

When you are trying to find out information about a population it seems sensible to take random samples and then consider the values obtained from them. It is therefore useful to know how these sample values are distributed.

## THE DISTRIBUTION OF THE SAMPLE MEAN

Imagine carrying out the following procedure:

- Take a random sample of $n$ independent observations from a population. Note that from a finite population, sampling should be with replacement to ensure that the observations are independent.
- Calculate the mean of these $n$ sample values. This is known as the sample mean.
- Now repeat the procedure until you have taken all possible samples of size $n$, calculating the sample mean of each one.
- Form a distribution of all the sample means.

The distribution that would be formed is called the **sampling distribution of means**.

# The mean and variance of the sampling distribution of means

It is possible to work out the mean and variance of this sampling distribution using expectation algebra.

Consider a population $X$ in which $E(X) = \mu$ and $\text{Var}(X) = \sigma^2$.
Take $n$ independent observations $X_1, X_2, ..., X_n$, from $X$.

Since $\quad E(X) = \mu$,
$$E(X_1) = \mu, \quad E(X_2) = \mu, \quad ..., \quad E(X_n) = \mu$$

Since $\quad \text{Var}(X) = \sigma^2$,
$$\text{Var}(X_1) = \sigma^2, \quad \text{Var}(X_2) = \sigma^2, \quad ..., \quad \text{Var}(X_n) = \sigma^2$$

The sample mean,

$$\overline{X} = \frac{X_1 + X_2 + \cdots + X_n}{n}$$

$$= \frac{1}{n} X_1 + \frac{1}{n} X_2 + \cdots + \frac{1}{n} X_n$$

$$E(\overline{X}) = E\left(\frac{1}{n} X_1 + \frac{1}{n} X_2 + \cdots + \frac{1}{n} X_n\right)$$

$$= E\left(\frac{1}{n} X_1\right) + E\left(\frac{1}{n} X_2\right) + \cdots + E\left(\frac{1}{n} X_n\right)$$

$$= \frac{1}{n} E(X_1) + \frac{1}{n} E(X_2) + \cdots + \frac{1}{n} E(X_n) \quad \text{using } E(aX) = aE(X) \quad \text{(page 246)}$$

$$= \frac{1}{n} \mu + \frac{1}{n} \mu + \cdots + \frac{1}{n} \mu$$

$$= n \times \frac{1}{n} \mu$$

$$= \mu$$

$$\text{Var}(\overline{X}) = \text{Var}\left(\frac{1}{n} X_1 + \frac{1}{n} X_2 + \cdots + \frac{1}{n} X_n\right)$$

$$= \text{Var}\left(\frac{1}{n} X_1\right) + \text{Var}\left(\frac{1}{n} X_2\right) + \cdots + \text{Var}\left(\frac{1}{n} X_n\right)$$

$$= \left(\frac{1}{n}\right)^2 \text{Var}(X_1) + \left(\frac{1}{n}\right)^2 \text{Var}(X_2) + \cdots + \left(\frac{1}{n}\right)^2 \text{Var}(X_n) \quad \text{using } \text{Var}(aX) = a^2\text{Var}(X) \quad \text{(page 250)}$$

$$= \frac{1}{n^2} \sigma^2 + \frac{1}{n^2} \sigma^2 + \cdots + \frac{1}{n^2} \sigma^2$$

$$= n \times \frac{\sigma^2}{n^2}$$

$$= \frac{\sigma^2}{n}$$

$$E(\overline{X}) = \mu \quad \text{and} \quad \text{Var}(\overline{X}) = \frac{\sigma^2}{n}$$

The standard deviation of the sampling distribution is $\sqrt{\dfrac{\sigma^2}{n}}$, usually written $\dfrac{\sigma}{\sqrt{n}}$. This is known as the **standard error of the mean**.

The mean of the sampling distribution is the same as the mean of the population. The standard deviation of the sampling distribution is much smaller than that of the population since $\sigma^2$ has been divided by $n$. This implies that the sample means are much more clustered around $\mu$ than the population values are. In fact, the larger the sample size, the more clustered they are.

The following diagrams help to illustrate the shape of the sampling distribution of means resulting from different sized samples from given populations.

## (a) The distribution of $\overline{X}$ when the population of X is normal

Distriubtion of $X$ when $X \sim N(100, 64)$

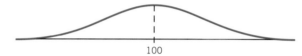

100

Distribution of $\overline{X}$ when $n = 2$, 5 and 25.

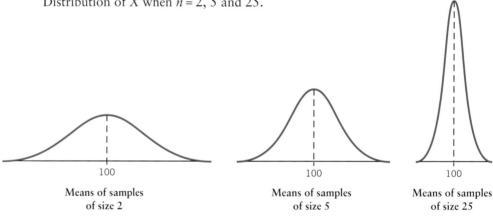

| 100 | 100 | 100 |
|:---:|:---:|:---:|
| Means of samples of size 2 | Means of samples of size 5 | Means of samples of size 25 |

From the diagrams, you can see that if samples are taken from a normal population, the sampling distribution of means is normal for any sample size.

$$\text{If } X \sim N(\mu, \sigma^2) \text{ then } \overline{X} \sim N\!\left(\mu, \frac{\sigma^2}{n}\right)$$

## Example 9.13

At a college the masses of the male students can be modelled by a normal distribution with mean mass 70 kg and standard deviation 5 kg.
Four male students are chosen at random. Find the probability that their mean mass is less than 65 kg.

## Solution 9.13

$X$ is the mass, in kilograms, of a male student at the college,
and $X \sim N(\mu, \sigma^2)$, with $\mu = 70$ and $\sigma = 5$.

Since the distribution of $X$ is normal, the distribution of $\bar{X}$ is also normal and

$$\bar{X} \sim N\left(\mu, \frac{\sigma^2}{n}\right) \quad \text{with } \mu = 70, \, \sigma^2 = 25, \, n = 4.$$

i.e. $\bar{X} \sim N\left(70, \frac{25}{4}\right)$

so $\bar{X} \sim N(70, 6.25)$

$$P(\bar{X} < 65) = P\left(Z < \frac{65 - 70}{\sqrt{6.25}}\right)$$
$$= (Z < -2)$$
$$= 1 - \Phi(2)$$
$$= 1 - 0.9772$$
$$= 0.0228$$

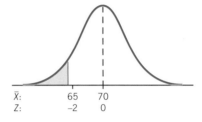

**The probability that the mean mass is less than 65 kg is 0.023 (2 s.f.).**

The diagram below shows the distributions of $X$ and $\bar{X}$ drawn to scale.

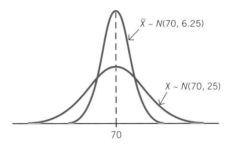

---

## Example 9.14

The distribution of the random variable $X$ is $N(25, 340)$. The mean of a random sample of
size $n$ drawn from this distribution is $\bar{X}$. Find the value of $n$, correct to two significant figures,
given that $P(\bar{X} > 28)$ is approximately 0.005. (C)

## Solution 9.14

$X \sim N(25, 340)$

For samples of size $n$, $\bar{X} \sim N\left(25, \frac{340}{n}\right)$

$$\therefore \quad P(\bar{X} > 28) = P\left(Z > \frac{28 - 25}{\sqrt{\frac{340}{n}}}\right)$$

$$= P\left(Z > \frac{3\sqrt{n}}{\sqrt{340}}\right)$$

You are given that $P(\overline{X} > 28) = 0.005$,

so $P\left(Z > \dfrac{3\sqrt{n}}{\sqrt{340}}\right) = 0.005$

$\therefore\quad P\left(Z < \dfrac{3\sqrt{n}}{\sqrt{340}}\right) = 1 - 0.005 = 0.995$

$\dfrac{3\sqrt{n}}{\sqrt{340}} = \Phi^{-1}(0.995)$

$= 2.576$

$3\sqrt{n} = 2.576 \times \sqrt{340}$

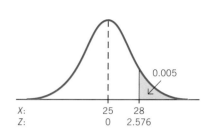

Squaring both sides

$9n = 2.576^2 \times 340$

$n = 250.68\ldots$

so $\qquad n = 250 \quad \textbf{(2 s.f.)}.$

---

## (b) The distribution of $\overline{X}$ when $X$ is not normally distributed

The following diagrams illustrate the distribution of $\overline{X}$ for samples of different sizes taken from a population $X$:

(i)  Distribution of $X$ when $X \sim B\,(10, 0.25)$

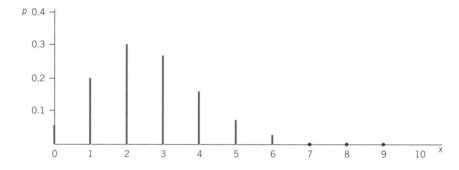

Distribution of $\overline{X}$ for samples of size 10, 15 and 30

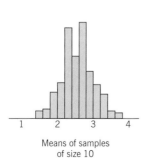

Means of samples
of size 10

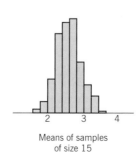

Means of samples
of size 15

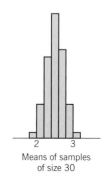

Means of samples
of size 30

(ii) Distribution of $X$ when $X \sim \text{Po}\,(4)$

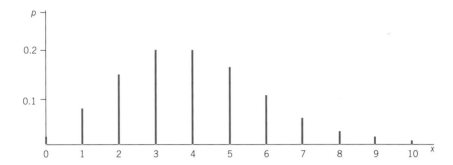

Distribution of $\overline{X}$ for samples of size 10, 15 and 30

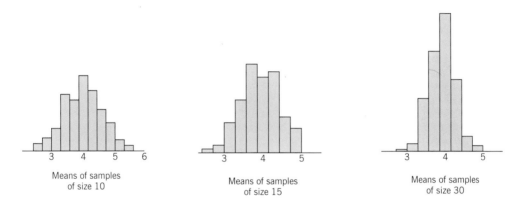

Means of samples
of size 10

Means of samples
of size 15

Means of samples
of size 30

(iii) Distribution of $X$ when $X \sim R\,(3, 7)$

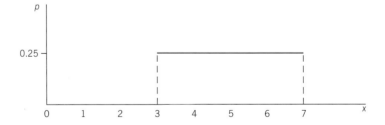

Distribution of $\overline{X}$ for samples of size 10, 15 and 30

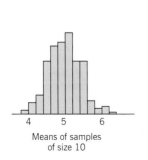

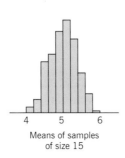

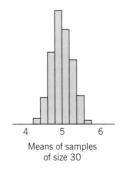

Means of samples
of size 10

Means of samples
of size 15

Means of samples
of size 30

# CENTRAL LIMIT THEOREM

From the diagrams you can see that when samples are taken from a population that is **not normally distributed,** the sampling distribution takes on the characteristic normal shape as the sample size increases. **For large *n*** the distribution of the sample mean is **approximately normal.**

This result is known as the **central limit theorem.** It is somewhat surprising, since it holds when the population of $X$ is discrete (as in the binomial and Poisson distributions) and when $X$ is continuous (as in the uniform distribution).

For samples taken from a non-normal population with mean $\mu$ and variance $\sigma^2$, by the central limit theorem, $\bar{X}$ is approximately normal

and $\qquad \bar{X} \sim N\!\left(\mu, \dfrac{\sigma^2}{n}\right)$

provided that the sample size, $n$, is large ($n \geqslant 30$ say).

## Example 9.15

Thirty random observations are taken from each of the following distributions and the sample mean calculated. Find, in each case, the probability that the sample mean exceeds 5.

(a) $X$ is the number of telephone calls made in an evening to a counselling service, where $X \sim \text{Po}(4.5)$.
(b) $X$ is the number of heads obtained when an unbiased coin is tossed nine times.
(c) $X$ is distributed uniformly throughout the range $2 \leqslant x \leqslant 7$.

## Solution 9.15

(a) $X \sim \text{Po}(4.5)$

$\mu = \lambda = 4.5, \qquad \sigma^2 = \lambda = 4.5$     (see page 293).

By the central limit theorem, since $n$ is large, $\bar{X}$ is approximately normal,

so $\bar{X} \sim N\!\left(\mu, \dfrac{\sigma^2}{n}\right)$ with $n = 30$

i.e. $\bar{X} \sim N\!\left(4.5, \dfrac{4.5}{30}\right)$

$\bar{X} \sim N(4.5, 0.15)$

$P(\bar{X} > 5) = P\!\left(Z > \dfrac{5 - 4.5}{\sqrt{0.15}}\right)$

$\qquad\qquad = P(Z > 1.291)$
$\qquad\qquad = 1 - \Phi(1.291)$
$\qquad\qquad = 1 - 0.9017$
$\qquad\qquad = 0.098 \ (\textbf{2 s.f.})$

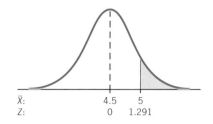

$\bar{X}$:    4.5   5
$Z$:    0   1.291

(b) $X \sim B(9, 0.5)$

$\mu = np = 9 \times 0.5 = 4.5$     (see page 286)
$\sigma^2 = npq = 9 \times 0.5 \times 0.5 = 2.25$

By the central limit theorem, since $n$ is large, $\bar{X}$ is approximately normal and

$$\bar{X} \sim N\left(\mu, \frac{\sigma^2}{n}\right) \text{ with } n = 30$$

i.e. $\bar{X} \sim N\left(4.5, \frac{2.25}{30}\right)$

$\bar{X} \sim N(4.5, 0.075)$

$$P(\bar{X} > 5) = P\left(Z > \frac{5 - 4.5}{\sqrt{0.075}}\right)$$

$= P(Z > 1.826)$

$= 1 - 0.9660$

$= \mathbf{0.034} \quad \textbf{(2 s.f.)}$

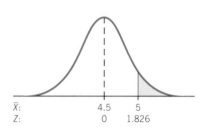

(c) When $X$ is uniformly distributed and $a \leqslant x \leqslant b$,

$E(X) = \frac{1}{2}(a + b)$ and $\text{Var}(X) = \frac{1}{12}(b - a)^2$ (see page 363)

Since $X$ is valid for $2 \leqslant x \leqslant 7$, $a = 2$ and $b = 7$

$\mu = E(X) = \frac{1}{2}(2 + 7) = 4.5 \qquad \sigma^2 = \text{Var}(X) = \frac{1}{12}(7 - 2)^2 = \frac{25}{12}$

By the central limit theorem, since $n$ is large, $\bar{X}$ is approximately normal and

$$\bar{X} \sim N\left(\mu, \frac{\sigma^2}{n}\right) \text{ with } n = 30$$

i.e. $\bar{X} \sim N\left(4.5, \frac{\frac{25}{12}}{30}\right)$

$\bar{X} \sim N(4.5, 0.0694\ldots)$

$$P(\bar{X} > 5) = P\left(Z > \frac{5 - 4.5}{\sqrt{0.0694\ldots}}\right)$$

$= P(Z > 1.897)$

$= 1 - 0.9711$

$= \mathbf{0.029} \quad \textbf{(2 s.f.)}$

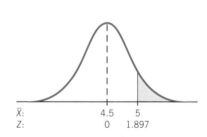

## Exercise 9c   The distribution of the sample mean, $\bar{X}$, for

- samples of any size from a normal population
- large samples from a non-normal population

1. The volumes of wine in bottles are normally distributed with a mean of 758 ml and a standard deviation of 12 ml. A random sample of 10 bottles is taken and the mean volume found.
   Calculate the probability that the sample mean is less than 750 ml.

2. The heights of a new variety of sunflower can be modelled by a normal distribution with mean 2 m and standard deviation of 40 cm.

(a) A random sample containing 50 sunflowers is taken and the mean height calculated. What is the probability that the sample mean lies between 195 cm and 205 cm?

(b) A hundred such samples, each of 50 observations, are taken. In how many of these would you expect the sample mean to be greater than 210 cm?

3. In an examination taken by a large number of students the mean mark was 64.5 and the variance was 64. The mean mark in a random sample of 100 scripts is denoted by $\bar{X}$. Find

(a) $P(\bar{X} > 65.5)$
(b) $P(63.8 < \bar{X} < 64.5)$

4. The mean of 50 observations of $X$, where $X \sim B(12, 0.4)$, is $\bar{X}$.

(a) State the approximate distribution of $\bar{X}$.
(b) Hence find $P(\bar{X} < 5)$

5. A normal variable $X$ has standard deviation $\sigma$. The mean of 20 independent observations of $X$ is $\bar{X}$.

(a) Given that $\text{Var}(\bar{X}) = 3.2$, find the value of $\sigma$.
(b) Would your answer be different if the variable was not normal?

6. Independent observations are taken from a normal distribution with mean 30 and variance 5.

(a) Find the probability that the average of 10 observations exceeds 30.5.
(b) Find the probability that the average of 40 observations exceeds 30.5.
(c) Find the probability that the average of 100 observations exceeds 30.5.
(d) Find the least value of $n$ such that the probability that the average of $n$ observations exceeds 30.5 is less than 1%.

7. The standard deviation of the masses of articles in a large population is 4.55 kg. Random samples of size 100 are drawn from the population. Find the probability that a sample mean will differ from the population mean by less than 0.8 kg.

8. The variable $X$ is such that $X \sim N(\mu, 4)$. A random sample of size $n$ is taken from the population. Find the least $n$ such that $P(|\bar{X} - \mu| < 0.5) > 0.95$.

9. (a) A large number of random samples of size $n$ are taken from $B(20, 0.2)$. Approximately 90% of the sample means are less than 4.354. Estimate $n$.
   (b) A large number of random samples of size $n$ are taken from Po(2.9). Approximately 1% of the sample means are greater than 3.41. Estimate $n$.

10. The random variable $X$ has standard deviation $\sigma$. The mean of 40 observations of $X$ is $\bar{X}$. Given that $\text{Var}(\bar{X}) = 0.625$, find the value of $\sigma$.

11. The mean of a sample of 100 observations of the random variable $X$ is denoted by $\bar{X}$. The mean of $\bar{X}$ is 20 and the standard deviation of $\bar{X}$ is 0.3. Find the mean and the standard deviation of $X$.

12. A sample of $n$ independent observations is taken from a normal population with mean 74 and standard deviation 6. The sample mean is denoted by $\bar{X}$.

(a) Find $n$ if $P(\bar{X} > 75) = 0.282$.
(b) Find $n$ if $P(\bar{X} < 70.4) = 0.0037$.

13. To estimate the mean and standard deviation of the life of a certain brand of car tyre a large number of random samples of size 50 were tested. The mean and standard error of the sampling distribution obtained were 20 500 km and 250 km respectively. Estimate the mean and standard deviation of the life of this brand of car tyre. Explain what part the use of the central limit theorem has played in the calculations.

14. The diameters, $x$, of 110 steel rods were measured in centimetres and the results were summarised as follows:

$$\Sigma x = 36.5, \qquad \Sigma x^2 = 12.49.$$

Find the mean and standard deviation of these measurements.
Assuming these measurements are a sample from a normal distribution with this mean and this variance, find the probability that the mean diameter of a sample of size 110 is greater than 0.345 cm. $\qquad$ (O & C)

15. In a certain nation, men have heights distributed normally with mean 1.70 m and standard deviation 10 cm. Find the probability that a man chosen randomly has height not less than 1.83 m.
What is the probability that the average height of 3 men chosen randomly is greater than 1.78 m and the probability that all three will have heights greater than 1.83 m? $\qquad$ (MEI)

16. Two red balls and 2 white balls are placed in a bag. Balls are drawn one by one, at random and without replacement. The random variable $X$ is the number of white balls drawn before the first red ball is drawn.

(a) Show that $P(X = 1) = \frac{1}{3}$, and find the rest of the probability distribution of $X$.
(b) Find $E(X)$ and show that $\text{Var}(X) = \frac{5}{9}$.
(c) The sample mean for 80 independent observations of $X$ is denoted by $\bar{X}$. Using a suitable approximation, find $P(\bar{X} > 0.75)$. $\qquad$ (C)

# THE DISTRIBUTION OF THE SAMPLE PROPORTION, $p$

Suppose a random sample of $n$ observations is taken from a population in which the proportion of successes is $p$ and the proportion of failures is $q = 1 - p$.

If $X$ is the number of successes in the sample, then $X$ follows a binomial distribution i.e. $X \sim B(n, p)$ and $E(X) = np$, $Var(X) = nqp$ (see page 286).

The random variable for the proportion of success in the sample is $\dfrac{X}{n}$.

This can be written $P_s$ where $P_s = \dfrac{X}{n} = \dfrac{1}{n} X$

It is possible to work out the mean and the variance of $P_s$ using expectation algebra as follows:

$$E(P_s) = E\left(\frac{1}{n} X\right) \qquad\qquad Var(P_s) = Var\left(\frac{1}{n} X\right)$$

$$= \frac{1}{n} E(X) \qquad\qquad\qquad = \left(\frac{1}{n}\right)^2 \times Var(X)$$

$$= \frac{1}{n} \times np \qquad\qquad\qquad = \frac{1}{n^2} \times npq$$

$$= p \qquad\qquad\qquad\qquad\quad = \frac{pq}{n}$$

The distribution of $P_s$ has mean $p$ and variance $\dfrac{pq}{n}$.

When $n$ is large, the distribution of $P_s$ is approximately normal, and

$$P_s \sim N\left(p, \frac{pq}{n}\right)$$

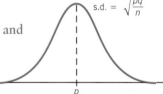

s.d. $= \sqrt{\dfrac{pq}{n}}$

The larger the sample size, $n$, the better the approximation.

The distribution of $P_s$ is known as the **sampling distribution of proportions**. The standard deviation of this distribution is $\sqrt{\dfrac{pq}{n}}$ and it is known as the **standard error of proportion**.

*NOTE:* When considering the normal approximation to the binomial distribution, a continuity correction of $\pm \frac{1}{2}$ is needed (see page 383).

Since $P_s = \dfrac{1}{n} \times X$, use a continuity correction $\dfrac{1}{n} \times \left(\pm \dfrac{1}{2}\right)$ i.e. $\pm \dfrac{1}{2n}$.

## Example 9.16

It is known that 3% of frozen pies delivered to a canteen are broken. What is the probability that, on a morning when 500 pies are delivered, 5% or more are broken?

## Solution 9.16

Let $p$ be the probability that a pie is broken, so $p = 0.03$.

Let $P_s$ be the proportion of pies in the sample that are broken.

Then $P_s \sim N\left(p, \dfrac{pq}{n}\right)$  with $p = 0.03$, $q = 0.97$, $n = 500$.

$$P_s \sim N\left(0.03, \dfrac{0.03 \times 0.97}{500}\right)$$

i.e.  $P_s \sim N(0.03, 0.000\ 058\ 2)$

To find the probability that 5% or more are broken,

find $P(P_s \geqslant 0.05) \rightarrow P\left(P_s > 0.05 - \dfrac{1}{2 \times 500}\right)$  (continuity correction)

$= P(P_s > 0.049)$

$= P\left(Z > \dfrac{0.049 - 0.03}{\sqrt{0.000\ 058\ 2}}\right)$

$= P(Z > 2.491)$

$= 1 - \Phi(2.491)$

$= 1 - 0.9936$

$= 0.0064$

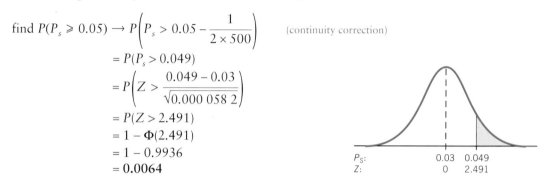

$P_S$:    $\quad$ 0.03  0.049
$Z$:    $\quad$ 0      2.491

### Alternative method for Solution 9.16

Instead of considering $p$, the proportion of broken pies, you could consider $X$, the number of broken pies in the sample.
In this case, $X \sim B(n, p)$ with $n = 500$, $p = 0.03$, $q = 1 - p = 0.97$.

Now $np = 500 \times 0.03 = 15$ and $nq = 500 \times 0.97 = 485$.

Since $n$ is large such that $np > 5$ and $nq > 5$, use the normal approximation for the binomial distribution (see page 382),
where  $X \sim N(np, npq)$ with $np = 15$ and $npq = 500 \times 0.03 \times 0.97 = 14.55$
i.e.    $X \sim N(15, 14.55)$.

You want the probability that 5% or more are broken.
5% of 500 = 25, so find the probability that 25 or more are broken.

$P(X \geqslant 25) \rightarrow P(X > 24.5)$  (continuity correction)

$= P\left(Z > \dfrac{24.5 - 15}{\sqrt{14.55}}\right)$

$= P(Z > 2.491)$

$= 0.0064$  (as above)

$X$:    $\quad$ 15    24.5
$Z$:    $\quad$ 0      2.491

NOTE: Since the same underlying theory has been used, probabilities of this type can be found either by considering $P_s$, the distribution of sample proportions, or by considering $X$, the distribution of the number of successes, and applying the normal approximation to the binomial distribution. In either case, the sample size, $n$, must be large.

Note that if the continuity correction is used in both cases, or omitted in both cases, the standardized $z$ values will agree exactly.

# Exercise 9d  Distribution of sample proportions (large samples)

1. 2% of the trees in a plantation are known to have a certain disease. A random sample of 300 trees is checked. Find the probability that the proportion of diseased trees in the sample is

   (a) less than 1%,
   (b) more than 4%.

2. A fair coin is tossed 150 times. Find the probability that

   (a) fewer than 40% of the tosses will result in heads,
   (b) between 40% and 50% (inclusive) of the tosses will result in heads,
   (c) at least 55% of the tosses will result in heads.

3. A fair coin is tossed 300 times.
   Work through part (c) as in question 2.
   Explain why your answer is different from that obtained in question 2.

4. Mr Hand gained 48% of the votes in the District Council elections.

   (a) Find the probability that a poll of 100 randomly selected voters would show over 50% in favour of Mr Hand.
   (b) Find the corresponding probability if the sample consists of 1000 randomly selected voters.

5. Three-quarters of the households in a particular area are connected to the internet. Find the probability that at least 73 of a random sample of 100 households are connected to the internet.

6. A die is biased so that 1 in 5 throws results in a six. Find the probability that, when the die is thrown 300 times, the number that result in a six

   (a) is more than 70,
   (b) is at least 70,
   (c) is less than 57.

7. 70% of the strawberry plants of a particular variety produce more than ten strawberries per plant. Find the probability that a random sample of 50 plants of this variety consists of more than 37 plants which produce more than ten strawberries per plant.

## UNBIASED ESTIMATES OF POPULATION PARAMETERS

In order to define a binomial distribution you need to know $n$ and $p$; to define a Poisson distribution you need to know $\lambda$ and to define a normal distribution you need to know $\mu$ and $\sigma^2$. These are known as the **population parameters** of the distributions.

Suppose that you do not know the value of a particular parameter of a distribution, for example the mean or the variance or the proportion of successes. It seems sensible that you would take a random sample from the distribution and use it in some way to make an **estimate** of the value of your unknown parameter.

This estimate is **unbiased** if the average (or expectation) of a large number of values taken in the same way is the true value of the parameter. There may be several ways of obtaining an unbiased estimate but the best (most efficient) estimate is the one with the smallest variance.

## POINT ESTIMATES

If the random sample taken is of size $n$,

- the best unbiased estimate of $p$, the proportion of successes in the population, is $\hat{p}$ where

   $$\hat{p} = p_s \qquad\qquad p_s \text{ is the proportion of successes in the sample}$$

- the best unbiased estimate of $\mu$, the population mean, is $\hat{\mu}$ where

   $$\hat{\mu} = \bar{x} = \frac{\sum x}{n} \qquad\qquad \bar{x} \text{ is the mean of the sample}$$

- the best unbiased estimate of $\sigma^2$, the population variance, is $\hat{\sigma}^2$ where

   $$\hat{\sigma}^2 = \frac{n}{n-1} \times s^2 \qquad\qquad s^2 \text{ is the variance of the sample}$$

There are alternative formats for $\hat{\sigma}^2$:

$$\hat{\sigma}^2 = \frac{n}{n-1} \times \frac{\Sigma(x-\bar{x})^2}{n} = \frac{\Sigma(x-\bar{x})^2}{n-1}$$

or

$$\hat{\sigma}^2 = \frac{n}{n-1}\left(\frac{\Sigma x^2}{n} - \left(\frac{\Sigma x}{n}\right)^2\right) = \frac{1}{n-1}\left(\Sigma x^2 - \frac{(\Sigma x)^2}{n}\right)$$

NOTE: that if you are using your calculator in SD mode, it is possible to find the value of $\hat{\sigma}$ directly. Look for a key marked $\boxed{x_{\sigma_{n-1}}}$. On some models this is obtained by pressing $\boxed{\text{SHIFT}}$ $\boxed{3}$. Find the key on your model.

## Example 9.17

A railway enthusiast simulates train journeys and records the number of minutes, $x$, to the nearest minute, trains are late according to the schedule being used. A random sample of 50 journeys gave the following times.

```
17   5   3  10   4   3  10   5   2  14
 3  14   5   5  21   9  22  36  14  34
22   4  23   6   8  15  41  23  13   7
 6  13  33   8   5  34  26  17   8  43
24  14  23   4  19   5  23  13  12  10
```

Given that $\Sigma x = 738$ and $\Sigma x^2 = 16\ 526$, calculate to two decimal places, unbiased estimates of the mean and the variance of the population from which this sample was drawn. (L)

## Solution 9.17

$X$ is the number of minutes that the train is late.
Let $E(X) = \mu$ and $\text{Var}(X) = \sigma^2$.

Unbiased estimate of $\mu$

$$\hat{\mu} = \bar{x} = \frac{\Sigma x}{n} = \frac{738}{50} = 14.76$$

Unbiased estimate of $\sigma^2$

$$\hat{\sigma}^2 = \frac{1}{n-1}\left(\Sigma x^2 - \frac{(\Sigma x)^2}{n}\right)$$

$$= \frac{1}{49}\left(16\ 526 - \frac{(738)^2}{50}\right)$$

$$= 114.961\ldots$$

$$= 114.96 \quad \text{(2 d.p.)}$$

Try this using the raw data and your calculator in SD mode (see page 40). Input the data as follows:

| Casio 570W/85W/85WA | |
|---|---|
| Set SD mode | MODE MODE 1 or MODE 2 |
| Clear memories | SHIFT Scl = |
| Input data | 17 DT |
| | 5 DT |
| | 3 DT |
| | ⋮ |
| | 10 DT |
| To obtain | |
| $\hat{\sigma}^2 = 114.961 \ldots$ | SHIFT 3 $x^2$ = |
| $\bar{x} = 14.76$ | SHIFT 1 = |
| You can also check | |
| $\Sigma x = 738$ | RCL B |
| $\Sigma x^2 = 16\ 526$ | RCL A |
| $n = 50$ | RCL C |
| To clear SD mode | MODE 1 |

## Example 9.18

For the data given in Example 9.17, estimate the proportion of trains that are more than 25 minutes late.

## Solution 9.18

Number in sample that are more than 25 minutes late = 7
Proportion in sample, $p_s = \frac{7}{50} = 0.14$
Unbiased estimate of population proportion, $p$, is $\hat{p}$, where $\hat{p} = p_s = \mathbf{0.14}$.

## INTERVAL ESTIMATES

Another way of using a sample value to give a good idea of an unknown population parameter is to construct an **interval**, known as a **confidence interval**.

In general terms, this is an interval that has a specified probability of **including** the parameter. The interval is usually written $(a, b)$ and the end-values, $a$ and $b$, are known as **confidence limits**. The probabilities most often used in confidence intervals are 90%, 95% and 99%.

Suppose you do not know the mean $\mu$ of a particular population and you want to work out a 95% confidence interval for it. You would need to construct an interval $(a, b)$ so that

$$P(a < \mu < b) = 0.95.$$

In this case, the probability that the interval *includes* $\mu$ is 0.95 or 95%.

The interval that you construct uses the value of the mean of a random sample of size $n$ taken from the population. This mean is denoted by $\bar{x}$.

Before constructing your interval for $\mu$, it is essential to ask the following questions.

- Is the distribution of the population normal or not?
- Do you know the variance of the population?
- Is the sample size large or small?

Your answers will then determine how to proceed. The following theory illustrates various situations.

## (a) Confidence interval for $\mu$, the population mean

- **of a normal population,**
- **with known variance $\sigma^2$**
- **using any size sample, $n$ large or small**

Consider first how to calculate the end-values of the most commonly used interval, the **95% confidence interval**. The method can then be adapted for other levels of confidence.

Note that it is useful to be able to follow the theory for the derivation of the end-points, but in practice you will probably only need to be able to apply the formula.

As you saw on page 438, for random samples of size $n$,

$$\text{if } X \sim N(\mu, \sigma^2), \quad \text{then } \bar{X} \sim N\left(\mu, \frac{\sigma^2}{n}\right)$$

Standardising, $\quad Z = \dfrac{\bar{X} - \mu}{\sigma/\sqrt{n}}$, where $Z \sim N(0, 1)$

Consider the distribution of $Z$.
For a 95% confidence interval you need to find the values of $z$ between which the central 95% of the distribution lies. This means that the upper tail probability is 0.025 and the lower tail is 0.975.

$P(Z < z) = 0.975$

$\qquad z = \Phi^{-1}(0.975)$

$\qquad = 1.96$

The values of $z$ are $\pm 1.96$.

So $P(-1.96 < Z < 1.96) = 0.95$

i.e. $P\left(-1.96 < \dfrac{\bar{X} - \mu}{\sigma/\sqrt{n}} < 1.96\right) = 0.95$

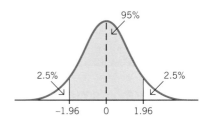

Now consider the inequality in two parts:

$-1.96 < \dfrac{\bar{X} - \mu}{\dfrac{\sigma}{\sqrt{n}}}$

$-1.96 \dfrac{\sigma}{\sqrt{n}} < \bar{X} - \mu$

$\mu < \bar{X} + 1.96 \dfrac{\sigma}{\sqrt{n}}$

$\dfrac{\bar{X} - \mu}{\dfrac{\sigma}{\sqrt{n}}} < 1.96$

$\bar{X} - \mu < 1.96 \dfrac{\sigma}{\sqrt{n}}$

$\bar{X} - 1.96 \dfrac{\sigma}{\sqrt{n}} < \mu$

Writing these two inequalities in one statement gives

$$\bar{X} - 196 \frac{\sigma}{\sqrt{n}} < \mu < \bar{X} + 1.96 \frac{\sigma}{\sqrt{n}}$$

Therefore the probability statement is

$$P\left(\bar{X} - 1.96 \frac{\sigma}{\sqrt{n}} < \mu < \bar{X} + 1.96 \frac{\sigma}{\sqrt{n}}\right) = 0.95$$

This enables you to construct the 95% confidence interval for $\mu$.
Comparing this with $P(a < \mu < b) = 0.95$, if the mean obtained from *your* sample is $\bar{x}$, then the end-values, or confidence limits are

$$\bar{x} - 1.96 \frac{\sigma}{\sqrt{n}} \quad \text{and} \quad \bar{x} + 1.96 \frac{\sigma}{\sqrt{n}}.$$

These are sometimes written $\bar{x} \pm 1.96 \frac{\sigma}{\sqrt{n}}$.

If $\bar{x}$ is the mean of a random sample of any size $n$ taken from a normal population with known variance $\sigma^2$,
then a 95% confidence interval for $\mu$ is given by

$$\left(\bar{x} - 1.96 \frac{\sigma}{\sqrt{n}}, \bar{x} + 1.96 \frac{\sigma}{\sqrt{n}}\right)$$

## Example 9.19

The mass of vitamin E in a capsule manufactured by a certain drug company is normally distributed with standard deviation 0.042 mg. A random sample of five capsules was analysed and the mean mass of vitamin E was found to be 5.12 mg. Calculate a symmetric 95% confidence interval for the population mean mass of vitamin E per capsule. Give the values of the end-points of the interval correct to three significant figures. (C)

## Solution 9.19

$X$ is the mass, in milligrams of a vitamin E capsule.
$X \sim N(\mu, \sigma^2)$ with $\sigma = 0.042$.

$\bar{X} \sim N\left(\mu, \dfrac{\sigma^2}{n}\right)$ with $n = 5$.   $X$ is normally distributed so any size random sample is acceptable.

The 95% confidence interval for $\mu$ is $\left(\bar{x} - 1.96 \frac{\sigma}{\sqrt{n}}, \quad \bar{x} + 1.96 \frac{\sigma}{\sqrt{n}}\right)$.

$$\bar{x} \pm 1.96 \frac{\sigma}{\sqrt{n}} = 5.12 \pm 1.96 \times \frac{0.042}{\sqrt{5}}$$

$$= 5.12 \pm 0.0368 \ldots$$
$$\uparrow$$

Store this value in your calculator

Lower confidence limit = 5.12 − 0.0368 ... = 5.08   (3 s.f.)
Upper confidence limit = 5.12 + 0.0368 ... = 5.16   (3 s.f.)

**So the 95% confidence interval for $\mu$, based on the sample mean, is (5.08 mg, 5.16 mg).**

NOTE: The probability that the interval (5.08 mg, 5.16 mg) includes, or has trapped, $\mu$ is 0.95, i.e. 95%. If you took another random sample of the same size, you would probably get a different interval. If you took lots of samples in a similar way then, on average, 95% of these intervals would include the true population mean $\mu$.

---

The following computer simulation illustrates the intervals obtained when 100 confidence intervals are constructed, each with 95% confidence. On average, 5% do *not* include $\mu$. In practice, you would only construct *one* interval. Remember that there is a 5% chance that *your* interval does not include $\mu$.

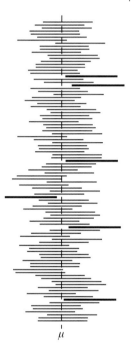

The intervals shown in bold are the ones which do **not** include $\mu$. You will see that in this case just six of the 100 do not include $\mu$. On **average** 95% of intervals constructed in this way will include the true population mean.

### Critical z-values in confidence intervals

The $z$-value in the confidence interval is known as the **critical value** and is obtained for different levels of confidence as follows:

In a **90% confidence interval,**
    the upper tail probability is 0.05
so the lower tail probability is 0.95.

$$P(Z < z) = 0.95$$
i.e.    $$\Phi(z) = 0.95$$
$$z = \Phi^{-1}(0.95)$$
$$= 1.645$$

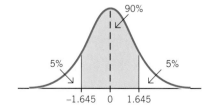

In a **95% confidence interval**,
the upper tail probability is 0.025
so the lower tail probability is 0.975.

$P(Z < z) = 0.975$

i.e. $\Phi(z) = 0.975$

$z = \Phi^{-1}(0.975)$

$= 1.96$

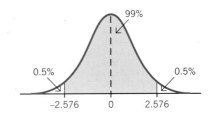

In a **99% confidence interval**,
the upper tail probability is 0.005
so the lower tail probability is 0.995.

$P(Z < z) = 0.995$

i.e. $\Phi(z) = 0.995$

$z = \Phi^{-1}(0.995)$

$= 2.576$

### Summary

A symmetric 90% confidence interval for $\mu$ is

$$\left(\bar{x} - 1.645 \frac{\sigma}{\sqrt{n}}, \quad \bar{x} + 1.645 \frac{\sigma}{\sqrt{n}}\right)$$

A symmetric 95% confidence interval for $\mu$ is

$$\left(\bar{x} - 1.96 \frac{\sigma}{\sqrt{n}}, \quad \bar{x} + 1.96 \frac{\sigma}{\sqrt{n}}\right)$$

A symmetric 99% confidence interval for $\mu$ is

$$\left(\bar{x} - 2.576 \frac{\sigma}{\sqrt{n}}, \quad \bar{x} + 2.576 \frac{\sigma}{\sqrt{n}}\right)$$

### Table of critical values

In some tables, the most commonly used critical $z$-values are summarised as follows:

If $Z \sim N(0, 1)$ then, for each value of $p$, the table gives the value of $z$ such that $P(Z \leqslant z) = p$.

| $p$ | 0.75 | 0.90 | 0.95 | 0.975 | 0.99 | 0.995 | 0.9975 | 0.999 | 0.9995 |
|-----|------|------|------|-------|------|-------|--------|-------|--------|
| $z$ | 0.674 | 1.282 | 1.645 | 1.960 | 2.326 | 2.576 | 2.807 | 3.090 | 3.291 |

– for a 90% confidence interval, $P(Z < z) = 0.95$; $p = 0.95$ gives $z = 1.645$,

– for a 95% confidence interval, $P(Z < z) = 0.975$; $p = 0.975$ gives $z = 1.96$,

– for a 99% confidence interval, $P(Z < z) = 0.995$; $p = 0.995$ gives $z = 2.576$.

If you want a 98% confidence interval, this implies an upper tail probability of 0.01.
$P(Z < z) = 0.99$ gives $z = 2.326$.

If the $p$-value for the confidence interval that you require is not in this summary table, you will need to work from the main body of the normal distribution table.

## (b) Confidence interval for $\mu$, the population mean

- of a non-normal population,
- with a known variance $\sigma^2$
- using a large sample, $n \geqslant 30$ say

In this case, since the sample size is large, the **central limit theorem** can be used.

$\overline{X}$ is approximately normal and $\overline{X} \sim N\left(\mu, \dfrac{\sigma^2}{n}\right)$ (see page 442).

If $\bar{x}$ is the mean of a random sample of size $n$, where $n$ is large ($n \geqslant 30$), taken from a non-normal population with known variance $\sigma^2$,
then a 95% confidence interval for $\mu$ is given by

$$\left(\bar{x} - 1.96 \frac{\sigma}{\sqrt{n}}, \bar{x} + 1.96 \frac{\sigma}{\sqrt{n}}\right)$$

### Example 9.20

The heights of men in a particular district are distributed with mean $\mu$ cm and the standard deviation $\sigma$ cm.
On the basis of the results obtained from a random sample of 100 men from the district, the 95% confidence interval for $\mu$ was calculated and found to be (177.22 cm, 179.18 cm).

Calculate
(a) the value of the sample mean,
(b) the value of $\sigma$,
(c) a symmetric 90% confidence interval for $\mu$.

### Solution 9.20

Let $X$ be the height, in centimetres, of a man in the district.
The distribution of $X$ is not known, but $E(X) = \mu$ and $Var(X) = \sigma^2$.

Since the sample size, $n$, is large ($n = 100$), using the **central limit theorem**, the distribution of $\overline{X}$ is approximately normal with mean $\mu$ and variance $\dfrac{\sigma^2}{n}$.

(a) A 95% confidence interval for $\mu$ is given by

$$\left(\bar{x} - 1.96 \frac{\sigma}{\sqrt{n}}, \bar{x} + 1.96 \frac{\sigma}{\sqrt{n}}\right) \quad \text{with } n = 100, \text{ so } \sqrt{n} = 10$$

Since the interval is (177.22, 179.18)

$$\bar{x} - 1.96 \frac{\sigma}{10} = 177.22 \ \ldots \ \text{①}$$

$$\bar{x} + 1.96 \frac{\sigma}{10} = 179.18 \ \ldots \ \text{②}$$

Adding ① and ②
$$2\bar{x} = 356.4$$
$$\bar{x} = 178.2$$

**The sample mean is 178.2 cm.**

(b) Subtracting: ② – ①     $2 \times 1.96 \dfrac{\sigma}{10} = 1.96$

$$\sigma = 5$$

(c) A symmetric 90% confidence interval for $\mu$ is

$$\left( \bar{x} - 1.645 \frac{\sigma}{\sqrt{n}} , \bar{x} + 1.645 \frac{\sigma}{\sqrt{n}} \right)$$

$$\bar{x} \pm 1.645 \frac{\sigma}{\sqrt{n}} = 178.2 \pm 1.645 \times \frac{5}{10}$$

$$= 178.2 \pm 0.8225$$

So the 90% confidence interval = $(178.2 - 0.8225, 178.2 + 0.8225)$

$$= (177.38 \text{ cm}, 179.02 \text{ cm}) \quad \textbf{(2 d.p.)}$$

---

## Example 9.21

A plant produces steel sheets whose weights are known to be normally distributed with a standard deviation of 2.4 kg. A random sample of 36 sheets had a mean weight of 31.4 kg. Find 99% confidence limits for the population mean. *(L)*

## Solution 9.21

$X$ is the weight, in kilograms, of a steel sheet. Then $X \sim N(\mu, 2.4^2)$.

A sample of size 36 is taken, so $n = 36$ and the sample mean $\bar{x} = 31.4$.

The end-values of a 99% confidence interval for $\mu$ are

$$\bar{x} \pm 2.576 \frac{\sigma}{\sqrt{n}} = 31.4 \pm 2.576 \times \frac{2.4}{\sqrt{36}}$$

$$= 31.4 \pm 1.0304$$

so the 99% confidence interval is $(31.4 - 1.0304, 31.4 + 1.0304) = (30.3696, 32.4304)$

$$= (30.4 \text{ kg}, 32.4 \text{ kg}) \quad \textbf{(3 s.f.)}$$

---

## Width of a confidence interval

In Example 9.21,

width of the 99% confidence interval = $2 \times 2.576 \times \dfrac{\sigma}{\sqrt{n}}$

$$= 2.0608 \text{ kg.}$$

$\bar{x} - 2.576 \dfrac{\sigma}{\sqrt{n}} \qquad \bar{x} \qquad \bar{x} + 2.576 \dfrac{\sigma}{\sqrt{n}}$

width

$2 \times 2.576 \dfrac{\sigma}{\sqrt{n}}$

*For the same data*
The width of the 95% confidence interval

$$= 2 \times 1.96 \frac{\sigma}{\sqrt{n}}$$

$$= 2 \times 1.96 \times \frac{2.4}{\sqrt{36}}$$

$$= 1.568 \text{ kg}$$

For a given sample size $n$,

the greater the level of confidence, the wider the confidence interval.

[diagram: $\bar{x} - 1.96 \frac{\sigma}{\sqrt{n}}$    $\bar{x}$    $\bar{x} + 1.96 \frac{\sigma}{\sqrt{n}}$; width $= 2 \times 1.96 \frac{\sigma}{\sqrt{n}}$]

## Determination of sample size

### Example 9.22

The result $X$ of a stress test is known to be a normally distributed random variable with mean $\mu$ and standard deviation 1.3. It is required to have a 95% symmetrical confidence interval for $\mu$ with total width less than 2. Find the least number of tests that should be carried out to achieve this.

(L)

### Solution 9.22

$X \sim N(\mu, 1.3^2)$, and for samples of size $n$, $\bar{X} \sim N\left(\mu, \frac{1.3^2}{n}\right)$

The 95% confidence interval for $\mu$ is

$$\left(\bar{x} - 1.96 \frac{\sigma}{\sqrt{n}}, \bar{x} + 1.96 \frac{\sigma}{\sqrt{n}}\right)$$

Interval width $= 2 \times 1.96 \frac{\sigma}{\sqrt{n}}$

$$= 2 \times 1.96 \times \frac{1.3}{\sqrt{n}} \quad \ldots \text{①}$$

$$= \frac{5.096}{\sqrt{n}}$$

The width of the interval must be less than 2,

$$\therefore \frac{5.096}{\sqrt{n}} < 2 \quad \ldots \text{②}$$

$$\sqrt{n} > \frac{5.096}{2}$$

$$\sqrt{n} > 2.548$$

$$n > 2.548^2$$

i.e. $\quad n > 6.49\ldots$

**The least number of tests that should be carried out is 7.**

Now suppose that, in Example 9.22, the 95% confidence interval for $\mu$ must have a width less than 1. Will the sample size $n$ be larger or smaller than when the total width was less than 2?

To answer this, look at equation ②. This now becomes

$$\frac{5.096}{\sqrt{n}} < 1$$

i.e. $\sqrt{n} > 5.096$

$\quad n > 25.96$

So the least number of tests would be 26. The sample size has increased.

For a given confidence level,

the smaller the interval width, the larger the sample size required.

Now consider the situation as in Example 9.22, where the total width must be less than 2, but the confidence level is increased to 99%.
Will the sample size $n$ be larger or smaller than that required for the 95% confidence interval?

The calculations needed to find the width are similar to those given in Solution 9.22, equation ① but the value 1.96 will be replaced by 2.576, the $z$-value for the 99% confidence interval.

So interval width $= 2 \times 2.576 \times \dfrac{1.3}{\sqrt{n}}$

$$= \frac{6.6976}{\sqrt{n}}$$

Therefore $\quad \dfrac{6.6976}{\sqrt{n}} < 2$

$$\sqrt{n} > \frac{6.6976}{2}$$

$$n > 11.21 \ldots$$

For the 99% confidence interval, the least number of tests required is 12, whereas for the 95% confidence interval it was 7, so the sample size must be larger.

For a given interval width,

the greater the level of confidence, the larger the sample size required.

## (c) Confidence interval for $\mu$, the population mean

- **of a normal or non-normal population,**
- **with unknown variance** $\sigma^2$
- **using a large sample,** $n$

When calculating confidence intervals it is often the case that the population variance, $\sigma^2$, is not known. **Provided that the sample size, $n$, is large, ($n \geqslant 30$ say) it is permissible to use $\hat{\sigma}^2$,** the best unbiased estimate for $\sigma^2$ (see page 447).

Ideally the distribution of $X$ should be normal, but an approximate confidence interval can also be given when the distribution of $X$ is not normal. Remember that in both cases, $n$ must be large.

Provided that $n$ is large, ($n \geqslant 30$ say),
a 95% confidence interval for $\mu$ is

$$\left( \bar{x} - 1.96 \frac{\hat{\sigma}}{\sqrt{n}}, \quad \bar{x} + 1.96 \frac{\hat{\sigma}}{\sqrt{n}} \right)$$

where $\hat{\sigma}^2 = \dfrac{n}{n-1} s^2$      $s^2$ is the sample variance

or     $\hat{\sigma}^2 = \dfrac{1}{n-1} \left( \Sigma x^2 - \dfrac{(\Sigma x)^2}{n} \right)$.

## Example 9.23

The fuel consumption of a new model of car is being tested. In one trial, 50 cars chosen at random, were driven under identical conditions and the distances, $x$ km, covered on 1 litre of petrol were recorded. The results gave the following totals:

$$\Sigma x = 525, \quad \Sigma x^2 = 5625.$$

Calculate a 95% confidence interval for the mean petrol consumption, in kilometres per litre, of cars of this type.

## Solution 9.23

$$\bar{x} = \frac{\Sigma x}{n} = \frac{525}{50} = 10.5$$

$\sigma^2$ is unknown, so use $\hat{\sigma}^2$ where $\hat{\sigma}^2 = \dfrac{1}{n-1} \left( \Sigma x^2 - \dfrac{(\Sigma x)^2}{n} \right)$

$$= \frac{1}{49} \left( 5625 - \frac{525^2}{50} \right)$$

$$= 2.2959\ldots$$

$$\hat{\sigma} = 1.515\ldots$$

95% confidence limits for $\mu$ are

$$\bar{x} \pm 1.96 \frac{\hat{\sigma}}{\sqrt{n}} = 10.5 \pm 1.96 \times \frac{1.515\ldots}{\sqrt{50}}$$

$$= 10.5 \pm 0.42$$

**95% confidence interval for $\mu$ = (10.08 km/litre, 10.92 km/litre)**

## Example 9.24

The height, $x$ cm, of each man in a random sample of 200 men living in the UK was measured. The following results were obtained:

$\Sigma x = 35\,050, \Sigma x^2 = 6\,163\,109.$

(a) Calculate unbiased estimates of the mean and variance of the heights of men living in the UK.

(b) Determine an approximate 90% confidence interval for the mean height of men living in the UK. Name the theorem that you have assumed.

(NEAB)

## Solution 9.24

(a) $\hat{\mu} = \bar{x} = \dfrac{\sum x}{n} = \dfrac{35\,050}{200} = 175.25$

$\hat{\sigma}^2 = \dfrac{n}{n-1} s^2$       see page 447

$\phantom{\hat{\sigma}^2} = \dfrac{n}{n-1}\left(\dfrac{\sum x^2}{n} - \bar{x}^2\right)$

$\phantom{\hat{\sigma}^2} = \dfrac{200}{199}\left(\dfrac{6\,163\,109}{200} - 175.25^2\right)$

$\phantom{\hat{\sigma}^2} = 103.5$

Alternatively

$\hat{\sigma}^2 = \dfrac{1}{n-1}\left(\sum x^2 - \dfrac{(\sum x)^2}{n}\right)$       see page 448

$\phantom{\hat{\sigma}^2} = \dfrac{1}{199}\left(6163\,109 - \dfrac{35\,050^2}{200}\right)$

$\phantom{\hat{\sigma}^2} = 103.5$

(b) The confidence limits for 90% confidence interval for $\mu$ are

$\bar{x} \pm 1.645\,\dfrac{\hat{\sigma}}{\sqrt{n}} = 175.25 \pm 1.645 \times \dfrac{\sqrt{103.5}}{\sqrt{200}}$

$\phantom{\bar{x} \pm 1.645\,\dfrac{\hat{\sigma}}{\sqrt{n}}} = 175.25 \pm 1.1833$

So 90% confidence interval is $(175.25 - 1.1833\ldots, 175.25 + 1.1833\ldots)$
$= (174.07\text{ cm}, 176.43\text{ cm})$    (2 d.p.)

The **central limit theorem** has been used to give an approximate distribution for $\bar{X}$, the sample mean where $\bar{X} \sim N\!\left(\mu, \dfrac{\sigma^2}{n}\right)$.

# Exercise 9e    Point estimates and confidence intervals for $\mu$ (using normal distribution)

1. The concentrations, in milligrams per litre, of a trace element in 7 randomly chosen samples of water from a spring were

   240.8, 237.3, 236.7, 236.6, 234.2, 233.9, 232.5.

   Determine the unbiased estimates of the mean and the variance of the concentration of the trace element per litre of water from the spring.    (L)

2. Find the best unbiased estimates of the mean $\mu$ and variance $\sigma^2$ of the population from which each of the following samples is drawn. It is a good idea to do parts (a) to (c) both with and without a calculator.

   (a)  46,  48,  51,  50,  45,  53,  50,  48

   (b)  1.684,  1.691,  1.687,  1.688,  1.689, 1.688,  1.690,  1.693,  1.685

   (c)

   | $x$ | 20 | 21 | 22 | 23 | 24 | 25 |
   |-----|----|----|----|----|----|----|
   | $f$ | 4  | 14 | 17 | 26 | 20 | 9  |

   (d)  $\Sigma x = 120$,    $\Sigma x^2 = 2\ 102$,    $n = 8$

   (e)  $\Sigma x = 100$,    $\Sigma x^2 = 1\ 028$,    $n = 10$

   (f)  $n = 34$,    $\Sigma x = 330$,    $\Sigma x^2 = 23\ 700$

3. A measuring rule was used to measure the length of a rod of stated length 1 m. On 8 successive occasions the following results, in millimetres, were obtained.

   1000, 999, 999, 1002, 1001, 1000, 1002, 1001.

   Calculate unbiased estimates of the mean and, to two significant figures, the variance of the errors occurring when the rule is used for measuring a 1 m length.    (L)

4. Cartons of orange are filled by a machine. A sample of 10 cartons selected at random from the production contained the following quantities (in millilitres)

   201.2   205.0   209.1   202.3   204.6
   206.4   210.1   201.9   203.7   207.3

   Calculate unbiased estimates of the mean and variance of the population from which the sample was taken.    (L)

5. A certain type of tennis ball is known to have a height of bounce which is normally distributed with standard deviation 2 cm. A sample of 60 tennis balls is tested and the mean height of bounce of the sample is 140 cm.

   (a)  Find a 95% confidence interval for the mean height of bounce of this type of tennis ball.
   (b)  State any assumptions made in calculating your interval.

6. A random sample of 6 items taken from a normal population with mean $\mu$ and variance 4.5 cm$^2$ gave the following data:
   Sample values: 12.9 cm, 13.2 cm, 14.6 cm, 12.6 cm, 11.3 cm, 10.1 cm.

   (a)  Find the 95% confidence interval for $\mu$.
   (b)  What is the width of this confidence interval?

7. A factory produces cans of meat whose masses are normally distributed with standard deviation 18 g. A random sample of 25 cans is found to have a mean mass of 458 g.

   (a)  Obtain the 99% confidence interval for the population mean mass of a can of meat produced at the factory.
   (b)  Explain what the interval means.
   (c)  Would the interval be wider if a 90% confidence interval was calculated? Explain your reasoning.

8. A random sample of 100 observations from a normal population with mean $\mu$ gave the following data: $\Sigma x = 8200$, $\Sigma x^2 = 686\ 800$.

   (a)  Find a 98% confidence interval for $\mu$.
   (b)  Find a 99% confidence interval for $\mu$.
   (c)  Would your answers have been different if the population was not normal? Explain your answer.

9. Eighty employees at an insurance company were asked to measure their pulse rates when they woke up in the morning. The researcher then calculated the mean and the standard deviation of the sample and found these to be 69 beats and 4 beats respectively. Calculate a 97% confidence interval for the mean pulse rate of all the employees at the company, stating any assumptions that you have made.

10. One hundred and fifty bags of flour are taken from a production line and found to have a mean mass of 748 g and standard deviation of 3.6 g.

    (a)  Calculate an unbiased estimate of the standard deviation of a bag of flour produced on this production line.
    (b)  Calculate a 98% confidence interval for the mean mass of a bag of flour produced on this production line.
    (c)  State any assumptions you have made.

11. (a)  A 95% confidence interval for the mean length of life of a particular brand of light bulb was calculated and the confidence limits were 1023.3 hours and 1101.7 hours. The interval was based on the results of a random sample of 36 light bulbs. Find the 99% confidence interval for $\mu$, the mean length of life of this brand of light bulb.

(b) Forty random samples of 36 light bulbs are taken and a 90% confidence interval for $\mu$ is calculated for each sample. Find the expected number of intervals that contain $\mu$.

12. An efficiency expert wishes to determine the mean time taken to drill a number of holes in a metal sheet. Determine how large a random sample is needed so that the expert can be 95% certain that the sample mean will differ from the true mean time by less than 15 seconds. Assume that it is known from previous studies that the population standard deviation is 40 seconds. (L)

13. A random sample of 60 loaves is taken from a population whose masses are normally distributed with mean $\mu$ and standard deviation 10 g.
   (a) Calculate the width of a symmetric 95% confidence interval for $\mu$ based on this sample.
   (b) Find the confidence level of a symmetric 95% confidence interval having the same width as before but based on a random sample of 40 loaves.

14. The distribution of measurements of thicknesses of a random sample of yarns produced in a textile mill is shown in the following table.

| Yarn thickness in microns (mid-interval values) | Frequency |
|---|---|
| 72.5 | 6 |
| 77.5 | 18 |
| 82.5 | 32 |
| 87.5 | 57 |
| 92.5 | 102 |
| 97.5 | 51 |
| 102.5 | 25 |
| 107.5 | 9 |

Illustrate these data on a histogram.
Estimate, to two decimal places, the mean and standard deviation of yarn thickness. Hence estimate the standard error of the mean to two decimal places, and use it to determine approximate symmetric 95% confidence limits, giving your answer to one decimal place. (MEI)

15. The age, $X$, in years at last birthday, of 250 mothers when their first child was born is given in the following table:

| $x$ | No. of mothers |
|---|---|
| 18– | 14 |
| 20– | 36 |
| 22– | 42 |
| 24– | 57 |
| 26– | 48 |
| 28– | 26 |
| 30– | 17 |
| 32– | 7 |
| 34– | 2 |
| 36– | 0 |
| 38– | 1 |

(The notation implies that, for example in row 1, there are 14 mothers for whom the continuous variable $X$ satisfies $18 \leqslant X < 20$.)
Calculate, to the nearest 0.1 of a year, estimates of the mean and the standard deviation of $X$.

If the 250 mothers are a random sample from a large population of mothers, find 95% confidence limits for the mean age, $\mu$, of the total population. (C)

16. The lifetimes of 200 electrical components were recorded to the nearest hour and classified in the frequency tabulation.

| Lifetime | Frequency | Lifetime | Frequency |
|---|---|---|---|
| 0– | 80 | 600– | 4 |
| 100– | 48 | 700– | 3 |
| 200– | 30 | 800– | 2 |
| 300– | 18 | 900– | 0 |
| 400– | 10 | 1000– | 0 |
| 500– | 5 | | |

Draw a histogram of the data and estimate the mean and standard deviation of the distribution. Calculate a symmetric 90% confidence interval for the population means, using a suitable normal approximation for the distribution of the sample mean. (MEI)

## (d) Confidence interval for $\mu$ when

- the population is normal
- $\sigma^2$ is unknown,
- sample size $n$ is small,

When calculating confidence intervals, you have already encountered the situation when **large samples** ($n \geqslant 30$) are taken from a **normal** population with **unknown variance** $\sigma^2$.

**For large samples,**

$$\frac{\overline{X} - \mu}{\hat{\sigma}/\sqrt{n}} = Z \quad \text{where } Z \sim N(0, 1)$$

But if the **sample size is small** ($n < 30$), $\dfrac{\overline{X} - \mu}{\hat{\sigma}/\sqrt{n}}$ no longer has a normal distribution.

**For small samples,**

$$\frac{\overline{X} - \mu}{\hat{\sigma}/\sqrt{n}} = T \quad \text{where } T \text{ has a } t\text{-distribution.}$$

Before looking at confidence intervals $\mu$ when the sample size is small, consider further the $t$-distribution.

## THE $t$-DISTRIBUTION

The distribution of $T$ is a member of a family of $t$-distributions. All $t$-distributions are symmetric about zero and have a single parameter $\nu$ (pronounced new) which is a positive integer.

$\nu$ is known as the **number of degrees of freedom** of the distribution and if, for example, $T$ has a $t$-distribution with five degrees of freedom, you would write $T \sim t(5)$.

The diagram below shows two curves, $t(2)$ and $t(5)$.
Note that as $\nu$ increases, the corresponding $t(\nu)$ curve resembles the standardised normal distribution $N(0, 1)$. In fact when $\nu \geqslant 30$, the difference between the $t(\nu)$ distribution and the normal distribution is negligible.
For samples of size $n$, it can be shown that

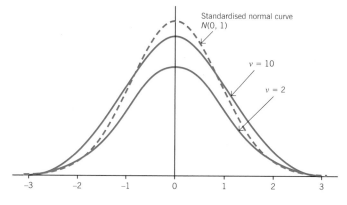

$$T = \frac{\overline{X} - \mu}{\hat{\sigma}/\sqrt{n}} \text{ follows a } t\text{-distribution with } (n - 1) \text{ degrees of freedom.}$$

For example, for a sample of size 8,
$T$ follows a $t$-distribution with 7 degrees of freedom. You would write $T \sim t(7)$.

The 95% confidence interval for $\mu$ is obtained as follows:

If $\bar{x}$ and $s^2$ are the mean and variance of a small sample ($n < 30$) from a normal population with unknown mean $\mu$ and unknown variance $\sigma^2$,
then a 95% confidence interval for $\mu$ is given by

$$\left( \bar{x} - t\,\frac{\hat{\sigma}}{\sqrt{n}},\ \bar{x} + t\,\frac{\hat{\sigma}}{\sqrt{n}} \right) \quad \text{where } \hat{\sigma}^2 = \frac{n}{n-1}\,s^2$$

and $t$ is the value from a $t(n-1)$ distribution such that $P(-t \leqslant T \leqslant t) = 0.95$,

i.e. $(-t, t)$ encloses 95% of the $t(n-1)$ distribution.

To find the required value of $t$, known as the **critical value**, you will need to use $t$-distribution tables. These give the $t$-value such that $P(T \leqslant t) = p$, for various values of $\mu$. The tables are printed on page 650 and an extract is reproduced here.

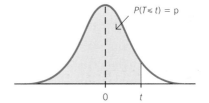

| $p$ | 0.75 | 0.90 | 0.95 | 0.975 | 0.99 | 0.995 | 0.9975 | 0.999 | 0.9995 |
|---|---|---|---|---|---|---|---|---|---|
| $v = 1$ | 1.000 | 3.078 | 6.314 | 12.71 | 31.82 | 63.66 | 127.3 | 318.3 | 636.6 |
| 2 | 0.816 | 1.886 | 2.920 | 4.303 | 6.965 | 9.925 | 14.09 | 22.33 | 31.60 |
| 3 | 0.765 | 1.638 | 2.353 | 3.182 | 4.541 | 5.841 | 7.453 | 10.21 | 12.92 |
| 4 | 0.741 | 1.533 | 2.132 | 2.776 | 3.747 | 4.604 | 5.598 | 7.173 | 8.610 |
| 5 | 0.727 | 1.476 | 2.015 | 2.571 | 3.365 | 4.032 | 4.773 | 5.893 | 6.869 |
| 6 | 0.718 | 1.440 | 1.943 | 2.447 | 3.143 | 3.707 | 4.317 | 5.208 | 5.959 |
| 7 | 0.711 | 1.415 | 1.895 | 2.365 | 2.998 | 3.499 | 4.029 | 4.785 | 5.408 |
| 8 | 0.706 | 1.397 | 1.860 | 2.306 | 2.896 | 3.355 | 3.833 | 4.501 | 5.041 |
| 9 | 0.703 | 1.383 | 1.833 | 2.262 | 2.821 | 3.250 | 3.690 | 4.297 | 4.781 |
| 10 | 0.700 | 1.372 | 1.812 | 2.228 | 2.764 | 3.169 | 3.581 | 4.144 | 4.587 |
| 11 | 0.697 | 1.363 | 1.796 | 2.201 | 2.718 | 3.106 | 3.497 | 4.025 | 4.437 |
| ⋮ | ⋮ | ⋮ | ⋮ | ⋮⋮ | ⋮ | ⋮ | ⋮ | ⋮ | ⋮ |
| 30 | 0.683 | 1.310 | 1.697 | 2.042 | 2.457 | 2.750 | 3.030 | 3.385 | 3.646 |
| 40 | 0.681 | 1.303 | 1.684 | 2.021 | 2.423 | 2.704 | 2.971 | 3.307 | 3.551 |
| 60 | 0.679 | 1.296 | 1.671 | 2.000 | 2.390 | 2.660 | 2.915 | 3.232 | 3.460 |
| 120 | 0.677 | 1.289 | 1.658 | 1.980 | 2.358 | 2.617 | 2.860 | 3.160 | 3.373 |
| ∞ | 0.674 | 1.282 | 1.645 | 1.960 | 2.326 | 2.576 | 2.807 | 3.090 | 3.291 |

You will see that as $v$ increases, the corresponding $t$-distribution becomes more and more like the normal distribution. Compare the last row, $v = \infty$, with the critical values for the normal distribution, printed on page 649.

You will find that you use the *t*-distribution tables in a slightly different way from the normal tables, so you need to ensure that you can use them correctly.

In this extract, the highlighted values are referred to in the text.

| $p$ | 0.75 | 0.90 | 0.95 | 0.975 | 0.99 | 0.995 | 0.9975 | 0.999 | 0.9995 |
|---|---|---|---|---|---|---|---|---|---|
| $v = 1$ | 1.000 | 3.078 | 6.314 | 12.71 | 31.82 | 63.66 | 127.3 | 318.3 | 636.6 |
| 2 | 0.816 | 1.886 | 2.920 | 4.303 | 6.965 | 9.925 | 14.09 | 22.33 | 31.60 |
| ⋮ | ⋮ | ⋮ | ⋮ | ⋮ | ⋮ | ⋮ | ⋮ | ⋮ | ⋮ |
| 10 | 0.700 | 1.372 | 1.812 | 2.228 | 2.764 | 3.169 | 3.581 | 4.144 | 4.587 |
| (a) 11 | 0.697 | 1.363 | 1.796 | 2.201 | 2.718 | 3.106 | 3.497 | 4.025 | 4.437 |
| 12 | 0.695 | 1.356 | 1.782 | 2.179 | 2.681 | 3.055 | 3.428 | 3.930 | 4.318 |
| 13 | 0.694 | 1.350 | 1.771 | 2.160 | 2.650 | 3.012 | 3.372 | 3.852 | 4.221 |
| (b) 14 | 0.692 | 1.345 | 1.761 | 2.145 | 2.624 | 2.977 | 3.326 | 3.787 | 4.140 |

## Example 9.25(a)

Consider $T$ following a *t*-distribution with 11 degrees of freedom, i.e. $v = 11$ and $T \sim t(11)$.

Find    (i)    $P(T < 1.796)$
         (ii)    $P(T > 3.106)$
        (iii)    $P(|T| < 2.201)$

## Solution 9.25(a)

(i)    $v = 11$, so find row 11 and go across to 1.796, then up to the top of the column. This gives 0.95.
     $P(T < 1.796) = 0.95 = 95\%$

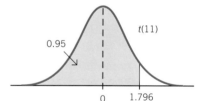

(ii)    Find row 11, go across to 3.106, which is in column 0.995.
     So   $P(T < 3.106) = 0.995$
     $\therefore$    $P(T > 3.106) = 1 - 0.995 = 0.005$
     i.e.   $P(T > 3.106) = 0.5\%$

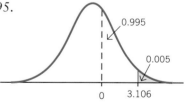

(iii) You need $P(|T| < 2.201)$ i.e. $P(-2.201 < T < 2.201)$
Find row 11, go across to 2.201 which is in column 0.975.

So $\quad P(T < 2.201) = 0.975$
$\quad\quad P(T > 2.201) = 1 - 0.975 = 0.025$

It follows that

$P(T < -2.201) = 0.025$ also
$\therefore \quad P(-2.201 < T < 2.201) = 1 - 0.025 - 0.025$
$\quad\quad\quad\quad\quad\quad\quad\quad\quad = 0.95$
$\quad\quad\quad\quad\quad\quad\quad\quad\quad = 95\%$

The *t*-values that enclose the central 95% are ±2.201.

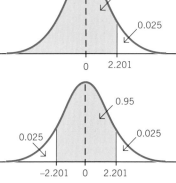

## Example 9.25(b)

The random variable $T$ has a *t*-distribution with 14 degrees of freedom, i.e. $T \sim t(14)$.
Find the value of $t$ for which
(i) $P(T < t) = 0.90$ $\quad$ (ii) $P(|T| < t) = 0.98$

## Solution 9.25(b)

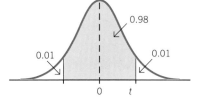

(i) $\quad v = 14$, row 14, column 0.90 gives $t = \textbf{1.345}$.
(ii) $\quad$ The required value for $t$ corresponds to an upper tail
probability of 0.01, so the *p*-value must be 0.99.
Row 14, column 0.99 gives $t = \textbf{2.624}$.

# Critical *t*-values for a 95% confidence interval

For a 95% confidence interval, you want $t$ such that
$P(|T| < t) = 0.95$, i.e. $P(-t < T < t) = 0.95$.

This corresponds to an upper tail probability of 0.025,
so $P(T < t) = 0.975$.

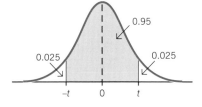

| $p$ | 0.75 | 0.90 | 0.95 | 0.975 |
|---|---|---|---|---|
| $v = 1$ | 1.000 | 3.078 | 6.314 | 12.71 |
| 2 | 0.816 | 1.886 | 2.920 | 4.303 |
| 3 | 0.765 | 1.638 | 2.353 | 3.182 |
| 4 | 0.741 | 1.533 | 2.132 | 2.776 |
| 5 | 0.727 | 1.476 | 2.015 | 2.571 |
| 6 | 0.718 | 1.440 | 1.943 | 2.447 |
| 7 | 0.711 | 1.415 | 1.895 | 2.365 |
| 8 | 0.706 | 1.397 | 1.860 | 2.306 |
| 9 | 0.703 | 1.383 | 1.833 | 2.262 |

So if $T \sim t(9)$ for example,

$P(T < t) = 0.975$   gives $t = 2.262$

and the critical values of $t$ for the 95% confidence interval are ±2.262

NOTE: This value will be used in Example 9.26.

To find the critical values of $t$
for a 95% confidence interval, look under column 0.975.

Similarly
for a 90% confidence interval, look under column 0.95,
for a 98% confidence interval, look under column 0.99,
for a 99% confidence interval, look under column 0.995.

The following examples illustrate how to calculate confidence intervals when critical values are found using a $t$-distribution.

## Example 9.26

The mass, in grams, of a packet of biscuits of a particular brand, follows a normal distribution with mean $\mu$. Ten packets of biscuits are chosen at random and their masses noted. The results, in grams, are

397.3, 399.6, 401.0, 392.9, 396.8, 400.0, 397.6, 392.1, 400.8, 400.6

These can be summarised as follows: $\Sigma x = 3978.8$, $\Sigma x^2 = 1\,583\,098.3$.
Calculate a 95% confidence interval for $\mu$.

## Solution 9.26

$X$ is the mass, in grams, of a packet of biscuits.
$X \sim N(\mu, \sigma^2)$ with both $\mu$ and $\sigma^2$ unknown.

Since $\sigma^2$ is unknown, find $\hat{\sigma}^2$ (see page 447)

$$\hat{\sigma} = \frac{1}{n-1}\left(\Sigma x^2 - \frac{(\Sigma x)^2}{n}\right) \quad \text{with } n = 10$$

$$= \frac{1}{9}\left(1\,583\,098.3 - \frac{3978.7^2}{10}\right)$$

$$= 10.325\ldots$$

$$\hat{\sigma} = 3.213\ldots$$

The sample mean, $\bar{x} = \dfrac{\Sigma x}{n} = \dfrac{3978.7}{10} = 397.87$.

Since $n$ is small, a $t(n-1)$ distribution is required.
$n = 10$, so use a $t(9)$ distribution.

The 95% confidence limits for $\mu$ are

$$\bar{x} \pm t\frac{\hat{\sigma}}{\sqrt{n}} \quad \text{where } (-t, t) \text{ enclose 95\% of the } t(9) \text{ distribution.}$$

From tables, as illustrated on page 466, the critical value $t$ is 2.262

$$\therefore \text{ confidence limits are } 397.87 \pm 2.262 \times \frac{3.213\ldots}{\sqrt{10}}$$

$$= 397.87 \pm 2.298 \ldots$$

95% confidence interval for $\mu = (397.87 - 2.298 \ldots, 397.87 + 2.298 \ldots)$

$$= \textbf{(395.6 g, 400.3 g)} \quad \textbf{(1 d.p.)}$$

---

Note that in Example 9.27, the value of $\hat{\sigma}$ can be obtained directly by using the calculator in standard deviation mode. Look for the key $\boxed{x_{\sigma_{n-1}}}$ (see page 448).
You should practise finding $\hat{\sigma}$ using the data of Example 9.27.

## Example 9.27

A student, studying the height of a particular plant, knows that it follows a normal distribution with mean $\mu$ and variance $\sigma^2$, but he does not know the value of either of these parameters. He selects 15 plants at random, measures their heights and calculates that the mean height of the sample is 12.2 cm and the standard deviation is 1.4 cm. Using these values, calculate a 90% confidence interval for $\mu$. Calculate also the width of this interval.

## Solution 9.27

$X$ is the height, in centimetres, of a plant, where $X \sim N(\mu, \sigma^2)$
Sample values: $\bar{x} = 12.2$ and $s = 1.4$ where $s$ is the standard deviation.

$$\hat{\sigma}^2 = \frac{n}{n-1}s^2 \quad \text{with } n = 15$$

$$= \tfrac{15}{14} \times 1.4 = 1.5$$

$$\hat{\sigma} = \sqrt{1.5} = 1.22 \ldots$$

Since $n = 15$, the $t(14)$ distribution is considered.

| $p$ | 0.75 | 0.90 | 0.95 |
|---|---|---|---|
| $\nu = 1$ | 1.000 | 3.078 | 6.314 |
| 2 | 0.816 | 1.886 | 2.920 |
| $\vdots$ | $\vdots$ | $\vdots$ | $\vdots$ |
| 12 | 0.695 | 1.356 | 1.782 |
| 13 | 0.694 | 1.350 | 1.771 |
| 14 | 0.692 | 1.345 | 1.761 |

For a symmetric 90% confidence interval, column $p = 0.95$ is required in order to find the critical value of $t$.
When $\nu = 14$, $t = 1.761$,
so $(-1.761, 1.761)$ encloses the central 90% of the $t(14)$ distribution.

(Extract from tables on page 650)

The 90% confidence limits for $\mu$ are

$$\bar{x} \pm t\frac{\hat{\sigma}}{\sqrt{n}} = 12.2 \pm 1.761 \times \frac{1.22\ldots}{\sqrt{15}}$$
$$= 12.2 \pm 0.556\ldots$$

90% confidence interval $= (12.2 - 0.556\ldots, 12.2 + 0.556\ldots)$
$$= (\textbf{11.64 cm, 12.76 cm}) \quad \textbf{(2 d.p.)}$$

Width of interval $= 2 \times 0.556\ldots$
$$= \textbf{1.11 cm} \quad \textbf{(2 d.p.)}$$

## Exercise 9f  Confidence intervals involving small samples requiring the *t*-distribution

1. The heights, in metres, of a random sample of 6 policemen from a particular station were as follows:

   1.80, 1.76, 1.79, 1.81, 1.83, 1.79.

   Assuming that the heights of policemen from that station are normally distributed with mean $\mu$,

   (a) calculate a 95% confidence interval for $\mu$,
   (b) state the width of this interval.

2. A sample of 8 independent observations of a normally distributed variable gave the following values:

   3.6, 3.9, 4.5, 3.8, 4.4, 4.9, 4.2, 3.8.

   (a) Determine a 99% confidence interval for the population mean $\mu$.
   (b) Find the difference between the widths of a 90% confidence interval for $\mu$ and a 95% confidence interval for $\mu$.

3. Twenty measurements of $x$, the life, in hours, of a particular make of candle gave the following data:

   $\Sigma x = 172$, $\Sigma x^2 = 1495.5$.

   Assuming that the length of life is modelled by a normal distribution with mean $\mu$, find a 98% confidence interval for $\mu$.

4. A random sample of 8 independent observations of a normal variable gave

   $\Sigma x = 261.2$, $\Sigma (x - \bar{x})^2 = 3.22$.

   Calculate a 95% confidence interval for the population mean.
   If 400 such samples were taken, how many of these would you expect not to include the population mean?

5. A random sample of 7 independent observations of a normal variable gave

   $\Sigma x = 35.9$, $\Sigma x^2 = 186.19$.

   Calculate

   (a) an unbiased estimate of the population mean,
   (b) an unbiased estimate of the population standard deviation,
   (c) a 90% confidence interval for the population mean.

6. The masses, in grams, of 13 washers selected from a production line at random are:

   15.4, 15.2, 14.6, 16.1, 14.8,
   15.3, 15.9, 16.0, 15.4, 14.6,
   15.0, 15.5, 16.1.

   Calculate 98% confidence limits for the mean mass of the washers on this particular production line, assuming that the mass can be modelled by a normal distribution.

7. Fifteen pupils performed experiments to find the value of $g$, the acceleration due to gravity. Their results were as follows:

   9.806, 9.807, 9.810, 9.802, 9.805,
   9.806, 9.804, 9.811, 9.801, 9.804,
   9.805, 9.808, 9.803, 9.809, 9.807.

   Assuming that these are taken from a normal population, calculate 95% confidence limits for the value of $g$ based on these results.

# CONFIDENCE INTERVALS FOR THE POPULATION PROPORTION, $p$

Imagine that you want to find $p$, the proportion of successes in a particular population. To obtain an idea of its value, you could take a random sample of size $n$ and calculate $p_s$, the proportion of successes in your sample. This would give the best unbiased estimate $\hat{p}$, where $\hat{p} = p_s$ (see page 447). You could also use this value of $p_s$ to obtain an interval estimate of $p$, known as a **confidence interval for $p$**.

The theory needed to derive the confidence interval for $p$ is based on the sampling distribution of proportions, $P_s$, described on page 445.

This states that, **provided the sample size $n$ is large, $(n \geqslant 30)$,**

the distribution of $P_s$ is **normal**, so $P_s \sim N\left(p, \dfrac{pq}{n}\right)$ where $q = 1 - p$.

The standard deviation of the sampling distribution of proportions, $\sqrt{\dfrac{pq}{n}}$ is needed in the

calculation of the limits for the confidence interval. The difficulty is, however, that its value isn't known, since $p$ isn't known!

To overcome this, use $\hat{p} = p_s$. Writing $1 - p_s$ as $q_s$, the standard deviation of the sampling

distribution is approximately $\sqrt{\dfrac{p_s q_s}{n}}$.

You are then able to find *approximate* confidence intervals for $p$ as follows:

| | Confidence limits | Confidence interval | Width |
|---|---|---|---|
| 90% | $p_s \pm 1.645 \sqrt{\dfrac{p_s q_s}{n}}$ | $\left(p_s - 1.645 \sqrt{\dfrac{p_s q_s}{n}},\ p_s + 1.645 \sqrt{\dfrac{p_s q_s}{n}}\right)$ | $2 \times 1.645 \sqrt{\dfrac{p_s q_s}{n}}$ |
| 95% | $p_s \pm 1.96 \sqrt{\dfrac{p_s q_s}{n}}$ | $\left(p_s - 1.96 \sqrt{\dfrac{p_s q_s}{n}},\ p_s + 1.96 \sqrt{\dfrac{p_s q_s}{n}}\right)$ | $2 \times 1.96 \sqrt{\dfrac{p_s q_s}{n}}$ |
| 99% | $p_s \pm 2.576 \sqrt{\dfrac{p_s q_s}{n}}$ | $\left(p_s - 2.576 \sqrt{\dfrac{p_s q_s}{n}},\ p_s + 1.96 \sqrt{\dfrac{p_s q_s}{n}}\right)$ | $2 \times 2.576 \sqrt{\dfrac{p_s q_s}{n}}$ |

Remember that the sample size, $n$, should be large $(n \geqslant 30$ say), since the normal approximation to the binomial distribution is used in obtaining the distribution of sample proportions. Also, since a continuous distribution has been used as an approximation for a discrete distribution, continuity corrections should be used. These are usually omitted, however, when calculating confidence intervals.

## Example 9.28

A manufacturer wants to assess the proportion of defective items in a large batch produced by a particular machine. He tests a random sample of 300 items and finds that 45 items are defective.

(a) Calculate an approximate 95% confidence interval for the proportion of defective items in the batch.
(b) If 200 such tests are performed and a 95% confidence interval calculated for each, how many would you expect to include the proportion of defective items in the batch?

## Solution 9.28

(a) $p_s = \dfrac{45}{300} = 0.15$, $q_s = 1 - p_s = 0.85$, $n = 300$.

The 95% confidence limits for $p$ are

$$p_s \pm 1.96 \sqrt{\dfrac{p_s q_s}{n}} = 0.15 \pm 1.96 \sqrt{\dfrac{0.15 \times 0.85}{300}}$$

$$= 0.15 \pm 0.0404$$

95% confidence interval $= (0.15 - 0.0404, 0.15 + 0.0404)$
$$= \mathbf{(0.1096, 0.1904)}.$$

(b) The expected number of tests that include the proportion of defective items in the batch
$$= 200 \times 0.95 = \mathbf{190}.$$

---

## Example 9.29

In a random sample of 400 carpet shops, it was discovered that 136 of them sold carpets at below the list prices recommended by the manufacturer.

(a) Estimate the percentage of all carpet shops selling below list price.
(b) Calculate an approximate 90% confidence interval for the proportion of shops that sell below list price and explain briefly what this means.
(c) What size sample would have to be taken in order to estimate the percentage to within ±2%, with 90% confidence?

## Solution 9.29

(a) $p_s = \dfrac{136}{400} = 0.34$

An estimate of $p$, the proportion of all carpet shops selling below list price, is $\hat{p}$ where $\hat{p} = p_s = 0.34$. So an estimate of the percentage of shops is **34%**.

(b) An approximate 90% confidence interval for $p$ is given by

$$p_s \pm 1.645 \sqrt{\dfrac{p_s q_s}{n}} = 0.34 \pm 1.645 \times \sqrt{\dfrac{0.34 \times 0.66}{400}}$$

$$= 0.34 \pm 0.039$$
$$= (0.301, 0.379)$$
$$= (30.1\%, 37.9\%)$$

The probability that the interval (30.1%, 37.9%) includes the true population percentage is 0.90. If a large number of intervals are calculated in the same way, 90% of them would include the true percentage.

(c) In part (b) the percentage of shops was estimated to within ±3.9%.
You now require $n$ such that the percentage is to within ±2%,

so that $p_s \pm 1.645 \sqrt{\dfrac{p_s q_s}{n}} = p_s \pm 0.02$

Taking the + sign on both sides

$$p_s + 1.645 \sqrt{\dfrac{p_s q_s}{n}} = p_s + 0.02$$

$$\therefore \quad 1.645 \sqrt{\dfrac{p_s q_s}{n}} = 0.02$$

i.e $\quad 1.645 \times \sqrt{\dfrac{0.34 \times 0.66}{n}} = 0.02$

$$\dfrac{1.645 \times \sqrt{0.34 \times 0.66}}{0.02} = \sqrt{n}$$

$$\sqrt{n} = 38.96 \ldots$$
$$n = 1520 \quad (3 \text{ s.f.})$$

**1520 shops would have to be sampled.**

## Exercise 9g  Confidence intervals for $p$

1. In a survey of a random sample of
250 households in a large city, 170 households
owned at least one pet.

   (a) Find an approximate 95% confidence
   interval for the proportion of households in
   the city that own at least one pet.
   (b) Explain why the interval is approximate.

2. In order to assess the probability of a successful
outcome, an experiment was performed 200
times. The number of successful outcomes was
72.

   (a) Find a 95% confidence interval for $p$, the
   probability of a successful outcome.
   (b) Find a 99% confidence interval for $p$.

3. A survey was undertaken of the use of the
internet by residents in a large city and it was
discovered that in a random sample of 150
residents, 45 logged on to the internet at least
once a day.

   (a) Calculate an approximate 90% confidence
   interval for $p$, the proportion of residents in
   the city that log on to the internet at least
   once a day.
   (b) One hundred similar surveys are carried out
   and the 90% confidence interval calculated
   for each survey. State the expected number
   of intervals that include $p$.

4. Recruits are issued with boots when they join the
army. The last 50 pairs of boots issued were the
following sizes:

| 8 | 9 | 8 | 10 | 11 | 8 | 7 | 12 | 12 | 9 |
|----|----|----|----|----|----|----|----|----|----|
| 9 | 8 | 11 | 8 | 9 | 7 | 11 | 12 | 11 | 10 |
| 9 | 10 | 10 | 10 | 8 | 8 | 7 | 12 | 9 | 9 |
| 10 | 13 | 7 | 8 | 9 | 9 | 10 | 10 | 8 | 12 |
| 9 | 9 | 10 | 10 | 11 | 12 | 9 | 9 | 10 | 9 |

   (a) Find the proportion in the sample requiring
   size 9.
   (b) Assuming that the last 50 recruits can be
   regarded as a random sample of all recruits,
   calculate an approximate 90% confidence
   interval for the proportion, $p$, of all recruits
   requiring size 9 boots.
   (c) Explain why the interval is approximate.

5. In a market research survey, 25 people out of a
random sample of 100 in a certain town said
that they regularly used a particular brand of
soap. Find approximate 97% confidence limits
for the proportion of people in the town who
regularly use this brand of soap.

6. A college principal decides to consult the
students about a proposed change in the times of
lectures. She finds that, out of a random sample
of 80 students, 57 are in favour of the change.

   (a) Find an approximate 90% confidence
   interval for the proportion of students who
   are not in favour of the change.
   (b) State the effect on the width of such a
   confidence interval when the confidence
   level is increased.

7. In an opinion poll, 2000 people were interviewed and 527 said that they preferred white chocolate to milk chocolate.

   (a) Calculate an approximate 95% confidence interval for the proportion of the population who prefer white chocolate. State any assumptions you have made.

   (b) The $\alpha\%$ confidence limits for the proportion preferring white chocolate, based on a sample of size 500, are 0.2278 and 0.2922. Calculate
      (i) the proportion of people in the sample of 500 who preferred white chocolate,
      (ii) the value of $\alpha$.

8. The results of a survey showed that 3600 out of 10 000 families regularly purchased a specific weekly magazine.

   (a) Find approximate 95% confidence limits for the proportion of families buying the magazine.

   (b) Estimate the additional number of families to be contacted if the probability that the estimated proportion is in error by more than 0.01 is to be at most 1%. *(AEB)*

9. The probability of success in each of a long series of $n$ independent trials is constant and equal to $p$. Explain how an approximate 95% confidence interval for $p$ may be obtained.
   In an opinion poll carried out before a local election, 501 people out of a random sample of 925 voters declare that they will vote for a particular one of the two candidates contesting the election. Find approximate 95% confidence limits for the proportion of all voters in favour of this candidate. *(AEB)*

## Summary

- Point estimates: unbiased estimates for
  - population mean $\mu$                  $\hat{\mu} = \bar{x}$, the sample mean
  - population proportion $p$             $\hat{p} = p_s$, the sample proportion
  - population variance $\sigma^2$

$$\hat{\sigma}^2 = \frac{n}{n-1} s^2 \; (s^2 \text{ is the sample variance})$$

$$= \frac{n}{n-1}\left(\frac{\Sigma x^2}{n} - \bar{x}^2\right)$$

$$= \frac{1}{n-1}\left(\Sigma x^2 - \frac{(\Sigma x)^2}{n}\right)$$

$\hat{\sigma}$ is given by $\boxed{\sigma_{x_{n-1}}}$ on your calculator.

- Distribution of the sample mean

  If a number of random samples, each of the same size $n$, is taken from a parent population and the mean $\bar{x}$, is calculated for each sample, then these means form a distribution called the **sampling distribution of means**:

  (a) when the samples are taken from a **normal** population $X \sim N(\mu, \sigma^2)$, with $\sigma^2$ known, then for samples of any size $n$, the distribution $\bar{X}$ is also normal such that $\bar{X} \sim N\left(\mu, \dfrac{\sigma^2}{n}\right)$.

  $\dfrac{\sigma}{\sqrt{n}}$ is called the **standard error of the mean**.

  (b) when the samples are taken from a **non-normal** population with known variance $\sigma^2$ then for **large values of $n$**, the distribution $\bar{X}$ is approximately normal such that $\bar{X} \sim N\left(\mu, \dfrac{\sigma^2}{n}\right)$.

  This is known as the **central limit theorem**.

  In both these cases, if the population variance, $\sigma^2$, is unknown, then $\hat{\sigma}^2$ can be used instead, provided that $n$ is large.

- Distribution of the sample proportion:

  If a number of random samples, each of the same size $n$, is taken from a parent population and the proportion of successes, $p_s$, calculated for each sample, then these proportions form a distribution called the **sampling distribution of proportions**.

  Provided that **$n$ is large**, the sampling distribution of proportions is approximately normal such that $P_s \sim N\left(p, \dfrac{pq}{n}\right)$ where $q = 1 - p$.

  $\sqrt{\dfrac{pq}{n}}$ is called the standard error of proportion.

- Interval estimates:

  Confidence interval for the population mean $\mu$

| Conditions | 95% confidence interval for $\mu$ |
|---|---|
| Normal population <br> – with known variance $\sigma^2$ <br> – sample size $n$ large or small <br> – sample mean $\bar{x}$ | $\left(\bar{x} - 1.96\,\dfrac{\sigma}{\sqrt{n}}, \bar{x} + 1.96\,\dfrac{\sigma}{\sqrt{n}}\right)$ |
| Non-normal population <br> – with known variance $\sigma^2$ <br> – sample size $n$ large ($n \geqslant 30$) <br> – sample mean $\bar{x}$ | $\left(\bar{x} - 1.96\,\dfrac{\sigma}{\sqrt{n}}, \bar{x} + 1.96\,\dfrac{\sigma}{\sqrt{n}}\right)$ |
| Non-normal population <br> – with unknown variance $\sigma^2$ <br> – sample size $n$ large ($n \geqslant 30$) <br> – sample mean $\bar{x}$ <br> – sample variance $s^2$ | $\left(\bar{x} - 1.96\,\dfrac{\hat{\sigma}}{\sqrt{n}}, \bar{x} + 1.96\,\dfrac{\hat{\sigma}}{\sqrt{n}}\right)$ <br><br> where $\hat{\sigma}^2 = \dfrac{n}{n-1}s^2$ |
| Normal population <br> – with unknown variance <br> – sample size $n$ small ($n < 30$) <br> – sample mean $\bar{x}$ <br> – sample variance $s^2$ | $\left(\bar{x} - t\,\dfrac{\hat{\sigma}}{\sqrt{n}}, \bar{x} + t\,\dfrac{\hat{\sigma}}{\sqrt{n}}\right)$ <br><br> where $\hat{\sigma}^2 = \dfrac{n}{n-1}s^2$ <br> and $(-t, t)$ encloses 95% of the $t(n - 1)$ distribution |

Note that the width of the confidence interval can be reduced in one of these ways:

- by increasing the sample size (making $n$ larger),

- by decreasing the percentage confidence (eg choosing a confidence level of 90% instead of 95%),

- by reducing the size of the population variance if possible (making $\sigma$ smaller).

Confidence interval for the population proportion $p$

| Conditions | 95% confidence interval for $p$ |
|---|---|
| – sample size $n$ large <br> – sample proportion $p_s$ | $\left(p_s - 1.96\,\sqrt{\dfrac{p_s q_s}{n}}, p_s + 1.96\,\sqrt{\dfrac{p_s q_s}{n}}\right)$ <br><br> where $q_s = 1 - p_s$ |

## Miscellaneous Worked Examples

### Example 9.30

Each of a random sample of 50 one-pound coins was weighed and their masses, $x$ grams, are summarised by

$$\Sigma x = 474.51, \quad \Sigma x^2 = 4503.8276.$$

(a) Use an unbiased estimate of variance to calculate an approximate 90% confidence interval for the mean mass (in grams) of all one-pound coins, giving the end-values of the interval to two decimal places.

(b) Estimate the size of a random sample of one-pound coins that would be required to give a 95% confidence interval whose width is half that of the interval calculated in (a).

(c) It was found later that the scales were consistently underweighing by 0.05 grams. State which of the results of (a) and (b) should be amended and which should not. Give the amended values. (C)

### Solution 9.30

$X$ is the mass, in grams of a one-pound coin.

(a) $\hat{\sigma}^2 = \dfrac{1}{n-1}\left(\Sigma x^2 - \dfrac{(\Sigma x)^2}{n}\right)$

$\qquad = \dfrac{1}{49}\left(4503.8276 - \dfrac{474.51^2}{50}\right)$

$\qquad = 0.01291\ldots$

$\hat{\sigma} = \sqrt{0129\ldots} = 0.1136\ldots$

$\bar{x} = \dfrac{\Sigma x}{n} = \dfrac{474.51}{50} = 9.4902$

By the central limit theorem, and using $\hat{\sigma}$ since $\sigma$ is unknown, 90% confidence limits for $\mu$ are

$$\bar{x} \pm 1.645\,\frac{\hat{\sigma}}{\sqrt{n}} = 9.4902 \pm 1.645 \times \frac{0.1136\ldots}{\sqrt{50}}$$

$$= 9.4902 \pm 0.0264\ldots$$

90% confidence interval = **(9.46 g, 9.52 g)** **(2 d.p.)**.

(b) Width of 90% confidence interval $= 2 \times 1.645\,\dfrac{\hat{\sigma}}{\sqrt{n}}$

$\qquad\qquad\qquad\qquad\qquad = 2 \times 0.0264\ldots$

$\qquad\qquad\qquad\qquad\qquad = 0.05287\ldots$

Width of required interval $= \frac{1}{2} \times 0.05287\ldots$

$\qquad\qquad\qquad\qquad = 0.0264\ldots$

$z$-value required for 95% confidence interval = 1.96

$$\therefore 2 \times 1.96 \times \frac{0.1136\ldots}{\sqrt{n}} = 0.0264\ldots$$

$$\sqrt{n} = \frac{0.445\ldots}{0.0264\ldots}$$

$$\sqrt{n} = 16.85\ldots$$

$$n = 283.9\ldots$$

The sample size required is **280**   **(2 s.f.)**.

(c) When the scales are underweighing by 0.05 g,
- the confidence interval in part (a) would be amended. It would be shifted 0.05 units to the right. The new confidence interval would be (9.51, 9.57).
- the confidence interval in part (b) would remain the same, since this uses the estimate of the variance which would not be altered if all the readings were increased by 0.05 g.

## Example 9.31

Out of a random sample of 1000 French people interviewed during Autumn 1996, 410 supported a single European Currency.

(a) Calculate an approximate 99% confidence interval for the population proportion, $p$, of French people who supported a single European Currency.
(b) Estimate the size of a sample that would have provided a 99% confidence interval of width 0.04 for $p$.
(c) Give one reason (other than rounding) why your answer to (b) is only an estimate.     (C)

## Solution 9.31

(a) $p_s = \dfrac{410}{1000} = 0.41$ and $q_s = 1 - p_s = 0.59$.

In a sample of size 1000, the 99% confidence limits for $p$ are

$$p_s \pm 2.576 \sqrt{\frac{p_s q_s}{n}} = 0.41 \pm 2.576 \times \sqrt{\frac{0.41 \times 0.59}{1000}}$$

$$= 0.41 \pm 0.4006\ldots$$

99% confidence interval = **(0.37, 0.45)**   **(2 s.f.)**.

(b) For a width of 0.04, confidence interval would be $0.41 \pm 0.02$

i.e.   $$2.576 \sqrt{\frac{0.41 \times 0.59}{n}} = 0.02$$

$$\frac{1.154\ldots}{\sqrt{n}} = 0.02$$

$$\sqrt{n} = \frac{1.154\ldots}{0.02}$$

$$= 57.73\ldots$$

$$n = 3332.8\ldots$$

```
0.41 – 0.02    0.41    0.41 + 0.02

         width = 0.04
```

Sample size = **3330**   **(3 s.f.)**.

(c) The answer is only an estimate because the estimate for $p$, $\hat{p} = p_s$, was used to obtain an approximate value for the standard deviation of the sampling distribution $\sqrt{\dfrac{pq}{n}}$.

Also in the sampling distribution of proportions (from which the confidence interval is obtained) a normal approximation is used for a binomial distribution.

## Example 9.32

It may be assumed that the breaking strength of paving slabs laid in public areas is normally distributed with mean 50 units and standard deviation 8 units. Random samples of $n$ paving slabs are taken. The mean breaking strength for a sample is denoted by $\bar{X}$.

(a) State the distribution of $\bar{X}$, giving its mean and variance.
(b) Find the probability that $\bar{X}$ exceeds 54 units in the case $n = 25$.
(c) Find the smallest possible sample size if the probability that $\bar{X}$ exceeds 54 units is less than 0.01.

Suppose that the breaking strength of paving slabs laid in public areas has mean 50 units and standard deviation 8 units, but that the form of the distribution of breaking strengths is not known. Random samples of $n$ paving slabs are taken. What can be said about the form of the distribution of the mean breaking strength of these samples in the case when $n$ is large, and also in the case when $n$ is small? (C)

## Solution 9.32

$X$ is the breaking strength, $X \sim N(50, 8^2)$.

(a) $\bar{X} \sim N\left(50, \dfrac{8^2}{n}\right)$,

so $\bar{X}$ follows a normal distribution with mean 50 and variance $\dfrac{64}{n}$.

(b) When $n = 25$, $\bar{X} \sim N\left(50, \dfrac{8^2}{25}\right)$.

Standard deviation of $\bar{X}$ is $\dfrac{8}{5}$.

$$P(\bar{X} > 54) = P\left(Z > \frac{54 - 50}{8/5}\right)$$
$$= P(Z > 2.5)$$
$$= 0.0062$$

(c) $P(\bar{X} > 54) < 0.01$

$$P\left(Z > \frac{54 - 50}{8/\sqrt{n}}\right) < 0.01$$

So $P\left(Z < \dfrac{54 - 50}{8/\sqrt{n}}\right) < 0.99$

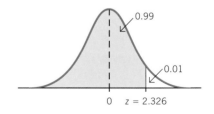

Now $\Phi^{-1}(0.99) = 2.326$, so a $z$-value of 2.326 gives an upper tail probability of 0.01.

$$\therefore \quad \frac{54 - 50}{8/\sqrt{n}} \text{ must lie to the } \textbf{right} \text{ of } 2.326.$$

$$\text{So} \quad \frac{54 - 50}{8/\sqrt{n}} > 2.326$$

$$4 > 2.326 \times \frac{8}{\sqrt{n}}$$

$$\sqrt{n} > 4.652 \dots$$

$$n > 21.64 \dots$$

**Smallest sample size is 22.**

When $n$ is large, by the central limit theorem

$$\overline{X} \text{ is approximately normal and } \overline{X} \sim N\left(50, \frac{8^2}{n}\right)$$

When $n$ is small, you cannot say what the distribution of $\overline{X}$ is. You only know that its mean is 50 and its variance is $\frac{8^2}{n}$.

---

## Example 9.33

The 'reading age' of children about to start secondary school is a measure of how good they are at reading and understanding printed text. A child's reading age, measured in years, is denoted by the random variable $X$. The distribution of $X$ is assumed to be $N(\mu, \sigma^2)$. The reading ages of a random sample of 20 children were measured, and the data obtained is summarised by $\Sigma x = 232.6$, $\Sigma x^2 = 2756.22$.

(a) Calculate unbiased estimates of $\mu$ and $\sigma^2$, giving your answers to correct to two decimal places.
(b) Calculate a symmetric 95% confidence interval for $\mu$. (C)

## Solution 9.33

(a) $\hat{\mu} = \bar{x} = \dfrac{\Sigma x}{n} = \dfrac{232.6}{20} = \mathbf{11.63}$

$$\hat{\sigma}^2 = \frac{1}{n-1}\left(\Sigma x^2 - \frac{(\Sigma x)^2}{n}\right)$$

$$= \frac{1}{19}\left(2756.22 - \frac{232.6^2}{20}\right)$$

$$= 2.688 \dots$$

$$= \mathbf{2.69} \quad \text{(2 d.p.).}$$

(b) Since the population is normal, with variance unknown, and the sample size is small, use the $t$-distribution.

The 95% confidence limits for $\mu$ are $\bar{x} \pm t \dfrac{\hat{\sigma}}{\sqrt{n}}$

where $(-t, t)$ encloses the central 95% of the $t(n-1)$ distribution.
Since $n = 20$, consider $t(19)$.

| $p$ | 0.75 | 0.90 | 0.95 | 0.975 |
|---|---|---|---|---|
| $v = 1$ | 1.000 | ... | ... | ... |
| $\vdots$ | $\vdots$ | $\vdots$ | $\vdots$ | $\vdots$ |
| 17 | 0.689 | 1.333 | 1.740 | 2.110 |
| 18 | 0.688 | 1.330 | 1.734 | 2.101 |
| 19 | 0.688 | 1.328 | 1.729 | 2.093 |

From tables, the critical $t$-value is found from column 0.975, $v = 19$ and it is $t = 2.093$.

$$95\% \text{ confidence limits are } 11.63 \pm 2.093 \times \frac{\sqrt{2.688\ldots}}{\sqrt{20}}$$
$$= 11.63 \pm 0.767\ldots$$

$$95\% \text{ confidence interval for } \mu = (11.63 - 0.767\ldots, 11.63 + 0.767\ldots)$$
$$= (\mathbf{10.9 \text{ years}, 12.4 \text{ years}}) \quad (\mathbf{1 \text{ d.p.}}).$$

# Miscellaneous exercise 9h

1. The mass of a certain brand of chocolate bar has a normal distribution with mean $\mu$ grams and standard deviation 0.85 grams. The masses, in grams, of 5 randomly chosen bars are

   124.31,  125.14,  124.23,  125.41,  125.76.

   Calculate a symmetric 90% confidence interval for $\mu$, giving the end-points correct to two decimal places.
   Forty random samples of 5 bars are taken, and a 90% confidence interval for $\mu$ is calculated for each sample. Find the expected number of intervals that do not contain $\mu$. (C)

2. A telephone company selected a random sample of size 150 from those customers who had not paid their bills one month after they had been sent out. The mean amount owed by the customers in the sample was £97.50 and the standard deviation was £29.00.
   Calculate a 90% confidence interval for the mean amount owed by all customers who had not paid their bills one month after they had been sent out. (AEB)

3. A catering company asked 50 randomly selected college students to state the amount of money, $x$, which they spent daily on lunch, and the results were summarised by $\Sigma x = 56.50$ and $\Sigma x^2 = 66.80$. Calculate unbiased estimates of the mean and the variance of the amount spent daily on lunch by students at the college, giving your answers correct to three significant figures.
   Hence find a symmetric 90% confidence interval for the mean amount spent daily on lunch, giving the end-points correct to the nearest $0.01. Justify the use of the normal distribution in constructing the confidence interval. (C)

4. A random sample of 250 adult men undergoing a routine medical inspection had their heights ($x$ cm) measured to the nearest centimetre, and the following data were obtained:

   $\Sigma x = 43\ 205$,   $\Sigma x^2 = 7\ 469\ 107$.

   Calculate an unbiased estimate of the population variance. Calculate also a symmetric 99% confidence interval for the population mean. (C)

5. A random sample of 600 was chosen from the adults living in a town in order to investigate the number $x$ of days of work lost through illness. Before taking the sample it was decided that certain categories of people would be excluded from the analysis of the number of working days lost although they would not be excluded from the sample. In the sample 180 were found to be from these categories. For the remaining 420 members of the sample:

$\Sigma x = 1260$   $\Sigma x^2 = 46\,000$.

(a) Estimate the mean number of days lost through illness, for the restricted population, and give a 95% confidence interval for the mean.

(b) Estimate the percentage of people in the town who fall into the excluded categories, and give a 99% confidence interval for this percentage.

(c) Give two examples, with reasons, of people who might fall into the excluded categories.

(O)

6. The proportion of bruised apricots in a large consignment is denoted by $p$. A sample of 100 apricots is examined and 11 apricots are found to be bruised.

(a) Give an assumption under which it would be valid to calculate an approximate confidence interval for $p$.

(b) Given that the assumption in part (a) is justified, calculate an approximate 90% confidence interval for $p$. Give the end-points correct to two decimal places.   (C)

7. The lifetimes of light bulbs of a certain type have standard deviation 25.3 hours. Each bulb in a randomly chosen box of 12 was tested to failure and the mean lifetime was found to be 1785.7 hours.

(a) State two assumptions which are required so that a symmetric 90% confidence interval for the population mean lifetime of the bulbs can be calculated.

(b) Calculate a symmetric 90% confidence interval, given the validity of the assumptions. The values of the end-points should be given to the nearest integer.   (C)

8. A consumer group wishes to estimate the proportion, $p$, of packages of sausages whose fat content is greater than that stated on the label. A random sample of 40 packages was tested and nine packages were found to contain more fat than stated on the label.

(a) Estimate the number of packages that would have to be tested in order that a 95% confidence interval for $p$ should have a width of 0.1.

(b) State, giving a reason, whether the number of packages to be tested would be larger or smaller than the answer in (a) if the confidence level were changed to 90%.   (C)

9. In June 1996, 150 randomly chosen people aged sixteen or more were asked whether they smoked cigarettes and 34 said that they did. Assuming that the responses were truthful, calculate an approximate 99% confidence interval for the population proportion of people aged sixteen or more who smoked cigarettes.
Give a reason why this interval might not contain the true population proportion.   (C)

10. A certain type of yarn is known to have a breaking strength with a mean of 25 newtons. In an attempt to increase its breaking strength the yarn is treated with a chemical. Each piece of yarn in a random sample of 80 treated pieces has its breaking strength, $x$ newtons, measured, producing the following summarised data:

$\Sigma x = 2122$   $\Sigma x^2 = 56\,384$

(a) Obtain unbiased estimates of the mean, $\mu$, and variance $\sigma^2$, of the breaking strengths of pieces of yarn treated with the chemical.

(b) Construct a symmetric 99% confidence interval for $\mu$.

(c) Hence state, with a reason, whether or not the manufacturer of the yarn is justified in claiming that the treatment increases the mean breaking strength of this type of yarn.

(d) Explain why you were able to construct your confidence interval without knowing the form of the distribution of the breaking strength of a piece of yarn.   (NEAB)

11. Shoe shop staff routinely measure the length of their customers' feet. Measurements of the length of one foot (without shoes) from each of 180 adult male customers yielded a mean length of 29.2 cm and a standard deviation of 1.47 cm.

(a) Calculate a 95% confidence interval for the mean length of male feet.

(b) Why was it not necessary to assume that the lengths of feet are normally distributed in order to calculate the confidence interval in part (a)?

(c) What assumption was it necessary to make in order to calculate the confidence interval in part (a)?

(d) Given that the lengths of male feet may be modelled by a normal distribution, and making any other necessary assumptions, calculate an interval within which 90% of the lengths of male feet will lie.

(e) In the light of your calculations in parts (a) and (d), discuss, briefly, the question 'is a foot a foot long?' (One foot is 30.5 cm.)

(AEB)

12. Before its annual overhaul, the mean operating time of an automatic machine was 103 seconds. After the annual overhaul, the following random sample of operating times (in seconds) was obtained:

    90, 97, 101, 92, 101, 95, 95, 98, 96, 95.

    Assuming that the time taken by the machine to perform the operation is a normally distributed random variable with a known standard deviation of 5 seconds, find 98% confidence limits for the mean operating time after the overhaul.
    Comment on the magnitude of these limits relative to the mean operating time before the overhaul. (AEB)

13. Packets of baking powder have a nominal weight of 200 g. The distribution of weights is normal and the standard deviation is 7g. Average quantity system legislation states that, if the nominal weight is 200 g,

    (i) the average weight must be at least 200 g.
    (ii) not more than 2.5% of packages may weigh less than 191 g.
    (iii) not more than 1 in 1000 packages may weigh less than 182 g.

    A random sample of 30 packages had the following weights:

        218, 207, 214, 189, 211, 206, 203, 217,
        183, 186, 219, 213, 207, 214, 203, 204,
        195, 197, 213, 212, 188, 221, 217, 184,
        186, 216, 198, 211, 216, 200.

    (a) Calculate a 95% confidence interval for the mean weight.
    (b) Find the proportion of packets in the sample weighing less than 191 g and use your result to calculate an approximate 95% confidence interval for the proportion of all packets weighing less than 191 g. (AEB)

14. A company manufactures bars of soap. In a random sample of 70 bars, 18 were found to be mis-shaped. Calculate an approximate 99% confidence interval for the proportion of mis-shaped bars of soap.
    Explain what you understand by a 99% confidence interval by considering

    (a) intervals in general based on the above method,
    (b) the interval you have calculated.

    The bars of soap are either pink or white in colour and differently shaped according to colour. The masses of both types of soap are known to be normally distributed, the mean mass of the white bars being 176.2 g. The standard deviation for both bars is 6.46 g. A sample of 12 of the pink bars of soap had masses, measured to the nearest gram, as follows:

    174, 164, 182, 169, 171, 187,
    176, 177, 168, 171, 180, 175.

    Find a 95% confidence interval for the mean mass of pink bars of soap.
    Calculate also an interval within which approximately 90% of the masses of the white bars of soap will lie. (AEB)

15. An experimental physicist needs to estimate the true viscosity, $\mu$ Pascal seconds (Pa s), of a light machine oil. Using the same apparatus he takes 12 independent measurements, $x$ Pa s, of the viscosity of the oil, obtaining the values below:

    25.8, 25.2, 24.7, 25.5, 25.3, 25.4,
    25.2, 25.3, 25.8, 25.9, 25.2, 24.9.

    $(\Sigma x = 304.2 \quad \Sigma x^2 = 7712.9)$

    When using this apparatus, measurements of the oil's viscosity are distributed with mean $\mu$ and variance $\sigma^2$.
    Obtain unbiased estimates of $\mu$ and $\sigma^2$. Hence obtain a symmetric 95% confidence interval for $\mu$. State any distributional assumptions you have made in obtaining your confidence interval.
    The physicist explained the meaning of his confidence interval by saying there was a probability of 0.95 that $\mu$ lay between the limits of the interval. Explain why this interpretation is wrong and provide a correct explanation of 95% *confidence* as used in this context.
    The manufacturer of the oil quotes a viscosity of 25.5 Pa s for the oil. With reference to your confidence interval, state any conclusion you can come to regarding the validity of this figure.
    (NEAB)

16. Three weeks before an election in a certain constituency an opinion poll was conducted using a random sample of 800 voters selected from the electoral roll. The numbers of persons who said they would vote for parties A, B, C are recorded below; the remainder were categorised as 'Don't know'.

| Party A | Party B | Party C | Don't know |
|---------|---------|---------|------------|
| 264     | 256     | 144     | 136        |

    (a) Calculate an approximate 90% symmetric confidence interval for the proportion of the total electorate in the constituency that will vote for party A in the election.
    (b) Give a *very brief* description of how the sample might have been selected, to ensure that it was random.
    (c) In the actual election, 41% of the total electorate voted for party A. Give two possible explanations for the fact that this value is not contained within the confidence interval calculated in (a). (NEAB)

17. In an investigation to assess the difference in use between a credit card and a store card a random sample of 20 people, each using both cards, was selected. They supplied information from which, in 1994, the difference between each person's mean monthly spending on the credit and store cards, £d, was calculated. The following summary data were then calculated.

$$\Sigma d = 1664 \quad \text{and} \quad \Sigma d^2 = 426\ 445.$$

Stating all necessary distributional assumptions, calculate a symmetric 90% confidence interval for the mean difference between the mean monthly spending for all users of the two cards.

(NEAB)

18. The mass, $x$ millgrams, of each of 10 randomly selected units of a new cancer drug was measured and the following results obtained:

35.9, 35.2, 35.0, 34.9, 35.4,
34.8, 35.0, 35.1, 35.3, 35.1.

Assuming that the masses are normally distributed with mean $\mu$, calculate an 80% confidence interval for $\mu$.

19. Ten random samples of nylon fibre were tested for the amount of stretching under tension. Each fibre had the same length and diameter and was stretched by applying a standard load.

The increase in length, in millimetres, were as follows:

13.52, 14.06, 13.19, 14.77, 12.80,
12.06, 15.12, 14.39, 15.81, 13.38.

Calculate a 95% confidence interval for the mean increase in length of the population of fibres, assuming that the increase in length can be modelled by a normal distribution.

20. During a particular evening, 10 babies were born on a particular maternity ward in a large hospital. The lengths, in centimetres, of the babies were noted:

50, 51, 45, 47, 49, 48, 54, 53, 45, 50.

Assuming that the sample came from an underlying normal population, calculate a 95% confidence interval for the mean of the population.

21. The external diameters (measured in units of 0.01 mm above a nominal value) of a sample of piston rings produced on the same machine were:

11, 9, 32, 18, 29, 1, 21, 19, 6.

Assuming a normal distribution, calculate a 95% confidence interval for the population mean.

(AEB)

# Mixed test 9A

1. A random sample of 40 nails is drawn from a population whose lengths are normally distributed with mean $\mu$ mm and standard deviation 0.48 mm.

   (a) Calculate the width of a symmetric 99% confidence interval for $\mu$ based on this sample.
   (b) Find the confidence level of a symmetric confidence interval having the same width as before, but based on a random sample of 20 nails. (C)

2. From time to time a firm manufacturing pre-packed furniture needs to check the mean distance between pairs of holes drilled by machine in pieces of chipboard to ensure that no change has occurred. It is known from experience that the standard deviation of the distance is 0.43 mm. The firm intends to take a random sample of size $n$, and to calculate a 99% confidence interval for the mean of the population. The width of this interval must be no more than 0.60 mm. Calculate the minimum value of $n$. (L)

3. Out of 248 cars parked in a car park, 72 were fitted with an anti-theft device on the steering wheel. Assuming that the cars form a random sample of parked cars, calculate an approximate 95% confidence interval for the population proportion of parked cars fitted with an anti-theft device on the steering wheel. Give a reason, other than rounding in the calculations, why the interval is approximate. Give a reason why the assumption of randomness might not be valid. (C)

4. The fat content of a well-known brand of beefburger was investigated by measuring the percentage of fat, $X$, in each of 12 randomly selected beefburgers. The results were summarised as follows:

$$\Sigma x = 228, \quad \Sigma x^2 = 4448.$$

Assuming the percentage fat content to be normally distributed, find a 90% confidence interval for the population mean $\mu$.

# Mixed test 9B

1. A group of 65 students is asked to guess the length of a particular object and their answers are recorded as $x$ cm, with the following results:

$$\Sigma x = 6019.0 \quad \text{and} \quad \Sigma x^2 = 557\,733.8.$$

   (a) Show that the estimated standard error of the sample mean is 0.3 cm.
   (b) Determine an approximate symmetric 95% confidence interval for the mean of the population of all such guesses, giving your limits correct to two decimal places.
   (c) State one assumption which you have made in your calculations. *(NEAB)*

2. A survey was carried out by a County Meals Service in order to gauge the response to a new 'healthy eating' menu. A random sample of 200 schoolchildren was selected from schools using the menu and it was found that 84 children approved of it. Calculate an approximate 95% confidence interval for the population proportion, $p$, who approve of the new menu. It is given that $p = 0.38$. Use a suitable approximation to calculate the probability that, in a random sample of 200 children, the proportion who approve of the new menu will be at least 0.42. *(C)*

3. A researcher is designing a study to standardise a new intelligence test. It is known that scores on this type of test are normally distributed with a standard deviation of 15.0.

   (a) Write down in terms of $\bar{x}$, the sample mean, and $n$, the sample size, an expression for a 99% symmetric confidence interval for the mean test score.
   (b) Calculate, to the nearest 100, the value of $n$ such that the width of this confidence interval will be less than 1.0. *(NEAB)*

4. In Tesbury's supermarket, economy packs of butter are marked 250 g. An inspector takes a random sample of 12 packs and weighs them. Correct to the nearest 0.1 g, the weights, in grams, were:

   246.5, 240.9, 245.3, 250.5, 248.7, 249.1, 251.0, 249.8, 249.8, 247.6, 246.2, 241.4.

   (a) Making any necessary assumptions, which should be stated, calculate a 99% confidence interval for the mean weight of the packs of butter.
   (b) Calculate the width of the 99% confidence interval.
   (c) How is the width affected when calculating a 90% confidence interval?

# 10

# Hypothesis tests: discrete distributions

*In this chapter you will learn about*

- the language of hypothesis testing
- how to perform a test
  - for the parameter $p$ of a binomial distribution (small sample)
  - for the mean $\lambda$ of a Poisson distribution
- Type I and Type II errors associated with hypothesis tests

**Background knowledge**
*You will need to be able to*

- recognise the conditions needed for a situation to be modelled by a binomial distribution or a Poisson distribution.
- find related probabilities by direct evaluation or by using cumulative probability tables.

## HYPOTHESIS TEST FOR A BINOMIAL PROPORTION, $p$
## (small sample size)

Sid says that he has psychic powers and can read people's thoughts. To test this claim, a volunteer from the audience sits on the stage while Sid sits in a separate room off stage. The volunteer chooses a card from a well-shuffled pack and concentrates on the card for five seconds. At the same time, Sid writes down the suit of the card, either hearts, diamonds, spades or clubs. The card is replaced in the pack, the pack is shuffled and another card drawn. The procedure is repeated until 20 cards have been drawn.

There are four suits, so Sid has a one in four chance of writing down the correct suit if he guesses the answer. If he isn't guessing, you would expect him to get more than one in four correct. So if he gets five (or fewer) correct answers out of the 20, you would definitely say that he is just guessing but if he gets as many as 19 or 20 correct you would have no hesitation in saying that he could read people's thoughts.

But what about other values? If he gets 12 correct answers, would this be very unusual? What would you say if he got 10 correct? What about 8 correct?

Somehow you have to decide on a cut-off point, $c$. This would be the *least* value you could find such that the probability of getting $c$ or more correct answers would be very small. It would be considered a rare event to get $c$ or more correct answers.

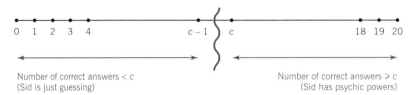

To decide on the value of $c$, you could choose a number that seemed reasonable. If however you perform a **hypothesis (or significance) test** you will be able to back up your argument and conclusion with statistical theory.

Suppose that $X$ is the number of correct answers that Sid writes down for the suits of the 20 cards. If you assume that Sid is just guessing, the probability that he writes down the correct suit is 0.25. The experiment is performed 20 times, so there are 20 independent trials, each with a probability of 0.25 of success. This suggests a binomial situation (see page 279). In fact, on the assumption that Sid is guessing, $X$ can be modelled by a binomial distribution with $n = 20$ and $p = 0.25$, i.e. $X \sim B(20, 0.25)$.

You now need to look for a value, $c$, in this distribution such that $P(X \geqslant c)$ is very small. Binomial probabilities can be calculated directly (see page 279) or found from cumulative probability tables which give $P(X \leqslant r)$ for various values of $n$ and $p$, where $X \sim B(n, p)$. The extract here relates to $X \sim B(20, 0.25)$ and has been reproduced from page 646.

The tables give probabilities to four decimal places indicating that, to four decimal places, $P(X \leqslant 13) = 1.0000$. This implies that $P(X \geqslant 14) = 1 - P(X \leqslant 13)$ tends to 0. If he is just guessing, it would be almost *impossible* for Sid to give 14, 15, 16, 17, 18, 19 or 20 correct answers. So if he gives, for example, 14 correct answers you would certainly have to conclude that he is able to read people's thoughts in some way!

Similarly $P(X \geqslant 13) = 1 - P(X \leqslant 12) = 1 - 0.9998 = 0.0002$.
Getting 13 or more correct answers would be a *very rare event*.

$P(X \geqslant 12) = 1 - P(X \leqslant 11) = 1 - 0.9991 = 0.0009$.
Getting 12 or more correct answers is still a *very rare event*.

$P(X \geqslant 11) = 1 - P(X \leqslant 10) = 1 - 0.9961 = 0.0039$.
On about four occasions in every thousand Sid might give 11 or more correct answers. This is still a *rare event*.

$P(X \geqslant 10) = 1 - P(X \leqslant 9) = 1 - 0.9861 = 0.0138 \approx 1\%$.
It would be *very unlikely* for Sid to give 10 or more correct answers but it could happen on about one occasion in every hundred.

$P(X \geqslant 9) = 1 - P(X \leqslant 8) = 1 - 0.99591 = 0.0409 \approx 4\%$.
It would still be *unlikely* for Sid to give 9 or more correct answers.

$P(X \geqslant 8) = 1 - P(X \leqslant 7) = 1 - 0.8952 = 0.1018 \approx 10\%$.
This probability is not that small. If Sid is just guessing, on 10% of occasions he could give 8 or more correct answers.

| | $P(X \leqslant r)$ for $X \sim B(20, 0.25)$ |
|---|---|
| | $p = 0.25$ |
| $n = 20$  $r = 0$ | 0.0032 |
| 1 | 0.0243 |
| 2 | 0.0913 |
| 3 | 0.2252 |
| 4 | 0.4148 |
| 5 | 0.6172 |
| 6 | 0.7858 |
| 7 | 0.8982 |
| 8 | 0.9591 |
| 9 | 0.9861 |
| 10 | 0.9961 |
| 11 | 0.9991 |
| 12 | 0.9998 |
| 13 | 1.0000 |
| 14 | |
| 15 | |

You have to make a decision about the value of the probability that is considered to imply an unlikely or rare event. This probability is called the **significance level** of the test. As a guide, events that have a probability of 5% or less are regarded as *unlikely* and events having a probability of 1% or less are regarded as *very unlikely*. Often a significance test at the 5% level is carried out.

The cut-off point $c$ is known as the **critical value** and the group of observations that are considered to be unusual or unlikely (rare) events is called the **critical (or rejection) region**. The critical value and critical region depend on the significance level chosen.

Suppose you choose a significance level of 5% to test Sid's claim. From the working above, $P(X \geqslant 8) \approx 10\%$. This is greater than 5%, so $x \geqslant 8$ is not the critical region; getting eight correct answers would not be considered an unlikely or rare event.

But $P(X \geqslant 9) \approx 4\%$, which is less than 5%, so getting nine correct answers would be considered an unlikely or rare event. Therefore the critical value for a 5% level of significance is 9 and the critical region is $x \geqslant 9$, i.e. 9, 10, 11, 12, ..., 19 or 20 correct answers.

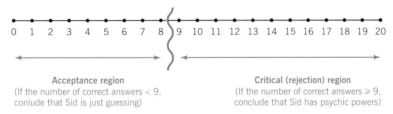

Acceptance region
(If the number of correct answers < 9, conclude that Sid is just guessing)

Critical (rejection) region
(If the number of correct answers ⩾ 9, conclude that Sid has psychic powers)

## The language used in hypothesis testing

The assumption that Sid is guessing is called the **null hypothesis** and it is written $H_0$. The null hypothesis is very important as it provides the model for the calculations. You would write

$$H_0: p = 0.25$$

null hypothesis    assumes Sid is just guessing, in which case he has a one in four chance of guessing correctly

If Sid has psychic powers, then he should get more than one in four correct and the probability that he gives the correct suit will be more than 0.25. This is called the **alternative hypothesis** and is denoted by $H_1$. You would write

$$H_1: p > 0.25$$

alternative hypothesis    assumes Sid has psychic powers

Since you are interested in whether the probability is **greater than** 0.25, the critical region in this example is at the right-hand end of the distribution. This is known as the **upper tail**.

The variable $X$, the number of correct answers, is the **test statistic**. The number of correct answers that Sid gives in the experiment is the **test value**. To perform the hypothesis test you need to work out whether the test value lies in the critical region or not.

If the test value lies in the critical (or rejection) region, **reject $H_0$ in favour of $H_1$**. This means that you will reject that Sid is guessing in favour of the alternative hypothesis that he has psychic powers.

If the test value does not lie in the critical region, do not reject $H_0$. There is not enough evidence to say that he has psychic powers. Writing this another way, if the test value lies in the acceptance region, **accept $H_0$** and conclude that Sid is guessing.

Suppose that in the experiment Sid gives seven correct answers and he says that this proves that he has psychic powers. Is this enough evidence statistically? From the critical region diagram, you can see that, at the 5% level of significance, $x = 7$ does not lie in the critical region. Therefore $H_0$ is not rejected. This means that there is not enough evidence to say that $p > 0.25$, i.e. to say Sid has psychic powers. You would conclude that he is just guessing.

## PROCEDURE FOR CARRYING OUT A HYPOTHESIS TEST

To find whether the test value is in the critical region you can work out the critical region as described above. This is a useful method as it gives a lot of information, but its disadvantage is that it can be rather time-consuming.

In this example, it may be quicker instead to calculate the probability that $X$ is greater than the test value. If this probability is less than 5%, this means that the test value is in the upper tail 5% of the distribution, i.e. it is in the critical region.

This method is illustrated in the working below which tests the sample value $x = 7$ and assumes that you have not found the critical region first. Note that the stages of the test are shown in the margin and additional commentary is given in italics.

1. Define the variable.

Let $X$ be the number of correctly identified suits out of the 20 trials. Assuming that the pack is well shuffled between each trial and the trials are independent, $X$ can be modelled by a binomial distribution, where $X \sim B(20, p)$.

2. State $H_0$ and $H_1$.

$H_0$: $p = 0.25$ (Sid is guessing)
$H_1$: $p > 0.25$ (Sid has psychic powers)

3. State the distribution according to $H_0$.

If $H_0$ is true, then $X \sim B(20, 0.25)$
$\uparrow$
*null hypothesis value for $p$*

4. State level and type of test.

Use a one-tailed (upper tail) test, at the 5% level.

5. State the rejection criterion.

*The test value, $x$, will lie in the critical region, (the upper tail 5% of the distribution), if $P(X \geqslant x) < 5\%$.*
Reject $H_0$ if $P(X \geqslant x) < 5\%$, where $x$ is the test value.

6. Calculate the required probability.

Sid gives 7 correct answers, so test $x = 7$ and find $P(X \geqslant 7)$.
From cumulative binomial tables

$P(X \geqslant 7) = 1 - P(X \leqslant 6)$

$= 1 - 0.7858$

$= 0.2141$

$\approx 21\%$

$P(X \leqslant r)$ for $X \sim B(20, 0.25)$

| | $p = 0.25$ | |
|---|---|---|
| $n = 20$ $r = 0$ | 0.0032 | |
| 1 | 0.0243 | |
| 2 | 0.0913 | |
| 3 | 0.2252 | |
| 4 | 0.4148 | |
| 5 | 0.6172 | |
| 6 | 0.7858 | $\leftarrow P(X \leqslant 6)$ |
| 7 | 0.8982 | |
| 8 | 0.9591 | |

If you do not have cumulative tables, work out the binomial probabilities as follows:

$$P(X \geqslant 7) = 1 - P(X \leqslant 6)$$
$$= 1 - (0.75^{20} + 20 \times 0.75^{19} \times 0.25 + {}^{20}C_2 \times 0.75^{18} \times 0.25^2$$
$$+ {}^{20}C_3 \times 0.75^{17} \times 0.25^3 + {}^{20}C_4 \times 0.75^{16} \times 0.25^4$$
$$+ {}^{20}C_5 \times 0.75^{15} \times 0.25^5 + {}^{20}C_6 \times 0.75^{14} \times 0.25^6)$$
$$= 0.2142$$
$$\approx 21\%$$

7. Make your
conclusion.

Since $P(X \geqslant 7) > 5\%$, the test value $x = 7$ is *not* in the critical region. There is not enough evidence to reject $H_0$.

You would conclude that Sid is just guessing; he does not have psychic powers.

*NOTE:* when you are testing the value $x = 7$, it may seem strange that you have to work out $P(X \geqslant 7)$ rather than just $P(X = 7)$. Remember that this is necessary as you are essentially looking for the critical *region* to see whether the test value lies in this region or not.

The probabilities and critical region can be illustrated diagrammatically. Below is the probability distribution for $X \sim B(20, 0.25)$. Note that it is positively skewed and the probabilities for 12 to 30 correct answers are so small that they cannot be shown on the diagram. The test value has been circled.

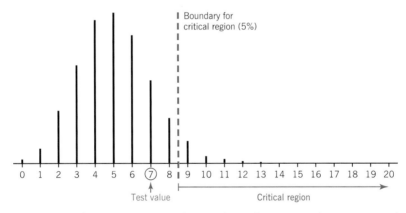

Since $P(X \geqslant 8) \approx 10\%$ and $P(X \geqslant 9) \approx 4\%$, the 5% boundary comes between 8 and 9. Note that with discrete distributions you will probably not get a perfect 5% in your calculations.

## Example 10.1

A drugs company produced a new pain-relieving drug for migraine sufferers and its advertisements stated that the drug had a 90% success rate. A doctor doubted whether the drug would be as successful as the company claimed. She prescribed the drug for 15 of her patients. After six months, 11 of these patients said that their migraine symptoms had been relieved by the drug.

(a) Test the drug company's claim, at the 5% level of significance.
(b) Should the doctor continue to prescribe the drug?

## Solution 10.1

1. Define the variable.

(a) Let $X$ be the number of patients in 15 whose symptoms are relieved by the drug. Assuming that the effect of the drug on a patient is independent of the effect on other patients, $X$ can be modelled by a binomial distribution, where $X \sim B(15, p)$.

2. State $H_0$ and $H_1$.

$H_0$: $p = 0.9$ (The success rate of the drug is 90%.)
$H_1$: $p < 0.9$ (The success rate is less than 90% and drug is not as successful as the company claims.)

3. State the distribution according to $H_0$.

If $H_0$ is true, then $X \sim B(15, 0.9)$.

$\uparrow$

value of $p$ specified in $H_0$

4. State level and type of test.

Since the alternative hypothesis is $p < 0.9$, the critical region is in the *lower tail* of the distribution, so use a one-tailed (lower tail) test, at the 5% level.

5. State the rejection criterion.

*The test value, $x$, will lie in the critical region, (the lower tail 5% of the distribution), if $P(X \leqslant x) < 5\%$.*

Reject $H_0$ if $P(X \leqslant x) < 5\%$, where $x$ is the test value.

6. Calculate the required probability.

The test value is $x = 11$, so find $P(X \leqslant 11)$.
Using cumulative binomial tables, if the tables give only values of $p$ up to 0.5, you need to use symmetry properties as illustrated on page 284.

$$P(X \leqslant 11 \mid p = 0.9) = P(X \geqslant 4 \mid p = 0.1)$$
$$= 1 - P(X \leqslant 3)$$
$$= 1 - 0.9444$$
$$= 0.0556$$
$$\approx 5.6\%$$

| $P(X \leqslant r)$ for $X \sim B(15, 0.1)$ | |
| --- | --- |
| | $p = 0.1$ |
| $n = 15 \quad r = 0$ | 0.2059 |
| 1 | 0.5490 |
| 2 | 0.8159 |
| 3 | 0.9444 $\quad \leftarrow P(X \leqslant 3)$ |
| 4 | 0.9873 |
| 5 | 0.9978 |

If you are calculating the probabilities directly:

$$P(X \leqslant 11) = 1 - P(X \geqslant 12)$$
$$= 1 - ({}^{15}C_{12} \times 0.1^3 \times 0.9^{12} + {}^{15}C_{13} \times 0.1^2 \times 0.9^{13}$$
$$+ 15 \times 0.1 \times 0.9^{14} + 0.9^{15})$$
$$= 1 - 0.944\ldots$$
$$= 0.0556 \text{ (4 d.p.)}$$
$$\approx 5.6\%$$

7. Make your conclusion.

$P(X \leqslant 11)$ is greater than 5%. *This means that boundary for the critical region (the lower tail 5% of the distribution) will be slightly to the left of $x = 11$. So $x = 11$ is not in the critical region.*
$H_0$ is not rejected and the drugs company's claim of a 90% success rate is upheld.

(b) $P(X \leqslant 11)$ is only just greater than 5%. With safety in mind, it would be wise to suggest that the doctor errs on the side of caution and carries out further tests before accepting that the success rate is 90%.

It is time-consuming to draw a probability diagram when carrying out the hypothesis test, but it can be helpful in illustrating the results. The distribution $X \sim B(15, 0.9)$ is very negatively skewed with the probabilities in the lower tail being too small to show in the diagram!

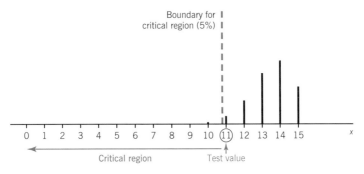

## ONE-TAILED AND TWO-TAILED TESTS

### One-tailed test

In the examples so far, one-tailed tests have been considered, with either the upper or lower tail being used for the critical region, depending on the alternative hypothesis.

In general, for a significance level of $\alpha\%$, null hypothesis $H_0$: $p = p_0$ and a test value $x$,

- if $H_1$ involves a > sign, indicating that you are looking for an **increase** in $p$, use the **upper tail** to find whether $P(X \geqslant x) < \alpha\%$,

- if $H_1$ involves a < sign, indicating that you are looking for a **decrease** in $p$, use the **lower tail** to find whether $P(X \leqslant x) < \alpha\%$.

Remember that in both cases you use $P(X \ldots) \; < \; \alpha\%$.

### Two-tailed test

A two-tailed test is carried out when the alternative hypothesis looks for a change in $p$, not specifically an increase or a decrease. If the significance level is $\alpha\%$, then the critical region is in two parts, half in the lower tail and half in the upper tail,

- if $H_1$ involves a $\neq$ sign, indicating that you are looking for a change in $p$,

in the lower tail, the critical region consists of values **less than or equal to** $c_1$ such that $P(X \leqslant c_1) < \frac{1}{2}\alpha\%$.

in the upper tail, the critical region consists of values **greater than or equal to** $c_2$ such that $P(X \geqslant c_2) < \frac{1}{2}\alpha\%$.

For example, for a 5% significance level (two-tailed test) the probability distribution and critical region might look like this:

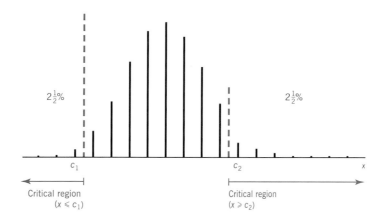

## Example 10.2

In last year's local elections, the Purple party gained 35% of the vote. Prior to this year's election, the party asked a researcher to find out whether support of the party had changed. Out of twelve voters selected at random, one said that he would vote for the Purple party.

(a) Test, at the 10% level, whether support for the Purple party has changed.
(b) Find the critical region for the test.

## Solution 10.2

1. Define the variable.

(a) Let $X$ be the number of voters in 12 who say that they will vote for the Purple party. Assuming that each person votes independently, $X$ can be modelled by a binomial distribution, where $X \sim B(12, p)$.

2. State $H_0$ and $H_1$.

$H_0$: $p = 0.35$ (The support has not changed)
$H_1$: $p \neq 0.35$ (The support has changed)

3. State the distribution according to $H_0$.

If $H_0$ is true, then $X \sim B(12, 0.35)$.

4. State level and type of test.

Since the alternative hypothesis is $p \neq 0.35$, consider both tails of the distribution and perform a two-tailed test at the 10% level
*In this case the 10% for the significance level is distributed evenly between the upper and lower tails, with 5% at each tail.*

5. State the rejection criterion.

*The test value, x will lie in the critical region, (the lower tail 5% or the upper tail 5%), if $P(X \leqslant x) < 5\%$, or $P(X \geqslant x) < 5\%$.*

Reject $H_0$ if $P(X \leqslant x) < 5\%$, or $P(X \geqslant x) < 5\%$.

6. Calculate the required probability.

*The test value is x = 1, so you need to look at the lower tail part of the critical region and find $P(X \leqslant 1)$.*

$$P(X \leqslant 1) = P(X = 0) + P(X = 1)$$
$$= 0.65^{12} + 12 \times 0.65^{11} \times 0.35$$
$$= 0.04244\dots$$
$$\approx 4.2\%$$

*(You can use cumulative binomial tables if they are available.)*

7. Make your
conclusion.

Since $P(X \leqslant 1) < 5\%$, the sample value $x = 1$ lies in the critical region, so $H_0$ is rejected in favour of $H_1$. **At the 10% significance level, there is evidence that support for the Purple party has changed.**

(b) To find the critical region, consider separately the upper and lower tails of the distribution.

*Critical region in lower tail:*
Find the maximum value of $c$ such that $P(X \leqslant c) < 0.05$.
You already know that $P(X \leqslant 1) < 0.05$, i.e. that 0 and 1 lie in the critical region, so try $c = 2$.

$$P(X \leqslant 2) = 0.0422\ldots + P(X = 2)$$
$$= 0.0422\ldots + {}^{12}C_2 \times 0.65^{10} \times 0.35^2$$
$$= 0.1512\ldots$$
$$\approx 15\%$$

$P(X \leqslant 2)$ is greater than 5%, indicating that $x = 2$ is not in the critical region. So the lower tail part of the critical region is $x = 0, 1$.

*Critical region in upper tail:*
Find the minimum value of $c$ that $P(X \geqslant c) < 0.05$. By guesswork, try $c = 8$.

$$P(X \geqslant 8) = {}^{12}C_8 \times 0.65^4 \times 0.35^8 + {}^{12}C_9 \times 0.65^3 \times 0.35^9 + {}^{12}C_{10} \times 0.65^2 \times 0.35^{10}$$
$$+ 12 \times 0.65 \times 0.35^{11} + 0.35^{12}$$
$$= 0.0255\ldots < 5\%$$

This indicates that $x \geqslant 8$ is in the critical region.

*But is 8 the smallest value in the critical region? To check this, try $c = 7$:*

$$P(X \geqslant 7) = {}^{12}C_7 \times 0.65^5 \times 0.35^7 + 0.0255\ldots$$
$$= 0.084\ldots > 5\%$$

This indicates that $x = 7$ is not in the critical region.
So the upper tail part of the critical region consists of $x \geqslant 8$.

**Therefore the critical region is $x = 0, 1, 8, 9, 10, 11, 12$.**

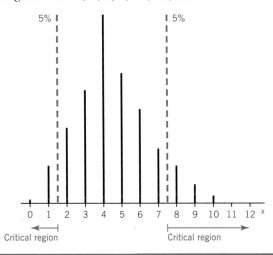

## Summary of the procedure for a significance test for the proportion, $p$ of a binomial distribution (small sample)

- Define $X$, the binomial variable being considered and the general form of its distribution, for example $X \sim B(12, p)$.

- State the null hypothesis $H_0$ and the alternative hypotheses $H_1$ concerning $p$ for example
$H_0: p = p_0$
$H_1: p < p_0$

- State the distribution of $X$ assuming that the null hypothesis is true, i.e. assuming that $X$ does follow a binomial distribution with the value of $p$ specified in $H_0$, for example $X \sim B(12, p_0)$

- State the type of test (one-tailed or two-tailed). This depends on whether the alternative hypothesis looks for an increase or a decrease (one-tailed) or a change (two-tailed) in $p$, for example,

| $H_0: p = p_0$ $H_1: p < p_0$ | $H_0: p = p_0$ $H_1: p > p_0$ | $H_0: p = p_0$ $H_1: p \neq p_0$ |
|---|---|---|
| indicates one-tailed test (*lower tail* considered for critical region) | indicates one-tailed test (*upper tail* considered for critical region) | indicates two-tailed test (*both tail ends* considered for critical region) |

- State the significance level of the test, $\alpha\%$. Remember that this defines the critical region.

- State the criterion for rejection of $H_0$, for example, for test value, $x$,

| Reject $H_0$ if $P(X \leqslant x) < \alpha\%$, i.e. if $x$ lies in the lower tail $\alpha\%$ of the distribution | Reject $H_0$ if $P(X \geqslant x) < \alpha\%$, i.e. if $x$ lies in the upper tail $\alpha\%$ of the distribution | Reject $H_0$ if $P(X \leqslant x) < \frac{1}{2}\alpha\%$, or if $P(X \geqslant x) < \frac{1}{2}\alpha\%$, i.e. if $x$ lies in the lower or upper tail $\frac{1}{2}\alpha\%$ of the distribution |
|---|---|---|

- Obtain the test value, $x$.

- Calculate the required probability to see whether $x$ lies in the critical (rejection) region.

- Make your conclusion by rejecting $H_0$ or not. Then relate this to the context of the situation being tested.

NOTE: The method is essentially the same for a **large sample**, but, in the case of large samples, use is made of the application of the **normal approximation to the binomial distribution**. This test is described on page 528.

# TYPE I AND TYPE II ERRORS

When you perform a significance test, there are four possible conclusions, and these are shown in the table below.

Two of the conclusions lead to correct decisions and the other two lead to wrong ones. The errors associated with wrong decisions are called **Type I and Type II errors**.

The outcomes and errors are summarised as follows:

(a) $H_0$ is true and your test leads you to accept $H_0$ – correct decision
(b) $H_0$ is true but your test leads you to reject $H_0$ – wrong decision – Type I error
(c) $H_0$ is false but your test leads you to accept $H_0$ – wrong decision – Type II error
(d) $H_0$ is false and your test leads you to reject $H_0$ – correct decision

It can be helpful to see these on a diagram:

|  |  | Test decision | |
|---|---|---|---|
|  |  | Accept $H_0$ | Reject $H_0$ |
| **Actual** | $H_0$ is true | ✓correct | Type I error |
| **situation** | $H_0$ is false | Type II error | ✓correct |

A Type I error is made when you reject $H_0$ when it is true,

i.e.  P(Type I error) = P(reject $H_0$ when $H_0$ is true)

You reject $H_0$ if the test value lies in the critical region. This region is fixed according to the level of significance of the test, so the probability that the test value lies in the critical region is the same as the significance level. So for a test carried out at the $a\%$ level of significance,

  P(Type I error) = $a\%$

A Type II error is made when you accept $H_0$ when it is false,

i.e.  P(Type II error) = P(accept $H_0$ when $H_0$ is false).

To calculate the probability of making a Type II error, **a specific value for $H_1$ is stated.** The error is then calculated as follows:

  P(Type II error) = P(accept $H_0$ when $H_1$ is true).

This is illustrated in the following example.

## Example 10.3

A random observation is taken from a binomial distribution $X \sim B(20, p)$ and used to test the null hypothesis $p = 0.8$ against the alternative hypothesis $p > 0.8$.

The critical region is chosen to be $x \geqslant 19$.

(a)  What is the significance level of the test?
(b)  What is the probability of making a Type I error?
(c)  Find the probability of making a Type II error if, in fact, $p = 0.85$.

## Solution 10.3

You are given that $X \sim B(20, p)$.

(a) $H_0: p = 0.8$
$H_1: p > 0.8$

The critical region is $x \geqslant 19$, so to find the significance level of the test, find $P(X \geqslant 19)$.

$$P(X \geqslant 19) = P(X = 19) + P(X = 20)$$
$$= {}^{20}C_{19} \times 0.2 \times 0.8^{19} + 0.8^{20}$$
$$= 0.0691...$$
$$\approx 7\%$$

**The significance level is approximately 7%.**

(b) $P(\text{Type I error}) = P(\text{reject } H_0 \text{ when } H_0 \text{ is true})$

$H_0$ is rejected if $x \geqslant 19$, so
$\qquad P(\text{Type I error}) = P(X \geqslant 19 \text{ when } p = 0.8) \approx 7\%$ (found above).

*Note that this could have been stated directly from part (a), since the probability of a Type I error is the same as the significance level of the test.*

(c) You make a Type II error when you accept $H_0$ (which you will do if $x < 19$) when $p$ is the value specified in $H_1$ (not the value given by $H_0$).
The hypotheses are now

$H_0: p = 0.8$
$H_1: p = 0.85$

So $P(\text{Type II error}) = P(\text{accept } H_0 \text{ when } H_1 \text{ is true})$
$\qquad\qquad\qquad\qquad = P(X < 19 \text{ when } p = 0.85)$
$\qquad\qquad\qquad\qquad = P(X < 19 \text{ when } X \sim B(20, 0.85))$

$$P(X \leqslant 18) = 1 - P(X = 19) - P(X = 20)$$
$$= 1 - {}^{20}C_{19} \times 0.15 \times 0.85^{19} - 0.85^{20}$$
$$= 0.8244...$$
$$\approx 82\%$$

**The probability of making a Type II error is 82%.**

This is a very high probability. To make this smaller, you could increase the significance level of the test. But this would of course increase the probability of making a Type I error!

## Exercise 10a – Testing $p$ in a binomial distribution (small samples)

1. A certain type of seed has a germination rate of 70%. The seeds undergo a new treatment after which 9 germinate in a packet of 10 seeds. Stating suitable null and alternative hypotheses, test, at the 5% level, whether this is evidence of an increase in the germination rate.

2. Hester suspected that a die was biased in favour of a four occurring. She decided to carry out a hypothesis test.

   (a) State suitable null and alternative hypotheses for the test.

   When she threw the die 15 times, she obtained a four on 6 occasions.

   (b) Carry out the test, at the 5% level, stating your conclusion clearly.

3. The random variable $X$ can be modelled by a binomial distribution with $n = 10$.
A random observation, $x$, is taken from the distribution.
Test, at the 8% level, the hypothesis that $p = 0.45$ against the alternative hypothesis $p \neq 0.45$, (a) when $x = 7$, (b) when $x = 1$.

4. Records kept in a hospital show that 3 out of every 10 casualties who come to the casualty department have to wait more than half an hour before receiving medical attention. The hospital decided to increase the staffing in the department by one person and it was then found that, of the next 20 casualties, 2 had to wait for more than half an hour for medical attention.
Test whether the new staffing has decreased the number of casualties who have to wait more than half an hour for medical attention.
Perform the test (a) at the 5% level, (b) at the 2% level. *(L)*

5. The random variable $X$ can be modelled by a binomial distribution with parameters $n = 9$ and $p$, whose value is unknown.

   (a) Find, at the 10% level of significance, the critical region to test the null hypothesis that $p = 0.3$ against the alternative hypothesis that $p > 0.3$.
   (b) Explain what is meant by a Type I error.
   (c) State the probability of making a Type I error in the test described in (a).

6. In each of the following, a random observation $x$ is taken from a binomial distribution $X \sim B(n, p)$. Test the given hypotheses at the significance level stated.

| | $x$ | $n$ | Hypotheses | Level of significance |
|---|---|---|---|---|
| (a) | 6 | 8 | $H_0: p = 0.45$, $H_1: p > 0.45$ | 5% |
| (b) | 1 | 10 | $H_0: p = 0.45$, $H_1: p < 0.45$ | 5% |
| (c) | 9 | 15 | $H_0: p = 0.35$, $H_1: p > 0.35$ | 5% |
| (d) | 9 | 15 | $H_0: p = 0.35$, $H_1: p \neq 0.35$ | 5% |
| (e) | 2 | 9 | $H_0: p = 0.45$, $H_1: p < 0.45$ | 5% |
| (f) | 16 | 20 | $H_0: p = 0.45$, $H_1: p > 0.45$ | 1% |
| (g) | 5 | 7 | $H_0: p = 0.4$, $H_1: p > 0.4$ | 10% |
| (h) | 2 | 20 | $H_0: p = 0.3$, $H_1: p < 0.3$ | 1% |

7. A driving instructor claims that 95% of his pupils pass their driving test at the first attempt. Tom is considering having lessons with this instructor but wonders whether 95% is an overestimate. He decides to conduct a significance test at the 5% level and discovers that last month, out of the 15 pupils who took the test for the first time, 11 passed.

   (a) What would Tom decide about the driving instructor's claim?
   (b) Find the critical region for the number of failures last month.

8. In a test of ten true-false questions, Siân got 8 correct. Test at the 5% level whether she could have obtained this score by guessing all the answers.

9. At a particular hospital it was found from past records that the probability that a patient does not turn up for an appointment is 0.3. Following a campaign to make patients more aware of the problems caused by missed appointments, a significance test at the 10% level was carried out to decide whether the campaign had been effective in reducing the number of patients who did not turn up for an appointment. A random sample of 16 patients was surveyed.

   (a) Find the critical region for the test.
   (b) Find the probability of making a Type II error in the test described in (a) if in fact the probability that a patient does not turn up for an appointment is now 0.25.

10. Jessica is trying to find out whether a particular coin is biased, so she performs a significance test. She decides that she will say that the coin is biased in favour of heads if, when she tosses it 15 times, at least two-thirds of the tosses result in heads.

    (a) What significance level did she use for her test?
    (b) What is the probability that she makes a Type I error?
    (c) If, in fact, the coin is biased, with probability of 0.7 of obtaining a head, what is the probability that she makes a Type II error?

11. A random observation is taken from a binomial distribution $X \sim B(25, p)$ and used to test the null hypothesis $p = 0.4$ against the alternative hypothesis $p < 0.4$.
The critical region is chosen to be $x \leq 6$.

    (a) At what significance level is the test carried out?
    (b) What is the probability of making a Type I error?
    (c) Find the probability of making a Type II error if, in fact, $p = 0.3$.

# SIGNIFICANCE TEST FOR A POISSON MEAN $\lambda$

## Summary of the procedure for a hypothesis test for the mean $\lambda$ of a Poisson distribution

This follows the same pattern as that for the binomial parameter $p$ as follows:

- Define $X$, the Poisson variable being considered and the general form of its distribution, for example $X \sim Po(\lambda)$.

- State the null hypothesis $H_0$ and the alternative hypothesis $H_1$ concerning $\lambda$, for example
  $H_0: \lambda = \lambda_0$
  $H_1: \lambda > \lambda_0$

- State the distribution of $X$ assuming that the null hypothesis is true, i.e. assuming that $X$ does follow a Poisson distribution with the value of $\lambda$ specified in $H_0$, for example $X \sim Po(\lambda_0)$

- State the type of test (one-tailed or two-tailed), for example,

| $H_0: \lambda = \lambda_0$ <br> $H_1: \lambda < \lambda_0$ | $H_0: \lambda = \lambda_0$ <br> $H_1: \lambda > \lambda_0$ | $H_0: \lambda = \lambda_0$ <br> $H_1: \lambda \neq \lambda_0$ |
|---|---|---|
| one-tailed test (lower tail considered for critical region) | one-tailed test (upper tail considered for critical region) | two-tailed test (both tail ends considered for critical region) |

- State the significance level of the test, for example $\alpha\%$. This defines the critical region.

- State the criterion for rejection of $H_0$, for example

| Reject $H_0$ if $P(X \leqslant x) < \alpha\%$, i.e. if $x$ lies in the lower tail $\alpha\%$ of the distribution | Reject $H_0$ if $P(X \geqslant x) < \alpha\%$, i.e. if $x$ lies in the upper tail $\alpha\%$ of the distribution | Reject $H_0$ if $P(X \leqslant x) < \frac{1}{2}\alpha\%$ or if $P(X \geqslant x) < \frac{1}{2}\alpha\%$, i.e. if $x$ lies in the lower or upper tail $\frac{1}{2}\alpha\%$ of the distribution |
|---|---|---|

- Obtain your test value, $x$.

- Calculate the required probability to see whether $x$ lies in the critical (rejection) region.

- Make your conclusion by rejecting $H_0$ or not. Then relate this to the context of the situation being tested.

When $\lambda$ is large, a normal approximation to the Poisson distribution can be applied. The test is similar to the one for normal approximation to the binomial described on page 528.

## Example 10.4

The number of misprints in the classified advertisements pages of the Daily Informer is found to have a Poisson distribution with average 6.5 misprints per page. A new proof reader is employed and the number of misprints on a page was found to be 12. The editor said that the average number of misprints had increased. Test this claim at the 5% level.

## Solution 10.4

| | |
|---|---|
| 1. Define the variable. | Let $X$ be the number of misprints on the classified advertisements page. Assuming that misprints occur randomly, $X$ can be modelled by a Poisson distribution, where $X \sim \text{Po}(\lambda)$. |
| 2. State $H_0$ and $H_1$. | $H_0$: $\lambda = 6.5$<br>$H_1$: $\lambda > 6.5$ (the number of misprints has increased) |
| 3. State the distribution according to $H_0$. | If $H_0$ is true, then $X \sim \text{Po}(6.5)$. |
| 4. State level and type of test. | Since the alternative hypothesis is $\lambda > 6.5$, use a one-tailed test at the 5% level and consider the upper tail of the distribution for the critical region. |

5. State the rejection criterion.

*At the 5% level, the sample value x will lie in the critical region if $P(X \geqslant x) < 5\%$.*
Reject $H_0$ if $P(X \geqslant x) < 5\%$, where $x$ is the test value.

6. Calculate the required probability.

The test value is $x = 12$, so find $P(X \geqslant 12)$.
Using cumulative Poisson tables (page 648)

$$P(X \geqslant 12) = 1 - P(X \leqslant 11)$$
$$= 1 - 0.9661$$
$$= 0.0339$$
$$= 3.39\%$$

| $P(X \leqslant r)$ where $X \sim \text{Po}(6.5)$ | |
|---|---|
| | $\lambda = 6.5$ |
| $r = 0$ | 0.0015 |
| 1 | 0.0113 |
| 2 | 0.0430 |
| 3 | 0.1118 |
| 4 | 0.2237 |
| 5 | 0.3690 |
| 6 | 0.5265 |
| 7 | 0.6728 |
| 8 | 0.7916 |
| 9 | 0.8774 |
| 10 | 0.9332 |
| 11 | 0.9661 |
| 12 | 0.9840 |
| 13 | 0.9929 |
| 14 | 0.9970 |

7. Make your conclusion.

Since $P(X \geqslant 12) < 5\%$, the sample value of 12 misprints lies in the critical region, so reject $H_0$ in favour of $H_1$.
**There is evidence, at the 5% level, that the average number of misprints has increased.**

*NOTE:* by further calculation you will find that
$P(X \geqslant 11) = 6.68\% > 5\%$, so the boundary for the critical region comes between 11 and 12. The critical region is therefore $x \geqslant 12$ and the null hypothesis will be rejected if 12 or more misprints are found in the sample.

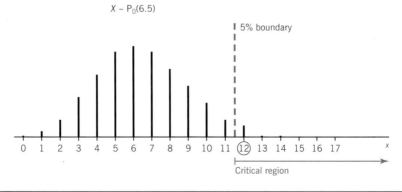

$X \sim \text{Po}(6.5)$

## Example 10.5

The number of breakdowns per week of an office photocopier can be modelled by a Poisson distribution with mean 4.5. The photocopier was serviced and during the next week it broke down just twice. There is no evidence, at the 10% level, of an improvement in the reliability of the photocopier.

## Solution 10.5

**1.** Define the variable.

Let $X$ be the number of breakdowns in a week, where $X \sim \text{Po}(\lambda)$.

**2.** State $H_0$ and $H_1$.

$H_0: \lambda = 4.5$
$H_1 \; \lambda < 4.5$ (the number of breakdowns has decreased implying that the photocopier is more reliable)

**3.** State the distribution according to $H_0$.

If $H_0$ is true, then $X \sim \text{Po}(4.5)$.

**4.** State level and type of test.

Since the alternative hypothesis is $\lambda < 4.5$, use a one-tailed test and consider the lower tail of the distribution for the critical region.

**5.** State the rejection criterion.

*At the 10% level, the sample value $x$ will lie in the critical region if*
$P(X \leqslant x) < 10\%$.
Reject $H_0$ if $P(X \leqslant x) < 10\%$, where $x$ is the test value.

**6.** Calculate the required probability.

*The test value is 2, so find $P(X \leqslant 2)$.*
Using cumulative Poisson tables (page 647)

$P(X \leqslant 2) = 0.1736 = 17.36\%$

NOTE: if you do not have access to tables, then calculate $P(X \leqslant 2)$ as follows (see page 292):

$P(X \leqslant 2) = P(X = 0) + P(X = 1) + P(X = 2)$

$$= e^{-4.5}\left(1 + 4.5 + \frac{4.5^2}{2!}\right)$$

$$= 0.17357\ldots$$
$$\approx 17.36\%$$

| $P(X \leqslant r)$ where $X \sim \text{Po}(4.5)$ $\lambda = 4.5$ | |
|---|---|
| $r = 0$ | 0.0111 |
| 1 | 0.0611 |
| 2 | 0.1736 |
| 3 | 0.3423 |
| 4 | 0.5321 |
| 5 | 0.7029 |
| 6 | 0.8311 |

**7.** Make your conclusion.

Since $P(X \leqslant 2) > 10\%$, the test value of two breakdowns does not lie in the critical region, so $H_0$ is not rejected.
There is no evidence, at the 10% level, of an improvement in the reliability of the photocopier.

## Example 10.6

The number of flaws per metre of fabric follows a Poisson distribution with mean 2. With the aim of reducing the number of flaws, the fabric is subjected to a different treatment process. After this treatment a significance test is devised to gauge whether it has been successful. The test states that the number of flaws has decreased if a randomly selected 4-metre length of cloth contains fewer than five flaws.

(a) State the null and alternative hypotheses for this significance test.
(b) Find the probability of making a Type I error when the test is carried out.

(c) State the significance level of the test.

(d) If, in fact, the treatment has reduced the number of flaws to one per metre, find the probability of making a Type II error when applying the test described above.

## Solution 10.6

*See page 493 for the definitions of Type I and Type II errors.*

(a) Let $X$ be the number of flaws in a 4-metre length of fabric. Assuming that the flaws occur randomly, $X \sim \text{Po}(\lambda)$.

In a 4-metre length, the expected number of flaws is eight. The hypotheses for the significance test would be

$H_0: \lambda = 8$
$H_1: \lambda < 8$   (the mean number of flaws in 4 metres is less than eight and the treatment has been successful in reducing the number per metre.)

(b) If $H_0$ is true, then $X \sim \text{Po}(8)$.

You are told that values of $x < 5$ are in the critical region. So a test value of $x < 5$ would lead to the null hypothesis being rejected.

$P(\text{Type I error}) = P(\text{reject } H_0 \text{ when } H_0 \text{ is true})$
$\qquad\qquad\qquad\quad = P(X < 5 \text{ when } X \sim \text{Po}(8))$

From tables

$P(X < 5) = P(X \leqslant 4)$
$\qquad\qquad = 0.0996$
$\qquad\qquad \approx 10\%$

**$P(\text{Type I error}) = 0.10$ (2 d.p.)**

| $P(X \leqslant r)$ where $X \sim \text{Po}(8)$ $\lambda = 8.0$ | |
|---|---|
| $r = 0$ | 0.0003 |
| 1 | 0.0030 |
| 2 | 0.0138 |
| 3 | 0.0424 |
| 4 | 0.0996 |
| 5 | 0.1912 |

(c) The significance level of the test is the same as the probability of making a Type I error, so **the significance level is 10%.**

(d) If, in fact, the number of flaws per metre has been reduced to 1, then the expected number of flaws in a 4-metre length is four. The hypotheses become

$H_0: \lambda = 8$
$H_1: \lambda = 4$

$P(\text{Type II error}) = P(\text{accept } H_0 \text{ when } H_1 \text{ is true})$

You reject $H_0$ when $x < 5$, so accept $H_0$ when $x \geqslant 5$.

$P(\text{Type II error}) = P(X \geqslant 5 \text{ when } \lambda = 4)$
$\qquad\qquad\qquad\quad = 1 - P(X \leqslant 4 \text{ when } X \sim \text{Po}(4))$
$\qquad\qquad\qquad\quad = 1 - 0.6288$
$\qquad\qquad\qquad\quad = 0.3712$
$\qquad\qquad\qquad\quad \approx 37\%$

| $P(X \leqslant r)$ where $X \sim \text{Po}(4)$ $\lambda = 4.0$ | |
|---|---|
| $r = 0$ | 0.0183 |
| 1 | 0.0916 |
| 2 | 0.2381 |
| 3 | 0.4335 |
| 4 | 0.6288 |
| 5 | 0.7851 |
| 6 | 0.8893 |

**The probability of making a Type II error is approximately 37%.**

## Exercise 10b – Testing $\lambda$ in a Poisson distribution

1. The number of white corpuscles on a slide has a Poisson distribution with mean 3.5. After treatment, a sample was taken and the number of white corpuscles was found to be 8. Test at the 5% level of significance, whether the number of white corpuscles has increased.

2. The number of telephone calls to an office on a weekday follows a Poisson distribution with a mean number of six per hour.
   (a) On Monday there were 5 calls between 10.00 a.m. and 10.30 a.m. Test, at the 5% level, whether the number of calls has increased.
   (b) On Wednesday there were 3 calls between 11.00 a.m. and 12.30 p.m. Test, at the 5% level whether the number of calls has decreased.

3. Over a long period of time, Jane has found that the bus taking her to school arrives late on average 9 times a month. In the month following the start of the new summer schedules, Jane finds that her bus arrives late 13 times. Assuming that the number of times the bus is late can be modelled by a Poisson distribution, test, at the 5% level of significance, whether the new schedules have in fact increased the number of times on which the bus is late. State clearly your null and alternative hypotheses. (L)

4. A single observation is to be taken from a Poisson distribution with mean $\lambda$ and used to test the null hypothesis $\lambda = 8$ against the alternative hypothesis $\lambda < 8$.
   The critical region is chosen to be $x \leqslant 3$.
   (a) Find the probability of making a Type I error.
   (b) Find the probability of making a Type II error if, in fact, $\lambda = 6$.

5. For each of the following, an observation, $x$, is taken from a Poisson distribution, where $X \sim Po(\lambda)$.
   Test the hypotheses at the level of significance stated.

| | $x$ | Hypotheses | Level of significance |
|---|---|---|---|
| (a) | 11 | $H_0: \lambda = 7$ <br> $H_1: \lambda > 7$ | 5% |
| (b) | 12 | $H_0: \lambda = 7$ <br> $H_1: \lambda \neq 7$ | 5% |
| (c) | 4 | $H_0: \lambda = 10$ <br> $H_1: \lambda < 10$ | 1% |
| (d) | 18 | $H_0: \lambda = 10$ <br> $H_1: \lambda > 10$ | 5% |
| (e) | 2 | $H_0: \lambda = 6.5$ <br> $H_1: \lambda \neq 6.5$ | 5% |
| (f) | 2 | $H_0: \lambda = 6.5$ <br> $H_1: \lambda < 6.5$ | 5% |

6. In a particular city it was found, over a period of time, that $X$, the number of cases of a certain medical condition reported in a month, has a Poisson distribution with mean 3.5. During the month of August, seven cases were reported. Stating a necessary assumption, perform a significance test at the 5% level to decide whether or not this number of reported cases suggests that the number of occurrences of the medical condition has increased. State your hypotheses and conclusions clearly.

## Summary of stages of a hypothesis (significance) test

The following summary shows the stages of the hypothesis test. For details relating to the particular distributions, see page 492 for the binomial test and page 496 for the Poisson test:

1. State the variable being considered.

2. State the null hypothesis $H_0$ and the alternative hypothesis $H_1$.

   If you are looking for

   an increase, then $\quad\quad H_1: \ldots > \ldots$ (one-tailed test, upper tail)

   a decrease, then $\quad\quad H_1: \ldots < \ldots$ (one-tailed test, lower tail)

   a change, then $\quad\quad H_1: \ldots \neq \ldots$ (two-tailed test, upper and lower tails)

3. Consider the appropriate distribution if the null hypothesis is true.

4. Decide on the significance level of the test. This fixes the critical (rejection) region.

5. Decide on your rejection criterion.

*Now consider the value of the test statistic.*

6. Perform any calculations necessary to find out whether the test statistic is in the critical region.
7. Make your conclusion:
   - If the test statistic is in the critical region, reject $H_0$ in favour of $H_1$.
   - If the test statistic is not in the critical region, do not reject $H_0$.

   Then relate this to the context of the situation being tested.

## Summary of Type I and Type II errors

| | | Test decision | |
|---|---|---|---|
| | | Accept $H_0$ | Reject $H_0$ |
| Actual | $H_0$ is true | ✓ | Type I error |
| situation | $H_0$ is false | Type II error | ✓ |

$P(\text{Type I error}) = P(\text{reject } H_0 \text{ given that } H_0 \text{ is true})$
$= \alpha\% \text{ (where } \alpha\% \text{ is the significance level)}$

$P(\text{Type II error}) = P(\text{accept } H_0 \text{ given that } H_0 \text{ is false})$
$= P(\text{accept } H_0 \text{ given that } H_1 \text{ is true})$

# Miscellaneous worked examples

## Example 10.7

When I used to play darts regularly I scored a bull's-eye on average on 40% of attempts. After a break of three months, I play darts one evening and score two bull's-eyes in 12 attempts. I wish to test whether the percentage of attempts on which I score a bull's-eye has decreased.

(a) Stating a necessary assumption, use an exact binomial distribution to carry out the test, using a 10% significance level.
(b) Comment on the validity of the assumptions made in (a). (C)

## Solution 10.7

(a) Let $X$ be the number of bull's eyes scored in 12 attempts.
   Assuming that the result of an attempt is independent of the results of all other attempts, $X$ can be modelled by a binomial distribution, where $X \sim B(12, p)$.

   $H_0$: $p = 0.4$ (I score a bull's eye on 40% of the attempts.)
   $H_1$: $p < 0.4$ (The percentage of attempts on which I score a bull's eye has decreased.)

   If $H_0$ is true, then $X \sim B(12, 0.4)$.

   Carry out a one-tailed (lower tail) test at the 10% level.

   Reject $H_0$ if $P(X \leq x) < 10\%$ where $x$ is the test value.

The test value is $x = 2$.
Now $P(X \leqslant 2) = P(X = 0) + P(X = 1) + P(X = 2)$
$$= 0.6^{12} + 12 \times 0.6^{11} \times 0.4 + {}^{12}C_2 \times 0.6^{10} \times 0.4^2$$
$$= 0.08344 \ldots$$
$$\approx 8.3\%$$

*(You can use cumulative binomial tables if they are available.)*

Since $P(X \leqslant 2) < 10\%$, the sample value $x = 2$ lies in the critical region, so $H_0$ is rejected in favour of $H_1$. **At the 10% significance level, there is evidence that the percentage of attempts on which I score a bull's eye has decreased.**

(b) One usually improves when making several attempts, so it is unlikely that the attempts are independent. If they are not independent, then a binomial model is not suitable and the test is not valid.

## Example 10.8

A die is suspected of bias towards showing more sixes than would be expected of an ordinary die. In order to test this, it is decided to throw the die 12 times. The null hypothesis $p = \frac{1}{6}$, where $p$ is the probability of the die showing a six, will be rejected in favour of the alternative hypothesis $p > \frac{1}{6}$ if the number of sixes obtained is 4 or more. Calculate, to three decimal places, the probability of making

(a) a Type I error,
(b) a Type II error if, in fact $p = \frac{1}{2}$.                      (C)

## Solution 10.8

(a) Let $X$ be the number of sixes obtained when the die is thrown 12 times.
Then $X \sim B(12, p)$.

$H_0$: $p = \frac{1}{6}$ (The die is fair)
$H_1$: $p > \frac{1}{6}$ (The die is biased in favour of sixes)

$P(\text{Type I error}) = P(\text{reject } H_0 \text{ when } H_0 \text{ is true})$

If $H_0$ is true, then $X \sim B(12, \frac{1}{6})$.

Also $H_0$ is rejected if $x \geqslant 4$,
so $P(\text{Type I error}) = P(X \geqslant 4 \text{ when } X \sim B(12, \frac{1}{6}))$
$$= 1 - P(X < 4)$$
$$= 1 - \left( (\tfrac{5}{6})^{12} + 12 \times (\tfrac{5}{6})^{11} \times \tfrac{1}{6} + {}^{12}C_2 \times (\tfrac{5}{6})^{10} \times (\tfrac{1}{6})^2 + {}^{12}C_3 \times (\tfrac{5}{6})^9 \times (\tfrac{1}{6})^3 \right)$$
$$= 0.1251 \ldots$$
$$= 0.13 \quad (\textbf{2 s.f.})$$

(b) The hypotheses now become

$H_0$: $p = \frac{1}{6}$
$H_1$: $p = \frac{1}{2}$

$P(\text{Type II error}) = P(\text{accept } H_0 \text{ when } H_1 \text{ is true})$

If $H_1$ is true, then $p = \frac{1}{2}$ and $X \sim B(12, \frac{1}{2})$
$H_0$ is accepted if $x < 4$,
so $P(\text{Type II error}) = P(X < 4 \text{ when } X \sim B(12, \frac{1}{2}))$
$$= (\tfrac{1}{2})^{12} + 12 \times (\tfrac{1}{2})^{12} + {}^{12}C_2 \times (\tfrac{1}{2})^{12} + {}^{12}C_3(\tfrac{1}{2})^{12}$$
$$= 0.073 \quad (\textbf{2 s.f.})$$

## Example 10.9

A manufacturer of windows has used a process which produced flaws in the glass randomly at a rate of 0.5 per $m^2$. In an attempt to reduce the number of flaws produced, a new process is tried out. A randomly chosen window produced using this new process has an area of 8 $m^2$ and contains only one flaw.

(a) Stating your hypotheses clearly, test at the 10% level of significance whether or not the rate of occurrence of flaws using the new procedure has decreased.

The new procedure actually produces flaws at a rate of 0.3 per $m^2$.

(b) Find the probability of making a Type II error using the test in part (a).   (L)

## Solution 10.9

Let $X$ be the number of flaws in an 8 $m^2$ window. Then $X \sim Po(\lambda)$.

(a) $H_0: \lambda = 4$
$H_1: \lambda < 4$ (the number of flaws has decreased)

If $H_0$ is true, then $X \sim Po(4)$.
Use a one-tailed test and consider the lower tail for the critical region.
At the 10% level the value $x = 1$ will be in the critical region if $P(X \leqslant 1) < 10\%$, therefore reject $H_0$ if $P(X \leqslant 1) < 10\%$.

$$P(X \leqslant 1) = P(X = 0) + P(X = 1)$$
$$= e^{-4}(1 + 4)$$
$$= 0.0915\ldots$$

Since $P(X \leqslant 1) < 10\%$, reject $H_0$ in favour of $H_1$.

**The rate of occurrence of flaws using the new procedure has decreased.**

(b) The hypotheses now become

$H_0: \lambda = 4$
$H_1: \lambda = 2.4$

If $H_1$ is true, then $X \sim Po(2.4)$.

$P(\text{Type II error}) = P(\text{accept } H_0 \text{ when } H_1 \text{ is true})$
The critical region is $x \leqslant 1$, so you would accept $H_0$ if $x > 1$.
$P(\text{Type II error}) = P(X > 1 \text{ when } X \sim Po(2.4))$
$$= 1 - e^{-2.4}(1 + 2.4)$$
$$= 0.6916$$
$$\approx 69\%$$

# Miscellaneous exercise 10c – Binomial and Poisson tests

1. Before I sat an examination, my teacher told me that I had a 60% chance of obtaining a grade A, but I thought I had a better chance than that. In preparation for the examination, we did seven tests each of the same standard as the examination. Assuming my teacher is right, find the probability that I would get a grade A on

   (a) all 7 tests,
   (b) exactly 6 tests out of 7,
   (c) exactly 5 tests out of 7.

   In fact I got a grade A on 6 tests out of 7. State suitable null and alternative hypotheses and carry out a statistical test to determine whether or not there is evidence that my teacher is underestimating my chances of a grade A. (*MEI*)

2. Harry Hotspur is a footballer who likes to take penalty kicks. On past performance he reckons that on average he scores 7 times out of 10. Assume that Harry is correct, and consider the next 8 penalty kicks he takes.

   (a) Find the probability that he will score at least 6 times.
   (b) Find the modal score and state its probability.
   (c) What further assumption have you made in calculating the probabilities in (a) and (b)?

   After a period of intense practice, Harry reckons that he has improved his penalty taking.

   (d) Write down suitable null and alternative hypotheses for testing the value of $p$, the probability that Harry scores from a penalty kick.

   He takes 15 penalty kicks and scores from 13 of them.

   (e) Carry out the hypothesis test, at the 10% level of significance, stating your conclusion clearly.
   (f) Harry takes a further set of 15 penalty kicks. Out of the total of 30 kicks he scores from 26. Without further calculation explain carefully whether this additional information strengthens Harry's case or not. (*MEI*)

3. The manufacturers of a certain type of microwave oven claim that at least 95% of their ovens will not fail during the first two years of use. In order to test this claim, a Consumer Agency purchased a random sample of 15 ovens and ran them under similar conditions over a two-year period. It was found that 12 ovens had not failed during that period.
   Test the manufacturer's claim using an exact binomial distribution. The significance level should be as close as possible to 5%.
   Explain why an exact 5% significance level is not possible. (*C*)

4. The ABC School of Motoring claim that at least 80% of their pupils pass the driving test first time. The XYZ School of Motoring suspect that more than 20% of ABC's pupils fail first time. They test this suspicion by checking the results of a random sample of 25 former ABC pupils, finding out how many failed first time.

   (a) State suitable null and alternative hypotheses to be used in the test.
   (b) Identify the model that should be used for the distribution of the number of failures.
   (c) Find the smallest number of failures which would allow ABC's claim to be rejected at the 5% level of significance. (*NEAB*)

5. For most small birds, the ratio of males to females may be expected to be about 1 : 1. In one ornithological study birds are trapped by setting fine-mesh nets. The trapped birds are counted and then released. The catch may be regarded as a random sample of the birds in the area. The ornithologists want to test whether the sex ratio of blackbirds is, in fact, 1 : 1.

   (a) Assuming that the sex ratio of blackbirds is 1 : 1, find the probability that a random sample of 16 blackbirds contains
   (i) 12 males   (ii) at least 12 males
   (iii) at least 12 of the same sex.
   (b) State the null and alternative hypotheses the ornithologists should use, clearly indicating why the alternative hypothesis takes the form it does.

   In one sample of 16 blackbirds there are 12 males and 4 females.

   (c) Carry out a suitable test using these data at the 5% significance level, stating your conclusion clearly. Find the critical region for the test.
   (d) Another ornithologist points out that, because female birds spend much time sitting on the nest, females are less likely to be caught than males. Explain how this would affect your conclusions. (*MEI*)

6. Over many years it has been found that at a particular station 20% of trains arrive late. A consumer group wishes to test whether the percentage of trains arriving late has increased recently. It decides to observe 20 trains. If more than four of the trains arrive late it will claim that the percentage of trains arriving late has increased.

   (a) In the case where the percentage of trains arriving late has remained at 20%, find the probability that the consumer group makes a Type I error.
   (b) In the case where the percentage of trains arriving late has increased to 25%, find the probability that the consumer group makes a Type II error.

(c) (i) Comment on your answer to part (a).
    (ii) Suggest an improvement to the procedure used by the consumer group.

    (NEAB)

7  A firm producing mugs has a quality control scheme in which a random sample of 10 mugs from each batch is inspected. For 50 such samples, the numbers of defective mugs are as follows:

| Number of defective mugs | 0 | 1 | 2 | 3 | 4 | 5 | 6+ |
|---|---|---|---|---|---|---|---|
| Number of samples | | 5 | 13 | 15 | 12 | 4 | 1 | 0 |

(a) Find the mean and standard deviation of the number of defective mugs per sample.
(b) Show that a reasonable estimate for $p$, the probability that a mug is defective, is 0.2. Use this figure to calculate the probability that a randomly chosen sample will contain exactly 2 defective mugs. Comment on the agreement between this value and the observed data.

The management is not satisfied with 20% of mugs being defective and introduces a new process to reduce the proportion of defective mugs.

(c) A random sample of 20 mugs, produced by the new process, contains just one which is defective. Test, at the 5% level, whether it is reasonable to suppose that the proportion of defective mugs has been reduced, stating your null and alternative hypotheses clearly.
(d) What would the conclusion have been if the management had chosen to conduct the test at the 10% level?    (MEI)

8. In a certain country, 90% of letters are delivered the day after posting.
A resident posts 8 letters on a certain day.
Find the probability that:

(a) all 8 letters are delivered the next day,
(b) at least 6 letters are delivered the next day,
(c) exactly half the letters are delivered the next day.

It is later suspected that the service has deteriorated as a result of mechanisation. To test this, 17 letters are posted and it is found that only 13 of them arrive the next day. Let $p$ denote the probability, after mechanisation, that a letter is delivered the next day.

(d) Write down suitable null and alternative hypotheses for the value of $p$. Explain why the alternative hypothesis takes the form it does.
(e) Carry out the hypothesis test, at the 5% level of significance, staring your results clearly.
(f) Write down the critical region for the test, giving a reason for your choice.    (MEI)

9. It is known that the number of defects in a one-metre length of steel pipe has mean 2.4. It has been suggested that a Poisson distribution would be a reasonable model for the number of defects in a randomly chosen one-metre length of this steel pipe.

(a) State two assumptions that would need to be made for a Poisson distribution to be an appropriate model in this case.
(b) Using this Poisson model, calculate the probability that in a randomly chosen one-metre length of steel pipe there are:
    (i) exactly 3 defects,
    (ii) more than 3 defects.
(c) Determine the probability that there are exactly 6 defects in a randomly chosen two-metre length of the same type of steel pipe.
(d) It is believed that the manufacturing process may now be producing more defects than before. In a quality control experiment a one-metre length of the steel pipe is chosen and is found to have 7 defects. Test, at the 5% level of significance, the hypothesis that the number of defects in this type of steel pipe has increased. State your hypotheses clearly.    (O)

10. (a) The number, $X$, of breakdowns per day of the lifts in a large block of flats has a Poisson distribution with mean 0.2. Find, to three decimal places, the probability that on a particular day
    (i) there will be at least one breakdown,
    (ii) there will be at most two breakdowns.
(b) Find, to three decimal places, the probability that, during a 20-day period, there will be no lift breakdowns.
(c) The maintenance contract for the lifts is given to a new company. With this company it is found that there are two breakdowns over a period of 30 days. Perform a significance test at the 5% level to decide whether or not the number of breakdowns has decreased.    (L)

11. The number, $X$, of emergency telephone calls to a gas board office in $t$ minutes at weekends is known to follow a Poisson distribution with mean $\frac{1}{90}t$. Given that the telephone in that office is unmanned for 10 minutes, calculate, to two significant figures, the probability that there will be at least 2 emergency telephone calls to the office during that time.
Find, to the nearest minute, the length of time that the telephone can be left unmanned for there to be a probability of 0.9 that no emergency telephone call is made to the office during the period the telephone is unmanned.
During a week of very cold weather it was found that there had been 10 emergency telephone calls to the office in the first 12 hours of the weekend. Using tables, or otherwise, determine whether the increase in the average number of emergency telephone calls to that office is significant at the 5% level.    (L)

# Mixed test 10A (Binomial)

1. The random variable, $R$, can be modelled by a binomial distribution with parameters $n = 10$ and $p$, whose value is unknown.
   Find the critical region for the test of

   $H_0$: $p = 0.5$ against $H_1$: $p \neq 0.5$

   at the 10% level of significance. (*NEAB*)

2. A large college introduced a new procedure to try to ensure that staff arrived on time for the start of lectures. A recent survey by the students had suggested that in 15% of cases the staff arrived late for the start of a lecture. In the first week following the introduction of this new procedure a random sample of 35 lectures was taken and in only one case did the member of staff arrive late.

   (a) Stating your hypotheses clearly test, at the 5% level of significance, whether or not there is evidence that the new procedure has been successful.

A student complained that this sample did not give a true picture of the effectiveness of the new procedure.

   (b) Explain briefly why the student's claim might be justified and suggest how a more effective check on the new procedure could be made. (*L*)

3. An enthusiastic gardener claimed that she could never work in the garden at the weekend because 'It always rains on Saturday and Sunday when I'm at home and it's always fine on weekdays when I'm not!' She noted the weather for the next month and recorded that, out of 10 wet days, five were either a Saturday or a Sunday. The gardener's claim may be modelled by regarding her observation as a single sample from a $B(10, p)$ distribution. Given that one would expect 2 out of every 7 wet days to be either a Saturday or a Sunday, the null hypothesis, $p = \frac{2}{7}$, may be tested against the alternative hypothesis, $p > \frac{2}{7}$. Carry out a hypothesis test to test her claim at the 10% significance level. (*C*)

# Mixed test 10B (Poisson)

1. The mean number of serious accidents at a motorway interchange is 2.1 per week.

   (a) State the probability distribution which may reasonably be used to model the weekly number of serious accidents at this motorway interchange, and give its parameter.

   (b) Use an appropriate distribution to determine the probability that he number of serious accidents is:
   (i) two or fewer in a randomly selected week:
   (ii) exactly one on a randomly selected **day**.

   (c) Given that there were 6 serious accidents during one wet winter week, test, at the 5% level of significance, the hypothesis that the accident rate is higher in wet weather. (*O*)

2. (a) The number of bacterial colonies that develop in dishes of nutrient exposed to an infected environment has a Poisson distribution with mean 7.5.
   (i) Calculate the probability that, in one such dish, the number of bacterial colonies that develop will be greater than 10.

   (ii) Calculate the probability that, in two such dishes, the total number of bacterial colonies that develop will be between 10 and 20 inclusive.

   (b) Experiments were conducted to determine the effectiveness of an antibiotic spray in reducing the number of bacterial colonies that develop.
   In one experiment in which one dish was sprayed, the number of bacterial colonies that developed was 3. Stating suitable null and alternative hypotheses, determine whether or not this result provides significant evidence at the 5% level that the spray is effective. (*NEAB*)

3. A single observation is taken from a Poisson distribution with mean $\mu$ and used to test the hypothesis $\mu = 6$ against the alternative hypothesis $\mu > 6$.
   The critical region is chosen to be $x \geqslant 11$.

   (a) At what significance level is the test carried out?

   (b) Find the probability of making a Type II error if, in fact, $\mu = 8.5$.

# 11

# Hypothesis testing (z-tests and t-tests)

*In this chapter you will*

- be reminded about the language of hypothesis (significance) testing introduced in Chapter 10
- be reminded about Type I and Type II errors
- learn how to perform the following hypothesis tests:

    **Test 1**: Testing $\mu$, the mean
    - 1a: of a normal distribution with known variance, any size sample (z-test)
    - 1b: of a distribution with known variance, large sample (z-test)
    - 1c: of a distribution with unknown variance, large sample (z-test)
    - 1d: of normal distribution with unknown variance, small sample (t-test)

    **Test 2**: Testing $p$, the proportion of a binomial distribution, $n$ large (z-test)

    **Test 3**: Testing $\mu_1 - \mu_2$, the difference between means of two normal distributions
    - 3a: when population variances are known (z-test)
    - 3b: when there is a known common population variance (z-test)
    - 3c: when the common population variance is unknown,
        - – large samples (z-test)
        - – small samples (t-test)

Background knowledge:

*For the z-tests you will need to be familiar with*
- *the normal distribution and the use of the standard normal tables (see page 362)*
- *the distribution of the sample mean (see page 436)*
- *the unbiased estimate for the population variance (see page 447)*
- *the normal approximation to the binomial distribution (see page 382)*

*For the t-tests you will need to be familiar with*
- *the use of the t-distribution tables (see page 463)*

## HYPOTHESIS TESTING

If you have worked through Chapter 10 you will be familiar with the terminology and methods used to carry out hypothesis tests relating to discrete distributions. For those new to the topic, these are described again in the following text, but this time in relation to continuous distributions. The following example illustrates the hypothesis test for the mean of a normal distribution.

In the production of ice packs for use in cool boxes, a machine fills packs with liquid and the packs are then frozen. Since space is needed in the packs for the liquid to expand, it is important that they are not over-filled. The volume of liquid in the packs follows a normal distribution with mean 524 ml and standard deviation 3 ml.

The machine breaks down and is repaired. In the next batch of production, there is a suspicion that the mean volume of liquid dispensed by the machine into the packs has increased and is greater than 524 ml. In order to investigate this, the supervisor takes a random sample of 50 packs and finds that the mean volume of liquid in these is 524.9 ml. Does this provide evidence that the machine is over-dispensing?

The mean volume of the sample, 524.9 ml, is higher than the established mean of 524 ml. But is it high enough to say that the mean volume of *all* the packs filled by the machine has increased? Perhaps the mean is still 524 ml and this higher value has occurred just because of sampling variation. A **hypothesis** (or **significance**) **test** will enable a decision to be made that is backed by statistical theory, not just based on a suspicion.

Let $X$ be the volume, in millilitres, of liquid dispensed into a pack after the machine has been repaired and let the mean of $X$ be $\mu$, where $\mu$ is unknown. Assuming that the standard deviation remains unchanged, $X \sim N(\mu, \sigma^2)$ with $\sigma = 3$.

The hypothesis is made that $\mu$ is 524 ml, i.e. the mean has remained the same as it was prior to the repair. This is known as the **null hypothesis, $H_0$** and is written

$H_0: \mu = 524$

Since it is suspected that the mean volume has *increased*, the **alternative hypothesis, $H_1$**, is that the mean is *greater than* 524 ml. This is written

$H_1: \mu > 524$

To carry out the test, the focus moves from $X$, the volume of liquid in a pack, to the distribution of $\overline{X}$, the *mean volume of a sample of 50 packs*. In this test, $\overline{X}$ is known as the **test statistic** and its distribution is needed. The distribution of $\overline{X}$ is known as the sampling distribution of means.

In Chapter 9 you saw that if $X \sim N(\mu, \sigma^2)$,

then, for samples of size $n$, $\overline{X} \sim N\left(\mu, \dfrac{\sigma^2}{n}\right)$.

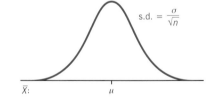

The hypothesis test starts by assuming that the value stated in the null hypothesis is true, so $\mu = 524$.

Since $\sigma = 3$ and $n = 50$,

$$\overline{X} \sim N\left(524, \frac{3^2}{50}\right), \quad \text{i.e. } \overline{X} \sim N(524, 0.18).$$

The sampling distribution of means, therefore, follows a normal distribution with mean 524 ml and variance 0.18 ml². The standard deviation is $\sqrt{0.18}$ ml.

NOTE: This is sometimes left in the form $\sqrt{\dfrac{\sigma^2}{n}} = \dfrac{\sigma}{\sqrt{n}} = \dfrac{3}{\sqrt{50}}$.

The result of the test depends on the whereabouts in the sampling distribution of the **test value** of 524.9 ml, the mean volume of the sample of 50 packs taken by the supervisor. She would need to find out whether 524.9 is close to 524 or far away from 524.

If it is *close to* 524 then it is likely to have come from a distribution with mean 524 ml and there would not be enough evidence to say that the mean volume has increased.

If it is *far away from* 524, i.e. in the right-hand (upper tail) of the distribution, then it is unlikely to have come from a distribution with a mean of 524 ml. The mean is likely to be higher than 524 ml.

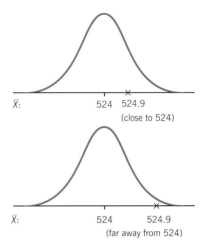

Note that the *upper tail* is being used because the supervisor suspects that there is an *increase* in $\mu$. This type of test is called a **one-tailed (upper tail) test**.

A decision needs to be taken about the cut off point, $c$, known as the **critical value**, which indicates the boundary of the region where values of $\bar{x}$ would be considered to be *too far away* from 524 ml and therefore would be *unlikely to occur*. The region is known as the **critical region** or **rejection region**.

The critical value and region are fixed using probabilities linked to the **significance level** of the test. In general, for an upper tail test at the $\alpha\%$ level, the critical value $c$ is fixed so that $P(\overline{X} > c) = \alpha\%$ and the critical (rejection) region is $\bar{x} > c$.

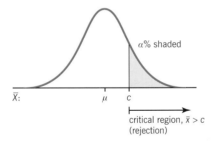

The hypothesis test involves finding whether or not the sample value, $\bar{x}$, lies in the critical region of the sampling distribution of means, $\overline{X}$.

In this example, **if $\bar{x}$ lies in the critical region**, then a decision is taken that it is too far away from 524 ml to have come from a distribution with this mean. In statistical language, you would **reject the null hypothesis, $H_0$** (that the mean is 524 ml), **in favour of the alternative hypothesis, $H_1$** (that the mean is greater than 524 ml).

**If $\bar{x}$ does not lie in the critical region,** there is not have enough evidence to reject $H_0$, so $H_0$ is **accepted**. In this example, $\bar{x} < c$ is known as the **acceptance region**.

For a significance level of $\alpha\%$, if the sample mean lies in the critical (or rejection) region, then the result is said to be **significant at the $\alpha\%$ level**.

Note that if a result is significant at, say, the 1% level, then it is automatically significant at any level greater than 1%, for example 5% or 10%.

Say that the supervisor chooses a significance level of 5%. She will then reject $H_0$ if the test value (i.e. the mean volume of the sample of 50 cans) lies in the upper tail 5% of the distribution of sample means.

Since this distribution is normal, instead of finding $c$, the critical $\bar{x}$ value, it is possible to work in standardised values and find the $z$-value that gives 5% in the upper tail. Using standard normal tables (page 649),

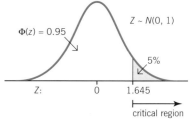

if $\quad P(Z > z) = 0.05$

then $\quad P(Z < z) = 1 - 0.05 = 0.95$

i.e. $\quad \Phi(z) = 0.95$

$\quad\quad z = \Phi^{-1}(0.95)$

$\quad\quad\quad = 1.645$

So $z$-values that are greater than 1.645 lie in the upper tail 5% of the distribution.

This enables a statement to be made, known as the **rejection criterion**, which tells you when to reject the null hypothesis:

Reject $H_0$ if $z > 1.645$, where $z$ is the standardised value of the mean of the sample of 50 packs,

i.e. $\quad z = \dfrac{\bar{x} - \mu}{\sigma/\sqrt{n}} = \dfrac{\bar{x} - 524}{3/\sqrt{50}}$

Note that to avoid being influenced by sample readings, it is important that the rejection criterion is decided upon *before* any sample values are taken.

When the sample was taken, it was found that $\bar{x} = 524.9$,

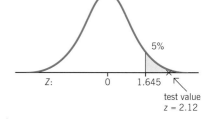

so $\quad z = \dfrac{524.9 - 524}{3/\sqrt{50}}$

$\quad\quad\quad = 2.12 \text{ (2 d.p.)}$

The result of the test is now stated in statistical terms and then related to the context of the test, as follows:

Since $z > 1.645$, $H_0$ is rejected in favour of $H_1$. The supervisor would conclude that the mean volume of liquid being dispensed by the machine is not 524 ml, but has increased, she would be wise therefore to stop production so that the setting on the machine could be adjusted.

Note that the critical $\bar{x}$-value, $c$, can be found by de-standardising the critical $z$-value of 1.645, where

$\dfrac{c - 524}{3/\sqrt{50}} = 1.645$

$\quad\quad c = 524 + 1.645 \times \dfrac{3}{\sqrt{50}}$

$\quad\quad\quad = 524.7$

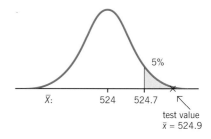

Since the test value of 524.9 is greater than 524.7, it lies in the critical region, confirming the result obtained above.

If you want even more information, you can find out *exactly* where the sample mean lies in the distribution of $\bar{X}$. Note that this is the method used in Chapter 10 for discrete variables.

$$\bar{X} \sim N\left(524, \frac{3^2}{50}\right)$$

so $\quad P(\bar{X} > 524.9) = P\left(Z > \dfrac{524.9 - 524}{3/\sqrt{50}}\right)$

$\qquad\qquad\qquad = P(Z > 2.1213 \ldots)$
$\qquad\qquad\qquad = 1 - 0.9831$
$\qquad\qquad\qquad = 0.0169$
$\qquad\qquad\qquad \approx 1.7\%$

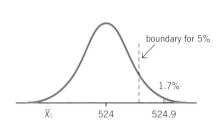

This probability is less than 5%, implying that the boundary for the critical region must lie to the left of the sample value of 524.9 and confirming that 524.9 lies in the critical region. This method also tells you that the test value of 524.9 will lie in the critical region for any level of significance above 1.7%.

This probability method can be used, if preferred, in the hypothesis test to find whether the sample value lies in the rejection region. In this example, for a 5% level of significance, the rejection criterion would be to **reject $H_0$** if $P(\bar{X} > \bar{x}) < 0.05$, where $\bar{x}$ is the sample mean.

## ONE-TAILED AND TWO-TAILED TESTS

Say that the null hypothesis is $\mu = \mu_0$.

In a **one-tailed test**, the alternative hypothesis $H_1$ looks for an **increase or a decrease** in $\mu$:

– for an increase, $H_1$ is $\mu > \mu_0$ and the critical region is in the **upper tail**,

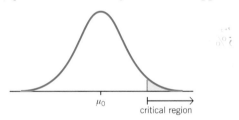

– for a decrease, $H_1$ is $\mu < \mu_0$ and the critical region is in the **lower tail**.

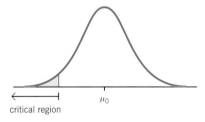

In a **two-tailed test**, the alternative hypothesis $H_1$ looks for a **change** in $\mu$ without specifying whether it is an increase or a decrease and $H_1$ is $\mu \neq \mu_0$. The critical region is in two parts:

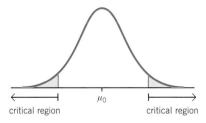

## CRITICAL z-VALUES

Critical values depend on the significance level and also whether the test is one- or two-tailed. The method of working in standardised values is widely used for tests involving the normal distribution because the critical z-values can be found easily from standard normal tables, as described on page 529. Sometimes the most commonly needed values are summarised in a critical value table. One such table is shown below and it is also printed in the appendix at the bottom of page 649. It gives the z-values for various values of $p$, where $p = P(Z < z) = \Phi(z)$.

| $p$ | 0.75 | 0.90 | 0.95 | 0.975 | 0.99 | 0.995 | 0.9975 | 0.999 | 0.9995 |
|---|---|---|---|---|---|---|---|---|---|
| $z$ | 0.674 | 1.282 | 1.645 | 1.960 | 2.326 | 2.576 | 2.807 | 3.090 | 3.291 |

For example, *for a one-tailed test, at 1% level:* you want to find $z$ such that $\Phi(z) = 0.99$, so look up $p = 0.99$.
From the table, $z = 2.326$. Therefore the upper tail critical value is 2.326. By symmetry, the lower tail critical value is –2.326.

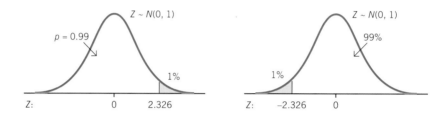

*For a two-tailed test, at the 1% level:* the 1% in the tails is split evenly between the upper and lower tails with 0.5% in each. There are two critical values.

To find the upper tail value, you need to find $z$ such that $\Phi(z) = 0.995$, so look up $p = 0.995$.
From the table $z = 2.576$.
So the upper tail critical value is 2.576 and the lower tail value (by symmetry) is –2.576.

Critical values:

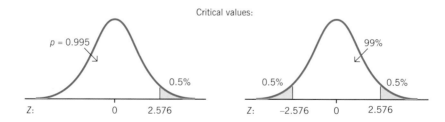

# SUMMARY OF CRITICAL VALUES AND REJECTION CRITERIA

The summary below shows the critical $z$-values and rejection criteria for the most commonly used levels of significance: 10%, 5% and 1%.

| | One-tailed test (lower tail) $H_0 : \mu = \mu_0$ $H_1 : \mu < \mu_0$ | One-tailed test (upper tail) $H_0 : \mu = \mu_0$ $H_1 : \mu > \mu_0$ | Two-tailed test $H_0 : \mu = \mu_0$ $H_1 : \mu \neq \mu_0$ |
|---|---|---|---|
| 10% significance level | Reject $H_0$ if $z < -1.282$ | Reject $H_0$ if $z > 1.282$ | Reject $H_0$ if $z < -1.645$ or $z > 1.645$ (written $\lvert z \rvert > 1.645$) |
| 5% significance level | Reject $H_0$ if $z < -1.645$ | Reject $H_0$ if $z > 1.645$ | Reject $H_0$ if $z < -1.96$ or $z > 1.96$ (written $\lvert z \rvert > 1.96$) |
| 1% significance level | Reject $H_0$ if $z < -2.326$ | Reject $H_0$ if $z > 2.326$ | Reject $H_0$ if $z < -2.576$ or $z > 2.576$ (written $\lvert z \rvert > 2.576$) |

# STAGES IN THE HYPOTHESIS TEST

When carrying out a hypothesis test, it is useful to work through the following stages. This is essentially the same procedure as in the tests for parameters of discrete distributions described in Chapter 10.

1. State the variable being considered.

2. State the null hypothesis $H_0$ and the alternative hypothesis $H_1$.

   Remember that if you are looking for
   a decrease, then $\quad H_1 : \ldots < \ldots \quad$ (one-tailed test, lower tail)
   an increase, then $\quad H_1 : \ldots > \ldots \quad$ (one-tailed test, upper tail)
   a change, then $\quad H_1 : \ldots \neq \ldots \quad$ (two-tailed test, upper and lower tails)

3. Consider the distribution of the test statistic, assuming that the null hypothesis is true. If you are testing a sample mean, then the test statistic is $\overline{X}$, and the sampling distribution of means is considered.

4. State the type of test (i.e. whether it is one-tailed or two-tailed) and decide the significance level of the test.

5. Decide on your rejection criterion, remembering that you will reject $H_0$ if the test value lies in the critical (or rejection) region fixed by the significance level.

*Now consider the value of the test statistic*

6. Perform any calculations necessary to find out whether the test value is in the critical region.

7. Make your conclusion in statistical terms:

- If the test value is in the critical region, reject $H_0$ in favour of $H_1$.
- If the test value is not in the critical region, do not reject $H_0$.

Then relate your conclusion to the situation being tested.

There are several hypothesis tests involving continuous distributions and some of these are illustrated in the following text.

# HYPOTHESIS TEST 1: TESTING $\mu$ (THE MEAN OF A POPULATION)

Consider a population $X$ with unknown mean $\mu$ and variance $\sigma^2$.
A value for $\mu$, call it $\mu_0$, is specified in the hypotheses, for example

$H_0 : \mu = \mu_0$
$H_1 : \mu < \mu_0$ (or $\mu > \mu_0$ or $\mu \neq \mu_0$)

To test the hypotheses, take a sample of size $n$ from the population and calculate the sample mean, $\bar{x}$. The test statistic is $\bar{X}$, and the sampling distribution of means is considered.

There are now several cases that may occur, depending on whether the population is normal or not, whether the sample size is large or small and whether the population variance is known or not.

## Test 1a: Testing $\mu$ when the population $X$ is normal and the variance $\sigma^2$ is known (any size sample)

Since the population is normally distributed, $X \sim N(\mu, \sigma^2)$. The sampling distribution of means, $\bar{X}$ is also normal for *all* sample sizes, with mean $\mu_0$ (as specified in the null hypothesis).

When testing the mean of a normal population $X$ with known variance $\sigma^2$ for samples of size $n$, the test statistic is

$\bar{X}$, where $\bar{X} \sim N\!\left(\mu_0, \dfrac{\sigma^2}{n}\right)$.

In standardised form, the test statistic is

$Z = \dfrac{\bar{X} - \mu_0}{\sigma/\sqrt{n}}$ where $Z \sim N(0, 1)$.

## Example 11.1

Each year trainees throughout the country sit a test. Over a period of time it has been established that the marks can be modelled by a normal distribution with mean 70 and standard deviation 6.

This year it was thought that trainees from a particular county did not perform as well as expected. The marks of a random sample of 25 trainees from the county were scrutinised and it was found that their mean mark was 67.3.

Does this provide evidence, at the 5% significance level, that trainees from this county did not perform well as expected?

## Solution 11.1

The stages of the hypothesis test are shown in the margin and additional comments are given in italics.

**1. Define the variable.**

Let $X$ be the mark of a trainee from the particular county and let the population mean mark be $\mu$.

Assuming that the standard deviation has not changed, then $X \sim N(\mu, \sigma^2)$ with $\sigma = 6$.

**2. State $H_0$ and $H_1$.**

$H_0 : \mu = 70$ (The trainees have performed as expected)
$H_1 : \mu < 70$ (The trainees have not performed as well as expected)

**3. State the distribution of the test statistic according to $H_0$.**

*The test is carried out based on the value of the sample mean, $\bar{x}$. The test statistic is $\overline{X}$ and you need to consider the sampling distribution of means.*

For samples of size $n$, $\overline{X} \sim N\left(\mu, \dfrac{\sigma^2}{n}\right)$ with $\sigma = 6$ and $n = 25$.

*You now use the value of $\mu$ given by the null hypothesis.*

If $H_0$ is true, then $\mu = 70$, so $\overline{X} \sim N\left(70, \dfrac{6^2}{25}\right)$, i.e. $\overline{X} \sim N(70, 1.44)$.

Note that the standard deviation is $\sqrt{1.44} = 1.2$.

The standard deviation is sometimes left in its uncalculated form:

$$\sqrt{\frac{\sigma^2}{n}} = \frac{\sigma}{\sqrt{n}} = \frac{6}{\sqrt{25}}$$

**4. State the level of the test.**

Use a one-tailed (lower tail) test at the 5% level.

*Note that the test is one-tailed (lower tail) since you are looking for a decrease in $\mu$.*

**5. Decide on your rejection criterion.**

*You need to find out whether the sample mean of 67.3 (known as the test value) lies in the critical region. To state your rejection criterion, find the critical z-value for the 5% lower tail. This is $-1.645$ (see page 513).*

Reject $H_0$ if $z < -1.645$ where $z = \dfrac{\bar{x} - \mu_0}{\sigma/\sqrt{n}} = \dfrac{\bar{x} - 70}{6/\sqrt{25}}$.

**6. Perform the required calculation.**

*The test value is $\bar{x} = 67.3$, so*

$$z = \frac{67.3 - 70}{6/\sqrt{25}}$$
$$= -2.25$$

**7. Make your conclusion.**

*State the conclusion statistically (either reject $H_0$, or do not reject $H_0$) and then relate it to the context of the question.*

Since $z < -1.645$, $H_0$ is rejected in favour of $H_1$.

**There is evidence, at the 5% level, that the trainees in this area have not performed as well as expected.**

NOTE 1:
To find the critical region, calculate the critical $\bar{x}$ value, $c$, as follows:

$$\frac{c - 70}{6/\sqrt{25}} = -1.645$$

$$c = 70 - 1.645 \times \frac{6}{\sqrt{25}}$$

$$\therefore \quad c = 68.026$$

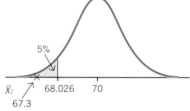

So the critical region is $\bar{x} < 68.026$. This means that any test value less than 68.026 would result in the null hypothesis being rejected.

NOTE 2:
If you prefer to use the probability method to decide whether $\bar{x}$ lies in the critical region, then, since the significance level is 5%, the rejection criterion would be to reject $H_0$ if $P(\bar{X} < 67.3) < 0.05$.

$$\text{Now} \quad P(\bar{X} < 67.3) = P\left(Z < \frac{67.30 - 70}{6/\sqrt{25}}\right)$$

$$= P(Z < -2.25)$$

$$= 0.013$$

$$= 1.3\%$$

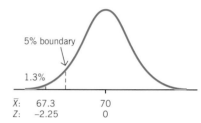

Since $P(\bar{X} < 67.3) < 0.05$, reject $H_0$ (as before).

This method also tells you that $H_0$ would be rejected at any significance level above 1.3%.

## Example 11.2

A sample of size 16 is taken from the distribution of $X \sim N(\mu, 3^2)$ and a hypothesis test is carried out at the 1% level of significance. On the basis of the value of the sample mean $\bar{x}$, the null hypothesis $\mu = 100$ is rejected in favour of the alternative hypothesis $\mu > 100$.

What can be said about the value of $\bar{x}$?

## Solution 11.2

*In this question you are being asked to find the critical region in terms of $\bar{x}$.*

It is given that $X \sim N(\mu, \sigma^2)$ with $\sigma = 3$.

The hypotheses are $\quad H_0 : \mu = 100$
$\qquad\qquad\qquad\qquad H_1 : \mu > 100$

Considering the sampling distribution of means for samples of size $n$,

$$\bar{X} \sim N\left(\mu, \frac{\sigma^2}{n}\right) \text{ with } \sigma = 3 \text{ and } n = 16.$$

If $H_0$ is true, then $\mu = 100$, so $\overline{X} \sim N\left(100, \dfrac{3^2}{16}\right)$

Note that the standard deviation is $\dfrac{\sigma}{\sqrt{n}} = \dfrac{3}{\sqrt{16}}\ (= 0.75)$.

Performing a one-tailed (upper tail) test at the 1% level, you are told that $H_0$ is rejected.

This means that the sample mean, $\bar{x}$ must lie in the critical region, i.e. $\bar{x}$ must be greater than the critical value, $c$.

Working first in standardised values, the critical $z$-value that gives 1% in the upper tail is $z = 2.576$ (see page 649).

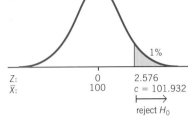

De-standardising to give $c$:  $\dfrac{c - 100}{3/\sqrt{16}} = 2.576$

$$c = 100 + 2.576 \times \dfrac{3}{\sqrt{16}}$$

$$\therefore \quad c = 101.932$$

So the critical region is $\bar{x} > 101.932$.
Since the null hypothesis is rejected, **the sample mean, $\bar{x}$, is greater than 101.932.**

---

# Test 1b: Testing $\mu$ when the population X is not normal, the variance $\sigma^2$ is known and the sample size $n$ is large

Since the population is not normal, you cannot say that the distribution of $\overline{X}$ is normal for all sample sizes. If the sample size $n$ is large, however, you can apply the **central limit theorem (see page 442)**. This states that for **large samples taken from a non-normal population**, the sampling distribution of means $\overline{X}$ is *approximately* normal, whatever the distribution of the parent population.

When testing the mean of a non-normal population $X$ with known variance $\sigma^2$, provided that the sample size $n$ is large,

the test statistic is $\overline{X}$, where $\overline{X}$ is approximately normal, $\overline{X} \sim N\left(\mu_0, \dfrac{\sigma}{n}\right)$.

In standardised form,

the test statistic is $Z = \dfrac{\overline{X} - \mu_0}{\sigma/\sqrt{n}}$ where $Z \sim N(0, 1)$.

## Example 11.3

The management of a large hospital states that the mean age of its patients is 45 years. Records of a random sample of 100 patients give a mean age of 48.4 years. Using a population standard deviation of 18 years, test at the 5% significance level whether there is evidence that the management's statement is incorrect. State clearly your null and alternative hypotheses.

(C)

## Solution 11.3

1. Define the variable.

Let $X$ be the age, in years, of a patient and let the population mean age be $\mu$. The population standard deviation $\sigma = 18$.

2. State $H_0$ and $H_1$.

$H_0 : \mu = 45$ (The management's claim is correct)
$H_1 : \mu \neq 45$ (The management's claim is incorrect)

3. State the distribution of the test statistic according to $H_0$.

*You are performing a test based on a sample mean, $\bar{x}$, so you need to consider the sampling distribution of means, $\bar{X}$.*

The sample size is 100. Since $n$ is large, by the central limit theorem, $\bar{X}$ is approximately normal, so

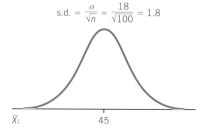

s.d. $= \dfrac{\sigma}{\sqrt{n}} = \dfrac{18}{\sqrt{100}} = 1.8$

$$\bar{X} \sim N\left(\mu, \frac{\sigma^2}{n}\right) \text{ with } \sigma = 18 \text{ and } n = 100.$$

If $H_0$ is true, then $\mu = 45$,

so $\quad \bar{X} \sim N\left(45, \dfrac{18^2}{100}\right).$

$\bar{X}$:  45

4. State the level of the test.

Use a two-tailed test, at the 5% level.

*The test is two-tailed since you are looking for a change in $\mu$, not specifically an increase or a decrease.*

5. Decide on your rejection criterion.

*The critical z-values for a two-tailed test at the 5% level are ±1.96 (see page 649). Remember that the 5% is shared evenly between the tails.*

Reject $H_0$ if $z < -1.96$ or $z > 1.96$, i.e. if $|z| > 1.96$,

where $\quad z = \dfrac{\bar{x} - \mu}{\sigma/\sqrt{n}} = \dfrac{\bar{x} - 45}{18/\sqrt{100}}$

6. Perform the required calculation.

The test value is $\bar{x} = 48.4$,

so $\quad z = \dfrac{48.4 - 45}{18/\sqrt{100}}$

$\quad\quad = 1.888 \ldots$

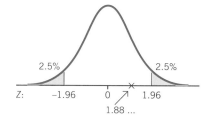

2.5%  2.5%

$Z$:  −1.96  0  1.96
1.88 ...

7. Make your conclusion.

Since $|z| < 1.96$, do not reject $H_0$.

**There is not sufficient evidence, at the 5% level of significance, to reject the management's claim that the mean age is 45 years.**

# Test 1c: Testing the mean $\mu$ of a population $X$ where the variance $\sigma^2$ is unknown and the sample size $n$ is large

The variance of the population, $\sigma^2$, is unknown, but, as you saw in on page 447, an unbiased estimate, $\hat{\sigma}^2$ can be used instead,

where $\hat{\sigma}^2 = \dfrac{n}{n-1} \times s^2$   ($s^2$ is the sample variance)

Alternative formats:

$$\hat{\sigma}^2 = \frac{1}{n-1}\Sigma(x-\bar{x})^2 \quad \text{or} \quad \hat{\sigma}^2 = \frac{1}{n-1}\left(\Sigma x^2 - \frac{(\Sigma x)^2}{n}\right)$$

Ideally the population distribution should be normal, but if it is not, then the central limit theorem can be applied, since the sample size is large.

When testing the mean of a population $X$ with unknown variance $\sigma^2$, provided the sample size $n$ is large,

the test statistic is $\bar{X}$ where $\bar{X} \sim N\left(\mu_0, \dfrac{\hat{\sigma}^2}{n}\right)$.

In standardised form,

the test statistic $Z = \dfrac{\bar{X}-\mu_0}{\hat{\sigma}/\sqrt{n}}$, where $Z \sim N(0, 1)$.

## Example 11.4

The packaging on an electric light bulb states that the average length of life of bulbs is 1000 hours. A consumer association thinks that this is an overestimate and tests a random sample of 64 bulbs, recording the life $x$ hours, of each bulb.

The results are summarised as follows:

$\Sigma x = 63\ 910.4$,     $\Sigma x^2 = 63\ 824\ 061$.

(a) Calculate the sample mean, $\bar{x}$.
(b) Calculate an unbiased estimate for the standard deviation of the length of life of all light bulbs of this type.
(c) Is there evidence, at the 10% significance level, that the statement on the packaging is overestimating the length of life of this type of light bulb?

## Solution 11.4

(a) $\bar{x} = \dfrac{\Sigma x}{n} = \dfrac{63\ 910.4}{64} = \mathbf{998.6\ hours}$

(b) $\hat{\sigma}^2 = \dfrac{1}{n-1}\left(\Sigma x^2 - \dfrac{(\Sigma x)^2}{n}\right)$

$\qquad = \dfrac{1}{63}\left(63\ 824\ 061 - \dfrac{63\ 910.4^2}{64}\right)$

$\qquad = 49.77\ \ldots$

$\hat{\sigma} = \sqrt{49.77\ \ldots}$

$\qquad = 7.055$ (3 d.p.)

(c) Perform the hypothesis test as follows:

**1. Define the variable.**

Let $X$ be the lifetime, in hours, of a light bulb.
Let the population mean be $\mu$ and the population standard deviation be $\sigma$.

**2. State $H_0$ and $H_1$.**

$H_0 : \mu = 1000$ (The statement on the packaging is correct)
$H_1 : \mu < 1000$ (The statement on the packaging is overestimating the length of life)

**3. State the distribution of the test statistic according to $H_0$.**

For samples of size $n$, where $n$ is large, by the central limit theorem and using $\hat{\sigma}$ for $\sigma$,

$$\overline{X} \sim N\left(\mu, \frac{\hat{\sigma}^2}{n}\right) \quad \text{with } n = 64 \text{ and } \hat{\sigma} = 7.055.$$

If $H_0$ is true, $\mu = 1000$, so $\overline{X} \sim N\left(1000, \frac{7.055^2}{64}\right)$.

**4. State the level of the test.**

Use a one-tailed test, at the 10% level.

**5. Decide on your rejection criterion.**

*The critical z-value for a one-tailed 10% test (lower tail) is $-1.282$. (see page 649).*

Reject $H_0$ if $z < -1.282$, where $z = \dfrac{\bar{x} - \mu_0}{\hat{\sigma}/\sqrt{n}} = \dfrac{\bar{x} - 1000}{7.055/\sqrt{64}}$

**6. Perform the required calculation.**

From the sample, $\bar{x} = 998.6$.

$$\therefore \quad z = \frac{998.6 - 1000}{7.055/\sqrt{64}}$$
$$= -1.587 \ldots$$

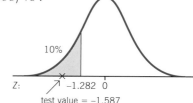

10%

Z:    $-1.282$   0

test value = $-1.587$

**7. Make your conclusion.**

Since $z < -1.282$, reject $H_0$ (the mean is 1000 hours) in favour of $H_1$ (the mean is less than 1000 hours).

**There is evidence, at the 10% level, that the statement on the packaging overestimates the length of life of this type of bulb.**

## TYPE I AND TYPE II ERRORS

When you make your decision about whether or not to reject $H_0$ there are two types of error that could be made. These were described in Chapter 10 (page 493) and are called Type I and Type II errors:

– a Type I error is made when you wrongly reject a true hypothesis,
– a Type II error is made when you wrongly accept a false hypothesis.

These can be summarised in a table:

| | | Test decision | |
| --- | --- | --- | --- |
| | | Accept $H_0$ | Reject $H_0$ |
| Actual situation | $H_0$ is true | ✓(correct) | Type I error |
| | $H_0$ is false | Type II error | ✓ (correct) |

A Type I error is made if $H_0$ is rejected when $H_0$ is true.
This is written

$P(\text{Type I error}) = P(H_0 \text{ is rejected} \mid H_0 \text{ is true}).$

The | notation was introduced in Chapter 3 to describe conditional probability and means 'given that'.

If the significance level is $a\%$ then the probability of rejecting $H_0$ is $a\%$, so the significance level of the test and the probability of making a Type I error are both the same.

$\therefore P(\text{Type I error}) = a\%$

A Type II error is made if $H_0$ is accepted when $H_0$ is false.
This is written

$P(\text{Type II error}) = P(H_0 \text{ is accepted} \mid H_0 \text{ is false}).$

To calculate the probability of a Type II error, a **particular value** must be specified in the alternative hypothesis $H_1$.

So $P(\text{Type II error}) = P(H_0 \text{ is accepted} \mid H_1 \text{ is true})$

## POWER OF A TEST

The power of a test $= P(\text{reject } H_0 \text{ when } H_1 \text{ is true})$
$= 1 - P(\text{Type II error})$

### Example 11.5

A random variable has a normal distribution with mean $\mu$ and standard deviation 3.

The null hypothesis $\mu = 20$ is to be tested against the alternative hypothesis $\mu > 20$ using a random sample of size 25. It is decided that the null hypothesis will be rejected if the sample mean is greater than 21.4.

(a) Calculate the probability of making a Type I error.
(b) Calculate the probability of making a Type II error, when in fact $\mu = 21$.

### Solution 11.5

(a) You are given that $X \sim N(\mu, 3^2)$

and $\quad H_0: \mu = 20$
$\quad\quad H_1: \mu > 20$

For samples of size 25, $\bar{X} \sim N\left(\mu, \dfrac{3^2}{n}\right)$ with $n = 25$.

If the null hypothesis is true, $\mu = 20$ and $\bar{X} \sim N\left(20, \dfrac{3^2}{25}\right)$

$P(\text{Type I error}) = P(H_0 \text{ is rejected when } H_0 \text{ is true})$
$= P(X > 21.4 \text{ when } \mu = 20)$
$= P\left(Z > \dfrac{21.4 - 20}{3/\sqrt{25}}\right)$
$= P(Z > 2.333)$
$= 1 - \Phi(2.333)$
$= 1 - 0.9902$
$= 0.0098$
$\approx 1\%$

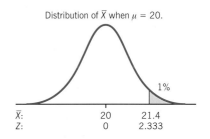

Distribution of $\bar{X}$ when $\mu = 20$.

1%

| $\bar{X}$: | 20 | 21.4 |
| $Z$: | 0 | 2.333 |

**The probability of making a Type I error = 1%.**

NOTE: This gives the significance level of the test, so if values of the sample mean greater than 21.4 are rejected, the significance level of the test is 1%.

(b) If, in fact, $\mu = 21$, the hypotheses become

$H_0: \mu = 20$

$H_1: \mu = 21$

$P(\text{Type II error}) = P(\text{accept } H_0 \text{ when } H_0 \text{ is false})$
$= P(\text{accept } H_0 \text{ when } H_1 \text{ is true})$

You are given that $H_0$ is rejected if $\bar{x} > 21.4$, so $H_0$ **will be accepted if** $\bar{x} < 21.4$.

If $H_1$ is true, then $\mu = 21$ and $\bar{X} \sim N\left(21, \dfrac{3^2}{25}\right)$,

so $P(\text{Type II error}) = P(\bar{X} < 21.4 \text{ when } \mu = 21)$

$= P\left(Z < \dfrac{21.4 - 21}{3/\sqrt{25}}\right)$

$= P(Z < 0.667)$

$= \Phi(0.667)$

$= 0.7477$

$\approx 75\%$

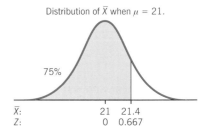

Distribution of $\bar{X}$ when $\mu = 21$.

75%

$\bar{X}$:  21  21.4
$Z$:  0  0.667

**The probability of making a Type II error is 75%.**

---

## Exercise 11a   z-tests for a normal population or large sample size (Tests 1a, 1b and 1c)

1. For each of the following, $X$ follows a normal distribution with unknown mean $\mu$ and known standard deviation $\sigma$. A random sample of size $n$ is taken from the population of $X$ and the sample mean, $\bar{x}$, is calculated.
   Test the hypotheses stated, at the significance level indicated.

|   | $n$ | $\bar{x}$ | $\sigma$ | Hypotheses | | Level of significance |
|---|---|---|---|---|---|---|
| (a) | 30 | 15.2 | 3 | $H_0: \mu = 15.8,$ | $H_1: \mu \neq 15.8$ | 5% |
| (b) | 10 | 27 | 1.2 | $H_0: \mu = 26.3,$ | $H_1: \mu > 26.3$ | 5% |
| (c) | 49 | 125 | 4.2 | $H_0: \mu = 123.5,$ | $H_1: \mu > 123.5$ | 1% |
| (d) | 100 | 4.35 | 0.18 | $H_0: \mu = 4.40,$ | $H_1: \mu < 4.40$ | 2% |

2. A machine fills cans with soft drinks so that their contents have a nominal volume of 330 ml. Over a period of time it has been established that the volume of liquid in the cans follows a normal distribution with mean 335 ml and standard deviation 3 ml. A setting on the machine is altered, following which the operator suspects that the mean volume of liquid discharged by the machine into the cans has decreased. He takes a random sample of 50 cans and finds that the mean volume of liquid in these cans is 334.6 ml. Does this confirm his suspicion? Perform a significance test at the 5% level and assume that the standard deviation remains unchanged.

3. In a significance (hypothesis) test of the mean of a population, a null hypothesis $\mu = 103.5$ is tested against an alternative hypothesis $\mu < 103.5$, where $\mu$ is the mean of a normally distributed variable with known variance. A sample is taken from the population and a standardised normal test statistic $z = -1.35$ is calculated.
   What conclusion, at the 5% level of significance, can be reached about the mean of the population?

4. A machine packs flour into bags. A random sample of 11 filled bags was taken and the masses of the bags, to the nearest 0.1 g were

   1506.8, 1506.6, 1506.7, 1507.2, 1506.9,

   1506.8, 1506.6, 1507.0, 1507.5, 1506.3, 1506.4

Filled bags are supposed to have a mass of 1506.5 g. Assuming that the mass of a bag has normal distribution with variance $0.16 \text{ g}^2$, test whether the sample provides significant evidence, at the 5% level, that the machine produces overweight bags. (C)

5. A variable with known variance 32 is thought to have a mean of 55. A random sample of 81 observations of the variable has a mean of 56.2. Does this provide evidence at the 10% level of significance that the mean is not 55?

   Explain what part the central limit theorem has played in your calculations.

6. A random sample of 75 eleven-year-olds performed a simple task and the time taken, $t$ minutes, noted for each. The results were summarised as follows:

   $\Sigma t = 1215, \quad \Sigma t^2 = 21\ 708.$

   Test, at the 1% level, whether there is evidence that the mean time taken to perform the task is greater than 15 minutes.

7. Cassette tapes manufactured by a particular firm are such that the playing time of the tapes can be modelled by a normal distribution with standard deviation 1.8 minutes. The tapes are advertised as having a playing time of 90 minutes, but the manufacturer claims that they actually have a mean playing time of 92 minutes. An investigator selected 36 tapes at random and checked the playing times. He calculated the mean playing time of the tapes in the sample and, on the basis of the value obtained, he rejected the manufacturer's claim at the 5% level, saying that the mean time was less than 92 minutes.

   (a) What can be said about the value of the sample mean for this decision to be taken?

   The mean playing time of the firm's cassettes is in fact 90.8 minutes.

   (b) Find the probability of making a Type II error.

   (c) Find the probability that a cassette tape, picked at random, has a playing time less than the stated value on the pack of 90 minutes.

8. A sample of 40 observations from a normal distribution $X$ gave $\Sigma x = 24, \quad \Sigma x^2 = 596$. Performing a two-tailed test at the 5% level, test whether the mean of the distribution is zero.

9. The masses of components produced at a particular workshop are normally distributed with standard deviation 0.8 g. It is claimed that the mean mass is 6 g.

To test this claim the mean mass of a random sample of 50 components is calculated and a significance test at the 5% level carried out. On the basis of the test, the claim is accepted. Between what values did the mean mass of the 50 components in the sample lie?

10. For each of the following distributions, $X$, a random sample of size $n$ is taken and the values of $\Sigma x$, $\Sigma x^2$ or $\Sigma(x - \bar{x})^2$ summarised as shown. Test the hypotheses stated at the significance level indicated.

| | $n$ | $\Sigma x$ | $\Sigma x^2$ | $\Sigma(x - \bar{x})^2$ | Hypotheses | Level of significance |
|---|---|---|---|---|---|---|
| (a) | 65 | 6500 | 650 842.4 | | $H_0$: $\mu = 99.2$, $H_1$: $\mu \neq 99.2$ | 5% |
| (b) | 65 | 6500 | 650 842.4 | | $H_0$: $\mu = 99.2$, $H_1$: $\mu > 99.2$ | 5% |
| (c) | 80 | 6824 | | 2508.8 | $H_0$: $\mu = 86.2$, $H_1$: $\mu < 86.2$ | 10% |
| (d) | 100 | 685 | 4728.25 | | $H_0$: $\mu = 7$, $H_1$: $\mu \neq 7$ | 1% |

11. A large random sample was taken from a population with mean $\mu$ and known variance. The null hypothesis $\mu = 52$ was tested against the alternative hypothesis $\mu \neq 52$ at the 4% significance level. The calculated value of the standardised test statistic was 2.19.

    (a) Carry out a significance test for $\mu$ based on this result, stating your conclusion clearly.

    (b) State the probability of making a Type I error.

12. A sample of size 15 is taken from the distribution of $X$ where $X \sim N(\mu, 4)$. If the sample mean is greater than 10.72, the null hypothesis $\mu = 10$ is rejected in favour of the alternative hypothesis $\mu > 10$.

    (a) Find the probability of making a Type I error.

    (b) Find the probability of making a Type II error if $\mu = 10.5$.

13. An IQ test is developed such that the mean quotient is 100 and standard deviation is 12. It is given to a random sample of 50 children in one area. The average mark was 105. Does this provide evidence, at the 5% level, that children from this area are generally more intelligent?

14. Boxes of a certain breakfast cereal have contents whose masses are normally distributed with mean $\mu$ g and standard deviation 15 g. A test of the null hypothesis $\mu = 375$ against the alternative hypothesis $\mu > 375$ is carried out at the $2\frac{1}{2}\%$ significance level using a random sample of 16 boxes.

(a) Show that the alternative hypothesis is accepted when $\bar{x} > 382.35$, where $\bar{x}$ g is the sample mean mass.

(b) Given that the actual value of $\mu$ is 385, find the probability of making a Type II error.
(c) Find the range of values of $\mu$ for which the probability of making a Type II error is less than 0.025.
(d) The test is carried out, independently, on two different occasions. Find the probability that at least one Type I error is made.　(C)

## Test 1d: Testing the mean $\mu$ when the population $X$ is normal but the variance $\sigma^2$ is unknown and the sample size $n$ is small

In this case, the population is normal, so $X \sim N(\mu, \sigma^2)$. Since $\sigma^2$ is unknown, $\hat{\sigma}^2$ is used instead (as in Test 1c on page 519).

Consider the distribution of the sample mean $\bar{X}$. When the sample size is small, $\bar{X}$ *does not* follow a normal distribution. As you saw in Chapter 9 (page 462), the standardised statistic is called $T$ and it follows a $t$-distribution with $(n-1)$ degrees of freedom.

When testing the mean of a normal population $X$ with unknown variance $\sigma^2$, when the sample size $n$ is small,

the test statistic is $T$ where $T = \dfrac{\bar{X} - \mu_0}{\hat{\sigma}/\sqrt{n}}$ and $T \sim t(n-1)$.

When finding the critical $t$-values, **$t$-distribution tables** are needed and these are printed on page 650. You may need to remind yourself how to use them by reading again the notes on page 464.

### Example 11.6

Five readings of the resistance $X$, in ohms, of a piece of wire gave the following results:

1.51, 1.49, 1.54, 1.52, 1.54

These are summarised by $\Sigma x = 7.6$, $\Sigma x^2 = 11.5538$.

If the wire is pure, the resistance is 1.50 ohms. If the wire is impure, its resistance is higher than 1.50 ohms. Assuming that the resistance can be modelled by a normal variable with mean $\mu$, and standard deviation $\sigma$, calculate

(a) the sample mean, $\bar{x}$,
(b) an unbiased estimate of $\sigma$,

Is there evidence, at the 5% level of significance, that the wire is impure?

### Solution 11.6

1. Define the variable.

Let $X$ be the resistance, in ohms, of the wire.
Let the population mean be $\mu$, and the population standard deviation be $\sigma$.
Then $X \sim N(\mu, \sigma^2)$.

(a) $\bar{x} = \dfrac{\Sigma x}{n} = \dfrac{7.6}{5} = 1.52$ ohms

(b) $\hat{\sigma}^2 = \dfrac{1}{n-1}\left(\Sigma x^2 - \dfrac{(\Sigma x)^2}{n}\right)$

$\qquad = \dfrac{1}{4}\left(11.5538 - \dfrac{7.6^2}{5}\right)$

$\qquad = 0.000\ 45$

so $\quad \hat{\sigma} = 0.0212$ (3 s.f.)

**2.** State $H_0$ and $H_1$.

(c) $H_0$: $\mu = 1.50$ (the wire is pure silver)
$\qquad H_1$: $\mu > 1.50$ (the wire is impure)

**3.** State the distribution of the test statistic according to $H_0$.

If $H_0$ is true, $\mu = 1.50$. Since $n$ is small and $\sigma^2$ is unknown, the test statistic is $T$,

where $\quad T = \dfrac{\overline{X} - 1.50}{\hat{\sigma}/\sqrt{n}} \quad$ and $T \sim t(n-1)$,

i.e. $\quad T = \dfrac{\overline{X} - 1.50}{0.0212/\sqrt{5}} \quad$ and $T \sim t(4)$.

**4.** State the level of the test.

Use a one-tailed test (upper tail) at the 5% level.

**5.** Decide on your rejection criterion.

From the tables on page 650, *the critical value for t is found from row v = 4, p = 0.95 giving 2.132.*

Reject $H_0$ if the test value of $t$ is greater than 2.132.

| $p$ | 0.75 | 0.90 | 0.95 | 0.975 |
|---|---|---|---|---|
| $v = 1$ | 1.000 | 3.078 | 6.314 | 12.71 |
| 2 | 0.816 | 1.886 | 2.920 | 4.303 |
| 3 | 0.765 | 1.638 | 2.353 | 3.182 |
| 4 | 0.741 | 1.533 | 2.132 | 2.776 |

**6.** Perform the required calculation.

From the sample, $\bar{x} = 1.52$.

$\therefore \quad t = \dfrac{1.52 - 1.50}{0.0212/\sqrt{5}} = 2.109$

**7.** Make your conclusion.

Since $t < 2.132$, $H_0$ is not rejected.

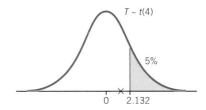

There is not enough evidence, at the 5% level, to indicate that the wire is impure.

## Example 11.7

A machine is supposed to produce paper with a mean thickness of 0.05 mm. Eight random measurements of the paper gave a mean of 0.047 mm with a standard deviation of 0.002 mm. Assuming that the thickness of the paper produced by the machine is normally distributed, test at the 1% level whether the output from the machine is different from expected.

## Solution 11.7

**1.** Define the variable.

Let $X$ be the thickness, in millimetres, of paper produced by the machine.
Let the population mean be $\mu$ and the population standard deviation be $\sigma$.
Then $X \sim N(\mu, \sigma^2)$.

Since $\sigma^2$ is unknown, the unbiased estimate $\hat{\sigma}^2$ is used, where

$$\hat{\sigma}^2 = \frac{n}{n-1} \times s^2 \qquad \text{(where } s^2 \text{ is the sample variance).}$$

$$\hat{\sigma}^2 = \frac{8}{7} \times 0.002^2$$

$$= 4.57 \ldots \times 10^{-6}$$

$$\hat{\sigma} = 0.002\ 14 \text{ (3 s.f.)}$$

**2. State $H_0$ and $H_1$.**

$H_0$: $\mu = 0.05$ (the thickness is as expected)
$H_1$: $\mu \neq 0.05$ (the thickness is different from that expected)

**3. State the distribution of the test statistic according to $H_0$.**

If $H_0$ is true, $\mu = 0.05$.
Since $n$ is small and $\sigma^2$ is unknown, the test statistic is $T$ where

$$T = \frac{\overline{X} - 0.05}{\hat{\sigma}/\sqrt{n}} \qquad \text{and} \qquad T \sim t(n-1)$$

i.e. $\quad T = \dfrac{\overline{X} - 0.05}{0.00214/\sqrt{8}} \qquad$ and $\quad T \sim t(7)$.

**4. State the level of the test.**

Use a two-tailed test at the 1% level.

**5. Decide on your rejection criterion.**

*The critical value for t is found from row $v = 7$, p = 0.995 (because you want 0.5% in the each tail) giving ±3.499.*

| $p$ | 0.75 | 0.90 | 0.95 | 0.975 | 0.99 | 0.995 |
|---|---|---|---|---|---|---|
| $v-1$ | 1.000 | 3.078 | 6.314 | 12.71 | 31.82 | 63.66 |
| 2 | 0.816 | 1.886 | 2.920 | 4.303 | 6.965 | 9.925 |
| ⋮ | ⋮ | ⋮ | ⋮ | ⋮ | ⋮ | ⋮ |
| 7 | 0.711 | 1.415 | 1.895 | 2.365 | 2.998 | 3.499 |

Reject $H_0$ if $t < -3.499$ or $t > 3.499$ i.e. if $|t| > 3.499$.

**6. Perform the required calculation.**

From the sample, $\bar{x} = 0.047$.

$$\therefore \quad t = \frac{0.047 - 0.05}{0.002\ 14/\sqrt{8}} = -3.96 \ldots$$

**7. Make your conclusion.**

Since $|t| > 3.499$, $H_0$ is rejected.

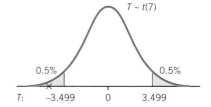

**There is evidence, at the 1% level, that the output from the machine is different from that expected.**

# Exercise 11b   $t$-tests for a normal population, small sample size (Test 1d)

1. For each of the following, $X$ follows a normal distribution with unknown mean $\mu$ and known standard deviation $\sigma$. A random sample of size $n$ is taken from the population of $X$ and the sample mean, $\bar{x}$, is calculated.
   Test the hypotheses stated, at the significance level indicated.

   | | $n$ | $\Sigma x$ | $\Sigma(x - \bar{x})^2$ | $\Sigma x^2$ | Hypotheses | Level |
   |---|---|---|---|---|---|---|
   | (a) | 12 | 298.8 | | 7542.42 | $H_0: \mu = 24.1$, $H_1\ \mu > 24.1$ | 5% |
   | (b) | 17 | 605.2 | | 23016.92 | $H_0: \mu = 40$, $H_1: \mu \neq 40$ | 5% |
   | (c) | 6 | 9034.8 | 50.8 | | $H_0: \mu = 1503$, $H_1: \mu \neq 1503$ | 10% |
   | (d) | 10 | 1298 | 97.6 | | $H_0: \mu = 133.0$, $H_1: \mu < 133.0$ | 1% |

2. An athlete finds that her times for running a race are normally distributed with mean 10.6 seconds. She trains intensively for a week and then records her time in the next 6 races. Her times, in seconds, are

   10.70, 10.65, 10.75, 10.80, 10.60

   Is there evidence, at the 5% level, that training intensively has improved her times?

3. Family packs of bacon slices are sold in 1.5 kg packs. A sample of 12 packs was selected at random and their masses, measured in kilograms, noted. The following results were obtained:

   $\Sigma x = 17.81$,   $\Sigma x^2 = 26.4357$

   Assuming that the masses of packs follow a normal distribution with variance $\sigma^2$, test at the 1% level whether the packs are underweight,
   (a) if $\sigma^2$ is unknown,   (b) if $\sigma^2 = 0.0003$.

4. It is thought that a normal population has mean 1.6. A random sample of 10 observations gives a mean of 1.49 and standard deviation of 0.3. Does this provide evidence, at the 5% level, that the population mean is less than 1.6?

5. A random sample of 8 observations of a normal variable gave

   $\Sigma x = 36.5$,   $\Sigma(x - \bar{x})^2 = 0.74$

   Test, at the 5% level, the hypothesis that the mean of the distribution is 4.3 against the alternative hypothesis that the mean is greater than 4.3.

6. The cholesterol levels of 8 women were measured, with the following results.

   3.1, 2.8, 1.5, 1.7, 2.4, 1.9, 3.3, 1.6

   Making any necessary assumptions,
   (a) test, at the 5% level, whether the sample has been drawn from a distribution with mean cholesterol level 3.1,
   (b) calculate a symmetric 95% confidence interval for the mean cholesterol level.

7. A marmalade manufacturer produces thousands of jars of marmalade each week. The mass of marmalade in a jar is an observation from a normal distribution having mean of 455 g and standard deviation 0.8 g.

   Following a slight adjustment to the filling machine, a random sample of 10 jars is found to contain the following masses, in grams, of marmalade:

   454.8, 453.8, 455.0, 454.4, 455.4, 454.4, 454.4, 455.0, 455.0, 453.6

   (a) Assuming that the variance of the distribution is unaltered by the adjustment, test at the 5% significance level the hypothesis that there has been no change in the mean of the distribution.
   (b) Assuming that the variance of the distribution may have been altered, obtain an unbiased estimate of the new variance and, using this estimate, test at the 5% level of significance the hypothesis that there has been no change in the mean of this distribution.   (C)

8. Six observations of a continuous random variable $X$ gave the following values:

   120.3, 122.4, 119.8, 121.0, 122.5, 119.6

   State any conditions that are necessary for the valid use of a $t$-test to test a hypothesis about the mean of $X$.
   Assuming that the use of the $t$-test is valid, test the null hypothesis that the mean of $X$ is 120 against the alternative hypothesis that the mean is not 120, using a 5% significance level.

9. A random sample of 12 independent observations of a normally distributed random variable $X$ is taken from a population and a test statistic, $t = 2.9$, calculated. It is thought that the population mean $\mu$ is 27. Write down suitable null and alternative hypotheses to carry out a two-tailed significance test for $\mu$ and use a $t$-test to test your hypotheses at the 1% level.

10. A firm of solicitors claims that, on average, interviews with clients last 50 minutes. A random sample of 15 interviews is chosen, and the time taken for each interview, $x$ minutes, is noted. The results are summarised by $\Sigma x = 746$ and $\Sigma x^2 = 37\ 180$. Assuming that the time for an interview has a normal distribution, use a $t$-test to determine, at the 5% significance level, whether the firm is overstating the average interview time. Give null and alternative hypotheses, full details of your procedure and a conclusion. (C)

# HYPOTHESIS TEST 2: TESTING A BINOMIAL PROPORTION $p$ WHEN $n$ IS LARGE

Consider the situation when independent trials are carried out, each with a probability $p$ of success, where $p$ is constant. If $X$ is the number of successes in $n$ trials, then $X$ follows a binomial distribution i.e. $X \sim B(n, p)$ (page 279).

In Chapter 10 (hypothesis tests for discrete variables, page 483) you learnt how to carry out a hypothesis test for an unknown binomial proportion $p$. This involved calculating binomial probabilities which are relatively easy to find when $n$ is small.

When $n$ is large, however, the calculations can become very cumbersome and in such cases it is useful to use the **normal approximation to the binomial distribution**:

If $n$ is sufficiently large such that $np > 5$ and $nq > 5$, then the binomial distribution
$X \sim B(n, p)$
can be approximated by a normal distribution
$X \sim N(np, npq)$, where $q = 1 - p$.

When performing the hypothesis test, you are able to work in standardised $z$-values. Since the normal distribution is continuous and the binomial is discrete, you will need to use a **continuity correction** (see page 383) and this involves amending your test value by adding or subtracting 0.5. Further details are given in the following examples. The stages of the test are the same as in the general procedure outlined on pages 513.

When testing the proportion $p$, of a binomial population, the test statistic is $X$, the number of successes in $n$ trials, where $X \sim B(n, p)$.

When $n$ is large such that $np > 5$, and $nq > 5$, $X$ is approximately normal and $X \sim N(np, npq)$.

In standardised form,
the test statistic is $Z = \dfrac{X - np}{\sqrt{npq}}$ where $Z \sim N(0, 1)$.

s.d. $= \sqrt{npq}$

$X$:  $np$

## Example 11.8

Caroline was asked to test whether a coin is biased in favour of heads, using a 5% level of significance. She tossed the coin 100 times and obtained 57 heads. What should she have concluded?

## Solution 11.8

1. Define the variable.

Let $X$ be the number of heads in 100 tosses and let the probability of obtaining a head be $p$. Then $X \sim B(100, p)$.

2. State $H_0$ and $H_1$.

$H_0$: $p = 0.5$ (the coin is not biased and heads or tails are equally likely to occur)
$H_1$: $p > 0.5$ (the coin is biased in favour of heads)

**3.** State the distribution of the test statistic according to $H_0$.

If $H_0$ is true, then $p = 0.5$, so $X \sim B(100, 0.5)$.

Now $n$ (the number of tosses) is large and $np = 100 \times 0.5 = 50 > 5$, $nq = 50 > 5$.

Since $np > 5$ and $nq > 5$, use the normal approximation,

$X \sim N(np, npq)$ with $np = 50$ and $npq = 100 \times 0.5 \times 0.5 = 25$,

i.e. $X \sim N(50, 25)$

s.d. $= \sqrt{npq} = \sqrt{25} = 5$

**4.** State the level of the test.

Use a one-tailed test at the 5% level.

**5.** Decide on your rejection criterion.

The critical $z$-value for the 5% critical region (upper tail) is 1.645.

Reject $H_0$ if $z > 1.645$, where $z$ is the sample value when standardised,

**6.** Perform the required calculation.

*When standardising the sample value of 57 heads you have to use a continuity correction. Think of the discrete value of 57 being represented by a rectangle over the continuous values from 56.5 to 57.5.*

*In order to reject* $H_0$, *the complete rectangle must lie in the critical region. Therefore take as the test value the lower boundary, 56.5.*

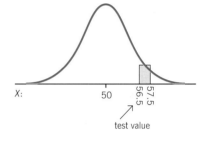

So $z = \dfrac{56.5 - np}{\sqrt{npq}}$

$= \dfrac{56.5 - 50}{5}$

$= 1.3$

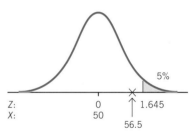

**7.** Make your conclusion.

Since $z < 1.645$, the sample value of 57 heads is not in the critical region and $H_0$ is not rejected.

**On the statistical evidence, Caroline should have concluded that the coin is not biased in favour of heads.**

It is interesting to work out how many heads would need to be obtained to conclude that the coin is biased in favour of heads. This can be done as follows:

The standardised test value lies in the critical region if $z > 1.645$.

If the number of heads is $x$, then, applying the continuity correction, you need to consider $x - 0.5$ when standardising the test value.

Therefore $\dfrac{(x - 0.5) - np}{\sqrt{npq}} > 1.645$

i.e. $\dfrac{(x - 0.5) - 50}{5} > 1.645$

$x > 50 + 0.5 + 1.645 \times 5$

$x > 58.725$

Since $x$ is an integer, the least value of $x$ is 59. So if Caroline had obtained 59 or more heads when she tossed the coin 100 times, she would have concluded that the coin was biased in favour of heads.

*NOTE*: This result is perhaps surprising. Would you have thought that more heads would be needed?

## Example 11.9

A manufacturer claims that a particular brand of seeds has a 90% germination rate. To test this claim, 150 randomly selected seeds are planted and it is noted that 124 germinate. Does this provide evidence, at the 1% significance level, that the manufacturer is overstating the germination rate of the seeds?

## Solution 11.9

**1.** Define the variable.

Let $X$ be the number of seeds that germinate in 150 and let the probability that a seed germinates be $p$. Then $X \sim B(150, p)$.

**2.** State $H_0$ and $H_1$.

$H_0$: $p = 0.9$ (the germination rate is 90%)
$H_1$: $p < 0.9$ (the germination rate is less than 90% and the manufacturer is overstating the rate)

**3.** State the distribution of the test statistic according to $H_0$.

If $H_0$ is true, then $p = 0.9$, so $X \sim B(150, 0.9)$.

*Since n is large, check whether the normal approximation can be used.*

Now $np = 150 \times 0.9 = 135 > 5$ and $nq = 150 \times 0.1 = 15 > 5$.

Since $np > 5$ and $nq > 5$, use the normal approximation,

$$X \sim N(np, npq) \text{ with } npq = 150 \times 0.9 \times 0.1 = 13.5$$

i.e.    $X \sim N(135, 13.5)$

**4.** State the level of the test.

Use a one-tailed test at the 1% level.

**5.** Decide on your rejection criterion.

The critical $z$-value for the critical region (lower tail) is $-2.326$.

*So $H_0$ is rejected if your test value, when standardised, is less than $-2.326$. The test value is 124, but when you consider the continuity correction in this lower tail test you want to see whether the complete rectangle for 124 lies in the critical region, so you need to standardise 124.5.*

Reject $H_0$ if the standardised value of 124.5 is less than $-2.326$.

**6.** Perform the required calculation.

$$z = \frac{124.5 - np}{\sqrt{npq}}$$
$$= \frac{124.5 - 135}{\sqrt{13.5}}$$
$$= -2.857 \ldots$$

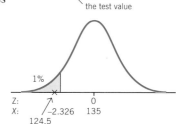

**7.** Make your conclusion.

Since $z < -2.326$, the sample value is in the critical region and so $H_0$ (the germination rate is 90%) is rejected in favour of $H_1$ (the germination rate is less than 90%).

**There is evidence that the manufacturer is overstating the germination rate.**

## Example 11.10

The random variable $X$ can be modelled by a binomial distribution with parameters $n = 200$ and $p$, whose value is unknown. A significance test is performed, based on a sample value $x$, to test the null hypothesis $p = 0.4$ against the alternative hypothesis $p < 0.4$. The probability of making a Type I error when performing this test is 0.05.

(a) Find the critical region for $x$.

(b) Find the probability of making a Type II error in the case when $p = 0.3$.

## Solution 11.10

(a) You are given that $X \sim B(200, p)$.

The hypotheses are $H_0$: $p = 0.4$
$H_1$: $p < 0.4$

If $H_0$ is true, then $p = 0.4$, so $X \sim B(200, 0.4)$.

Now $np = 200 \times 0.4 = 80$ and $nq = 200 \times 0.8 = 160$.
Since $np > 5$ and $nq > 5$, use the normal approximation,

$\qquad X \sim N(np, npq)$ with $np = 80$, $npq = 200 \times 0.4 \times 0.6 = 48$,

so $\quad X \sim N(80, 48)$.

You are given that $P(\text{Type I error}) = 0.05$.

Since $P(\text{Type I error}) = P(H_0 \text{ is rejected when } H_0 \text{ is true})$, this is the same as the significance level of the test.

So the significance level of the test is 5%.

Using a one-tailed test, at the 5% level, the critical $z$-value for the critical region (lower tail) is $-1.645$. So reject $H_0$ if the sample value, when standardised, is less than $-1.645$.

*To find the critical region for $x$, remember that a continuity correction is needed. Since you are considering values in the lower tail and you want to include the complete rectangle representing $x$, use $x + 0.5$.*

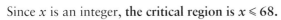

$\dfrac{x + 0.5 - 80}{\sqrt{48}} < -1.645$

$\therefore \quad x < 80 - 0.5 - 1.645 \times \sqrt{48}$
$\qquad x < 68.10 \ldots$

Since $x$ is an integer, **the critical region is $x \leqslant 68$.**

Check:

When $x = 68$, $z = \dfrac{68.5 - 80}{\sqrt{48}} = -1.659 < -1.645$, so 68 is in the critical region.

When $x = 69$, $z = \dfrac{69.5 - 80}{\sqrt{48}} = -1.515 > -1.645$, so 69 is not in the critical region.

(b) If $p = 0.3$, the hypotheses become
$H_0: p = 0.4$
$H_1: p = 0.3$

From part(a), the critical region is $X \le 68$,
so $H_0$ is accepted when $X > 68$.

$P(\text{Type II error}) = P(H_0 \text{ is accepted when } H_1 \text{ is true})$
$\qquad\qquad\qquad\quad = P(X > 68 \text{ when } p = 0.3)$

When $p = 0.3$, $np = 200 \times 0.3 = 60$ and $npq = 200 \times 0.3 \times 0.7 = 42$

Therefore $X \sim N(60, 42)$.

Note that the conditions $np > 5$, $nq > 5$ are satisfied so the normal approximation can be applied.

Now $P(X > 68) \rightarrow P(X > 68.5)$ (continuity correction),

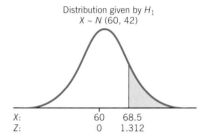

Distribution given by $H_1$
$X \sim N(60, 42)$

$P(\textbf{Type II error}) = P(X > 68.5 \text{ when } X \sim N(60, 42))$
$$= P\left(Z > \frac{68.5 - 60}{\sqrt{42}}\right)$$
$$= P(Z > 1.312)$$
$$= 1 - 0.6224$$
$$\approx \textbf{38\%}$$

$X$:     60    68.5
$Z$:     0    1.312

---

# Exercise 11c   Testing a binomial proportion, large $n$

1. In the following, $X \sim B(n, p)$ with $n$ as shown. $p$ is unknown and $x$ is the number of successes in the sample.
   Test the hypotheses stated at the level of significance indicated.

   |  | $n$ | $x$ | Hypotheses | Level |
   |---|---|---|---|---|
   | (a) | 50 | 45 | $H_0: p = 0.8,$ $H_1: p > 0.8$ | 5% |
   | (b) | 60 | 42 | $H_0: p = 0.55,$ $H_1: p \neq 0.55$ | 2% |
   | (c) | 120 | 21 | $H_0: p = 0.25,$ $H_1: p \neq 0.25$ | 5% |
   | (d) | 300 | 213 | $H_0: p = 0.65,$ $H_1: p \neq 0.65$ | 1% |
   | (e) | 90 | 56 | $H_0: p = 0.76,$ $H_1: p < 0.76$ | 1% |

2. A manufacturer claims that 8 out 10 dogs prefer its brand of dog food to any other. In a random sample of 120 dogs, it was found that 85 appeared to prefer that brand. Test, at the 5% level, whether you would accept the manufacturer's claim.

3. In a survey it was found that 3 out of every 10 people supported a particular political party. A month later a party representative claimed that the popularity of the party had increased. Would you accept that the number who supported the party was still 3 out of 10 if a further survey revealed that in a random sample of 100 people, 38 supported the party? Test at the 3% level.

4. A large college claims that it admits equal numbers of men and women. In a random sample of 500 students at the college there were 267 males. Is this evidence, at the 5% level, that the college population is not evenly divided between males and females?

5. A theory predicts that the probability of an event is 0.4. The theory is tested experimentally and in 400 independent trials, the event occurred 140 times. Is the number of occurrences significantly less than that predicted by the theory? Test at the 1% level.

6. It is thought that the proportion of defective items produced by a particular machine is 0.1. A random sample of 100 items is inspected and found to contain 15 defective items. Does this provide evidence, at the 5% level, that the machine is producing more defective items than expected?

7. In an investigation into the ownership of mobile phones amongst school children, 200 randomly chosen school children were interviewed and 142 owned a mobile phone. Test, at the 5% level of significance, the hypothesis that 65% of school children own a mobile phone against the alternative hypothesis that more than 65% own a mobile phone.

8. (a) A gardener sows 150 Special cabbage seeds and knows that the germination rate is 75%. By using a suitable approximation find the probability that:
   (i)  more than 122 seeds germinate
   (ii) fewer than 106 seeds germinate

   (b) The gardener also sows 120 Everyday cabbage seeds and finds that 81 germinate. Test whether the Everyday seeds have a germination rate less than 75%. Perform a significance test at the 4% level.

9. A government report states that a third of teenagers in Great Britain belong to a youth organisation. A survey, conducted among a random sample of 1000 teenagers from a certain city revealed that 370 belonged to a youth organisation. Does this provide evidence, at the 2% level, that the proportion of teenagers in this city who belong to a youth organisation is greater than the national average?

10. A questionnaire was sent to a large number of people, asking for their opinions about a proposal to alter an examination syllabus. Of the 180 replies received, 134 were in favour of the proposal. Stating a necessary assumption,

    (a) test, at the 5% level, the hypothesis that the population proportion in favour of the proposal is 0.7 against the alternative that it is more than 0.7,
    (b) find a symmetric 95% confidence interval for the population proportion in favour of the proposal. (C)

11. Over a long period of time it has been found that in Enrico's restaurant the ratio of non-vegetarian to vegetarian meals ordered is 3 to 1.
    During one particular day at Enrico's restaurant, a random sample of 20 people contained two who ordered a vegetarian meal.

    (a) Carry out a significance test to determine whether or not the proportion of vegetarian meals ordered that day is lower than usual. State clearly your hypotheses and use a 10% significance level. *Use an exact binomial test.*

In Manuel's restaurant, of a random sample of 100 people ordering meals, 31 ordered vegetarian meals.

    (b) Set up null and alternative hypotheses and, using a suitable approximation, test whether or not the proportion of people eating vegetarian meals at Manuel's restaurant is different from that at Enrico's restaurant. Use a 5% level of significance. (L)

12. When a drawing pin is dropped on to the floor, the probability that it lands point up is $p$.

    (a) A teacher drops a drawing pin 900 times and observes that it lands point up 315 times. Test, at the 1% level, the hypothesis that $p = 0.4$ against the alternative hypothesis $p < 0.4$.
    (b) A student drops a drawing pin 600 times and observes that it lands point up 251 times. Using the student's results, find a symmetric 95% confidence interval for $p$.

    As part of a statistics investigation, 1500 students carry out similar experiments and they each calculate (correctly) their own symmetric 95% confidence interval for $p$. Find the expected number of these intervals that do not contain the true value of $p$. (C)

13. After carrying out a survey, a market research company asserted that 75% of TV viewers watched a certain programme. Another company interviewed 75 viewers and found that 51 had watched the programme and 24 had not. Does this provide evidence, at the 5% significance level, that the first company's figure of 75% was incorrect?

14. The Paper Engineering Company has traditionally supplied 85% of the retail outlets for origami products. With the onset of increased competition they feared that this proportion might have fallen. They examined a random sample of 500 retail outlets and found that 405 of them sold Paper Engineering Company products. Use a normal approximation to the binomial distribution to carry out a hypothesis test at the 1% significance level to test whether or not their proportion of the retail outlets has fallen. Give suitable null and alternative hypotheses and state your conclusion clearly. (C)

# HYPOTHESIS TEST 3: TESTING $\mu_1 - \mu_2$, THE DIFFERENCE BETWEEN MEANS OF TWO NORMAL POPULATIONS

This test is used when you have two **normal** populations $X_1$ and $X_2$ with unknown means, $\mu_1$ and $\mu_2$, and you want to test the difference between the means of these populations. Consider $X_1 \sim N(\mu_1, \sigma_1^2)$ and $X_2 \sim N(\mu_2, \sigma_2^2)$.

The hypotheses might be:

$H_0: \mu_1 - \mu_2 = \cdots$
$H_1: \mu_1 - \mu_2 > \cdots$ (or $\mu_1 - \mu_2 < \cdots$ or $\mu_1 - \mu_2 \neq \cdots$)

Often the test involves the null hypothesis that the means are the same, i.e. $\mu_1 = \mu_2$ or $\mu_1 - \mu_2 = 0$, so the null hypothesis would be $H_0: \mu_1 - \mu_2 = 0$.

To test the difference between the means, take a random sample of size $n_1$ from $X_1$ and work out its sample mean, $\bar{x}_1$. Also take a random sample of size $n_2$ from $X_2$ and work out its sample mean $\bar{x}_2$.

The test statistic is $\bar{X}_1 - \bar{X}_2$, and you need to consider the sampling distribution of the difference between means. The mean and variance of this distribution can be found as follows:

$$E(\bar{X}_1 - \bar{X}_2) = E(\bar{X}_1) - E(\bar{X}_2) = \mu_1 - \mu_2 \qquad \text{(see pages 403 and 437)}$$

$$\text{Var}(\bar{X}_1 - \bar{X}_2) = \text{Var}(\bar{X}_1) + \text{Var}(\bar{X}_2) = \frac{\sigma_1^2}{n_1} + \frac{\sigma_2^2}{n_2}$$

$$\uparrow$$
Remember the + sign here

The distribution of $\bar{X}_1 - \bar{X}_2$ depends on various factors and careful analysis of a given situation is required in order to decide which test to use. In each situation described below, the underlying distributions are normal.

Note that, for reference, the 95% confidence intervals for $\mu_1 - \mu_2$ are also given.

## Test 3a: The population variances $\sigma_1^2$ and $\sigma_2^2$ are known

If the variances $\sigma_1^2$ and $\sigma_2^2$ are known,

the test statistic is $\bar{X}_1 - \bar{X}_2$ where $\bar{X}_1 - \bar{X}_2 \sim N\left(\mu_1 - \mu_2, \frac{\sigma_1^2}{n_1} + \frac{\sigma_2^2}{n_2}\right)$.

In standardised form,

the test statistic is $Z = \dfrac{\bar{X}_1 - \bar{X}_2 - (\mu_1 - \mu_2)}{\sqrt{\dfrac{\sigma_1^2}{n_1} + \dfrac{\sigma_2^2}{n_2}}}$ where $Z \sim N(0, 1)$.

Note that the 95% confidence limits for $\mu_1 - \mu_2$ are $(\bar{x}_1 - \bar{x}_2) \pm 1.96 \sqrt{\dfrac{\sigma_1^2}{n_1} + \dfrac{\sigma_2^2}{n_2}}$.

# Test 3b: The populations have a common variance, $\sigma^2$, which is known

If there is a common population variance $\sigma^2 (= \sigma_1^2 = \sigma_2^2)$ and $\sigma^2$ is known, then

the test statistic is $\bar{X}_1 - \bar{X}_2$ where $\bar{X}_1 - \bar{X}_2 \sim N\left(\mu_1 - \mu_2, \sigma^2\left(\dfrac{1}{n_1} + \dfrac{1}{n_2}\right)\right)$.

In standardised form,

the test statistic is $Z = \dfrac{\bar{X}_1 - \bar{X}_2 - (\mu_1 - \mu_2)}{\sigma\sqrt{\dfrac{1}{n_1} + \dfrac{1}{n_2}}}$, where $Z \sim N(0, 1)$.

Note that the 95% confidence limits for $\mu_1 - \mu_2$ are $(\bar{x}_1 - \bar{x}_2) \pm 1.96\sigma\sqrt{\dfrac{1}{n_1} + \dfrac{1}{n_2}}$.

# Test 3c: The populations have a common population variance, $\sigma^2$, which is unknown

If the common population variance, $\sigma^2$, is unknown, then an unbiased estimate, $\hat{\sigma}^2$, is used instead. This is sometimes known as a **pooled two-sample estimate**, where

$\hat{\sigma}^2 = \dfrac{n_1 s_1^2 + n_2 s_2^2}{n_1 + n_2 - 2}$     ($s_1^2$ and $s_2^2$ are the sample variances)

An alternative format for $\hat{\sigma}^2$ is

$\hat{\sigma}^2 = \dfrac{\Sigma(x_1 - \bar{x}_1)^2 + \Sigma(x_2 - \bar{x}_2)^2}{n_1 + n_2 - 2}$

The distribution of $\bar{X}_1 - \bar{X}_2$ depends on the whether the samples taken are large or small.

### Large samples

For large samples the distribution of $\bar{X}_1 - \bar{X}_2$ is approximately normal.

The test statistic is $\bar{X}_1 - \bar{X}_2$ where $\bar{X}_1 - \bar{X}_2 \sim N\left(\mu_1 - \mu_2, \hat{\sigma}^2\left(\dfrac{1}{n_1} + \dfrac{1}{n_2}\right)\right)$.

In standardised form,

the test statistic $Z = \dfrac{\bar{X}_1 - \bar{X}_2 - (\mu_1 - \mu_2)}{\hat{\sigma}\sqrt{\dfrac{1}{n_1} + \dfrac{1}{n_2}}}$ where $Z \sim N(0, 1)$.

Note that the 95% confidence limits for $\mu_1 - \mu_2$ are $\bar{x}_1 - \bar{x}_2 \pm 1.96\hat{\sigma}\sqrt{\dfrac{1}{n_1} + \dfrac{1}{n_2}}$.

### Small samples

For small samples the standardised form of the distribution of $\bar{X}_1 - \bar{X}_2$ follows a *t*-distribution.

The test statistic is $T = \dfrac{\bar{X}_1 - \bar{X}_2 - (\mu_1 - \mu_2)}{\hat{\sigma}\sqrt{\dfrac{1}{n_1} + \dfrac{1}{n_2}}}$, where $T \sim t(n_1 + n_2 - 2)$.

Note that the 95% confidence limits for $\mu_1 - \mu_2$ are $(\bar{x}_1 - \bar{x}_2) \pm t\hat{\sigma}\sqrt{\dfrac{1}{n_1} + \dfrac{1}{n_2}}$ where $t$ is such that $P(T < t) = 0.975$ for $t(n_1 + n_2 - 2)$.

## Example 11.11

Due to differences in the environment, the masses of a certain species of small animal are believed to be greater in Region A than in Region B. It is known that the masses in both regions are normally distributed, with masses in Region A having a standard deviation of 0.04 kg and masses in region B having a standard deviation of 0.09 kg.

To test the theory, random samples are taken: 60 animals from Region A had a mean mass of 3.03 kg and 50 animals from Region B had a mean mass of 3.00 kg.

Does this provide evidence, at the 1% level that the animals of this species in Region A have a greater mass than those in Region B?

## Solution 11.11

1. Define the variables.

Let $X_1$ be the mass, in kilograms, of an animal in Region A and let the population mean be $\mu_1$. Then $X_1 \sim N(\mu_1, 0.04^2)$.
Let $X_2$ be the mass, in kilograms, of an animal in Region B and let the population mean be $\mu_2$. Then $X_2 \sim N(\mu_2, 0.09^2)$.

2. State $H_0$ and $H_1$.

$H_0: \mu_1 - \mu_2 = 0$ (there is no difference in the masses between the regions)
$H_1: \mu_1 - \mu_2 > 0$ (the animals in Region A have greater mass)

3. State the distribution of the test statistic according to $H_0$.

Consider the distribution of the difference between the means, $\bar{X}_1 - \bar{X}_2$.

$$\bar{X}_1 - \bar{X}_2 \sim N\left(\mu_1 - \mu_2, \frac{\sigma_1^2}{n_1} + \frac{\sigma_2^2}{n_2}\right) \text{ with } n_1 = 60, n_2 = 50$$

If $H_0$ is true then $\mu_1 - \mu_2 = 0$,

$$\text{so} \quad \bar{X}_1 - \bar{X}_2 \sim N\left(0, \frac{0.04^2}{60} + \frac{0.09^2}{50}\right).$$

4. State the level of the test.

Use a one-tailed test (upper tail) at the 1% level.

5. Decide on your rejection criterion.

The critical $z$-value is 2.326, so reject $H_0$ if $z > 2.326$

6. Perform the required calculation.

$$\text{where} \quad z = \frac{\bar{x}_1 - \bar{x}_2}{\sqrt{\dfrac{0.04^2}{60} + \dfrac{0.09^2}{50}}}$$

$$= \frac{3.03 - 3.00}{0.0137 \ldots}$$

$$= 2.184 \ldots$$

7. Make your conclusion.

Since $z < 2.326$, do not reject $H_0$.

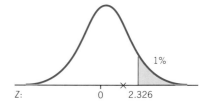

There is no evidence, at the 1% level, that the animals in region A have a greater mass than those in region B.

## Example 11.12

The same physical fitness test was given to a group of 100 scouts and to a group of 144 guides. The maximum score was 30. The guides obtained a mean score of 26.81 and the scouts obtained a mean score of 27.53. Assuming that the fitness scores are normally distributed with a common population standard deviation of 3.48, test at the 5% level of significance whether the guides did not do as well as the scouts in the fitness test.

## Solution 11.12

**1.** Define the variables.

Let $X_1$ be a guide's score and let the population mean be $\mu_1$.
Then $X_1 \sim N(\mu_1, \sigma^2)$ with $\sigma = 3.48$.

Let $X_2$ be a scout's score and let the population mean be $\mu_2$.
Then $X_2 \sim N(\mu_2, \sigma^2)$ with $\sigma = 3.48$.

**2.** State $H_0$ and $H_1$.

$H_0: \mu_1 - \mu_2 = 0$ (there is no difference in the performance)
$H_1: \mu_1 - \mu_2 < 0$ (the guides did not perform as well as the scouts)

**3.** State the distribution of the test statistic according to $H_0$.

Consider the distribution of the difference between the means $\bar{X}_1 - \bar{X}_2$.

$$\bar{X}_1 - \bar{X}_2 \sim N\left(\mu_1 - \mu_2, \sigma^2\left(\frac{1}{n_1} + \frac{1}{n_2}\right)\right) \text{ with } n_1 = 144, n_2 = 100$$

If $H_0$ is true then $\mu_1 - \mu_2 = 0$,

$$\text{so} \quad \bar{X}_1 - \bar{X}_2 \sim N\left(0, 3.48^2\left(\frac{1}{144} + \frac{1}{100}\right)\right)$$

**4.** State the level of the test.

Use a one-tailed test (lower tail) at the 5% level.

**5.** Decide on your rejection criterion.

The critical $z$-value is $-1.645$, so reject $H_0$ if $z < -1.645$, where $z = \dfrac{\bar{x}_1 - \bar{x}_2 - 0}{\hat{\sigma}\sqrt{\dfrac{1}{n_1} + \dfrac{1}{n_2}}}$

**6.** Perform the required calculation.

The sample values are $\bar{x}_1 = 26.81, \bar{x}_2 = 27.53$.

$$z = \frac{\bar{x}_1 - \bar{x}_2 - 0}{3.48\sqrt{\dfrac{1}{144} + \dfrac{1}{100}}}$$
$$= \frac{26.81 - 27.53}{0.452\ldots}$$
$$= -1.589$$

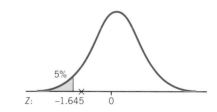

**7.** Make your conclusion.

Since $z > -1.645$, do not reject $H_0$.

**There is no evidence, at the 5% level, that the guides did not perform as well as the scouts in the fitness tests.**

## Example 11.13

An investigation was carried out to assess the effects of adding certain vitamins to the diet. A group of two-week old rats was given a vitamin supplement in their diet for a period of one month, after which time their masses were noted. A control group of rats of the same age was fed on an ordinary diet and their masses were also noted after one month.

The results are summarised in the table:

| | Number in sample | Mean | Standard deviation |
|---|---|---|---|
| With vitamin supplement | 64 | 89.6 g | 12.96 g |
| Without vitamin supplement | 36 | 83.5 g | 11.41 g |

Treating the samples as large samples from normal distributions with the same variance, $\sigma^2$, test at the 5% level whether the results provide evidence that rats given the vitamin supplement have a greater mass, at age six weeks, than those not given the vitamin supplement.

## Solution 11.13

**1. Define the variables.**

Let $X_1$ be the mass of a rat given a vitamin supplement and let the population mean be $\mu_1$. Then $X_1 \sim N(\mu_1, \sigma^2)$ with $\sigma$ unknown.

Let $X_2$ be the mass of a rat in the control group and let the population mean be $\mu_2$. Then $X_2 \sim N(\mu_2, \sigma^2)$ with $\sigma$ unknown

Since the common population variance $\sigma^2$ is unknown, use $\hat{\sigma}^2$ where

$$\hat{\sigma}^2 = \frac{n_1 s_1^2 + n_2 s_2^2}{n_1 + n_2 - 2}$$

$$= \frac{64 \times 12.96^2 + 36 \times 11.41^2}{64 + 36 - 2}$$

$$= 157.5 \ldots$$

$$\hat{\sigma} = 12.55 \ldots$$

**2. State $H_0$ and $H_1$.**

$H_0$: $\mu_1 - \mu_2 = 0$ (there is no difference in the masses of the two groups)
$H_1$: $\mu_1 - \mu_2 > 0$ (the rats given vitamin supplements are heavier)

**3. State the distribution of the test statistic according to $H_0$.**

Consider the distribution of the difference between the means, $\overline{X}_1 - \overline{X}_2$.

$$\overline{X}_1 - \overline{X}_2 \sim N\left(\mu_1 - \mu_2, \hat{\sigma}^2\left(\frac{1}{n_1} + \frac{1}{n_2}\right)\right) \text{ with } \hat{\sigma} = 12.55, n_1 = 64, n_2 = 36$$

If $H_0$ is true then $\mu_1 - \mu_2 = 0$,

so $\overline{X}_1 - \overline{X}_2 \sim N\left(0, 12.55^2\left(\frac{1}{64} + \frac{1}{36}\right)\right)$.

**4. State the level of the test.**
**5. Decide on your rejection criterion.**

Use a one-tailed test (upper tail) at the 5% level.

The critical $z$-value is 1.645, so reject $H_0$ if $z > 1.645$

**6. Perform the required calculation.**

where $z = \dfrac{\bar{x}_1 - \bar{x}_2 - 0}{\hat{\sigma}\sqrt{\dfrac{1}{n_1} + \dfrac{1}{n_2}}}$

$$= \frac{89.6 - 83.5}{12.55\sqrt{\dfrac{1}{64} + \dfrac{1}{36}}}$$

$$= 2.332 \ldots$$

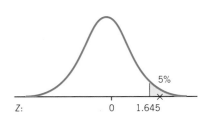

7. Make your conclusion.

Since $z > 1.645$, reject $H_0$.

**There is evidence, at the 5% level, that the rats given the vitamin supplement have a greater mass than the rats not given the supplement.**

## Example 11.14

Two statistics teachers, Mr Chalk and Mr Talk, argue about their abilities at golf. Mr Chalk claims that with a number 7 iron he can hit the ball, on average, at least 10 m further than Mr Talk. They conducted an experiment, measuring the distances for several shots.
Denoting the distance Mr Chalk hits the ball by $x$ metres, the following results were obtained:
$n_1 = 40$, $\Sigma x = 4080$, $\Sigma(x - \bar{x})^2 = 1132$.

Denoting the distance Mr Talk hits the ball by $y$ metres, the following results were obtained:
$n_2 = 35$, $\Sigma y = 3325$, $\Sigma(y - \bar{y})^2 = 1197$.

Assuming that the populations have a common variance, test whether there is evidence, at the 1% level, to support Mr Chalk's claim.

## Solution 11.14

1. Define the variables.

Let $X$ be the distance, in metres, for Mr Chalk and let the population mean be $\mu_1$.
Then $X \sim N(\mu_1, \sigma^2)$ with $\sigma$ unknown.

Let $Y$ be the distance, in metres, for Mr Talk and let the population mean be $\mu_2$.
Then $Y \sim N(\mu_2, \sigma^2)$ with $\sigma$ unknown.

An unbiased estimate for $\sigma^2$ is $\hat{\sigma}^2$ where

$$\hat{\sigma}^2 = \frac{\Sigma(x - \bar{x})^2 + \Sigma(y - \bar{y})^2}{n_1 + n_2 - 2}$$
$$= \frac{1132 + 1197}{40 + 35 - 2}$$
$$= 31.904 \ldots$$
$$\hat{\sigma} = 5.6483 \ldots$$

The unbiased estimate of the common population standard deviation is 5.648 (3 d.p.).

$H_0: \mu_1 - \mu_2 = 10$ (Mr Chalk hits the ball 10 m further than Mr Talk)

2. State $H_0$ and $H_1$.

*Mr Chalk claims that he can hit the ball at least 10 m further than Mr Talk. Mr Talk wants to refute this, so take as alternative hypothesis that Mr Chalk hits the ball less than 10 m further than Mr Talk.*

$H_1: \mu_1 - \mu_2 < 10$ (Mr Chalk hits the ball less than 10 m further than Mr Talk)

Consider the distribution of the difference between the means $\bar{X} - \bar{Y}$.

3. State the distribution of the test statistic according to $H_0$.

$$\bar{X} - \bar{Y} \sim N\left(\mu_1 - \mu_2, \hat{\sigma}^2\left(\frac{1}{n_1} + \frac{1}{n_2}\right)\right) \text{ with } \hat{\sigma} = 5.648, \, n_1 = 40, \, n_2 = 35$$

If $H_0$ is true then $\mu_1 - \mu_2 = 10$,

$$\text{so} \quad \bar{X} - \bar{Y} \sim N\left(10, 5.648^2\left(\frac{1}{40} + \frac{1}{35}\right)\right).$$

| 4. State the level of the test. | Use a one-tailed test (lower tail) at the 1% level. |
| --- | --- |

5. Decide on your rejection criterion.

The critical $z$-value is $-2.326$, so reject $H_0$ if $z < -2.326$.

where $\quad z = \dfrac{\bar{x}_1 - \bar{x}_2 - 10}{\hat{\sigma}\sqrt{\dfrac{1}{n_1} + \dfrac{1}{n_2}}} = \dfrac{\bar{x}_1 - \bar{x}_2 - 10}{5.648\sqrt{\dfrac{1}{40} + \dfrac{1}{35}}}$

6. Perform the required calculation.

From the samples, $\bar{x} = \dfrac{\Sigma x}{n_1} = \dfrac{4080}{40} = 102$

$\bar{y} = \dfrac{\Sigma y}{n_2} = \dfrac{3325}{35} = 95$

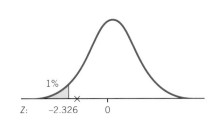

So $\quad z = \dfrac{102 - 95 - 10}{5.648\sqrt{\dfrac{1}{40} + \dfrac{1}{35}}}$

$= -2.29\ldots$

7. Make your conclusion.

Since $z > -2.326$, do not reject $H_0$.

**There is not sufficient evidence, at the 1% level, to reject Mr Chalk's claim that he hits the ball, on average, at least 10 m further than Mr Talk.**

## Example 11.15

An investigation was conducted into the dust content in the flue gases of two types of solid fuel boilers.

Thirteen boilers of type A and 9 boilers of type B were used under identical fuelling and extraction conditions. Over a similar period, the following quantities, in grams, of dust were deposited in similar traps inserted in each of the 22 flues.

Dust deposit (g) in Type A boilers:

73.1, 56.4, 82.1, 67.2, 78.7, 75.1, 48.0, 53.3, 55.5, 61.5, 60.6, 55.2, 63.1

Dust deposit (g) in Type B boilers:

53.0, 39.3, 55.8, 58.8, 41.2, 66.6, 46.0, 56.4, 58.9

(a) Find the mean and variance of each of the samples.

(b) Assuming that these independent samples came from normal populations with the same variances:
  (i)  use a two-sample $t$-test at the 5% level of significance to determine whether there is any difference between the two samples as regards the mean dust deposit,
  (ii) test at the 5% level of significance whether there is any difference between the two samples as regards the mean dust deposit where this time you should also assume that the population variances are both known to be 196.0.

(c) Explain the apparent contradiction in your results. *(AEB)*

## Solution 11.15

(a) Using a calculator in standard deviation mode, the following values were obtained. Check them using your calculator.

|  | Mean | Variance |
|---|---|---|
| Type A boiler | 63.83 | 104.32 |
| Type B boiler | 52.89 | 72.07 |

**1. Define the variables.**

(b) (i) Let $X_1$ be the mass of dust deposited in a type A boiler and let the population mean be $\mu_1$ and population variance be $\sigma^2$.

Then $X_1 \sim N(\mu_1, \sigma^2)$ with $\sigma$ unknown.

Let $X_2$ be the mass of dust deposited in a type B boiler and let the population mean be $\mu_2$ and population variance be $\sigma^2$.

Then $X_2 \sim N(\mu_2, \sigma^2)$ with $\sigma$ unknown.

Since $\sigma^2$ is unknown, use $\hat{\sigma}^2$ where

$$\hat{\sigma}^2 = \frac{n_1 s_1^2 + n_2 s_2^2}{n_1 + n_2 - 2}$$
$$= \frac{13 \times 104.32 + 9 \times 72.07}{13 + 9 - 2}$$
$$= 100.23 \ldots$$
$$\hat{\sigma} = 10.01 \ (2 \text{ d.p.})$$

**2. State $H_0$ and $H_1$.**

$H_0: \mu_1 - \mu_2 = 0$ (there is no difference in the masses deposited)
$H_1: \mu_1 - \mu_2 \neq 0$ (there is a difference in the masses deposited)

**3. State the distribution of the test statistic according to $H_0$.**

Consider the distribution of the difference between the means, $\overline{X}_1 - \overline{X}_2$.

Since the samples are small and the common population variance is unknown, the test statistic is $T$ where $T \sim t(n_1 + n_2 - 2)$ and

$$T = \frac{\overline{X}_1 - \overline{X}_2 - (\mu_1 - \mu_2)}{\hat{\sigma} \sqrt{\dfrac{1}{n_1} + \dfrac{1}{n_2}}}, \quad \text{with } \hat{\sigma} = 10.01,\ n_1 = 13,\ n_2 = 9$$

If $H_0$ true then $\mu_1 - \mu_2 = 0$,

so $T = \dfrac{\overline{X}_1 - \overline{X}_2 - 0}{10.01 \sqrt{\dfrac{1}{13} + \dfrac{1}{9}}}$ and $T \sim t(20)$

**4. State the level of the test.**

Use a two-tailed test at the 5% level.

**5. Decide on your rejection criterion.**

*Because you want 2.5% in the each tail, the critical value for $t$ is found from row $v = 20$, p = 0.975 giving $\pm 2.086$*

Critical values for $t$ (see page 650)

| $p$ | 0.75 | 0.90 | 0.95 | 0.975 |
|---|---|---|---|---|
| $v = 1$ | 1.000 | 3.078 | 6.314 | 12.71 |
| 2 | 0.816 | 1.886 | 2.920 | 4.303 |
| 19 | 0.688 | 1.328 | 1.729 | 2.093 |
| 20 | 0.687 | 1.325 | 1.725 | 2.086 |

Reject $H_0$ if $t < -2.086$ or $t > 2.086$, i.e. if $|t| > 2.086$.

6. Perform the required calculation.

From the samples, $\bar{x}_1 = 63.83$, $\bar{x}_2 = 52.89$

$$t = \frac{63.83 - 52.89}{10.01\sqrt{\dfrac{1}{13} + \dfrac{1}{9}}}$$

$$= 2.52 \ldots$$

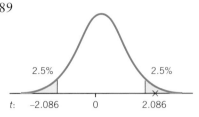

2.5%          2.5%

$t:$    $-2.086$    $0$    $2.086$

7. Make your conclusion.

since $t > 2.086$, reject $H_0$.

**There is a difference between the samples with regard to the mean dust deposit.**

(b) (ii)  This time the population variances are known to be 196 and so a $z$-test is performed, rather than a $t$-test.

$X_1 \sim N(\mu_1, \sigma^2)$ with $\sigma = \sqrt{196} = 14$
$X_2 \sim N(\mu_2, \sigma^2)$ with $\sigma = 14$

The hypotheses are as before, but the test statistic, the difference between the means, is distributed

$$\bar{X}_1 - \bar{X}_2 \sim N\left(\mu_1 - \mu_2, \sigma^2\left(\frac{1}{n_1} + \frac{1}{n_2}\right)\right) \text{ with } \sigma = 14, \, n_1 = 13, \, n_2 = 9.$$

i.e.    $$\bar{X}_1 - \bar{X}_2 \sim N\left(0, 14^2\left(\frac{1}{13} + \frac{1}{9}\right)\right)$$

A two-tailed test at the 5% level gives critical $z$-values of $\pm 1.96$ (see page 649).
So reject $H_0$ if $z < -1.96$ or $z > 1.96$, i.e. if $|z| < 1.96$

where    $$z = \frac{\bar{x}_1 - \bar{x}_2 - 0}{\sigma\sqrt{\dfrac{1}{n_1} + \dfrac{1}{n_2}}}$$

$$z = \frac{63.83 - 52.89}{14\sqrt{\dfrac{1}{13} + \dfrac{1}{9}}}$$

$$= 1.802 \ldots$$

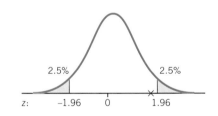

2.5%          2.5%

$z:$    $-1.96$    $0$    $1.96$

since $|z| < 1.96$, $H_0$ is not rejected.

**There is no difference between the samples with regard to the mean dust deposit.**

(c)  Considering the variances of the samples, it would seem that the common population variance of 196.0 given in part (b) is suspect. The value of just over 100 given by the unbiased estimate appears more reasonable and so result (a) is likely to be more accurate.

# Exercise 11d   Testing the difference between means of two normal populations

## Section A: z-tests

1. In each of the following, a random sample of size $n_1$ is taken from population $X$ and a random sample of size $n_2$ is taken from population $Y$.

   Use the information given to test the hypotheses stated at the level of significance indicated.

   (a)  $X \sim N(\mu_1, \sigma_1^2)$, $X \sim N(\mu_2, \sigma_1^2)$

| | $n_1$ | $\Sigma x$ | $\sigma_1^2$ | $n_2$ | $\Sigma y$ | $\sigma_2^2$ | Hypotheses | Level |
|---|---|---|---|---|---|---|---|---|
| (i) | 100 | 4250 | 30 | 80 | 3544 | 35 | $H_0: \mu_1 - \mu_2 = 0$ $H_1: \mu_1 - \mu_2 \neq 0$ | 5% |
| (ii) | 20 | 95 | 2.3 | 25 | 135 | 2.5 | $H_0: \mu_1 = \mu_2$ $H_1: \mu_1 < \mu_2$ | 2% |
| (iii) | 50 | 1545 | 6.5 | 50 | 1480 | 7.1 | $H_0: \mu_1 = \mu_2$ $H_1: \mu_1 > \mu_2$ | 1% |

   (b)  $X$ and $Y$ have a common variance $\sigma^2$, so $X \sim N(\mu_1, \sigma^2)$, $Y \sim N(\mu_2, \sigma^2)$.

| | $n_1$ | $\Sigma x$ | $n_2$ | $\Sigma y$ | Common population standard deviation ($\sigma$) | Hypotheses | Level |
|---|---|---|---|---|---|---|---|
| (i) | 50 | 2480 | 40 | 1908 | 4.5 | $H_0: \mu_1 = \mu_2$ $H_1: \mu_1 \neq \mu_2$ | 2% |
| (ii) | 100 | 12 730 | 100 | 12 410 | 10.9 | $H_0: \mu_1 = \mu_2$ $H_1: \mu_1 > \mu_2$ | 5% |
| (iii) | 30 | 192 | 45 | 315 | 1.25 | $H_0: \mu_1 = \mu_2$ $H_1: \mu_1 < \mu_2$ | 1% |
| (iv) | 200 | 18 470 | 300 | 27 663 | 0.86 | $H_0: \mu_1 = \mu_2$ $H_1: \mu_1 \neq \mu_2$ | 10% |

| | $n_1$ | $\Sigma x$ | $\Sigma(x-\bar{x})^2$ | $n_2$ | $\Sigma y$ | $\Sigma(y-\bar{y})^2$ | Hypotheses | Level |
|---|---|---|---|---|---|---|---|---|
| (v) | 40 | 2128 | 810 | 50 | 2580 | 772 | $H_0: \mu_1 = \mu_2$ $H_1: \mu_1 > \mu_2$ | 5% |
| (vi) | 80 | 6824 | 2508 | 100 | 8740 | 3969 | $H_0: \mu_1 - \mu_2$ $H_1: \mu_1 \neq \mu_2$ | 2% |
| (vii) | 65 | 5369 | 8886 | 80 | 4672 | 5026 | $H_0: \mu_1 - \mu_2 = 20$ $H_1: \mu_1 - \mu_2 > 20$ | 1% |

2. A large group of sunflowers is growing in the shady side of a garden. A random sample of 36 of these sunflowers is measured. The sample mean height is found to be 2.86 m, and the sample standard deviation is found to be 0.60 m. A second group of sunflowers is growing in the sunny side of the garden. A random sample of 26 of these sunflowers is measured. The sample mean height is found to be 3.29 m and the sample standard deviation is found to be 0.9 m. Treating the samples as large samples from normal distributions having the same variance but possibly different means, obtain a pooled estimate of the variance and test whether the results provide significant evidence (at the 5% level) that the sunny-side sunflowers grow taller, on average, than the shady-side sunflowers.   (C)

3. The lengths, in millimetres, of 9 screws selected at random from a large consignment are found to be:

   8.00,  8.02,  8.03,  7.99,  8.00,
   8.01,  8.01,  7.99,  8.01.

   From a second large consignment, 16 screws are selected at random and their mean length is found to be 7.992 mm.
   Assuming that both samples are from normal populations with variance 0.0001, test, at the 5% significance level, the hypothesis that the second population has the same mean as the first population, against the alternative hypothesis that the second population has a smaller mean that the first population.   (C)

4. Hischi and Taschi are two makes of video tapes. They are both advertised as having a recording time of 3 hours. A sample of 49 Hischi tapes was tested and denoting the actual recording time by $h$ minutes, the following results were obtained:

$$\Sigma h = 8673, \quad \Sigma(h - \bar{h})^2 = 12\,720$$

A sample of 81 Taschi tapes was also tested. Denoting the actual recording time by $t$ minutes, the results obtained were:

$$\Sigma t = 14\,904, \quad \Sigma(t - \bar{t})^2 = 33\,488$$

If the recording times for the two makes are normally distributed and have a common variance, show that the unbiased estimate of this common variance is 361. Test whether there is significant evidence, at the 5% level, of a difference in the mean recording times. Is the difference significant at the 4% level?

5. A large number of tomato plants are grown under controlled conditions. Half of the plants, chosen at random, are treated with a new fertiliser, and the other half of the plants are treated with a standard fertiliser. Random samples of 100 plants are selected from each half, and records are kept of the total crop mass of each plant. For those treated with the new fertiliser, the crop masses (in suitable units) are summarized by the figures

$$\Sigma x = 1030.0, \quad \Sigma x^2 = 11\,045.59.$$

The corresponding figures for those plants treated with the standard fertiliser are

$$\Sigma y = 990.0, \quad \Sigma y^2 = 10\,079.19.$$

Treating the sample as a large sample from a normal distribution, and assuming that the population variances of both distributions are equal, obtain a two-sample pooled estimate of the common population variance.
Assuming that it is impossible for the new fertiliser to be less efficacious than the old fertiliser and assuming that both distributions are normal, test whether the results provide significant evidence (at the 3% level) that the new fertiliser is associated with a greater mean crop mass, stating clearly your null and alternative hypotheses. (C)

6. Mr Brown and Mr Green work at the same office and live next door to each other.
Each day they leave for work together but travel by different routes. Mr Brown maintains that his route is quicker, on average, by at least four minutes. Both men time their journeys in minutes over a period of ten weeks. The results obtained were:

Mr Brown: $n_1 = 50, \quad \bar{x}_1 = 21, \quad s_1^2 = 10.24$
Mr Green: $n_2 = 50, \quad \bar{x}_2 = 24, \quad s_2^2 = 7.84$

Assuming that the times are normally distributed and that they have a common population variance, test at the 5% level whether Mr Brown's claim can be accepted.

7. A random sample of size 100 is taken from a normal population with variance $\sigma_1^2 = 40$. The sample mean $\bar{x}_1$ is 38.3. Another random sample, of size 80, is taken from a normal population with variance $\sigma_2^2 = 30$. The sample mean $\bar{x}_2$ is 40.1. Test, at the 5% level, whether there is a significant difference in the population means $\mu_1$ and $\mu_2$.

8. A certain political group maintains that girls reach a higher standard in single-sex classes than in mixed classes. To test this hypothesis 140 girls of similar ability are split into two groups, with 68 attending classes containing only girls and 72 attending classes with boys. All the classes follow the same syllabus and after a specified time the girls are given a test. The test results are summarised thus:

Girls in the mixed classes:
$\Sigma x = 7920, \quad \Sigma x^2 = 879\,912$
Girls in single-sex classes:
$\Sigma y = 7820, \quad \Sigma y^2 = 904\,808$

Treating both samples as large samples from normal distributions having the same variance, obtain a two-sample pooled estimate of the common population variance. Test whether the results provide significant evidence, at the 1% level, that girls reach a higher standard in single-sex classes.

9. The mean height of 50 male students of a college who took an active part in athletic activities was 178 cm with a standard deviation of 5 cm while 50 male students who showed no interest in such activities had a mean height of 176 cm with a standard deviation of 7 cm. Test the hypothesis that male students who take an active part in athletic activities have the same mean height as the other male students.
If both samples had been of size $n$, instead of 50, find the least value of $n$ which would ensure that the observed difference of 2 cm in the mean height would be significant at the 1% level. (Assume that the samples continue to have the same means and standard deviations.) (C)

10. A random sample of 27 individuals from the population of young men aged 18 and of high intelligence have foot lengths (in centimetres, to the nearest centimetre) as summarised below.

| Foot length (in cm) | 24 | 25 | 26 | 27 | 28 | 29 | 30 |
|---|---|---|---|---|---|---|---|
| Number with this foot length | 1 | 2 | 3 | 9 | 6 | 5 | 1 |

Obtain the sample mean and show that the unbiased estimate of the population variance, based on this sample, is 2.00. Obtain a 96% confidence interval for the mean foot length of this type of person.

A random sample of 48 individuals from the population of young men aged 18 and of moderate intelligence have foot lengths summarised by $\bar{x} = 26.6$, $\Sigma(x - \bar{x})^2 = 123.20$. A complex genetic theory suggests that persons of high intelligence have a greater foot length than do those of moderate intelligence. The two samples described above may be assumed to have been drawn at random from independent normal distributions having a common variance. Obtain an unbiased two-sample estimate of this common variance. Treating the samples as large samples, test this genetic theory, using a significance test at the 1% significance level and stating clearly the hypotheses under comparison. (C)

## Section B: t-tests

1. A random sample of size $n_1$ is taken from population $X \sim N(\mu_1, \sigma^2)$ and a random sample of size $n_2$ is taken from population $Y \sim N(\mu_2, \sigma^2)$.

   (a) Obtain an unbiased estimate of $\sigma^2$ by pooling the results from the two samples.
   (b) Test the hypotheses stated at the level of significance indicated.

|       | $n_1$ | $\Sigma x$ | $\Sigma(x-\bar{x})^2$ | $n_2$ | $\Sigma y$ | $\Sigma(y-\bar{y})^2$ | Hypotheses | Level |
|-------|-------|------------|------------------------|-------|------------|------------------------|------------|-------|
| (i)   | 6     | 171        | 83                     | 7     | 164.5      | 112                    | $H_0: \mu_1 = \mu_2$ $H_1: \mu_1 > \mu_2$ | 5% |
| (ii)  | 5     | 678.5      | 562.3                  | 7     | 971.6      | 308.6                  | $H_0: \mu_1 = \mu_2$ $H_1: \mu_1 \neq \mu_2$ | 5% |
| (iii) | 8     | 238.4      | 296                    | 10    | 206        | 145                    | $H_0: \mu_1 - \mu_2 = 4$ $H_1: \mu_1 - \mu_2 > 4$ | 1% |
| (iv)  | 12    | 116.16     | 45.1                   | 18    | 156.96     | 72                     | $H_0: \mu_1 = \mu_2$ $H_1: \mu_1 \neq \mu_2$ | 10% |

2. The heights (measured to the nearest centimetre) of a random sample of six policemen from a certain force in Wales were found to be:

   176, 180, 179, 181, 183, 179.

   The heights (measured to the nearest centimetre) of a random sample of 11 policemen from a certain force in Scotland gave the following data:

   $\Sigma y = 1991$, $\Sigma(y - \bar{y})^2 = 54$.

   Test at the 5% level, the hypothesis that Welsh policemen are shorter than Scottish policemen. Assume that the heights of policemen in both forces are normally distributed and have a common population variance.

3. An expert golfer wishes to discover whether the average distances travelled by two different brands of golf ball differ significantly. He tests each ball by hitting it with his driver and measuring the distance $X$ (in metres) that it travels. The distribution of $X$ may be assumed to be normal.
   His results for a random sample of 9 'Farfly' golf balls were $\bar{x} = 214$ and $\Sigma(x - \bar{x})^2 = 2048$.
   His results for a random sample of 16 'Gofar' golf balls were
   $\bar{x} = 224$ and $\Sigma(x - \bar{x})^2 = 2460$.

   Assuming that the variance of $X$ is the same for both types of golf ball, obtain a pooled (two sample) estimate of this variance and, test at the 5% level whether his results for 'Gofar' golf balls differ significantly from those for 'Farfly' golf balls. (C)

4. Mr Mean notes the time, in minutes, that it takes him to drive to work in the mornings. The results are:

   $n_1 = 8$, $\Sigma x_1 = 120$, $\Sigma x_1^2 = 1827$.

   For his return journey in the rush hour, Mr Mean notes that:

   $n_2 = 10$, $\Sigma x_2 = 230$, $\Sigma x_2^2 = 5436$.

   He maintains that, on average, it takes him at least ten minutes longer to drive home.

   (a) Using the results from the two samples, find an unbiased estimate of the common population variance.
   (b) Assuming that the times of all journeys are normally distributed, use the two-sample t-test at the 5% level to test Mr Mean's claim.

5. Random samples of year 10 pupils at two schools are given the same mathematics test. The results are summarised thus:

   School A:  $n_1 = 20$,  $\bar{x} = 43$,  $\Sigma(x - \bar{x})^2 = 1296$
   School B:  $n_2 = 17$,  $\bar{y} = 36$,  $\Sigma(y - \bar{y})^2 = 1388$

Assuming that the distributions of marks are normal with a common population variance, test at the 2% level whether there is a significant difference in the mathematical ability of the Year 10 pupils at the two schools.

6. A random sample of size $n_1$ is taken from a population $P_1$ whose mean is $\mu_1$ and variance $\sigma_1^2$ and a random sample of size $n_2$ is taken from population $P_2$ with mean $\mu_2$ and variance $\sigma_2^2$. Under what circumstances is it valid to test the hypothesis $\mu_1 - \mu_2 = 0$ using a two-sample $t$-test?

   A machine fills bags of sugar and a random sample of 20 bags selected from a week's production yielded a mean weight of 499.8 g with standard deviation 0.63 g. A week later a sample of 25 bags yielded a mean weight of 500.2 g with standard deviation 0.48 g.

   Assuming that your stated conditions are satisfied, perform a test to determine whether the mean has increased significantly during the second week.

   Test whether the mean during the second week could be 500 g. (Use a 5% significance level for both tests.)

7. A liquid product is sold in containers. The containers are filled by a machine. The volumes of liquid (in millilitres) in a random sample of 6 containers were found to be:

   497.8,  501.4,  500.2,  500.8,  498.3,  500.0.

   After overhaul of the machine, the volumes (in millilitres) in a random sample of 11 containers were found to be

   501.1,  499.6,  500.3,  500.9,  498.7,  502.1, 500.4,  499.7,  501.0,  500.1,  499.3.

It is desired to examine whether the average volume of liquid delivered to a container by the machine is the same after overhaul as it was before.
   (a) State the assumptions that are necessary for the use of the customary $t$-test.
   (b) State formally the null and alternative hypotheses that are to be tested.
   (c) Carry out the $t$-test, using a 5% level of significance.
   (d) Discuss briefly which of the assumptions in (a) is least likely to be valid in practice and why.  (MEI)

8. The performances of trainee actors who have passed through a drama school are rated by a panel of experienced actors who assign an overall mark for each trainee. The drama school has recently introduced a new training method which, it is claimed, will lead to better performances.
   The marks for a random sample of 6 trainees using the old training method were

   243,  228,  220,  206,  230,  198.

   and the marks for a random sample of 8 using the new method were

   235,  259,  227,  242,  238,  253,  221,  217.

   Use an appropriate $t$-test to examine, at the 5% level of significance, whether there is evidence that the new method has led, on average, to higher scores. State carefully the assumptions on which this procedure is based.
   Provide a two-sided 95% confidence interval for the true difference in mean scores between the old and new methods. State carefully the interpretation of this interval.  (MEI)

## Summary

## Hypothesis test 1 (z-tests and *t*-tests)

- For stages in a hypothesis test, see page 513
- For critical values and rejection criteria for a *z*-test see page 513

- Standardised test statistics:

**Test 1: Testing an unknown population mean $\mu$, $H_0$: $\mu = \mu_0$.**

When $\sigma^2$ is known.

1a    $X$ is normally distributed, $X \sim N(\mu, \sigma^2)$

For samples of size $n$ (**any size**),

$$\overline{X} \sim N\left(\mu_0, \frac{\sigma^2}{n}\right)$$

Test statistic $Z = \dfrac{\overline{X} - \mu_0}{\sigma/\sqrt{n}}$ where $Z \sim N(0, 1)$.

1b    $X$ is **not normally distributed**

For **large** samples of size $n$, by the central limit theorem,

$$\overline{X} \sim N\left(\mu_0, \frac{\sigma^2}{n}\right)$$

Test statistic $Z = \dfrac{\overline{X} - \mu_0}{\sigma/\sqrt{n}}$ where $Z \sim N(0, 1)$.

When $\sigma^2$ is unknown,                      $\hat{\sigma}^2 = \dfrac{n}{n-1} s^2$

1c    $X$ is preferably **normally distributed**. For **large** $n$,

$$\overline{X} \sim N\left(\mu_0, \frac{\hat{\sigma}^2}{n}\right)$$

Test statistic $Z = \dfrac{\overline{X} - \mu_0}{\hat{\sigma}/\sqrt{n}}$ where $Z \sim N(0, 1)$.

1d    $X$ is normally distributed, $X \sim N(\mu, \sigma^2)$. For **small** $n$,

Test statistic $T = \dfrac{\overline{X} - \mu_0}{\hat{\sigma}/\sqrt{n}}$ where $T \sim t(n - 1)$.

**Test 2: Testing a binomial proportion $p$, where $X \sim B(n, p)$.**

$X$ is the number of successes in $n$ trials.

If $n$ is **large** such that $np > 5$ and $nq > 5$, then $X \sim N(np, npq)$.

Test statistic $Z = \dfrac{X - np}{\sqrt{npq}}$ where $Z \sim N(0, 1)$.      Remember to use a continuity correction ($\pm 0.5$).

**Test 3: Testing $\mu_1 - \mu_2$, the difference between means of two normal distributions**

3a $\sigma_1^2, \sigma_2^2$ known

$$\bar{X}_1 - \bar{X}_2 \sim N\left(\mu_1 - \mu_2, \frac{\sigma_1^2}{n_1} + \frac{\sigma_2^2}{n_2}\right)$$

Test statistic $Z = \dfrac{\bar{X}_1 - \bar{X}_2 - (\mu_1 - \mu_2)}{\sqrt{\dfrac{\sigma_1^2}{n_1} + \dfrac{\sigma_2^2}{n_2}}}$ where $Z \sim N(0, 1)$.

3b **Common population variance $\sigma^2$ known**

$$\bar{X}_1 - \bar{X}_2 \sim N\left(\mu_1 - \mu_2, \sigma^2\left(\frac{1}{n_1} + \frac{1}{n_2}\right)\right)$$

Test statistic $Z = \dfrac{\bar{X}_1 - \bar{X}_2 - (\mu_1 - \mu_2)}{\sigma\sqrt{\dfrac{1}{n_1} + \dfrac{1}{n_2}}}$ where $Z \sim N(0, 1)$.

3c **Common population variance $\sigma^2$ unknown**

Use $\hat{\sigma}^2 = \dfrac{n_1 s_1^2 + n_2 s_2^2}{n_1 + n_2 - 2}$    ($s_1^2, s_2^2$ sample variances)

$$= \dfrac{\Sigma(x_1 - \bar{x}_1)^2 + \Sigma(x_2 - \bar{x}_2)^2}{n_1 + n_2 - 2}$$

**When $n$ is large**

Test statistic $Z = \dfrac{\bar{X}_1 - \bar{X}_2 - (\mu_1 - \mu_2)}{\hat{\sigma}\sqrt{\dfrac{1}{n_1} + \dfrac{1}{n_2}}}$ where $Z \sim N(0, 1)$.

**When $n$ is small**

Test statistic $T = \dfrac{\bar{X}_1 - \bar{X}_2 - (\mu_1 - \mu_2)}{\hat{\sigma}\sqrt{\dfrac{1}{n_1} + \dfrac{1}{n_2}}}$ where $T \sim t(n_1 + n_2 - 2)$.

# Miscellaneous worked examples

## Example 11.16

An inspector of items from a production line takes, on average, 21.75 seconds to check each item. After the installation of a new lighting system the times, $t$ seconds, to check each of 50 randomly chosen items from the production line are summarised by $\Sigma t = 1107$, $\Sigma t^2 = 24\,592.35$.

(a) Calculate an unbiased estimate of the population variance of the time taken to check an item under the new lighting system.

(b) Test at the 5% significance level whether there is evidence that the population mean time has changed from 21.75 seconds.

A technician who carried out the above test concluded with the following incorrect statement.
Give a corrected version.

'It is not necessary for the population to be normal since the sample size is large and the central limit theorem states that any sufficiently large sample is normal.' (C)

## Solution 11.16

Let $T$ be the time, in seconds, to check an item.

(a) $\hat{\sigma}^2 = \dfrac{1}{n-1}\left(\Sigma t^2 - \dfrac{(\Sigma t)^2}{n}\right)$

$= \dfrac{1}{49}\left(24\,592.35 - \dfrac{1107^2}{50}\right)$

$= 1.7014 \ldots$

$= 1.70 \text{ (3 s.f.)}$

(b) Let $\mu$ be the population mean time.

$H_0: \mu = 21.75$ (the population mean has not changed)
$H_1: \mu \neq 21.75$ (the population mean has changed)

Since $n$ is large, by the central limit theorem, $\overline{T}$ is approximately normal, and

$\overline{T} \sim N\left(\mu, \dfrac{\sigma^2}{n}\right).$

According to $H_0$, $\mu = 21.75$.

Since $\sigma^2$ is unknown, $\hat{\sigma}^2$ is used instead.

$\therefore \quad \overline{T} \sim N\left(21.75, \dfrac{1.70}{50}\right)$

Carry out a two-tailed test at the 5% level and reject $H_0$ if $|z| > 1.96$ where $z = \dfrac{\bar{t} - \mu}{\hat{\sigma}/\sqrt{n}}$.

From the sample, $\bar{t} = \dfrac{\Sigma t}{n} = \dfrac{1107}{50} = 22.14$

$\therefore \qquad z = \dfrac{22.14 - 21.75}{\sqrt{1.70}/\sqrt{50}} = 2.114 \ldots$

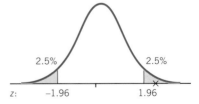

Since $|z| > 1.96$, reject $H_0$.

**There is evidence, at the 5% level, that the population mean time has changed from 21.75 seconds.**

The central limit theorem states that the distribution of means is approximately normal for large sample size $n$.

NOTE: the variable in Example 11.16 was given as $T$. Do not confuse with the standardised statistic in the $t$-distribution.

## Example 11.17

The random variable $X$ is distributed $N(\mu, 3.5^2)$. A test of the null hypothesis $\mu = 15$ against the alternative hypothesis $\mu > 15$ is required and the probability of a Type I error should be 0.05. A random sample of 30 observations on $X$ is taken.

(a) Find the critical region for the sample mean $\bar{X}$.

The mean of the sample was 16.00.

(b) Find a 95% confidence interval for $\mu$.
(c) Find $P$(Type II error) for the test in part (a) when $\mu = 17$.

The size of the sample is increased but $P$(Type I error) is still 0.05.

(d) State what effect this change will have on the critical value for $\bar{X}$ and on $P$(Type II error).

$(L)$

## Solution 11.17

$P$(Type I error) = $P(H_0$ is rejected when $H_0$ is true) = 0.05.

So the significance level of the test is 5%.

$X \sim N(\mu, 3.5^2)$

$H_0: \mu = 15$
$H_1: \mu > 15$

According to $H_0$, $X \sim N(15, 3.5^2)$

So, for a sample of size 30,

$$\bar{X} \sim N\left(15, \frac{3.5^2}{30}\right).$$

(a) Using a one-tailed (upper tail) test at the 5% level, reject $H_0$ if $z > 1.645$

where $\quad z = \dfrac{\bar{x} - 15}{3.5/\sqrt{30}}$.

So the critical (rejection) region for $\bar{x}$ is given by

$$\frac{\bar{x} - 15}{3.5/\sqrt{30}} > 1.645$$

i.e. $\quad \bar{x} > 15 + 1.645 \times \dfrac{3.5}{\sqrt{30}}$

$\qquad \bar{x} > 16.05$ **(2 d.p.)**

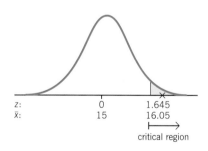

z:    0    1.645
x̄:    15    16.05
critical region

(b) 95% confidence limits for $\mu$:

$$\bar{x} \pm 1.96 \frac{\sigma}{\sqrt{n}} = 16.00 \pm 1.96 \times \frac{3.5}{\sqrt{30}}$$

$$= 16.00 \pm 1.252 \ldots$$

95% confidence interval $= (16.00 - 1.252 \ldots, 16.00 + 1.252 \ldots)$

$$= \mathbf{(14.7, 17.3)} \textbf{ (1 d.p.)}.$$

(c)  From part (a), $H_0$ is accepted if $\bar{x} < 16.05$.

$P(\text{Type II error}) = P(H_0 \text{ is accepted when } H_1 \text{ is true})$

$H_1: \mu = 17$, so under $H_1$, $\bar{X} \sim N\left(17, \dfrac{3.5^2}{30}\right)$.

$\therefore \quad P(\text{Type II error}) = P\left(\bar{X} < 16.05 \text{ when } \bar{X} \sim N\left(17, \dfrac{3.5^2}{30}\right)\right)$

$$= P\left(Z < \dfrac{16.05 - 17}{3.5/\sqrt{30}}\right)$$
$$= P(Z < -1.487)$$
$$= 1 - 0.9316$$
$$= 0.0684$$

So  **$P(\text{Type II error}) \approx 7\%$.**

If *n* is increased, but $P(\text{Type I error}) = 0.05$, the critical value for $\bar{X}$ will decrease. $P(\text{Type II error})$ will also decrease.

This is illustrated in the following diagrams:

When $n = 30$:

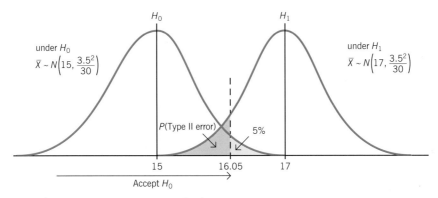

When $n > 30$, the curves are more squashed:

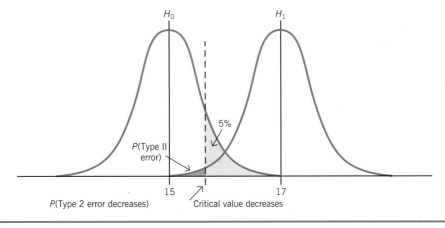

## Example 11.18

When cars arrive at a certain T-junction they turn either right or left. Part of a study of road usage involved deciding between the following alternatives.

Cars are equally likely to turn right or left.

Cars are more likely to turn right than left.

(a) State suitable null and alternative hypotheses, involving a probability, for a significance test.
(b) Out of a random sample of 40 cars, $n$ turned right. Use a suitable approximation to find the least value of $n$ for which the null hypothesis will be rejected at the 2% significance level.
(c) For the test described in (b), calculate the probability of making a Type II error when, in fact, 80% of all cars arriving at the junction turn right. (C)

## Solution 11.18

Let $X$ be the number of cars in 40 that turn right. Then $X \sim B(40, p)$.

(a) $H_0$: $p = 0.5$
$H_1$: $p > 0.5$

(b) According to $H_0$, $X \sim B(40, 0.5)$
Now $n$ is large and $np = 40 \times 0.5 = 20 > 5$, $nq = 40 \times 0.5 = 20 > 5$.
Since $np > 5$, $nq > 5$, use the normal approximation

$$X \sim N(np, npq), \quad \text{with } npq = 40 \times 0.5 \times 0.5 = 10$$
i.e. $X \sim N(20, 10)$.

Use a one-tailed (upper tail) test at the 2% level and reject $H_0$ if $P(X \geqslant n) < 2\%$. With the continuity correction this becomes $P(X > n - 0.5) < 2\%$.

i.e. reject $H_0$ if $P\left(Z > \dfrac{n - 0.5 - 20}{\sqrt{10}}\right) < 0.02$.

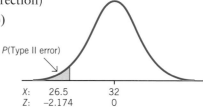

From tables, $P(Z > 2.055) = 0.02$,

so $\dfrac{n - 0.5 - 20}{\sqrt{10}} > 2.055$

$$n > 20.5 + 2.055 \times \sqrt{10}$$
$$n > 26.998 \dots$$

Since $n$ is an integer, **least value of $n = 27$.**

$\therefore$   $H_0$ is rejected if $X \geqslant 27$.

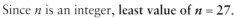

(c) $H_0$: $p = 0.5$
$H_1$: $p = 0.8$
$X \sim B(40, 0.8)$
$np = 40 \times 0.8 = 32 > 5$, $nq = 40 \times 0.2 = 8 > 5$,
so, using the normal approximation, $X \sim N(np, npq)$ with $npq = 40 \times 0.8 \times 0.2 = 6.4$
$X \sim N(32, 6.4)$
$P(\text{Type II error}) = P(H_0 \text{ is accepted when } H_1 \text{ is true})$
$H_0$ is accepted if $x < 27$, i.e. if $x < 26.5$ (continuity correction)
so $P(\text{Type II error}) = P(X < 26.5 \text{ when } X \sim N(32, 6.4))$

$$= P\left(Z < \frac{26.5 - 32}{\sqrt{6.4}}\right)$$
$$= P(Z < -2.174)$$
$$= 1 - 0.9852$$
$$= 0.0148$$
$$\approx \mathbf{1.5\%}$$

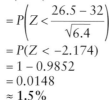

## Example 11.19

The volume of paint, in litres, in a randomly chosen two-litre can is denoted by the random variable $V$. Ten random observations of $V$ are taken, with the following results.

2.12   2.03   2.07   1.99   1.95   2.01   2.00   2.08   1.94   1.99

Assuming that $V$ has a normal distribution, use a single-sample *t*-test, at the 10% significance level, to test the claim that the mean volume of paint in two-litre cans is not 2 litres.

After installation of a new machine for filling the cans, there were worries that, on average, the new machine was dispensing less paint than the old machine. The volumes of paint dispensed by the new machine in a random sample of 20 cans were measured, and the results are summarised by

$$\Sigma w = 38.1 \qquad \text{and} \qquad \Sigma(w - \overline{w})^2 = 0.060\ 40,$$

where $w$ is the volume of paint, in litres, in a can. Use a two-sample *t*-test to test whether there is evidence, at the 10% significance level, that the new machine is dispensing less paint than the old machine.

State one condition required of the two populations for the two-sample *t*-test to be valid.   (C)

## Solution 11.19

Let $V$ be the volume (in litres) of paint in a can.

$V \sim N(\mu_1, \sigma^2)$ with $\mu_1$ and $\sigma^2$ unknown.

Sample readings: $n = 10$,   $\overline{v} = 2.018$,   $s_1^2 = 0.002976$ (using calculator)

An unbiased estimate of $\sigma^2 = \hat{\sigma}^2 = \dfrac{n}{n-1}\, s_1^2 = \dfrac{10}{9}\,(0.002976) = 0.0033$

$\therefore\ \hat{\sigma} = 0.0575$

$H_0: \mu_1 = 2$
$H_1: \mu_1 \neq 2$

According to $H_0$, the standardised statistic $T$ is such that

$$T = \frac{\overline{V} - 2}{\hat{\sigma}/\sqrt{n}} \quad \text{and} \quad T \sim t(9).$$

Carry out a two-tailed test at 10% level.
Reject $H_0$ if $|t| > 1.833$ (see tables on page 650, $v = 9$, $p = 0.95$).

$$t = \frac{2.018 - 2}{0.0575/\sqrt{10}} = 0.989 \ldots$$

Since $|t| < 1.833$, do not reject $H_0$.
**There is not enough evidence to say that the mean volume is not two litres.**

Let $W$ be the volume, in litres, dispensed by the new machine.
Assume that $W$ has a normal distribution and the two samples have a common population variance $\sigma^2$. This time the two samples are considered to give a two-sample estimate of $\sigma^2$.

$$\hat{\sigma}^2 = \frac{n_1 s_1^2 + n_2 s_2^2}{n_1 + n_2 - 2} \qquad (s_1^2 \text{ and } s_2^2 \text{ are the sample variances.})$$

where $s_1^2 = 0.002\ 976$ (from first part of question)

and $\quad s_2^2 = \dfrac{\Sigma(w - \overline{w})^2}{n_2}$

$\therefore \quad n_2 s_2^2 = \Sigma(w - \overline{w})^2 = 0.060\ 40$

so $\quad \hat{\sigma}^2 = \dfrac{10 \times 0.002\ 976 + 0.060\ 40}{10 + 20 - 2}$

$\qquad = 0.003\ 22$

$\qquad \hat{\sigma} = 0.0567$ (3 s.f.)

$H_0: \mu_1 - \mu_2 = 0$ (the machines dispense the same amount)
$H_1: \mu_1 - \mu_2 > 0$ (the new machine is dispensing less than the old)

The test statistic $T = \dfrac{\overline{V} - \overline{W} - 0}{\hat{\sigma}\sqrt{\dfrac{1}{n_1} + \dfrac{1}{n_2}}}$ $\qquad$ where $\qquad T \sim t(n_1 + n_2 - 2)$

$\therefore \quad T = \dfrac{\overline{V} - \overline{W}}{0.0567\sqrt{\dfrac{1}{10} + \dfrac{1}{20}}}$ $\qquad$ where $\qquad T \sim t(28)$.

Carry out a one-tailed test, at the 10% level. Reject $H_0$ if $t > 1.313$, where $\quad t = \dfrac{\overline{v} - \overline{w}}{\hat{\sigma}\sqrt{\dfrac{1}{n_1} - \dfrac{1}{n_2}}}$.

From the sample, $\overline{v} = 2.018$, $\overline{w} = \dfrac{\Sigma w}{n_2} = \dfrac{38.1}{20} = 1.905$

$t = \dfrac{2.018 - 1.905}{0.0567\sqrt{\dfrac{1}{10} + \dfrac{1}{20}}} = 5.145 \ldots$

Since $t > 1.313$, reject $H_0$.

**There is evidence at the 10% level that the new machine is dispensing less paint than the old machine.**

The condition required: **The two populations must be normal with common variance.**

# Miscellaneous exercise 11e

1. The amount of nicotine, in milligrams, in a cigarette of a certain brand is normally distributed with mean $\mu$ and standard deviation 2.5. A random sample of 10 cigarettes yielded a mean nicotine value of 18.4. Obtain a symmetric 90% confidence interval for $\mu$, giving values to three significant figures.
   Give a reason why the value of $\mu$ might not be inside this interval.
   Test the null hypothesis $\mu = 17.8$ against the alternative hypothesis $\mu \neq 17.8$ at the 10% significance level. (C)

2. A study is made of the numbers of boys and girls in families. A random sample of families is chosen. The total number of children is 500, of whom 261 are girls. It is desired to test the null hypothesis that boys and girls are equally likely in the population against the alternative hypothesis that they are not equally likely.

   (a) State an assumption necessary for these 500 children to be considered as a random sample of the population of all children.
   (b) Test at the 10% significance level whether the data indicate that boys and girls are not equally likely in the population. (C)

3. A resident of an urban road claims that the average speed of vehicles using the road is greater than the 30 m.p.h. speed limit. To investigate this claim the police time a randomly selected sample of 25 vehicles over a measured mile on the road. It is assumed that the speeds calculated from their observations come from a normal distribution with mean $\mu$ m.p.h. and standard deviation 12 m.p.h.

   (a) State appropriate null and alternative hypotheses for a significance test.

(b) Find a critical region for a 5% significance test in the form,

sample mean $\overline{X} > k$,

where the value of the constant $k$ is given correct to two decimal places.

(c) State, with a reason, your conclusion for the test when the mean speed calculated from the sample was 35 m.p.h.

(d) Calculate the power of the test when, in fact, $\mu = 40$. (NEAB)

4. A supermarket's statistician reports that, over the past three months, the mean amount spent per customer has been £43 with a standard deviation of £20.

The supermarket carries out a promotion for one week by offering 'buy two ... get one free' on a range of products which it sells. The management hopes that this will increase the mean amount spent per customer; you may assume that the standard deviation remains unchanged.

A random sample of 50 customers visiting the supermarket that week spent a total of £2400.

(a) Write down suitable null and alternative hypotheses in order to test whether or not the promotion has increased the average level of spending per customer.

(b) Explain carefully the use of the Central limit theorem in carrying out this hypothesis test.

(c) Carry out the hypothesis test at the 5% significance level, clearly stating your conclusion.

(d) Find a 90% confidence interval for the mean amount spent by customers during the period of the promotion. State, giving a reason, whether this is consistent with your conclusion in (c). (MEI)

5. The process of manufacturing a certain kind of dinner plate results in a proportion 0.13 of faulty plates. An alteration is made to the process which is intended to reduce the proportion of faulty plates. State suitable null and alternative hypotheses for a statistical test of the effectiveness of the alteration.

In order to carry out the test, the quality control department count the number of faulty plates in a random sample of 2500. If 290 or fewer faulty plates are found then it will be accepted that the alteration does result in a reduction in the proportion of faulty plates. Calculate the significance level of this test, using a suitable normal approximation.

Calculate the probability of making a Type II error in the above test, given that the alteration results in a decrease in the proportion of faulty plates to 0.11. (C)

6. Water from a cooling tower at a power station is discharged into a river. In order to test whether the mean temperature of discharged water is greater than the permitted maximum of 65 °C, the temperature ($x$ °C) of 40 randomly selected samples of water will be taken and the sample

mean $\bar{x}$ used to test the null hypothesis $\mu = 65$ against the alternative hypothesis $\mu > 65$, where $\mu$ °C is the population mean temperature of discharged water. It may be assumed that the population standard deviation of $x$ is 5.0.

(a) State, in the context of the question, what you understand by
  (i) a Type I error,
  (ii) a Type II error.

(b) The probability of a Type I error is fixed at 0.1. Show that the range of values of $\bar{x}$ for which the null hypothesis is rejected is given by $\bar{x} > 66.01$, correct to two decimal places.

(c) State the conclusion of the test when $\bar{x} = 65.7$, and the type of error that might be made in this case.

(d) Calculate the probability of making a Type II error when, in fact, $\mu = 68$.

What can be deduced about the probability of making a Type II error when, in fact, $\mu > 68$? (C)

7. The manager of a large supermarket wishes to judge the effect of a new layout on the customers. On the day that the layout was introduced the first 200 customers in the store were asked whether or not they approved of the new layout.

Comment on the manner in which the sample was chosen, and suggest a way of obtaining a more suitable sample.

Out of a suitably chosen sample of 200 customers, 148 approved of the new layout. Calculate an approximate 95% confidence interval for the population percentage of customers who approve of the new layout.

The supermarket manager claims that 80% of customers approve of the new layout. Show that the data provide evidence at the $2\frac{1}{2}$% significance level that the population percentage is less than 80%. (C)

8. The random variable $X$ has a normal distribution with mean $\mu$ (unknown) and variance $\sigma^2$ (known). To test the null hypothesis $H_0 : \mu = \mu_0$, a random sample of $n$ observations of $X$ is taken, and the sample mean is $\bar{x}$. Find, in terms of $\mu_0$, $\sigma$ and $n$, the set of values of $\bar{x}$ which will result in each of the following:

(a) $H_0$ being rejected in favour of $H_1 : \mu \neq \mu_0$ at the 5% level of significance,

(b) $H_0$ not being rejected in favour of $H_1 : \mu < \mu_0$ at the 1% level of significance.

9. The masses of components used in making a model car are being checked. Each of a random sample of 200 components is weighed and the masses, $x$ g, are summarised by

$n = 200$, $\quad \Sigma x = 1484.2$, $\quad \Sigma x^2 = 11\ 098.19$.

(a) Calculate an unbiased estimate of the population variance.

(b) State what you understand by 'unbiased estimate'.

The components are produced in large batches. It is desired that the mean mass of components in a batch should be at least 7.40 g. In order to decide whether to accept or reject a batch each of a random sample of 50 components from the batch is weighed. The sample data is used to perform a test of the null hypothesis $\mu = 7.40$ against the alternative hypothesis $\mu < 7.40$, where $\mu$ g is the mean mass of components in the batch. For the test the population variance is taken to have the value found in part (a). The batch is rejected if the null hypothesis is rejected using a $2\frac{1}{2}\%$ significance level. Show that the batch will be rejected if the sample mean mass is less than 7.22 g.
For one such batch the sample data is summarised by $n = 50$, $\quad \Sigma x = 366.0$.

Determine whether or not this batch is rejected. Calculate the probability of making a Type II error in carrying out the above test for a batch whose mean mass is actually 7.10 g.    (C)

10. A box of dice contains some which are unbiased and some which are biased in such a way that the probability of throwing a six with one of these dice is $\frac{1}{4}$. One die is selected at random from the box and, in order to decide whether it is biased, it is thrown 240 times and the number of sixes, $N$, is counted. The probability of throwing a six with this die is denoted by $p$. The null hypothesis $p = \frac{1}{6}$ is tested against an appropriate alternative hypothesis at the 5% significance level.

(a) State an appropriate alternative hypothesis.
(b) Find the set of values of $N$ for which it is accepted that the die is biased.
(c) Find the probability of making a Type II error in the test. [$\Phi(3.355) = 0.9996$]    (C)

11. A manufacturer makes two grades of squash ball: 'slow' and 'fast'. Slow balls have a 'bounce' (measured under standard conditions) which is known to be a normal variable with mean 10 cm and standard deviation 2 cm. The 'bounce' of fast balls is a normal variable with mean 15 cm and standard deviation 2 cm. A box of balls is unlabelled so that it is not known whether they are all slow or all fast.
Devise a test, based on an observation of the mean bounce of a sample of four balls from the box such that the Type I error is 0.05 and state the magnitude of the Type II error for this test.    (C)

12. An ambulance station serves an area which includes more than 10 000 houses. It has been decided that if the mean distance of the houses from the ambulance station is greater than ten miles then a new ambulance station will be necessary. The distance, $x$ miles, from the station of each of a random sample of 200 houses was

measured, the results being summarised by:
$\Sigma x = 2092.0$ and $\Sigma x^2 = 24\ 994.5$.

(a) Calculate, to four significant figures, unbiased estimates of
   (i) the population mean distance, $\mu$ miles, of the houses from the station,
   (ii) the population variance of the distances of the houses from the station.
   State what you understand by the term 'unbiased estimate'.
(b) Using the sample data, a significance test of the null hypothesis $\mu = 10$ against the alternative hypothesis $\mu > 10$ is carried out at the $\alpha\%$ significance level. In the test, the sample mean is compared with the critical value of 10.65; as the sample mean is less than 10.65 the null hypothesis is not rejected. Calculate the value of $\alpha$.
(c) Give a reason why it is not necessary for the distances to be normally distributed for the test to be valid.    (C)

13. A particular investigation concentrated on people recently re-employed following a first period of unemployment. Each of a random sample of 50 such persons was asked the duration, in months, of this period of unemployment. A summary of the results is as follows.

mean = 16.7 months,  variance = 193.21 month$^2$

Investigate at the 5% level of significance the claim that, for people re-employed after a first period of unemployment, the mean duration of unemployment is more than 12 months. Indicate why, in carrying out your test, **no** assumption regarding the distribution of the duration of the first period of unemployment is necessary.    (NEAB)

14. The error in the readings made on a measuring instrument can be modelled by the continuous random variable $X$ which has mean $\mu$ and standard deviation $\sigma$. If the instrument is correctly calibrated then $\mu = 0$.
In order to check the calibration of the instrument, the errors in a random sample of 40 readings were determined. These data are summarised by:
$\Sigma x = 120, \quad \Sigma x^2 = 3285$.

(a) Estimate $\sigma^2$.
(b) Carry out a hypothesis test, at the 5% level of significance, to test whether the machine is, or is not, correctly calibrated. You should state your hypotheses and conclusions carefully.
(c) Obtain a symmetric 95% confidence interval for $\mu$, explaining why it is only approximate.
(d) Suppose the data from the 40 readings had been such that the estimate of $\sigma^2$ as found in part (a) was larger, but without changing the sample mean. State the effect this would have on the value of the test statistic in part (b). Explain why this might affect the conclusion to part (b).    (MEI)

15. A study of the annual rainfall, $x$ cm to the nearest centimetre, over the last 20 years for a small town gave the following results:

$\Sigma x = 1325, \quad \Sigma x^2 = 90\ 316.$

(a) Find unbiased estimates of the mean and the variance of the annual rainfall for this town.

Archive records show that the annual rainfall for this town, prior to this period, had a mean value of 62.50 cm and a standard deviation of 11.45 cm.

(b) Assuming that the standard deviation remains unchanged at 11.45 cm, test at the 5% level of significance whether or not there is evidence of an increase in mean annual rainfall over the last 20 years. State your hypotheses clearly. *(L)*

16. In 1978 the Borsetshire County Council tree officer did a survey of a random sample of 64 separate areas, each 1 km square, and found an average of 19.5 diseased trees per square. The following year, to test whether the disease had spread, she took a new random sample of 36 separate areas, also each 1 km square, and found an average of 21.7 diseased trees per square.

(a) Assume that, in both years, the number of diseased trees per 1 km square had a normal distribution with population variance 18.2. Test, at the 1% significance level, the hypothesis that the mean number of diseased trees per 1 km square in 1979 was the same as in 1978, against the hypothesis that the mean number had increased.

(b) Further evidence suggests that the number of diseased trees is not normally distributed. Say what changes you might have to make, if any, to the test you have carried out, explaining the reasons for your answer. Do not carry out any further tests. *(C)*

17. When watching games of men's basketball, I have noticed that the players are often tall. I am interested to find out whether or not men who play basketball really are taller than men in general.

I know that the heights, in metres, of men in general have the distribution $N(1.73, 0.08^2)$. I make the assumption that the heights, $X$ metres, of male basketball players are also normally distributed, with the same variance as the heights of men in general, but possibly with a larger mean.

(a) Write down the null and alternative hypotheses under test.

I propose to base my test on the heights of eight male basketball players who recently appeared for our local team, and I shall use a 5% level of significance.

(b) Write down the distribution of the sample mean, $\overline{X}$, for samples of size 8 drawn from the distribution of $X$ assuming that the null hypothesis is true.

(c) Determine the critical region for my test, illustrating your answer with a sketch.

(d) Carry out the test, given that the mean height of the eight players is 1.765 m. You should present your conclusions carefully, stating any additional assumption you need to make.

In fact, the distribution of $X$ is $N(1.80, 0.06^2)$.

(e) Find the probability that a test based on a random sample of size 8 and using the critical region in part (c) will lead to the conclusion that male basketball players are *not* taller than men in general. *(MEI)*

18. The ingredients for concrete are mixed together to obtain a mean breaking strength of 2000 newtons. If the mean breaking strength drops below 1800 newtons then the composition must be changed The distribution of the breaking strength is normal with standard deviation 200 newtons.

Samples are taken in order to investigate the hypotheses:

$H_0 : \mu = 2000$ newtons
$H_1 : \mu = 1800$ newtons

How many samples must be tested so that $P(\text{Type I error}) = 0.05$ and $P(\text{Type II error}) = 0.1$?

# Test 11A (z-tests)

1. Cans of lemonade are filled by a machine which is set to dispense a mean amount of 330 ml into each can. The manufacturer suspects that the machine is tending to over-dispense and, in order to test the suspicion, measures the contents, $x$ ml, of a random sample of 30 cans. The results are summarised by:

   $$\Sigma x = 9925, \quad \Sigma x^2 = 3\ 284\ 137.$$

   (a) Calculate an unbiased estimate of the population variance of the amount dispensed into each can. Give four significant figures in your answer.
   (b) Test the manufacturer's suspicion at the 10% significance level.
   (c) Indicate where the central limit theorem is used in the test, and state why the use of the central limit theorem is necessary. (C)

2. The proportion of patients who suffer an allergic reaction to a certain drug used to treat a particular medical condition is assumed to be 0.045.
   When 400 patients were treated, 25 suffered an allergic reaction. Using a normal approximation, test at the 5% significance level whether the quoted figure of 0.045 is an underestimate. (C)

3. (a) A null hypothesis $H_0$ is to be tested against an alternative hypothesis $H_1$. Explain what is meant by:
   (i) a Type I error,
   (ii) a Type II error.

   (b) The tar yields in cigarettes of a particular brand are distributed normally with mean $\mu$ mg and standard deviation 0.8 mg. In order to test $H_0 : \mu = 17.5$ against $H_1 : \mu > 17.5$ at the 1% level of significance, a random sample of 10 cigarettes of this brand is to be obtained and the sample mean $\bar{X}$ calculated.
   (i) In the case when the yields were recorded, in milligrams, as:

      17.1, 18.3, 18.9, 17.8, 16.9, 19.2, 17.8, 18.3, 18.5, 18.2

      carry out the required significance test.
   (ii) Determine a critical region for the test in the form, $\bar{X} > c$, where $c$ is a constant whose value is to be determined.
   (iii) Calculate the size of the Type II error for this test when, in fact, $\mu = 18.0$. (NEAB)

4. The random variable $X$ is distributed as $N(\mu, 16)$. A random sample of size 25 is available. The null hypothesis $\mu = 0$ is to be tested against the alternative hypothesis $\mu \neq 0$. The null hypothesis will be accepted if $-1.5 < \bar{x} < 1.5$, where $\bar{x}$ is the value of the sample mean, otherwise it will be rejected. Calculate the probability of a Type I error. Calculate the probability of a Type II error if in fact $\mu = 0.5$; comment on the value of this probability. (MEI)

# Test 11B (z-tests)

1. A certain brand of mineral water comes in bottles. The amount of water in a bottle, in millilitres, follows a normal distribution of mean $\mu$ and standard deviation 2. The manufacturer claims that $\mu$ is 125. In order to maintain standards the manufacturer takes a sample of 15 bottles and calculates the mean amount of water per bottle to be 124.2 millilitres. Test, at the 5% level, whether or not there is evidence that the value of $\mu$ is lower than the manufacturer's claim. State your hypotheses clearly. (L)

2. A newspaper headline stated 'Majority would vote for Prime Minister'. The article explained that in a survey of 70 randomly selected people, 38 had said that they would vote for the Prime Minister. A spokesman for the opposition party said that such evidence was inconclusive, and, according to standard statistical techniques, the result was consistent with only 40% of the whole population voting for the Prime Minister.

   A spokesman for the government stated that the results showed that 40% was too low. Stating the null and alternative hypotheses, test at the 5% level which of the spokesmen was justified in his assertion. (L)

3. The playing times of a particular brand of audio tape are normally distributed with mean $\mu$ minutes and standard deviation 0.24 minutes. The manufacturer states that $\mu = 60$. A large batch of these tapes is delivered to a store and, in order to check the manufacturer's statement, the playing times of a random sample of ten tapes were measured. The null hypothesis $\mu = 60$ is tested against the alternative hypothesis $\mu < 60$ at the 1% significance level.

   (a) Find the range of values of the sample mean $\bar{X}$ for which the null hypothesis is rejected, giving two decimal places in your answer.

(b) State what 'a Type II error has occurred' means in the context of the playing times of tapes.

(c) Calculate the probability of making a Type II error when, in fact, $\mu = 59.7$. (C)

4. The top 40 chart of the Recorded Music Association has been compiled every week for some years, and the standard deviation of the number of weeks which a record spends at Number One in the chart has been found to be 0.87 weeks. The number of weeks which the last ten Number One records featuring female singers spent in the Number One position are:

3, 1, 1, 2, 1, 2, 3, 2, 1, 1.

For the last 15 Number One records featuring male singers, the data are:

1, 1, 2, 2, 2, 3, 4, 2, 1, 2, 3, 5, 1, 2, 3.

A music industry producer wished to test whether there was any difference in the time spent at Number One between female and male singers. She assumes that both the distributions from which the two samples are drawn are normal with standard deviation 0.87 weeks.

(a) State the null and alternative hypotheses she must use.

(b) Carry out the test at the 5% level of significance.

(c) Give a reason why her assumption of normality may be invalid. (L)

# Test 11C (*t*-tests)

1. Six cleaning firms were selected at random and asked about their hourly rates of pay, $x$, with the following results:

7.00,  6.80,  6.62,  6.94,  7.48,  7.04
[$\Sigma x = 41.88$,  $\Sigma x^2 = 292.74$.]

Carry out a *t*-test at the 1% significance level to establish whether the mean hourly rate of pay, paid by cleaning firms, falls below a proposed minimum of $7.40.
State clearly an assumption made in applying the *t*-test in this context. (C)

2. On a certain day in July the maximum temperature, $m$ °C, was recorded at 11 points chosen at random on the island of San Marco. On the same day the maximum temperature, $p$ °C, was recorded at 20 points chosen at random on the island of San Polo. The results are summarised by:

$\bar{m} = 25.30$,   $\bar{p} = 26.45$,
$\Sigma(m - \bar{m})^2 = 16.74$,   $\Sigma(p - \bar{p})^2 = 15.29$.

Test, at the 2.5% significance level, the claim that San Marco was cooler than San Polo on that day, giving your null and alternative hypotheses and stating any assumptions necessary for your test to be valid. (C)

3. The contents of a packet of crisps are marked as 30 g. The manufacturer believes that one of their machines is faulty and is issuing too many crisps per packet. A sample of 10 packets is selected at random from this machine and the masses of their contents were:

31.5,  28.9,  30.5,  32.2,  35.5,
34.2,  31.8,  32.8,  29.1,  32.1.

(a) Calculate an unbiased estimate of the population variance.

(b) Is there evidence at the 10% level that the machine is issuing too many crisps per packet? State any distributional assumptions made.

(c) How would the test procedure in part (b) have differed if the population variance was known? (AQA)

4. The customers of a local branch of a bank are invited to comment on various aspects of the service. Their comments are translated into an overall 'satisfaction score'. This score can be taken as normally distributed over the whole population of customers.
A staff training programme has recently been completed. A random sample of scores before the programme was as follows:

126,   93,   114,   107,   98,   112.

A separate random sample of scores after the programme was as follows:

124,   107,   117,   136,   120,   122.

Test at the 5% level of significance the null hypothesis that the mean score is the same after the training programme as it was before against the alternative that the new mean score is higher, stating your assumption concerning the underlying variances. Provide a two-sided 99% confidence interval for the true difference in mean scores. (MEI)

# 12

# The $\chi^2$ significance test

*In this chapter you will learn*

- about the $\chi^2$ distribution
- how to use $\chi^2$ tables and work out the number of degrees of freedom, $\nu$
- how to perform a $\chi^2$ goodness-of-fit test for the following
  - Test 1: a uniform distribution
  - Test 2: a distribution in a given ratio
  - Test 3: a binomial distribution
  - Test 4: a Poisson distribution
  - Test 5: a normal distribution
- how to perform a $\chi^2$ test for independence between variables, using a contingency table
- about applying Yates' continuity correction when $\nu = 1$

**Background knowledge**
*For the goodness of fit tests you will need to know how to calculate probabilities for the uniform, binomial, Poisson and normal distributions.*

## THE $\chi^2$ SIGNIFICANCE TEST

There are two main situations when a $\chi^2$ significance test is used:

($\chi^2$ is pronounced 'kye squared' and is sometimes written chi-squared)

### 1. A $\chi^2$ goodness-of-fit test
This is used when you have some practical data and you want to know how well a particular statistical distribution, such as a binomial or a normal, models that data. The null hypothesis $H_0$ is that the particular distribution does provide a model for the data; the alternative hypothesis $H_1$ is that it does not.

### 2. A $\chi^2$ test for independence (or for association).
This is used when you have some practical data concerning two variables and you want to know whether they are independent or whether there is an association between them. The null hypothesis $H_0$ is that the factors are independent; the alternative hypothesis $H_1$ is that they are not.

Following the pattern for hypothesis tests established in earlier chapters, you assume that the null hypothesis is true and then calculate the frequencies that you would expect to occur based on this assumption. These are denoted by $E$ or $f_e$. These expected frequencies are then compared with the actual (or observed) frequencies, denoted by $O$ or $f_o$.

A **test statistic** involving $O$ and $E$ is calculated. This is often written $X^2$ and, subject to certain conditions, it can be approximated by a $\chi^2$ distribution. Before looking in detail at how the test statistic is calculated and how to perform the test, consider some of the features of the $\chi^2$ distribution.

# The $\chi^2$ distribution

The $\chi^2$ distribution has one parameter, $\nu$, pronounced 'new', and the shape of the distribution is different for different values of $\nu$. Here are some examples.

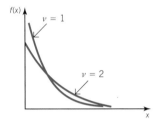
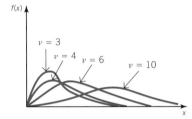

Some features of the $\chi^2$ distribution are:

(a)  It is reverse J-shaped for $\nu = 1$ and $\nu = 2$.
(b)  It is positively skewed for $\nu > 2$.
(c)  The larger the value of $\nu$, the more symmetric the distribution becomes.
(d)  When $\nu$ is large, the distribution is approximately normal.

# Degrees of freedom, $\nu$

The parameter $\nu$ is known as the number of **degrees of freedom** and it is the number of independent variables used in calculating the test statistic. Details of how to find $\nu$ are given in the following text and in the summary table on pages 579 and 590.

# Critical values and levels of significance

The $\chi^2$ test is conducted as a one-tailed (upper tail) test. When carrying out the test, you will want to know whether the calculated value of the test statistic lies in the main bulk of the $\chi^2$ distribution or whether it is in the upper tail **critical (or rejection) region**. The boundary of the critical region is called the **critical value**.

The critical value depends on the **level of significance** of the test. Often a 5% or a 1% level of significance is used and the critical values can be found from $\chi^2$ tables. For example, for a 5% level of significance, the critical value is such that 5% of the area is in the upper tail and the critical value is written $\chi^2_{5\%}(\nu)$, for a particular value of $\nu$.

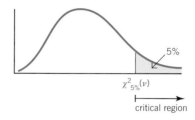

If the test value lies in the critical region, then the null hypothesis $H_0$ is rejected in favour of the alternative hypothesis $H_1$.

# $\chi^2$ tables

$\chi^2$ tables are usually set in one of two formats and you must make sure that you are familiar with the format that you will be given in an examination. In each case the rows refer to the value of $\nu$.

## Format 1: $\chi^2$ tables giving lower-tail probabilities

In this format the column headings indicate the area in the lower tail. For a 5% significance level, since there would be 5% in the upper tail, there would be 95% in the lower tail, so look for the column headed 0.95.

| | | | | | 5% level $\downarrow$ | | 1% level $\downarrow$ | | |
|---|---|---|---|---|---|---|---|---|---|
| $p$ | 0.01 | 0.025 | 0.05 | 0.9 | 0.95 | 0.975 | 0.99 | 0.995 | 0.999 |
| $\nu = 1$ | 0.0001571 | 0.0009821 | 0.003932 | 2.706 | 3.841 | 5.024 | 6.635 | 7.879 | 10.83 |
| 2 | 0.02010 | 0.05064 | 0.1026 | 4.605 | 5.991 | 7.378 | 9.210 | 10.60 | 13.82 |
| 3 | 0.1148 | 0.2158 | 0.3518 | 6.251 | 7.815 | 9.348 | 11.34 | 12.84 | 16.27 |
| 4 | 0.2971 | 0.4844 | 0.7107 | 7.779 | 9.488 | 11.14 | 13.28 | 14.86 | 18.47 |

Examples (highlighted in the extract)

(a) For a significance level of 5% and 4 degrees of freedom, the critical value, $\chi^2_{5\%}(4) = 9.488$.

(b) For a significance level of 1% and 2 degrees of freedom, the critical value $\chi^2_{1\%}(2) = 9.210$.

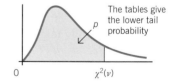

The tables give the lower tail probability

## Format 2: $\chi^2$ tables giving upper-tail probabilities

(This is the format printed on page 651.)

In this format the columns indicate the area in the upper tail, i.e. they give the significance level of the test.

For example, the column headed 0.05 gives the critical value such that 5% of the area lies in the upper tail.

| $v$ | 0.990 | 0.975 | 0.950 | 0.100 | 0.050 | 0.025 | 0.010 | 0.005 |
|---|---|---|---|---|---|---|---|---|
| | | | | 10% level | 5% level | | 1% level | |
| 1 | 0.000 | 0.001 | 0.004 | 2.705 | 3.841 | 5.024 | 6.635 | 7.879 |
| 2 | 0.020 | 0.051 | 0.103 | 4.605 | 5.991 | 7.378 | 9.210 | 10.597 |
| 3 | 0.115 | 0.216 | 0.352 | 6.251 | 7.815 | 9.348 | 11.345 | 12.838 |
| 4 | 0.297 | 0.484 | 0.711 | 7.779 | 9.488 | 11.143 | 13.277 | 14.860 |

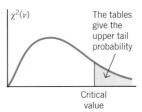

The tables give the upper tail probability

Critical value

Examples (highlighted in the extract)

(c) For a significance level of 5% and 3 degrees of freedom, $\chi^2{}_{5\%}(3) = 7.815$.

(d) For a significance level of 10% and 2 degrees of freedom, $\chi^2{}_{10\%}(2) = 4.605$.

# The test statistic $X^2$

The test statistic $X^2$ uses the values of the observed ($O$) and expected ($E$) frequencies.

$X^2$ is defined as $\sum \dfrac{(O - E)^2}{E}$.

The $X^2$ distribution can be used as an approximation for the distribution of $X^2$, provided none of the expected frequencies ($E$) falls below 5.

You would write $X^2 \sim \chi^2(v)$.

$X^2$ is calculated as follows:

1. Find the difference, $O - E$, for each set of values.

2. Square each difference to obtain $(O - E)^2$. This gives due weight to any particularly large differences and also means that all the values are positive.

3. Divide $(O - E)^2$ by $E$ for each set of values to obtain $\dfrac{(O - E)^2}{E}$.

   This has the effect of standardising that element. In this way, for example, a small difference will be more important when the expected frequency is small than when the expected frequency is large.

4. Finally, find the sum, $\sum \dfrac{(O - E)^2}{E}$. **The smaller this quantity is, the better the fit.**

The following example shows how to calculate $X^2$ for given data in a goodness-of-fit test.

# PERFORMING A $\chi^2$ GOODNESS-OF-FIT TEST

Random numbers consist of lists of the ten digits 0, 1, 2, 3, 4, 5, 6, 7, 8, 9 and are such that each digit has an equal chance of appearing at any stage. Each digit, therefore, has a probability of 0.1 of occurring, i.e. $P(X = x) = 0.1$. This is the discrete uniform distribution (see page 270).

By pressing the random number key ⟨Ran#⟩ on a calculator it is possible to generate a random three-digit number between 0.000 and 0.999. For example

⟨Ran#⟩  0.593  ⟨Ran#⟩ 0.194  ⟨Ran#⟩ 0.106 and so on.

In this case the random digits are 5, 9, 3, 1, 9, 4, 1, 0, 6

Here are 100 digits generated on a calculator.

| 4 | 9 | 8 | 3 | 3 | 3 | 7 | 1 | 3 | 9 |
|---|---|---|---|---|---|---|---|---|---|
| 9 | 9 | 6 | 1 | 8 | 1 | 3 | 6 | 1 | 6 |
| 0 | 3 | 7 | 7 | 3 | 3 | 5 | 4 | 7 | 3 |
| 3 | 8 | 1 | 4 | 2 | 8 | 8 | 6 | 1 | 9 |
| 4 | 5 | 3 | 4 | 9 | 4 | 3 | 8 | 5 | 5 |
| 8 | 6 | 6 | 7 | 5 | 9 | 2 | 6 | 3 | 3 |
| 3 | 8 | 2 | 4 | 8 | 4 | 1 | 9 | 8 | 4 |
| 1 | 4 | 2 | 2 | 1 | 7 | 0 | 8 | 2 | 5 |
| 7 | 5 | 8 | 0 | 4 | 7 | 6 | 9 | 1 | 2 |
| 9 | 7 | 7 | 5 | 3 | 7 | 4 | 0 | 6 | 6 |

A $\chi^2$ goodness of fit test is used to test whether the numbers generated on the calculator are random enough. To make it easier to analyse the data, arrange the digits in a frequency table:

| Digit | 0 | 1 | 2 | 3 | 4 | 5 | 6 | 7 | 8 | 9 | |
|-------|---|---|---|---|---|---|---|---|---|---|---|
| Frequency | 4 | 10 | 7 | 16 | 12 | 8 | 10 | 11 | 12 | 10 | Total 100 |

Make null and alternative hypotheses as follows:

$H_0$: the digits are random
$H_1$: the digits are not random.

Then calculate the frequencies that you would expect if the digits are random:

Expected frequency for each digit is $100 \times 0.1 = 10$

Add another row to the table so that the observed frequencies ($O$) and the expected frequencies ($E$) can be compared.

| Digit | 0 | 1 | 2 | 3 | 4 | 5 | 6 | 7 | 8 | 9 | |
|-------|---|---|---|---|---|---|---|---|---|---|---|
| Observed frequency ($O$) | 4 | 10 | 7 | 16 | 12 | 8 | 10 | 11 | 12 | 10 | Total 100 |
| Expected frequency ($E$) | 10 | 10 | 10 | 10 | 10 | 10 | 10 | 10 | 10 | 10 | Total 100 |

The frequencies can be illustrated by a vertical line graph:

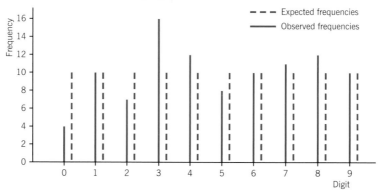

Distribution of 100 random digits generated on a calculator

At first glance the observed frequency for the digit 3 seems much too high and that for the digit 0 seems much too low.

The $\chi^2$ test compares each observed frequency with the corresponding expected frequency.

For each pair, calculate $\dfrac{(O-E)^2}{E}$, then calculate the sum to give the test statistic

$$X^2 = \sum \frac{(O-E)^2}{E}$$

If $X^2 = 0$, then there is exact agreement between the observed and expected frequencies. If $X^2 > 0$, then $O$ and $E$ do not agree exactly; the larger the value of $X^2$ the greater the discrepancy. A low value of $X^2$ implies a good fit, whereas a high value of $X^2$ implies a poor fit.

For the above data,

$$X^2 = \frac{(4-10)^2}{10} + \frac{(10-10)^2}{10} + \frac{(7-10)^2}{10} + \cdots + \frac{(10-10)^2}{10} = 9.4$$

The calculations are usually summarised in a table:

| $O$ | $E$ | $\dfrac{(O-E)^2}{E}$ |
|---|---|---|
| 4 | 10 | 3.6 |
| 10 | 10 | 0 |
| 7 | 10 | 0.9 |
| 16 | 10 | 3.6 |
| 12 | 10 | 0.4 |
| 8 | 10 | 0.4 |
| 10 | 10 | 0 |
| 11 | 10 | 0.1 |
| 12 | 10 | 0.4 |
| 10 | 10 | 0 |
| $\Sigma O = 100$ | $\Sigma E = 100$ | 9.4 |

$$X^2 = \sum \frac{(O-E)^2}{E}$$
$$= 9.4$$

To decide whether the data give a good fit, you need to know whether 9.4 lies in the main body of the distribution or whether it is in the critical (rejection) region in the upper tail. If it lies in the critical region, reject $H_0$.

The boundary of the critical region is found from the appropriate $\chi^2$ distribution which depends on the number of degrees of freedom, $\nu$, the number of independent variables used in calculating the test statistic. It is found as follows:

$\nu$ = number of classes – number of restrictions.

The number of classes is 10 and there is one restriction (that the total of the expected frequencies is 100), so $\nu = 10 - 1 = 9$. Consider the $\chi^2(9)$ distribution.

Say that the test is to be carried out at the 5% significance level. The critical value, $\chi^2_{5\%}(9)$ is found from tables (see page 651).

From tables, $\chi^2_{5\%}(9) = 16.919$

So $H_0$ will be rejected if $X^2 > 16.919$.

Since the calculated value of the test statistic, 9.4, is less than 16.919, it does not lie in the critical region and $H_0$ is not rejected.

On the evidence obtained, you would accept that the digits are true random digits.

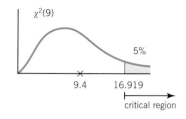

# Summary of the procedure for performing a $\chi^2$ goodness-of-fit test:

For a set of data with observed frequencies $O$:

1.  Make the **null hypothesis** $H_0$ that the data are distributed in a particular way and the **alternative hypothesis** $H_1$ that they are not.

2.  Calculate $E$, the frequencies **expected** if the distribution follows the one given in $H_0$. Note that when calculating $X^2$, small values of $E$ tend to give a large value of $X^2$, so it is advisable to adopt the rule that **expected frequencies below 5 should not be used**. If $E < 5$ for any class, combine adjacent classes to form a class that is sufficiently large. Combine corresponding frequencies in the observed data also and make a revised table.

3.  Work out the number of degrees of freedom, $\nu$, where

    $$\nu = \text{number of classes} - \text{number of restrictions}$$

    Decide on the level of the test and look up the appropriate critical value in tables. For example, for a 5% significance level, look up $\chi^2_{5\%}(\nu)$.

    Use it to state the rejection criteria:

    **If $X^2 > \chi^2_{5\%}(\nu)$** then the test value lies in the critical (rejection) region.
    The discrepancy between the observed and expected frequencies is considered to be too great and $H_0$ is rejected.

    **If $X^2 \leqslant \chi^2_{5\%}(\nu)$** the test value does not lie in the critical region and $H_0$ is not rejected.

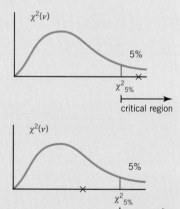

4.  Now calculate $X^2 = \sum \dfrac{(O - E)^2}{E}$.

    Note, however, that if $\nu = 1$, it is advisable to use Yates' continuity correction. In this case the formula is

    $$X^2 = \sum \frac{(|O - E| - 0.5)^2}{E}$$

5. Compare the calculated value of $X^2$ with the critical value. Make your conclusion ($H_0$ is rejected or $H_0$ is not rejected) and relate it to the context of the situation being investigated.

Note that when the value of $X^2$ is *very small*, it is wise to query the reliability of the observed data. This is where the lower tail (left-hand) probabilities might be useful.

For example, suppose that the test involves a $\chi^2(4)$ distribution and that the calculated value of the test statistic is $X^2 = 0.7$.

You can see from the tables on page 651 that $\chi^2_{95\%}(4) = 0.711$, which means that if the null hypothesis is true you would expect a value less than 0.711 from at most 5% of samples, so this would be quite rare. You might wonder whether the observed data have been fiddled.

# TEST 1: GOODNESS-OF-FIT TEST FOR A UNIFORM DISTRIBUTION

## Example 12.1

The table shows the number of employees absent for just one day during a particular period of time.

| Day of the week | Mon | Tues | Wed | Thurs | Fri | |
|---|---|---|---|---|---|---|
| Number of absentees | 121 | 87 | 87 | 91 | 114 | Total 500 |

(a) Find the frequencies expected according to the hypothesis that the number of absentees is independent of the day of the week.

(b) Test at the 5% level whether the differences in the observed and expected data are significant.

## Solution 12.1

1. State $H_0$ and $H_1$.

$H_0$: The number of absentees is independent of the day of the week.
$H_1$: The number of absentees is not independent of the day of the week.

2. Calculate $E$ and check that expected frequencies are greater than 5.

*If the number of absentees is independent of the day of the week then you would expect the total of 500 to be spread uniformly throughout the week.*

Expected number of absentees for any day is 100.

| | Mon | Tues | Wed | Thurs | Fri | |
|---|---|---|---|---|---|---|
| Observed frequencies ($O$) | 121 | 87 | 87 | 91 | 114 | $\Sigma O = 500$ |
| Expected frequencies ($E$) | 100 | 100 | 100 | 100 | 100 | $\Sigma E = 500$ |

3. Work out $v$

*Degrees of freedom $v$.*
There are five classes and there is one restrictions ($\Sigma E = 100$).
Therefore $v = 5 - 1 = 4$, so consider the $\chi^2(4)$ distribution.

**4.** State the level of the test and the rejection criterion.

**5.** Calculate $X^2$.

Perform the test at the 5% level.
From tables $\chi^2_{5\%}(4) = 9.488$, so reject $H_0$ if $X^2 > 9.488$.

| $O$ | $E$ | $\dfrac{(O-E)^2}{E}$ |
|---|---|---|
| 121 | 100 | 4.41 |
| 87 | 100 | 1.69 |
| 87 | 100 | 1.69 |
| 91 | 100 | 0.81 |
| 114 | 100 | 1.96 |
| $\Sigma\, O = 500$ | $\Sigma\, E = 500$ | 10.56 |

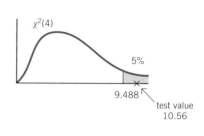

$$X^2 = \sum \frac{(O-E)^2}{E} = 10.56$$

**6.** Make your conclusion.

Since $X^2 > 9.488$, reject $H_0$. **There is evidence, at the 5% level, that the number of absentees on a day is not independent of the day of the week.**

Note that the test does not indicate what the relationship might be. The observed frequencies, however, suggest a tendency towards a greater number of absentees on Mondays and Fridays.

## TEST 2: GOODNESS-OF-FIT TEST FOR A DISTRIBUTION IN A GIVEN RATIO

### Example 12.2

According to a particular genetic theory the number of colour strains, pink, white and blue, in a certain flower should appear in the ratio 3 : 2 : 5. In 100 randomly chosen plants, the corresponding numbers of each colour were 24, 14 and 62. Test at the 1% level whether the differences between the observed and expected frequencies are significant.

### Solution 12.2

**1.** State $H_0$ and $H_1$.

$H_0$: The colours are in the ratio 3 : 2 : 5.
$H_1$: The colours are not in the ratio 3 : 2 : 5.

**2.** Calculate $E$ and check that expected frequencies are greater than 5.

According to the null hypothesis, the colours should be in the ratio 3 : 2 : 5, so the expected frequencies are

pink $\quad \frac{3}{10} \times 100 = 30$ $\qquad$ white $\quad \frac{2}{10} \times 100 = 20$ $\qquad$ blue $\quad \frac{5}{10} \times 100 = 50$

| Colour | Pink | White | Blue | |
|---|---|---|---|---|
| Observed frequency ($O$) | 24 | 14 | 62 | $\Sigma\, O = 100$ |
| Expected frequency ($E$) | 30 | 20 | 50 | $\Sigma\, E = 100$ |

**3.** Work out $v$.

*Degrees of freedom $v$*
There are three classes and there is one restriction ($\Sigma\, E = 100$).
Therefore $v = 3 - 1 = 2$, so consider the $\chi^2(2)$ distribution.

4. State the level of the test and the rejection criterion.

5. Calculate $X^2$.

Perform the test at the 1% level.
From tables $\chi^2{}_{1\%}(2) = 9.210$, so reject $H_0$ if $X^2 > 9.210$.

| $O$ | $E$ | $\dfrac{(O-E)^2}{E}$ |
|---|---|---|
| 24 | 30 | 1.2 |
| 14 | 20 | 1.8 |
| 62 | 50 | 2.88 |
| $\Sigma O = 100$ | $\Sigma E = 100$ | 5.88 |

$$X^2 = \sum \frac{(O-E)^2}{E} = 5.88$$

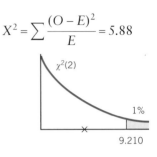

6. Make your conclusion.

Since $X^2 < 9.210$, do not reject $H_0$.
**The differences between the observed and expected frequencies are not significant at the 1% level. The colours are in the ratio $3:2:5$.**

# Exercise 12a  Goodness-of-fit test (uniform and given ratio)

1. A tetrahedral die is thrown 120 times and the number on which it lands is noted.

| Number | 1 | 2 | 3 | 4 | |
|---|---|---|---|---|---|
| Frequency | 35 | 32 | 25 | 28 | Total 120 |

Test, at the 5% level whether the die is fair.

2. From a list of 500 digits, the occurrence of each digit is noted.

| Digit | 0 | 1 | 2 | 3 | 4 | 5 | 6 | 7 | 8 | 9 |
|---|---|---|---|---|---|---|---|---|---|---|
| Frequency | 40 | 58 | 49 | 53 | 38 | 56 | 61 | 53 | 60 | 32 |

Test, at the 1% level, whether the sequence is a random sample from a uniform distribution.

3. The outcomes, $A$, $B$ and $C$, of a certain experiment are thought to occur in the ratio $1:2:1$. The experiment is performed 200 times and the observed frequencies of $A$, $B$ and $C$ are 36, 115 and 49 respectively. Is the difference in the observed and expected results significant? Test at the 5% level.

4. According to genetic theory the number of colour strains, red, yellow, blue and white, in a certain flower should appear in proportions $4:12:5:4$. Observed frequencies of red, yellow, blue and white strains amongst 800 plants were 110, 410, 150, 130 respectively. Are these differences from the expected frequencies significant at the 5% level? If the number of plants had been 1600 and the observed frequencies 220, 820, 300, 260, would the difference have been significant at the 5% level?

(C)

5. It is thought that each of the 8 outcomes of an experiment is equally likely to occur. When the experiment is performed 400 times, the observed frequencies are 45, 42, 55, 53, 40, 62, 47 and 56. Perform a test at the 1% level to investigate the validity of the theory.

6. In a particular subject students are set multiple choice questions each of which contain five alternatives A, B, C, D and E. A teacher suggests that when students do not know the correct answer they are twice as likely to choose one of B, C or D than to choose A or E. For 160 questions where it was known that the student answered without knowing the correct answer, A, B, C, D, E were chosen 23, 45, 36, 43 and 13 times respectively. Is there evidence, at the 5% level, to support the teacher's theory?

7. For a given set of data the observed and expected frequencies are shown:

| Result | 1 | 2 | 3 | 4 | 5 |
|---|---|---|---|---|---|
| Observed frequency | 30 | 31 | 42 | 40 | 57 |
| Expected frequency | 38 | 45 | 36 | 36 | 45 |

Are the differences between the observed and expected frequencies significant at the 1% level?

8. The random variable $Y$ has a $\chi^2$ distribution with eight degrees of freedom. Find $y$, such that $P(Y > y) = 0.05$.

(L)

9. During the course of one year a tutor marked 111 assignments. The grades he awarded and the comparable national proportions are given in the table:

| Grade | A | B | C | D |
|---|---|---|---|---|
| Number he awarded | 86 | 18 | 6 | 1 |
| National proportion | 71% | 16% | 7% | 6% |

Calculate the expected numbers (to one decimal place) based on the national proportions. The $\chi^2$ goodness of fit test requires the summation of terms of the form

$$\frac{(O - E)^2}{E}$$

where $O$ and $E$ are observed and expected frequencies. Suggest reasons why

(a) the difference between $O$ and $E$ is used
(b) this difference is squared, and
(c) the squared difference is divided by $E$.

Test, at the 5% level, whether there is any difference between the tutor's and the national awarding of grades. State your conclusions clearly. (O)

10. A calibrated instrument is used over a wide range of values. To assess the operator's ability to read the instrument accurately, the final digit in each of 700 readings was noted. The results are tabulated below.

| Final digit | Frequency |
|---|---|
| 0 | 75 |
| 1 | 63 |
| 2 | 50 |
| 3 | 58 |
| 4 | 73 |
| 5 | 95 |
| 6 | 96 |
| 7 | 63 |
| 8 | 46 |
| 9 | 81 |

Use an approximate $\chi^2$ statistic to test whether there is any evidence of bias in the operator's reading of the instrument. Use a 5% significance level and state your null and alternative hypotheses. (L)

11. The grades in a statistics examination for a group of students were as follows.

| Grade | A | B | C | D | E |
|---|---|---|---|---|---|
| Number of students | 14 | 18 | 32 | 20 | 16 |

Test the hypothesis that the distribution of grades is uniform. Use a 5% level of significance. (L)

12. An ordinary die is thrown 120 times and each time the number on the uppermost face is noted. The results are as follows:

| Number on die | 1 | 2 | 3 | 4 | 5 | 6 |
|---|---|---|---|---|---|---|
| Frequency | 14 | 16 | 24 | 22 | 24 | 20 |

Is the die fair? Test at the 10% level.

13. In a certain town an investigation was carried out into accidents in the home to children under 12 years of age. The numbers of reported accidents and the ages of the children concerned are summarised in the table.

| Group | Age of child (yrs) | No. of accidents |
|---|---|---|
| A | 0 to < 2 | 42 |
| B | 2 to < 4 | 52 |
| C | 4 to < 6 | 28 |
| D | 6 to < 8 | 20 |
| E | 8 to < 10 | 18 |
| F | 10 to < 12 | 16 |

(a) State the modal class.
(b) Calculate, to the nearest month, the mean age and the standard deviation of the distribution of ages.
(c) Draw a cumulative frequency curve, and from it estimate, to the nearest month, the median, and the interquartile range for the ages of all children under 12 years of age concerned in reported accidents in the home. State, giving a reason, whether you consider the mean, the mode or the median best represents the average age for accidents in the home to children under 12 years of age.
(d) An investigator believes that children in the groups A, B, C, D, E, F are likely to have accidents in the home in the ratios 2 : 2 : 1 : 1 : 1 : 1 respectively. Use a $\chi^2$ test at a 5% significance level to decide whether or not this belief is justified. (L)

# TEST 3: GOODNESS-OF-FIT TEST FOR A BINOMIAL DISTRIBUTION

## Example 12.3

A farmer kept a record of the number of heifer calves born to each cow during the first five years of breeding of the cow. The results are summarised in the table:

| Number of heifers | 0 | 1 | 2 | 3 | 4 | 5 |
|---|---|---|---|---|---|---|
| Number of cows | 4 | 19 | 41 | 52 | 26 | 8 |

(a) Test, at the 5% level of significance, whether or not the binomial distribution with parameters $n = 5$, $p = 0.5$, is an adequate model for these data.
(b) Explain briefly what changes you would make in your analysis if you were testing whether or not the binomial distribution with $n = 5$ and unspecified $p$ fitted the data. (AEB)

## Solution 12.3

1. State $H_0$ and $H_1$.

(a) Let $X$ be the number of heifer calves born to a cow in the first five years of breeding.

$H_0$: $X \sim B(5, 0.5)$
$H_1$: $X$ is not distributed in this way.

2. Calculate $E$ and check that expected frequencies are greater than 5.

To calculate the binomial probabilities, use cumulative probability tables which give $P(X \leqslant x)$.

Alternatively, calculate the probabilities using

$$P(X = x) = {}^5C_x(0.5)^{5-x}(0.5)^x = {}^5C_x(0.5)^5$$

The total number of cows is 150, so the expected frequencies are found by multiplying $P(X = x)$ by 150.

*Note on accuracy: When calculating it is often necessary to approximate, say to the nearest integer or to one decimal place. If you have memory facilities on your calculator for retaining several numbers then you may prefer to do so.*

Using tables:                                    (Extract from page 645)

| $P(X = x)$ | $E = 150 \times P(X = x)$ | $X \sim B(5, 0.5)$ | |
|---|---|---|---|
| | | $n = 5$ | $P(X \leqslant r)$ |
| $P(X = 0) = 0.0313$ | 4.7 | $r = 0$ | 0.0313 |
| $P(X = 1) = 0.1875 - 0.0313 \doteq 0.1562$ | 23.4 | 1 | 0.1875 |
| $P(X = 2) = 0.5 - 0.1875 = 0.3125$ | 46.9 | 2 | 0.5000 |
| $P(X = 3) = 0.8125 - 0.5 = 0.3125$ | 46.9 | 3 | 0.8125 |
| $P(X = 4) = 0.9688 - 0.8125 = 0.1563$ | 23.4 | 4 | 0.9688 |
| $P(X = 5) = 1 - 0.9688 = 0.0312$ | 4.7 | 5 | 1.0000 |

*Check on size of expected frequencies:*
Since the expected frequencies for the first and last classes are less than 5, combine them with the next classes.

Do a revised table for the expected frequencies and also show the corresponding observed frequencies:

| Number of heifers $x$ | 0 or 1 | 2 | 3 | 4 or 5 | |
|---|---|---|---|---|---|
| Observed frequencies (O) | 23 | 41 | 52 | 34 | $\Sigma O = 150$ |
| Expected frequencies (E) | 28.1 | 46.9 | 46.9 | 28.1 | $\Sigma E = 150$ |

**3. Work out $\nu$.**

*Degrees of freedom $\nu$*
There are four classes and there is one restriction ($\Sigma E = 150$).
Therefore $\nu = 4 - 1 = 3$, so consider the $\chi^2(3)$ distribution.

**4. State the level of the test and the rejection criterion.**

Perform the test at the 5% level.
From tables $\chi^2{}_{5\%}(3) = 7.815$, so reject $H_0$ if $X^2 > 7.815$.

**5. Calculate $X^2$.**

| $O$ | $E$ | $\dfrac{(O-E)^2}{E}$ |
|---|---|---|
| 23 | 28.1 | 0.925 ... |
| 41 | 46.9 | 0.742 ... |
| 52 | 46.9 | 0.554 ... |
| 34 | 28.1 | 1.238 ... |
| $\Sigma O = 150$ | $\Sigma E = 150$ | 3.461 ... |

$$X^2 = \sum \frac{(O-E)^2}{E} = 3.46 \quad \text{(2 d.p.)}$$

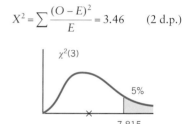

$\chi^2(3)$

5%

7.815

**6. Make your conclusion.**

Since $X^2 < 7.815$, do not reject $H_0$.
**The binomial distribution with $n = 5$ and $p = 0.5$ is an adequate model for the data.**

(b) If you want to test whether the distribution $B(5, p)$ provides an adequate model, *with p unspecified*, you would need to *estimate p* from the data using the fact that the mean of a binomial distribution is $np$.

From the data

$$\bar{x} = \frac{\Sigma fx}{\Sigma f} = \frac{401}{105} = 2.673\ldots$$

Since

$$\bar{x} = np$$
$$2.673 = 5p$$
$$p = 0.535 \quad \text{(3 d.p.)}$$

So the null hypothesis would be $H_0: X \sim B(5, 0.535)$ and the expected frequencies would be calculated using $p = 0.535$.

When working out $\nu$, the number of degrees of freedom, you would take into account that there are now *two* restrictions, one is that $\Sigma E = 150$ (as before) and the other is that $p$ is estimated from the sample.

Try working through this test. You should find that $\nu = 5 - 2 = 3$, $X^2 \approx 0.85$ (depending on degree of approximation in your calculations) and $H_0$ not rejected.

# TEST 4: GOODNESS OF FIT FOR A POISSON DISTRIBUTION

## Example 12.4

An analysis if the number of goals scored by the local football team gave the following results:

| Goals per match (x) | 0 | 1 | 2 | 3 | 4 | 5 | 6 | 7 | |
|---|---|---|---|---|---|---|---|---|---|
| Number of matches | 14 | 18 | 29 | 18 | 10 | 7 | 3 | 1 | Total 100 |

Carry out a $\chi^2$ goodness of fit test at the 10% significance level to determine whether or not the above distribution can be reasonably modelled by a Poisson distribution with parameter 2.

## Solution 12.4

1. State $H_0$ and $H_1$.

Let X be the number of goals scored in a match.

$H_0$: $X \sim \text{Po}(2)$
$H_1$: $X$ is not distributed in this way.

2. Calculate $E$ and check that expected frequencies are greater than 5.

To calculate the Poisson probabilities, use cumulative probability tables which give $P(X \leqslant x)$.

Alternatively, calculate the probabilities using

$$P(X = x) = e^{-2} \frac{2^x}{x!}.$$

The total number of matches is 100, so the expected frequencies are found by multiplying $P(X = x)$ by 100.

Using tables:

(Extract from page 647)

$X \sim \text{Po}(2)$

| $P(X = x)$ | $E = 100P(X = x)$ |
|---|---|
| $P(X = 0) = 0.1353$ | 13.53 |
| $P(X = 1) = 0.4060 - 0.1353 = 0.2707$ | 27.07 |
| $P(X = 2) = 0.6767 - 0.4060 = 0.2707$ | 27.07 |
| $P(X = 3) = 0.8571 - 0.6767 = 0.1804$ | 18.04 |
| $P(X = 4) = 0.9473 - 0.8571 = 0.0902$ | 9.02 |
| $P(X = 5) = 0.9834 - 0.9473 = 0.0361$ | 3.61 |
| $P(X = 6) = 0.9955 - 0.9834 = 0.0121$ | 1.21 |
| $P(X = 7 \text{ or more}) = 1 - 0.9955 = 0.0045$ | 0.45 |
| | $\Sigma E = 100$ |

| $\lambda = 2.0$ | $P(X \leqslant x)$ |
|---|---|
| $r = 0$ | 0.1353 |
| 1 | 0.4060 |
| 2 | 0.6767 |
| 3 | 0.8571 |
| 4 | 0.9473 |
| 5 | 0.9834 |
| 6 | 0.9955 |
| 7 | 0.9989 |
| 8 | 0.9998 |
| 9 | 1.0000 |
| 10 | |
| 11 | |

*Check on size of expected frequencies:*
The $x^2$ test is not valid for expected frequencies less than 5, so combine the last three classes to give 5 or more goals.

*Revised table:*

| $x$ | 0 | 1 | 2 | 3 | 4 | 5 or more | |
|---|---|---|---|---|---|---|---|
| $O$ | 14 | 18 | 29 | 18 | 10 | 11 | $\Sigma O = 100$ |
| $E$ | 13.53 | 27.07 | 27.07 | 18.04 | 9.02 | 5.27 | $\Sigma E = 100$ |

3. Work out $v$.

*Degrees of freedom $v$*
There are six classes and there is one restriction ($\Sigma E = 100$).
Therefore $v = 6 - 1 = 5$, so consider the $\chi^2(5)$ distribution.

4. State the level of the test and the rejection criterion.

Perform the test at the 10% level.
From tables $\chi^2_{10\%}(5) = 9.236$, so reject $H_0$ if $X^2 > 9.236$

5. Calculate $X^2$.

| $O$ | $E$ | $\dfrac{(O-E)^2}{E}$ |
|---|---|---|
| 14 | 13.53 | 0.016 ... |
| 18 | 27.07 | 3.038 ... |
| 29 | 27.07 | 0.137 ... |
| 18 | 18.04 | 0.000 ... |
| 10 | 9.02 | 0.106 ... |
| 11 | 5.27 | 6.230 ... |
| $\Sigma O = 100$ | $\Sigma E = 100$ | 9.529 ... |

$$X^2 = \sum \frac{(O-E)^2}{E} = 9.53 \quad \text{(2 d.p.)}$$

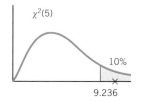

6. Make your conclusion.

Since $X^2 > 9.236$, reject $H_0$.
**The number of goals per match cannot be modelled by a Poisson distribution with parameter 2.**

---

## Example 12.5

Can the data of Example 12.4 be modelled by a Poisson distribution having the same mean as the observed data? Test at the 10% level.

## Solution 12.5

For the observed data, $\bar{x} = \dfrac{\Sigma fx}{\Sigma f} = \dfrac{230}{100} = 2.3$.

The null hypothesis is that the distribution is Poisson, with parameter 2.3,

i.e. $H_0: X \sim \text{Po}(2.3)$
$H_1: X$ is not distributed in this way.

The probabilities are found from cumulative tables or by calculating using

$$P(X = x) = e^{-2.3} \frac{(2.3)^x}{x!}, \quad x = 0, 1, 2, ...$$

The expected frequencies are given by $100 \times P(X = x)$.

Using the formula:

| $P(X \doteq x)$ to 4 d.p. | $E = 100P(X = x)$ |
|---|---|
| $P(X = 0) = 0.10025 \ldots$ | 10.03 |
| $P(X = 1) = 0.2306$ | 23.06 |
| $P(X = 2) = 0.2652$ | 26.52 |
| $P(X = 3) = 0.2033$ | 20.33 |
| $P(X = 4) = 0.1169$ | 11.69 |
| $P(X = 5) = 0.0538$ | 5.38 |
| $P(X = 6) = 0.0206$ | 2.06 |
| $P(X = 7$ or more$) = 0.0099$ | 0.99 |
| | $\Sigma E = 100$ |

The last three classes are combined to give a class with expected frequency greater than 5.

*Revised table:*

| $x$ | 0 | 1 | 2 | 3 | 4 | 5 or more | |
|---|---|---|---|---|---|---|---|
| $O$ | 14 | 18 | 29 | 18 | 10 | 11 | $\Sigma O = 100$ |
| $E$ | 10.03 | 23.06 | 26.52 | 20.33 | 11.69 | 8.43 | $\Sigma E = 100$ |

*Degrees of freedom v:*
There are six classes.

There are two restrictions:

- $\Sigma E = 100$,
- the mean of the Poisson distribution has to be estimated from the data.

Therefore $v = 6 - 2 = 4$, so consider the $\chi^2(4)$ distribution.

Perform the test at the 10% level.
From tables $\chi^2_{10\%}(4) = 7.779$, so reject $H_0$ if $X^2 > 7.779$.

Calculating $X^2$

| $O$ | $E$ | $\dfrac{(O - E)^2}{E}$ |
|---|---|---|
| 14 | 10.03 | 1.571 ... |
| 18 | 23.06 | 1.110 ... |
| 29 | 26.52 | 0.231 ... |
| 18 | 20.33 | 0.267 ... |
| 10 | 11.69 | 0.244 ... |
| 11 | 8.43 | 0.783 ... |
| $\Sigma O = 100$ | $\Sigma E = 100$ | 4.208 ... |

$$X^2 = \sum \frac{(O - E)^2}{E} = 4.208 \quad (3 \text{ d.p.})$$

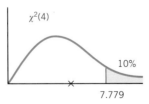

$\chi^2(4)$

10%

7.779

Since $X^2 < 7.779$, do not reject $H_0$.
**The number of goals per match can be modelled by a Poisson distribution with the same mean as the observed data.**

# TEST 5: GOODNESS-OF-FIT TEST FOR A NORMAL DISTRIBUTION

### Example 12.6

The height, in centimetres, gained by a conifer in its first year of planting is denoted by the random variable X. The value of X is measured for a random sample of 86 conifers and the results obtained are summarised in the table:

| X | <35 | 35–45 | 45–55 | 55–65 | >65 |
|---|---|---|---|---|---|
| Observed frequency | 10 | 18 | 28 | 18 | 12 |

(a) Assuming that X is modelled by a $N(50, 15^2)$ distribution, calculate the expected frequencies for each of the five classes.
(b) Carry out a $\chi^2$ goodness of fit analysis to test, at the 5% level, the hypothesis that X can be modelled as in (a). (C)

### Solution 12.6

(a) $X \sim N(50, 15^2)$
Standardise each X value,
e.g. when $x = 35$
$$z = \frac{x - \mu}{\sigma} = \frac{35 - 50}{15} = -1$$

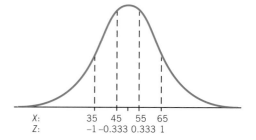

| X: | 35 | 45 | 55 | 65 |
|---|---|---|---|---|
| Z: | −1 | −0.333 | 0.333 | 1 |

Notice that there is symmetry in the diagram.

| Probabilities | $E$ = probability × 86 |
|---|---|
| $P(X < 35) = P(Z < -1) = 1 - 0.8413 = 0.1587$ | 13.7 |
| $P(35 < X < 45) = P(-1 < Z < -0.333) = 0.8413 - 0.6304 = 0.2109$ | 18.1 |
| $P(45 < X < 55) = P(-0.333 < Z < 0.333) = 2 \times 0.6304 - 1 = 0.2608$ | 22.4 |
| $P(55 < X < 65) = 0.2109$  (by symmetry) | 18.1 |
| $(P > 65) = 0.1587$  (by symmetry) | 13.7 |

$$\Sigma E = 86$$

Note that the expected frequencies have been given to 1 d.p.

(b) $\chi^2$ goodness-of-fit test:

1. State $H_0$ and $H_1$.

$H_0$: $X \sim N(50, 15^2)$
$H_1$: $X$ is not distributed in this way.

2. Calculate $E$ and check that expected frequencies are greater than 5.

| x | <35 | 35–45 | 45–55 | 55–65 | >65 | |
|---|---|---|---|---|---|---|
| O | 10 | 18 | 28 | 18 | 12 | $\Sigma O = 86$ |
| E | 13.7 | 18.1 | 22.4 | 18.1 | 13.7 | $\Sigma E = 86$ |

(Note that all expected frequencies are greater than 5 so there is no need to combine classes)

3. Work out $\nu$.

*Degrees of freedom $\nu$*
There are five classes and one restriction ($\Sigma E = 86$).
Therefore $\nu = 5 - 1 = 4$, so consider the $\chi^2(4)$ distribution.

**4.** State the level of the test and the rejection criterion.

Perform the test at the 5% level.
From tables $\chi^2_{5\%}(4) = 9.488$, so reject $H_0$ if $X^2 > 9.488$

**5.** Calculate $X^2$.

| $O$ | $E$ | $\dfrac{(O - E)^2}{E}$ |
|-----|-----|-----|
| 10 | 13.7 | 0.999 ... |
| 18 | 18.1 | 0.0005 ... |
| 28 | 22.4 | 1.4 ... |
| 18 | 18.1 | 0.0005 ... |
| 12 | 13.7 | 0.210 ... |
| $\Sigma O = 86$ | $\Sigma E = 86$ | 2.611 ... |

$$X^2 = \sum \frac{(O - E)^2}{E} = 2.61 \quad \text{(2 d.p.)}$$

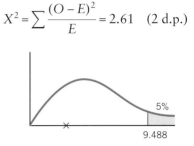

**6.** Make your conclusion.

Since $X^2 < 9.488$, do not reject $H_0$.
**The normal model $N(50, 15^2)$ is a suitable model.**

## Example 12.7

A weaving mill sells lengths of cloth with a nominal length of 70 m. A customer measured 100 lengths and obtained the following frequency distribution:

| Length (m) | 61–67 | 67–69 | 69–71 | 71–73 | 73–75 | 75–81 |
|-----------|-------|-------|-------|-------|-------|-------|
| Frequency | 1 | 16 | 26 | 19 | 20 | 18 |

Use a $\chi^2$ test at the 5% level of significance to show that the normal distribution is not an adequate model for the data. (*AEB*)

## Solution 12.7

The null hypothesis is that the distribution is normal, but since neither the mean nor the variance is given, they have to be estimated from the data.

| Mid-interval value $x$ | 64 | 68 | 70 | 72 | 74 | 78 |
|-----------------------|----|----|----|----|----|----|
| Frequency | 1 | 16 | 26 | 19 | 20 | 18 |

From the calculator
$\hat{\mu} = \bar{x} = 72.24$     (see page 32)
$\hat{\sigma}^2 = 11.578$   (3 d.p.)   (see page 449)

$\chi^2$ goodness-of-fit test

**1.** State $H_0$ and $H_1$.

$H_0$: $X \sim N(72.24, 11.578)$
$H_1$: $X$ is not distributed in this way.

**2.** Calculate $E$ and check that expected frequencies are greater than 5.

Standardise the boundary values of the intervals (to 3 d.p.) using $z = \dfrac{x - \mu}{\sigma} = \dfrac{x - 72.24}{\sqrt{11.578}}$

when $x = 61$, $z = \dfrac{61 - 72.24}{\sqrt{11.578}} = -3.303$.

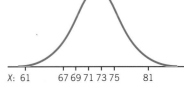

NOTE: $P(X < 61) = P(Z < -3.303) \to 0$, so take the first class as $X < 67$.

| Probabilities | $E = \text{prob} \times 100$ |
|---|---|
| $P(X < 67) = P(Z < -1.540) = 1 - 0.9382 = 0.0618$ | 6.18 |
| $P(67 < X < 69) = P(-1.540 < z < -0.952) = 0.9382 - 0.8294 = 0.1088$ | ·10.88 |
| $P(69 < X < 71) = P(-0.952 < Z < -0.364) = 0.8294 - 0.6421 = 0.1873$ | 18.73 |
| $P(71 < X < 73) = P(-0.364 < Z < 0.223) = 0.6421 + 0.5883 - 1 = 0.2254$ | 22.54 |
| $P(73 < X < 75) = P(0.223 < Z < 0.811) = 0.7913 - 0.5883 = 0.208$ | 20.8 |
| $P(75 < X < 81) = P(0.811 < Z < 2.574) = 0.995 - 0.7913 = 0.2037$ | 20.37 |
| $P(X > 81) = P(Z > 2.574) = 1 - 0.995 = 0.005$ | 0.05 |

↑
Combine the last two classes to give $X > 75$.

| $x$ | <67 | 67–69 | 69–71 | 71–73 | 73–75 | 75 and over | |
|---|---|---|---|---|---|---|---|
| $O$ | 1 | 16 | 26 | 19 | 20 | 18 | $\Sigma O = 100$ |
| $E$ | 6.18 | 10.88 | 18.73 | 22.54 | 20.8 | 21.42 | $\Sigma E = 100$ |

3. Work out $\nu$.

*Degrees of freedom $\nu$*
There are six classes.
There are three restrictions:

- $\Sigma E = 100$

- The mean of the normal distribution has been estimated from the data.

- The variance of the normal distribution has been estimated from the data.

Therefore $\nu = 6 - 3 = 3$, so consider the $\chi^2(3)$ distribution.

4. State the level of the test and the rejection criterion.

Perform the test at the 5% level.
From tables $\chi^2_{5\%}(3) = 7.815$, so reject $H_0$ if $X^2 > 7.815$

5. Calculate $X^2$.

| $O$ | $E$ | $\dfrac{(O-E)^2}{E}$ |
|---|---|---|
| 1 | 6.18 | 4.341 … |
| 16 | 10.88 | 2.409 … |
| 26 | 18.73 | 2.821 … |
| 19 | 22.54 | 0.555 … |
| 20 | 20.8 | 0.030 … |
| 18 | 21.42 | 0.546 … |
| $\Sigma O = 100$ | $\Sigma E = 100$ | 10.705 … |

$$X^2 = \sum \frac{(O-E)^2}{E} = 10.7 \quad (1 \text{ d.p.})$$

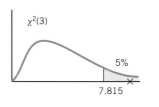

6. Make your conclusion.

Since $X^2 > 7.815$, reject $H_0$.
**The normal distribution is not an adequate model for the data.**

# Summary of the number of degrees of freedom for goodness-of-fit tests

| Distribution | | $\nu$ |
|---|---|---|
| Uniform | | $\nu = n - 1$ |
| Given ratio | | $\nu = n - 1$ |
| Binomial | (a) if $p$ is known | $\nu = n - 1$ |
| | (b) if $p$ is unknown and it is estimated from the observed frequencies using $\bar{x} = np$ | $\nu = n - 2$ |
| Poisson | (a) if $\lambda$ is known | $\nu = n - 1$ |
| | (b) if $\lambda$ is unknown and it is estimated from the observed frequencies using $\bar{x} = \lambda$ | $\nu = n - 2$ |
| Normal | (a) if $\mu$ and $\sigma^2$ are known | $\nu = n - 1$ |
| | (b) if $\mu$ and $\sigma^2$ are unknown and are estimated from the observed frequencies | $\nu = n - 3$ |

## Exercise 12b $\quad \chi^2$ goodness-of-fit tests for binomial, Poisson and normal distributions

1. Perform a $\chi^2$ test to investigate whether the following is drawn from a binomial distribution with $p = 0.3$. Use a 5% level of significance.

| $x$ | 0 | 1 | 2 | 3 | 4 | 5 |
|---|---|---|---|---|---|---|
| $f$ | 12 | 39 | 27 | 15 | 4 | 3 |

2. A six-sided die with faces numbered as usual from 1 to 6 was thrown five times and the number of sixes was recorded. The experiment was performed 200 times, with the following results:

| $x$ | 0 | 1 | 2 | 3 | 4 | 5 |
|---|---|---|---|---|---|---|
| Frequency | 66 | 82 | 40 | 10 | 2 | 0 |

On this evidence, would you consider the die to be biased? Fit a suitable distribution to the data and test and comment on the goodness of fit.

(*MEI*)

3. Under what circumstances would you expect a variable, $X$, to have a binomial distribution? What is the mean of $X$ if it has a binomial distribution with parameters $n$ and $p$?

A new fly spray is applied to 50 samples each of five flies and the number of living flies counted after one hour. The results were as follows:

| Number living | 0 | 1 | 2 | 3 | 4 | 5 |
|---|---|---|---|---|---|---|
| Frequency | 7 | 20 | 12 | 9 | 1 | 1 |

Calculate the mean number of living flies per sample and hence an estimate for $p$, the probability of a fly surviving the spray. Using your estimate calculate the expected frequencies (each correct to one decimal place) corresponding to a binomial distribution and perform a $\chi^2$ goodness-of-fit test using a 5% significance level.

4. Two dice were thrown 216 times, and the number of sixes at each throw were counted. The results were:

| No. of sixes | 0 | 1 | 2 | |
|---|---|---|---|---|
| Frequency | 130 | 76 | 10 | Total 216 |

Test the hypothesis that the distribution is binomial with the parameter $p = \frac{1}{6}$. Explain how the test would be modified if the hypothesis to be tested is that the distribution is binomial with the parameter $p$ unknown. (Do not carry out the test.)

(*O*)

5. Smallwoods Ltd. run a weekly football pools competition. One part of this involves a fixed-odds contest where the entrant has to forecast correctly the result of each of five given matches. In the event of a fully correct forecast the entrant is paid out at odds of 100 to 1. During the last two years Miss Fortune has entered this fixed-odds contest 80 times. The table below summarises her results.

| Number of matches correctly forecast per entry ($x$) | Number of entries with $x$ correct forecasts ($f$) |
|---|---|
| 0 | 8 |
| 1 | 19 |
| 2 | 25 |
| 3 | 22 |
| 4 | 5 |
| 5 | 1 |

(a) Find the frequencies of the number of matches correctly forecast per entry given by a binomial distribution having the same mean and total as the observed distribution.
(b) Use the $\chi^2$ distribution and a 10% level of significance to test the adequacy of the binomial distribution as a model for these data.
(c) On the evidence before you, and assuming that the point of entering is to win money, would you advise Miss Fortune to continue with this competition and why? (AEB)

6. Samples of size 5 are selected regularly from a production line and tested. During one week 500 samples are taken and the number of defective items in each sample is recorded.

| Number of defectives, $x$ | 0 | 1 | 2 | 3 | 4 | 5 |
|---|---|---|---|---|---|---|
| Frequency, $f$ | 170 | 180 | 120 | 20 | 8 | 2 |

(a) It is suggested that a binomial model, with mean the same as the observed data, can be used. Find the frequencies expected by this model.
(b) Test whether this binomial model is a good one. Use a 5% level of significance.

7. A group of students are performing an experiment where 20 drawing pins are dropped randomly on to the floor and the number landing point down is counted. The procedure is then repeated several times. Describe the assumptions you would need to make in order to be satisfied with modelling this situation by a binomial distribution. The experiment was carried out until the students had 50 observations; their results are given in the table:

| Number landing point down | Frequency |
|---|---|
| 3 | 2 |
| 4 | 2 |
| 5 | 5 |
| 6 | 7 |
| 7 | 17 |
| 8 | 8 |
| 9 | 6 |
| 10 | 1 |
| 11 | 2 |

(a) Calculate the mean number landing point down. Hence show that an estimate for the probability of a drawing pin landing point down is 0.35.
(b) What are the parameters of the appropriate binomial distribution for these data? Calculate the probability of exactly eight landing point down, and hence write down, accurate to one decimal place, its expected frequency.
(c) Using appropriate tables, find, making your method clear, the expected number of times five or fewer pins would land point down.
(d) The chi-squared goodness-of-fit test can be used to judge how well data follow a distribution. Group the above data in the following manner and evaluate the missing expected or observed frequencies:

| Number of pins | ≤5 | 6 | 7 | 8 | ≥9 |
|---|---|---|---|---|---|
| Expected | | 8.6 | | | 11.8 |
| Observed | 9 | 7 | 17 | | |

Calculate the value of the chi-squared statistic for this data.
(e) How many degrees of freedom does your test have? By referring to your tables carry out the test and make your findings clear. (O)

8. A local council has records of the number of children and the number of households in its area. It is therefore known that the average number of children per household is 1.40. It is suggested that the number of children per household can be modelled by the Poisson distribution with parameter 1.40. In order to test this, a random sample of 1000 households is taken, giving the following data.

| Number of children | 0 | 1 | 2 | 3 | 4 | 5+ |
|---|---|---|---|---|---|---|
| Number of households | 273 | 361 | 263 | 78 | 21 | 4 |

(a) Find the corresponding expected frequencies obtained from the Poisson distribution with parameter 1.40.

(b) Carry out a $\chi^2$ test, at the 5% level of significance, to determine whether or not the proposed model should be accepted. State clearly the null and alternative hypotheses being tested and the conclusion which is reached. *(MEI)*

9. The numbers of cars passing a check-point during 100 intervals, each of time 5 minutes, were noted:

| Number of cars | Frequency |
|---|---|
| 0 | 5 |
| 1 | 23 |
| 2 | 23 |
| 3 | 25 |
| 4 | 14 |
| 5 | 10 |
| 6 or more | 0 |

Fit a Poisson distribution to these data and test the goodness of fit.

10. During the weaving of cloth the thread sometimes breaks. 147 lengths of thread of equal length were observed during weaving and the table records the number of these threads for which the indicated number of breaks occurred.

| Number of breaks per thread | 0 | 1 | 2 | 3 | 4 | 5 |
|---|---|---|---|---|---|---|
| Number of threads | 48 | 46 | 30 | 12 | 9 | 2 |

Fit a Poisson distribution to the data and examine whether the deviation between theory and experiment is significant. *(MEI)*

11. A shop that repairs television sets keeps a record of the number of sets brought in for repair each day. The numbers brought in during a random sample of 40 days were as follows.

4 0 0 0 2 1 1 0 0 0   0 1 1 0 3 0 0 0 1 0
4 0 0 0 0 0 2 0 1 0   0 0 0 1 1 1 0 2 0 0

Test, at the 5% significance level, the hypothesis that these numbers are observations from a Poisson distribution. *(C)*

12. The table gives the distribution for the number of heavy rainstorms reported by 330 weather stations in the United States of America over a one-year period.

| Number of rainstorms ($x$) | Number of stations ($f$) reporting $x$ rainstorms |
|---|---|
| 0 | 102 |
| 1 | 114 |
| 2 | 74 |
| 3 | 28 |
| 4 | 10 |
| 5 | 2 |
| more than 5 | 0 |

(a) Find the expected frequencies of rainstorms given by the Poisson distribution having the same mean and total as the observed distribution.

(b) Use the $\chi^2$ distribution to test the adequacy of the Poisson distribution as a model for these data. *(AEB)*

13. Over a period of 50 weeks the numbers of road accidents reported to a police station are shown in the table below.

| No. of accidents | 0 | 1 | 2 | 3 |
|---|---|---|---|---|
| No. of weeks | 23 | 13 | 10 | 4 |

Find the mean number of accidents per week. Use this mean, a 5% level of significance, and your table of $\chi^2$ to test the hypothesis that these data are a random sample from a population with a Poisson distribution. *(O&C)*

14. (a) The data in the following table are the result of counting radioactive events in five-second intervals:

| Number of events | 0 | 1 | 2 | $\geqslant 3$ |
|---|---|---|---|---|
| Number of observations | 5 | 14 | 13 | 8 |

Show that the mean number of events in a five-second interval is 1.7 (taking the group with frequency 8 to have a mean of 3.5).

(b) Write down the probability of 0, 1, 2, $\geqslant 3$ events for a Poisson distribution with mean 1.7. Hence obtain to one decimal place the expected frequencies.

(c) Use the chi-squared goodness of fit test to assess whether it is reasonable to claim that the data come from a Poisson distribution. Make your method clear and conduct your test at the 10% level.

(d) A student conducting a similar experiment found the chi-squared statistic for his results was 0.015. What conclusions do you draw from this value? *(O)*

15. For a period of six months 100 similar hamsters were given a new type of feedstuff. The gains in mass are recorded in the table below:

| Gain in mass (g) $x$ | Observed frequency $f$ |
|---|---|
| $-\infty < x \leqslant -10$ | 3 |
| $-10 < x \leqslant -5$ | 6 |
| $-5 < x \leqslant 0$ | 9 |
| $0 < x \leqslant 5$ | 15 |
| $5 < x \leqslant 10$ | 24 |
| $10 < x \leqslant 15$ | 16 |
| $15 < x \leqslant 20$ | 14 |
| $20 < x \leqslant 25$ | 8 |
| $25 < x \leqslant 30$ | 3 |
| $30 < x \leqslant \infty$ | 2 |

It is thought that these data follow a normal distribution, with mean 10 and variance 100. Use the $\chi^2$ distribution at the 5% level of significance to test this hypothesis. Describe briefly how you would modify this test if the mean and variance were unknown. (*AEB*)

16. The following data give the heights in centimetres of 100 male students.

| Height (cm) | Frequency |
|---|---|
| 155–160 | 5 |
| 161–166 | 17 |
| 167–172 | 38 |
| 173–178 | 25 |
| 179–184 | 9 |
| 185–190 | 6 |

(a) Test, at the 5% level, whether the data follow a normal distribution with mean 173.5 cm and standard deviation 7 cm.
(b) Find the expected frequencies for a normal distribution having the same mean and variance as the data given, and test the goodness of fit, using a 5% level of significance.

17. In a European country registration for military service is compulsory for all eighteen-year-old males. All males must report to a barracks where, after an inspection some people, including all those less than 1.6 m tall, are excused service. The heights of a sample of 125 eighteen-year-olds measured at the barracks were as follows:

| Height, m | 1.2– | 1.4– | 1.6– | 1.8– | 2.0–2.2 |
|---|---|---|---|---|---|
| Frequency | 6 | 34 | 31 | 42 | 12 |

(a) Use a $\chi^2$ test and a 5% significance level to confirm that the normal distribution is not an adequate model for this data.
(b) Show that, if the second and third classes (1.4– and 1.6–) are combined, the normal distribution does appear to fit the data. Comment on this apparent contradiction in the light of the information at the beginning of the question. (*AEB*)

# THE $\chi^2$ SIGNIFICANCE TEST FOR INDEPENDENCE

Sometimes situations arise when data are classified according to two different factors or attributes and these are often displayed in a table, known as a **contingency table**, for example

(a) examination grades for Mathematics in three further education colleges

| | | College | | |
|---|---|---|---|---|
| | | Bradley | Cooper | Dunstan |
| Examination Grade | A | 27 | 35 | 17 |
| | B | 52 | 36 | 28 |
| | C | 63 | 31 | 64 |
| | D | 31 | 43 | 21 |
| | E | 16 | 17 | 12 |
| | N | 5 | 12 | 8 |

This is a 6 by 3 contingency table (6 rows and 3 columns).

(b) age of voter and voting preference

|  |  | Candidate | |
|---|---|---|---|
|  |  | A | B |
| Age of voter | 18–25 | 373 | 62 |
|  | 25–40 | 484 | 187 |
|  | 40–60 | 167 | 563 |
|  | Over 60 | 100 | 492 |

This is a 4 by 2 contingency table (4 rows and 2 columns).

You can use a $\chi^2$ test to investigate whether the two factors are independent or whether there is an association between them. The test follows a similar pattern to the goodness of fit test, but this time the null hypothesis $H_0$ is that the two factors are independent and the alternative hypothesis $H_1$ is that there is an association between them.

The following example explains how to calculate the expected frequencies for data given in a contingency table.

## Example 12.8

The members of a sports team are interested in whether the weather has an effect on their results. They play 50 matches, with the following results.

|  |  | Weather | | |
|---|---|---|---|---|
|  |  | Good | Bad | Total |
| Result | Win | 12 | 4 | 16 |
|  | Draw | 5 | 8 | 13 |
|  | Lose | 7 | 14 | 21 |
|  | Total | 24 | 26 | 50 |

Formulate suitable null and alternative hypotheses, and use a $\chi^2$ test to test the claim, at the 1% significance level, that the weather has no effect on the team's results. State your conclusion clearly. (C)

## Solution 12.8

*Note that the factors are the result of the match and the type of weather and they have been linked in a 3 by 2 contingency table.*

1. State $H_0$ and $H_1$.

The hypotheses are:

$H_0$: The weather has no effect on the team's results
$H_1$: The weather has an effect on the team's results

2. Calculate $E$ and check that expected frequencies are greater than 5.

When calculating the expected frequencies, the row and column totals must remain the same.

Consider the cell linking a win with good weather:

Total number of
wins = 16, therefore

$P(\text{result is a win}) = \dfrac{16}{50}$.

Total number of matches
in good weather = 24,
therefore

$P(\text{good weather}) = \dfrac{24}{50}$.

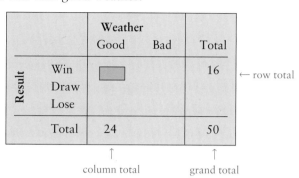

|  |  | Weather | | |
|---|---|---|---|---|
|  |  | Good | Bad | Total |
| Result | Win | ▢ |  | 16 | ← row total |
|  | Draw |  |  |  |
|  | Lose |  |  |  |
|  | Total | 24 |  | 50 |

↑ column total        ↑ grand total

According to the null hypothesis, the events 'the result is a win' and 'the weather is good' are independent, so, using the **multiplication rule for independent events** (see page 198)

$P(\text{win and good weather}) = P(\text{win}) \times P(\text{good weather})$

$$= \frac{16}{50} \times \frac{24}{50}$$

Expected number of wins in good weather $= \cancel{50}^{1} \times \dfrac{16}{\cancel{50}^{1}} \times \dfrac{24}{50}$

$$= \frac{16 \times 24}{50}$$

$$= 7.68$$

Note that the calculation $\dfrac{16 \times 24}{50}$ gives a clue to the quick way of working out the expected frequency:

$$\text{Expected frequency} = \frac{\text{row total} \times \text{column total}}{\text{grand total}}$$

So, for example, the expected number of draws in bad weather is calculated as follows:

|  |  | Weather | | |
|---|---|---|---|---|
|  |  | Good | Bad | Total |
| Result | Win |  |  |  |
|  | Draw |  | ▢ | 13 |
|  | Lose |  |  |  |
|  | Total |  | 26 | 50 |

Expected frequency

$$= \frac{\text{row total} \times \text{column total}}{\text{grand total}}$$

$$= \frac{13 \times 26}{50}$$

$$= 6.76$$

The completed table for the expected frequencies is:

|  |  | Weather | | |
|---|---|---|---|---|
|  |  | Good | Bad | Total |
| Result | Win | 7.68 | 8.32 | 16 |
|  | Draw | 6.24 | 6.76 | 13 |
|  | Lose | 10.08 | 10.92 | 21 |
|  | Total | 24 | 26 | 50 |

Note that all the expected frequencies are greater than five, so cells do not need to be combined.

**3.** Work out $\nu$.

*Degrees of freedom, $\nu$*

Notice that in this table once two of the expected frequencies in different rows have been calculated (for example those in bold type), the others are known automatically. This is because the row and column totals must agree with those in the observed data, for example

if expected number of wins in good weather = 7.68,

then expected number of wins in bad weather = 16 − 7.68 = 8.32

Number of degrees of freedom, $\nu = 2$ and the $\chi^2(2)$ distribution is considered.

**4.** State the level of the test and the rejection criterion.

Test at the 1% level.

From tables $\chi^2_{1\%}(2) = 9.21$, so reject $H_0$ if $X^2 > 9.21$.

**5.** Calculate $X^2$.

| $O$ | $E$ | $\dfrac{(O-E)^2}{E}$ |
|---|---|---|
| 12 | 7.68 | 2.43 |
| 5 | 6.24 | 0.246 ... |
| 7 | 10.08 | 0.941 ... |
| 4 | 8.32 | 2.243 ... |
| 8 | 6.76 | 0.227 ... |
| 14 | 10.92 | 0.868 ... |
| $\Sigma O = 50$ | $\Sigma E = 50$ | 6.956 ... |

$$X^2 = \sum \frac{(O-E)^2}{E} = 6.96 \quad (2 \text{ d.p.})$$

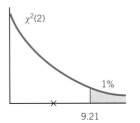

**6.** Make your conclusion.

**Since $X^2 < 9.21$ do not reject $H_0$,** and conclude that the team's results are independent of the weather.

**At the 1% level conclude that the weather has no effect on the team's result.**

# Finding the number of degrees of freedom, $\nu$, in an $h$ by $k$ contingency table

There is a general rule for calculating $\nu$ for data in a contingency table. In each of the tables shown below, it is possible to work out all the expected frequencies once the values indicated with a × have been found.

**4 by 3 table**

$$\nu = (4-1) \times (3-1)$$
$$= 3 \times 2$$
$$= 6$$

2 by 4 table

$$v = (2 - 1) \times (4 - 1)$$
$$= 1 \times 3$$
$$= 3$$

3 by 2 table

$$v = (3 - 1) \times (2 - 1)$$
$$= 2 \times 1$$
$$= 2$$

2 by 2 table

$$v = (2 - 1) \times (2 - 1)$$
$$= 1 \times 1$$
$$= 1$$

In general, if there are $h$ rows, then once $(h - 1)$ expected frequencies in a row have been calculated, the last value in the row is known because the row total must agree.

Similarly, if there are $k$ columns, once $(k - 1)$ expected frequencies in a column have been calculated, the last value in the column is known because the column total must agree.

For an $h$ by $k$ contingency table,

number of degrees of freedom $= (h - 1) \times (k - 1)$.

# Yates' correction for a 2 by 2 contingency table

In particular, for a 2 by 2 contingency table, $v = 1$ and the $\chi^2(1)$ distribution is considered. In this case, Yates' correction should be applied when calculating $X^2$, where

$$X^2 = \sum \frac{(|O - E| - 0.5)^2}{E}$$

### Example 12.9

A driving school examined the results of 100 candidates who took their test for the first time. It was found that out of the 40 men, 28 passed and out of the 60 women, 34 passed. Do these results indicate, at the 5% significance level, a relationship between the sex of candidate and the ability to pass the driving test at the first attempt?

## Solution 12.9

Displaying the results in a contingency table:

|  | **Result of driving test** Pass | Fail | Total |
|---|---|---|---|
| Male | 28 | 12 | 40 |
| Female | 34 | 26 | 60 |
| Totals | 62 | 38 | 100 |

1. State $H_0$ and $H_1$.

The hypotheses are:

$H_0$: There is no relationship between the sex of a candidate and the ability to pass at the first attempt.
$H_1$: There is a relationship.

2. Calculate $E$ and check that expected frequencies are greater than 5.

To calculate expected frequencies, use

$$\text{Expected frequency} = \frac{\text{row total} \times \text{column total}}{\text{grand total}}$$

So expected number of males who pass $= \dfrac{40 \times 62}{100} = 24.8$

Use the fact that row and column totals agree with the observed data to work out all the remaining frequencies:

|  | **Result of driving test** Pass | Fail | Total |
|---|---|---|---|
| Male | 24.8 | 15.2 | 40 |
| Female | 37.2 | 22.8 | 60 |
| Totals | 62 | 38 | 100 |

$40 - 24.8 = 15.2$

$60 - 37.2 = 22.8$

$62 - 24.8 = 37.2$

*Note that there are no expected frequencies that are less than 5.*

3. Work out $v$.

*Degrees of freedom, $v$*
$v = (2 - 1)(2 - 1) = 1$, so use the $\chi^2(1)$ distribution.

4. State the level of the test and the rejection criterion.

Test at the 5% level.
From tables $\chi^2_{5\%}(1) = 3.841$, so reject $H_0$ if $X^2 > 3.841$.

5. Calculate $X^2$.     Using Yates' correction,

| $O$ | $E$ | $\dfrac{(|O-E|-0.5)^2}{E}$ |
|---|---|---|
| 28 | 24.8 | 0.293 ... |
| 34 | 37.2 | 0.195 ... |
| 12 | 15.2 | 0.479 ... |
| 26 | 22.8 | 0.319 ... |
| $\Sigma O = 100$ | $\Sigma E = 100$ | 1.289 ... |

$$X^2 = \sum \frac{(|O-E|-0.5)^2}{E}$$
$$= 1.29 \text{ (2 d.p.)}$$

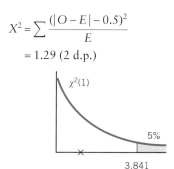

6. Make your conclusion.

Since $X^2 < 3.841$, do not reject $H_0$.

**The driving test results do not indicate a relationship between the sex of a candidate and the ability to pass the driving test at the first attempt.**

## Exercise 12c   Contingency tables

1. Two schools enter their pupils for a particular public examination and the results obtained are shown below.

|  | Credit | Pass | Fail |
|---|---|---|---|
| School $A$ | 51 | 10 | 19 |
| School $B$ | 39 | 10 | 21 |

By using an approximate $\chi^2$ statistic, assess at the 5% level whether or not there is a significant difference between the two schools with respect to the proportions of pupils in the three grades. State your null and alternative hypotheses.    (L)

2. Students in the Sociology department of a university decided to conduct a survey into the roles of married couples in performing tasks of housework and child care. They designed questionnaire for this purpose.
They then contacted 240 married couples who were willing to take part in the survey. Each of the participating couples was randomly allocated to one of two groups. In the first group the wife was asked to complete the questionnaire and in the second group the husband was asked to complete it.
Four response categories were available for a question which asked how the work of cleaning the house was shared between husband and wife. The following table shows the numbers of husbands and wives choosing each category.

| Response category | Husbands | Wives |
|---|---|---|
| Wife does it all | 21 | 30 |
| Wife does most of it | 63 | 58 |
| Shared half and half | 28 | 25 |
| Husband does all or most of it | 8 | 7 |

Carry out a $\chi^2$ test to investigate whether there is an association between the sex of the respondent and the respondent's view of how the work is shared.
Comment on any differences revealed by this survey between the opinions of husbands and wives about who does the household cleaning.
(NEAB)

3. The following are data on 150 chickens, divided into two groups according to breed, and into three groups according to yield of eggs:

|  | Yield | | |
|---|---|---|---|
|  | High | Medium | Low |
| Rhode Island Red | 46 | 29 | 28 |
| Leghorn | 27 | 14 | 6 |

Are these data consistent with the hypothesis that the yield is not affected by the type of breed?

4. A research worker studying the ages of adults and the number of credit cards they possess obtained the results shown in the table.

|  |  | Number of cards possessed | |
|---|---|---|---|
|  |  | ⩽3 | >3 |
| Age | <30 | 74 | 20 |
|  | ⩾30 | 50 | 35 |

Use the $\chi^2$ statistic and a significance test at the 5% level to decide whether or not there is an association between age and number of credit cards possessed.    (L)

5. An investigation into colourblindness and the sex of a person gave the following results:

| | | Colourblindness | |
| | | Colourblind | Not colourblind |
|---|---|---|---|
| Sex | Male | 36 | 964 |
| | Female | 19 | 981 |

Is there evidence, at the 5% level, of an association between the sex of a person and whether or not they are colourblind?

6. In a small survey 350 car owners from four districts P, Q, R, S were found to have cars in price ranges A, B, C, D, the frequencies of the prices being as shown in the table.

| | | P | Q | R | S |
|---|---|---|---|---|---|
| Price of car | A | 9 | 10 | 12 | 19 |
| | B | 13 | 20 | 18 | 29 |
| | C | 24 | 29 | 12 | 25 |
| | D | 34 | 41 | 18 | 37 |

Find the expected frequencies on the hypothesis that there is no association between the district and the price of the car. Use the $\chi^2$ distribution to test this hypothesis. (AEB)

7. A random sample of 100 shoppers was asked by a market research team whether or not they used Sudsey Soap. 58 said yes and 42 said no. In a second random sample of 80 shoppers, 62 said yes and 18 said no. By considering a suitable 2 × 2 contingency table, test whether these two samples are consistent with each other. (O & C)

8. The table summarises the incidence of cerebral tumours in 141 neurosurgical patients.

| | | Type of tumour | | |
| | | Benign | Malignant | Others |
|---|---|---|---|---|
| Site of tumour | Frontal lobes | 23 | 9 | 6 |
| | Temporal lobes | 21 | 4 | 3 |
| | Elsewhere | 34 | 24 | 17 |

Find the expected frequencies on the hypothesis that there is no association between the type and site of a tumour. Use the $\chi^2$ distribution to test this hypothesis. (AEB)

9. In an examination 37 out of 47 boys passed and 27 out of 41 girls passed. By considering a suitable 2 × 2 contingency table, test whether boys and girls differ in their ability in this subject.

10. In an investigation into eye colour and left- or right-handedness the following results were obtained:

| | | Handedness | |
| | | Left | Right |
|---|---|---|---|
| Eye colour | Blue | 15 | 85 |
| | Brown | 20 | 80 |

Is there evidence, at the 5% level, of an association between eye colour and left- or right-handedness?

11. In 1988 the number of new cases of insulin-dependent diabetes in children under the age of 15 years was 1495. The table below breaks down this figure according to age and sex.

| Age (yrs) | 0–4 | 5–9 | 10–14 | Total |
|---|---|---|---|---|
| Boys | 205 | 248 | 328 | 781 |
| Girls | 182 | 251 | 281 | 714 |
| Total | 387 | 499 | 609 | 1495 |

Perform a suitable test, at the 5% significance level, to determine whether age and sex are independent factors. (C)

12. When analysing the results of a 3 × 2 contingency table it was found that

$$\sum_{i=1}^{6} \frac{(O_i - E_i)^2}{E_i} = 2.38.$$

Write down the number of degrees of freedom and the critical value appropriate to these data in order to carry out a $\chi^2$ test of significance at the 5% level. (L)

13. In a college, three different groups of students sit the same examination. The results of the examination are classified as Credit, Pass or Fail. In order to test whether or not there is a difference between the groups with respect to the proportion of students in the three grades the statistic

$$\sum \frac{(O - E)^2}{E}$$

is evaluated and found to be equal to 10.28.

(a) Explain why there are four degrees of freedom in this situation.
(b) Using a 5% level of significance, carry out the test and state your conclusions. (L)

14. The personnel manager of a large firm is investigating whether there is any association between the length of service of the employees and the type of training they receive from the firm. A random sample of 200 employee records is taken from the last few years and is classified according to these criteria. Length of service is classified as short (meaning less than 1 year), medium (1–3 years) and long (more than 3 years). Type of training is classified as being merely an initial 'induction course', proper initial on-the-job training but little if any more, and regular and continuous training. The data are as follows:

|  |  | Length of service | | |
|---|---|---|---|---|
|  |  | Short | Medium | Long |
| Type of training | Induction course | 14 | 23 | 13 |
|  | Initial on-the-job | 12 | 7 | 13 |
|  | Continuous | 28 | 32 | 58 |

Examine at the 5% level of significance whether these data provide evidence of association between length of service and type of training, stating clearly your null and alternative hypotheses.
Discuss your conclusions. *(MEI)*

15. A market research organisation interviewed a random sample of 120 users of launderettes in London and found that 37 preferred brand $X$ washing powder, 66 preferred brand $Y$ and the remainder preferred brand $Z$. A similar survey was carried out in Birmingham. In this survey, of 80 people interviewed, 19 preferred brand $X$, 40 preferred brand $Y$ and the remainder preferred brand $Z$. Test whether these results provide significant evidence, at the 5% level, of different preferences in the two cities. *(C)*

16. The results obtained by 200 students in chemistry and biology are shown in the table. Test, at the 5% level, whether the performances in both subjects are related.

|  |  | Chemistry | |
|---|---|---|---|
|  |  | Pass | Fail |
| Biology | Pass | 102 | 45 |
|  | Fail | 21 | 32 |

## Summary

### $\chi^2$ significance test

- The test statistic is $X^2$ where

  where $\quad X^2 = \sum \dfrac{(O-E)^2}{E} \quad$ and $\quad X^2 \sim \chi^2(v)$.

  When $v = 1$, use Yates' correction,

  where $\quad X^2 = \sum \dfrac{(|O-E|-0.5)^2}{E}$.

  Remember to combine classes if $E \leqslant 5$.

- Degrees of freedom, $v$

  $\chi^2$ goodness of fit tests

  $v$ = number of classes – number of restrictions (see table on page 579)

  $\chi^2$ tests for independence

  For an $h$ by $k$ contingency table, $v = (h-1)(k-1)$.

# Miscellaneous worked examples

### Example 12.10

In experiments in pea breeding Gregor Mendel obtained the following data relating to 556 peas.

| Round and Yellow | Wrinkled and Yellow | Round and Green | Wrinkled and Green |
|---|---|---|---|
| 315 | 101 | 108 | 32 |

According to Mendel's theoretical results, the expected figures are in the ratios $9 : 3 : 3 : 1$. Calculate the value of $\chi^2$ for these data on the assumption that the theory is correct.

Test at the 10% significance level whether the theory is contradicted.

It has been suggested that Mendel's results are suspect in that they are unlikely to have been obtained from random observations. Comment on this suggestion in relation to the value of $\chi^2$ calculated.　　　　(C)

### Solution 12.10

$H_0$: The different types of peas occur in the ratio $9 : 3 : 3 : 1$.
$H_1$: The different types of peas do not occur in this ratio.

Expected frequencies, according to $H_0$:

Round and yellow     $\frac{9}{16} \times 556 = 312.75$
Wrinkled and yellow     $\frac{3}{16} \times 556 = 104.25$
Round and green     $\frac{3}{16} \times 556 = 104.25$
Wrinkled and green     $\frac{1}{16} \times 556 = 34.75$

| | Round and yellow | Wrinkled and yellow | Round and green | Wrinkled and green | |
|---|---|---|---|---|---|
| Observed ($O$) | 315 | 101 | 108 | 32 | $\Sigma O = 556$ |
| Expected ($E$) | 312.75 | 104.25 | 104.25 | 34.75 | $\Sigma E = 556$ |

There are four classes and one restriction ($\Sigma E = 556$)
Therefore $v = 4 - 1 = 3$ and the $\chi^2(3)$ distribution is considered.

Perform the test at the 10% level.
From tables $\chi^2_{10\%}(3) = 6.251$, so reject $H_0$ if $X^2 > 6.251$.

| $O$ | $E$ | $\frac{(O-E)^2}{E}$ |
|---|---|---|
| 315 | 312.75 | 0.0161 ... |
| 101 | 104.25 | 0.101 ... |
| 108 | 104.25 | 0.134 ... |
| 32 | 34.75 | 0.217 ... |
| $\Sigma O = 556$ | $\Sigma E = 556$ | 0.470 ... |

$$X^2 = \sum \frac{(O-E)^2}{E} = 0.47 \text{ (2 d.p.)}$$

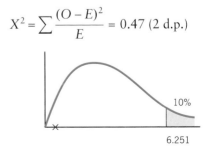

Since $X^2 < 6.25$, accept $H_0$ and conclude that **the types are in the ratio 9 : 3 : 3 : 1.**

The calculated value of $X^2$ is very small indeed, suggesting very little discrepancy between the observed and expected frequencies.

From $\chi^2$ tables, $P(X^2 < 0.352) = 5\%$ so on only just over 5% of occasions would you expect to have a test value this low. This could suggest that the data are not random observations.

## Example 12.11

Mr and Mrs Smith live in a small town with two primary schools $A$ and $B$. They are trying to decide which school would provide the better learning environment for their children. They have available the results of recent national tests in mathematics, English and science. Each child in the final year took three tests, one in each subject, and they either passed or failed each test. These results are summarised in the table below.

|  | 3 passes | 1 or 2 passes | No passes |
|---|---|---|---|
| School $A$ | 15 | 6 | 5 |
| School $B$ | 10 | 14 | 13 |

(a) Stating your hypotheses clearly test, at the 5% level of significance, whether or not there is evidence of an association between school and test results.

Mr and Mrs Smith also have available the results of a questionnaire about the annual family income $x$, in thousands of pounds, of the families of the children taking these tests. The results are summarised in the table below.

|  | $x > 30$ | $20 < x \leqslant 30$ | $15 < x \leqslant 20$ | $x \leqslant 15$ |
|---|---|---|---|---|
| School $A$ | 7 | 5 | 9 | 5 |
| School $B$ | 6 | 13 | 8 | 10 |

A $\chi^2$ test for association between school and family income using this information gave a test statistic of 3.545. There was no pooling of classes.

(b) Using a 5% level of significance, interpret this statistic stating the critical value used.

(c) In the light of parts (a) and (b) state, giving reasons, which of the two schools Mr and Mrs Smith might choose for their children. (L)

## Solution 12.11

(a) $H_0$: The two factors 'school' and 'results' are independent.
$H_1$: The factors are not independent and there is an association between school and results.

Observed data:

|  | 3 passes | 1 or 2 passes | No passes | Totals |
|---|---|---|---|---|
| School $A$ | 15 | 6 | 5 | 26 |
| School $B$ | 10 | 14 | 13 | 37 |
| Totals | 25 | 20 | 18 | 63 |

Expected data:
For school $A$ and three passes

$$\text{expected frequency} = \frac{\text{row total} \times \text{column total}}{\text{grand total}}$$

$$= \frac{26 \times 25}{63}$$

$$= 10.32 \ (2 \text{ d.p.})$$

The complete table is as follows:

| | 3 passes | 1 or 2 passes | No passes | Totals |
|---|---|---|---|---|
| School $A$ | 10.32 | 8.25 | 7.43 | 26 |
| School $B$ | 14.68 | 11.75 | 10.57 | 37 |
| Totals | 25 | 20 | 18 | 63 |

The table has 2 rows and 3 columns
so $\nu = (2-1)(3-1) = 1 \times 2 = 2$ and the $\chi^2(2)$ distribution is considered.

Test at the 5% level.
From tables, $\chi^2_{5\%}(2) = 5.991$, so reject $H_0$ if $X^2 > 5.991$.

| $O$ | $E$ | $\dfrac{(O-E)^2}{E}$ |
|---|---|---|
| 15 | 10.32 | 2.122 ... |
| 6 | 8.25 | 0.613 ... |
| 5 | 7.43 | 0.794 ... |
| 10 | 14.68 | 1.491 ... |
| 14 | 11.75 | 0.430 ... |
| 13 | 10.57 | 0.558 ... |
| $\Sigma O = 63$ | $\Sigma E = 63$ | 6.012 ... |

$$X^2 = \sum \frac{(O-E)^2}{E} = 6.01 \ (2 \text{ d.p.})$$

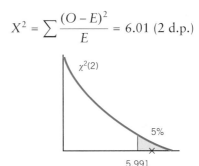

Since $X^2 > 5.991$, reject $H_0$ and conclude that **there is evidence of an association between the school and the test results.**

(b) The table has 2 rows and 4 columns
so $\nu = (2-1)(4-1) = 1 \times 3 = 3$ and the $\chi^2(3)$ distribution is considered.

$H_0$: The two factors 'school' and 'family income' are independent.
$H_1$: There is an association between school and family income.

From tables $\chi^2_{5\%}(3) = 7.815$, so reject $H_0$ if $X^2 > 7.815$.

It is given that $X^2 = 3.545$

Since $X^2 < 7.815$, do not reject $H_0$.
**There is no association between school and family income.**

(c) As there is no association between school and family income, Mr and Mrs Smith are likely to base their choice on the results of the national tests. Since 57% of the pupils in school $A$ obtained three passes, as compared with only 27% with three passes in school $B$, Mr and Mrs Smith might conclude that school $A$ provides the better learning environment.

# Miscellaneous exercise 12d

1. It is suggested that preferences for three proposed routes for a town by-pass are associated with where people live. Each person in a random sample of 150 people, chosen from the inhabitants of the town and surrounding villages, was asked which route he or she preferred. The results are given in the following table.

|         | Town | Surrounding villages |
|---------|------|----------------------|
| Route 1 | 50   | 25                   |
| Route 2 | 28   | 22                   |
| Route 3 | 16   | 9                    |

State appropriate null and alternative hypotheses, and use a $\chi^2$ test to test, at the 5% significance level, the suggestion that there is an association between preferred route and where people live. (C)

2. (a) A random sample of supermarkets was sent a questionnaire on which they were asked to report the number of cases of shoplifting they had dealt with in each month of the previous year. The totals for each month were as follows.

| J  | F  | M  | A  | M | J  | J  | A  | S  | O  | N  | D  |
|----|----|----|----|---|----|----|----|----|----|----|----|
| 16 | 12 | 10 | 17 | 6 | 18 | 16 | 17 | 10 | 22 | 14 | 16 |

Carry out a chi-squared test at an appropriate level of significance to determine whether or not shoplifting is more likely to occur in some months than others. (You may take all months to be of the same length.) Make clear your null and alternative hypotheses, the level of significance you are using, and your conclusion.

You may, if you wish, use the fact that, when all the values of $f_e$ are equal, the usual chi-squared test statistic may be written as

$$\frac{1}{f_e}\Sigma f_o{}^2 - \Sigma f_o$$

(b) Prove the result given at the end of part (a). (MEI)

3. It is thought that there is an association between the colour of a person's eyes and the reaction of the person's skin to ultraviolet light. In order to investigate this each of a random sample of 120 people was subjected to a standard dose of ultraviolet light. The degree of their reaction was noted, '−' indicating no reaction, '+' indicating slight reaction and '++' indicating strong reaction. The results are shown in the table below.

|          |      | Blue | Eye colour<br>Grey or<br>green | Brown |
|----------|------|------|------|------|
| Reaction | −    | 7    | 8    | 18   |
|          | +    | 29   | 10   | 16   |
|          | ++   | 21   | 9    | 2    |

Perform an appropriate test at the 5% significance level, stating your null and alternative hypotheses.

4. Describe briefly how the number of degrees of freedom is calculated in a $\chi^2$ goodness-of-fit test. The following set of grouped data from 100 observations has mean 1.03. The data are thought to come from a normal distribution with known variance 1 but unknown mean. Using an appropriate $\chi^2$-distribution, test this hypothesis at the 1% significance level.

| Lower value of<br>grouping interval | Number of<br>Observations |
|---------------------|----------|
| $-\infty$           | 0        |
| −2.0                | 1        |
| −1.5                | 0        |
| −1.0                | 6        |
| −0.5                | 10       |
| 0.0                 | 12       |
| 0.5                 | 15       |
| 1.0                 | 23       |
| 1.5                 | 16       |
| 2.0                 | 13       |
| 2.5                 | 3        |
| 3.0                 | 1        |
| 3.5                 | 0        |

(C)

5. A farmers' cooperative decided to test three new brands of fertiliser, A, B and C, allocating them at random to 75 plots. The yield of the crop was classified as high, medium or low. The results are summarised in the table below.

|       |        | Fertiliser | | | Total |
|-------|--------|---|---|---|-------|
|       |        | A | B | C |       |
| Yield | High   | 12 | 15 | 3  | 30 |
|       | Medium | 8  | 8  | 8  | 24 |
|       | Low    | 5  | 7  | 9  | 21 |
| Total |        | 25 | 30 | 20 | 75 |

(a) Stating your hypotheses clearly, test at the 5% level of significance whether or not there is any evidence of an association between brand of fertiliser and yield.

Fertilisers A and B are produced by *Quickgrow* whereas C is produced by *Bumpercrops*. The farmers wanted to decide from which company to purchase fertiliser and combined the figures for A and B to give a 3 × 2 table. The statistic

$$\sum \frac{(O-E)^2}{E}$$

for this new table was calculated and gave the value 7.622.

(b) By carrying out a suitable test at the 5% level of significance, advise the farmers whether or not there is any evidence of an association between the choice of company and yield.

(c) Giving your reason, advise the farmers which company they should use. *(L)*

6. A statistician, who is suspected to be suffering from asthma, is asked to record his peak flow measurement four times each day for a period of four weeks.
He groups by value the 112 recorded measurements into seven classes giving observed frequencies, $o_i$, $i = 1, 2, ..., 7$. He then calculates correctly corresponding expected frequencies, $e_i$, using a normal distribution having mean and variance estimated from the original measurements.

The value of the test statistic

$$\sum_{i=1}^{7} \frac{(o_i - e_i)^2}{e_i}$$

is then calculated correctly by the statistician as 5.624.

(a) Using a 1% level of significance and stating the null hypothesis, complete the test.

(b) Give the usual requirement made on each of the values $e_i$ prior to calculating the test statistic, and indicate how a failure to meet the requirement may be overcome. *(NEAB)*

7. A random sample of 100 people was asked for their opinions about the amount of sport shown on TV. Each person had to say whether there was too much sport shown, about the right amount, or not enough. The numbers of men and women making each response are shown in the table.

|  | Men | Women |
|---|---|---|
| Too much sport | 13 | 26 |
| About right | 22 | 22 |
| Not enough sport | 12 | 5 |

The null hypothesis is that a person's opinion about the amount of sport shown on TV is independent of the person's sex.

(a) Construct a table showing the expected frequencies, assuming that the null hypothesis is true.

(b) Use a $\chi^2$ test to test this null hypothesis, using a 5% significance level. Show full details of your method and state your conclusion clearly. *(C)*

8. The Director of Studies at a College of Further Education believed that there was a connection between candidates' grades in mathematics and physics at A-level. For a set of candidates who had taken both examinations, she recorded the number of candidates in each of four categories, as shown in the table.

|  | Mathematics grades A–C | Mathematics grades D–U |
|---|---|---|
| Physics grades A–C | 22 | 9 |
| Physics grades D–U | 8 | 15 |

(a) Test the Director's belief at the 2.5% level of significance, stating your null and alternative hypotheses.

Her colleague said that she was losing accuracy by combining the grades A to C in one group, and grades D to U in another. He suggested that she should create a 7 × 7 table showing all possible combinations of grades.

(b) State why his suggestion might lead to a problem in performing the test. *(L)*

9. During a working day a machine requires occasional adjustments which appear to be randomly distributed throughout the day. A factory foreman records the number of adjustments made to the machine each day for a period of 200 working days, obtaining the data displayed in the table.

| Number of adjustments | 0 | 1 | 2 | 3 | 4 | 5 |
|---|---|---|---|---|---|---|
| Number of days | 34 | 78 | 61 | 20 | 5 | 2 |

Previous experience has suggested that the daily number of adjustments to this machine follows a Poisson distribution with mean 1.5.

(a) Perform a $\chi^2$ goodness of fit test to decide whether the data in the table can reasonably be considered as conforming to a Poisson distribution with mean 1.5.

(b) Outline, without detailed calculation, the necessary modifications to your test if the Poisson mean is not assumed to be 1.5.

(c) The distribution B(5, 0.3) is a very good fit to the data in the table. Without further calculation, explain why, despite this good fit, the binomial model is *not* appropriate. *(NEAB)*

10. A department store has five doorways, each for entrance and exit. It is claimed that the proportion of shoppers entering or leaving the store is the same for each of the five doorways. The number of customers entering or leaving the store is counted at each doorway for three randomly selected days with the following results.

| Doorway | Number of customers |
|---------|---------------------|
| A | 601 |
| B | 673 |
| C | 626 |
| D | 618 |
| E | 702 |

Test whether or not these data support the claim.

The same store also records the daily number of sales charged to stolen credit cards. The results for the first four months of 1990 are as follows.

| Number of sales | Number of days |
|-----------------|----------------|
| 0 | 31 |
| 1 | 39 |
| 2 | 19 |
| 3 | 11 |
| $\geqslant 4$ | 0 |

Explain why a Poisson distribution may be appropriate as a model for the daily number of sales charged to stolen credit cards. Test the hypothesis that the daily number of sales does follow a Poisson distribution.     (NEAB)

11. In the mathematics department of a college, candidates in an examination are graded A, B, C, D or E. Records from previous years show that examiners have awarded a grade A to 15% of candidates, B to 20%, C to 35%, D to 25% and E to 5%. A new syllabus is examined by a new board of examiners who award the grades to 200 candidates as follows:

A, 33;  B, 37;  C, 81;  D, 36;  E, 13

(a) Stating clearly your hypotheses and using a 5% level of significance investigate whether or not the new board of examiners awards grades in the same proportions as the previous one.

In addition to being classified by examination grade, these 200 students are classified as male or female and the results summarised in a contingency table. Assuming all expected values are 5 or more, the statistic

$$\sum_{i=1}^{10} \frac{(O_i - E_i)^2}{E_i} \text{ was } 14.27.$$

(b) Stating your hypotheses and using a 1% significance level, investigate whether or not sex and grade are associated.     (L)

12. A factory operates four production lines. Maintenance records show that the daily number of stoppages due to mechanical failure were as shown in the table below (it is possible for a production line to break down more than once on the same day). You may assume that $\Sigma f = 1400$, $\Sigma fx = 1036$.

| Number of stoppages, $x$ | 0 | 1 | 2 | 3 | 4 | 5 | 6 or more |
|--------------------------|---|---|---|---|---|---|-----------|
| Number of days, $f$ | 728 | 447 | 138 | 48 | 26 | 13 | 0 |

(a) Use a $\chi^2$ distribution and a 1% significance level to determine whether the Poisson distribution is an adequate model for the data.

(b) The maintenance engineer claims that breakdowns occur at random and that the mean rate has remained constant throughout the period. State, giving a reason, whether your answer to (a) is consistent with this claim.

(c) Of the 1036 breakdowns which occurred 230 were on production line A, 303 on B, 270 on C and 233 on D. Test at the 5% significance level whether these data are consistent with breakdowns occurring at an equal rate on each production line.     (AEB)

13. A group of students studying A-level statistics was set a paper, to be attempted under examination conditions, containing four questions requiring the use of the $\chi^2$ distribution. The following table shows the type of question and the number of students who obtained good (14 or more out of 20) and bad (fewer than 14 out of 20) marks.

| | Type of question | | | |
|---|-------------------|-----------|------------|-------------|
| | Contingency table | Binomial fit | Normal fit | Poisson fit |
| Good mark | 25 | 12 | 12 | 11 |
| Bad mark | 4 | 11 | 3 | 12 |

(a) Test at the 5% significance level whether the mark obtained (by the students who attempted the question) is associated with the type of question.

(b) Under some circumstances it is necessary to combine classes in order to carry out a test. If it had been necessary to combine the binomial fit question with another question, which question would you have combined it with and why?

(c) Given that a total of 30 students sat the paper, test, at the 5% significance level whether the number of students attempting a particular question is associated with the type of question.

(d) Compare the difficulty and popularity of the different types of question in the light of your answers to (a) and (c). (*AEB*)

14. (a) The number of books borrowed from a library during a certain week were 518 on Monday, 431 on Tuesday, 485 on Wednesday, 443 on Thursday and 523 on Friday.

Is there any evidence that the number of books borrowed varies between the five days of the week? Use a 1% level of significance. Interpret fully your conclusions.

(b) Analysis of the rate of turnover of employees by a personnel manager produced the following table showing the length of stay of 200 people who left the company for other employment.

|  |  | Length of employment (years) | | |
|  |  | 0–2 | 2–5 | > 5 |
|---|---|---|---|---|
| Grade | Managerial | 4 | 11 | 6 |
|  | Skilled | 32 | 28 | 21 |
|  | Unskilled | 25 | 23 | 50 |

Using a 1% level of significance, analyse this information and state fully the conclusions from your analysis. (*AEB*)

15. Over a long period of time, a research team monitored the number of car accidents which occurred in a particular county. Each accident was classified as being trivial (minor damage and no personal injuries), serious (damage to vehicles and passengers, but no deaths) or fatal (damage to vehicles and loss of life). The colour of the car which, in the opinion of the research team, caused the accident was also recorded, together with the day of the week on which the accident occurred. The following data were collected.

| Colour | Trivial | Serious | Fatal |
|---|---|---|---|
| White | 50 | 25 | 16 |
| Black | 35 | 39 | 18 |
| Green | 28 | 23 | 13 |
| Red | 25 | 17 | 11 |
| Yellow | 17 | 20 | 16 |
| Blue | 24 | 33 | 10 |

Analyse these data for evidence of association between the colour of the car and the type of accident.

State the condition which sometimes necessitates the amalgamation of rows or columns in contingency tables. Explain why amalgamation might not be appropriate for this table.

The following table summarises the data relating to the day of the week on which the accident occurred.

| Day | Number of accidents |
|---|---|
| Monday | 60 |
| Tuesday | 54 |
| Wednesday | 48 |
| Thursday | 53 |
| Friday | 53 |
| Saturday | 75 |
| Sunday | 77 |

Investigate the hypothesis that these data are a random sample from a uniform distribution. (*AEB*)

16. (a) The number of accidents per day on a stretch of motorway was recorded for 100 days and the following results obtained.

| Number of accidents | Frequency |
|---|---|
| 0 | 44 |
| 1 | 32 |
| 2 | 9 |
| 3 | 10 |
| 4 | 5 |
| 5 or more | 0 |

Examine whether or not a Poisson model is suitable to represent the number of accidents per day on this stretch of road. Use a 1% level of significance.

(b) The results of a survey to establish the attitude of individuals to a particular political proposal showed that three-quarters of those interviewed were house owners. Of the 44 interviewed, only 6 of the 35 in favour of the proposal were not house owners.

Does the survey indicate that a person's opinion on the proposal is independent of house ownership? Use a 1% level of significance. (*AEB*)

# Mixed test 12A

1. The number of telephone calls received per day over a period of 150 days is shown in the table below.

   | Number of calls | 0 | 1 | 2 | 3 |
   |---|---|---|---|---|
   | Number of days | 50 | 54 | 36 | 10 |

   (a) Estimate the mean number of calls per day.
   (b) What must be assumed for a Poisson model to be appropriate in this case?
   (c) Carry out a $\chi^2$ goodness of fit analysis to test the null hypothesis that the number of telephone calls received per day has a Poisson distribution with mean 0.95 at the 5% significance level. Give full details of your method. (C)

2. A University Sociology Department believes that students with a good grade in A level General Studies tend to do well on sociology degree courses. To check this it collected information on a random sample of 100 students who had just graduated and had also taken General Studies at A level. The students' performance in General Studies was divided into two categories, those with grade A or B, and 'others'. Their degree classes were recorded as Class I, Class II, Class III, Fail. The data is given in the table below.

   | | | Class of Degree | | | | |
   |---|---|---|---|---|---|---|
   | | | Class I | Class II | Class III | Fail | Total |
   | General | Grade | | | | | |
   | Studies | A or B | 11 | 22 | 6 | 1 | 40 |
   | Grade | Others | 4 | 28 | 24 | 4 | 60 |
   | | Total | 15 | 50 | 30 | 5 | 100 |

   Use this data to test, at the 1% significance level, the hypothesis that degree class is independent of General Studies A level performance. State your conclusion clearly. (C)

3. The heights ($x$) of 100 police officers recruited to a police force in a particular year are summarised in the following table. The mean and standard deviation of the original data are 180 cm and 3 cm respectively.

   | Height (cm) | Frequency |
   |---|---|
   | $x < 175$ | 2 |
   | $175 \leqslant x < 177$ | 15 |
   | $177 \leqslant x < 179$ | 29 |
   | $179 \leqslant x < 181$ | 25 |
   | $181 \leqslant x < 183$ | 12 |
   | $183 \leqslant x < 185$ | 10 |
   | $185 \leqslant x$ | 7 |

   Fit an appropriate normal distribution to the above data, and test the goodness of fit at the 5% level. (C)

4. A survey in a college was commissioned to investigate whether or not there was any association between gender and passing a driving test. A group of 50 male and 50 female students was asked whether they passed or failed their driving test at the first attempt. All the students asked had taken the test. The results were as follows.

   | | Pass | Fail |
   |---|---|---|
   | Male | 23 | 27 |
   | Female | 32 | 18 |

   Stating your hypotheses clearly test, at the 10% level, whether or not there is any evidence of an association between gender and passing a driving test at the first attempt. (L)

# Mixed test 12B

1. It is claimed that when homing pigeons are disorientated harmlessly they will exhibit no particular preference for any direction of flight after take-off. To test this, 128 pigeons, from lofts in a particular region, were disorientated harmlessly and then all released from a position 100 miles south of the region. The direction of flight of each pigeon was recorded with the following results.

| Flight direction | 0°–90° | 90°–180° | 180°–270° | 270°–360° |
|---|---|---|---|---|
| Number of pigeons | 30 | 35 | 36 | 27 |

Use the $\chi^2$ goodness of fit test to determine whether or not these data can be used to discredit the claim. (NEAB)

2. An increasing number of people are spending their working hours in front of a visual display unit (VDU). Sixty-five workers using non-adjustable screens and 66 workers using adjustable screens were asked if they experienced annoying reflections from the screens. The resulting responses are given in the table below.

|  |  | Annoying reflection | |
|---|---|---|---|
|  |  | No | Yes |
| Screen type | Non-adjustable | 15 | 50 |
|  | Adjustable | 28 | 38 |

Test the claim that there is no association between screen type and a worker's experience of annoying reflections. (NEAB)

3. A six-sided die is believed to be biased in the following way:

the probabilities of throwing a one, a two, a three or a four are equal;
the probability of throwing a five is twice the probability of throwing a one;
the probability of throwing a six is three times the probability of throwing a one.

The die is thrown 150 times, and the results are recorded in the table below.

| Score | 1 | 2 | 3 | 4 | 5 | 6 |
|---|---|---|---|---|---|---|
| Frequency | 18 | 15 | 19 | 20 | 39 | 39 |

Test, at the 5% significance level, the belief that the die is biased in the way described. (C)

4. A student of botany believed that *multifolium uniflorum* plants grow in random positions in grassy meadowland. He recorded the number of plants in one square metre of grassy meadow, and repeated the procedure to obtain the 148 results in the table.

| Number of plants | 0 | 1 | 2 | 3 | 4 | 5 | 6 | 7 or greater |
|---|---|---|---|---|---|---|---|---|
| Frequency | 9 | 24 | 43 | 34 | 21 | 15 | 2 | 0 |

(a) Show that, to two decimal places, the mean number of plants in one square metre is 2.59.
(b) Give a reason why the Poisson distribution might be an appropriate model for these data.

Using the Poisson model with mean 2.59, expected frequencies corresponding to the given frequencies were calculated, to two decimal places, and are shown in the table below.

| Number of plants | Expected frequencies |
|---|---|
| 0 | 11.10 |
| 1 | 28.76 |
| 2 | s |
| 3 | 32.15 |
| 4 | 20.82 |
| 5 | 10.78 |
| 6 | 4.65 |
| 7 or greater | t |

(c) Find the values of $s$ and $t$ to two decimal places.
(d) Stating clearly your hypotheses, test at the 5% level of significance, whether or not this Poisson model is supported by these data.

(L)

# 13

## Significance tests for correlation coefficients

*In this chapter you will learn about*

- a significance test for $r$, the product moment correlation coefficient
- a significance test for $r_s$, Spearman's coefficient of rank correlation

**Background knowledge**
*You need to be familiar with the ideas associated with correlation (see Chapter 2 page 119) and the methods for calculating the product-moment correlation coefficient (page 139) and Spearman's coefficient of rank correlation (page 146).*

## SIGNIFICANCE TESTS FOR CORRELATION COEFFICIENTS

*Before tackling this section you need to review the work covered in Chapter 2 on the product-moment correlation coefficient and Spearman's rank correlation coefficient.*

When a correlation coefficient has been calculated it is usual to make an assessment of the degree of correlation. You might say, for example, that there is good positive correlation between the variables or that there is weak negative correlation. There is a significance test that allows you to decide whether there is a correlation between the variables, backed by statistical theory rather than just a suspicion.

## TEST FOR THE PRODUCT-MOMENT CORRELATION COEFFICIENT, *r*

In Chapter 2 (page 139) you learnt how to calculate $r$, the **product-moment correlation coefficient** between two sets of data $X$ and $Y$.

Using small $s$ format:

$$r = \frac{s_{xy}}{s_x s_y} \quad \text{where} \quad s_{xy} = \frac{1}{n} \Sigma xy - \bar{x}\bar{y} = \frac{\Sigma xy}{n} - \bar{x}\bar{y}$$

$$s_x = \sqrt{s_{xx}} = \sqrt{\frac{1}{n} \Sigma x^2 - \bar{x}^2} = \sqrt{\frac{\Sigma x^2}{n} - \bar{x}^2}$$

$$s_y = \sqrt{s_{yy}} = \sqrt{\frac{1}{n} \Sigma y^2 - \bar{y}^2} = \sqrt{\frac{\Sigma y^2}{n} - \bar{y}^2}$$

Using big $S$ format:

$$r = \frac{S_{xy}}{S_x S_y} \quad \text{where} \quad S_{xy} = \Sigma xy - \frac{\Sigma x \, \Sigma y}{n}$$

$$S_x = \sqrt{S_{xx}} = \sqrt{\Sigma x^2 - \frac{(\Sigma x)^2}{n}}$$

$$S_y = \sqrt{S_{yy}} = \sqrt{\Sigma y^2 - \frac{(\Sigma y)^2}{n}}$$

Remember that $r$ is such that $-1 \leqslant r \leqslant 1$, where

$r = -1$    indicates perfect negative correlation

$r = 0$     indicates no correlation

$r = 1$     indicates perfect positive correlation.

If $r$ is very close to zero, then you would probably say that the two variables $X$ and $Y$ are not related at all. If $r$ is very close to 1, for example $r = 0.992$, then you would probably say that there is a strong positive linear correlation between $X$ and $Y$. But what about a value for $r$ of 0.694? Would you be able to claim that this indicates positive correlation? What about a value of $-0.5$? Does this indicate negative correlation between the variables? A significance test is needed!

In order to carry out a significance test, assume that $X$ and $Y$ are jointly normally distributed with correlation coefficient $\rho$, referred to as the population correlation coefficient. Data must be collected so that they constitute a random sample from the whole population values of $X$ and $Y$.

# The null hypothesis, $H_0$

The null hypothesis is always that the correlation coefficient is zero, i.e. there is no correlation between the variables. This is written $H_0: \rho = 0$.

# The alternative hypothesis, $H_1$

The alternative hypothesis depends on whether the test is one-tailed or two-tailed.

**One-tailed tests**

If you think there is a positive correlation between the variables $X$ and $Y$, the alternative hypothesis is $H_1: \rho > 0$ (there is a positive correlation between the variables).

If you think there is a negative correlation between the variables $X$ and $Y$, the alternative hypothesis is $H_1: \rho < 0$ (there is a negative correlation between the variables).

**Two-tailed tests**

If you are looking for a correlation but not specifying whether it is positive or negative, then the alternative hypothesis is $\rho \neq 0$ (there is some correlation between the variables).

The calculated value of $r$, the product-moment correlation coefficient, is compared with the critical value which is found from tables. An extract is given below and the tables are printed on page 652.

**Critical values for product-moment correlation coefficient**

| | 0.10 | 0.05 | Level 0.025 | 0.01 | 0.005 | Sample size |
|---|---|---|---|---|---|---|
| | 0.8000 | 0.9000 | 0.9500 | 0.9800 | 0.9900 | 4 |
| | 0.6870 | 0.8054 | 0.8783 | 0.9343 | 0.9587 | 5 |
| (i) | 0.6084 | 0.7293 | 0.8114 | 0.8822 | 0.9172 | 6 |
| | 0.5509 | 0.6694 | 0.7545 | 0.8329 | 0.8745 | 7 |
| (ii) | 0.5067 | 0.6215 | 0.7067 | 0.7887 | 0.8343 | 8 |
| | 04716 | 0.5822 | 0.6664 | 0.7498 | 0.7977 | 9 |
| (iii) | 0.4428 | 0.5494 | 0.6319 | 0.7155 | 0.7645 | 10 |

The tables are easy to use. The highlighted values are referred to in the following illustrations:

(i)   Consider hypotheses

$H_0$: $\rho = 0$ (there is no correlation between the variables)
$H_1$: $\rho > 0$ (there is a positive correlation between the variables).

This is a one-tailed (upper tail) test. At the 5% level, the critical value is found under column 0.05. If $r$ has been calculated from, say, six pairs of data, i.e. sample size 6, the critical value is 0.7293.

This means that in random samples from a distribution in which $\rho = 0$, only 5% of these samples will give a value of $r$ greater than 0.7293. So, at the 5% level of significance, you would reject $H_0$ (that there is no correlation) in favour of $H_1$ (that there is positive correlation) if $r > 0.7293$

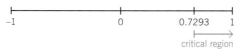

(ii)   The same tables are used when testing for a negative correlation. Consider hypotheses

$H_0$: $\rho = 0$ (there is no correlation between the variables)
$H_1$: $\rho < 0$ (there is a negative correlation between the variables).

This test is one-tailed (lower tail). At the 1% level, look up the value in the column headed 0.01. For a sample size of eight pairs of data, the value given in the table is 0.7887, indicating that the critical value is −0.7887. At the 1% level, you would reject $H_0$ if $r < -0.7887$.

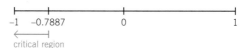

(iii)   Now consider hypotheses
$H_0$: $\rho = 0$ (there is no correlation between the variables)
$H_1$: $\rho \neq 0$ (there is some correlation between the variables).

This test is two-tailed. At the 5% level of significance, you want critical values that give 2.5% in each tail, so look under the column headed 0.025. For a sample size of 10, the critical value given in the table is 0.6319. This means that you would reject $H_0$ in favour of $H_1$ if $r > 0.6319$ or $r < -0.6319$ i.e. if $|r| > 0.6319$.

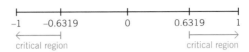

## Example 13.1

The scatter diagram illustrating ten pairs of values $(x, y)$ is shown below.

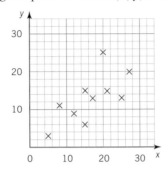

(a) Comment on the diagram.

(b) Calculate the value of $r$, the product-moment correlation coefficient for the pairs of data shown in the diagram.

(c) Assuming that $X$ and $Y$ are jointly normally distributed with correlation coefficient $\rho$, and the data constitutes a random sample, test, at the 5% level, whether there is a positive correlation between $X$ and $Y$.

(d) Would your conclusion be the same at the 1% level?

## Solution 13.1

(a) From the scatter diagram, there appears to be some positive linear correlation but it does not appear to be very strong.

(b) In the diagram, the data points are

| $x$ | 5 | 8 | 12 | 15 | 15 | 17 | 20 | 21 | 25 | 27 |
|---|---|---|---|---|---|---|---|---|---|---|
| $y$ | 3 | 11 | 9 | 6 | 15 | 13 | 25 | 15 | 13 | 20 |

Using the calculator in LR mode, it can be shown that $r = 0.6954$ (4d.p.).

*(See page 140 if you need to review how to calculate r, with or without a calculator.)*

(c) The significance test is carried out as follows:

1. State $H_0$ and $H_1$.

$H_0: \rho = 0$ (there is no correlation between $X$ and $Y$)
$H_1: \rho > 0$ (there is positive correlation between $X$ and $Y$)

2. State level and type of test.

Perform a one-tailed (upper tail) test at the 5% level.

3. State the rejection criterion.

The sample size is 10.
From tables, the critical value is 0.5494, so reject $H_0$ if $r > 0.5494$.

4. Calculate $r$.

From the calculations in (b), $r = 0.6954$.

5. Make conclusion.

Since $r > 0.5494$, $H_0$ is rejected in favour of $H_1$.
**There is evidence of positive correlation between $X$ and $Y$.**

(d) For a test at the 1% level, the critical value is 0.7155 so $H_0$ is rejected if $r > 0.7155$. Since $r = 0.6954 < 0.7155$, do not reject $H_0$.

**At the 1% level, there is not enough evidence to say that there is positive correlation between X and Y.**

## Exercise 13a    Significance test for product-moment correlation coefficient

1. In each of the following significance tests for the product-moment correlation coefficient the calculated value of $r$ is as shown. Use tables of critical values to decide whether $H_0$ is rejected or not.

|     | $n$ | $r$ | Hypotheses | Level of significance |
|-----|-----|-----|------------|------------------------|
| (a) | 7   | 0.893  | $H_0: \rho = 0, H_1: \rho \neq 0$ | 2%  |
| (b) | 14  | 0.499  | $H_0: \rho = 0, H_1: \rho > 0$    | 1%  |
| (c) | 28  | 0.324  | $H_0: \rho = 0, H_1: \rho \neq 0$ | 10% |
| (d) | 28  | 0.324  | $H_0: \rho = 0, H_1: \rho > 0$    | 1%  |
| (e) | 16  | -0.419 | $H_0: \rho = 0, H_1: \rho < 0$    | 5%  |
| (f) | 12  | -0.689 | $H_0: \rho = 0, H_1: \rho \neq 0$ | 10% |
| (g) | 12  | 0.689  | $H_0: \rho = 0, H_1: \rho > 0$    | 1%  |
| (h) | 10  | 0.733  | $H_0: \rho = 0, H_1: \rho > 0$    | 1%  |

2. A small bus company provides a service for a small town and some neighbouring villages. In a study of their service a random sample of 20 journeys was taken and the distances $x$, in kilometres, and journey times $t$, in minutes, were recorded. The average distance was 4.535 km and the average journey time was 15.15 minutes.

   (a) Using $\Sigma x^2 = 493.77$, $\Sigma t^2 = 4897$, $\Sigma xt = 1433.8$, calculate the product-moment correlation coefficient for these data.

   (b) Stating your hypotheses clearly, test, at the 5% level, whether or not there is evidence of a positive correlation between journey time and distance.

   (c) State any assumptions that have to be made to justify the test in (b).            (L)

3. In order to investigate the strength of the correlation between the value of a house and the value of the householder's car, a random sample of householders was questioned. The resulting data are shown in the table, the units being thousands of pounds.

   $\Sigma x = 762$      $\Sigma x^2 = 68\ 088$      $\Sigma y = 64.5$
   $\Sigma y^2 = 606.63$      $\Sigma xy = 6067.4$

   (a) Represent the data graphically.
   (b) Calculate the product-moment correlation coefficient.
   (c) Carry out a hypothesis test, at a suitable level of significance, to determine whether or not it is reasonable to suppose that the value of a house is positively correlated with the value of the householder's car.

| Value of house, $x$ | Value of car, $y$ |
|---------------------|-------------------|
| 110 | 12  |
| 106 | 9.5 |
| 51  | 2.4 |
| 94  | 4.2 |
| 66  | 4.1 |
| 26  | 0.3 |
| 72  | 3.2 |
| 51  | 6.0 |
| 53  | 7.8 |
| 133 | 15  |

   (d) A student argues that when two variables are correlated one must be the cause of the other. Briefly discuss this view with regard to the data in this question.            (MEI)

4. For the sets of data given, test the hypotheses indicated. Then draw a scatter diagram and comment on whether this reinforces your conclusion.
   [$\rho$ is the population product-moment correlation coefficient.]

   (a)

| $x$ | 7  | 12 | 13 | 17 | 23 | 25 | 30 | 20 |
|-----|----|----|----|----|----|----|----|----|
| $y$ | 23 | 22 | 18 | 15 | 7  | 13 | 8  | 27 |

   $H_0: \rho = 0$, $H_1: \rho < 0$; 5% significance level

   (b)

| $x$ | $y$ |
|------|------|
| 5.1  | 5.3  |
| 5.4  | 10.2 |
| 5.5  | 15.7 |
| 10   | 5    |
| 10.2 | 10.9 |
| 10.4 | 15.1 |
| 15   | 5.3  |
| 15.4 | 10.9 |
| 15.6 | 15.3 |
| 30   | 25.1 |
| 20.2 | 20   |

   $H_0: \rho = 0$, $H_1: \rho > 0$; 1% significance level

## SPEARMAN'S COEFFICIENT OF RANK CORRELATION, $r_s$

Spearman's coefficient of rank correlation is calculated using the ranks of the data. As you saw on page 146, for $n$ data points, if $d$ is the difference between the ranks for a data point, then

$$r_s = 1 - \frac{6 \Sigma d^2}{n(n^2 - 1)}.$$

Remember that $-1 \leqslant r_s \leqslant 1$, with $r_s = 1$ indicating perfect agreement between the rankings, $r_s = -1$ indicating that the rankings are in exact reverse order (complete disagreement) and $r_s = 0$ indicating no correlation between the rankings.

Writing $\rho_s$ for the population rank correlation coefficient, the null hypothesis is always

$$H_0: \rho_s = 0 \text{ (there is no correlation between the rankings)}$$

The alternative hypothesis is either

$$H_1: \rho_s > 0 \text{ and there is positive correlation (agreement) between the rankings}$$
(one-tailed (upper tail) test)

or $\quad H_1: \rho_s < 0$ and there is negative correlation (disagreement) between the rankings (one-tailed (lower tail) test)

or $\quad H_1: \rho_s \neq 0$ and there is correlation between the rankings (two-tailed test).

Note that the test for Spearman's coefficient of rank correlation does not make any assumptions about the population parameters. It is known as a **non-parametric test**.

The critical values for Spearman's rank correlation coefficient are found from tables which are very similar in format to those for the product-moment correlation coefficient. An extract is shown below and the tables are printed on page 652.

## Critical values for Spearman's rank correlation coefficient

| Sample size | Level 0.05 | 0.025 | 0.01 |
|---|---|---|---|
| 4 | 1.0000 | — | — |
| 5 | 0.9000 | 1.0000 | 1.0000 |
| 6 | 0.8286 | 0.8857 | 0.9429 |
| 7 | 0.7143 | 0.7857 | 0.8929 |
| 8 | 0.6429 | 0.7381 | 0.8333 |
| 9 | 0.6000 | 0.7000 | 0.7833 |
| 10 | 0.5636 | 0.6485 | 0.7455 |
| 11 | 0.5364 | 0.6182 | 0.7091 |

For a **one-tailed** test at the 5% level, sample size 7, look under column 0.05. This gives the value 0.7143 and means that

- for $H_1: \rho_s > 0$, $H_0$ is rejected if $r_s > 0.7143$

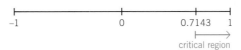

- for $H_1$: $\rho_s < 0$, $H_0$ is rejected if $r_s < -0.7143$.

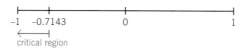

For a **two-tailed test** at the 5% level, sample size 9, look under column 0.025 (half of 5%). This gives the value 0.7000 and means that

- for $H_1$: $\rho_s \neq 0$, $H_0$ is rejected if $r_s > 0.7000$ or $r_s < -0.7000$, i.e. if $|r_s| > 0.7000$.

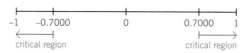

## Example 13.2

A teacher selects one boy and one girl at random from her class, and asks them to arrange 11 types of food in order of preference. The food types are labelled $A$ to $K$ and the results are given below.

| Boy's order: | E | K | F | C | B | I | D | A | G | J | H |
|---|---|---|---|---|---|---|---|---|---|---|---|
| Girl's order: | F | K | E | C | B | I | H | D | A | J | G |

(a) Calculate Spearman's rank correlation coefficient for these data.

(b) Stating your hypotheses clearly test, at the 1% level of significance, whether or not there is evidence of a positive correlation.

(c) Interpret your conclusion to the test in part (b). (L)

## Solution 13.2

(a)

| Food type | A | B | C | D | E | F | G | H | I | J | K |
|---|---|---|---|---|---|---|---|---|---|---|---|
| Boy's order, $x$ | 8 | 5 | 4 | 7 | 1 | 3 | 9 | 11 | 6 | 10 | 2 |
| Girl's order, $y$ | 9 | 5 | 4 | 8 | 3 | 1 | 11 | 7 | 6 | 10 | 2 |
| $d$ | -1 | 0 | 0 | -1 | -2 | 2 | -2 | 4 | 0 | 0 | 0 |
| $d^2$ | 1 | 0 | 0 | 1 | 4 | 4 | 4 | 16 | 0 | 0 | 0 |

$\Sigma d^2 = 30$ and $n = 11$, so $r_s = 1 - \dfrac{6\Sigma d^2}{n(n^2 - 1)}$

$$= 1 - \frac{6 \times 30}{11 \times 120}$$

$$= 0.8636 \dots$$

(b) The significance test is carried out as follows:

| | |
|---|---|
| 1. State $H_0$ and $H_1$. | $H_0$: $\rho_s = 0$ (there is no correlation) <br> $H_1$: $\rho_s > 0$ (there is positive correlation between the boy's and girl's preferences) |
| 2. State level and type of test. | Perform a one-tailed (upper tail) test at the 1% level. |
| 3. State the rejection criterion. | The sample size is 11. <br> From tables (page 652), the critical value is 0.7091, so reject $H_0$ if $r_s > 0.7091$. |
| 4. Calculate $r_s$. | From part (a), $r_s = 0.8636 \ldots$ |
| 5. Make conclusion. | Since $r_s > 0.7091$, $H_0$ is rejected in favour of $H_1$. <br> **There is evidence of positive correlation between the boy's and girl's preferences.** |

(c) **The boy and girl agree in their preferences.**

---

# Exercise 13b  Significance test for Spearman's coefficient of rank correlation

1. In each of the following significance tests for Spearman's rank correlation coefficient, the value of $\Sigma d^2$ obtained when calculating $r_s$ is as shown. Use tables of critical values to decide whether $H_0$ is rejected or not.

| | $n$ | $\Sigma d^2$ | Hypotheses | Level of significance |
|---|---|---|---|---|
| (a) | 9 | 212 | $H_0$: $\rho_s = 0$, $H_1$: $\rho_s < 0$ | 1% |
| (b) | 8 | 30 | $H_0$: $\rho_s = 0$, $H_1$: $\rho_s > 0$ | 5% |
| (c) | 8 | 30 | $H_0$: $\rho_s = 0$, $H_1$: $\rho_s \neq 0$ | 5% |
| (d) | 10 | 78 | $H_0$: $\rho_s = 0$, $H_1$: $\rho_s > 0$ | 5% |
| (e) | 10 | 252 | $H_0$: $\rho_s = 0$, $H_1$: $\rho_s < 0$ | 5% |
| (f) | 10 | 274 | $H_0$: $\rho_s = 0$, $H_1$: $\rho_s \neq 0$ | 5% |
| (g) | 7 | 18 | $H_0$: $\rho_s = 0$, $H_1$: $\rho_s \neq 0$ | 10% |
| (h) | 7 | 106 | $H_0$: $\rho_s = 0$, $H_1$: $\rho_s < 0$ | 1% |
| (i) | 7 | 14 | $H_0$: $\rho_s = 0$, $H_1$: $\rho_s \neq 0$ | 5% |

2. An expert on porcelain is asked to place 7 china bowls in date order of manufacture assigning the rank 1 to the oldest bowl. The actual dates of manufacture and the order given by the expert are shown.

| Bowl | Date of manufacture | Order given by expert |
|---|---|---|
| A | 1920 | 7 |
| B | 1857 | 3 |
| C | 1710 | 4 |
| D | 1896 | 6 |
| E | 1810 | 2 |
| F | 1690 | 1 |
| G | 1780 | 5 |

Find, to three decimal places, the Spearman rank correlation coefficient between the order of manufacture and the order given by the expert.

Refer to tables of critical values to comment on the significance of your result. State clearly the null hypothesis which is being tested. (L)

3. Applicants for a job with a company are interviewed by two of the personnel staff. After the interviews each applicant is awarded a mark by each of the interviewers. The marks are given below.

| | Candidate | | | | | | | |
|---|---|---|---|---|---|---|---|---|
| | A | B | C | D | E | F | G | H |
| Interviewer 1 | 22 | 27 | 24 | 17 | 20 | 22 | 16 | 13 |
| Interviewer 2 | 28 | 23 | 25 | 14 | 26 | 17 | 20 | 15 |

(a) Calculate, to two decimal places, the Spearman rank correlation coefficient between these two sets of marks.
(b) stating your hypotheses and using a 5% level of significance, interpret your result. (L)

4. Ten architects each produced a design for a new building and two judges, A and B, independently awarded marks, x and y respectively, to the 10 designs, as given in the table below.

| Design | Judge A (x) | Judge B (y) |
|--------|-------------|-------------|
| 1 | 50 | 46 |
| 2 | 35 | 26 |
| 3 | 55 | 48 |
| 4 | 60 | 44 |
| 5 | 85 | 62 |
| 6 | 25 | 28 |
| 7 | 65 | 30 |
| 8 | 90 | 60 |
| 9 | 45 | 34 |
| 10 | 40 | 42 |

Calculate Spearman's rank correlation coefficient for the data and test, at the 5% level, the hypothesis that there is no correlation between the marks awarded by the two judges. (C)

5. In a ski-jumping contest each competitor made two jumps. The orders of merit for the 10 competitors who completed both jumps are shown in the table.

| Ski jumper | First jump | Second jump |
|------------|-----------|-------------|
| A | 2 | 4 |
| B | 9 | 10 |
| C | 7 | 5 |
| D | 4 | 1 |
| E | 10 | 8 |
| F | 8 | 9 |
| G | 6 | 2 |
| H | 5 | 7 |
| I | 1 | 3 |
| J | 3 | 6 |

(a) Calculate, to two decimal places, a rank correlation coefficient for the performances of the ski-jumpers in the two jumps.
(b) Using a 5% level of significance and quoting from tables of critical values, interpret your result. State clearly your null and alternative hypotheses. (L)

6. The positions in a league of 8 hockey clubs at the end of a season are shown in the table. Shown also are the average attendances (in hundreds) at home matches during that season.
Calculate a coefficient of rank correlation between position in the league and average home attendance.

| Club | Position | Average attendance |
|------|----------|--------------------|
| A | 1 | 30 |
| B | 2 | 32 |
| C | 3 | 12 |
| D | 4 | 19 |
| E | 5 | 27 |
| F | 6 | 18 |
| G | 7 | 15 |
| H | 8 | 25 |

Refer to the appropriate table of critical values to comment on the significance of your result, stating clearly the null hypothesis being tested. (L)

## Summary

● **Significance test for the product-moment correlation coefficient, r**

*The assumptions are that X and Y are jointly normally distributed and the sample must constitute a random sample from the whole populations of X and Y.*

1. State $H_0: \rho = 0$ (there is no correlation between X and Y)
   State $H_1$ as follows

   $H_1: \rho > 0$ (there is positive correlation between X and Y)    $H_1: \rho < 0$ (there is negative correlation between X and Y)    $H_1: \rho \neq 0$ (there is correlation between X and Y)

2. State the level and type of test, e.g. one-tailed test at the 5% level.

3.  State the rejection criterion, obtaining the critical value from tables.

| Reject $H_0$ if | Reject $H_0$ if | Reject $H_0$ if |
|---|---|---|
| $r >$ critical value | $r < -$ critical value | $|r| >$ critical value |

4.  Calculate $r$ and compare with the critical value.
5.  Make your conclusion.

## Significance test for Spearman's rank correlation coefficient, $r_s$

Note that this is a non-parametric test.

1.  State $H_0: \rho_s = 0$ (there is no correlation between the ranks of $X$ and $Y$)
    State $H_1$ as follows

| $H_1: \rho_s > 0$ (there is agreement between the ranks of $X$ and $Y$) | $H_1: \rho_s < 0$ (there is disagreement between the ranks of $X$ and $Y$) | $H_1: \rho_s \neq 0$ (there is correlation between the ranks of $X$ and $Y$) |
|---|---|---|

2.  State the level and type of test, e.g. one-tailed test at the 5% level.
3.  State the rejection criterion, obtaining the critical value from tables:

| Reject $H_0$ if | Reject $H_0$ if | Reject $H_0$ if |
|---|---|---|
| $r_s >$ critical value | $r_s < -$ critical value | $|r_s| >$ critical value |

4.  Calculate $r_s$ and compare with the critical value.
5.  Make your conclusion.

# Miscellaneous worked example

## Example 13.3

During the lambing season 8 ewes and the lambs they bore were weighed at the time of birth with the following results:

| Ewe | A | B | C | D | E | F | G | H |
|---|---|---|---|---|---|---|---|---|
| Weight of ewe, $x$ kg | 44 | 41 | 43 | 40 | 41 | 37 | 38 | 35 |
| Weight of lamb, $y$ kg | 3.5 | 2.8 | 3.2 | 2.7 | 2.9 | 2.5 | 2.8 | 2.6 |

You may assume $\Sigma x = 319$, $\Sigma y = 23.0$, $\Sigma x^2 = 12\ 785$, $\Sigma y^2 = 66.88$, $\Sigma xy = 923.2$.

Calculate the product-moment correlation coefficient between $X$ and $Y$.
Making any necessary assumptions, test whether the data could have come from a population with correlation coefficient $\rho = 0$. Use a 5% level of significance. (*AEB*)

## Solution 13.3

Using small $s$ formula to calculate $r$:

$$s_{xy} = \frac{\Sigma xy}{n} - \bar{x}\bar{y} = \frac{923.2}{8} - \frac{319}{8} \times \frac{23.0}{8} = 0.759 \ldots$$

$$s_{xx} = \frac{\Sigma x^2}{n} - \bar{x}^2 = \frac{12\,785}{8} - \left(\frac{319}{8}\right)^2 = 8.1093 \ldots$$

$$s_{yy} = \frac{\Sigma y^2}{n} - \bar{y}^2 = \frac{66.88}{8} - \left(\frac{23.0}{8}\right)^2 = 0.0943 \ldots$$

$$r = \frac{s_{xy}}{s_x s_y} = \frac{0.759 \ldots}{\sqrt{8.1093 \ldots} \times \sqrt{0.0943 \ldots}} = 0.868 \text{ (3 s.f.)}$$

Using big $S$ formula to calculate $r$:

$$S_{xy} = \Sigma xy - \frac{\Sigma x \Sigma y}{n} = 923.2 - \frac{319 \times 23.0}{8} = 6.075$$

$$S_{xx} = \Sigma x^2 - \frac{(\Sigma x)^2}{n} = 12\,785 - \frac{319^2}{8} = 64.875$$

$$S_{yy} = \Sigma y^2 - \frac{(\Sigma y)^2}{n} = 66.88 - \frac{23.0^2}{8} = 0.755$$

$$r = \frac{S_{xy}}{S_x S_y} = \frac{6.075}{\sqrt{64.875} \times \sqrt{0.755}} = 0.868 \text{ (3 s.f.)}$$

**The product moment correlation coefficient between the weight of a ewe and the weight of its lamb is 0.868.**

---

Assume that $X$ and $Y$ are jointly normally distributed with product-moment correlation coefficient $\rho$ and the data form a random sample from the populations of $X$ and $Y$. The significance test is carried out as follows:

1. State $H_0$ and $H_1$.

$H_0: \rho = 0$ (there is no correlation between the weight of a ewe and its lamb)
$H_1: \rho \neq 0$ (there is correlation between the weight of a ewe and its lamb)

2. State level and type of test.

Perform a two-tailed test at the 5% level.

3. State the rejection criterion.

*The sample size is 8. From tables, the critical value for a two-tailed test at the 5% level is 0.7067 (page 652, row n = 8, column 0.025).*

$H_0$ is rejected if $|r| > 0.7067$.

4. Calculate $r$.

For the data, $r = 0.868$.

5. Make conclusion.

Since $|r| > 0.7067$, $H_0$ is rejected in favour of $H_1$. There is evidence of correlation between the weight of a ewe and its lamb.

**It is unlikely that the data came from a population with correlation coefficient $\rho = 0$.**

Note that the conclusion would have been the same if you had chosen to carry out a one-tailed test. In this case $H_1$ is $\rho > 0$, the critical value of $r$ is 0.6215 and $H_0$ is rejected since $r > 0.6215$.

## Example 13.4

The coursework grades, *A* highest to *G* lowest, and examination marks of 8 candidates are given below.

| Coursework Grade | Examination Mark |
|:---:|:---:|
| A | 92 |
| C | 75 |
| D | 63 |
| B | 54 |
| F | 48 |
| C | 45 |
| G | 34 |
| E | 18 |

(a) Calculate the value of an appropriate measure of correlation between these two sets of data.

(b) Test whether this value indicates evidence of correlation between coursework grades and examination grades at a 5% significance level.

(c) Give a practical interpretation of your value.                    (*AQA*)

## Solution 13.4

(a) Calculating Spearman's rank correlation coefficient, $r_s$

| Coursework | A | C | D | B | F | C | G | E |
|---|---|---|---|---|---|---|---|---|
| Examination mark | 92 | 75 | 63 | 54 | 48 | 45 | 34 | 18 |
| Coursework rank | 1 | 3.5 | 5 | 2 | 7 | 3.5 | 8 | 6 |
| Examination mark rank | 1 | 2 | 3 | 4 | 5 | 6 | 7 | 8 |
| $\lvert d \rvert$ | 0 | 1.5 | 2 | 2 | 2 | 2.5 | 1 | 2 |
| $d^2$ | 0 | 2.25 | 4 | 4 | 4 | 6.25 | 1 | 4 |

$$\Sigma d^2 = 25.5, \quad n = 8, \quad \text{therefore } r_s = 1 - \frac{6 \, \Sigma d^2}{n(n_2 - 1)}$$

$$= 1 - \frac{6 \times 25.5}{8 \times 63}$$

$$= 0.696 \quad (3 \text{ s.f.})$$

(b) $H_0$: $\rho_s = 0$ (there is no correlation)
$H_1$: $\rho_s \neq 0$ (there is evidence of correlation)

Perform a two-tailed test at 5% level.
From tables (page 651), critical value is 0.7381 ($n = 8$, column 0.025).

Reject $H_0$ if $\lvert r_s \rvert > 0.7381$.

Since $r_s = 0.696 < 0.7381$, do not reject $H_0$. **There is no evidence of correlation.**

(c) **Performance in the examination does not reflect on performance in coursework.**

# Miscellaneous exercise 13c

1  To test the belief that milder winters are followed by warmer summers, meteorological records are obtained for a random sample of 10 years. For each year the mean temperatures are found for January and July. The data, in degrees Celsius, are given below.

| Jan | July |
|-----|------|
| 8.3 | 16.2 |
| 7.1 | 13.1 |
| 9.0 | 16.7 |
| 1.8 | 11.2 |
| 3.5 | 14.9 |
| 4.7 | 15.1 |
| 5.8 | 17.7 |
| 6.0 | 17.3 |
| 2.7 | 12.3 |
| 2.1 | 13.4 |

(a)  Rank the data and calculate Spearman's rank correlation coefficient.

(b)  Test, at the 2.5% level of significance, the belief that milder winters are followed by warmer summers. State clearly the null and alternative hypotheses under test.

(c)  Would it be more appropriate, less appropriate or equally appropriate to use the product-moment correlation coefficient to analyse these data? Briefly explain why.
(MEI)

2.  Bird abundance may be assessed in several ways. In one long-term study in a nature reserve, two independent surveys (A and B) are carried out. The data show the number of wren territories recorded (survey A) and the numbers of adult wrens trapped in a fine mesh net (survey B) over a number of years.

| Survey A | Survey B |
|----------|----------|
| 16 | 11 |
| 19 | 12 |
| 27 | 15 |
| 50 | 18 |
| 60 | 22 |
| 70 | 35 |
| 79 | 35 |
| 79 | 71 |
| 84 | 46 |
| 85 | 53 |
| 97 | 52 |

(a)  Plot a scatter diagram to compare results for the two surveys.

(b)  Calculate Spearman's coefficient of rank correlation.

(c)  Perform a significance test, at the 5% level, to determine whether there is any association between the results of the two surveys. Explain what your conclusion means in practical terms.

(d)  Would it be more appropriate, less appropriate or equally appropriate to use the product moment correlation coefficient to analyse these data? Briefly explain why.
(MEI)

3.  The data below shows the height above sea level, $x$ metres, and the temperature, $y$ °C, at 7.00 a.m., on the same day in summer at nine places in Europe.

| Height, $x$ | Temperature, $y$ |
|-------------|------------------|
| 1400 | 6 |
| 400 | 15 |
| 280 | 18 |
| 790 | 10 |
| 390 | 16 |
| 590 | 14 |
| 540 | 13 |
| 1250 | 7 |
| 680 | 13 |

(a)  Plot these data on a scatter diagram.

(b)  Calculate the product–moment correlation coefficient between $x$ and $y$.
(Use $\Sigma x^2 = 5\,639\,200$, $\Sigma y^2 = 1524$, $\Sigma xy = 66\,450$)

(c)  Give an interpretation of your coefficient. On the same day the number of hours of sunshine was recorded and Spearman's rank correlation between hours of sunshine and temperature, based on $\Sigma d^2 = 28$ was 0.767.

(d)  Stating clearly your hypotheses and using a 5% two-tailed test, interpret this rank correlation coefficient.
(L)

4.  At the end of a season a league of eight ice hockey clubs produced the following table showing the position of each club in the league and the average attendances (in hundreds) at home matches.

| Club | A | B | C | D | E | F | G | H |
|------|---|---|---|---|---|---|---|---|
| Position | 1 | 2 | 3 | 4 | 5 | 6 | 7 | 8 |
| Average | 37 | 38 | 19 | 27 | 34 | 26 | 22 | 32 |

(a)  Calculate the Spearman rank correlation coefficient between position in the league and average home attendance.

(b) Stating clearly your hypotheses and using a 5% two-tailed test, interpret your rank correlation coefficient.

Many sets of data include tied ranks.

(c) Explain briefly how tied ranks can be dealt with. *(L)*

5. At an agricultural show ten Shetland sheep were ranked by a qualified judge and by a trainee judge. Their rankings are shown in the table.

| Qualified judge | Trainee judge |
|:---:|:---:|
| 1 | 1 |
| 2 | 2 |
| 3 | 5 |
| 4 | 6 |
| 5 | 7 |
| 6 | 8 |
| 7 | 10 |
| 8 | 4 |
| 9 | 3 |
| 10 | 9 |

Calculate a rank correlation coefficient for these data.
Using one of the tables provided and a 5% significance level, state your conclusions as to whether there is some degree of agreement between the two sets of ranks. *(L)*

6. A teacher recorded the following data which refer to the marks gained by 13 children in an aptitude test and a statistics examination.

| Child | Aptitude Test, $x$ | Statistics Examination, $y$ |
|:---:|:---:|:---:|
| A | 54 | 84 |
| B | 52 | 68 |
| C | 42 | 71 |
| D | 31 | 37 |
| E | 43 | 79 |
| F | 23 | 58 |
| G | 32 | 33 |
| H | 49 | 60 |
| I | 37 | 47 |
| J | 13 | 60 |
| K | 13 | 44 |
| L | 36 | 64 |
| M | 39 | 49 |

(a) Draw a scatter diagram to represent these two sets of marks.

(b) Calculate, to three decimal places, the product–moment correlation coefficient between the test mark and the examination mark.

(You may use $\Sigma x^2 = 18\,672$, $\Sigma y^2 = 46\,626$, $\Sigma xy = 28\,234$)

(c) Comment on your result.

(d) The teacher decided that, on the basis of the scatter diagram, children $F$, $J$ and $K$ performed differently from the rest of the group. Suggest why the teacher might have come to that decision.

(e) The teacher decides to analyse the data ignoring these three children. Calculate, to three decimal places, the Spearman rank correlation coefficient between the other ten pairs of observations.

(f) Using a 5% level of significance and quoting from tables of critical values interpret the rank correlation coefficient. Use a one-tailed test.
State clearly the null and alternative hypotheses. *(L)*

7. The yield (per hectare) of a crop, $c$, is believed to depend on the May rainfall, $m$. For 9 regions records are are kept of the average values of $c$ and $m$, and these are recorded below.

| $c$ | $m$ |
|:---:|:---:|
| 8.3 | 14.7 |
| 10.1 | 10.4 |
| 15.2 | 18.8 |
| 6.4 | 13.1 |
| 11.8 | 14.9 |
| 12.2 | 13.8 |
| 13.4 | 16.8 |
| 11.9 | 11.8 |
| 9.9 | 12.2 |

($\Sigma c = 99.2$, $\Sigma m = 126.5$, $\Sigma c^2 = 1150.16$, $\Sigma m^2 = 1832.07$, $\Sigma mc = 1427.15$)

(a) Find the equation of the appropriate regression line.

(b) Find $r$, the linear (product–moment) correlation coefficient between $c$ and $m$.

(c) In a tenth region the average May rainfall was 14.6. Estimate the average yield of the crop for that region, giving your answer correct to one decimal place.

(d) Calculate the value of $r_s$, Spearman's rank correlation coefficient, for the above data and determine whether it is significantly greater than zero at the 5% level.

(e) State, with a reason, which of $r$ and $r_s$ you regard as being more appropriate for these data. *(C)*

8. In a random sample of 8 areas, residents were asked to express their approval or disapproval of the services provided by the local authority. A score of 0 represented complete dissatisfaction, and 10 represented complete satisfaction. The

table below shows the mean score for each local authority together with the authority's level of community charge.

| Authority | Community Charge (£) | Approval rating |
|---|---|---|
| A | 485 | 3.0 |
| B | 490 | 4.4 |
| C | 378 | 5.0 |
| D | 451 | 4.6 |
| E | 384 | 4.1 |
| F | 352 | 5.5 |
| G | 420 | 5.8 |
| H | 212 | 6.1 |

Calculate Spearman's rank correlation coefficient for these data.
Carry out a significance test at the 5% level using the value of the correlation coefficient which you have calculated. State carefully the null and alternative hypotheses under test and the conclusion to be drawn. (*MEI*)

9. A local historian was studying the number of births in a town and found the following figures relating to the years 1925 to 1934.

| Male births, $x$ | Female births, $y$ |
|---|---|
| 223 | 219 |
| 218 | 205 |
| 223 | 209 |
| 223 | 239 |
| 242 | 252 |
| 278 | 256 |
| 299 | 254 |
| 256 | 257 |
| 255 | 259 |
| 292 | 323 |

(a) Draw a scatter diagram to illustrate this information.

The historian calculated the following summary statistics from the data:

$S_{xx} = 8276.9$,   $S_{yy} = 10\,230.1$,   $S_{xy} = 7206.3$.

(b) Calculate the product–moment correlation coefficient.

The historian believed these data gave strong evidence of a positive correlation between male and female births.

(c) Stating your hypotheses clearly, test at the 1% level of significance whether or not there is evidence to support the historian's belief.

(d) State an assumption required for the validity of the test in part (c) and comment on whether or not you consider it to be met.

In 1924 there were 249 male and 177 female births.

(e) Without carrying out any further calculations state, giving a reason, what effect the inclusion of these figures would have on the value of the product moment correlation coefficient. (*L*)

10. Seven rock samples taken from a particular locality were analysed. The percentages, C and M, of two oxides contained in each sample were recorded. The results are shown in the table.

| Sample | C | M |
|---|---|---|
| 1 | 0.60 | 1.06 |
| 2 | 0.42 | 0.72 |
| 3 | 0.51 | 0.94 |
| 4 | 0.56 | 1.04 |
| 5 | 0.31 | 0.84 |
| 6 | 1.04 | 1.16 |
| 7 | 0.80 | 1.24 |

Given that

$\Sigma CM = 4.459$,   $\Sigma C^2 = 2.9278$,   $\Sigma M^2 = 7.196$,

find, to three decimal places, the product–moment correlation coefficient of the percentages of the two oxides. Calculate also, to three decimal places, a rank correlation coefficient. Using tables state any conclusions which you draw from the value of your rank correlation coefficient. State clearly the null hypothesis being tested. (*L*)

11. In the table below, $x$ is the average weekly household income in £ and $y$ the infant mortality per 1000 live births in 11 regions of the UK in 1985.

| Region | $x$ | $y$ |
|---|---|---|
| A | 170.4 | 8.4 |
| B | 183.2 | 9.4 |
| C | 172.9 | 10.3 |
| D | 187.1 | 10.5 |
| E | 203.2 | 8.3 |
| F | 204.8 | 9.4 |
| G | 208.8 | 8.5 |
| H | 248.0 | 9.0 |
| I | 198.3 | 9.4 |
| J | 187.1 | 9.8 |
| K | 179.1 | 9.6 |

It is hypothesised that a high value of $x$ will be associated with a low value of $y$. Calculate a rank correlation coefficient and test its significance.

It appears that region $A$ is exceptional. What would your findings be if this region were omitted from the analysis?

# Mixed test 13A   Correlation Coefficients

1. The bivariate sample illustrated in the scatter diagram below shows the heights, $x$ cm, and masses, $y$ kg, of a random sample of 20 students.

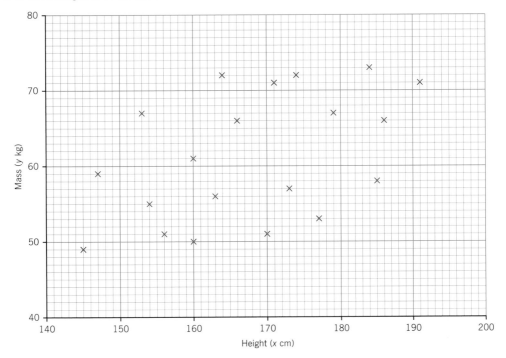

Height (x cm)

You are given that $\Sigma x = 3358$, $\quad \Sigma x^2 = 567\,190$, $\Sigma y = 1225$, $\quad \Sigma y^2 = 76\,357$, $\quad \Sigma xy = 206\,680$.

(a) Calculate the product-moment correlation coefficient.
(b) Carry out a hypothesis test, at the 5% level of significance, to determine whether or not there is evidence that the height of a student is positively correlated with his or her mass. What feature of the scatter diagram suggests that this test is appropriate?
(c) A statistics student suggests that a positive correlation between height and mass implies that 'the taller a student is the heavier he or she will be'. Comment on this statement with reference to your conclusions in part (b).                                              (MEI)

2. Two judges give marks for artistic impression (out of a maximum 6.0) to 10 ice skaters.

| Skater | Judge 1 | Judge 2 |
|--------|---------|---------|
| A | 5.3 | 5.4 |
| B | 4.9 | 5.0 |
| C | 5.6 | 5.8 |
| D | 5.2 | 5.6 |
| E | 5.7 | 5.2 |
| F | 4.8 | 4.5 |
| G | 5.2 | 4.7 |
| H | 4.6 | 4.8 |
| I | 5.1 | 5.3 |
| J | 4.9 | 4.9 |

(a) Calculate the value of Spearman's rank correlation coefficient for the marks of the two judges.
(b) Use your answer to (a) to test, at the 5% level of significance, whether it appears that there is some overall agreement between the judges. State your hypotheses and your conclusions carefully.

(c) For these marks the product-moment correlation coefficient is 0.6705. Use this to test, at the 5% level, whether there is any correlation between assessments of the two Judges.

(d) Comment on which is the more appropriate test to use in this situation. (MEI)

3. It is hypothesised that there is a positive correlation between the population of a country and its area. The following table gives a random sample of 13 countries with their area $x$, in thousands of square kilometres, and population $y$, in millions.

| Country | $x$ | $y$ |
|---------|------|------|
| 1 | 2.5 | 0.5 |
| 2 | 28 | 5 |
| 3 | 30 | 2 |
| 4 | 72 | 4 |
| 5 | 98 | 42 |
| 6 | 121 | 21 |
| 7 | 128 | 16 |
| 8 | 176 | 3 |
| 9 | 239 | 14 |
| 10 | 313 | 37 |
| 11 | 407 | 6 |
| 12 | 435 | 17 |
| 13 | 538 | 22 |

Plot a scatter diagram and comment on its implication for the hypothesis.
Calculate a suitable correlation coefficient and test its significance at the 5% level. (C)

4. A psychologist was studying the relationship between short term memory and ability in mathematics. A sample of 8 students was shown a tray of objects for 5 seconds and the students were asked to recall as many of the objects as they could. The number of objects recalled correctly was recorded and compared with their mark (percentage) in a recent mathematics examination. The results are given below.

| Student | A | B | C | D | E | F | G | H |
|---------|----|----|----|----|----|----|----|----|
| No. of objects | 3 | 5 | 12 | 8 | 7 | 11 | 4 | 9 |
| Maths % | 56 | 64 | 75 | 69 | 48 | 63 | 52 | 84 |

(a) Calculate Spearman's rank correlation coefficient for these data.

(b) Using a 5% level of significance, and stating your hypotheses clearly, interpret your result.

(c) Give a reason why it may be more appropriate to use Spearman's rank correlation coefficient for the hypothesis test than the product-moment correlation coefficient. (L)

# ICT STATISTICS SUPPLEMENT

## Contents

## INTRODUCTION

The intention of this supplement is to explore the use of ICT in the teaching and learning of statistics and probability. It is well established now that dynamic and interactive computer images can bring subjects to life in a way that was impossible to imagine before. The principle benefit is that lessons can now have variety. The same topic can be presented in the traditional way on the board, explored in a practical simulation, investigated using a spreadsheet or illustrated using a graph plotter. There is also quite likely to be a useful JAVA applet or some interesting real data on it from the internet.

Furthermore students can now present their findings electronically, and teachers can store, share and continually refine their lesson plans. And if you add to all this the obvious benefit of the computer's ability to carry out calculations without effort, ICT methods are almost guaranteed to enhance the enjoyment of those teaching and studying this subject.

## USING A SPREADSHEET

A spreadsheet, such as Excel, has enormous power; it can effortlessly analyse huge data sets, conduct simple simulations. Getting familiar with all this can come only with practice, and there are plenty of spreadsheet tutorials around, both in print and on the net. For starters, here is a summary of some of the more important features that relate to statistics, and in each of the following pages, features that relate to specific topics are listed.

## Entering formulae and functions

Enter '=', or click on the '=' button

Use the Functions menu that is now inserted to the left of the cross and tick to find the formula you want, or type it in if you know it, then click on the '='. This way you get a helpful entry box, explaining what each parameter is, and a useful HELP link. Details of appropriate formulae are given under the chapter headings.

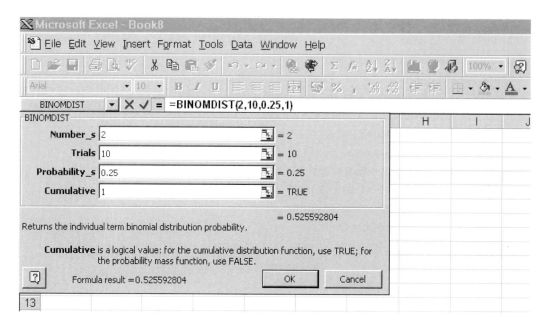

## Array

This word is used for a continuous selection of cells in a spreadsheet. Arrays are referred to by the top-left and bottom-right coordinates, e.g. A2:B16

## The edit menu

To insert a data set from some other source it is sometimes useful to try
Edit ⟹ Paste Special ⟹ select "text"

To generate a series, e.g. 1, 2, 3, ... 1000: First enter the start value in a cell and highlight it.
Edit ⟹ Fill ⟹ Series

## The format menu

To control the appearance of the numbers:
Format ⟹ cells ⟹ Format Cells ⟹ Number tab ⟹ Number ⟹ Decimal places

To control the appearance of a selected array:
Format ⟹ column ⟹ autofit
Format ⟹ autoformat (fancy styles)

To control format conditionally:
Format ⟹ conditional formatting
e.g. if = 0 then put a frame round it.

## The tools menu

To solve equations:
Tools ⟹ solver: this finds the value of a cell that makes the target cell a max, a min or = a value, e.g. to solve $x^3 - 2 = 0$, set A1 = 1 and B1 = A1^3 − 2. Select B1 then Tools ⟹ Solver. B1 is the 'target cell', "Equal to Value of" (0), by changing cell A1. Solve!

To show formulae instead of results:
Tools ⟹ options ⟹ View ⟹ Window options
TICK 'show formulas' (or use Ctrl-`)

To hide the grid lines:
Tools ⟹ options ⟹ View ⟹ Window options
UNTICK 'show gridlines'

To ensure auto-recalculation on pressing F9:
Tools ⟹ options ⟹ Calculations ⟹ Automatic

For iterative solving, e.g. $x = g(x)$:
Tools ⟹ options ⟹ Iteration
⟹ Max iterations (100) ⟹ Max change 0.001
e.g. Enter A1 = 1
    Select A1
    Enter = COS(A1)
    Press  F9

For a cell formula referring to itself without iteration:
Tools ⟹ options ⟹ Iteration
⟹ Max iterations (1) (leave Max change)
(e.g. D1 = D1 + 1 to increment by 1)

To draw a histogram (applicable only to equal classes):
Tools ⟹ Data Analysis ⟹ Histogram
(see Chapter 1)

For random number generation (see Chapter 9):
Tools ⟹ Data Analysis
    ⟹ Random Number generation
Continuous Uniform $(a, b)$, normal $(m, s)$, Bernouilli $(p)$, binomial $(n, p)$, Poisson$(m)$, Discrete, in 2 columns: $x$, $P(X = x)$

For sampling $n$ items from an array:
Tools ⟹ Data Analysis ⟹ Sampling

For t-tests:
Tools ⟹ Data Analysis
    ⟹ t-test (paired 2 samples for means)
2-sample assuming equal variances
2-sample assuming unequal variances

For the forms toolbar:
Tools ⇒ Customize ⇒ Forms
Label, Group-box, check-box, option button, list-box, combo-box, scroll-bar, spinner

## The data menu

Click first on the top left hand corner of the data.

*Sorting and filtering:*
Data ⇒ sort (up to 3 columns deep)
Data ⇒ filter ⇒ auto-filter

This is useful if a dataset has pasted incorrectly into a single column. Converting text into columns:
Data ⇒ Text to Columns

*For re-interpreting multiple data sets:*
Data ⇒ PivotTable report

## Charts

Excel was never written as an educational tool, so it is worth getting to know the charts option to sort out what is going to be useful and what is purely decorative. Of the chart options available 'Column', 'Bar', 'Line', and 'XY (Scatter)' all work well, but Excel is not good at Histograms, or Time Series Moving Averages.

## USING INTERNET RESOURCES AND WORD

There is an ever-growing amount of useful information on the net for the study of statistics and probability. Mixed in with all that is the less useful, and the task of sifting out quality resources is getting harder. Listed here are some authenticated sites, but the pace of change being what it is there is no guarantee that they will still be there and still be useful by the time this is read!

## Data sets

*http://lib.stat.cmu.edu/DASL/*
DASL: The Data and Story Library (USA), categorised by topic.

*http://www.maths.uq.edu.au/~gks/data*
OzDASL (University of Queensland, Australia). Australian version of the above.

*http://forum.swarthmore.edu/workshops/sum96/data.collections/datalibrary/*
The Data Library (from the Math Forum, USA)

*http://www.ni.com.au/mercury/mathguys/mercindx.htm*
Chance and Data (from Tasmania, Australia)

*http://lottery.merseyworld.com/*
The UK Lottery Web Site – includes statistics from all the draws.

*http://www.fa-premier.com/results/*
UK Premier Football results and statistics

*http://www.stats.ox.ac.uk/links/schoenfield.htm*
Schoenfield's List of Data Archives (Oxford University)

*http://sunsite.unc.edu/lunarbin/worldpop*
Demographic statistics (including up to the minute world population)

*http://www.nist.gov/itl/div898/strd/*
US Statistics Reference Database

*http://www.statistics.gov.uk*
UK National Statistics

*http://www.un.org/Pubs/CyberSchoolBus/infonation/e_infonation.htm*
Data from the UN, by country

## Teaching statistics

*http://www.rss.org.uk*
Royal Statistical Society

*http://science.ntu.ac.uk/rsscse/TS/*
Teaching Statistics magazine – home page

*http://www.stats.gla.ac.uk:80/cti/*
CTI Statistics (changing to LTSN-CMSOR)

*http://www.kuleuven.ac.be/ucs/java/index.htm*
JAVA Statistics – some fascinating 'applets' from Belgium

*http://surfstat.newcastle.edu.au/surfstat/main/surfstat.html*
Australia – online text. An introductory course by Annett Dobson *et al.*, Newcastle University

*http://cast.massey.ac.nz*
CAST: Computer Assisted Statistics Teaching [registration required] – a complete course, by Doug Stirling, Massey University, Palmerston North, New Zealand

*http://193.61.107.61/volume*
DISCUSS statistics teaching resources
An important contribution to the understanding of probability and statistics from the team at Coventry University.

*http://www.mis.coventry.ac.uk/~styrrell/resource.htm*
Personal selection of statistical web resources from Sidney Tyrrell, Coventry University, co-author of DISCUSS

## Getting all this into word

Any **text or graphics** can be copied straight into Word. Simply

1. Mark any text you want, or hover over any graphic you want.

2. Right-click 'Copy' (or Ctrl-C)

3. Click where you want to insert it in Word

4. Right-click 'Paste' (or Ctrl-V)

It is often better to paste into a text box, so you have more control over the layout and positioning. Note: any internet links (underlined in blue) will be copied too.

## Copying a data set:

If data is presented on the web page in columns, it should copy and paste in TAB-separated (.tsv) or COMMA-separated (.csv) format.

Pasting into Word should therefore also put it in columns. You may need to adjust the TABS settings to suite the data once it is in.

Copying into EXCEL can be less successful, with all the data often ending up in the first column. If this happens, use the 'text to columns' feature in the 'DATA' menu, and try 'Tab', 'Comma' or 'Space' until it works. Alternatively try
Edit $\Rightarrow$ Paste special and select 'Text'.

## USING AUTOGRAPH

Autograph is a dynamic graphing package that operates in both bivariate and single variable modes. In the bivariate mode, as well as a full range of equations and coordinate geometry operations, data sets can be represented as scatter diagrams. In the single-variable mode, data can be displayed in all the usual diagrams, and probability distributions can be drawn. A variety of on-screen calculations are available.

Many of these operations can also be created very effectively on a spreadsheet, and throughout this supplement both approaches are explored.

## Bivariate data

In Autograph, the word 'cursor' is used to describe a coordinate point that is added by the user, either by 'point and click' or entering coordinates directly.

Most operations are available on the button bar, or through the right-click menu. This is dependent on the selection of objects that has been made, and standard rules for object selection are used.

A bivariate data set can be created in various ways:

(a) By adding 'cursors', perhaps in a pattern that will help make a particular teaching point (e.g. mostly in a well-correlated line, but with one outlier). Cursors can of course be moved around at will subsequently.

Use the right-click options: 'Select all cursors' and 'Convert to data set'.

This will change the cursors into a single data object, though you can still move an individual cursor around if you hold down Ctrl.

You can double-click on any one cursor in the data set to open the 'Edit data set' dialogue box.

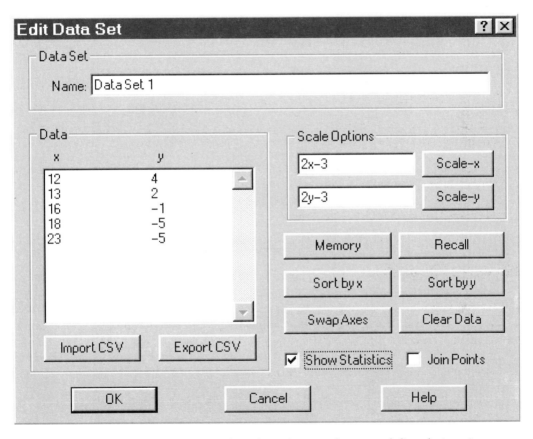

(b) By using the Edit Data Set dialogue box: here data can be entered directly in pairs, imported by loading a CSV file (comma-separated), or pasted in from two columns in a spreadsheet.

Data can then be sorted (by *x* or by *y*), scaled by any formula, or swapped over. Tick 'Show Statistics' to create a dynamically linked set of results, which change if any points are dragged around (while holding down Ctrl).

## Single variable statistics

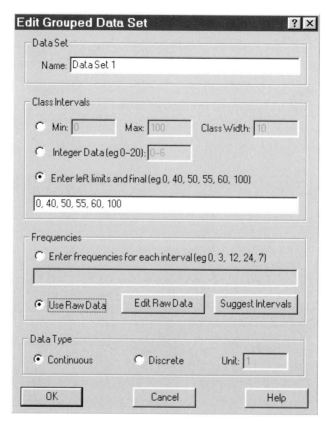

A **Grouped Data Set** can be defined either by its class intervals and frequencies, or by an underlying a raw data set. Plotting can treat the data either as continuous or discrete.

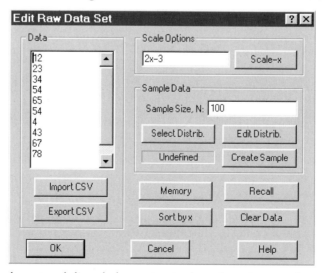

A raw data set can be entered directly by typing in the values, imported or pasted from a spreadsheet, or created by sampling from a probability distribution.

# CHAPTER 1   REPRESENTATION AND SUMMARY OF DATA

Data sets come in all shapes and sizes these days. Computers can make light work of presenting data in a digestible form, but users need to take care to use the right tool for the job.

## Using a spreadsheet

The following functions are relevant:

| | |
|---|---|
| $\Sigma x$ | SUM(array) |
| $\Sigma fx$ | SUMPRODUCT(array, array) |
| $\Sigma fx^2$ | SUMPRODUCT(array, array, array) |
| $m = (1/n)\,\Sigma x$ | AVERAGE(array) |
| $(1/n)\,\Sigma(x^2) - m^2$ | STDEV(array) |
| $n$ | COUNT(array) |
| | COUNTIF(array, test) |
| $k$th smallest | SMALL(array, $k$) |
| $k$th largest | LARGE(array, $k$) |
| Minimum | MIN(array) |
| Maximum | MAX(array) |
| Mode | MODE(array) |
| Median | MEDIAN(array) |
| Quartiles | QUARTILE(array, q) |

$q = 0 \Rightarrow$ Min, $q = 1$: LQ, $q = 2$: Median,
$q = 3$: UQ, $q = 4$: Max

## Calculating frequencies:

{ = FREQUENCY(array1,array2)}

To get this to work you need:

(a)  array1 with all the data in a single column

(b)  array2 called the 'bin' array, listing the right hand ends of the classes, e.g. 20, 40, 60, 80

(c)  array3 (empty) marked where you want the frequencies to go.

This operation then returns array of frequencies for $\leqslant 20$, $\leqslant 40$, $\leqslant 60$, $\leqslant 80$ and also $> 80$, but you need to have marked an array first ready to receive this information. (Note, it is one more cell than the 'bin' array-2). This last cell is optional.

*NOTE*: this formula is generating an array. Excel requires that you press SHIFT-CTRL-ENTER when you have finished editing the formula: this puts curly brackets round the formula.

## To draw a histogram in Excel:

This is related to the frequencies function above, but can also run without it.

This is not really a histogram as it only works if the classes are even.

Tools ⟹ Data Analysis ⟹ Histogram
Input Range = raw data array (which can run over several columns)
Bin Range = array of upper class interval limits
Output range = array to place the resulting frequency column

TICK 'Chart Output' to draw the histogram
Double-click on the shaded histogram section 'Format Data Series' ⟹ Options ⟹ Set Gap Width to zero.

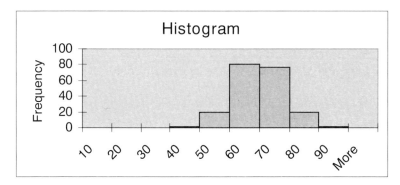

## Using Autograph

With a grouped data set entered (with or without underlying raw data) you can draw:

● Histogram (see next page)

● Cumulative frequency diagram (frequency or percentile scale)

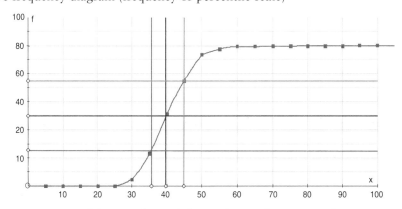

On-screen measurements enable quartiles and the median to be measured

- Box and whisker diagram
- Dot plot

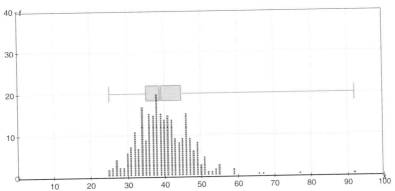

The dot plot is useful for showing where the raw data points actually are, especially when drawn at the same time as another diagram, e.g. a box and whisker or a histogram.

- Numerical statistics can be generated as text:

(a) summary statistics (mean, mode, quartiles, SD, range, etc, for raw data and for grouped)

(b) tabulated results, including mid-interval value

(c) stem and leaf diagram (really only works for discrete integer data)

An example of a stem and leaf diagram, generated as text in the 'Results Box'.

```
 0:
10:  1  3  5  5  6
20:  2  2  4  4  4  5  5  5  5  6  6  9  9  9  9
30:  1  1  1  1  1  2  4  4  4  4  4  4  4  4  4  4  4  5  5  6  8  9  9  9  9  9  9
40:  0  0  1  2  4  5  5  6  6  8  8  8  8
50:  0  0  1  2  3  5
60:  1  5  6  6
70:  2
80:
90:  2
```

# Illustrating discrete and continuous data on a histogram

The x-axis on most statistical diagrams is a continuous scale. Therefore it is import that discrete data is represented correctly.

The grouped data entry box in Autograph requires that the data is represented as continuous or discrete. The 'unit' should be set = 1 for integers, or 0.1 if data is 'to the nearest 0.1', etc.

A discrete data item, e.g. 53 (unit = 1) will be represented on the histogram as a region from 52.5–53.5. Similarly a class interval 20–29 is represented by the region 19.5–29.5, i.e. 20–30 shifted to the left by 0.5 (half the unit).

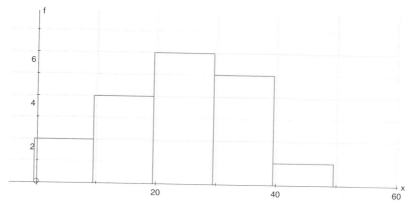

Cumulative frequency, frequency polygon and box and whisker diagrams are similarly displaced.

In the **table of values option,** the mid-interval value is used to calculate the mean and SD, and will take account of the nature of the data:

*Continuous*:

| Class Interval | Mid-interval Value (x) | Class Width | Frequency (f) | Cumulative Frequency |
|---|---|---|---|---|
| 0–20 | 10.0 | 20 | 0 | 0 |
| 20–30 | 25.0 | 10 | 10 | 10 |
| 30–40 | 35.0 | 10 | 130 | 140 |
| 40–45 | 42.5 | 5 | 90 | 230 |
| 45–50 | 47.5 | 5 | 55 | 285 |
| 50–100 | 75.0 | 50 | 75 | 360 |

$\Sigma f = 360$   $\Sigma fx = 16\ 863$   $\Sigma fx^2 = 874\ 031$
Mean = 46.84 SD = 15.29

*Discrete*:

| Class Interval | Mid-interval Value (x) | Class Width | Frequency (f) | Cumulative Frequency |
|---|---|---|---|---|
| 0–19 | 9.5 | 20 | 0 | 0 |
| 20–29 | 24.5 | 10 | 10 | 10 |
| 30–39 | 34.5 | 10 | 130 | 140 |
| 40–44 | 42.0 | 5 | 90 | 230 |
| 45–49 | 47.0 | 5 | 55 | 285 |
| 50–99 | 74.5 | 50 | 75 | 360 |

$\Sigma f = 360$   $\Sigma fx = 16\ 683$   $\Sigma fx^2 = 857\ 259$
Mean = 46.34 ( = Continuous mean – 0.5) SD = 15.29 (unchanged)

# Illustrating frequency and frequency density on a histogram

This can be a difficult concept to get across, and a visual approach can be very effective.

*Example*: Consider a grouped data set entered into Autograph defined by these variable-width class intervals:

0, 20, 30, 40, 45, 50, 100

and the following associated frequencies:

0, 10, 130, 90, 55, 75

If you select to draw a histogram from this data, the dialogue box asks you to choose 'frequency' or 'frequency density'

**Frequency:**

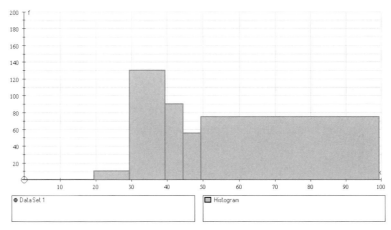

Here, a data set is clearly being mis-represented: the mode is wrong and there is undue weight to the final class.

**Frequency Density:**

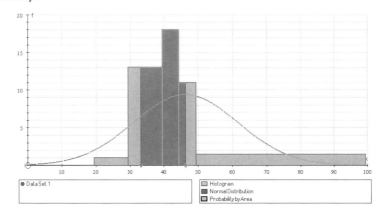

When selecting frequency density, you need also to specify the 'per' unit. The default value = 1 so that the area under the histogram is a direct measurement of the frequency.

*Example*: Enter the following discrete raw data:

1, 2, 3, 4, 5, 6, 7, 8, 9, 10, 12, 23, 24, 24, 25, 25, 27, 31 and draw a histogram using the unequal classes 0, 10, 40 (i.e. 0–9 and 10–39). There are nine items in each class, so plotting 'frequency' gives two equal frequencies. Here is the same data plotted as a frequency density.

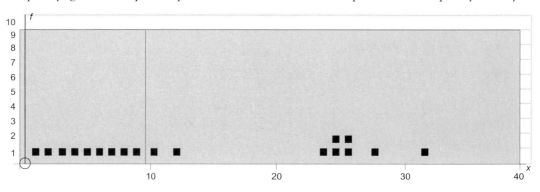

# CHAPTER 2   CORRELATION AND REGRESSION

## Using a spreadsheet

The following functions are relevant to bivariate statistics:

PEARSON(array-y, array-x)
returns correlation coefficient, r [PMCC]

CORREL(array-y, array-x)
Spearman's rank correlation coefficient

FORECAST(x, array-y, array-x)
estimates $y(x)$ given the 2 arrays

INTERCEPT(array-y, array-x)
returns $c$ in $y$-on-$x$ regression line $y = mx + c$

SLOPE(array-y, array-x) $\Rightarrow$ 'm'
returns $m$ in $y$-on-$x$ regression line $y = mx + c$

DEVSQ(array-y, array-x)
returns $R^2$, the sum of the squares of deviations from the sample mean.

There is also a sophisticated facility to create a full regression analysis, including residuals and normal probability plots. Use: Tools $\Rightarrow$ Data Analysis $\Rightarrow$ Regression

### The scatter diagram chart option:

If you select 'XY (Scatter)' and create the diagram, there are a number of useful options available:

Double-click on the either axis to reformat that axis
Double-click a data point to open the 'Format Data Series' dialogue box
Click on a data-point (they should all turn yellow), then right-click $\Rightarrow$ 'Add Trend Line'

Choose Type: linear, and
Options:    Forecast forward/back
            display equation, display $R^2$

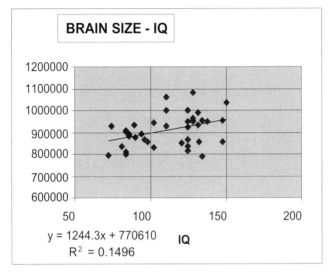

An example of a data set copied off an internet page into Excel. There were many columns of data: to select to two for this plot, press Ctrl as you select the second (non-adjacent) column.

## Using Autograph

With a bivariate data set in place, there are a number of options to help illustrate its properties.

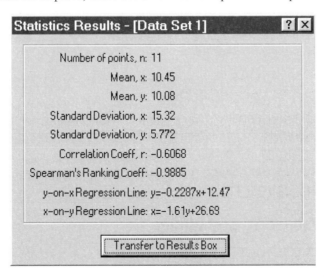

The statistics box (tick the option in the Edit Data Set dialogue box) gives all the standard results, and these will change dynamically if a point in the data set is altered (hold down Ctrl and drag).

If the 'Junior' option is chosen (from View $\Rightarrow$ Preferences), this box is simplified, and gives only the means and the 'line of best fit'.

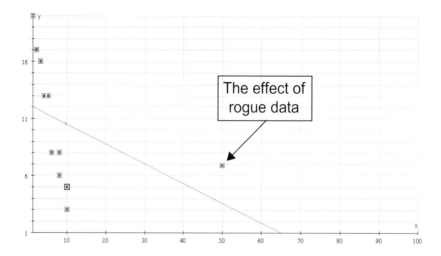

Here a set of data has a rogue point (which can be moved around by pressing Ctrl and dragging with the mouse), and the centroid and line of best fit are also shown.

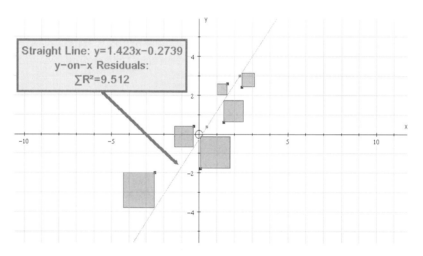

Here the principle of least squares regression is being illustrated, relating to a variable straight line through the centroid.

## CHAPTER 3   PROBABILITY

## Using a spreadsheet for probability

There are various ways to create random numbers in Excel:

| | |
|---|---|
| RAND() | Random number $0 \leqslant x \leqslant 1$ |
| INT(6 * RAND() + 1) | Random integer $1 \leqslant x \leqslant 6$ |
| RANDBETWEEN(a, b) | Random integer $a \leqslant x \leqslant b$ |

*NOTE*: If RANDBETWEEN does not work, go to Tools $\Rightarrow$ Add-Ins and tick 'Analysis Tool Pack'.

*Example*: Simulation of the sum of two dice:

|    | A      | B    | C                      |
|----|--------|------|------------------------|
| 1  | DICE 1 |      | = INT(6 * RAND()) + 1  |
| 2  | DICE 2 |      | = INT(6 * RAND()) + 1  |
| 3  | SUM    |      | = C1 + C2              |
| 4  |        | 'x'  | 'f'                    |
| 5  | Score  | 2    | = C5 + (B5 = $C$3)     |
| 6  | Score  | 3    |                        |
| 7  | Score  |      |                        |
| 21 | Score  | 12   | Fill down, mark B4:(2) and plot |

The logical statement in cell C5, '(B5 = $C$3)' equals 1 if TRUE, otherwise equals 0. This is a simple way to add 1 to a total if a condition is met.

However, a cell formula referring to itself is called a 'circular reference' and you need to set the following to make it work in this instance:

Tools     ⟹ Options ⟹ Calculation tab ⟹ Tick Iteration
            ⟹ Max iterations = 1 (leave Max change)

Then hold down F9 to run the simulation.

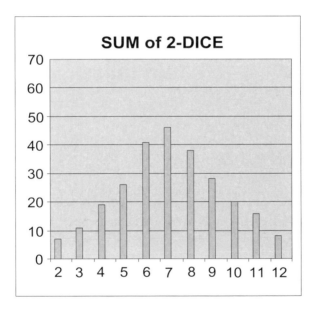

To make the *x*-axis work properly in this chart, proceed as follows:

Choose the 'Column' chart type and observe that it is plotting '*x*' and '*f*' against the row number.

Click 'next' ⟹ 'Chart Source Data' ⟹ 'Series'.

'*f*' is OK.

Select '*x*' (which is plotting on the wrong axis). Copy and paste its array into the Category X axis labels slot, then click 'Remove' to take it off the y-axis list.

## Internet resources for probability

### 1. The DISCUSS site

This is a growing set of teaching resources from Coventry University, covering many aspects of school level probability and statistics.

Simulations available include one on Buffon's needle.

### 2. The Chance and data site from Tasmania

This has an excellent probability section which links probability theory to stories in the newspapers.

## From the Autograph 'extras'

### 1. Monte-Carlo simulation

This simulation is based on the probability of a random point within a square (of side 2) also landing within the unit circle. The probability is $\frac{\pi}{4}$. This simulation is very slow to converge, but is a good illustration of randomness.

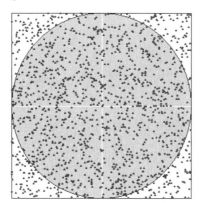
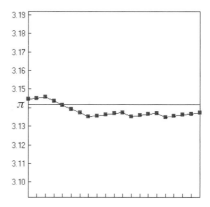

## 2. Dice throwing

This is similar to the spreadsheet example above, but more automatic to use. Options include

(a) the sum of 2 dice
(b) the difference between 2 dice
(c) throwing one die
(d) throwing $n$ dice

# CHAPTERS 4 AND 5   DISCRETE RANDOM VARIABLES

# Using a spreadsheet

The following formulae are available for generating discrete random variables in Excel:

BINOMDIST(r, n, p, T)
e.g.   T = 0: X ~ B(10, 0.5)
        BINOMDIST(2, 10, 0.5, 0) $\Rightarrow$ P(X = 2)

e.g.   T = 1: (cumulative)
        BINOMDIST(2, 10, 0.5, 1) $\Rightarrow$ P(X $\leqslant$ 2)

POISSON(x, m, T)
e.g. T = 0: X ~ Po(4)          POISSON(2, 4, 0) $\Rightarrow$ P(X = 2)
e.g. T = 1: (cumulative)       POISSON(2, 4, 1) $\Rightarrow$ P(X $\leqslant$ 2)

*Example*: To produce the distribution and the cumulative distribution for X ~ Bin(10, p)

Name A2 'n'
Name B2 'p'

Formula in D2 = binomdist(x, n, p, 0)
or for cumulative: = binomdist(x, n, p, 1)

Note 'x' is the column heading C1 and
this can be used in the formula.

Enter C2 = 0
To create 0–10 in C2–C12,
use Edit $\Rightarrow$ Fill $\Rightarrow$ Series
Fill down D2–D12 (double-click on the
D2 cell dot)

|   | A | B | C | D |
|---|---|---|---|---|
| 1 | n | p | x | Bin(n,p) |
| 2 | 10 | 0.60 | 0 | 0.0001 |
| 3 |  |  | 1 | 0.0016 |
| 4 | SLIDER |  | 2 | 0.0106 |
| 5 | 0 ≤ p ≤ 1 |  | 3 | 0.0425 |
| 6 | ◀ | ▶ | 4 | 0.1115 |
| 7 |  |  | 5 | 0.2007 |
| 8 |  |  | 6 | 0.2508 |
| 9 | Dummy: p = B10/100 |  | 7 | 0.2150 |
| 10 |  | 60 | 8 | 0.1209 |
| 11 |  |  | 9 | 0.0403 |
| 12 |  |  | 10 | 0.0060 |

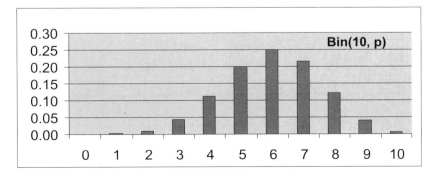

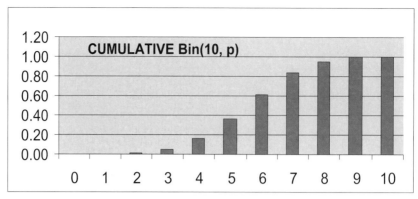

To put in a slider to control p:

Right-click over toolbar $\Rightarrow$ Forms $\Rightarrow$ Scroll Bar

Drag it into position. Right-click: Format control.

Unfortunately this slider only works with integers, so you need the dummy cell, B10: set the slider to vary from 0–100, and set p = B10/100.

## Using Autograph

The following discrete probability distributions are available in Autograph:

- Rectangular:  $X \sim R(a, b)$   $r = a, \ldots b$

  $P(X = r) = 1/(b - a + 1)$

  Mean, $\mu = (a + b)/2$

  Variance, $\sigma^2 = (b - a)(b - a + 2)/12$

- Binomial:  $X \sim B(n, p)$   $r = 0, 1, 2, \ldots n$

  $P(X = r) = nCr.p^r.q^{n-r}$

  Mean, $\mu = np$

  Variance, $\sigma^2 = npq$

- Poisson:  $X \sim Po(\lambda)$   $r = 0, 1, 2 \ldots$

  $P(X = r) = \lambda^r/r!.e^{-\lambda}$

  Mean, $\mu = \lambda$

  Variance, $\sigma^2 = \lambda$

  $P(X = r + 1) = P(X = r).\lambda/(r + 1)$

  also the distribution $Po(npq) \approx B(n, p)$

- Geometric:  $X \sim G(p)$   $r = 1, 2, 3, \ldots$

  $P(X = r) = q^{r-1}.p$

  Mean, $\mu = 1/p$

  Variance, $\sigma^2 = q/p^2$

  $P(X = r + 1) = P(X = r).q$

- User defined:

  Mean, $\mu = \sum r.P(x = r)$

  Variance, $\sigma^2 = \sum r^2.P(X = r) - \mu^2$

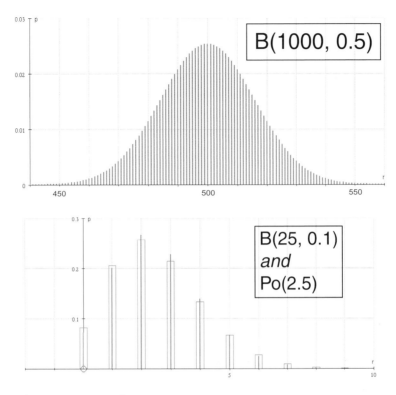

Table of Values of Po(2.5): $\mu = 2.5$, $\sigma^2 = 2.5$

| $r$ | $P(X = r)$ | $P(X \leqslant r)$ | $P(X \geqslant r)$ |
|---|---|---|---|
| 0 | 0.08208 | 0.08208 | 1 |
| 1 | 0.2052 | 0.2873 | 0.9179 |
| 2 | 0.2565 | 0.5438 | 0.7127 |
| 3 | 0.2138 | 0.7576 | 0.4562 |
| 4 | 0.1336 | 0.8912 | 0.2424 |
| 5 | 0.0668 | 0.958 | 0.1088 |
| 6 | 0.02783 | 0.9858 | 0.04202 |
| 7 | 0.009941 | 0.9958 | 0.01419 |
| 8 | 0.003106 | 0.9989 | 0.004247 |

# CHAPTERS 6, 7 AND 8   CONTINUOUS DISTRIBUTIONS AND THE NORMAL DISTRIBUTION

## Using a spreadsheet

The following normal distribution formulae are available in Excel:

NORMDIST$(x, m, s, T)$
$X \sim N(m, s^2)$
With $T = 0$ this returns the value of the pdf
With $T = 1$ this returns $P(X < x)$

NORMINV($p, m, s$)
For $X \sim N(m, s^2)$
this returns $x$ such that $P(X < x) = p$

NORMDIST(z)
For $Z \sim N(0, 1)$, this returns $P(Z \leqslant z)$
where $Z = (X - m)/s$

NORMINV(p)
returns $z$ such that $P(Z < z) = p$

STANDARDIZE $(x, m, s)$
returns $z = (x - m)./s$

*NOTE*: the parameter used in the normal distribution formula is standard deviation ($s$) and not variance ($s^2$).

| | A | B | C |
|---|---|---|---|
| 1 | X ~ N(m,s²): NORMAL CALCULATOR | | |
| 2 | Mean, m = | 500 | b2 named "m" |
| 3 | SD, s = | 100 | b3 named "s" |
| 4 | x = | 400 | b4 named "x" |
| 5 | z=(x–m)/s = | -1.000 | (x–m)/s |
| 6 | y = | 700 | b6 named "y" |
| 7 | z=(y–m)/s = | 2.000 | (y–m)/s |
| 9 | P(X<x) = | 0.159 | NORMDIST(x,m,s,1) |
| 10 | P(X>x) = | 0.841 | 1–b8 |
| 11 | P(X<y) = | 0.977 | NORMDIST(y,m,s,1) |
| 12 | P(X>y) = | 0.023 | 1–b10 |
| 13 | | | |
| 14 | P(x<X<y) = | 0.819 | b10–b8 |
| 15 | P(x>X>y) = | 0.181 | 1–b13 |
| 17 | p (%) = | 2.5 | b16 named "p" |
| 18 | x:P(X<x)=p% | 304.004 | NORMINV(p/100,m,s) |
| 19 | z1 = | -1.960 | (b18–m)/s) |
| 20 | x:P(X>x)=p% | 695.996 | NORMINV(1–p/100,m,s) |
| 21 | z2 = | 1.960 | (b20–m)/s) |

In the example, the NORMDIST and NORMINV formulae have been used to create a general Normal Distribution calculator.

The 'C' column explains the entries and formulae that have been used. Notice the extensive use of named cells – this makes all subsequent formulae so much more friendly.

# Using Autograph

The following continuous probability distributions are available in Autograph:

- Uniform:  $X \sim U(a, b)$                   $a \leqslant x \leqslant b$
  $f(x) = 1/(b - a)$
  Mean, $\mu = (a + b)/2$
  Variance, $\sigma^2 = (a - b)^2/12$

- Normal:  $X \sim N(\mu, \sigma^2)$
  $z = (x - \mu)/\sigma$     $f(x) = 1/(\sigma\sqrt{(2\pi)}).e^{(-\frac{1}{2}z^2)}$

- User defined:
  Mean, $\mu = \int x.f(x) \, dx$
  Variance, $\sigma^2 = \int x^2.f(x) \, dx - \mu^2$

*Example*: A continuous function $f(x) = x^2$, $-2 < x < 2$

The important principle to appreciate is that the total area must = 1. Therefore the function to be plotted must be $f(x) = kx^2$, where $k = \int x^2 \, dx$ over the range.

Autograph automatically converts any $f(x)$ entry to $k.f(x)$, and areas can be measured on-screen by entering limits. By dragging the limits around, it can be seen that the total area = 1, and so areas represent probability.

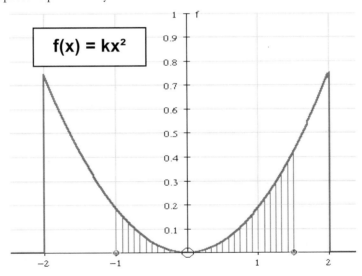

*Example*: $X \sim N(500, 100^2)$: here areas between limits and inverse calculations are possible. The parameters $\mu$ and $\sigma^2$ can also be varied dynamically.

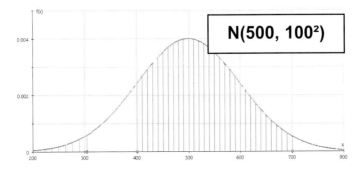

# CHAPTER 9   SAMPLING AND ESTIMATION

## Using a spreadsheet

Excel includes a feature which can generate a sample of random data from a number of distributions. The choice of distributions is:

Uniform $(a, b)$ [equivalent to RANDBETWEEN$(a, b)$]
Normal $(m, s)$
Bernouilli $(p)$
Binomial $(n, p)$
Poisson $(m)$,
User-defined Discrete [2 cols: $x$, $P(X = x)$]

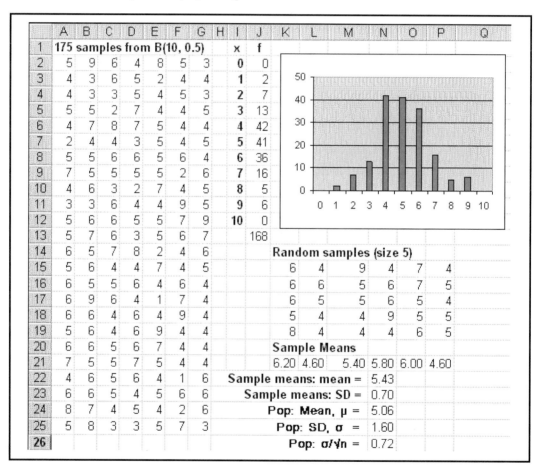

*Example*: To create a large sample data set from B(10, 0.5), and take samples of size 5 from it.

To achieve the above, use Tools ⟹ Data Analysis ⟹ Random Number Generation. Leave 'Number of Variables' ( = columns) and 'Number of Random Numbers' ( = rows) blank. Instead click on 'Output Range' and drag out the array you want to fill with random numbers. Choose 'Binomial', and enter 'p Value' and 'Number of trials'. Leave 'Random Seed' blank (this is used for creating exactly the same set of random numbers again if required).

Click 'OK'. Then to create a frequency chart first set up a 'bin array' as a column of figures 0–10, and use the formula { = FREQUENCY(data array, bin array} on a new array next to the bin array. Don't forget SHIFT-CTRL-ENTER! Select the bin array and the frequency array and draw a bar chart

To create a random sample from this sample, use Tools $\Rightarrow$ Data Analysis $\Rightarrow$ Sampling. The 'Input range' is the data. Use 'Random' with $n = 5$, and mark the output range. Unfortunately there is no easy way to create many such samples.

After five such samples it is useful to compare the mean and SD of the sample means with $\mu$ and $\sigma/\sqrt{n}$ calculated from the original data set.

## Using Autograph

Use New Statistics Page $\Rightarrow$ Add Grouped Data $\Rightarrow$ Use Raw Data $\Rightarrow$ Edit Raw Data $\Rightarrow$ Select Distribution. There is the option to create a set of random data from the following probability distributions:

Rectangular $(a, b)$ – discrete or continuous
Binomial $(n, p)$
Poisson $(\lambda)$
Geometric $(p)$
Normal $(\mu, \sigma^2)$
User defined continuous $f(x)$

['User defined discrete' is not yet implemented]
Use 'Edit Distrib.' to enter the parameters

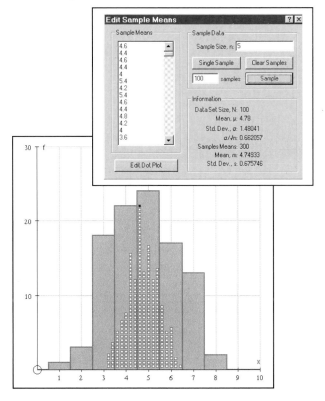

Enter the sample size N, and press 'Create Sample'. Click 'OK', then 'Suggest Intervals' (amend if necessary), then click 'Continuous' or 'Discrete' as appropriate, then 'OK'.

With a data set in place, select 'histogram' then 'autoscale'. Then choose 'Sample Means'.

In the 'Edit Sample Means' dialogue box, enter the sample size (e.g. $n = 5$). You can then

(a) take samples one at a time, in which case the actual samples are indicated on the diagram together with their mean.

(b) take many samples (e.g. 100), in which case a dot plot is created.

The **Central Limit Theorem** is very effectively demonstrated with almost any parent population.

# CHAPTERS 10 AND 11   HYPOTHESIS TESTS

## Using spreadsheets

Using Excel, the following formulae are useful when investigating hypothesis tests:

BINOMDIST($r, n, p, T$). e.g. for $X \sim B(10, 0.5)$
$T = 0$: BINOMDIST(2, 10, 0.5, 0) = $P(X = 2)$
$T = 1$: BINOMDIST(2, 10, 0.5, 1) = $P(X \leqslant 2)$

CRITBINOM($n, p$, test). e.g. for $X \sim B(n.p)$
This finds the smallest $x$ such that $P(X \leqslant x) \geqslant$ test

POISSON($x, m, T$). e.g. for $X \sim Po(4)$
$T = 0$: POISSON(2, 4, 0) = $P(X = 2)$
$T = 1$: POISSON(2, 4, 1) = $P(X \leqslant 2)$

NORMDIST($x, m, s, T$). e.g. for $X \sim N(m, s^2)$
$T = 0 \Rightarrow$ value of the pdf (for plotting the curve)
$T = 1 \Rightarrow P(X < x)$

NORMINV($p, m, s$). e.g. for $X \sim N(m, s^2)$
This returns the value $x$ such that $P(X < x) = p$

NORMSDIST($z$)
    returns probability $P(Z \leqslant z)$
NORMSINV($p$)
    returns z such that $P(Z \leqslant z) = p$
STANDARDIZE $(x, m, s) \Rightarrow z = (x - m)./s$

*Example*: A tough driving examiner claims to pass only 20% of his candidates. After 50 tests, what is the smallest number of passes required to refute this claim at 5%?
Answer: critbinom(25, 0.2, 0.05) = 2
This means $P(X \leqslant 2) > 5\%$
whereas $P(X \leqslant 1) < 5\%$

*Example*: CONFIDENCE($a$, $s$, $n$)
This returns the confidence intervals for a sample size of $n$ from a population with SD = $s$, at a significance level = $a$. Unfortunately, '$a$' is a measure of the probability outside the intervals. So for 95% confidence, $a = 1 - 95/100$.

e.g. CONFIDENCE$(1 - 95/100, 2.5, 50) = 0.69$
$\Rightarrow$ C.I. = sample mean $\pm 0.69$ at 95%

*Example*: CHITEST(array-1, array-2)
array-1 = actual frequencies
array-2 = expected frequencies
returns the $\chi^2$ calculation for the two arrays

This enables a set of actual data (frequencies) to be tested against frequencies calculated from an underlying probability distribution.

# Using Autograph

*Example*: Hypothesis testing on discrete probability distributions

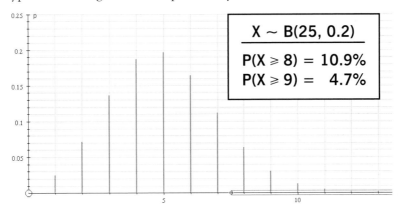

Here $H_0$ is $p = 0.2$ under B(25, 0.2) and $H_1$ is $p > 0.2$. If $x \geqslant 9$ $H_0$ is rejected. The boundary limits can be dragged up and down the $x$-axis.

*Example*: Hypothesis testing on continuous probability distributions

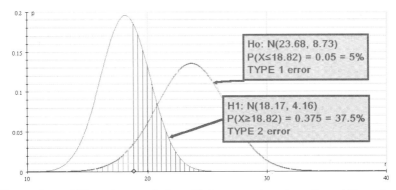

This is an illustration of Type 1 error (when $H_0$ is correctly rejected) and Type 2 error (when $H_0$ is accepted but $H_1$ is true) using two normal distributions.

*Example*: Fitting a probability distribution to a data set.

Autograph will find the best parameters to fit a **binomial, poisson** or **normal** distribution to a data set.

First draw a histogram, then a normal distribution (any parameters), then chose 'fit to data'. The probability distribution will only appear to be a good fit if the frequency density scale (unit = 1) is used – this way the total area on both diagrams = 1.

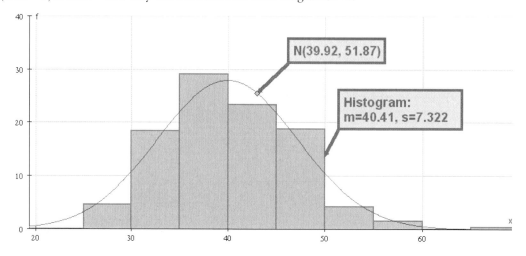

# Appendix

## CUMULATIVE BINOMIAL PROBABILITIES

The tabulated value is $P(X \leqslant r)$ where $X \sim B(n, p)$

| $p =$ | | 0.05 | 0.10 | 0.15 | 0.20 | 0.25 | 0.30 | 0.35 | 0.40 | 0.45 | 0.50 |
|---|---|---|---|---|---|---|---|---|---|---|---|
| $n = 2$ | $r = 0$ | 0.9025 | 0.8100 | 0.7225 | 0.6400 | 0.5625 | 0.4900 | 0.4225 | 0.3600 | 0.3025 | 0.2500 |
| | 1 | 0.9975 | 0.9900 | 0.9775 | 0.9600 | 0.9375 | 0.9100 | 0.8775 | 0.8400 | 0.7975 | 0.7500 |
| | 2 | 1.0000 | 1.0000 | 1.0000 | 1.0000 | 1.0000 | 1.0000 | 1.0000 | 1.0000 | 1.0000 | 1.0000 |
| $n = 3$ | $r = 0$ | 0.8574 | 0.7290 | 0.6141 | 0.5120 | 0.4219 | 0.3430 | 0.2746 | 0.2160 | 0.1664 | 0.1250 |
| | 1 | 0.9928 | 0.9720 | 0.9393 | 0.8960 | 0.8438 | 0.7840 | 0.7183 | 0.6480 | 0.5748 | 0.5000 |
| | 2 | 0.9999 | 0.9990 | 0.9966 | 0.9920 | 0.9844 | 0.9730 | 0.9571 | 0.9360 | 0.9089 | 0.8750 |
| | 3 | 1.0000 | 1.0000 | 1.0000 | 1.0000 | 1.0000 | 1.0000 | 1.0000 | 1.0000 | 1.0000 | 1.0000 |
| $n = 4$ | $r = 0$ | 0.8145 | 0.6561 | 0.5220 | 0.4096 | 0.3164 | 0.2401 | 0.1785 | 0.1296 | 0.0915 | 0.0625 |
| | 1 | 0.9860 | 0.9477 | 0.8905 | 0.8192 | 0.7383 | 0.6517 | 0.5630 | 0.4752 | 0.3910 | 0.3125 |
| | 2 | 0.9995 | 0.9963 | 0.9880 | 0.9728 | 0.9492 | 0.9163 | 0.8735 | 0.8208 | 0.7585 | 0.6875 |
| | 3 | 1.0000 | 0.9999 | 0.9995 | 0.9984 | 0.9961 | 0.9919 | 0.9850 | 0.9744 | 0.9590 | 0.9375 |
| | 4 | | 1.0000 | 1.0000 | 1.0000 | 1.0000 | 1.0000 | 1.0000 | 1.0000 | 1.0000 | 1.0000 |
| $n = 5$ | $r = 0$ | 0.7738 | 0.5905 | 0.4437 | 0.3277 | 0.2373 | 0.1681 | 0.1160 | 0.0778 | 0.0503 | 0.0313 |
| | 1 | 0.9774 | 0.9185 | 0.8352 | 0.7373 | 0.6328 | 0.5282 | 0.4284 | 0.3370 | 0.2562 | 0.1875 |
| | 2 | 0.9988 | 0.9914 | 0.9734 | 0.9421 | 0.8965 | 0.8369 | 0.7648 | 0.6826 | 0.5931 | 0.5000 |
| | 3 | 1.0000 | 0.9995 | 0.9978 | 0.9933 | 0.9844 | 0.9692 | 0.9460 | 0.9130 | 0.8688 | 0.8125 |
| | 4 | | 1.0000 | 0.9999 | 0.9997 | 0.9990 | 0.9976 | 0.9947 | 0.9898 | 0.9815 | 0.9688 |
| | 5 | | | 1.0000 | 1.0000 | 1.0000 | 1.0000 | 1.0000 | 1.0000 | 1.0000 | 1.0000 |
| $n = 6$ | $r = 0$ | 0.7351 | 0.5314 | 0.3771 | 0.2621 | 0.1780 | 0.1176 | 0.0754 | 0.0467 | 0.0277 | 0.0156 |
| | 1 | 0.9672 | 0.8857 | 0.7765 | 0.6554 | 0.5339 | 0.4202 | 0.3191 | 0.2333 | 0.1636 | 0.1094 |
| | 2 | 0.9978 | 0.9842 | 0.9527 | 0.9011 | 0.8306 | 0.7443 | 0.6471 | 0.5443 | 0.4415 | 0.3438 |
| | 3 | 0.9999 | 0.9987 | 0.9941 | 0.9830 | 0.9624 | 0.9295 | 0.8826 | 0.8208 | 0.7447 | 0.6563 |
| | 4 | 1.0000 | 0.9999 | 0.9996 | 0.9984 | 0.9954 | 0.9891 | 0.9777 | 0.9590 | 0.9308 | 0.8906 |
| | 5 | | 1.0000 | 1.0000 | 0.9999 | 0.9998 | 0.9993 | 0.9982 | 0.9959 | 0.9917 | 0.9844 |
| | 6 | | | | 1.0000 | 1.0000 | 1.0000 | 1.0000 | 1.0000 | 1.0000 | 1.0000 |
| $n = 7$ | $r = 0$ | 0.6983 | 0.4783 | 0.3206 | 0.2097 | 0.1335 | 0.0824 | 0.0490 | 0.0280 | 0.0152 | 0.0078 |
| | 1 | 0.9556 | 0.8503 | 0.7166 | 0.5767 | 0.4449 | 0.3294 | 0.2338 | 0.1586 | 0.1024 | 0.0625 |
| | 2 | 0.9962 | 0.9743 | 0.9262 | 0.8520 | 0.7564 | 0.6471 | 0.5323 | 0.4199 | 0.3164 | 0.2266 |
| | 3 | 0.9998 | 0.9973 | 0.9879 | 0.9667 | 0.9294 | 0.8740 | 0.8002 | 0.7102 | 0.6083 | 0.5000 |
| | 4 | 1.0000 | 0.9998 | 0.9988 | 0.9953 | 0.9871 | 0.9712 | 0.9444 | 0.9037 | 0.8471 | 0.7734 |
| | 5 | | 1.0000 | 0.9999 | 0.9996 | 0.9987 | 0.9962 | 0.9910 | 0.9812 | 0.9643 | 0.9375 |
| | 6 | | | 1.0000 | 1.0000 | 0.9999 | 0.9998 | 0.9994 | 0.9984 | 0.9963 | 0.9922 |
| | 7 | | | | | 1.0000 | 1.0000 | 1.0000 | 1.0000 | 1.0000 | 1.0000 |
| $n = 8$ | $r = 0$ | 0.6634 | 0.4305 | 0.2725 | 0.1678 | 0.1001 | 0.0576 | 0.0319 | 0.0168 | 0.0084 | 0.0039 |
| | 1 | 0.9428 | 0.8131 | 0.6572 | 0.5033 | 0.3671 | 0.2553 | 0.1691 | 0.1064 | 0.0632 | 0.0352 |
| | 2 | 0.9942 | 0.9619 | 0.8948 | 0.7969 | 0.6785 | 0.5518 | 0.4278 | 0.3154 | 0.2201 | 0.1445 |
| | 3 | 0.9996 | 0.9950 | 0.9786 | 0.9437 | 0.8862 | 0.8059 | 0.7064 | 0.5941 | 0.4770 | 0.3633 |
| | 4 | 1.0000 | 0.9996 | 0.9971 | 0.9896 | 0.9727 | 0.9420 | 0.8939 | 0.8263 | 0.7396 | 0.6367 |
| | 5 | | 1.0000 | 0.9998 | 0.9988 | 0.9958 | 0.9887 | 0.9747 | 0.9502 | 0.9115 | 0.8555 |
| | 6 | | | 1.0000 | 0.9999 | 0.9996 | 0.9987 | 0.9964 | 0.9915 | 0.9819 | 0.9648 |
| | 7 | | | | 1.0000 | 1.0000 | 0.9999 | 0.9998 | 0.9993 | 0.9983 | 0.9961 |
| | 8 | | | | | | 1.0000 | 1.0000 | 1.0000 | 1.0000 | 1.0000 |

# CUMULATIVE BINOMIAL PROBABILITIES

The tabulated value is $P(X \leqslant r)$ where $X \sim B(n, p)$

| $p =$ | | 0.05 | 0.10 | 0.15 | 0.20 | 0.25 | 0.30 | 0.35 | 0.40 | 0.45 | 0.50 |
|---|---|---|---|---|---|---|---|---|---|---|---|
| $n = 9$ | $r = 0$ | 0.6302 | 0.3874 | 0.2316 | 0.1342 | 0.0751 | 0.0404 | 0.0207 | 0.0101 | 0.0046 | 0.0020 |
| | 1 | 0.9288 | 0.7748 | 0.5995 | 0.4362 | 0.3003 | 0.1960 | 0.1211 | 0.0705 | 0.0385 | 0.0195 |
| | 2 | 0.9916 | 0.9470 | 0.8591 | 0.7382 | 0.6007 | 0.4628 | 0.3373 | 0.2318 | 0.1495 | 0.0898 |
| | 3 | 0.9994 | 0.9917 | 0.9661 | 0.9144 | 0.8343 | 0.7297 | 0.6089 | 0.4826 | 0.3614 | 0.2539 |
| | 4 | 1.0000 | 0.9991 | 0.9944 | 0.9804 | 0.9511 | 0.9012 | 0.8283 | 0.7334 | 0.6214 | 0.5000 |
| | 5 | | 0.9999 | 0.9994 | 0.9969 | 0.9900 | 0.9747 | 0.9464 | 0.9006 | 0.8342 | 0.7461 |
| | 6 | | 1.0000 | 1.0000 | 0.9997 | 0.9987 | 0.9957 | 0.9888 | 0.9750 | 0.9502 | 0.9102 |
| | 7 | | | | 1.0000 | 0.9999 | 0.9996 | 0.9986 | 0.9962 | 0.9909 | 0.9805 |
| | 8 | | | | | 1.0000 | 1.0000 | 0.9999 | 0.9997 | 0.9992 | 0.9980 |
| | 9 | | | | | | | 1.0000 | 1.0000 | 1.0000 | 1.0000 |
| $n = 10$ | $r = 0$ | 0.5987 | 0.3487 | 0.1969 | 0.1074 | 0.0563 | 0.0282 | 0.0135 | 0.0060 | 0.0025 | 0.0010 |
| | 1 | 0.9139 | 0.7361 | 0.5443 | 0.3758 | 0.2440 | 0.1493 | 0.0860 | 0.0464 | 0.0233 | 0.0107 |
| | 2 | 0.9885 | 0.9298 | 0.8202 | 0.6778 | 0.5256 | 0.3828 | 0.2616 | 0.1673 | 0.0996 | 0.0547 |
| | 3 | 0.9990 | 0.9872 | 0.9500 | 0.8791 | 0.7759 | 0.6496 | 0.5138 | 0.3823 | 0.2660 | 0.1719 |
| | 4 | 0.9999 | 0.9984 | 0.9901 | 0.9672 | 0.9219 | 0.8497 | 0.7515 | 0.6331 | 0.5044 | 0.3770 |
| | 5 | 1.0000 | 0.9999 | 0.9986 | 0.9936 | 0.9803 | 0.9527 | 0.9051 | 0.8338 | 0.7384 | 0.6230 |
| | 6 | | 1.0000 | 0.9999 | 0.9991 | 0.9965 | 0.9894 | 0.9740 | 0.9452 | 0.8980 | 0.8281 |
| | 7 | | | 1.0000 | 0.9999 | 0.9996 | 0.9984 | 0.9952 | 0.9877 | 0.9726 | 0.9453 |
| | 8 | | | | 1.0000 | 1.0000 | 0.9999 | 0.9995 | 0.9983 | 0.9955 | 0.9893 |
| | 9 | | | | | | 1.0000 | 1.0000 | 0.9999 | 0.9997 | 0.9990 |
| | 10 | | | | | | | | 1.0000 | 1.0000 | 1.0000 |
| $n = 15$ | $r = 0$ | 0.4633 | 0.2059 | 0.0874 | 0.0352 | 0.0134 | 0.0047 | 0.0016 | 0.0005 | 0.0001 | 0.0000 |
| | 1 | 0.8290 | 0.5490 | 0.3186 | 0.1671 | 0.0802 | 0.0353 | 0.0142 | 0.0052 | 0.0017 | 0.0005 |
| | 2 | 0.9638 | 0.8159 | 0.6042 | 0.3980 | 0.2361 | 0.1268 | 0.0617 | 0.0271 | 0.0107 | 0.0037 |
| | 3 | 0.9945 | 0.9444 | 0.8227 | 0.6482 | 0.4613 | 0.2969 | 0.1727 | 0.0905 | 0.0424 | 0.0176 |
| | 4 | 0.9994 | 0.9873 | 0.9383 | 0.8358 | 0.6865 | 0.5155 | 0.3519 | 0.2173 | 0.1204 | 0.0592 |
| | 5 | 0.9999 | 0.9978 | 0.9832 | 0.9389 | 0.8516 | 0.7216 | 0.5643 | 0.4032 | 0.2608 | 0.1509 |
| | 6 | 1.0000 | 0.9997 | 0.9964 | 0.9819 | 0.9434 | 0.8689 | 0.7548 | 0.6098 | 0.4522 | 0.3036 |
| | 7 | | 1.0000 | 0.9994 | 0.9958 | 0.9827 | 0.9500 | 0.8868 | 0.7869 | 0.6535 | 0.5000 |
| | 8 | | | 0.9999 | 0.9992 | 0.9958 | 0.9848 | 0.9578 | 0.9050 | 0.8182 | 0.6964 |
| | 9 | | | 1.0000 | 0.9999 | 0.9992 | 0.9963 | 0.9876 | 0.9662 | 0.9231 | 0.8491 |
| | 10 | | | | 1.0000 | 0.9999 | 0.9993 | 0.9972 | 0.9907 | 0.9745 | 0.9408 |
| | 11 | | | | | 1.0000 | 0.9999 | 0.9995 | 0.9981 | 0.9937 | 0.9824 |
| | 12 | | | | | | 1.0000 | 0.9999 | 0.9997 | 0.9989 | 0.9963 |
| | 13 | | | | | | | 1.0000 | 1.0000 | 0.9999 | 0.9995 |
| | 14 | | | | | | | | | 1.0000 | 1.0000 |
| $n = 20$ | $r = 0$ | 0.3585 | 0.1216 | 0.0388 | 0.0115 | 0.0032 | 0.0008 | 0.0002 | 0.0000 | 0.0000 | 0.0000 |
| | 1 | 0.7358 | 0.3917 | 0.1756 | 0.0692 | 0.0243 | 0.0076 | 0.0021 | 0.0005 | 0.0001 | 0.0000 |
| | 2 | 0.9245 | 0.6769 | 0.4049 | 0.2061 | 0.0913 | 0.0355 | 0.0121 | 0.0036 | 0.0009 | 0.0002 |
| | 3 | 0.9841 | 0.8670 | 0.6477 | 0.4114 | 0.2252 | 0.1071 | 0.0444 | 0.0160 | 0.0049 | 0.0013 |
| | 4 | 0.9974 | 0.9568 | 0.8298 | 0.6296 | 0.4148 | 0.2375 | 0.1182 | 0.0510 | 0.0189 | 0.0059 |
| | 5 | 0.9997 | 0.9887 | 0.9327 | 0.8042 | 0.6172 | 0.4164 | 0.2454 | 0.1256 | 0.0553 | 0.0207 |
| | 6 | 1.0000 | 0.9976 | 0.9781 | 0.9133 | 0.7858 | 0.6080 | 0.4166 | 0.2500 | 0.1299 | 0.0577 |
| | 7 | | 0.9996 | 0.9941 | 0.9679 | 0.8982 | 0.7723 | 0.6010 | 0.4159 | 0.2520 | 0.1316 |
| | 8 | | 0.9999 | 0.9987 | 0.9900 | 0.9591 | 0.8867 | 0.7624 | 0.5956 | 0.4143 | 0.2517 |
| | 9 | | 1.0000 | 0.9998 | 0.9974 | 0.9861 | 0.9520 | 0.8782 | 0.7553 | 0.5914 | 0.4119 |
| | 10 | | | 1.0000 | 0.9994 | 0.9961 | 0.9829 | 0.9468 | 0.8725 | 0.7507 | 0.5881 |
| | 11 | | | | 0.9999 | 0.9991 | 0.9949 | 0.9804 | 0.9435 | 0.8692 | 0.7483 |
| | 12 | | | | 1.0000 | 0.9998 | 0.9987 | 0.9940 | 0.9790 | 0.9420 | 0.8684 |
| | 13 | | | | | 1.0000 | 0.9997 | 0.9985 | 0.9935 | 0.9786 | 0.9423 |
| | 14 | | | | | | 1.0000 | 0.9997 | 0.9984 | 0.9936 | 0.9793 |
| | 15 | | | | | | | 1.0000 | 0.9997 | 0.9985 | 0.9941 |
| | 16 | | | | | | | | 1.0000 | 0.9997 | 0.9987 |
| | 17 | | | | | | | | | 1.0000 | 0.9998 |
| | 18 | | | | | | | | | | 1.0000 |

# CUMULATIVE POISSON PROBABILITIES

The tabulated value is $P(X \leqslant r)$ where $X \sim Po(\lambda)$

| $\lambda =$ | 0.2 | 0.4 | 0.5 | 0.6 | 0.8 | 1.0 | 1.2 | 1.4 | 1.5 |
|---|---|---|---|---|---|---|---|---|---|
| $r = 0$ | 0.8187 | 0.6703 | 0.6065 | 0.5488 | 0.4493 | 0.3679 | 0.3012 | 0.2466 | 0.2231 |
| 1 | 0.9825 | 0.9384 | 0.9098 | 0.8781 | 0.8088 | 0.7358 | 0.6626 | 0.5918 | 0.5578 |
| 2 | 0.9989 | 0.9921 | 0.9856 | 0.9769 | 0.9526 | 0.9197 | 0.8795 | 0.8335 | 0.8088 |
| 3 | 0.9999 | 0.9992 | 0.9982 | 0.9966 | 0.9909 | 0.9810 | 0.9662 | 0.9463 | 0.9344 |
| 4 | 1.0000 | 0.9999 | 0.9998 | 0.9996 | 0.9986 | 0.9963 | 0.9923 | 0.9857 | 0.9814 |
| 5 | | 1.0000 | 1.0000 | 1.0000 | 0.9998 | 0.9994 | 0.9985 | 0.9968 | 0.9955 |
| 6 | | | | | 1.0000 | 0.9999 | 0.9997 | 0.9994 | 0.9991 |
| 7 | | | | | | 1.0000 | 1.0000 | 0.9999 | 0.9998 |
| 8 | | | | | | | | 1.0000 | 1.0000 |

| $\lambda =$ | 1.6 | 1.8 | 2.0 | 2.2 | 2.4 | 2.5 | 2.6 | 2.8 | 3.0 |
|---|---|---|---|---|---|---|---|---|---|
| $r = 0$ | 0.2019 | 0.1653 | 0.1353 | 0.1108 | 0.0907 | 0.0821 | 0.0743 | 0.0608 | 0.0498 |
| 1 | 0.5249 | 0.4628 | 0.4060 | 0.3546 | 0.3084 | 0.2873 | 0.2674 | 0.2311 | 0.1991 |
| 2 | 0.7834 | 0.7306 | 0.6767 | 0.6227 | 0.5697 | 0.5438 | 0.5184 | 0.4695 | 0.4232 |
| 3 | 0.9212 | 0.8913 | 0.8571 | 0.8194 | 0.7787 | 0.7576 | 0.7360 | 0.6919 | 0.6472 |
| 4 | 0.9763 | 0.9636 | 0.9473 | 0.9275 | 0.9041 | 0.8912 | 0.8774 | 0.8477 | 0.8153 |
| 5 | 0.9940 | 0.9896 | 0.9834 | 0.9751 | 0.9643 | 0.9580 | 0.9510 | 0.9349 | 0.9161 |
| 6 | 0.9987 | 0.9974 | 0.9955 | 0.9925 | 0.9884 | 0.9858 | 0.9828 | 0.9756 | 0.9665 |
| 7 | 0.9997 | 0.9994 | 0.9989 | 0.9980 | 0.9967 | 0.9958 | 0.9947 | 0.9919 | 0.9881 |
| 8 | 1.0000 | 0.9999 | 0.9998 | 0.9995 | 0.9991 | 0.9989 | 0.9985 | 0.9976 | 0.9962 |
| 9 | | 1.0000 | 1.0000 | 0.9999 | 0.9998 | 0.9997 | 0.9996 | 0.9993 | 0.9989 |
| 10 | | | | 1.0000 | 1.0000 | 0.9999 | 0.9999 | 0.9998 | 0.9997 |
| 11 | | | | | | 1.0000 | 1.0000 | 1.0000 | 0.9999 |
| 12 | | | | | | | | | 1.0000 |

| $\lambda =$ | 3.2 | 3.4 | 3.5 | 3.6 | 3.8 | 4.0 | 4.5 | 5.0 | 5.5 |
|---|---|---|---|---|---|---|---|---|---|
| $r = 0$ | 0.0408 | 0.0334 | 0.0302 | 0.0273 | 0.0224 | 0.0183 | 0.0111 | 0.0067 | 0.0041 |
| 1 | 0.1712 | 0.1468 | 0.1359 | 0.1257 | 0.1074 | 0.0916 | 0.0611 | 0.0404 | 0.0266 |
| 2 | 0.3799 | 0.3397 | 0.3208 | 0.3027 | 0.2689 | 0.2381 | 0.1736 | 0.1247 | 0.0884 |
| 3 | 0.6025 | 0.5584 | 0.5366 | 0.5152 | 0.4735 | 0.4335 | 0.3423 | 0.2650 | 0.2017 |
| 4 | 0.7806 | 0.7442 | 0.7254 | 0.7064 | 0.6678 | 0.6288 | 0.5321 | 0.4405 | 0.3575 |
| 5 | 0.8946 | 0.8705 | 0.8576 | 0.8441 | 0.8156 | 0.7851 | 0.7029 | 0.6160 | 0.5289 |
| 6 | 0.9554 | 0.9421 | 0.9347 | 0.9267 | 0.9091 | 0.8893 | 0.8311 | 0.7622 | 0.6860 |
| 7 | 0.9832 | 0.9769 | 0.9733 | 0.9692 | 0.9599 | 0.9489 | 0.9134 | 0.8666 | 0.8095 |
| 8 | 0.9943 | 0.9917 | 0.9901 | 0.9883 | 0.9840 | 0.9786 | 0.9597 | 0.9319 | 0.8944 |
| 9 | 0.9982 | 0.9973 | 0.9967 | 0.9960 | 0.9942 | 0.9919 | 0.9829 | 0.9682 | 0.9462 |
| 10 | 0.9995 | 0.9992 | 0.9990 | 0.9987 | 0.9981 | 0.9972 | 0.9933 | 0.9863 | 0.9747 |
| 11 | 0.9999 | 0.9998 | 0.9997 | 0.9996 | 0.9994 | 0.9991 | 0.9976 | 0.9945 | 0.9890 |
| 12 | 1.0000 | 0.9999 | 0.9999 | 0.9999 | 0.9998 | 0.9997 | 0.9992 | 0.9980 | 0.9955 |
| 13 | | 1.0000 | 1.0000 | 1.0000 | 1.0000 | 0.9999 | 0.9997 | 0.9993 | 0.9983 |
| 14 | | | | | | 1.0000 | 0.9999 | 0.9998 | 0.9994 |
| 15 | | | | | | | 1.0000 | 0.9999 | 0.9998 |
| 16 | | | | | | | | 1.0000 | 0.9999 |
| 17 | | | | | | | | | 1.0000 |
| 18 | | | | | | | | | |

# CUMULATIVE POISSON PROBABILITIES

The tabulated value is $P(X \leqslant r)$ where $X \sim \text{Po}(\lambda)$

| $\lambda =$ | 6.0 | 6.5 | 7.0 | 7.5 | 8.0 | 8.5 | 9.0 | 9.5 | 10.0 |
|---|---|---|---|---|---|---|---|---|---|
| $r = 0$ | 0.0025 | 0.0015 | 0.0009 | 0.0006 | 0.0003 | 0.0002 | 0.0001 | 0.0001 | 0.0000 |
| 1 | 0.0174 | 0.0113 | 0.0073 | 0.0047 | 0.0030 | 0.0019 | 0.0012 | 0.0008 | 0.0005 |
| 2 | 0.0620 | 0.0430 | 0.0296 | 0.0203 | 0.0138 | 0.0093 | 0.0062 | 0.0042 | 0.0028 |
| 3 | 0.1512 | 0.1118 | 0.0818 | 0.0591 | 0.0424 | 0.0301 | 0.0212 | 0.0149 | 0.0103 |
| 4 | 0.2851 | 0.2237 | 0.1730 | 0.1321 | 0.0996 | 0.0744 | 0.0550 | 0.0403 | 0.0293 |
| 5 | 0.4457 | 0.3690 | 0.3007 | 0.2414 | 0.1912 | 0.1496 | 0.1157 | 0.0885 | 0.0671 |
| 6 | 0.6063 | 0.5265 | 0.4497 | 0.3782 | 0.3134 | 0.2562 | 0.2068 | 0.1649 | 0.1301 |
| 7 | 0.7440 | 0.6728 | 0.5987 | 0.5246 | 0.4530 | 0.3856 | 0.3239 | 0.2687 | 0.2202 |
| 8 | 0.8472 | 0.7916 | 0.7291 | 0.6620 | 0.5925 | 0.5231 | 0.4557 | 0.3918 | 0.3328 |
| 9 | 0.9161 | 0.8774 | 0.8305 | 0.7764 | 0.7166 | 0.6530 | 0.5874 | 0.5218 | 0.4579 |
| 10 | 0.9574 | 0.9332 | 0.9015 | 0.8622 | 0.8159 | 0.7634 | 0.7060 | 0.6453 | 0.5830 |
| 11 | 0.9799 | 0.9661 | 0.9467 | 0.9208 | 0.8881 | 0.8487 | 0.8030 | 0.7520 | 0.6968 |
| 12 | 0.9912 | 0.9840 | 0.9730 | 0.9573 | 0.9362 | 0.9091 | 0.8758 | 0.8364 | 0.7916 |
| 13 | 0.9964 | 0.9929 | 0.9872 | 0.9784 | 0.9658 | 0.9486 | 0.9261 | 0.8981 | 0.8645 |
| 14 | 0.9986 | 0.9970 | 0.9943 | 0.9897 | 0.9827 | 0.9726 | 0.9585 | 0.9400 | 0.9165 |
| 15 | 0.9995 | 0.9988 | 0.9976 | 0.9954 | 0.9918 | 0.9862 | 0.9780 | 0.9665 | 0.9513 |
| 16 | 0.9998 | 0.9996 | 0.9990 | 0.9980 | 0.9963 | 0.9934 | 0.9889 | 0.9823 | 0.9730 |
| 17 | 0.9999 | 0.9998 | 0.9996 | 0.9992 | 0.9984 | 0.9970 | 0.9947 | 0.9911 | 0.9857 |
| 18 | 1.0000 | 0.9999 | 0.9999 | 0.9997 | 0.9993 | 0.9987 | 0.9976 | 0.9957 | 0.9928 |
| 19 | | 1.0000 | 1.0000 | 0.9999 | 0.9997 | 0.9995 | 0.9989 | 0.9980 | 0.9965 |
| 20 | | | | 1.0000 | 0.9999 | 0.9998 | 0.9996 | 0.9991 | 0.9984 |
| 21 | | | | | 1.0000 | 0.9999 | 0.9998 | 0.9996 | 0.9993 |
| 22 | | | | | | 1.0000 | 0.9999 | 0.9999 | 0.9997 |
| 23 | | | | | | | 1.0000 | 0.9999 | 0.9999 |
| 24 | | | | | | | | 1.0000 | 1.0000 |

# THE STANDARD NORMAL DISTRIBUTION FUNCTION

If $Z$ has a normal distribution with mean 0 and variance 1 then, for each value of $z$, the table gives the value of $\Phi(z)$ where

$$\Phi(z) = P(Z \leqslant z).$$

For negative values of $z$ use $\Phi(-z) = 1 - \Phi(z)$.

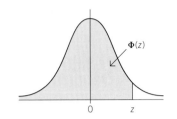

| z | 0 | 1 | 2 | 3 | 4 | 5 | 6 | 7 | 8 | 9 | 1 | 2 | 3 | 4 | 5 | 6 | 7 | 8 | 9 |
|---|---|---|---|---|---|---|---|---|---|---|---|---|---|---|---|---|---|---|---|
| | | | | | | | | | | | | | | | ADD | | | | |
| 0.0 | 0.5000 | 0.5040 | 0.5080 | 0.5120 | 0.5160 | 0.5199 | 0.5239 | 0.5279 | 0.5319 | 0.5359 | 4 | 8 | 12 | 16 | 20 | 24 | 28 | 32 | 36 |
| 0.1 | 0.5398 | 0.5438 | 0.5478 | 0.5517 | 0.5557 | 0.5596 | 0.5636 | 0.5675 | 0.5714 | 0.5753 | 4 | 8 | 12 | 16 | 20 | 24 | 28 | 32 | 36 |
| 0.2 | 0.5793 | 0.5832 | 0.5871 | 0.5910 | 0.5948 | 0.5987 | 0.6026 | 0.6064 | 0.6103 | 0.6141 | 4 | 8 | 12 | 15 | 19 | 23 | 27 | 31 | 35 |
| 0.3 | 0.6179 | 0.6217 | 0.6255 | 0.6293 | 0.6331 | 0.6368 | 0.6406 | 0.6443 | 0.6480 | 0.6517 | 4 | 7 | 11 | 15 | 19 | 22 | 26 | 30 | 34 |
| 0.4 | 0.6554 | 0.6591 | 0.6628 | 0.6664 | 0.6700 | 0.6736 | 0.6772 | 0.6808 | 0.6844 | 0.6879 | 4 | 7 | 11 | 14 | 18 | 22 | 25 | 29 | 32 |
| 0.5 | 0.6915 | 0.6950 | 0.6985 | 0.7019 | 0.7054 | 0.7088 | 0.7123 | 0.7157 | 0.7190 | 0.7224 | 3 | 7 | 10 | 14 | 17 | 20 | 24 | 27 | 31 |
| 0.6 | 0.7257 | 0.7291 | 0.7324 | 0.7357 | 0.7389 | 0.7422 | 0.7454 | 0.7486 | 0.7517 | 0.7549 | 3 | 7 | 10 | 13 | 16 | 19 | 23 | 26 | 29 |
| 0.7 | 0.7580 | 0.7611 | 0.7642 | 0.7673 | 0.7704 | 0.7734 | 0.7764 | 0.7794 | 0.7823 | 0.7852 | 3 | 6 | 9 | 12 | 15 | 18 | 21 | 24 | 27 |
| 0.8 | 0.7881 | 0.7910 | 0.7939 | 0.7967 | 0.7995 | 0.8023 | 0.8051 | 0.8078 | 0.8106 | 0.8133 | 3 | 5 | 8 | 11 | 14 | 16 | 19 | 22 | 25 |
| 0.9 | 0.8159 | 0.8186 | 0.8212 | 0.8238 | 0.8264 | 0.8289 | 0.8315 | 0.8340 | 0.8365 | 0.8389 | 3 | 5 | 8 | 10 | 13 | 15 | 18 | 20 | 23 |
| 1.0 | 0.8413 | 0.8438 | 0.8461 | 0.8485 | 0.8508 | 0.8531 | 0.8554 | 0.8577 | 0.8599 | 0.8621 | 2 | 5 | 7 | 9 | 12 | 14 | 16 | 19 | 21 |
| 1.1 | 0.8643 | 0.8665 | 0.8686 | 0.8708 | 0.8729 | 0.8749 | 0.8770 | 0.8790 | 0.8810 | 0.8830 | 2 | 4 | 6 | 8 | 10 | 12 | 14 | 16 | 18 |
| 1.2 | 0.8849 | 0.8869 | 0.8888 | 0.8907 | 0.8925 | 0.8944 | 0.8962 | 0.8980 | 0.8997 | 0.9015 | 2 | 4 | 6 | 7 | 9 | 11 | 13 | 15 | 17 |
| 1.3 | 0.9032 | 0.9049 | 0.9066 | 0.9082 | 0.9099 | 0.9115 | 0.9131 | 0.9147 | 0.9162 | 0.9177 | 2 | 3 | 5 | 6 | 8 | 10 | 11 | 13 | 14 |
| 1.4 | 0.9192 | 0.9207 | 0.9222 | 0.9236 | 0.9251 | 0.9265 | 0.9279 | 0.9292 | 0.9306 | 0.9319 | 1 | 3 | 4 | 6 | 7 | 8 | 10 | 11 | 13 |
| 1.5 | 0.9332 | 0.9345 | 0.9357 | 0.9370 | 0.9382 | 0.9394 | 0.9406 | 0.9418 | 0.9429 | 0.9441 | 1 | 2 | 4 | 5 | 6 | 7 | 8 | 10 | 11 |
| 1.6 | 0.9452 | 0.9463 | 0.9474 | 0.9484 | 0.9495 | 0.9505 | 0.9515 | 0.9525 | 0.9535 | 0.9545 | 1 | 2 | 3 | 4 | 5 | 6 | 7 | 8 | 9 |
| 1.7 | 0.9554 | 0.9564 | 0.9573 | 0.9582 | 0.9591 | 0.9599 | 0.9608 | 0.9616 | 0.9625 | 0.9633 | 1 | 2 | 3 | 4 | 4 | 5 | 6 | 7 | 8 |
| 1.8 | 0.9641 | 0.9649 | 0.9656 | 0.9664 | 0.9671 | 0.9678 | 0.9686 | 0.9693 | 0.9699 | 0.9706 | 1 | 1 | 2 | 3 | 4 | 4 | 5 | 6 | 6 |
| 1.9 | 0.9713 | 0.9719 | 0.9726 | 0.9732 | 0.9738 | 0.9744 | 0.9750 | 0.9756 | 0.9761 | 0.9767 | 1 | 1 | 2 | 2 | 3 | 4 | 4 | 5 | 5 |
| 2.0 | 0.9772 | 0.9778 | 0.9783 | 0.9788 | 0.9793 | 0.9798 | 0.9803 | 0.9808 | 0.9812 | 0.9817 | 0 | 1 | 1 | 2 | 2 | 3 | 3 | 4 | 4 |
| 2.1 | 0.9821 | 0.9826 | 0.9830 | 0.9834 | 0.9838 | 0.9842 | 0.9846 | 0.9850 | 0.9854 | 0.9857 | 0 | 1 | 1 | 2 | 2 | 2 | 3 | 3 | 4 |
| 2.2 | 0.9861 | 0.9864 | 0.9868 | 0.9871 | 0.9875 | 0.9878 | 0.9881 | 0.9884 | 0.9887 | 0.9890 | 0 | 1 | 1 | 1 | 2 | 2 | 2 | 3 | 3 |
| 2.3 | 0.9893 | 0.9896 | 0.9898 | 0.9901 | 0.9904 | 0.9906 | 0.9909 | 0.9911 | 0.9913 | 0.9916 | 0 | 1 | 1 | 1 | 1 | 2 | 2 | 2 | 2 |
| 2.4 | 0.9918 | 0.9920 | 0.9922 | 0.9924 | 0.9927 | 0.9929 | 0.9931 | 0.9932 | 0.9934 | 0.9936 | 0 | 0 | 1 | 1 | 1 | 1 | 1 | 2 | 2 |
| 2.5 | 0.9938 | 0.9940 | 0.9941 | 0.9943 | 0.9945 | 0.9946 | 0.9948 | 0.9949 | 0.9951 | 0.9952 | 0 | 0 | 0 | 1 | 1 | 1 | 1 | 1 | 1 |
| 2.6 | 0.9953 | 0.9955 | 0.9956 | 0.9957 | 0.9958 | 0.9960 | 0.9961 | 0.9962 | 0.9963 | 0.9964 | 0 | 0 | 0 | 0 | 1 | 1 | 1 | 1 | 1 |
| 2.7 | 0.9965 | 0.9966 | 0.9967 | 0.9968 | 0.9969 | 0.9970 | 0.9971 | 0.9972 | 0.9973 | 0.9974 | 0 | 0 | 0 | 0 | 0 | 1 | 1 | 1 | 1 |
| 2.8 | 0.9974 | 0.9975 | 0.9976 | 0.9977 | 0.9977 | 0.9978 | 0.9979 | 0.9979 | 0.9980 | 0.9981 | 0 | 0 | 0 | 0 | 0 | 0 | 0 | 1 | 1 |
| 2.9 | 0.9981 | 0.9982 | 0.9982 | 0.9983 | 0.9984 | 0.9984 | .09985 | 0.9985 | 0.9986 | 0.9986 | 0 | 0 | 0 | 0 | 0 | 0 | 0 | 0 | 0 |

# CRITICAL VALUES FOR THE NORMAL DISTRIBUTION

The table gives the value of $z$ such that $P(Z \leqslant z) = p$, where $Z \sim N(0, 1)$.

| p | 0.75 | 0.90 | 0.95 | 0.975 | 0.99 | 0.995 | 0.9975 | 0.999 | 0.9995 |
|---|---|---|---|---|---|---|---|---|---|
| z | 0.674 | 1.282 | 1.645 | 1.960 | 2.326 | 2.576 | 2.807 | 3.090 | 3.291 |

# CRITICAL VALUES FOR THE $t$-DISTRIBUTION

If $T$ has a $t$-distribution with $v$ degrees of freedom then, for each pair of values of $p$ and $v$, the table gives the value of $t$ such that $P(T \leqslant t) = p$.

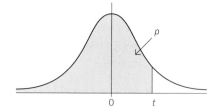

| $p$ | 0.75 | 0.90 | 0.95 | 0.975 | 0.99 | 0.995 | 0.9975 | 0.999 | 0.9995 |
|---|---|---|---|---|---|---|---|---|---|
| $v = 1$ | 1.000 | 3.078 | 6.314 | 12.71 | 31.82 | 63.66 | 127.3 | 318.3 | 636.6 |
| 2 | 0.816 | 1.886 | 2.920 | 4.303 | 6.965 | 9.925 | 14.09 | 22.33 | 31.60 |
| 3 | 0.765 | 1.638 | 2.353 | 3.182 | 4.541 | 5.841 | 7.453 | 10.21 | 12.92 |
| 4 | 0.741 | 1.533 | 2.132 | 2.776 | 3.747 | 4.604 | 5.598 | 7.173 | 8.610 |
| 5 | 0.727 | 1.476 | 2.015 | 2.571 | 3.365 | 4.032 | 4.773 | 5.893 | 6.869 |
| 6 | 0.718 | 1.440 | 1.943 | 2.447 | 3.143 | 3.707 | 4.317 | 5.208 | 5.959 |
| 7 | 0.711 | 1.415 | 1.895 | 2.365 | 2.998 | 3.499 | 4.029 | 4.785 | 5.408 |
| 8 | 0.706 | 1.397 | 1.860 | 2.306 | 2.896 | 3.355 | 3.833 | 4.501 | 5.041 |
| 9 | 0.703 | 1.383 | 1.833 | 2.262 | 2.821 | 3.250 | 3.690 | 4.297 | 4.781 |
| 10 | 0.700 | 1.372 | 1.812 | 2.228 | 2.764 | 3.169 | 3.581 | 4.144 | 4.587 |
| 11 | 0.697 | 1.363 | 1.796 | 2.201 | 2.718 | 3.106 | 3.497 | 4.025 | 4.437 |
| 12 | 0.695 | 1.356 | 1.782 | 2.179 | 2.681 | 3.055 | 3.428 | 3.930 | 4.318 |
| 13 | 0.694 | 1.350 | 1.771 | 2.160 | 2.650 | 3.012 | 3.372 | 3.852 | 4.221 |
| 14 | 0.692 | 1.345 | 1.761 | 2.145 | 2.624 | 2.977 | 3.326 | 3.787 | 4.140 |
| 15 | 0.691 | 1.341 | 1.753 | 2.131 | 2.602 | 2.947 | 3.286 | 3.733 | 4.073 |
| 16 | 0.690 | 1.337 | 1.746 | 2.120 | 2.583 | 2.921 | 3.252 | 3.686 | 4.015 |
| 17 | 0.689 | 1.333 | 1.740 | 2.110 | 2.567 | 2.898 | 3.222 | 3.646 | 3.965 |
| 18 | 0.688 | 1.330 | 1.734 | 2.101 | 2.552 | 2.878 | 3.197 | 3.610 | 3.922 |
| 19 | 0.688 | 1.328 | 1.729 | 2.093 | 2.539 | 2.861 | 3.174 | 3.579 | 3.883 |
| 20 | 0.687 | 1.325 | 1.725 | 2.086 | 2.528 | 2.845 | 3.153 | 3.552 | 3.850 |
| 21 | 0.686 | 1.323 | 1.721 | 2.080 | 2.518 | 2.831 | 3.135 | 3.527 | 3.819 |
| 22 | 0.686 | 1.321 | 1.717 | 2.074 | 2.508 | 2.819 | 3.119 | 3.505 | 3.792 |
| 23 | 0.685 | 1.319 | 1.714 | 2.069 | 2.500 | 2.807 | 3.104 | 3.485 | 3.768 |
| 24 | 0.685 | 1.318 | 1.711 | 2.064 | 2.492 | 2.797 | 3.091 | 3.467 | 3.745 |
| 25 | 0.684 | 1.316 | 1.708 | 2.060 | 2.485 | 2.787 | 3.078 | 3.450 | 3.725 |
| 26 | 0.684 | 1.315 | 1.706 | 2.056 | 2.479 | 2.779 | 3.067 | 3.435 | 3.707 |
| 27 | 0.684 | 1.314 | 1.703 | 2.052 | 2.473 | 2.771 | 3.057 | 3.421 | 3.690 |
| 28 | 0.683 | 1.313 | 1.701 | 2.048 | 2.467 | 2.763 | 3.047 | 3.408 | 3.674 |
| 29 | 0.683 | 1.311 | 1.699 | 2.045 | 2.462 | 2.756 | 3.038 | 3.396 | 3.659 |
| 30 | 0.683 | 1.310 | 1.697 | 2.042 | 2.457 | 2.750 | 3.030 | 3.385 | 3.646 |
| 40 | 0.681 | 1.303 | 1.684 | 2.021 | 2.423 | 2.704 | 2.971 | 3.307 | 3.551 |
| 60 | 0.679 | 1.296 | 1.671 | 2.000 | 2.390 | 2.660 | 2.915 | 3.232 | 3.460 |
| 120 | 0.677 | 1.289 | 1.658 | 1.980 | 2.358 | 2.617 | 2.860 | 3.160 | 3.373 |
| $\infty$ | 0.674 | 1.282 | 1.645 | 1.960 | 2.326 | 2.576 | 2.807 | 3.090 | 3.291 |

# CRITICAL VALUES FOR THE $\chi^2$ DISTRIBUTION

If $X$ has a $\chi^2$ distribution with $v$ degrees of freedom, then for each pair of values of $p$ and $v$, the table gives the value of $x$ such that $P(X \geqslant x) = p$

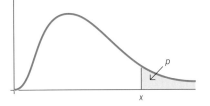

| $p$ | 0.990 | 0.975 | 0.950 | 0.100 | 0.050 | 0.025 | 0.010 | 0.005 |
|---|---|---|---|---|---|---|---|---|
| $v = 1$ | 0.000 | 0.001 | 0.004 | 2.705 | 3.841 | 5.024 | 6.635 | 7.879 |
| 2 | 0.020 | 0.051 | 0.103 | 4.605 | 5.991 | 7.378 | 9.210 | 10.597 |
| 3 | 0.115 | 0.216 | 0.352 | 6.251 | 7.815 | 9.348 | 11.345 | 12.838 |
| 4 | 0.297 | 0.484 | 0.711 | 7.779 | 9.488 | 11.143 | 13.277 | 14.860 |
| 5 | 0.554 | 0.831 | 1.145 | 9.236 | 11.070 | 12.832 | 15.086 | 16.750 |
| 6 | 0.872 | 1.237 | 1.635 | 10.645 | 12.592 | 14.449 | 16.812 | 18.548 |
| 7 | 1.239 | 1.690 | 2.167 | 12.017 | 14.067 | 16.013 | 18.475 | 20.278 |
| 8 | 1.646 | 2.180 | 2.733 | 13.362 | 15.507 | 17.535 | 20.090 | 21.955 |
| 9 | 2.088 | 2.700 | 3.325 | 14.684 | 16.919 | 19.023 | 21.666 | 23.589 |
| 10 | 2.558 | 3.247 | 3.940 | 15.987 | 18.307 | 20.483 | 23.209 | 25.188 |
| 11 | 3.053 | 3.816 | 4.575 | 17.275 | 19.675 | 21.920 | 24.725 | 26.757 |
| 12 | 3.571 | 4.404 | 5.226 | 18.549 | 21.026 | 23.337 | 26.217 | 28.300 |
| 13 | 4.107 | 5.009 | 5.892 | 19.812 | 22.362 | 24.736 | 27.688 | 29.819 |
| 14 | 4.660 | 5.629 | 6.571 | 21.064 | 23.685 | 26.119 | 29.141 | 31.319 |
| 15 | 5.229 | 6.262 | 7.261 | 22.307 | 24.996 | 27.488 | 30.578 | 32.801 |
| 16 | 5.812 | 6.908 | 7.962 | 23.542 | 26.296 | 28.845 | 32.000 | 34.267 |
| 17 | 6.408 | 7.564 | 8.672 | 24.769 | 27.587 | 30.191 | 33.409 | 35.718 |
| 18 | 7.015 | 8.231 | 9.390 | 25.989 | 28.869 | 31.526 | 34.805 | 37.156 |
| 19 | 7.633 | 8.907 | 10.117 | 27.204 | 30.144 | 32.852 | 36.191 | 38.582 |
| 20 | 8.260 | 9.591 | 10.851 | 28.412 | 31.410 | 34.170 | 37.566 | 39.997 |
| 21 | 8.897 | 10.283 | 11.591 | 29.615 | 32.671 | 35.479 | 38.932 | 41.401 |
| 22 | 9.542 | 10.982 | 12.338 | 30.813 | 33.924 | 36.781 | 40.289 | 42.796 |
| 23 | 10.196 | 11.689 | 13.091 | 32.007 | 35.172 | 38.076 | 41.638 | 44.181 |
| 24 | 10.856 | 12.401 | 13.848 | 33.196 | 36.415 | 39.364 | 42.980 | 45.558 |
| 25 | 11.524 | 13.120 | 14.611 | 34.382 | 37.652 | 40.646 | 44.314 | 46.928 |
| 26 | 12.198 | 13.844 | 15.379 | 35.563 | 38.885 | 41.923 | 45.642 | 48.290 |
| 27 | 12.879 | 14.573 | 16.151 | 36.741 | 40.113 | 43.194 | 46.963 | 49.645 |
| 28 | 13.565 | 15.308 | 16.928 | 37.916 | 41.337 | 44.461 | 48.278 | 50.993 |
| 29 | 14.256 | 16.047 | 17.708 | 39.088 | 42.557 | 45.722 | 49.588 | 52.336 |
| 30 | 14.953 | 16.791 | 18.493 | 40.256 | 43.773 | 46.979 | 50.892 | 53.672 |

# CRITICAL VALUES FOR CORRELATION COEFFICIENTS

These tables concern tests of the hypothesis that a population correlation coefficient $\rho$ is 0. The values in the tables are the minimum values which need to be reached by a sample correlation coefficient in order to be significant at the level shown, on a one-tailed test.

| Product-moment Coefficient | | | | | Sample size | Spearman's Coefficient | | |
|---|---|---|---|---|---|---|---|---|
| | | Level | | | | | Level | |
| 0.10 | 0.05 | 0.025 | 0.01 | 0.005 | | 0.05 | 0.025 | 0.01 |
| 0.8000 | 0.9000 | 0.9500 | 0.9800 | 0.9900 | 4 | 1.0000 | — | — |
| 0.6870 | 0.8054 | 0.8783 | 0.9343 | 0.9587 | 5 | 0.9000 | 1.0000 | 1.0000 |
| 0.6084 | 0.7293 | 0.8114 | 0.8822 | 0.9172 | 6 | 0.8286 | 0.8857 | 0.9429 |
| 0.5509 | 0.6694 | 0.7545 | 0.8329 | 0.8745 | 7 | 0.7143 | 0.7857 | 0.8929 |
| 0.5067 | 0.6215 | 0.7067 | 0.7887 | 0.8343 | 8 | 0.6429 | 0.7381 | 0.8333 |
| 0.4716 | 0.5822 | 0.6664 | 0.7498 | 0.7977 | 9 | 0.6000 | 0.7000 | 0.7833 |
| 0.4428 | 0.5494 | 0.6319 | 0.7155 | 0.7646 | 10 | 0.5636 | 0.6485 | 0.7455 |
| 0.4187 | 0.5214 | 0.6021 | 0.6851 | 0.7348 | 11 | 0.5364 | 0.6182 | 0.7091 |
| 0.3981 | 0.4973 | 0.5760 | 0.6581 | 0.7079 | 12 | 0.5035 | 0.5874 | 0.6783 |
| 0.3802 | 0.4762 | 0.5529 | 0.6339 | 0.6835 | 13 | 0.4835 | 0.5604 | 0.6484 |
| 0.3646 | 0.4575 | 0.5324 | 0.6120 | 0.6614 | 14 | 0.4637 | 0.5385 | 0.6264 |
| 0.3507 | 0.4409 | 0.5140 | 0.5923 | 0.6411 | 15 | 0.4464 | 0.5214 | 0.6036 |
| 0.3383 | 0.4259 | 0.4973 | 0.5742 | 0.6226 | 16 | 0.4294 | 0.5029 | 0.5824 |
| 0.3271 | 0.4124 | 0.4821 | 0.5577 | 0.6055 | 17 | 0.4142 | 0.4877 | 0.5662 |
| 0.3170 | 0.4000 | 0.4683 | 0.5425 | 0.5897 | 18 | 0.4014 | 0.4716 | 0.5501 |
| 0.3077 | 0.3887 | 0.4555 | 0.5285 | 0.5751 | 19 | 0.3912 | 0.4596 | 0.5351 |
| 0.2992 | 0.3783 | 0.4438 | 0.5155 | 0.5614 | 20 | 0.3805 | 0.4466 | 0.5218 |
| 0.2914 | 0.3687 | 0.4329 | 0.5034 | 0.5487 | 21 | 0.3701 | 0.4364 | 0.5091 |
| 0.2841 | 0.3598 | 0.4227 | 0.4921 | 0.5368 | 22 | 0.3608 | 0.4252 | 0.4975 |
| 0.2774 | 0.3515 | 0.4133 | 0.4815 | 0.5256 | 23 | 0.3528 | 0.4160 | 0.4862 |
| 0.2711 | 0.3438 | 0.4044 | 0.4716 | 0.5151 | 24 | 0.3443 | 0.4070 | 0.4757 |
| 0.2653 | 0.3365 | 0.3961 | 0.4622 | 0.5052 | 25 | 0.3369 | 0.3977 | 0.4662 |
| 0.2598 | 0.3297 | 0.3882 | 0.4534 | 0.4958 | 26 | 0.3306 | 0.3901 | 0.4571 |
| 0.2546 | 0.3233 | 0.3809 | 0.4451 | 0.4869 | 27 | 0.3242 | 0.3828 | 0.4487 |
| 0.2497 | 0.3172 | 0.3739 | 0.4372 | 0.4785 | 28 | 0.3180 | 0.3755 | 0.4401 |
| 0.2451 | 0.3115 | 0.3673 | 0.4297 | 0.4705 | 29 | 0.3118 | 0.3685 | 0.4325 |
| 0.2407 | 0.3061 | 0.3610 | 0.4226 | 0.4629 | 30 | 0.3063 | 0.3624 | 0.4251 |
| 0.2070 | 0.2638 | 0.3120 | 0.3665 | 0.4026 | 40 | 0.2640 | 0.3128 | 0.3681 |
| 0.1843 | 0.2353 | 0.2787 | 0.3281 | 0.3610 | 50 | 0.2353 | 0.2791 | 0.3293 |
| 0.1678 | 0.2144 | 0.2542 | 0.2997 | 0.3301 | 60 | 0.2144 | 0.2545 | 0.3005 |
| 0.1550 | 0.1982 | 0.2352 | 0.2776 | 0.3060 | 70 | 0.1982 | 0.2354 | 0.2782 |
| 0.1448 | 0.1852 | 0.2199 | 0.2597 | 0.2864 | 80 | 0.1852 | 0.2201 | 0.2602 |
| 0.1364 | 0.1745 | 0.2072 | 0.2449 | 0.2702 | 90 | 0.1745 | 0.2074 | 0.2453 |
| 0.1292 | 0.1654 | 0.1966 | 0.2324 | 0.2565 | 100 | 0.1654 | 0.1967 | 0.2327 |

# RANDOM NUMBERS

| | | | | | | | | | |
|---|---|---|---|---|---|---|---|---|---|
| 65 23 | 68 00 | 77 82 | 58 14 | 10 85 | 11 85 | 57 11 | 73 74 | 45 25 | 50 46 |
| 09 56 | 76 51 | 04 73 | 94 30 | 16 74 | 69 59 | 04 38 | 83 98 | 30 20 | 87 85 |
| 55 99 | 98 60 | 01 33 | 06 93 | 85 13 | 23 17 | 25 51 | 92 04 | 52 31 | 38 70 |
| 72 82 | 45 44 | 09 53 | 04 83 | 03 83 | 98 41 | 67 41 | 01 38 | 66 83 | 11 99 |
| 04 21 | 28 72 | 73 25 | 02 74 | 35 81 | 78 49 | 52 67 | 61 40 | 60 50 | 47 50 |
| 87 01 | 80 59 | 89 36 | 41 59 | 60 27 | 64 89 | 47 45 | 18 21 | 69 84 | 76 06 |
| 31 62 | 46 53 | 84 40 | 56 31 | 74 76 | 52 23 | 72 95 | 96 06 | 56 83 | 85 22 |
| 29 81 | 57 94 | 35 91 | 90 70 | 94 24 | 19 35 | 50 22 | 23 72 | 87 34 | 83 15 |
| 39 98 | 74 22 | 77 19 | 12 81 | 29 42 | 04 50 | 62 34 | 36 81 | 43 07 | 97 92 |
| 56 14 | 80 10 | 76 52 | 38 54 | 84 13 | 99 90 | 22 55 | 41 04 | 72 37 | 89 33 |
| 29 56 | 62 74 | 12 67 | 09 35 | 89 33 | 04 28 | 44 75 | 01 57 | 87 45 | 52 21 |
| 93 32 | 57 38 | 39 36 | 87 42 | 72 55 | 73 97 | 98 36 | 57 41 | 76 09 | 11 68 |
| 95 69 | 51 54 | 43 19 | 20 49 | 57 25 | 90 55 | 26 20 | 70 98 | 43 73 | 56 45 |
| 65 71 | 32 43 | 64 67 | 22 55 | 65 65 | 48 86 | 10 88 | 20 12 | 40 18 | 49 25 |
| 90 27 | 33 43 | 97 84 | 20 57 | 49 91 | 41 20 | 17 64 | 29 60 | 66 87 | 55 97 |
| 90 29 | 42 45 | 61 34 | 30 13 | 30 39 | 21 52 | 59 28 | 64 98 | 08 76 | 09 27 |
| 99 74 | 06 29 | 20 55 | 72 70 | 11 43 | 95 82 | 75 37 | 90 24 | 77 43 | 63 21 |
| 87 87 | 66 91 | 16 97 | 51 50 | 61 36 | 96 47 | 76 68 | 49 11 | 50 56 | 51 06 |
| 46 24 | 17 74 | 97 37 | 39 03 | 54 83 | 34 00 | 74 61 | 77 51 | 43 63 | 15 67 |
| 66 79 | 81 43 | 40 92 | 84 72 | 88 32 | 83 24 | 67 01 | 41 34 | 70 19 | 26 93 |
| 36 42 | 94 58 | 83 30 | 92 39 | 18 40 | 03 00 | 12 90 | 32 37 | 91 65 | 48 15 |
| 07 66 | 25 08 | 99 27 | 69 48 | 85 32 | 16 46 | 19 31 | 85 02 | 86 36 | 22 96 |
| 93 10 | 05 72 | 18 26 | 36 67 | 68 48 | 31 69 | 68 58 | 93 49 | 45 86 | 99 29 |
| 49 50 | 63 99 | 26 71 | 47 94 | 32 71 | 72 91 | 34 18 | 74 06 | 32 14 | 40 80 |
| 20 75 | 58 89 | 39 04 | 42 73 | 37 93 | 11 07 | 28 77 | 91 36 | 60 47 | 82 62 |
| 02 40 | 62 09 | 00 71 | 09 37 | 80 44 | 50 37 | 32 70 | 20 38 | 71 86 | 75 34 |
| 59 87 | 21 38 | 29 78 | 72 67 | 42 83 | 65 21 | 54 79 | 66 42 | 47 86 | 31 15 |
| 48 08 | 99 66 | 43 38 | 28 13 | 50 25 | 47 93 | 11 15 | 07 84 | 28 30 | 19 07 |
| 54 26 | 86 75 | 44 15 | 20 39 | 20 03 | 58 54 | 80 29 | 62 53 | 06 97 | 71 51 |
| 35 35 | 58 45 | 23 58 | 63 66 | 09 62 | 80 92 | 14 55 | 81 41 | 21 48 | 87 34 |
| 73 84 | 90 49 | 01 21 | 90 29 | 57 06 | 68 73 | 51 10 | 51 95 | 63 08 | 57 99 |
| 34 64 | 78 00 | 92 59 | 67 74 | 58 48 | 92 09 | 42 20 | 40 37 | 63 80 | 58 93 |
| 68 56 | 87 47 | 63 06 | 24 71 | 41 98 | 79 06 | 07 18 | 58 29 | 16 49 | 67 37 |
| 72 47 | 05 42 | 88 07 | 27 55 | 58 74 | 82 08 | 42 28 | 26 48 | 25 32 | 00 31 |
| 44 44 | 96 75 | 89 57 | 12 60 | 42 38 | 77 36 | 45 69 | 21 68 | 32 70 | 04 96 |
| 28 11 | 57 47 | 61 57 | 89 88 | 62 18 | 93 67 | 57 32 | 96 72 | 21 17 | 13 54 |
| 87 22 | 38 88 | 91 99 | 16 08 | 17 76 | 27 47 | 52 14 | 98 86 | 35 68 | 23 85 |
| 44 93 | 14 59 | 67 40 | 24 10 | 11 63 | 40 47 | 07 56 | 14 22 | 62 74 | 93 39 |
| 81 84 | 37 25 | 90 43 | 56 62 | 94 58 | 49 03 | 84 22 | 57 22 | 47 98 | 86 37 |
| 09 75 | 35 21 | 04 47 | 54 08 | 98 44 | 08 16 | 44 86 | 69 71 | 20 52 | 64 94 |
| 77 65 | 05 04 | 22 18 | 20 10 | 81 87 | 05 69 | 43 70 | 96 76 | 42 05 | 21 10 |
| 19 06 | 51 61 | 34 03 | 61 55 | 98 58 | 83 50 | 01 48 | 99 85 | 08 67 | 15 91 |
| 52 91 | 87 07 | 19 62 | 32 28 | 04 91 | 42 48 | 65 24 | 86 09 | 87 68 | 55 51 |
| 52 47 | 25 14 | 93 91 | 75 51 | 49 26 | 49 41 | 20 83 | 30 30 | 43 22 | 69 08 |
| 52 67 | 87 40 | 63 41 | 91 86 | 10 47 | 80 70 | 56 87 | 25 86 | 89 94 | 21 42 |
| 66 25 | 71 73 | 78 60 | 50 62 | 91 04 | 95 97 | 64 16 | 71 31 | 32 80 | 19 61 |
| 29 97 | 56 42 | 56 90 | 16 75 | 74 95 | 99 26 | 01 63 | 25 16 | 54 18 | 54 46 |
| 15 25 | 03 68 | 92 45 | 53 00 | 06 29 | 46 43 | 46 66 | 27 12 | 85 05 | 22 44 |
| 82 08 | 65 67 | 64 13 | 51 14 | 38 28 | 24 30 | 39 62 | 20 35 | 23 90 | 57 36 |
| 81 35 | 03 25 | 87 24 | 83 59 | 04 67 | 51 52 | 26 21 | 69 75 | 87 28 | 61 50 |

Each digit in this table is an independent sample from a population where each of the digits 0 to 9 has a probability of occurrence of 0.1. It should be noted that these digits have been computer generated, and are therefore 'pseudo' random numbers.

# ANSWERS

## Chapter 1

### Exercise 1a Stemplots (page 8)

NOTE: There are alternative formats

1. (a)

| | |
|---|---|
| 50 | 2 |
| 55 | 2 2 4 |
| 60 | 1 3 4 4 |
| 65 | 0 2 2 3 3 3 |
| 70 | 0 1 1 2 3 4 4 |
| 75 | 0 1 1 2 4 4 |
| 80 | 1 3 |
| 85 | 1 |

Key: 85 | 1 means 86

   (b) 68 kg

2.

| | |
|---|---|
| 3 | 2 2 2 2 |
| 3 | 0 1 1 1 |
| 2 | 9 9 |
| 2 | 6 6 6 6 7 7 7 |
| 2 | 4 5 5 |

Key: 2 | 7 means 27

3.

| | |
|---|---|
| 1 | 4 4 4 4 |
| 1 | 7 7 7 7 7 7 |
| 1 | 8 |
| 2 | 0 0 0 0 1 1 1 1 |
| 2 | 2 2 3 3 3 3 |
| 2 | 4 4 4 |
| 2 | 6 6 |

Key: 2 | 1 means 0.21 seconds

4.

| | |
|---|---|
| 3 | 9 |
| 4 | |
| 5 | 3 4 5 5 |
| 6 | 1 1 5 7 8 |
| 7 | 0 0 1 3 4 5 6 6 8 9 |
| 8 | 0 1 2 2 4 8 |
| 9 | 2 6 |
| 10 | 0 1 |

Key: 5 | 3 means 5.3 cm

5.

| | |
|---|---|
| 12 | 5 9 |
| 11 | 1 3 6 |
| 10 | 4 |
| 9 | 7 8 |
| 8 | 3 4 |
| 7 | 0 3 5 5 6 8 |
| 6 | 1 2 8 |
| 5 | 6 6 8 |
| 4 | 3 8 |
| 3 | |
| 2 | 4 6 |
| 1 | 6 |
| 0 | 0 2 6 8 |

Key: 7 | 3 means 7.3 hours

6. (a) 7.4 hours, 0.5 hrs
   (b) 0.074 g, 0.005 g

7. (a)

| Before | | After |
|---|---|---|
| 8 | 4 | |
| 7 3 1 1 0 | 5 | |
| 9 9 6 6 4 | 6 | 9 |
| 9 5 3 3 0 0 | 7 | 0 5 5 7 7 |
| 1 | 8 | 0 0 1 4 4 6 |
| 3 3 3 3 1 0 0 | 9 | 5 6 7 |
| 5 5 | 10 | 4 4 4 6 8 9 |
| 1 1 0 | 11 | 7 |
| | 12 | 5 |
| | 13 | 0 0 1 7 7 |
| | 14 | 3 5 |

Key: 9 | 7 means 79        Key: 8 | 4 means 84

Rate much faster after exercise

(b)

| School A | | School B |
|---|---|---|
| 9 8 7 5 3 3 | 2 | 3 5 9 |
| 9 9 9 7 7 7 4 3 3 1 1 | 3 | 4 6 6 8 8 |
| 8 8 8 8 6 6 5 5 5 0 0 | 4 | 0 1 2 2 3 4 5 5 6 7 7 9 |
| 9 4 4 3 3 1 1 | 5 | 0 0 2 2 4 4 6 6 6 7 8 8 9 9 9 |
| 1 | 6 | 0 |

Key: 9 | 5 means 59        Key: 5 | 9 means 59

Older teaching staff in School B

(c)

| Boys | | Girls |
|---|---|---|
| | 2 | 4 5 5 |
| 3 3 3 2 2 | 2 | 2 2 2 2 2 |
| 1 0 | 2 | 1 1 |
| 9 9 8 8 | 1 | 8 8 9 9 9 |
| 6 6 6 6 | 1 | 6 6 7 7 |
| 5 5 4 | 1 | |
| | 1 | |
| 1 | 1 | |
| 9 | 0 | |

Key: 8 | 1 means 0.18 s        Key: 1 | 8 means 0.18 s

Boys have faster reaction time.
Girls' reaction times more consistent.

### Exercise 1b Histograms and frequency polygons (page 21)

1. Boundary points 5, 10, 20, 25, 40, 45
   f.d. 0.4, 1.2, 1.4, 1, 0.4

2. (a)

| Mass (g) | Frequency | f.d. |
|---|---|---|
| 85–89 | 4 | 0.8 |
| 90–94 | 6 | 1.2 |
| 95–99 | 7 | 1.4 |
| 100–104 | 13 | 2.6 |
| 105–109 | 10 | 2 |
| 110–114 | 5 | 1 |
| 115–119 | 5 | 1 |

(b)

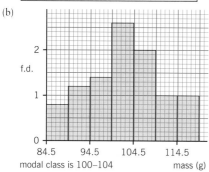

modal class is 100–104

(c)
```
 8 | 6 6 7 8
 9 | 2 2 2 2 3 3
 9 | 5 6 6 7 8 9 9
10 | 0 0 0 1 1 1 1 1 2 2 3 3 4
10 | 5 5 5 6 6 7 7 8 8 9
11 | 0 1 3 3 4
11 | 6 6 7 8 8
```
Key: 10 | 3 means 103

mode = 101 g

3. Boundary points 0, 25, 60, 80, 150, 300
f.d. 2.48, 2, 4.4, 4, 0.2

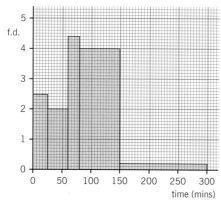

4. Boundary points 40.5, 50.5, 55.5, 60.5, 70.5, 75.5
f.d. 2.1, 12.4, 11, 5, 2.4

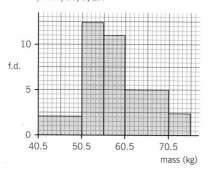

5.

| Speed | 0– | 20– | 24– | 30– | 32– | 38– | 48– | 60– |
|---|---|---|---|---|---|---|---|---|
| frequency | 20 | 24 | 24 | 16 | 12 | 10 | 6 | 0 |

6. Boundary points 176.5, 186.5, 191.5, 196.5, 201.5, 206.5, 216.5
f.d. 1.2, 1.6, 1.6, 1.8, 1.4, 0.6

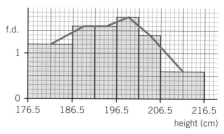

7. Plot polygon at $(0.75, 2)$, $(2.25, 4\frac{2}{3})$, $(4.5, 7\frac{1}{3})$, $(9, 3\frac{1}{3})$, $(13.5, 2)$, $(18, 1)$.
8.

| Number of occurrences of e | Frequency | Width | f.d. |
|---|---|---|---|
| 0–2 | 1 | 3 | $\frac{1}{3}$ |
| 3–5 | 5 | 3 | $1\frac{2}{3}$ |
| 6–8 | 6 | 3 | 2 |
| 9–11 | 3 | 3 | 1 |
| 12–14 | 5 | 3 | $1\frac{2}{3}$ |
| 15–17 | 4 | 3 | $1\frac{1}{3}$ |

Plot boundaries at −0.5, 2.5, 5.5, 8.5, 11.5, 14.5, 17.5
or       at 0, 3, 6, 9, 12, 15, 18
9. Plot polygon at (18, 17.5), (22.5, 94), (27.5, 107), (32.5, 56), (40, 11.8).
Modal class 25–30, skewed with a tail to the right.
(Other answers possible)
10. Boundary points −0.5, 9.5, 14.5, 19.5, 29.5, 39.5, 59.5 (say)
or       0, 10, 15, 20, 30, 40, 60 (say)
f.d. 0.5, 1.6, 6.4, 4.1, 1.6, (0.1).
11. Boundary points 9.5, 29.5, 39.5, 49.5, 59.5, 64.5, 69.5, 84.5
or       10, 30, 40, 50, 60, 65, 70, 85
or       9, 29, 39, 49, 59, 64, 69, 84
f.d. 1.1, 1.8, 2.2, 2.4, 2.8, 2.4, 1.6
12. 6, 8, 8, 6, 4, 10
13. Take boundary points 50, 100, 150, 200, 250, 300
Lucy: Plot (75, 0.12), (125, 0.28), (175, 0.2), (225, 0.12), (275, 0.08)
Jack: Plot (75, 0.04), (125, 0.12), (175, 0.2), (225, 0.32), (275, 0.12)

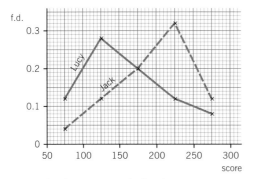

In general Jack's scores were higher than Lucy's scores.

14.

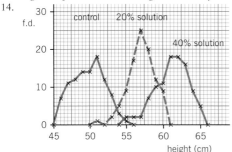

The maize seedlings showed a tendency to grow taller with the stronger solution.

15. (a) a = 20, b = 26, c = 12
(b) 88

## Exercise 1c Pie charts (page 26)

1. (a) 154°, 26°, 64°, 116°
(c) 5.51 cm
2. 208°, 46°, 38°, 36°, 32°; 5.25 cm
3. 66°, 156°, 24°, 42°, 72°; 5.5 cm, 6 cm; 50°
4. (a) £120 000 (b) 68 000 (c) 90°, 27°, 9°, 30°; 7.5 cm
5. (a) 42 (b) 40° (c) 91; 420, 30.0 cm
6. (a) 86°, 38°, 32°, 20°, 168°, 16° (b) 5.5 cm
7. (a) £2000, £8000 (b) £400 (c) 27° (d) 80°
8. 28.8°, 72°, 115.2°, 144°; 180
9. (a) £4500 (b) 1550, 1650 (c) 132°, 24°; 8 cm

## Exercise 1d The mean (page 34)

1. (a) 9.7 (b) 154.8 (c) 51.375 (d) $1775\frac{5}{7}$
(e) 0.908 (3 s.f.) (f) 4 (g) 29.54 (h) 122.82
2. 49.3
3. 45 (2 s.f.)
4. (a) Boundary points 0, 5, 10, 15, 20, 40
f.d. 2.4, 7.6, 8.4, 4, 0.4
(b) £11.92
5. Boundary points 0, 15, 30, 50, 70, 100
f.d. 3.6, 5.2, 6, 4.4, 2
43.35 years
6. 21.4 cm
7. (a) There should not be gaps between the bars. Heights should be adjusted so that area ∝ frequency
(b) Boundary points 4.5, 9.5, 12.5, 15.5, 18.5, 28.5
f.d. 2.8, 6, 5, $1\frac{1}{3}$, 0.8

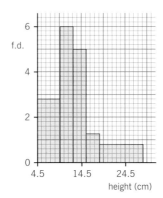

(c) 12.9 m (3 s.f.)
8. (a) Boundary points 9.5, 19.5, 24.5, 29.5, 30.5, 34.5, 39.5, 59.5
f.d. 2, 4, 3, 14, 4, 2, 0.5
(b) 28 seconds.

## Exercise 1e Weighted means (page 36)

1. 10.4
2. Class teacher, 1.65%
3. 40.6
4. 4
5. 5, 65.8

## Exercise 1f Mean and standard deviation (page 44)

1. (a) 5, 2 (b) 8.5, 1.80 (c) 18.8, 6.46 (d) $10\frac{5}{6}$, 4.10
(e) 3.42, 1.91 (f) 205, 3.16
2. (a) f.d. 0.2, 0.32, 0.625, 1.04, 0.08

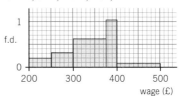

(b) £338.25, £59.60
3. 69.3, 1.7
4. 115.8, 7.58
5. 16.6 seconds, 2.63 seconds
6. 6.8, 1.11
7. (a) 2 min 38 sec, 1 min 54 sec
(b) Histogram f.d. 6, 10, 15, 2.5, 0.8
Frequency polygon: plot (0.5, 6), (1.5, 10), (2.5, 15), (4, 2.5), (7.5, 0.8)
8. 29, 5.9
9. 5.10
10. 5
11. (a) 10 (b) 11.7
12. (a) 121, 6.19 (b) 14, 1703.8 (c) 1716, 3.59
(d) 1026, 58 770
13. (a) Frequency = 5 + 18 + 22 + 28 + 22 + 18 + 5 = 118
(Area = f.d. × width)
(b) Symmetric. Midpoints of intervals have been taken to represent the interval.
(c) 3.5 mm (2 s.f.)
14. 28.15, 3.84
15. 5.3
16. 30.0 mph, 5.85 mph

## Exercise 1g Mean and standard deviation (page 50)

1. 19
2. 8
3. 7
4. 3.74
5. $a = 6, b = 4$
6. 15.6, 7.66
7. 12
8. 15, 7
9. 25.9, 1.99
10. 2.3, 1.41
11. 11.7%, 2.2%
12. (a) 4.6, 2 (b) 4.56, 2.04
13.

| 25 | 1 2 4 4 |
|----|---------|
| 30 | 0 1 1 2 2 2 3 3 3 4 |
| 35 | 0 1 2 3 3 3 3 4 |
| 40 | 0 2 2 4 |
| 45 | 0 4 |
| 50 | 2 |
| 55 | |
| 60 | 1 |

Key: 45 | 4 means 49

Features: modal class 30–34, skewed to the right, 61 extreme value (outlier), 36.87, 35.59
14. £195.45, £14.12
15. 11.87, 0.80
16. 4.44

## Exercise 1h Scaling sets of data (page 55)

1. (a) 6, 2.14 (b) 516, 2.14 (c) 78, 27.8
2. 50, 12
3. (a) $\mu + k, \sigma$ (b) $p\mu, p\sigma$; $3\mu + 5, 3\sigma$
4. (a) 2 (b) 200 (c) 2.02 (d) −4, −1, 2, 5, 8, 11, 14
5. (a) $a = \frac{3}{4}, b = 22$ (b) 70 (c) 76
6. (a) 38, 8.99 (b) 34, 77
7. $a = 1.6, b = 10$
8. $a = 0.8, b = -5$; 6.25
9. (b) Take mark intervals $0 \leqslant mark < 10, 10 \leqslant mark < 20$, etc.
   f.d. 0.1, 0.8, 1.9, 2.8, 2.5, 1.7, 0.7, 0.3, 0.1.
   boundaries 0, 10, 20, 30, 40, 50, 60, 70, 80
   (c) midpoints 5, 15, 25, etc, 40.4, 15.4;
   (d) $a = 24, b = 0.65$ (2 s.f.)
10. (a) 12.5 (b) 20; 80, 5.

## Exercise 1i Coding (page 58)

1. (a) 313.76, 5.19 (b) 431, 132 (c) 0.0171, 0.00818
2. 51.235, 0.927
3. 89.3275
4. 31.7 mins.
5. 71.2, 3.82
6. $46\frac{2}{3}$ secs.

## Exercise 1j Cumulative frequency, median and quartiles – ungrouped data (page 73)

1. (a) 9 (b) 207 (c) 1896 (d) 0.55
2. 4
3. (a) 61 (b) 52 (c) 73 (d) 21
4. (a) 46, 35 (b) 1.8, 1.2 (c) 20.5, 11.5
5. (a) 7, 2 (b) 14, 3

6. (a)

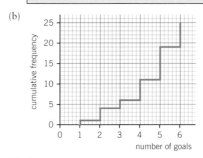

| number of goals | 0 | $\leqslant 1$ | $\leqslant 2$ | $\leqslant 3$ | $\leqslant 4$ | $\leqslant 5$ | $\leqslant 6$ |
|---|---|---|---|---|---|---|---|
| cumulative frequency | 0 | 1 | 4 | 6 | 11 | 19 | 25 |

(b)

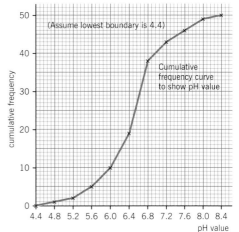

(c) 5
(d) 2
7. (a) 2 (b) 3 (c) 2.47 (d) 1.94
8. (a) 2, 3 (b) 2
   (c) It only considers the middle 50% and does not take account of large families.

## Exercise 1k Cumulative frequency, median and quartiles – grouped data (page 81)

Some answers are approximate and depend on the curve drawn

1. (a)

| mass (kg) | cumulative frequency |
|---|---|
| $\leqslant 39.5$ | 0 |
| $\leqslant 44.5$ | 3 |
| $\leqslant 49.5$ | 5 |
| $\leqslant 54.5$ | 12 |
| $\leqslant 59.5$ | 30 |
| $\leqslant 64.5$ | 48 |
| $\leqslant 69.5$ | 51 |
| $\leqslant 74.5$ | 52 |

Plot
(39.5, 0), (44.5, 3), (49.5, 5), (54.5, 12), (59.5, 30), (64.5, 48), (69.5, 51), (74.5, 52). Join with a smooth curve.
(b) 21 (c) 16 (d) 62 kg (e) 58.4 kg (f) 7.2 kg
2. (a)

(b) 82%

(c) 6.5, median

(d)

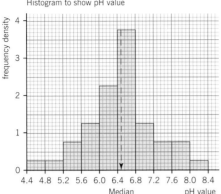

Histogram to show pH value

3.

| mass (g) | cumulative frequency |
|---|---|
| ⩽ 50 | 3 |
| ⩽ 54 | 5 |
| ⩽ 58 | 10 |
| ⩽ 62 | 22 |
| ⩽ 66 | 32 |
| ⩽ 70 | 38 |
| ⩽ 74 | 40 |

Plot (50, 3), (54, 5), (58, 10), (62, 22), (66, 32), (70, 38), (74, 40)

Median mass = 61.3 g

4. (a)

| time (minutes) | cumulative frequency |
|---|---|
| ⩽ 5 | 2 |
| ⩽ 10 | 4 |
| ⩽ 15 | 7 |
| ⩽ 20 | 13 |
| ⩽ 25 | 25 |
| ⩽ 30 | 41 |
| ⩽ 35 | 47 |
| ⩽ 40 | 50 |

(b) 24 (c) 26 (d) 23 (e) 25 mins (f) 4.5 mins

5. (a) 687.5 hours (b) 13.2 hours

6. (a) 80.75 g (b) 215

7. (a) Cumulative frequency curve to show maximum temperatures

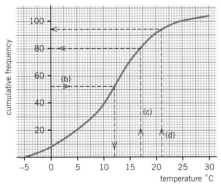

(b) 12°C (c) 80 (d) Approx. 10%

8.

| time (mins) | cumulative frequency |
|---|---|
| < 39.5 | 0 |
| < 44.5 | 8 |
| < 49.5 | 30 |
| < 54.5 | 64 |
| < 59.5 | 94 |
| < 64.5 | 120 |

For the curve, plot (39.5, 0), (44.5, 8), (49.5, 30), (54.5, 64), (59.5, 94), (64.5, 120) and join points with a smooth curve.

(a) 9 mins (b) Approx. 11%; 56 mins

9.

| distance (km) | cumulative frequency |
|---|---|
| 0 | 0 |
| < 4 | 1 |
| < 10 | 3 |
| < 20 | 9 |
| < 35 | 28 |
| < 60 | 40 |
| < 100 | 50 |

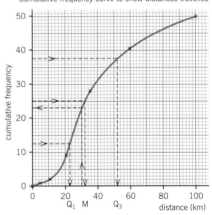

Cumulative frequency curve to show distances travelled

(a) 32 km (b) Approx. 30 km (c) Approx. 54%

10.

| price (£x) | cumulative frequency |
|---|---|
| ⩽ 75 | 0 |
| ⩽ 95 | 6 |
| ⩽ 100 | 16 |
| ⩽ 105 | 28 |
| ⩽ 110 | 41 |
| ⩽ 120 | 48 |
| ⩽ 135 | 53 |

Plot (75, 0), (95, 6), (100, 16), (105, 28), (110, 41), (120, 48), (135, 53)

(a) £104 (b) £13 (c) 47

11. $x = 25, y = 17$

12. Plot (405, 0), (415, 4), (425, 7), (435, 13), (445, 23), (455, 28), (465, 30).
437, 412.5, 453.

13. Plot (80, 0), (85, 6), (90, 18), (95, 40), (100, 71), (105, 86), (110, 93), (115, 97), (120, 99), (125, 100)
(a) 97 mins (b) 10 mins (c) 62

14. Plot (165, 0), (170, 18), (175, 55), (180, 115), (185, 180), (190, 228), (195, 250)
(a) 180.5 cm (b) 175.5 cm (c) 187 cm (d) 189.5 cm

15. (a) 57 mins (b) 71.5 mins (c) 32%.

16. Plot (69.5, 0), (74.5, 8), (79.5, 28), (84.5, 53), (89.5, 84), (94.5, 94), (99.5, 100).
    9.3 secs, 22, 75.5 secs.
17. 50p, £4.96, £5.96. Large amounts affect the mean but not the median
18. Histogram: frequency densities 0.2, 0.5, 0.9, 0.8, 0.1

| thickness (mm) | 0 | < 20 | < 30 | < 40 | < 50 | < 60 |
|---|---|---|---|---|---|---|
| cumulative number of strata | 0 | 2 | 7 | 16 | 24 | 25 |

Plot (0, 0), (20, 2), (30, 7), (40, 16), (50, 24), (60, 25)
36 mm, 15 mm, 0.24.

## Exercise 1l Skewness (page 90)

1. (a) 0.535  (b) −0.674
2. −2.4
3. 2
4. (a) Frequency densities: 0.8, 3, 5, 1.8, 1.2, 0.47, 0.2
   (b) Positively skewed
5. Vertical line graph, 2, 3, 3.53, 1.985, 0.801, 0.771
6. −0.482
7. (a) $B$  (b) $A$  (c) $C$
8. (a) (i) 0.75  (ii) 0.28
   (b) Frequency densities: 0.2, 1, 1.2, 1.8, 2.8, 0.6, 1.2, 1, 0.4, 0.2
9. (a) 9.6 mins, 1 min  (b) 0.33
   (c) (4.65 mins, 14.61 mins)
   (d) (4.3 mins, 15.27 mins)
10. (a) 0.143 ($Q_1 = 17$, $Q_2 = 26$, $Q_3 = 38$)
    (b) 0.0668 ($Q_1 = 11.9$, $Q_2 = 16.1$, $Q_3 = 20.9$)
    (c) 0.333 ($Q_1 = 9$, $Q_2 = 11$, $Q_3 = 15$).

## Exercise 1m Box plots (page 99)

1. (a) Plot (0, 0), (1, 8), (2, 19), (3, 36), (5, 44), (10, 50)
   (b) 2.35 mins, 1.4 mins, 3.4 mins
   (c) Positively skewed

Length of call (mins)

2. Group 1: $Q_1 = 0.17$, $Q_2 = 0.21$, $Q_3 = 0.23$; times from 0.14 to 0.26
   Group 2: $Q_1 = 0.16$, $Q_2 = 0.19$, $Q_3 = 0.22$; times from 0.09 to 0.25

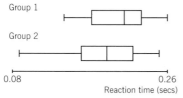

Reaction time (secs)

3. $Q_1 = 22$, $Q_2 = 35$ $Q_3 = 51$; whiskers from 16 to 97. Boundary for outliers 94.5; outlier 97

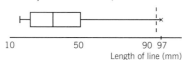

Length of line (mm)

4. (a) u.c.b. 0, 20, 30, 40, 50; c.f. 0, 20, 40, 65, 69; $Q_1 = 17.5$, $Q_2 = 27.5$, $Q_3 = 35$; 7.5, 10; negatively skewed
   (b) u.c.b. 0, 20, 40, 80, 100; c.f. 0, 4, 10, 34, 44; $Q_1 = 41.7$, $Q_2 = 60$, $Q_3 = 78.3$; 18.3, 18.3; negatively skewed, zero quartile skewness

(c) u.c.b. 0, 5, 10, 15, 20, 25, 35; c.f. 0, 1, 6, 9, 11, 12, 13; $Q_1 = 7.25$, $Q_2 = 10.8$, $Q_3 = 16.875$; 6.075, 3.55; positively skewed
(d) u.c.b. 0, 5, 10, 15, 20, 25, 30; c.f. 0, 5, 20, 45, 90, 140, 160; $Q_1 = 14$, $Q_2 = 18.9$, $Q_3 = 23$; 4.1, 4.9; negatively skewed

5. $\bar{x} = 63.9$, $s = 29.5$, outliers would be less than 4.9 mins, greater than 122.8 mins, outliers are 133, 144.
6. Compare median, quartiles, range, skewness
7. December: $Q_1 = 0.3$, $Q_2 = 1.8$, $Q_3 = 2.7$; July: $Q_1 = 4.1$, $Q_2 = 6.5$, $Q_3 = 9.8$

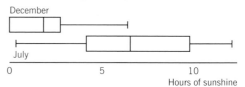

Hours of sunshine

8. (a)
```
0 | 1 2 2 5 9
1 | 0 0 2 3 5 7 9 9
2 | 2 5 9 9 9
3 | 0 1
4 | 5 7 8
5 | 3
```
*Key:* 2 | 5 means 9.25 a.m.

(b) 9.19 a.m.
(c) 9.10 a.m., $29\frac{1}{2}$ minutes past 9.
(d)

Time of delivery

9. (a) $Q_1 = 1$, $Q_2 = 1$, $Q_3 = 2$

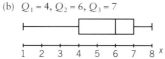

(b) $Q_1 = 4$, $Q_2 = 6$, $Q_3 = 7$

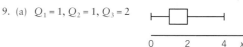

10. (a) 6, 5
    (b) More than 3 standard deviations from the mean
    (c) (i) older brother or sister also attended
        (ii) a mistake had been made
    (d) 5.5, 5
    (e) decrease
    (f) positive, less
11. (a)

| | height gain (grams) |
|---|---|
| 36 | 0 9 9 |
| 37 | 6 |
| 38 | |
| 39 | 1 7 7 9 |
| 40 | 2 3 7 |
| 41 | 0 0 |
| 42 | 0 5 7 |
| 43 | 0 4 |
| 44 | 5 |
| 45 | |
| 46 | 2 |

*Key:* 39 | 7 means 397

(b) Draw plots – New corn: whiskers from 360 to 462, $Q_1 = 397$, $Q_2 = 450$, $Q_3 = 426$; Standard corn: whiskers from 321 to 423; $Q_1 = 353$, $Q_2 = 368.5$, $Q_3 = 383$

12. (a)

| Stem | Leaf |
|---|---|
| 4 | 1 2 3 4 4 6 7 7 8 8 |
| 5 | 0 2 2 2 3 4 6 7 8 8 |
| 6 | 0 2 3 3 6 6 7 7 8 |
| 7 | 0 0 2 2 4 4 6 7 8 8 8 |
| 8 | 0 1 2 5 5 6 6 7 |
| 9 | 3 3 4 |

Key: 4 | 2 means 42

(b) $Q_2 = 66$ miles, $Q_1 = 52$ miles, $Q_3 = 78$ miles

(c)

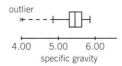

40    50    60    70    80    90
Distance (miles)

(d) (i) Gives a visual impression of the data whilst keeping the details.
(ii) Gives an immediate impression of an approximately symmetrical distribution with the middle 50% lying between 52 and 78 miles.

## Miscellaneous exercise 1n (page 110)

1. (a) $\bar{x} = 5.42$, $s = 0.33$; range = 1.79, $Q_2 = 5.46$, $Q_1 = 5.295$, $Q_3 = 5.615$, outlier = 4.07

(b) (i) 5.465
(ii) 5.47
(iii) 0.22

outlier

4.00    5.00    6.00
specific gravity

2. (a) Boundary points for histogram
689.5, 709.5, 719.5, 729.5, 739.5, 744.5, 749.5, 754.5, 759.5, 769.5, 789.5
First interval l.c.b. 689.5, u.c.b. 709.5
f.d. 0.15, 0.7, 1.5, 3.8, 8.2, 7, 4.2, 3.2, 1.4, 0.5

(b) Plot (689.5, 0), (709.5, 3), (719.5, 10), (729.5, 25), (739.5, 63), (744.5, 104), (749.5, 139), (754.5, 160), (759.5, 176), (769.5, 190), (789.5, 200).

(c) 744.24, 14.86

(d) 744.01, 736.08, 752.12

(e) 0.046

(f) 0.011

(g) In box plot, draw whiskers from 689.5 to 789.5, with median and quartiles as in (d).

3. 16, 6   (a) 5.86   (b) 15, 7

4. 35 yrs 1 month, 11 yrs 3 months.
(a) median = 33 yrs 9 months, IQR = 17 yrs 11 months
(b) 61.8%

5. (a) 44.5
(b) 51.75
(c)

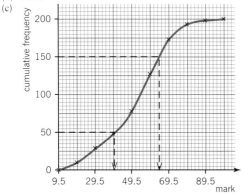

cumulative frequency

9.5    29.5    49.5    69.5    89.5
mark

$Q_1 = 40$, $Q_3 = 64$

(d) $a = 0.85$, $b = 1$ (dependent on values in (c))
(e) yes

6. (a) (i) 49.66   (ii) 433.97   (iii) 20.83
(b) c.f. 3, 9, 18, 28, 40, 58, 72, 83, 88
(c) Plot (0, 0), (10, 3), (20, 9), (30, 18), (40, 28), (50, 40), (60, 58), (70, 72), (80, 83), (90, 88)
(d) (i) 52   (ii) 32
(e) 11

7. c.f. 11, 39, 77, 111, 138, 150
Take as boundaries 0.90, 1.15, 1.30 etc. or 0.91, 1.16, 1.31, etc. or 0.905, 1.155, etc. Median ≈ £1.30.

8. (a)

| Stem (£) | Leaf (p) |
|---|---|
| 3 | 40  60  75  95 |
| 4 | 20  50  75 |
| 5 | ⑳  75 |
| 6 | 45  60 |
| 7 | 25 |
| 8 | 75 |
| 9 | 60 |
| 10 | |
| 11 | |
| 12 | 25 |

$Q_2 = £5.20$, range = £8.85

(b) $\bar{x} = £6$, $s = £2.47$
(c) A: $\bar{x} = £6.30$, $s = £2.47$
B: $\bar{x} = £6.30$, $s = £2.59$
(d) mean remains the same; lower paid workers do not benefit under scheme B.

9. (a) $8, 6, 4\frac{1}{6}, 3$

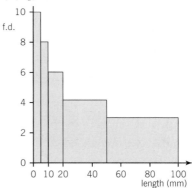

f.d.

0   10 20    40    60    80    100
length (mm)

(c) Approx. 2.5 mm (modal class is $0 \leqslant x < 5$)
(d) (i) 39.9 mm
(ii) 35 mm

10. (a) 275
(c) Comparative bar chart

11. 57,   (a) it becomes 39
(b) $x = 3x - 141$ does not have an integer solution.

12. 100.7 mm, 0.4 mm; machine B nearer 100 on average, less variation with machine A.

13. (a) 1, could be 1 or 2
    (b) Positive skew, possible outlier
    (c) 2, 1.7; more than 3 standard deviations from the mean
    (d) (A) a mistake.
        (B) could be correct.
    (e) 1.88, 1.48

14.

| Cost (£1000) | $\leqslant 50$ | $\leqslant 60$ | $\leqslant 70$ | $\leqslant 100$ | $\leqslant 150$ |
|---|---|---|---|---|---|
| c.f. | 540 | 1690 | 3010 | 3870 | 4320 |

Plot (20 000, 0), (50 000, 540), (60 000, 1690),
(70 000, 3010), (100 000, 3870), (150 000, 4320).
$Q_2 \approx £63\,000$, IQR: a value between £18 000 and £23 000 is acceptable

15. f.d. 0.93, 2.4, 1.4, 1.6, 0.9

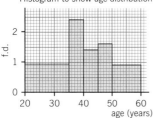

Histogram to show age distribution

(a) 40.15  (b) $35\frac{1}{2}$ yrs.

16.
| Time (mins) | Frequency | Frequency density |
|---|---|---|
| $0 \leqslant x \leqslant 1$ | 20 | 20 |
| $1 < x \leqslant 2$ | 47 | 47 |
| $2 < x < 2.5$ | 51 | 102 |
| $2.5 < x \leqslant 3$ | 59 | 118 |
| $3 < x \leqslant 5$ | 138 | 69 |
| $5 < x \leqslant 10$ | 85 | 17 |

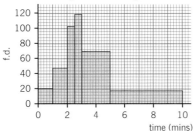

Histogram to show length of call

$3\frac{1}{3}$ mins, divides area in half.

17. (a) 8, 9.5 mins
    (b) Boundaries 0, 5, 10, 15, 20, 25, 30;
        f.d. 8, 11.2, 5.6, 4, 2.4, 0.8
    (c) 10
    (d) A False, B True, C False, D True

## Mixed test 1A (page 114)

1.
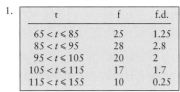

| t | f | f.d. |
|---|---|---|
| $65 < t \leqslant 85$ | 25 | 1.25 |
| $85 < t \leqslant 95$ | 28 | 2.8 |
| $95 < t \leqslant 105$ | 20 | 2 |
| $105 < t \leqslant 115$ | 17 | 1.7 |
| $115 < t \leqslant 155$ | 10 | 0.25 |

(a)
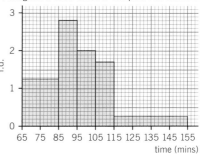

Histogram to show times to complete half-marathon

(b) 96.15 mins

2. (a) 7, 6, 4, 8
   (b) 6.55, 5.7, 8.1
   (c)
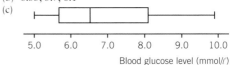

Blood glucose level (mmol/ℓ)

   (d) Positive skew.

3. (a) 4.5  (b) 1.5
   (c) No change to mean, standard deviation is increased.

4. (a) Pie chart, bar chart
   (b) Children in school, sample not representative.
   (c) f.d. 3.6, 6.4, 4.4, 1.4, 0.7

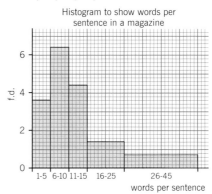

Histogram to show words per sentence in a magazine

NB: boundary points could be 0.5, 5.5, 10.5, 15.5, 25.5, 45.5

   (d) 13.8, 10.2
   (e) 9.11

5. (a) Histogram
   (b) Individual values are not known and mid points have been taken as representatives of the intervals.
   (c) 69.5, 7.6
   (d) Median – no effect, IQR – no effect, mean – increased.

6. (a) 7, 15, 35, 20, 13, 10
   (b) 9, 5.43, 14.5
   (c)

Male employees

Female employees

## Mixed test 1B (page 116)

1. (a) 1.15  (b) 1  (c) 1.09
2. (a) (i) Easier to see the spread

   (ii)
   ```
   1 | 1 2 2 3 4 4 4
   1 | 5 6 7 7 9
   2 | 1 1 1 2
   2 | 5 5 7
   3 | 1 2 2
   3 | 5 5 9
   4 | 1 3
   4 | 4 5
   ```
   Key: 1 | 5 is 15 cm

   (b) 24.6 cm

   (c) 21 cm

   (d) Median better; distribution not symmetrical.
3. (a) 44%

   (b) 33°
4. (a) Median same for both.

   B has 3 outliers; ignoring these, B's average waiting time would be lower.

   B's times are less variable than A's.

   A's times are positively skewed, B's are negatively skewed.

   (b) (i) If outliers are not the Post Office's fault, choose B for quicker service.

   (ii) If outliers are the Post Office's fault then the situation could happen again and there could be a long wait. A avoids long waits.
5. (a)
   ```
   00 | 6 7 8 8
   10 | 0 0 1 2 2 3 3 3 4 4
   10 | 5 6 6 6 6 7 7 7 8 8 9 9
   20 | 0 2 3
   20 | 7
   ```

   (b) $Q_2 = 15.5$ mins, $Q_1 = 12$ mins, $Q_3 = 18$ mins.

   (c)

   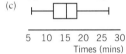

   5   10   15   20   25   30
   Times (mins)
6. (a) (i) 3 hrs 3 mins

   (ii) $Q_1 = 2$ hrs 42 mins, $Q_3 = 3$ hrs 42 mins.

   (b) 40, (200), 200, 60

   (c) (i) 3 hrs 20 mins  (ii) 54 mins

# Chapter 2

## Exercise 2a Equations of least squares regression lines (page 136)

1. Data set 1

   (a) $y = 4.50 + 0.64x$

   (b) $x = 4.42 + 0.75y$

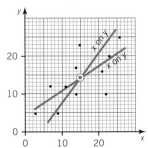

Good positive correlation

Data set 2

(a) $y = 90.31 - 1.78x$

(b) $x = 37.80 - 0.39y$

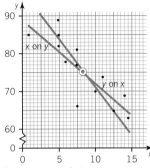

Good negative correlation

2. (a)

yield vs temperature graph

(b) $y = 0.614 + 0.0207x$
3. $y = -2.59 + 0.65x$; 36.5
4. $F = -6.33 + 0.90I$, $F = 20.8$
5. $y = 3.8 + 1.6x$, $x = -2.06 + 0.59y$
6. (a) $y = 15.83 + 0.72x$  (b) 66  (c) 59
7. (a)

total cost (£1000) vs units of output (1000's) graph

(b) $20.7 + 0.96x$

(c) 31 000 – 33 000 units  (d) Break-even point.
8. $y = 1.8 + 1.3x$
9. $y = -8 + 1.2x$
10. $c = 15, d = -5$

11. (a)

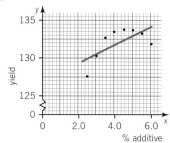

(b) $y = 3710 + 192x$
(c) Appears reasonably satisfactory apart from B and C who have earned substantially more than the equation suggests.
(d) (i) $y = 4210 + 192x$
(ii) $y = 4010 + 207x$
(iii) $y = 4160 + 200x$
(e) It would contain a term for employees who work away from home e.g. $y = a + bx + c$, where $c \approx £3000$ for employees who work away from home and zero otherwise.

12. 0.3, 0.6

13. (a)

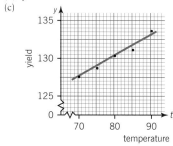

(b) $y = 127 + 1.17x$
(c)

(d) Argument invalid since relationship between yield and additive is not linear, yield declines above 4.5% additive; suggest additive 4.5%, temperature 90°.

## Exercise 2b Product-moment correlation coefficient (page 145)

1. (a) 0.930, strong positive correlation
(b) −0.828, strong negative correlation
(c) 0.867, strong positive correlation
(d) 0.742, positive correlation.
2. 0.82

3. (a) −0.558
(b) Low unemployment appears to be linked to high wage inflation, so suggestion justified.
4. 0.79
5. 0.73, $y = -25.4 + 0.53x$, $x = 94.4 + 1.01y$
6. 0.60, $W = -76 + 0.89$ h
7. 0.77
8. −0.415
9. (a) 0.954 (b) 2, 3
10. (a)

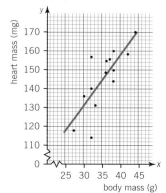

(b) $y = 48.35 + 2.75x$
(c) 0.787

## Exercise 2c Spearman's coefficient of rank correlation (page 151)

1. 0.26
2. (a) 0.43
(b) Some agreement between average attendance ranking and position in league, high position in league correlating with high attendance.
3. 0.033, little or no correlation.
4. 0.62, some agreement between the scores.
6. (a)

(b) (2.275, 38.375)
(c) Ranking both $p$ and $d$ from lowest to highest gives −0.839.
(d) In general the population density is greater nearer the centre of the town and less on the outskirts of the town.
(e) $H$, low population density and distance from centre of town.
7. (a) 0.3, 0.5, 0.7
(b) Mrs Brown and John; 1) Headrests 2) Heated rear window 3) Anti-rust treatment.

8. −0.036, no agreement.
9. 0.84, strong positive correlation between number of years smoking and extent of lung damage.
10. (a) 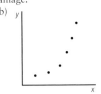 (b)

(c) −0.92 (d) −0.9
11. (a)

0.60, 0.60
12. (a) 0.7, good agreement between judges.
(b)

13. (a) (i) −0.976 (ii) −0.292 (or 0.292)
(b) The transport manager's order is more profitable for the seller, saleswomen is unlikely to try to dissuade.
(c) (i) No, maximum value is 1
(ii) Yes, higher performing cars generally do less mileage to the gallon.
(iii) No, the higher the engine capacity, the dearer the car.
(d) When only rankings are known; when relationship is non-linear.
14. 0.84; very good agreement between the rankings indicating strong positive correlation between the marks in English and the marks in History; E.

## Miscellaneous exercise 2d (page 160)

1. (a) $y = 3.07 + 1.17x$
(b) When the $y$ variable is the controlled or independent variable.
2. (a) $t$ on $w$ is required; $t = 18.8 - 0.853w$
(b) (i) −13.6° F (ii) −28.1° F
(c) −0.946, points lie close to the regression line.
(d) Good estimate for $w = 38$, since strong correlation. Estimate for $w = 55$ needs to be treated with care since extrapolation (outside range of data) is unreliable.
3. (a) Strong negative correlation
(b) $y = 6.85 - 0.0072x$
(c) pH = 6.85 at $t = 0°C$; for an increase of 10°C, pH drops by 0.07
(d) 6.71, reliable; 6.17, unreliable, outside range of data
(e) 48.6°C

4. (a)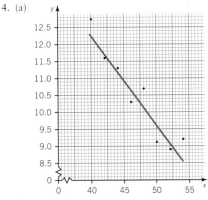

(b) $y = 23.0 - 0.267x$
(c) 7500. There isn't a wide degree of scatter, so estimate could be reliable, but in general it is unwise to extrapolate outside the range of data.
(d) No. The points do not lie in a line.
5. (a)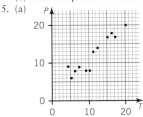

(b) 0.935
(c) b) indicates strong positive linear correlation and diagram confirms this is appropriate.
(d) $p = 2.58 + 0.88T$; 15
6. (a) $\Sigma m^2$, page 121 diagram 3
(b) $y = 7.77 - 0.005x$
(c) 5.77; treat with caution as outside range of data.
(d) The lower the percentage moisture content, the greater the heat output.
7. (a) −0.901, strong negative correlation, the greater the number of items finished, the lower the mean quality score.

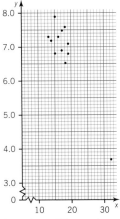

(b) Amend; possibly negative trend but not strong correlation, (32, 3.7) is an outlier
(c) Ignore outlier; weak negative correlation between number of items and quality score.

8. $y = 0.65 + 0.0157x$;
   Rate of about 1 hour per mile distance; 3 days 19 hours; out of range of data, travel across water required; 0.942, strong positive correlation, points close to regression line.

9. (a) $y = 12.033 - 0.009x$
   (b) 8.6 per 1000
   (c) Decreasing number of members of population per doctor not effective in reducing infant mortality rate.

10. (a) Spearman 0.613; grades given
    (b) Product-moment, 0.95; numerical data given
    (c) Students performed at a similar standard in the written and listening tests, but not in the oral test. Standard in oral test related more to listening performance than written performance.

11. (a)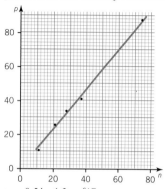

    (b) $p = -0.54 + 1.2$ n; £17
    (c) 0.998; the points will be close to a line with positive gradient.

12. (a) 0.96
    (b) points lie close to a straight line with positive gradient (strong positive correlation).
    (c) Equal to 0.986 since rankings will not change.

13. (a) 0.0705   (b) $y = 0.34 + 0.0085x$
    (c) 0.477   (d) unreliable since outside range of data

14. (a)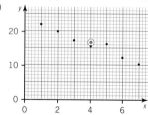

    Mean (4, 16.7).
    (b) average decrease of 1.80°C per month
    (c) $y = 23.9 - 1.80x$
    (d) 23.9°C; regression line is valid only within range of data.

15.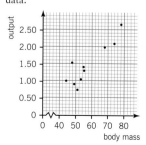

    0.91, strong positive correlation.

16. (a) 0.98   (b) $y = -7.42 + 1.115x$, $x = 6.89 + 0.862y$
    (c) 8.20 tons per acre   (d) 13.9 cm

17. (a) $y = 41.79 + 1.55x$   (b) 51
    (c) 43, but treat with caution as outside range of data.

18. (a) −0.9   (b) 0   (c) 0.9 (0.6 without outlier)

19. (a)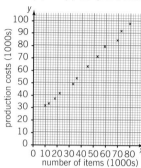

    (b) Diagram suggests a linear relationship
    (c) $y = 61.1 = 0.966 (x - 42.4)$
    (d) $y = 20.1 + 0.966x$
    (e) Initial costs are approx. £20 000, cost increases by approx. £1 per item

20. (a)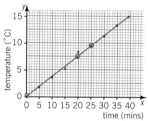

    (b) $y = -0.142 + 0.389x$
    (c) 23.2°C, outside range

## Mixed test 2A (page 166)

1. (a) $y = 3.667 + 0.038x$ (3 d.p.)
   (b) Mathematically $a = 3.667$ would indicate a yield of 3.667 tonnes with no water at all; in practice this would be nonsense, $b = 0.038$ indicates yield increases by 0.038 tonnes for every extra centimeter of water.
   (c) 4.7 tonnes (only just outside range, probably reliable), 9.3 tonnes (well outside range, unreliable).

2. (a) $y = 14.55 + 1.02$ t
   (b) Initial temperature of milk.
   (c) 19.14°C, 34.95°C
   (d) First; second outside range of data.
   (e)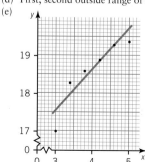

   (f) Temperature would stabilise at room temperature.
   (g) Points appear to lie on a curve, reaching a limit at room temperature.

3. 0.714

4. (a) −0.975; points lie close to a straight line with negative gradient (strong negative correlation).
   (b) −1, complete disagreement in the rankings.
   (c) (iii), data follow a non-linear relation.

## Mixed test 2B (page 167)

1. (a) −0.987, points lie close to a line with negative gradient.
   (b) $y$ on $x$, $y = 7.22 − 0.69x$; 4.45
   (c) Depreciation of £700 per year.
   (d) (i) No, outside range of data.
   
   (ii) Yes, since $x$ is controlled. Use $x = \dfrac{y − a}{b}$.

2.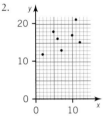

   0.515; 8.8 hours; regression line gives average value, points not that close to line as $r = 0.515$; $\Sigma m_i^2$ minimised where $m_i$ is vertical distance from point to line.

3. (a)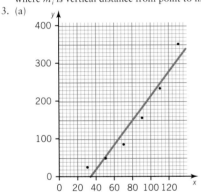

   (b) $y = −107 + 3.21x$
   (c) 214, overestimate (data fit a curve); 375, underestimate (outside range, also non-linear relationship), unreliable
   (d) no, better to use a curve.

4. (a) 0.714
   (b) Same, since there is no change in the rankings.
   (c) $d^2$ would decrease, therefore $1 − \dfrac{6\Sigma d^2}{n(n^2 − 1)}$ would increase.

# Chapter 3

## Exercise 3a Elementary probability (page 173)

1. (a) $\frac{1}{3}$  (b) 1  (c) $\frac{2}{3}$
2. (a) 0.375  (b) 0.625  (c) 0.75  (d) 0  (e) 0.8
3. (a) 0.3  (b) 0.75
4. (a) 0.4625  (b) 800
5. 0.73
6. (a) (i) $\frac{1}{52}$  (ii) $\frac{7}{26}$  (iii) $\frac{10}{13}$  (b) $\frac{13}{17}$
7. $\frac{4}{15}$

8. (a) $\frac{2}{7}$  (b) $\frac{3}{7}$
9. $\frac{4}{15}$
10. 0.27 (2 d.p.)
11. (a) (i) $\frac{75}{98}$  (ii) $\frac{15}{98}$  (b) $\frac{1}{5}$
12. 0.52
13. (a) $\frac{1}{18}$  (b) $\frac{1}{6}$  (c) $\frac{1}{6}$  (d) $\frac{1}{3}$
14. (a) (i) $\frac{1}{36}$  (ii) $\frac{1}{12}$  (iii) 0  (b) $t = 6$ or 12

## Exercise 3b (Probability) – combined events (page 181)

1. (a) $\frac{1}{2}$  (b) $\frac{1}{2}$  (c) $\frac{5}{6}$
2. $\frac{11}{30}$
3. (a) $\frac{4}{17}$  (b) $\frac{4}{51}$  (c) $\frac{5}{17}$  (d) $\frac{5}{17}$
4. $\frac{3}{4}$
5. (a) 0.5  (b) 0.4  (c) 0.2  (d) 0.1
6. (a) (i) $\frac{1}{3}$  (ii) $\frac{1}{9}$  (iii) $\frac{38}{45}$  (b) 0.2
7. (a) $\frac{3}{4}$  (b) 0  (c) $\frac{1}{4}$
8. 0.6
9. 0.7
10. (a) $\frac{7}{36}$  (b) $\frac{1}{6}$  (c) $\frac{5}{18}$  (d) $\frac{1}{12}$
11. $\frac{3}{4}$
12. (a) $\frac{11}{36}$  (b) $\frac{11}{36}$  (c) $\frac{5}{9}$
13. Yes
15. At least one tail is obtained; both coins show tails.
16. (a)

| | Fruit tree | Other tree | Total |
|---|---|---|---|
| Birds nest | 2 | 4 | 6 |
| No nest | 5 | 9 | 14 |
| Total | 7 | 13 | 20 |

   (b) 0.45  (c) $\frac{2}{7}$

## Exercise 3c Combined events (page 192)

1. (a) $\frac{1}{3}$  (b) 0
2. (a) 0.05  (b) 0.5
3. (a) 0.15  (b) 0.65; no
4. (a) $\frac{3}{10}$  (b) $\frac{1}{10}$
5. $\frac{1}{9}$
6. (a) $\frac{1}{4}$  (b) $\frac{1}{6}$
7. $\frac{1}{2}$
8. (a) 0.5  (b) 0.35  (c) 0.375  (d) 0.4
9. (a) $\frac{1}{2704}$  (b) $\frac{1}{16}$  (c) $\frac{1}{2}$  (d) $\frac{25}{169}$
10. (a)

| | B | G | Totals |
|---|---|---|---|
| Passed | 16 | 8 | 24 |
| Taken, failed | 7 | 6 | 13 |
| Learning | 10 | 8 | 18 |
| Too young | 2 | 3 | 5 |
| Totals | 35 | 25 | 60 |

   (b) $\frac{13}{60}$  (c) $\frac{14}{25}$  (d) $\frac{12}{35}$  (e) $\frac{1}{177}$  (f) $\frac{128}{875}$
11. (a) Independent; obtaining a head when a coin is tossed.
    (b) Mutually exclusive, 0.
12. $\frac{3}{8}$
13. 0.5
14. (a) $\frac{1}{3}$  (b) $\frac{2}{15}$  (c) $\frac{8}{15}$
15. (a) 0.2  (b) 0.03  (c) 0.32
16. (a) $\frac{1}{15}$  (b) $\frac{11}{15}$  (c) $\frac{1}{5}$
17. (a) (i) $\frac{1}{16}$  (ii) $\frac{24}{169}$  (iii) $\frac{1}{52}$  (b) (i) $\frac{32}{221}$  (ii) 0
18. (a) 0.5  (b) (i) 5p  (ii) 4p  (c) $\frac{1}{40}$
19. (a) 0.1  (b) 0.3  (c) 0.45
20. (a) $\frac{3}{7}$  (b) $\frac{3}{8}$
21. (a) $\frac{5}{21}$  (b) $\frac{2}{3}$  (c) $\frac{5}{12}$
22. (a) 0.15  (b) $\frac{7}{15}$  (c) $\frac{1}{75}$

## Exercise 3d Tree diagrams (page 200)

### Section A

1. (a) 0.0025  (b) 0.095
2. (a) $\frac{5}{14}$  (b) $\frac{17}{42}$
3. (a) 0.24  (b) 0.42
4. (a) (i) $\frac{8}{27}$  (ii) $\frac{4}{9}$  (iii) $\frac{7}{27}$
   (b) (i) $\frac{5}{21}$  (ii) $\frac{15}{28}$  (iii) $\frac{19}{84}$
5. 0.00599, 0.987
6. (a) $\frac{12}{49}$  (b) $\frac{20}{49}$
7. $\frac{7}{16}$
8. $\frac{25}{72}$
9. $\frac{5}{16}$
10. 0.35
11. 0.825
12. (a) 0.5  (b) 0.5  (c) 0.375
13. 0.788
14. (a) 0.02  (b) 0.64
15. (a) $\frac{3}{11}$  (b) $\frac{12}{55}$  (c) $\frac{3}{44}$
16. (a) 0.34  (b) 0.063  (c) 0.19  (d) 0.97; 3 white
17. 0.624
18. (a) $\frac{1}{4}$  (b) $\frac{1}{4}$  (c) $\frac{1}{16}$  (d) $\frac{1}{4}$  (e) $\frac{3}{4}$

### Section B

1. (a) $\frac{5}{12}$  (b) $\frac{3}{5}$
2. (a) $\frac{21}{38}$  (b) $\frac{15}{38}$  (c) $\frac{20}{83}$
3. (a) P(A occurs, given that B occurs)
   (i) mutually exclusive  (ii) independent
   (b) 0.88, 0.05
4. (a) 0.33  (b) $\frac{7}{11}$
5. (a) $\frac{1}{8}$  (b) $\frac{3}{8}$  (c) $\frac{8}{9}$
6. (b) $\frac{77}{95}$
7. (a) $\frac{7}{18}$  (b) (i) $\frac{5}{8}$  (ii) $\frac{8}{25}$
8. (a) $\frac{1}{25}$  (b) $\frac{106}{125}$  (c) $\frac{14}{19}$
9. (a) 0.096  (b) 0.156; $\frac{5}{13}$
10. (a) 0.7, 0.68  (b) 0.28  (c) 0.65625
11. (a) $\frac{6}{323}$  (b) $\frac{135}{323}$  (c) $\frac{1}{5}$  (d) $\frac{1}{5}$
    (i) Yes, no  (ii) No, yes
12. (a) 0.000877  (b) 0.421  (c) 0.65  (d) 0.642
13. (a) 0.042875  (b) 0.142  (c) 0.1215
    (d) 0.189  (e) 0.334125; 0.642
14. (a) (i) $\frac{9}{22}$  (ii) $\frac{6}{11}$  (iii) $\frac{2}{11}$  (iv) $\frac{4}{7}$
    (b) (i) 0.0303  (ii) 0.450  (iii) 0.0348
    (c) (i) 0.36  (ii) 0.848
15. $\frac{23}{45}, \frac{18}{23}$
16. (b) $\frac{9}{32}$  (c) $\frac{83}{128}$  (d) $\frac{17}{37}$
17. (a) 0.36  (b) 0.6875

## Exercise 3e Useful methods (page 206)

1. (a) 0.763  (b) 14
2. (a) 5  (b) 6
3. 0.5, 6
4. 0.999
5. 22
6. $\frac{5}{11}$
7. 1 : 8
8. 0.5  (a) $\frac{1}{6}$  (b) $\frac{25}{216}$  (c) $\frac{625}{7776}$; $\frac{6}{11}$
9. (a) (i) $\frac{1}{6}$  (ii) $\frac{1}{12}$  (iii) $\frac{2}{3}$  (b) $\frac{7}{12}$

## Exercise 3f Arrangements, permutations, combinations (page 219)

1. 9!, $\frac{1}{72}$
2. (a) 6!  (b) $\frac{1}{3}$

3. (a) 4! 9!  (b) $\frac{54}{55}$
4. $\frac{9}{11}$
5. $\frac{1}{126}$
6. (a) 8!  (b) $\frac{1}{28}$
7. (a) $\dfrac{12!}{(2!)^4}$  (b) $\frac{1}{66}$
8. $\frac{28}{153}$
9. $\frac{49}{143}$
10. $\frac{60}{143}$
11. (a) 210  (b) $\frac{2}{15}$  (c) $\frac{1}{30}$
12. 12
13. (a) $\frac{1}{14}$  (b) $\frac{3}{7}$  (c) $\frac{1}{30}$
14. (a) 65 268  (b) 4263
15. 510
16. $\frac{37}{42}$
17. 4608
18. (a) 1260  (b) 2520
19. (a) 420  (b) B 252, G 462  (c) 120  (d) $\frac{44}{133}$
20. (a) 5040  (b) 1680  (c) 672
21. 5005, 720, 72
22. 5040  (a) 144  (b) 120
23. (a) $2.5 \times 10^{-7}$  (b) 3 193 344
24. (a) $\frac{2}{7}$  (b) $\frac{2}{7}$
25. 130
26. (a) 360  (b) 6  (c) 12  (d) 1170
27. (a) 64  (b) 18  (c) $\frac{21}{32}$
28. (a) 9!  (b) $\frac{7}{36}$  (c) 1260  (d) $\frac{5}{9}$
29. (a) 75  (c) $\frac{181}{456}$  (d) (i) 6!  (ii) 72
30. 70, (a) 55
    (b) 30
    (c) 65
    (d) $\frac{2}{7}$
    (e) $\frac{1}{7}$
    (f) $\frac{1}{7}$

## Miscellaneous exercise 3g (page 228)

1. (a) 0.36  (b) 0.48  (c) 0.01024  (d) 0.98976
2. (a) C, C'  (b) C, D  (c) C, E
3. (a) 0.0902  (b) unsatisfactory test
4. 0.32, 0.467
5. (a) 0.325  (b) $\frac{51}{260}$  (c) $\frac{5}{13}$
6. (a) 0.28  (b) (i) 0.157  (ii) 0.363  (iii) 0.163
   (c) 0.0728  (d) 0.404
7. 0.166, 0.580
8. 5040  (a) 720  (b) 1440
9. (a) $\frac{1}{343}$  (b) $\frac{1}{49}$  (c) $\frac{30}{49}$  (d) $\frac{8}{343}$  (e) $\frac{1}{4}$  (f) 6
10. (a) $\frac{11}{24}$  (b) $\frac{11}{60}$  (c) $\frac{43}{120}$  (d) $\frac{49}{144}$
11. (a) (i) 0.005  (ii) 0.0955  (b) 0.999  (c) 0.136
12. (a) (i) $\frac{1}{3}$  (ii) $\frac{2}{9}$  (iii) $\frac{1}{3}$  (b) $\frac{2}{15}$  (c) $\frac{3}{10}$
13. 5005, 1960, 315  (a) $\frac{9}{56}$  (b) $\frac{27}{56}$
14. (a) 792  (b) 210  (c) $\frac{35}{132}$  (d) 120  (e) 0.1  (f) 0.1
15. (a) 40 320  (b) (i) 1440  (ii) 5760
    (c) (i) $\frac{1}{7}$  (ii) $\frac{6}{7}$  (d) 576  (e) $\frac{3}{35}$
16. (a) $\frac{1}{4}$  (b) $\frac{5}{14}$  (c) independent
    (d) $\frac{1}{7}$, $P(A|C) \neq P(A)$  (e) $\frac{7}{18}$

## Mixed test 3A (page 231)

1. (a) $\frac{1}{28}$ (b) $\frac{1}{13}$
2. (a)

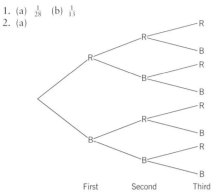

First draw    Second draw    Third draw

(b) $\frac{8}{15}$ (c) $\frac{2}{5}$ (d) $\frac{1}{2}$
3. (a) 0.4 (b) 0.2 (c) $\frac{13}{30}$
4. (a) $\frac{23}{30}$ (b) $\frac{119}{150}$ (c) $\frac{31}{150}$ (d) 30
   (e) The probability that a female employee is weekly paid. (f) 0.5
5. (a) $\frac{1}{16}$ (b) $\frac{7}{16}$ (c) $\frac{3}{32}$ (d) $\frac{29}{32}$

## Mixed test 3B (page 232)

1. (a) 0.64 (b) 0.75
2. (a) $q - 0.25$ (b) $\dfrac{2}{3(4q-1)}$ (c) $\frac{13}{20}$
3. (a) 0.857 (b) 0.135 (c) 0.13917 (d) 0.973
4. (a) 0.1792 (b) 0.1686 (c) 0.203
5. (a)

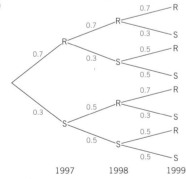

1997    1998    1999

(b) 0.372
(c) $\frac{49}{124}$
(d) 8

# Chapter 4

## Exercise 4a Probability distributions (page 236)

1 (a) 0.1

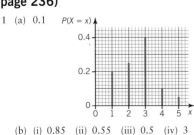

(b) (i) 0.85 (ii) 0.55 (iii) 0.5 (iv) 3

2.

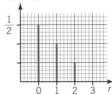

| $x$ | 12 | 13 | 14 |
|---|---|---|---|
| $P(X = x)$ | $12k$ | $13k$ | $14k$ |

, $k = \frac{1}{39}$

3. 0.1
4. (a) $\frac{7}{20}$ (b) $\frac{1}{4}$ (c) 0 (d) $\frac{13}{20}$
5. (a)

| $x$ | 0 | 1 | 2 |
|---|---|---|---|
| $P(X = x)$ | $\frac{1}{4}$ | $\frac{1}{2}$ | $\frac{1}{4}$ |

(b)

| $x$ | 0 | 1 | 2 | 3 |
|---|---|---|---|---|
| $P(X = x)$ | $\frac{1}{8}$ | $\frac{3}{8}$ | $\frac{3}{8}$ | $\frac{1}{8}$ |

6.

| $x$ | 0 | 1 | 2 | 3 |
|---|---|---|---|---|
| $P(X = x)$ | $\frac{1}{27}$ | $\frac{2}{9}$ | $\frac{4}{9}$ | $\frac{8}{27}$ |

7. (a) $\frac{1}{6}$ (b) $P(R = r)$

[graph of $P(R = r)$ with vertical axis marked $\frac{1}{2}$ and horizontal axis $r$ at 0, 1, 2, 3]

(c) $\frac{1}{2}$
8. $P(X = x) = 0.1$, $x = 0, 1, 2, ..., 9$
9. (a)

| $x$ | 0 | 1 | 2 | 3 |
|---|---|---|---|---|
| $P(X = x)$ | 0.216 | 0.432 | 0.288 | 0.064 |

(b) 0.648
10.

| $x$ | $-5$ | 5 | 15 |
|---|---|---|---|
| $P(X = x)$ | $\frac{25}{36}$ | $\frac{5}{18}$ | $\frac{1}{36}$ |

11.

| $x$ | 1 | 2 | 3 | 4 | 5 | 6 |
|---|---|---|---|---|---|---|
| $P(X = x)$ | $\frac{6}{72}$ | $\frac{7}{72}$ | $\frac{8}{72}$ | $\frac{9}{72}$ | $\frac{10}{72}$ | $\frac{11}{72}$ |

| $x$ | 7 | 8 | 9 | 10 | 11 | 12 |
|---|---|---|---|---|---|---|
| $P(X = x)$ | $\frac{6}{72}$ | $\frac{5}{72}$ | $\frac{4}{72}$ | $\frac{3}{72}$ | $\frac{2}{72}$ | $\frac{1}{72}$ |

$\frac{11}{18}$; Equally likely outcomes
12. For $x = 8$, draw vertical line to 0.2; for $x = 9$, draw vertical line to 0.3; symmetrical distribution.

## Exercise 4b Expectation (page 244)

1. 2.25
2. 7
3. (a) 0.3 (b) 2.9
4. 1
5. $\frac{12}{11}$
6. 0.75p
7.

| $x$ | 10 | 20 |
|---|---|---|
| $P(X = x)$ | 0.4 | 0.6 |

8. (a) 0.3 (b) 0.2
9. (a) 0.2 (b) 2.08
10. 2.75
11.

| $x$ | 4 | 6 | 8 | 9 | 11 | 14 |
|---|---|---|---|---|---|---|
| $P(X = x)$ | 0.16 | 0.32 | 0.16 | 0.16 | 0.16 | 0.04 |

Loss of £1.20.

12. $£\frac{3}{8}(7+x)$ (a) 5 (b) Loss of £3.75
13. (a) 24
(c)

| $x$ | 0 | 1 | 2 | 3 | 4 |
|---|---|---|---|---|---|
| $P(X = x)$ | $\frac{3}{8}$ | $\frac{1}{3}$ | $\frac{1}{4}$ | 0 | $\frac{1}{24}$ |

(d) 1
14. $\frac{11}{3}$
15. 2

## Exercise 4c Expectation and variance (page 251)

1. (a) 2.3 (b) 5.9 (c) 0.61
2. (a) 0.35 (b) 4.2
3. (a) 1.45 (b) 2.45 (c) 12.15
4. (a) 3.5 (b) $15\frac{1}{6}$ (c) 14.5 (d) $2\frac{11}{12}$
5. (a) 2.56
6. (a) 3.5 (b) 14 (c) 5.5 (d) 84 (e) 1.75
7. (a) 2 (b) 3 or −3
8. (a) $\frac{1}{32}$, 1, $1\frac{31}{32}$
9. (b)

| $x$ | 4 | 6 | 8 | 10 | 12 |
|---|---|---|---|---|---|
| $P(X = x)$ | $\frac{1}{16}$ | $\frac{1}{4}$ | $\frac{3}{8}$ | $\frac{1}{4}$ | $\frac{1}{16}$ |

(c) 4
10. (a) 4.2 (b) $7\frac{1}{3}$ (c) 3.67
11. (a) $\frac{7}{10}$ (b) $3\frac{1}{2}$ (c) $15\frac{7}{30}$ (d) $2\frac{59}{60}$ (e) $47\frac{11}{15}$
12. (a) $1\frac{2}{3}$ (b) $3\frac{1}{3}$ (c) $\frac{5}{9}$
13.

| $x$ | 0 | 1 | 2 |
|---|---|---|---|
| $P(X = x)$ | $\frac{1}{7}$ | $\frac{4}{7}$ | $\frac{2}{7}$ |

(a) $\frac{8}{7}$ (b) $\frac{12}{7}$ (c) $\frac{20}{49}$ (d) $\frac{180}{49}$
14. (a) 5 (b) 2.5 (c) 10 (d) 10
15. 144
16. (a) $\frac{6}{7}$ (b) 0.639
17. $P(X = x) = \dfrac{10 - x}{45}$, $x = 1, 2, ..., 9$; $3\frac{2}{3}$, 2.21, 1;
$P(X = x) = (\frac{4}{5})^{x-1}(\frac{1}{5})$, $x = 1, 2, ...$
18. (a) $\frac{1}{12}$ (b) 0 (c) 6 (d) 2.45
19. (a) 0.04 (b) 5 (c) 4 (d) 7 (e) 16
20. (a) Loss £3 (b) (i) $p = 0.12$, $q = 0.08$ (ii) 645, 8
21. (a) £2 (b) (i) 4 (ii) 17 (iii) 1

## Exercise 4d Cumulative distribution function (page 255)

1.

| $y$ | 0.1 | 0.2 | 0.3 | 0.4 | 0.5 |
|---|---|---|---|---|---|
| $P(Y \leqslant y)$ | 0.05 | 0.3 | 0.6 | 0.75 | 1 |

2. (a) 0.41 (b) 0.87 (c) 0.46 (d) 0.13 (e) 2.58
3. (a)

| $x$ | 0 | 1 | 2 |
|---|---|---|---|
| $F(x)$ | $\frac{25}{36}$ | $\frac{35}{36}$ | 1 |

(b)

| $x$ | 1 | 2 | 3 | 4 | 5 | 6 |
|---|---|---|---|---|---|---|
| $F(x)$ | $\frac{11}{36}$ | $\frac{5}{9}$ | $\frac{3}{4}$ | $\frac{8}{9}$ | $\frac{35}{36}$ | 1 |

(c)

| $x$ | 0 | 1 | 2 | 3 |
|---|---|---|---|---|
| $F(x)$ | $\frac{1}{8}$ | $\frac{1}{2}$ | $\frac{7}{8}$ | 1 |

4.

| $x$ | 3 | 4 | 5 | 6 | 7 |
|---|---|---|---|---|---|
| $P(X = x)$ | 0.01 | 0.22 | 0.41 | 0.22 | 0.14 |

; 0.9724

5. (a) $\frac{4}{9}$ (b) $\frac{1}{3}$ (c) $P(X = x) = \dfrac{2x - 1}{9}$, $x = 1, 2, 3$ (d) $\frac{17}{9}$
6. (a) $\frac{1}{3}$ (b) $\frac{2}{3}$ (c) $P(X = x) = \frac{1}{3}$, $x = 1, 2, 3$ (d) 0.816
7. (b)

| $x$ | 1 | 2 | 3 | 4 |
|---|---|---|---|---|
| $P(X = x)$ | $\frac{1}{4}$ | $\frac{1}{2}$ | $\frac{15}{64}$ | $\frac{1}{64}$ |

(c) $2\frac{1}{64}$, 0.547 (d) $\frac{1}{4}$
8. (a) 0.9900 (b) 0.1746 (c) 0.5886
(d) 0.5565 (e) 0.9785

## Exercise 4e Combinations of random variables (page 261)

1. (a) 26 (b) 15 (c) 17 (d) 59 (e) 59
2. (a) 0 or 12 or −12 (b) 294
3. (a) 1 (b) −1 (c) 34 (d) 14 (e) 14 (f) 30
4. (a) 1.3, 1, 1.01, 0.8
(b)

| $x + y$ | 0 | 1 | 2 |
|---|---|---|---|
| $P(X + Y = x + y)$ | 0.12 | 0.14 | 0.32 |

| $x + y$ | 3 | 4 | 5 |
|---|---|---|---|
| $P(X + Y = x + y)$ | 0.2 | 0.18 | 0.04 |

(e)

| $x - y$ | −2 | −1 | 0 |
|---|---|---|---|
| $P(X - Y = x - y)$ | 0.12 | 0.14 | 0.32 |

| $x - y$ | 1 | 2 | 3 |
|---|---|---|---|
| $P(X - Y = x - y)$ | 0.2 | 0.18 | 0.04 |

5. (a) 2.6, 0.24 (b) 5.2, 0.48 (c) 7.8, 0.72
6. $29\frac{1}{6}$
7. (a) 0.1 (b) 3 (c) 1 (d) 0.2 (e) 12 (f) 3

## Miscellaneous exercise 4f (page 266)

1. 0.1825, £1.75
2. (a) $\frac{1}{13}$ (b) 2, $\frac{12}{13}$
3. (a) 0.01 (b) 3.54, 0.4684 (c) 14.7, 11.71
4. $\frac{120}{49}$, 2.57
5. $\frac{1}{28}$, 3.5, 1.25, 12, 20
6. (a)

| $x$ | 1 | 2 | 4 | 5 |
|---|---|---|---|---|
| $P(X = x)$ | $\frac{1}{12}$ | $\frac{5}{12}$ | $\frac{1}{3}$ | $\frac{1}{6}$ |

(b)

| $y$ | 2 | 3 | 4 | 5 | 6 | 7 | 8 | 9 | 10 |
|---|---|---|---|---|---|---|---|---|---|
| $P(Y = y)$ | $\frac{1}{144}$ | $\frac{5}{72}$ | $\frac{25}{144}$ | $\frac{1}{18}$ | $\frac{11}{36}$ | $\frac{5}{36}$ | $\frac{1}{9}$ | $\frac{1}{9}$ | $\frac{1}{36}$ |

$6\frac{1}{6}$, $3\frac{35}{72}$
7. $\frac{1}{15}$, $\frac{2}{3}$, $\frac{34}{45}$, $\frac{1}{75}$, $\frac{4}{3}$, $\frac{68}{45}$
8. (a) $\frac{3}{8}$ (b)

| $d$ | 0 | 1 | 2 | 3 |
|---|---|---|---|---|
| $P(D = d)$ | $\frac{1}{4}$ | $\frac{3}{8}$ | $\frac{1}{4}$ | $\frac{1}{8}$ |

(c) 1.25
9. (a) 0.1248 (b) 2.8352, 236
10. (b)

| $b$ | 0 | 1 | 2 | 3 | 4 |
|---|---|---|---|---|---|
| $P(B = b)$ | $\frac{1}{3}$ | $\frac{2}{9}$ | $\frac{4}{27}$ | $\frac{8}{81}$ | $\frac{16}{81}$ |

(c) $1\frac{49}{81}$ (e) $\frac{16}{81}$
11.

| $y$ | 0 | 1 | 2 | 3 |
|---|---|---|---|---|
| $P(Y = y)$ | 0.3 | 0.34 | 0.2 | 0.16 |

(a) 1.22 (b) 1.0916 (c) 0.36

12.

| $x$ | 2 | 3 | 4 | 5 | 6 | 7 | 8 | 9 | 10 | 11 | 12 |
|---|---|---|---|---|---|---|---|---|---|---|---|
| $P(X=x)$ | $\frac{1}{25}$ | $\frac{2}{25}$ | $\frac{1}{25}$ | $\frac{2}{25}$ | $\frac{4}{25}$ | $\frac{4}{25}$ | $\frac{3}{25}$ | $\frac{2}{25}$ | $\frac{3}{25}$ | $\frac{2}{25}$ | $\frac{1}{25}$ |

7.2, £75

13. (a) $\frac{1}{36}$  (b) $\frac{5}{36}$  (c) $\frac{11}{36}$; $-\frac{1}{36}$, 7
14. (a) $\frac{1}{8}$, $\frac{5}{24}$  (b) 2.78  (c) 0.260
15. (a) 0.8  (b) $-0.24p$  (c) $3.34p^2$
16. (a) 1.7, 1.18  (b) 4.76
17. (a) 1, $\frac{4}{5}$  (b) $\frac{3}{5}$, $\frac{6}{25}$  (c) 11.2, 7.28

| $t$ | 0 | 1 | 2 | 3 | 4 |
|---|---|---|---|---|---|
| $P(T=t)$ | $\frac{2}{15}$ | $\frac{9}{25}$ | $\frac{8}{25}$ | $\frac{11}{75}$ | $\frac{1}{25}$ |

18. $P(X=x)=\frac{1}{6}$, $x=1, 2, 3, 4, 5$; $P(X=6)=0$; $P(X=x)=\frac{1}{36}$, $x=7, 8, ..., 12$; $4\frac{1}{12}$, $\frac{6}{17}$

## Mixed test 4A (page 269)

1. (a) 0.2  (b) 8  (c) 11.6
2. (b)

| $x$ | 0 | 1 | 2 | 3 | 4 |
|---|---|---|---|---|---|
| $P(X=x)$ | $\frac{1}{4}$ | $\frac{1}{3}$ | $\frac{5}{18}$ | $\frac{1}{9}$ | $\frac{1}{36}$ |

(c) $1\frac{1}{3}$  (e) 0, $\frac{20}{9}$
3. (a)

| $x$ | 0 | 1 | 2 | 3 | 4 | 5 |
|---|---|---|---|---|---|---|
| $P(X=x)$ | $\frac{1}{6}$ | $\frac{5}{18}$ | $\frac{2}{9}$ | $\frac{1}{6}$ | $\frac{1}{9}$ | $\frac{1}{18}$ |

(b) $1\frac{17}{18}$  (c) $\frac{4}{9}$

## Mixed test 4B (page 269)

1. (a) 0.4  (b) 0.8  (c) 2.6  (d) 1.44  (e) 15.6
2. (a)

| $x$ | 4 | 5 | 6 | 8 | 9 | 12 |
|---|---|---|---|---|---|---|
| $P(X=x)$ | $\frac{1}{9}$ | $\frac{1}{3}$ | $\frac{1}{4}$ | $\frac{1}{6}$ | $\frac{1}{6}$ | $\frac{1}{36}$ |

(b) $6\frac{1}{3}$, $43\frac{13}{18}$, $3\frac{11}{18}$  (c) Loss of £1  (d) $\frac{2}{19}$
3. (a)

| $s$ | 1 | 2 | 3 | 4 | 6 | 12 |
|---|---|---|---|---|---|---|
| $P(S=s)$ | $\frac{1}{8}$ | $\frac{1}{4}$ | $\frac{1}{8}$ | $\frac{1}{8}$ | $\frac{1}{4}$ | $\frac{1}{8}$ |

(b) 4.5  (c) 11

# Chapter 5

## Exercise 5a The uniform and geometric distributions (page 276)

1. (a) 0.2  (b) 8  (c) 0.4
2. (a) 0.096  (b) 0.179  (c) 0.725  (d) 2.86
3. (a) 0.9744  (b) 0.01024  (c) 1  (d) $1\frac{2}{3}$  (e) 2.5
4. (a) 1  (b) 0.7599
5. (a) 0.0226  (b) 0.00374
6. (a) (i) 0.6  (ii) 0.3  (iii) 4.5  (iv) 2.87
   (b) (i) 0.0531  (ii) 1  (iii) 10
7. (a) 1  (b) 2  (c) 1.41
8. (a) 0.128  (b) $X \sim \text{Geo}(0.2)$  (c) 0.512
9. (a) $P(X=4)=0.7^3 \times 0.3 = 0.1029$
   (b) The first success is at the $n$th attempt.
   (c) There are at least $n$ attempts before the first success is obtained.
10. 0.7225
11. 0.00026
12. 2
13. (a) 0.0864  (b) 2.5  (c) 1.94  (d) 1  (e) 0.028
14. £1.75
15. 0.0047, December 22nd
16. (a) $\frac{1}{6}$  (b) $\frac{25}{216}$  (c) $\frac{125}{216}$  (d) 1  (e) 6  (f) 17

## Exercise 5b The binomial distribution (page 285)

1. (a) 0.267  (b) 0.850
2. (a) 0.234  (b) 0.000107 (0.0001 from tables)
3. (a) 0.279  (b) 0.983  (c) 0.594
4. (a) 0.00549  (b) 0.157  (c) 0.503
5. 0.00200
6. (a) 0.318  (b) 0.671  (c) 0.647  (d) 0.0324
7. 0.344
8. 0.5
9. 0.3456
10. (a) (i) 0.0424  (ii) 0.623  (b) 12
11. 0.0963, improve with practice
12. (a) 0.0105  (b) 0.988  (c) 0.358
13. (a) 0.329  (b) 0.461
14.

| $x$ | $P(X=x)$ |
|---|---|
| 0 | 0.0156 |
| 1 | 0.0938 |
| 2 | 0.2344 |
| 3 | 0.3125 |
| 4 | 0.2343 |
| 5 | 0.0938 |
| 6 | 0.0156 |

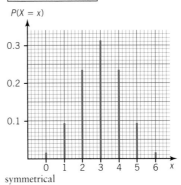

symmetrical
15. 9
16. 68; not strictly binomial as $p$ is not constant, but model can be used if there are a large number of bulbs in the box.
17. (a) 0.000416 (tables give 0.0004)  (b) 0.0197
18. 5
19.

| $x$ | $P(X=x)$ |
|---|---|
| 0 | 0.0000 |
| 1 | 0.0001 |
| 2 | 0.0011 |
| 3 | 0.0109 |
| 4 | 0.0617 |
| 5 | 0.2096 |
| 6 | 0.3960 |
| 7 | 0.3206 |

20. 4
21. Experiment 1 – no, 3 outcomes; Experiment 2 – yes, constant probability of obtaining black (or white), independent trials; Experiment 3 – no, trials not independent.

## Exercise 5c Expectation, variance and mode of the binomial distribution (page 290)

1. 2.5, 1.5
2. (a) 1.38  (b) 4
3. 8, 1.30
4. (a) 0.2  (b) 0.00551

5. (a) 0.25  (b) 2.5  (c) 0.282
6. 0.1, 0.23
7. (a) 10  (b) 0.000390
8. (a) 3  (b) 3  (c) 0.633
9 (a) 0.994  (b) 2
10. 0, 0, 3, 13, 30, 36, 18
11. 2500
12. 0.06; 293, 94, 12, 1, 0, 0
13. (a) 0.68  (b) 8, 1.6
14. (a) 0.25  (b) 1.5
15. 1, 0.894  (a) 5  (b) 0.2

### Exercise 5d The Poisson distribution (page 297)

1. (a) 0.180  (b) 0.0527  (c) 0.195  (d) 0.670
2. (a) 0.983  (b) 0.184  (c) 0.199
3. (a) 0.0821  (b) 0.560  (c) 0.0631
4. (a) 0.603  (b) 0.616  (c) 0.00246
5. (a) 0.0821  (b) 0.242  (c) 0.759
   (d) 0.0486  (e) 0.125
6. (a) 0.191  (b) 0.0498  (c) 2.45
7. 0.371
8. (a) 0.0382  (b) 0.122
9. 0.677
10. (a) 3  (b) 0.145
11. (a) 90, 72, 29, 8, 1, 0  (b) 44, 44, 22, 8, 2
12. Random events; 0.5, 0.481; 31, 16, 4, 1, 0
13. (a) 0.261  (b) 6

### Exercise 5e The Poisson approximation to the binomial (page 300)

1. (a) (i) 0.0476  (ii) 0.0498  (b) (i) 0.225  (ii) 0.224
   (c) (i) 0.171  (ii) 0.168
2. (a) (i) 0.184  (ii) 0.0190  (b) 0.271  (c) 0.0498
3. (a) 0.287  (b) 0.191
4. (a) $\frac{1}{36}$  (b) 0.713
5. (a) 0.647  (b) 0.185
6. 0.109, 185
7. (a) 0.677  (b) 0.017; 1498
8. (a) 0.468  (b) 0.703
9. 0.0150
10. (a) 0.47  (b) 0.041
    Poisson applies since $p < 0.1$ and $n = 50$. Events may not be independent. After mis-dialling, you are likely to be more careful.
11. Random sample, 0.305

### Exercise 5f Sums of Poisson variables (page 303)

1. 0.121
2. (a) 0.189  (b) 0.308  (c) 0.184
3. (a) 0.323  (b) 0.119
4. (a) 0.301  (b) 0.080  (c) 0.251

### Miscellaneous exercise 5g (page 307)

1. 0.752, 0.537
2. (a) 3  (b) 0.223  (c) 0.988
3. (a) 0.733  (b) 0.0703
4. (a) (i) 0.434  (ii) 0.378  (iii) 0.148  (iv) 0.0401
   (b) (i) 45  (ii) 111; $N > 20$
5. 0.507
6. (a) (i) 0.130  (ii) 0.271  (iii) 0.276; 65, 0.0159
   (b) 90, 3
7. (a) 0.270  (b) 0.350  (c) 0.182
   (d) 0.124  (e) £45

8. (a) 0.223  (b) 0.116  (c) 9.28, 2.86  (d) 18.9
   (e) Part (c) gives 223, part (d) gives 227, increase
9. (a) Large number of balls  (b) 0.799
10. 0.790, calls occur randomly
11. (a) 0.104  (b) 0.283  (c) 0.00113  (d) 9
12. 0.632, 0.069, 0.154
13. (a) (i) $X \sim B(28, 0.004)$  (ii) 0.00545  (b) 0.785
    (c) independence
14. (a) 0.311  (b) 0.959; 3.6, 1.2
15. (a) 0.253  (b) 3.6, 1.59
16. (a) (i) 0.201  (ii) 0.00637  (b) 2  (c) 5, 2  (d) 14
17. (a) 0.203  (b) (ii) 0.136  (c) 0.316
    (d) Assume $p$ constant; very unlikely in First World War
18. $P(X = x) = e^{-\lambda}\dfrac{\lambda^x}{x!}, \lambda, \lambda$
    (a) 0.082  (b) 0.242; 6.15
19. (a) 0.908  (b) 9
20. (a) 3, 7  (b) 20, 20
    Reason for (a) $E(Y - X) \neq \mathrm{Var}(Y - X)$
    Reason for (b) $2Y + 10$ could not take values less than 10.
21. 600 m, Po(2.5), 0.0821, 0.109, 0.779, 0.207
22. (a) (ii) 1.5  (b) 0.577  (c) 0.0249
23. 0.407, 0.366, 0.165, 0.0629, 0.816, 0.0518
24. (a) 22  (b) 19; 39
25. (a) 0.135  (b) 0.323; 0.81
26. (a) 0.387  (b) 0.929  (c) 0.893
    (d) 0.205  (e) 0.816; 0.029
27. (a) 0.0902  (b) 0.0613; 4

### Mixed test 5A (page 312)

1. (a) 0.159  (b) 0.766;
   Query independence: friends may have joint engagements.
2. (a) 0.152  (b) 0.567  (c) 0.285
3. (a) $X \sim B(150, \frac{1}{80})$, $\lambda = 1.875$, $p < 0.1$, $n > 50$
   (b) 0.559  (c) 369
4. (a) $X \sim Po(0.6)$, $X$ is number of boxes in a square km.
   (b) 0.549  (c) 0.0231
   (d) Probably not suitable; different scatter of telephone boxes in the city.
5. (a) 4.8, 0.98  (c) 0.737  (d) 0.388

### Mixed test 5B (page 313)

1. (a) $5(1 - p)p^4$  (b) $10(1 - p)^3p^2$; $\frac{2}{3}$
2. (a) $Y \sim \mathrm{Geo}(\frac{1}{6})$  (b) 30  (c) 0.233
3. (a) Binomial  (b) Poisson  (c) $e^{-\lambda}\dfrac{\lambda^3}{6}$
   (d) $1 - e^{-\lambda}\left(1 + \lambda + \dfrac{\lambda^2}{2}\right)$; 0.013, 0.014, 0.182
4. (a) 0.221  (b) 0.987
5. (a) 0.249  (b) 0.929  (c) 0.508; 0.542

# Chapter 6

### Exercise 6a Calculating probabilities (page 319)

1. (a) $\frac{3}{8}$  (b) $\frac{7}{8}$  (c) $\frac{13}{32}$
2. (a)

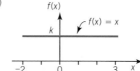

(b) 0.2  (c) 0.74

3. (a) 0.25

(b)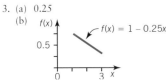

(c) 0.66

4. (a) $\frac{3}{56}$ (b) $\frac{19}{56}, \frac{37}{56}$

5. $c = 1, k = 4$

6. (a) 0.125 (b)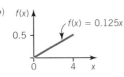

(c) 0.328

7. (a) 0.25

(b)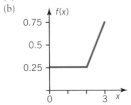

(c) 0.25 (d) 0.3125 (e) 0.3475

## Exercise 6b Expectation E(X) (page 323)

1. (a) $\frac{9}{16}$ (b) 1 (c) 2 (d) 1.6 (e) $2\frac{1}{24}$
2. (a)

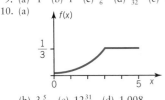

(b) $\frac{1}{3}$ (c) 2
3. 3
4. 6 m
5. (a) $\frac{2}{75}$ (b) $\frac{70}{9}$ (c) 0.48, money bond
6. 2, 0.124
7. 2.5, 0.803, 0.456
8. (a) 2.875 kg (b) £4.75, $\frac{3}{16}$
9. (a) 0.4 (b) 2.6 (c) 1.5

## Exercise 6c Standard deviation and variance (page 333)

1. (a) 1.5 (b) 2.4 (c) 0.15 (d) 0.387
2. (a) 0.5 (b) $2\frac{1}{3}$ (c) $2\frac{1}{12}$ (d) 1.44
3. (a) $1\frac{5}{6}$ (b) $3\frac{2}{3}$ (c) $\frac{11}{36}$ (d) 0.553
4. (a) $1\frac{3}{14}$ (b) $1\frac{27}{35}$ (c) $\frac{291}{980}$ (d) 0.545
5. (a) $\frac{4}{5}$ (b) $\frac{2}{3}$ (c) $\frac{2}{75}$ (d) 0.163
6. (a) $1\frac{19}{24}$ (b) $4\frac{1}{24}$ (c) $\frac{479}{576}$ (d) 0.912
7. (a) $\frac{5}{18}$ (b) $\frac{214}{405}$ (c) $\frac{731}{1620}$ (d) 0.672
8. (a) $\frac{3}{64}$ (b) $3, \frac{3}{5}$ (c) $\frac{7}{64}$
9. (a) 1 (b) 1 (c) $\frac{1}{6}$ (d) $\frac{19}{32}$ (e) 1
10. (a)

11. (b) $2, 4 - \dfrac{4}{\ln 3}$
12. $a = 2, k = 0.75$;

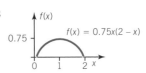

0.2
13. 0.6, 0.2

## Exercise 6d Cumulative distribution function (page 339)

1. (a) $F(x) = \begin{cases} \dfrac{x^3}{8} & 0 \leqslant x \leqslant 2 \\ 1 & x \geqslant 2 \end{cases}$

(b) 1.59

2. (a) $F(x) = \begin{cases} \dfrac{1}{8}(8x - x^2 - 7) & 1 \leqslant x \leqslant 3 \\ 1 & x \geqslant 3 \end{cases}$ (b) $\frac{9}{32}$

3. (a) $\frac{1}{5}$ (b) $F(x) = \begin{cases} \dfrac{1}{5}(x-1) & 1 \leqslant x \leqslant 6 \\ 1 & x \geqslant 6 \end{cases}$ (c) 2 (d) 2.5

4. (a) $F(x) = \begin{cases} \dfrac{x}{4} & 0 \leqslant x \leqslant 2 \\ \dfrac{1}{4}(x^2 - 3x + 4) & 2 \leqslant x \leqslant 3 \\ 1 & x \geqslant 3 \end{cases}$ (b) 2

5. (a) 0.1215 (b) 0.841 (c) 0.880

6. (a) 1.5 (b) 0.75 (c) $F(x) = \begin{cases} \dfrac{1}{3}x & 0 \leqslant x \leqslant 3 \\ 1 & x \geqslant 3 \end{cases}$

(d) 0.4 (e) 0.2

7. (a) $\frac{3}{7}$

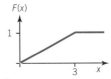

(b) 0.272 (c) $F(x) = \begin{cases} \dfrac{1}{7}(x^3 - 1) & 1 \leqslant x \leqslant 2 \\ 1 & x \geqslant 2 \end{cases}$ (d) 1.65

8. $\dfrac{3}{4}, \dfrac{19}{80}, F(x) = \begin{cases} \dfrac{3}{4}x - \dfrac{1}{16}x^3 & 0 \leqslant x \leqslant 2 \\ 1 & x \geqslant 2 \end{cases}$, 0.007

9. (a) $\frac{1}{3}, \frac{1}{3}$

(b) $F(x) = \begin{cases} \dfrac{x^2}{6} - \dfrac{2x}{3} + \dfrac{2}{3} & 2 \leqslant x \leqslant 3 \\ \dfrac{x}{3} - \dfrac{5}{6} & 3 \leqslant x \leqslant 5 \\ 2x - \dfrac{x^2}{6} - 5 & 5 \leqslant x \leqslant 6 \\ 1 & x \geqslant 6 \end{cases}$

(c) $\frac{1}{3}$ (d) $\frac{1}{24}$

(b) $3\frac{5}{12}$ (c) $12\frac{31}{45}$ (d) 1.008

10. (a)

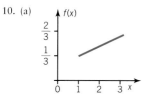

(b) $2\frac{1}{9}$ (c) $F(x) = \begin{cases} \frac{1}{6}x + \frac{1}{12}x^2 - \frac{1}{4} & 1 \leq x \leq 3 \\ 1 & x \geq 3 \end{cases}$ (d) 2.16

11. (a) 5 (b) $\frac{1}{6}$ (c) $\frac{5}{252}$; 543 tonnes

12. $F(x) = \begin{cases} \frac{1}{4}x & 0 \leq x \leq 1 \\ \frac{1}{5} + \frac{x^4}{20} & 1 \leq x \leq 2 \\ 1 & x \geq 2 \end{cases}$ 1.565, 0.821

13. (b) 1, 2 (c) 0 (d) $\frac{1}{\sqrt{2}}, -\frac{1}{\sqrt{2}}$

14. (a)

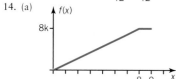

(c) $F(x) = \begin{cases} 0.0125x^2 & 0 \leq x \leq 8 \\ 0.2x - 0.8 & 8 \leq x \leq 9 \\ 1 & x \geq 9 \end{cases}$

(d) 0.55

15. (a) 0.75 (b) 0.2

(c) $F(x) = \begin{cases} 0.75x^2 - 0.25x^3 & 0 \leq x \leq 2 \\ 1 & x \geq 2 \end{cases}$ (d) 0.288

16. (a) 0.455, 3 (b) 3.64, 4.95

(c) $F(x) = \begin{cases} \frac{1}{\ln 9}\ln x & 1 \leq x \leq 9 \\ 1 & x \geq 9 \end{cases}$

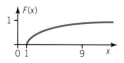

## Exercise 6e Obtaining $f(x)$ from $F(x)$ (page 343)

1. (a) $f(x) = \frac{1}{4}, 2 \leq x \leq 6$
(b)

(c) 4 (d) 2
2. (a) 0.794 (b) 0.75

3. (a) 0.25 (b) $f(x) = 1 - 0.5x, 0 \leq x \leq 2$
(c) 0.586 (d) $\frac{2}{9}$

4. (a) $\frac{1}{3}$ (b) $f(x) = \begin{cases} \frac{2}{3} & 0 \leq x < 1 \\ \frac{1}{3} & 1 \leq x \leq 2 \\ 0 & \text{otherwise} \end{cases}$

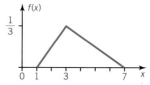

(c) $\frac{5}{6}$ (d) 0.553

5. (a) $f(x) = \begin{cases} \frac{1}{6}(x-1) & 1 \leq x \leq 3 \\ \frac{1}{12}(7-x) & 3 \leq x \leq 7 \\ 0 & \text{otherwise} \end{cases}$

(b) $3\frac{2}{3}$ (c) $1\frac{5}{9}$ (d) 3.45 (e) 0.595
6. (a)

(b) $1, \frac{5}{6}$ (c) 0.25

7. (a) $1, -\frac{1}{27}$ (b) $F(x) = \begin{cases} \frac{1}{27}x^3 & 0 \leq x \leq 3 \\ 1 & x \geq 3 \end{cases}$

(c) $f(x) = \frac{1}{9}x^2, 0 \leq x \leq 3$
8. (a) 2 (b) $f(x) = 2, 0 \leq x \leq 0.5$ (c) 0.25 (d) 0.144

## Exercise 6f Uniform distribution (page 349)

1. (a) $\frac{1}{3}$ (b) 4.5 (c) 0.75 (d) $\frac{1}{3}$
2. (a) 0.5 (b) −3.5 (c) 0.866
3. (a) 5 (b) 0.325 (c) 3 (d) $1\frac{1}{3}$
4. 0.4
5. 0.577
6. (a) 4.5 (b) $2\frac{1}{12}$
7. (a) $a = 3, b = 11$ (b) 0.125

(c) $F(x) = \begin{cases} \frac{1}{8}(x-3) & 3 \leq x \leq 11 \\ 1 & x \geq 11 \end{cases}$

8. (a) $f(x) = 0.2, -2 \leq x \leq 3$ (b) 1.44 (c) 2.5 (d) −1

## Miscellaneous exercise 6g (page 355)

1. (a) $1\frac{1}{3}$ (b) $6\frac{2}{3}$
2. (a) 2.4 (b) 20, $\frac{1}{3}$, 0.178

3. (a) $\frac{2}{3}$ (b) $f(x) = \begin{cases} \frac{2}{3}x & 0 \leq x \leq 1 \\ 1 - \frac{1}{3}x & 1 \leq x \leq 3 \\ 0 & \text{otherwise} \end{cases}$

$1\frac{1}{3}, \frac{7}{18}$
(c) 1.27; 0.875
4. 0.8, 0.16, £8

5. (b)

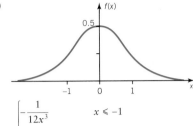

(c) $F(x) = \begin{cases} -\dfrac{1}{12x^3} & x \leqslant -1 \\ \dfrac{1}{2} + \dfrac{1}{2}x - \dfrac{1}{12}x^3 & -1 \leqslant x \leqslant 1 \\ 1 - \dfrac{1}{12x^3} & x \geqslant 1 \end{cases}$   (d) $0, \frac{11}{15}$

6. (a) −0.1875  (b) 0.2375  (d) 2
7. (b) 2.2  (c) 1.71  (d) 0.264  (e) 0.3645
8. (b) 0.5  (d) 0.36
9. (b) 30 hrs  (d) $\frac{65}{81}$  (e) 0.0390
  (f) The model does not allow for lifetimes over 90 hours.
10. (a) 3.8 hrs, 0.36 hrs$^2$  (b) 4 hrs  (c) Approx 60%
11. (a) 0  (b) 0.15625  (c) (i) symmetry  (ii) 0.05
  (d) The player might make a similar mistake each time, resulting in more hits above the line than below, or vice versa.
  (e) The range would reduce.
12. (b) 75 hours  (c) $\frac{5}{16}$
  (d) The model does not allow for $P(X > 2.5) > 0$, since $P(X > 2) = 0$
  (e) Change to exponential model for $x > 1.8$, say

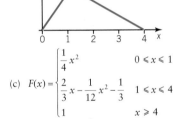

13. (a)

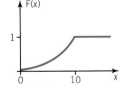

(c) $F(x) = \begin{cases} \dfrac{1}{4}x^2 & 0 \leqslant x \leqslant 1 \\ \dfrac{2}{3}x - \dfrac{1}{12}x^2 - \dfrac{1}{3} & 1 \leqslant x \leqslant 4 \\ 1 & x \geqslant 4 \end{cases}$

(d) £283.33  (e) $\frac{1}{3}$

14. 8, $\frac{1}{9}$, 39 litres

15. (a) 2.93  (b) $F(x) = \begin{cases} 1 - 0.01(x - 10)^2 & 0 \leqslant x \leqslant 10 \\ 1 & x \geqslant 10 \end{cases}$

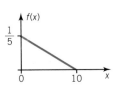

(c) $f(x) = \frac{1}{5} - \frac{1}{50}x, 0 \leqslant x \leqslant 10$

16. (b) $F(x) = \begin{cases} \dfrac{3}{17}x^2 & 0 \leqslant x \leqslant 1 \\ \dfrac{1}{17}(1 + 2x^3) & 1 \leqslant x \leqslant 2 \\ 1 & x \geqslant 2 \end{cases}$   (c) 1.55  (d) 0.89

17. $\lambda = \frac{1}{3}$

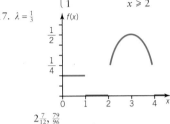

$2\frac{7}{12}, \frac{79}{96}$

18. (a) $\frac{1}{3}$  (b) $f(x) = \begin{cases} \alpha & -1 \leqslant x < 0 \\ 2\alpha & 0 \leqslant x < 1 \\ 0 & x \geqslant 1, x < -1 \end{cases}$

  (c) $\frac{1}{6}$  (d) 0.553  (e) $\frac{11}{18}$
19. (a) 2.1, 1.29  (b) 1, 0.5

## Mixed test 6A (page 358)

1. (a) f.d. 0.85, 0.76, 1.15, 0.8, 1, 0.9, 0.75, 0.36
  Histogram to show incomes

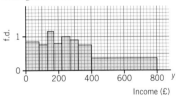

  (b) $\frac{1}{16}$  (c) 120
  (d) From original data, 106 have income in this range. In the model, $f(x) = 3k$, $0 \leqslant x \leqslant 4$ gives too high an estimate; perhaps $f(x) = 2.5k$, $0 \leqslant x \leqslant 4$ would be better.
2. 4, $\frac{8}{15}$, $\frac{11}{225}$, 0.541
3. (a) $F(w) = \begin{cases} \dfrac{w^4}{5^5}(25 - 4w) & 0 \leqslant w \leqslant 5 \\ 1 & w \geqslant 5 \end{cases}$

  (b) 0.650  (c) 0.794  (d) 3.75  (f) Negatively skewed

## Mixed test 6B (page 359)

1. (a)

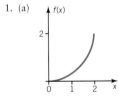

  (b) 1.6  (c) 0.327  (d) $F(m) = 0.5$, $F(\mu) < 0.5$ ∴ $m > \mu$
2. (b) $\frac{8}{15}$  (c) 0.577
3. (b) $1.25\left(1 - \dfrac{1}{m}\right) = 0.5$, $m = 1\frac{2}{3}$

  (c) 0.495  (d) $f(x) = \dfrac{1.25}{x^2}$, $1 \leqslant x \leqslant 5$

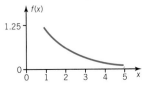

# Chapter 7

## Exercise 7a Finding probabilities, where $Z \sim N(0, 1)$ (page 367)

1. (a) 0.8089  (b) 0.8089  (c) 0.1911  (d) 0.1911
2. (a) 0.0359  (b) 0.2578  (c) 0.9931  (d) 0.9131
   (e) 0.0049  (f) 0.9911  (g) 0.9686  (h) 0.2343
   (i) 0.0312  (j) 0.9484  (k) 0.9803  (l) 0.0021
3. (a) 0.05  (b) 0.05  (c) 0.0999  (d) 0.025  (e) 0.005
   (f) 0.01  (g) 0.0025  (h) 0.075
4. (a) 0.044  (b) 0.8185  (c) 0.1336  (d) 0.3023
5. (a) 0.1703  (b) 0.5481  (c) 0.3639  (d) 0.4582
   (e) 0.4798  (f) 0.9624  (g) 0.0337  (h) 0.9082
   (i) 0.2729  (j) 0.030  (k) 0.925  (l) 0.4508
   (m) 0.9  (n) 0.02
6. 50%
7. (a) 0.9  (b) 0.7
8. (a) 0.55  (b) 0.15
9. (a) 0.9  (b) 0.1

## Exercise 7b Finding probabilities using $X \sim N(\mu, \sigma^2)$ (page 370)

1. (a) 0.0668  (b) 0.4013  (c) 0.1747
2. (a) 0.7054  (b) 0.0618  (c) 0.4621  (d) 0.0456
3. (a) 0.0548  (b) 0.1448  (c) 0.9544
4. (a) 0.0106  (b) 0.9857
5. (a) 0.3015  (b) 0.5231  (c) 0.3792
6. 740
7. 0.00003844
8. (a) 0.6554  (b) 8
9. (a) 0.0478  (b) 0.000817
10. (a) 0.9544  (b) 0.5784  (c) 0.0435
11. (a) 0.1056  (b) 0.7734  (c) 0.6678
12. 0.159, 0.775, 0.067, £37.56
13. 0.785, 0.397
14. 0.957

## Exercise 7c Using the standard normal tables in reverse (page 376)

1. (a) 0.015  (b) 0.796  (c) −1.887  (d) −0.454
   (e) −0.562  (f) 1.019  (g) 0.842
2. (a) 1.94  (b) −0.695  (c) −0.915  (d) 0.722
3. (a) 0.91  (b) 1.66  (c) 0.674  (d) 2.05
4. 0.674, −0.674; 0.524
5. (a) 70  (b) 4.65  (c) 190.742  (d) 1.468
6. (458.92, 546.52)
7. (a) 0.6247  (b) 629.52 g  (c) 3
8. 8, 1.158, (6.10, 9.90)
9. (a) (384.32, 415.68)  (b) (394.608, 405.392)
10. (a) 0.9332  (b) 0.383; 106.6, 137
11. (a) 0.0548  (b) 26  (c) 67.4  (d) 2183
12. (a) 37.8%  (b) (125.5, 194.5)  (c) 0.405

## Exercise 7d Finding $\mu$ or $\sigma$ or both, where $X \sim N(\mu, \sigma^2)$ (page 381)

1. 30
2. 10.7
3. 8.31, 35.9%
4. 35.5
5. 1.75
6. 52.73, 11.96
7. 2.74, 2.78
8. (a) 6.99, 0.324  (b) 0.0105
9. 39.5, 5.32
10. 53.87, 16.48
11. 0.203
12. 92.7%, 1.32, 1.7%
13. 4.299 g
14. 4.46
15. 2080, 236
16. (a) 0.4875  (b) 281, 5.00
17. 5.2007, 0.00346; 0.0269
18. (a) 0.1587  (b) 128.4  (c) 1.31
19. 0.0401  (a) 0.459  (b) 0.003
20. 490 g, 12.2 g
21. (a) 19.50  (b) not symmetrical  (c) 32

## Exercise 7e Continuity corrections (page 386)

1. $P(2.5 < X < 9.5)$
2. $P(3.5 < X < 8.5)$
3. $P(10.5 < X < 24.5)$
4. $P(1.5 < X < 7.5)$
5. $P(X > 54.5)$
6. $P(X > 75.5)$
7. $P(45.5 < X < 66.5)$
8. $P(X < 108.5)$
9. $P(X < 45.5)$
10. $P(55.5 < X < 56.5)$
11. $P(400.5 < X < 560.5)$
12. $P(66.5 < X < 67.5)$
13. $P(X > 59.5)$
14. $P(99.5 < X < 100.5)$
15. $P(33.5 < X < 42.5)$
16. $P(6.5 < X < 7.5)$
17. $P(X > 508.5)$
18. $P(X < 6.5)$
19. $P(26.5 < X < 28.5)$
20. $P(52.5 < X < 53.5)$

## Exercise 7f The normal approximation to the binomial (page 389)

1. 0.1958
2. (a) $np > 5, nq > 5$  (b) 0.0197  (c) 0.0968
3. (a) 0.0154  (b) 0.8145  (c) 0.02
4. (a) 0.657  (b) 0.2142
5. (a) 0.0318  (b) 0.8345
6. (a) 0.9474  (b) 0.6325  (c) 0.5914  (d) 0.0111
7. (a) 0.4502  (b) 0.0996  (c) 0.484
8. 20, 16, 0.0043
9. $P(R = r) = {}^nC_r(1 - p)^{n-r}p^r$, $np$, $np(1 - p)$
   (a) 0.2304  (b) 0.9222; 0.8531
10. 0.1432
11. 0.6886
12. $np > 5, nq > 5$  (a) 0.1853  (b) 0.1838  (c) 0.81%

## Exercise 7g The normal approximation to the Poisson distribution (page 390)

1. (a) 0.6201  (b) 0.39  (c) 0.5406
2. (a) 0.3998  (b) 0.2004  (c) 0.3661  (d) 0.0637
3. (a) 0.313  (b) 0.5078  (c) 0.8335  (d) 0.1101
4. (a) 0.2614  (b) 0.2343  (c) 0.0558
5. 0.8901
6. 0.6887, 4
7. (a) 0.4574  (b) 0.173  (c) 0.8312
8. (a) 0.4594  (b) 0.5363
9. (a) (i) 0.9815  (ii) 0.3486  (iii) 0.9244  (b) 0.0094
10. (a) 0.199  (b) 0.185; 0.870
11. (a) 0.927  (b) 0.0102; 0.297

12. (a) Weevils are randomly scattered in the grain, the grain is selected at random.
    (b) (i) 0.950  (ii) 0.105  (c) 0.158
13. (a) 0.953  (b) 0.745  (c) 0.19
14. (b) 0.133  (c) 11  (d) 0.7119

## Miscellaneous exercise 7h (page 398)

1. (a) 46.5%  (b) 0.532 m  (c) 1.00 M
2. (b) 0.0693  (c) 0.0746
3. (b) 11.5%
4. 50.154, 4
5. (a) (i) 0.0062  (ii) 0.5598  (b) 7.49 m  (c) 0.27
   (d) Brian, since $P(X \geqslant 8) = 0.0207$ whereas for Alan $P(X \geqslant 8) = 0.0062$.
6. (a) 0.886
   (b) Data not symmetric but showing a positive skew.
7. (a) 1.2  (b) 53.6  (c) 54.2; 0.066
8. (a) (i) 4.95%  (ii) 0, I, II
   (b) (i) 105.3  (ii) 106.45; 106.45
   (c) (i) 103.3, 3.98
   (ii) needs overhaul, standard deviation too high.
9. (a) 14.25 p  (b) 736 g  (c) 462 g
10. (a) (i) 0.250  (ii) 0.758  (iii) 0.00240  (b) 0.0433
11. (a) (i) 0.197  (ii) 0.820  (b) (ii) 19  (c) 0.2142
12. 0.360, 0.734
13. (a) 0.653  (b) 0.2224
14. (a) (i) 104  (ii) 33  (iii) 33  (b) 1000, 200
15. (a) 0.3154  (b) 0.3068; worse, 0.5245
16. 979.27, 17.27, 133
17. (a) random events, mean $\approx$ variance
    (b) 0.224  (c) 0.586  (e) 0.6201
18. (a) 0.988  (b) 0.855
    (c) 0.783 (Poisson), 0.784 (binomial)
19. (a) 0.649  (b) 0.965  (c) 0.371
20. (a) 0.988  (b) 0.624  (c) 0.828
21. (a) $np > 5$, $nq > 5$, $X \sim N(np, npq)$
    (b) $p < 0.1$, $n > 50$, $X \sim Po(np)$; 0.859
    (c) 0.204  (d) 0.034

## Mixed test 7A (page 401)

1. (a) 29%  (b) 402.62 ng/ml
2. (a) 25  (b) 0.673
3. (b) 0.0113  (c) 0.86
4. (a) 0.0548  (c) 0.356

## Mixed test 7B (page 402)

1. Luxibrite, 0.936
2. (a) 0.1056  (b) 0.8641  (c) 815.68
3. (a) (i) 0.8944  (ii) 0.4931
   (b) only able to stay for a maximum of 60 minutes
   (c) mean + $3\sigma$ gives 6.55 pm
4. (a) 7.5  (b) randomly scattered  (d) 0.901  (e) 0.2627
5. (a) (i) 0.0808  (ii) 0.1935
   (b) 0.295  (c) 0.0598

# Chapter 8

## Exercise 8a Sums and differences of normal variables (page 409)

1. (a) 210, 625  (b) $X \sim N(210, 625)$
   (c) 0.6554  (d) 0.7698
2. (a) 0.1319  (b) 0.0127
3. (b) 0.9324
4. 0.0745

5. (a) 0.5  (b) 0.8849  (c) 0.2779
6. (a) 0.0207  (b) (i) 0.0289  (ii) 0.0200  (iii) 0.6252
7. (a) 0.1247  (b) 0.6957
8. (a) 0.6298  (b) 0.1056
9. (a) 0.1728  (b) 0.6127  (c) 0.5
10. 0.2575
11. 0.1103, 0.753
12. 9.6, 0.522;  (a) 1.8%  (b) 22.2%
13. (a) (94.4, 105.6)  (b) 92.55%  (c) 22.14%
14. (a) 0.0787  (b) $3.02 \times 10^{-6}$

## Exercise 8b Multiples of normal variables (page 413)

1. (a) 0.8962  (b) 0.9386
2. (a) 0.2398  (b) 0.2523
3. (a) 0.244  (b) 0.659  (c) 0.409
4. (a) 6, $\sqrt{2}$  (b) 0.2074  (c) 0.7601  (d) 0.5143
5. 0.2762
6. (a) 0.3446  (b) 0.6915; 0.0033, 0.304

## Miscellaneous exercise 8c (page 417)

1. (a) 0.60  (b) 0.20  (c) 0.95  (d) 0.5
2. (a) 0.051  (b) 0.00155  (c) 0.9782
3. 1000, 172, 3000, 298, 0.16, 0.02
4. (a) 0.0888  (b) 0.6611
5. 0.0625, 0.2574, 0.5, 0.7123
6. (a) 0.0139  (b) 0.1587  (c) 0.9332
7. (a) 0.159  (c) 0.584
8. 12 kg, 57.0 g, 3.97%, 765 g
9. (a) (i) 0.1056  (ii) 0.8882  (b) 1028 g  (c) 0.0537
10. (a) (i) 0.0537  (ii) 0.144  (b) 0.0188
11. (a) 0.1416  (b) 0.5999  (c) 14.96 m  (d) 0.3043
12. (a) 0.798  (b) 0.323  (c) 0.132  (d) 0.228
13. (a) 0.252  (b) 0.0581  (c) 0.104

## Mixed test 8A (page 419)

1. (a) $S \sim N(600, 105.8)$, 0.0724  (b) 0.8392
   (c) 0.1606  (d) 30.54 g
2. (a) 0.733  (b) 0.984
3. (b) 0.0802  (c) 0.6729

## Mixed test 8B (page 420)

1. (a) 0.127  (b) (i) 0.0016  (ii) 0  (c) 0.1003
2. (a) 0.8413  (b) 0.5  (c) 0.4207; 0.9938
3. 0.84

# Chapter 9

## Exercise 9a Sampling methods (page 430)

2. (a) 6, 6, 6, 6, 6, 5, 5
4. (b) large : medium : small = 15 : 25 : 20

## Exercise 9b Simulating random samples from given distributions (page 435)

Some answers depend on the random numbers used and on the method of allocation. These are possible answers.
10. (a) 1, 1, 1, 0, 3  (b) 4
11. 33.134, 33.193, 28.712
12. (a) 3, 5  (b) 1, 5  (c) 1007.2, 1016.8
13. 1.52
14. means of sample means $\approx$ distribution mean; variance of sample means $\approx \frac{1}{2}$ variance of distribution
15. (a) 4  (b) 6.1826

## Exercise 9c The distribution of the sample mean, $\overline{X}$ (page 443)

1. 0.0176
2. (a) 0.6234  (b) Approx. 4
3. (a) 0.1056  (b) 0.3092
4. (a) $\overline{X} \sim N\left(4.8, \dfrac{2.88}{50}\right)$  (b) 0.7975
5. (a) 8  (b) no
6. (a) 0.2399  (b) 0.0787  (c) 0.0127  (d) $n = 109$
7. 0.9212
8. 62
9. (a) 42  (b) 60
10. 5
11. 20, 3
12. (a) 12  (b) 20
13. 20500, 1768; no
14. 0.332, 0.0587, 0.009
15. 0.4948, 0.4944, 0.1211
16. (a) $P(X = 0) = \frac{1}{2}$, $P(X = 1) = \frac{1}{3}$, $P(X = 2) = \frac{1}{6}$
    (b) $\frac{2}{3}$  (c) 0.159

## Exercise 9d Distribution of sample proportions (large samples) (page 447)

1. (a) 0.0745  (b) 0.0037
2. (a) 0.0057  (b) 0.527  (c) 0.1265
3. 0.0471
4. (a) 0.3085  (b) 0.0970
5. 0.7181
6. (a) 0.0648  (b) 0.0851  (c) 0.3068
7. (a) 0.22

## Exercise 9e Point estimates and confidence intervals for $\mu$ (page 460)

1. 236, 7.58
2. (a) 48.875, 6.98  (b) 1.69, $8 \times 10^{-6}$ (1 s.f.)
   (c) 22.79, 1.81
   (d) 15, 43.14  (e) 10, 3.11  (f) 9.71, 621.12
3. 0.5, 1.428
4. 205.16, 9.223
5. (a) (139.16, 140.5)  (b) random sample
6. (a) (10.75, 14.15)  (b) 3.4
7. (a) (448.7, 467.3)
   (b) The probability that this interval includes $\mu$ is 0.99.
   (c) No, $z$ value less
8. (a) (79.19, 84.81)  (b) (78.89, 85.11)
   (c) No, the central limit theorem can be used, since $n$ is large.
9. (68.0, 70.0), random sample, central limit theorem can be applied.
10. (a) 3.612  (b) (747.3 g, 748.7 g)
    (c) random sample, central limit theorem can be applied.
11. (a) (1011, 1114)  (b) 36
12. 28
13. (a) 5.06 g  (b) 89%
14. Histogram: frequency densities 1.2, 3.6, 6.4, 11.4, 20.4, 10.2, 5, 1.8; 91.32, 7.42, 0.43, (90.5, 92.2)
15. 25.3, 3.6, (24.9, 25.8)
16. Histogram: frequency densities 0.8, 0.48, 0.3, 0.18, 0.1, 0.05, 0.04, 0.03, 0.02; 194, 176, (173.5, 214.5)

## Exercise 9f Confidence intervals – small samples ($t$ – distribution) (page 468)

1. (a) (177.21 cm, 182.12 cm)  (b) 4.91 cm
2. (a) (3.59, 4.68)  (b) 0.146

3. (8.07, 9.13)
4. (32.08, 33.22), 380
5. (a) 5.13  (b) 0.588  (c) (4.70, 5.56)
6. (14.98 g, 15.78 g)
7. (9.804, 9.807)

## Exercise 9g Confidence intervals for $p$ (page 471)

1. (a) (0.622, 0.738)
   (b) The normal approximation to the binomial has been used in the underlying distribution.
2. (a) (0.293, 0.427)  (b) (0.273, 0.447)
3. (a) (0.238, 0.362)  (b) 90
4. (a) 0.28  (b) (0.176, 0.384)
5. (0.156, 0.344)
6. (a) (0.223, 0.352)  (b) wider
7. (a) Random sample (0.244, 0.283)
   (b) (i) 0.26  (ii) 90 approximately
8. (a) (0.351, 0.369)  (b) 5277
9. (0.509, 0.574)

## Miscellaneous exercise 9h (page 478)

1. (124.34, 125.60), 4
2. (£93.59, £101.48)
3. 1.13, 0.0603, ($1.07, $1.19)
4. 9.71, (172.3, 173.3)
5. (a) 3, (2.04, 3.96)  (b) 30%, (25.2%, 34.8%)
6. 0.059, 0.161
7. (a) Lifetime of bulb follows a normal distribution; the items in the box constitute a random sample.
   (b) (1774 hours, 1798 hours)
8. (a) 268  (b) smaller, critical $z$ value less
9. (0.139, 0.315); there is a 1% chance that the interval has not trapped $\mu$.
10. (a) 26.525, 1.24  (b) (26.20, 26.85)  (c) justified
    (d) $n$ large, use Central Limit theorem
11. (a) (28.98 cm, 29.42 cm)  (b) Large sample
    (c) $\overline{X}$ normally distributed, random sample
    (d) (26.78 cm, 31.62 cm)
    (e) no; 30.5 out of range of 95% confidence interval for $\mu$
12. (92.32, 99.68)
13. (a) (202.4, 207.4)  (b) 0.2, (0.057, 0.343)
14. (0.123, 0.392), (170.84, 178.16), (165.57, 186.83)
15. 25.35, 0.13, (25.15, 25.6), valid
16. (a) (0.303, 0.357)
    (c) 10% probability that interval did not trap $\mu$; people changed their minds at the last minute
17. (£35.60, £130.80)
18. (35.03 mg, 35.31 mg)
19. (13.10 mm, 14.72 mm)
20. (47.02 cm, 51.38 cm)
21. (0.0825 mm, 0.242 mm)

## Mixed test 9A (page 481)

1. (a) 0.391  (b) 93%
2. 14
3. (0.23, 0.35); the normal approximation to the binomial has been used in the underlying theory; only cars in the car park were sampled which may not constitute a random sample.
4. (18.51, 19.49)

## Mixed test 9B (page 482)

1. (b) (92.01, 93.19)
   (c) Central Limit theorem can be applied.

2. $(0.35, 0.49)$, $0.14$

3. (a) $\left(\bar{x} - \dfrac{38.64}{\sqrt{n}}, \bar{x} + \dfrac{38.64}{\sqrt{n}}\right)$  (b) 6000

4. (a) $(244.2\text{ g}, 250.2\text{ g})$  (b) 6.0 g  (c) smaller

# Chapter 10

## Exercise 10a Testing $p$ in a binomial distribution (small samples) (page 494)

1. $H_0: p = 0.7$, $H_1: p > 0.7$; no evidence
2. (a) $H_0: p = 1/6$, $H_1: p > 1/6$
   (b) There is no evidence that die is biased in favour of 4.
3. (a) Do not reject $H_0$  (b) Reject $H_0$
4. (a) Evidence to suggest decrease.
   (b) No evidence to suggest decrease.
5. (a) $x \geqslant 5$
   (b) The probability that $H_0$ is rejected when it is in fact true.  (c) 0.1
6. (a) Accept $H_0$  (b) Reject $H_0$  (c) Reject $H_0$
   (d) Accept $H_0$  (e) Accept $H_0$  (f) Reject $H_0$
   (g) Reject $H_0$  (h) Accept $H_0$
7. (a) Driving instructor is over-estimating pass rate.
   (b) $x \geqslant 3$
8. She could have been guessing.
9. (a) $x \leqslant 2$  (b) 0.803
10. (a) 15%  (b) 0.15(09)  (c) 28% (2 s.f.)
11. (a) 7.5% (2 s.f.)
    (b) same as significance level
    (c) 66% (2 s.f.)

## Exercise 10b Testing $\lambda$ in a Poisson distribution (page 500)

1. Increased
2. (a) Not increased  (b) Decreased
3. $H_0: \lambda = 9$, $H_1: \lambda > 9$, not increased
4. (a) 0.0424  (b) 0.849
5. (a) Accept $H_0$  (b) Accept $H_0$  (c) Accept $H_0$
   (d) Reject $H_0$  (e) Accept $H_0$  (f) Reject $H_0$
6. $H_0: \lambda = 3.5$, $H_1: \lambda > 3.5$, not increased

## Miscellaneous exercise 10c Binomial and Poisson tests (page 504)

1. (a) 0.028  (b) 0.131
   (c) 0.261; $H_0: p = 0.6$, $H_1: p > 0.6$; teacher is not underestimating
2. (a) 0.552  (b) 6, 0.296
   (c) The probability he scores a penalty kick remains constant at 0.7.
   (d) $H_0: p = 0.7$, $H_1: p > 0.7$
   (e) No evidence of improvement  (f) strengthened
3. Manufacturer's claim is not accepted; discrete distribution, $P(X \leqslant 12) = 3.6\%$, $P(X \leqslant 13) = 17.1\%$.
4. (a) $H_0: p = 0.2$, $H_1: p > 0.2$
   (b) $X \sim B(25, 0.2)$  (c) 9
5. (a) (i) 0.0278  (ii) 0.0384  (iii) 0.0768
   (b) $H_0: p = 0.5$, $H_1: p \neq 0.5$, no indication of whether looking for evidence of more males or more females.
   (c) Not evidence of more males than females, $x \geqslant 13$
6. (a) 37%  (b) 42%
   (c) (i) The consumer group has used a high value for the significance.
   (ii) Choose 5% or 10% significance level to maintain credibility.

7. (a) 2, 1.18  (b) 0.302
   (c) $H_0: p = 0.2$, $H_1: p < 0.2$, not reduced
   (d) reduced
8. (a) 0.430  (b) 0.962  (c) 0.00459
   (d) $H_0: p = 0.9$, $H_1: p < 0.9$, looking for a decrease
   (e) No evidence that service has deteriorated.
   (f) $x \leqslant 12$; $P(X \leqslant 12) < 0.05$, whereas $P(X \leqslant 13) > 0.05$
9. (a) Defects occur randomly and independently, with no two defects at the same spot.
   (b) (i) 0.209  (ii) 0.221
   (c) 0.140
   (d) $H_0: \lambda = 2.4$, $H_1: \lambda > 2.4$, evidence that number of defects has increased.
10. (a) (i) 0.181  (ii) 0.999  (b) 0.018
    (c) No evidence of decrease.
11. 0.0057, 9 mins, not significant

## Mixed test 10A (Binomial) (page 506)

1. 0, 1, 9, 10
2. (a) $H_0: p = 0.15$, $H_1: p < 0.15$, evidence that new procedure has been successful.
   (b) Staff making an effort during the first week, take sample over a longer period of time.
3. No evidence to support gardener's claim.

## Mixed test 10B (Poisson) (page 506)

1. (a) Poisson, 2.1
   (b) (i) 0.650  (ii) 0.222
   (c) Evidence suggests higher rate.
2. (a) (i) 0.138  (ii) 0.847
   (b) $H_0: \lambda = 7.5$, $H_1: \lambda < 7.5$, does not provide significant evidence.
3. (a) Nominally 5% (between 4.26% and 8.39%)
   (b) 76% (2 s.f.).

# Chapter 11

## Exercise 11a $z$-tests for a normal population or large sample size (page 522)

1. (a) $z = -1.095$, accept $H_0$  (b) $z = 1.845$, reject $H_0$
   (c) $z = 2.5$, reject $H_0$  (d) $z = -2.778$, reject $H_0$
2. $z = -0.943$, no
3. It could be 103.5
4. $z = 2.487$, yes
5. $z = 1.909$, distribution of the sample mean is approximately normal.
6. $z = 1.987$, no evidence
7. (a) $\bar{x} < 91.5065$ minutes  (b) 0.0093  (c) 0.3286
8. $z = 0.983$, accept mean is zero
9. $5.778 < \bar{x} < 6.222$
10. (a) $z = 1.778$, accept $H_0$  (b) $z = 1.778$, reject $H_0$
    (c) $z = -1.428$, reject $H_0$  (d) $z = -2.487$ accept $H_0$
11. (a) Reject $H_0$ and conclude mean is not 52.  (b) 0.04
12. (a) 0.0817  (b) 0.665
13. $z = 2.946$, yes
14. (b) 0.24  (c) $\mu > 389.7$  (d) 0.0494

## Exercise 11b $t$-tests for a normal population, small sample size (page 527)

1. (a) $t = 0.909$, accept $H_0$  (b) $t = -1.89$, accept $H_0$
   (c) $t = 2.15$, reject $H_0$  (d) $t = -3.07$, accept $H_0$
2. $t = 2.828$, evidence of improved times
3. (a) $t = -3.54$, underweight  (b) $z = -3.2$, underweight

4. $t = -1.1$, no
5. $t = 2.284$, mean greater than 4.3
6. (a) $t = -3.23$, no   (b) (1.69, 2.88)
7. (a) $z = -1.66$, no change in mean
   (b) 0.324, $t = -2.33$, change in mean
8. $X$ is normally distributed, $t = 1.80$, accept null hypothesis
9. $H_0: \mu = 27$, $H_1: \mu \neq 27$, $t = 2.9$, mean is 27
10. $H_0: \mu = 50$, $H_1: \mu < 50$, $t = -0.435$, not overstating

## Exercise 11c Testing a binomial proportion large $n$ (page 532)

1. (a) $z = 1.59$, accept $H_0$   (b) $z = 2.206$, accept $H_0$
   (c) $z = -1.79$, accept $H_0$   (d) $z = 2.118$, accept $H_0$
   (e) $z = -2.937$, reject $H_0$
2. $z = -2.40$, do not accept claim as there is evidence that proportion is less.
3. $z = 1.637$, yes
4. $z = 1.476$, no
5. $z = -1.990$, no
6. $z = 1.5$, no
7. $z = 1.705$, evidence that more than 65% own a mobile phone.
8. (a) (i) 0.0297   (ii) 0.0934
   (b) $z = -1.792$, germination rate less than 75% (only just – do further tests)
9. $z = 2.43$, yes
10. Replies were representative of the population.
    (a) $z = 1.220$, no evidence to suggest proportion in favour is more than 0.7.
    (b) (0.681, 0.808)
11. (a) Evidence that proportion is lower
    (b) No different
12. (a) $z = -3.03$, evidence that $p < 0.4$
    (b) (0.379, 0.458); 75
13. $z = -1.267$, no
14. $z = -2.44$, evidence that proportion has fallen

## Exercise 11d testing the difference between means of two normal populations
## Section A: $z$-tests (page 543)

1. (a) (i) $z = -2.096$, reject $H_0$   (ii) $z = -1.402$, accept $H_0$
   (iii) $z = 2.493$, reject $H_0$
   (b) (i) $z = 1.99$, accept $H_0$   (ii) $z = 2.076$, reject $H_0$
   (iii) $z = -2.036$, accept $H_0$   (iv) $z = 1.783$, reject $H_0$
   (v) $z = 1.779$, reject $H_0$   (vi) $z = -2.321$, accept $H_0$
   (only just)   (vii) $z = 2.55$, reject $H_0$
2. 0.567, $z = -2.219$, flowers on sunny side grow taller
3. $z = 3.52$, second population has smaller mean than first
4. $z = 2.036$, significant at 5% level, not significant at 4% level
5. 4.41, (9.87, 10.73), 3.61, $z = 1.49$, not significant evidence
6. $z = -1.646$, reject Mr Brown's claim (only just)
7. $z = -2.04$, evidence of difference
8. 103, $z = -2.913$, significant evidence
9. $z = 1.627$, accept; 124
10. 27.33 (26.77, 27.89), 2.4, $z = 1.97$, those of higher intelligence do not have greater foot length.

## Section B: $t$-tests (page 545)

1. (i) (a) 17.73   (b) $t = 2.135$, reject $H_0$
   (ii) (a) 87.09   (b) $t = -0.567$, accept $H_0$
   (iii) (a) 27.5625   (b) $t = 2.088$, accept $H_0$
   (iv) (a) 4.182   (b) $t = 1.260$, accept $H_0$

2. $t = -1.13$, no evidence that Welsh policemen are shorter than Scottish policemen.
3. 196, $z = -1.714$, do not differ significantly
4. (a) 10.8125   (b) $t = -1.282$, accept claim
5. $t = 2.423$, not significant difference
6. Normal populations with common variance, $t = 2.36$, evidence that mean has increased; $t = 2.041$, the mean could be 500 g.
7. (a) Normal populations with common variance
   (b) $H_0: \mu_1 = \mu_2$, $H_1: \mu_1 \neq \mu_2$   (c) $t = -0.942$, same
8. $t = 1.868$, evidence that new method has led to higher scores; (−2.60, 33.9)

## Miscellaneous exercise 11e (page 554)

1. (17.1, 19.7), there is a 10% chance that it hasn't trapped $\mu$: $z = 0.759$, $\mu$ could be 17.8
2. (a) Children within families selected are representative of all children.
   (b) $z = 0.939$, data do not indicate that boys and girls are not equally likely
3. (a) $H_0: \mu = 30$, $H_1: \mu > 30$   (b) $\bar{x} > 33.95$
   (c) Evidence that mean speed is greater than 39 mph ($\bar{x}$ is in critical region).
   (d) 0.9941
4. (a) $H_0: \mu = 43$, $H_1: \mu > 43$
   (b) Since $n$ is large, the distribution of the sample means is approximately normal
   (c) $z = 1.768$, mean amount has increased
   (d) (43.35, 52.65), consistent, 43 out of range of confidence interval
5. $H_0: p = 0.13$, $H_1: p < 0.13$, 2%, 0.161
6. (a) (i) $P(\bar{X} > \text{critical value} \mid \mu = 65)$
       (ii) $P(\bar{X} < \text{critical value} \mid \mu$ is value specified by the alternative hypothesis)
   (c) Accept $H_0$, Type II
   (d) 0.0059, Type II error would be less and tends to zero as $\mu$ increases
7. Not representative as it excludes people at work, school, etc; better to take random samples at random times during the day for a spread of days, (68%, 80%). $z = -2.03$, data provide significant evidence
8. (a) $\bar{x} > \mu_0 + 1.96\dfrac{\sigma}{\sqrt{n}}$ or $\bar{x} < \mu_0 - 1.96\dfrac{\sigma}{\sqrt{n}}$
   (b) $\bar{x} \geqslant \mu_0 - 2.326\dfrac{\sigma}{\sqrt{n}}$
9. (a) 0.422
   (b) E(unbiased estimate) = true value; batch not rejected, 9.6%
10. (a) $H_1: p > \frac{1}{6}$   (b) $N \geqslant 50$   (c) $0.059 \approx 6\%$
11. Accept as slow if mean bounce < 11.645, 0.0004
12. (a) (i) 10.46   (ii) 15.64, E(unbiased estimate) = true value
    (b) 1
    (c) Central Limit theorem holds when $n$ is large
13. $z = 3.367$, accept claim that mean duration is more than 12 months; $n$ large, use Central Limit theorem
14. (a) 75
    (b) $z = 2.19$, machine is not correctly calibrated
    (c) Unbiased estimate of standard deviation used, distribution of sample mean approximately normal; (0.316, 5.684)
    (d) Smaller, might lead to result that machine is correctly calibrated.
15. (a) 66.25, 133.40   (b) $H_0: \mu = 62.5$, $H_1 > 62.5$, $z = 1.465$, no evidence of increase
16. (a) $z = 2.475$, mean has increased

17. (a) $H_0: \mu = 1.73$, $H_1: \mu > 1.73$    (b) $\bar{X} \sim N(1.73, 0.0008)$
     (c) $\bar{x} > 1.777$
     (d) men who play basketball are not taller    (e) 0.14
18. 9

## Test 11A ($z$-tests) (page 558)

1. (a) 21.25
     (b) $z = 0.99$, no evidence to support manufacturer's suspicion
     (c) obtaining distribution of $\bar{X}$, distribution of $X$ not known
2. $z = 1.567$, not sufficient evidence to say that the quoted figure is an underestimate
3. (a) (i) P(reject $H_0$ when $H_0$ is true)
       (ii) P(accept $H_0$ when $H_1$ is true)
     (b) (i) $z = 2.372$, mean is greater than 17.5
       (ii) $\bar{x} > 18.09$
       (iii) 0.639
4. 0.0606 ($\approx 6\%$), 0.1118

## Test 11B ($z$-tests) (page 558)

1. $H_0: \mu = 125$, $H_1: \mu < 125$, $z = -1.549$, no evidence that $\mu$ is lower
2. $z = 2.318$, government spokesman
3. (a) $\bar{x} < 59.82$
     (b) It is accepted that the mean is 60 when in fact it is an alternative value (less than 60).
     (c) 0.057
4. (a) $H_0: \mu_1 - \mu_2 = 0$, $H_1: \mu_1 - \mu_2 \neq 0$
     (b) $z = 1.6$, no difference
     (c) Distribution likely to be skewed rather than symmetric

## Test 11C ($t$-tests) (page 559)

1. $t = -3.560$, evidence that mean falls below $7.40; normal distribution
2. $t = -2.915$, San Marco cooler
3. (a) 4.238    (b) Normal distribution, $t = 2.857$, yes
     (c) Perform $z$-test not $t$-test
4. $t = -2.046$, new score higher; $(-6.948, 32.282)$ or $(-32.282, 6.948)$

# Chapter 12

There will be variation in answers, depending on the degree of accuracy used in various stages of the working.

## Exercise 12a Goodness of fit test – uniform and given ratio (page 569)

1. $X^2 = 1.93$, $v = 3$, die is fair
2. $X^2 = 18.16$, $v = 9$, uniform distribution
3. $X^2 = 6.19$, $v = 2$, yes
4. $X^2 = 4.95$, $v = 3$, no; $X^2 = 9.90$, $v = 3$, yes
5. $X^2 = 8.24$, $v = 7$, accept theory
6. $X^2 = 4.15$, $v = 4$, yes
7. $X^2 = 10.68$, $v = 4$, no
8. 15.5
9. 78.81, 17.8, 7.8, 6.7 $X^2 = 5.92$, $v = 3$, no difference
10. $X^2 = 38.2$, $v = 9$, evidence of bias
11. $X^2 = 10$, $v = 4$, not uniform
12. $X^2 = 4.4$, $v = 5$, die is fair

13. (a) modal class 2 to <4
     (b) 4 years 8 months, 3 years 2 months
     (c) For cumulative frequency curve plot (0, 0), (2, 42), (4, 94), (6, 122), (8, 142), (10, 160), (12, 176); 3 years 9 months, 4 years 9 months
     (d) $X^2 = 5.73$, $v = 5$, justified

## Exercise 12b Goodness of fit tests – binomial, Poisson and normal distributions (page 579)

1. Combine last three classes, $X^2 = 4.09$, $v = 3$, accept $X \sim B(5, 0.3)$
2. $X \sim B(5, \frac{1}{6})$, E = 80.5, 80.5, 32, 7 (last 3 classes combined), $X^2 = 8.21$, $v = 3$, biased; $\bar{x} = 1$, $p = 0.2$, $X \sim B(5, 0.2)$, E = 66, 82, 41, 11 (last three classes combined) $X^2$ is very small, $v = 2$, too good a fit, query data.
3. $np$, 1.6, 0.32, E = 7.3, 17.1, 16.1, 7.5, 1.8, 0.2 (combine last 3 classes) $X^2 = 1.79$, $v = 2$, good fit
4. $X \sim B(2, \frac{1}{6})$, E = 150, 60, 6, $X^2 = 9.6$, $v = 2$, reject; use $\bar{x} = 0.444$, $p = 0.222$, find E, $v = 1$
5. (a) $\bar{x} = 2$, $p = 0.4$, E = 6, 21, 28, 18, 7 (combine last 2 classes)
     (b) $X^2 = 2.21$, $v = 3$    (c) yes, binomial adequate
6. (a) $X \sim B(5, 0.2088)$, E = 155, 205, 108, 28, 4, 0 (combine last 3 classes)
     (b) $X^2 = 5.959$, $v = 2$, binomial (but only just)
7. (a) 7
     (b) $n = 20$, $p = 0.35$; 0.16135, 8.1
     (c) 12.3
     (d) E = 12.3, 8.6, 9.2, 8.1, 11.8, O = 9, 7, 17, 8, 9, $X^2 = 8.46$    (e) 3, not good fit at 5% level
8. (a) E = 246.6, 345.2, 241.7, 112.8, 39.5, 14.2
     (b) $X^2 = 32.2$, $v = 5$, not accepted
9. $\bar{x} = 2.5$, E = 8, 21, 26, 21, 13, 11 (combine end classes), $X^2 = 2.59$, $v = 4$, good
10. $\bar{x} = 1.28$, E = 41, 52, 34, 14, 6 (combine end classes), $X^2 = 6.81$, $v = 3$, not significant
11. $\bar{x} = 0.65$, E = 20.88, 13.57, 5.55 (combine end classes), $X^2 = 1.85$, $v = 1$, accept
12. (a) $\bar{x} = 1.2$, E = 99, 119, 72, 29, 9, 2 (combine end classes)
     (b) $X^2 = 0.48$, $v = 3$, very good fit
13. $\bar{x} = 0.9$, E = 21, 18, 11 (combine last 3 classes), $X^2 = 1.80$, $v = 1$, yes, consistent
14. (b) E = 7.3, 12.4, 10.6, 9.7
     (c) $X^2 = 177$, $v = 2$, reasonable
     (d) very low, suspicious
15. E = 6.68, 9.19, 14.98, 19.15, 19.15, 14.98, 9.19, 6.68, $X^2 = 3.197$, $v = 7$, accept. If $\mu$, $\sigma^2$ unknown, $v = 5$
16. (a) E = 3, 13, 28, 32, 18, 6 (combine first 2 classes), $X^2 = 11.9$, $v = 4$, reject
     (b) $\bar{x} = 171.54$, $s = 7.11$, E = 6, 18, 32, 28, 13, 3 (combine last 2 classes), $X^2 = 1.73$, $v = 2$, accept normal
17. (a) $\bar{x} = 1.732$, $\hat{\sigma} = 0.216$ (3 d.p.), E = 7.78, 26.05, 44.12, 33.64, 13.41, $X^2 = 8.96$, $v = 2$
     (b) $X^2 = 2.42$, $v = 1$

## Exercise 12c Contingency tables (page 588)

1. E = 48, 10.67, 21.33, 42, 9.33, 18.67, $X^2 = 1.037$, $v = 2$, no difference
2. E = 25.5, 25.5, 60.5, 60.5, 26.5, 26.5, 7.5, 7.5, $X^2 = 2.03$, $v = 3$, independent
3. E = 50.1, 29.5, 23.4, 22.9, 13.5, 10.6, $X^2 = 4.00$, $v = 2$, yes
4. E = 65.1, 28.9, 58.9, 26.1, $X^2 = 7.43$, $v = 1$, yes
5. E = 27.5, 972.5, 27.5, 972.5, $X^2 = 4.79$, $v = 1$, yes

6. $E = 11.4, 14.3, 8.6, 15.7, 18.3, 22.9, 13.7, 25.1, 20.6,$
   $25.7, 15.4, 28.3, 29.7, 37.1, 22.3, 40.9, X^2 = 12.0, v = 9,$
   accept
7. $E = 66.7, 33.3, 53.3, 26.7, X^2 = 6.81, v = 1,$ no
8. $E = 21.0, 10.0, 7.0, 15.5, 7.5, 5, 41.5, 19.5, 14, X^2 = 7.86,$
   $v = 4,$ accept
9. $E = 34.2, 29.8, 12.8, 11.2, X^2 = 1.22, v = 1,$ no
10. $E = 17.5, 82.5, 17.5, 82.5, X^2 = 0.58, v = 1,$ no
11. $E = 202.2, 260.7, 318.1, 184.8, 238.3, 290.9, X^2 = 2.02,$
    $v = 2,$ independent
12. $v = 2, X^2 = 5.99$
13. (a) $v = (3 - 1)(3 - 1) = 4$   (b) difference
14. $E = 13.5, 15.5, 21, 8.64, 9.92, 13.44, 31.86, 36.58, 49.56,$
    $X^2 = 11.35, v = 4,$ yes
15. $E = 33.6, 22.4, 63.6, 42.4, 22.8, 15.2, X^2 = 4.775, v = 2,$
    no difference
16. $E = 90.405, 56.595, 35.595, 20.405, X^2 = 13.3, v = 1,$
    related

## Miscellaneous exercise 12d (page 594)

1. $H_0$: Preference for proposed route is independent of where
   people live (no association between them) $H_1$: There is an
   association between them, $E = 47, 28, 31.33, 18.67, 15.67,$
   $9.33, X^2 = 1.479, v = 2,$ no association
2. (a) $H_0$: Occurrence of shoplifting is uniformly distributed
       between the months, $H_1$: Shoplifting is more likely to
       occur in some months than others.
       $E = 14.5$ (all classes), $X^2 = 14.268, v = 11,$ no associa-
       tion
3. $H_0$: No association between reaction and eye colour,
   $H_1$: There is an association between them,
   $E = 15.675, 7.425, 9.9, 26.125, 12.375, 16.5, 15.2, 7.2,$
   $9.6, X^2 = 20.9, v = 4,$ association between reaction and eye
   colour
4. $E = 6.3$ (combines first 4 classes), $8.85, 14.66, 18.99,$
   $19.28, 15.32, 9.52, 7.05$ (combine last 3 classes),
   $X^2 = 4.908, v = 6,$ good fit
5. (a) $H_0$: No association between brand of fertiliser and
       yield, $H_1$: There is an association between them,
       $E = 10, 12, 8, 8, 9.6, 6.4, 7, 8.4, 5.6, X^2 = 7.811,$
       $v = 4,$ no association
   (b) $v = 2,$ there is an association between choice of
       company and yield
   (c) Quickgrow
6. (a) $H_0$: Peak flow measurements are normally distributed
       (with mean and variance as estimated from the data),
       accept
   (b) Expected frequency must be greater than 5, combine
       classes
7. (a) $E = 18.33, 20.67, 20.68, 23.32, 7.99, 9.01$
   (b) $X^2 = 6.88, v = 2,$ there is an association
8. (a) $H_0$: No association between candidates' grades in
       mathematics and physics, $H_1$: There is an association
       between them, $E = 17.2, 13.8, 12.8, 10.2, X^2 = 5.672,$
       $v = 1,$ there is an association
   (b) Expected frequency might drop below 5
9. (a) $E = 44.62, 66.94, 50.2, 25.12, 13.12$ (combine last
       two), $X^2 = 10.6, v = 4,$ at 5% level, no
   (b) Use mean from data for $\lambda, v = n - 2 = 3$
   (c) Do not have independent events with a constant
       probability of success.
10. $E = 644$ (all classes), $X^2 = 10.95, v = 4,$ data do not
    support claim, random events, $\bar{x} = 1.1, E = 33.3, 36.6,$
    $20.1, 10$ (combining last 2 classes), $X^2 = 0.48, v = 2,$
    accept

11. (a) $H_0$: Grades are in the ratio $15 : 20 : 35 : 25 : 5$, $H_1$:
        Grades are not in this ratio $E = 30, 40, 70, 50, 10,$
        $X^2 = 7.074, v = 4,$ same proportion
    (b) $H_0$: Sex and grade are not associated, $H_1$: There is an
        association between them, $v = 4,$ there is an
        association
12. (a) $\bar{x} = 0.74,$ combine 4 and over, $E = 667.96, 494.29,$
        $182.89, 45.11, 9.75, X^2 = 108.8, v = 3,$ not adequate
    (b) Not consistent, Poisson model was not adequate
    (c) $E = 259$ (all classes), $X^2 = 13.8, v = 3,$ not consistent
13. (a) $E = 19.33, 15.33, 10, 15.33, 9.67, 7.67, 5, 7.67,$
        $X^2 = 12.08, v = 3,$ mark is associated with type of
        question
    (b) Poisson, this is most similar question
    (c) $E = 22.5, 22.5, 22.5, 22.5, 7.5, 7.5, 7.5, 7.5,$
        $X^2 = 17.6, v = 3,$ yes it is
    (d) Contingency table – popular and well answered;
        Binomial and Poisson fits – average popularity,
        relatively badly answered, normal fit – unpopular but
        well answered by those who attempted it.
14. (a) $E = 480$ (all classes), $X^2 = 14.8, v = 4,$ there is evidence
    (b) $E = 6.405, 6.51, 8.085, 24.705, 25.11, 31.185, 29.89,$
        $30.38, 37.73, X^2 = 16.9, v = 4,$ length of employment
        is associated with grade
15. $E = 38.78, 34.02, 18.2, 39.21, 34.39, 18.4, 27.28, 23.92,$
    $12.8, 22.59, 19.81, 10.6, 22.59, 19.81, 10.6, 28.55,$
    $25.05, 13.4, X^2 = 16.0, v = 10,$ no association, expected
    frequency must be greater than 5, would not make sense;
    $E = 60$ (all classes), $X^2 = 13.2, v = 6,$ reject hypothesis
16 (a) $\bar{x} = 1, E = 36.79, 36.79, 18.39, 8.026,$ (3 or more),
        $X^2 = 12.9, v = 2,$ not suitable
    (b) $E = 26.25, 6.75, 8.75, 2.25, X^2 = 3.77, v = 1,$ yes

## Mixed test 12A (page 598)

1. (a) 1.04   (b) Calls occur at random
   (c) $E = 58.01, 55.11, 26.18, 10.70, X^2 = 4.86, v = 3,$
       $Po(0.95)$ is suitable
2. $E = 6, 20, 12, 2, 9, 30, 18, 3, X^2 = 13.11, v = 3,$ there is a
   link between General Studies performance and degree class
3. $E = 15.9, 21.1, 26.1, 21.1, 15.9$ (combine first 2 and last 2
   classes), $X^2 = 7.08, v = 4, N(180, 9)$ is suitable
4. $H_0$: There is no association between gender and passing a
   driving test.
   $H_1$: There is an association.
   $E = 27.5, 22.5, 27.5, 22.5, X^2 = 2.585, v = 1,$ results do
   not indicate link

## Mixed test 12B (page 599)

1. $E = 32$ (all classes), $X^2 = 1.6875, v = 3,$ no particular
   preference, data cannot be used to discredit claim
2. $E = 21.34, 43.66, 21.66, 44.34, X^2 = 4.72, v = 1,$ there is
   an association between the two factors
3. $E = 16.67, 16.67, 16.67, 16.67, 33.33, 50, X^2 = 4.65,$
   $v = 5,$ die is biased in the way described
4. (b) Random positions   (c) 37.24, 2.50
   (d) $H_0$: The distribution can be modelled by $Po(2.59)$, $H_1$:
       The distribution cannot be modelled by $Po(2.59)$,
       $X^2 = 7.55, v = 5,$ Poisson model is supported by data

# Chapter 13

## Exercise 13a Significance test for product – moment correlation coefficient (page 604)

1. Reject a, c, f, g, h: do not reject b, d, e
2. (a) 0.3755
   (b) $H_0: \rho = 0$, $H_1: \rho > 0$, reject $H_0$, no evidence
   (c) $X$ and $T$ are jointly normally distributed with correlation coefficient $\rho$ and that the data constitute a random sample from all values of $x$ and $t$.
3. (a) Scatter diagram
   (b) 0.834
   (c) reasonable
   (d) Student's view is wrong; correlation does not imply causation; in this case there may be a common underlying cause such as wealth.
4. (a) $-0.690$, reject $H_0$ in favour of $H_1$
   (b) 0.686, reject $H_0$ in favour of $H_1$

## Exercise 13b Significance test for Spearman's rank correlation coefficient (page 607)

1. Reject b, f, h; do not reject a, c, d, g, i, e
2. 0.714, no evidence of agreement (only just)
3. (a) 0.52
   (b) $H_0: \rho_s = 0$, $H_1: \rho_s > 0$, do not reject $H_0$, no evidence of agreement between the interviewers
4. 0.745, evidence of correlation
5. (a) 0.66
   (b) evidence of positive correlation
6. 0.4286, $H_0: \rho_s = 0$; no evidence of positive correlation

## Miscellaneous exercise 13c (page 612)

1. (a) 0.636
   (b) $H_0; \rho_s = 0$, $H_1: \rho_s > 0$. Accept $H_0$, no evidence of positive correlation.
2. (b) 0.916
   (c) Evidence of positive correlation between the number of wren territories recorded and the number of adult wrens trapped.

3. (b) $-0.975$   (c) strong negative correlation
   (d) There is evidence of correlation between hours of sunshine and temperature.
4. (a) 0.4286
   (b) $H_0: \rho_s = 0$, $H_1; \rho_s \neq 0$, no evidence of correlation between attendance and position in the league
5. 0.527, no evidence of agreement
6. (b) 0.535
   (c) some positive correlation
   (d) Low mark in $x$, high mark in $y$
   (e) 0.794
   (f) $H_0: \rho_s = 0$, $H_1: \rho_s > 0$, evidence of positive correlation
7. (a) $c = 2.48 + 0.61m$
   (b) 0.593
   (c) 11.4   (d) 0.516, no   (e) $r$
8. 0.690, $H_0: \rho_s = 0$, $H_1: \rho_s \neq 0$, no evidence of correlation
9. (b) 0.783
   (c) $H_0: \rho = 0$, $H_1: \rho > 0$, evidence of positive correlation
   (c) Data constitute a random sample of all values of $x$ and $y$, years selected may not be representative.
   (d) lower
10. 0.825, 0.929, evidence of positive correlation (1% level)
11. $-0.3341$, not significant (5%), $-0.6939$, significant (2.5%)

## Mixed test 13A Correlation coefficients (page 615)

   (a) 0.473   (b) evidence
2. (a) 0.667
   (b) $H_0: \rho_s = 0$, $H_1: \rho_s > 0$, judges in broad overall agreement
   (c) evidence of correlation
   (d) Spearman's rank
3. $r = 0.310$, not significant, no evidence of possible correlation
4. (a) 0.619
   (b) $H_0: \rho_s = 0$, $H_1: \rho_s > 0$, not evidence of positive correlation (just)
   (c) Two very different sets of data being compared

# INDEX